10.

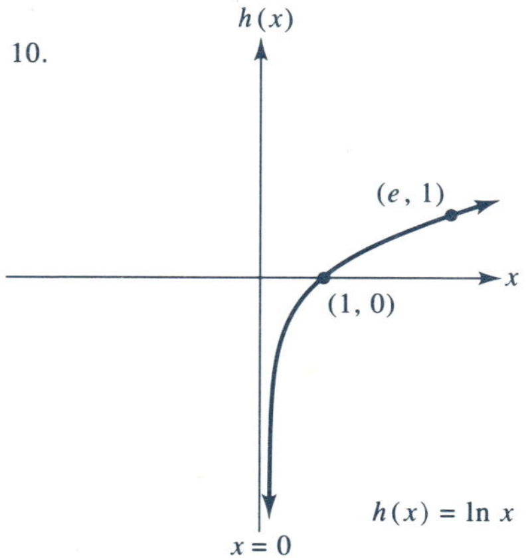

11.

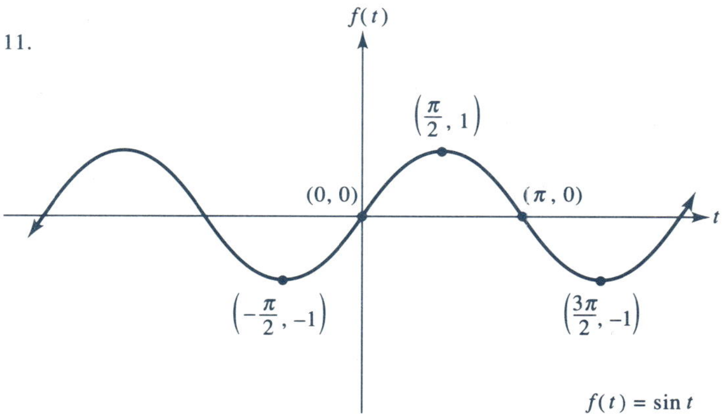

12.

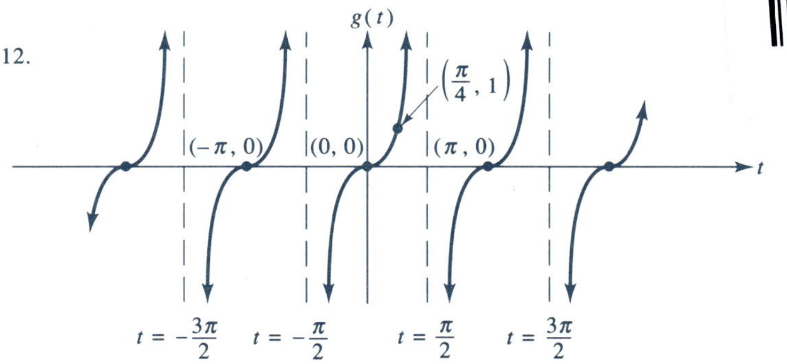

The graph continues to the left and right. $g(t) = \tan t$

13.

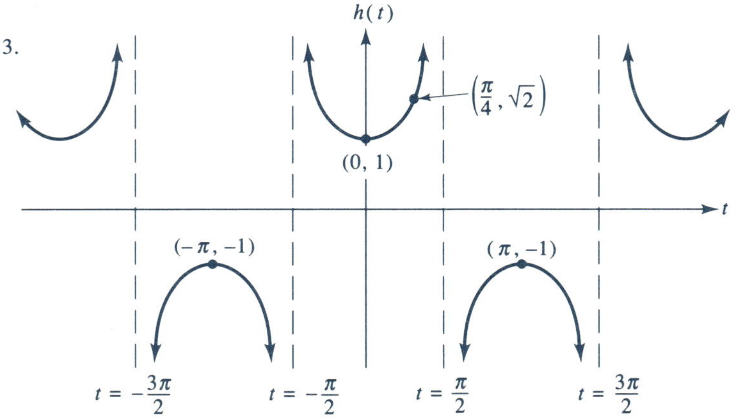

The graph continues to the left and right. $h(t) = \sec(t)$

Precalculus
Algebra and Trigonometry

Precalculus
Algebra and Trigonometry

Sharon Cutler Ross
DeKalb College

Linda Hawkins Boyd
DeKalb College

Brooks/Cole Publishing Company
Pacific Grove, California

Brooks/Cole Publishing Company
A Division of Wadsworth, Inc.

Printed in the United States of America

10 9 8 7 6 5 4 3 2 1

Library of Congress Cataloging-in-Publication Data
Ross, Sharon Cutler
Precalculus algebra and trigonometry / Sharon Cutler Ross, Linda Hawkins Boyd.
p. cm.
Includes index.
ISBN 0-534-14550-7
1. Algebra. 2. Trigonometry. I. Boyd, Linda Hawkins II. Title.
QA154.2.R67 1990
512′.13—dc20 90-2668
CIP

Sponsoring Editor: *Paula-Christy Heighton*
Editorial Assistants: *Sarah Wilson and Lainie Giuliano*
Production Editor: *Penelope Sky*
Production Assistant: *Dorothy Bell*
Manuscript Editor: *Linda L. Thompson*
Interior and Cover Design: *Sharon L. Kinghan*
Cover Photo: *Lee Hocker*
Art Coordinator: *Lisa Torri*
Interior Illustration: *Vantage Art, Inc.*
Typesetting: *Weimer Typesetting Company, Inc.*
Printing and Binding: *Arcata Graphics/Fairfield*

Preface

Precalculus Algebra and Trigonometry is designed to serve as the text for a one-semester or a two-quarter course used either to prepare for the calculus or to satisfy a college algebra requirement. Such courses are typically taught in the freshman year or whenever the student has a foundation in high school algebra. The text displays the diversity of mathematics without pretending to be a training course for prospective business, science, or other majors.

College algebra is a required course for all majors at many colleges and universities. This book fills the gap between precalculus texts whose focus is not appropriate for many students and the many "cookbook" college algebra-trigonometry texts. Revived interest in how and why required mathematics courses are taught at the freshman level encouraged us to present a fresh approach to this standard material. The book has a major emphasis on problem solving. Here in the Preface we discuss some of the ways in which students are encouraged to try different approaches to a problem, explore what-if questions, write to clarify their thoughts, search for patterns, and relate new problems to previously solved ones.

Our primary objective is to present mathematics in a unified, logical way, not as a sequence of isolated topics with steps for students to memorize. Functions and the solving of equations and inequalities are the text's twin themes. Graphing techniques are carefully developed with consistent labeling conventions used throughout. A conscious effort has been made to use a concept, once developed, in subsequent appropriate sections. This cycling helps the student not only in retention of the concept, but in putting ideas together in a truly mathematical way. The word problems included throughout the book serve as practice in problem solving and as a suggestion of the richness of mathematical applications. Such problems are not included as a spurious attempt at relevancy, but are written in a way that leads to a natural development of problem-solving skills. A special feature of the exercise sets is the inclusion of open-ended investigations and problems. These may preview a future topic, extend a current concept, or serve to generate discussion and creative thinking. Also, not all routine exercises have corresponding examples in the discussions, because we believe that too much mathematics at this level is done by rote with no real thinking taking place. This does not make the text "harder";

giving students something they can genuinely think about engages their interest, helps develop problem-solving skills, and provides the basis for a course both students and teachers can enjoy.

In summary, the text supports and integrates many of the current trends in mathematics education: an emphasis on problem solving, active participation by students, writing across the curriculum, geometric and analytic approaches to topics to accommodate different learning styles, use of the new technology such as graphing calculators, individual or group projects (see, in particular, the experiments in the *Instructor's Manual*), and the development of concepts.

Key Features

- Certain graphs are defined as standard graphs; each is a representative for a family of relations. Variations of these standard graphs by translation, expansion, and reflection are discussed throughout the book.
- All inequalities are treated by solving the corresponding equation to obtain "boundary numbers" or "boundary curves." Practice with the properties of inequalities appears in other settings.
- A small start is made on introducing discrete mathematics. In addition to the usual last-chapter topics, counting, mathematical induction, and so on, sets are defined and used often. The functions mod and div are also used in several places. Many of the extension exercises explore ideas in discrete mathematics.
- Consistency in the form of answers, in the treatment of approximate versus exact answers, and in graphing techniques is a matter of great concern to us. This is reflected in our treatment of these details.
- Careful instructions are given regarding exact and approximate answers throughout the exercises, encouraging the student to determine *when* as well as *how* to use the calculator. There are also selected graphing calculator problems. These are marked so that they may be easily identified.
- Along with the main threads of functions and equations and inequalities, many smaller strands are interwoven in the text. For example, writing explicit restrictions for reduced rational expressions leads into the discussion of domain and a method of finding vertical asymptotes and holes in graphs. The inclusion of explicit restrictions in trigonometric identities emphasizes this important concept in an area where it is generally ignored.
- Preview exercises lead the student to begin formulating more difficult concepts. For example, work with pseudo-trigonometric functions (Chapters 3 and 4) allows the student to work with a function whose definition is not algebraic, but given in terms of another function, before the trigonometric functions are developed.
- Pattern recognition and description are useful skills, and exercises for developing these appear in many sections. A number of opportunities to practice writing skills are also provided. Whenever possible, both geometric and algebraic definitions and explanations are given. Also, the student is asked to create as well as to analyze.

Structure

Many sections begin with a Toolkit—a summary of concepts needed for the subsequent discussion. Some topics are reviewed or defined in the Toolkit; for others, references to earlier chapters or the Appendix are given. The Appendix serves as both a review and a preview; even well-prepared students will find old ideas put to new uses here.

Each chapter summary contains a capsule review, a list of concepts covered in the chapter, a statement of skills to be mastered, and an exercise set designed to integrate the ideas of the chapter. Although it is sensible to break new material into small chunks for presentation, a chapter is built around a central theme. The forest is lost to the students if all they see are the isolated trees again. For this reason these summary exercises will often differ in ordering and form from those in the chapter's sections. A designated subset of the exercises in the summary may be used by the student as a self-test.

Each topic's inclusion had to be justified as appropriate for the courses described above and as directly related to our principal themes. Although Chapters 1, 2, and 3 contain the usual first-chapter topics, they do not come in the usual order. All the one- and two-variable inequalities are presented together to emphasize the generality of the solution method. The conics are developed naturally in a discussion of two-variable equations, instead of awkwardly fitting them in with functions as is often done. However, Sections 2.4 and 2.5 can be postponed. The discussion of translations, expansions, and reflections in Section 4.2 may need some amplification in class if Sections 2.4 and 2.5 have not been covered.

The first discussion of graphing rational functions is an outgrowth of the development of the hyperbola, $y = 1/x$. In Section 4.3, asymptotes are discussed as a significant feature which a graph may have; this means rational functions are graphed before the full development of polynomial functions in Chapter 8. Consequently, the discussion of the theory of polynomials can be more focused, and the idea of asymptote can be expanded in Section 8.4. Similarly, the natural exponential function, $y = e^x$, is introduced early as another function whose graph has an asymptote. Again, this gives the students some absorption time before a fuller treatment of the topic.

The real-valued versions of the trigonometric functions are presented first, and right-triangle trigonometry is treated as an application. Throughout Chapters 3 and 4, we have laid the groundwork for the wrapping function approach with preview exercises.

Chapters 8, 9, and 10 may be done any time after Chapter 5 has been covered. The examples and exercises using trigonometric functions may be safely omitted. A substantial piece of Section 9.4, and related exercises, which deals with rotation of graphs, does require Chapters 6 and 7.

In places we have included lots of detail, usually to emphasize a new approach; in others we have omitted details to prompt some thinking. Occasionally we even include a dead-end approach to show that mathematics is not always neat and tidy.

Acknowledgments

Our colleagues at DeKalb College and elsewhere have generously supported our efforts with encouragement, suggestions, and good humor. We work in a department whose members are deeply interested in mathematics and in the teaching of mathematics. The frequent discussions in the hallways and faculty lounge have been very important in the writing of this book. We had the good fortune of knowing a talented high school senior, Elisa Fiorenza, who not only worked nearly all the examples and exercises, but also read the whole book from a student's perspective. The text also benefited from the efforts of a number of dedicated reviewers: Patricia M. Bannantine, Marquette University; Franklin Cheek, University of Wisconsin, Platteville; David L. Fama, Germanna Community College; Marjorie Fitting, San Jose State University; Sandra C. McLaurin, University of North Carolina, Wilmington; George Morris, Sr., Parkland College; Catherine M. Murphy, Purdue University, Calumet; Dwayne S. Nymann, University of Tennessee, Chattanooga; Walter Peebles, Lansing Community College; Joanne Trimble, Marist College; Elaine Whittlesy, Sierra College; and Mary Winter, Michigan State University, East Lansing.

The entire staff at Brooks/Cole has contributed to making the development and production of this book a very pleasant experience. We are delighted to have worked with two fine editors, Jeremy Hayhurst and Paula-Christy Heighton.

And our husbands and children are delighted that we've finally returned from this writing adventure.

Linda H. Boyd
Sharon C. Ross

Contents

CHAPTER ONE — Equations of One Variable 1

1.1 Linear Equations 2
1.2 Quadratic Equations 13
1.3 Radical Equations 22
1.4 Miscellaneous Equations 29

CHAPTER TWO — Analytic Geometry and Equations of Two Variables 34

2.1 Introduction to Analytic Geometry 35
2.2 Linear Graphs 42
2.3 Parabolas 50
2.4 The Conic Sections 58
2.5 Translations, Expansions, and Reflections 73

CHAPTER THREE — Inequalities 85

3.1 One-Variable Inequalities I 86
3.2 One-Variable Inequalities II 91
3.3 Two-Variable Inequalities 98
3.4 Compound Inequalities 103

CHAPTER FOUR — Functions 111

4.1 Definition and Notation 112
4.2 Graphing Techniques I 121
4.3 Asymptotes 128
4.4 Properties of Graphs and Functions 136
4.5 Graphing Techniques II 143

CHAPTER FIVE ■ Algebra of Functions 151

5.1 New Functions from Old 152
5.2 Composition and Inverse Functions 157
5.3 Exponential and Logarithmic Functions 169
5.4 New Functions from Conditions 177

CHAPTER SIX ■ Trigonometric Functions 187

6.1 Preliminaries 188
6.2 The Trigonometric Functions 195
6.3 Graphs of the Trigonometric Functions 202
6.4 Graphing Variations of the Trigonometric Functions 208
6.5 The Inverse Trigonometric Functions 215
6.6 Equations Containing Trigonometric Expressions 220

CHAPTER SEVEN ■ Applications of the Trigonometric Functions 226

7.1 Right Triangles 227
7.2 Laws of Cosines and Sines 236
7.3 Basic Identities 246
7.4 Uses of the Basic Identities 257
7.5 Complex-Number Representation 264

CHAPTER EIGHT ■ Polynomial Functions 275

8.1 Synthetic Division and Its Uses 277
8.2 Complex and Rational Roots 282
8.3 Graphing Polynomial Functions 287
8.4 Rational and Other Functions Revisited 298

CHAPTER NINE ■ Systems of Equations 307

9.1 Linear Systems 308
9.2 Gauss-Jordan Elimination 315
9.3 Matrices 324
9.4 Applications of Matrices 332
9.5 Nonlinear Systems and Inequalities 343

CHAPTER TEN ■ Headstarts 354

10.1 Counting Techniques 355
10.2 Mathematical Induction 361
10.3 The Binomial Theorem 366

10.4 Sequences 372
10.5 Sigma Notation and Series 376

APPENDIX ■ Preview and Review 385

A.1 Introduction 386
A.2 Sets 388
A.3 Numbers 392
A.4 Exponents and Radicals 398
A.5 Expressions and Operations 402

■ Solutions to Odd-Numbered Exercises and Chapter Tests 413

■ Index 477

CHAPTER ONE

Equations of One Variable

- 1.1 Linear Equations
- 1.2 Quadratic Equations
- 1.3 Radical Equations
- 1.4 Miscellaneous Equations

Algebra gives us the means to represent symbolically relationships and facts about numbers. We will pursue two main threads of algebra in this book: solving equations and inequalities and working with functions. We expect that you have seen some algebra already.

Chapters and sections will sometimes begin with a toolkit, a brief list of skills and concepts needed for the discussion that follows. Many of these tools can be found in the appendix. It's a good place to check if you find a word or phrase about which you are unsure.

1.1 Linear Equations

Toolkit

The symbol $\equiv$ represents "is equivalent to." It is used to indicate that two quantities are identically the same and one may replace the other in any expression. For example, 1 foot $\equiv$ 12 inches and $x^2 - 1 \equiv (x + 1)(x - 1)$. There are also several formulas you will need. Among them are $d = rt$, relating distance, rate, and time; $i = prt$ for simple interest in terms of principal, interest rate, and time; and $C = 2\pi r$ for circumference of a circle in terms of its radius.

Sets are discussed in Section A.2; polynomials are introduced in Section A.5.

Let's begin with some basic definitions.

Definition An **equation** is a statement of equality between two numbers.

The more interesting equations are those where one or both numbers are represented by an expression involving variables.

Illustration Each of the following is an equation.

$$25 = 5^2$$
$$p^3 = -27$$
$$2x - 3y = 5$$
$$S = 2\pi r^2 + 2\pi rh$$
$$y - 14 = x^5 + 4x^4 - 12x^3 - x$$

□

In this chapter we will explore equations that involve only one variable. Our primary interest in equations is to find all replacements of the variable that make the equation true. For a one-variable equation, each replacement is a single number; the collection of replacements may be described conveniently using a set. Each such replacement is a **solution** of the equation. **To solve** a one-variable equation means to give the set of all solutions.

In later chapters, we will also find sets of solutions for different kinds of equations; what makes up a solution may differ in each case.

EXAMPLE **a.** Solve $2x = 5$. **a. The solution set is {2.5}.**

b. Solve $x^2 = 16$. **b. The solution set is {−4, 4}.** ■

A fundamental technique used to solve any equation is to transform it into a "simpler" equation without changing the solution set. Just what a simpler equation is may be hard to define precisely; usually it means fewer operations are used.

Definition Two equations that have the same solution set are called **equivalent equations.**

We use the symbol $\Leftrightarrow$ (read "is equivalent to") between two equations to state they are equivalent. We want to find an equivalent equation that we can solve by inspection—that is, just by looking at it.

Illustration $x^2 - 5 = 21$ and $x^2 = 26$ both have solution set $\{-\sqrt{26}, \sqrt{26}\}$ and are equivalent equations. We write $x^3 - 8 = 12 \Leftrightarrow x^3 = 20$ to show $x^3 - 8 = 12$ and $x^3 = 20$ are equivalent equations. □

We begin by restricting our one-variable equations to those with first-degree polynomials or constants as left- and right-hand sides.

Definition Equations whose left- and right-hand sides are first-degree polynomials or constants are called **linear equations.**

The reason for this name is discussed in Chapter 2.

Illustration

$6 = 8 + 7y$	Linear
$3x - 4 = 5(4 - 2x) + x$	Linear
$x^2 - 9 = 0$	Nonlinear; 2nd degree
$\sqrt{3w + 5} = 13$	Nonlinear; nonpolynomial

□

Sometimes the solution set for a linear equation is already obvious. We can see that the solution set for $x + 3 = 9$ is $\{6\}$. However, it is useful to have some techniques for handling more complex equations, such as $6.3y - 1.7(3 - 4.2y) = 9.5 - 4.8y$.

How can we be sure we have equivalent equations without producing the solution set for each? We have two rules from Section A.3 that give guaranteed ways to transform any equation into an equivalent one.

Rule 1.
Adding the same number to each side of an equation produces an equivalent equation.

Illustration

$$\begin{aligned} & 4x - 13 = 71 \\ \Leftrightarrow\quad & 4x - 13 + 13 = 71 + 13 \\ \Leftrightarrow\quad & 4x = 84 \end{aligned}$$

The solution set is $\{21\}$. □

Rule 2.
Multiplying both sides of an equation by the same nonzero number yields an equivalent equation.

Illustration

$$0.5x = -1.44$$
$$\Leftrightarrow \quad 2(0.5x) = 2(-1.44)$$

The solution set is $\{-2.88\}$. □

Note: Multiplying each side by 0 does not guarantee an equivalent equation.

Illustration

$3x = 9$ has solution set $\{3\}$.
$0(3x) = 0(9)$ has solution set $\mathbb{R}$, the reals. □

One common way to obtain a simpler equation is by undoing operations. For a linear equation this means using additive and multiplicative inverses. The order in which operations are evaluated requires that multiplications and divisions be done before additions and subtractions. When we undo, we generally work in reverse order; we undo additions before multiplications. As with evaluations it is possible and sometimes convenient to change the natural order, but reversing the natural order of operations always works.

EXAMPLE Solve $3.7x - 8 = 4.1x + 5$.
We have no preference for addition over subtraction, so we may choose either one to undo first. We add 8 to each side to undo the subtraction.

$$3.7x - 8 + \mathbf{8} = 4.1x + 5 + \mathbf{8}$$
$$\Leftrightarrow \quad 3.7x = 4.1x + 13$$

What action now would give us fewer operations? One is to subtract $4.1x$ from both sides in order to undo the addition. Remember, this is the same as adding its opposite, so Rule 1 applies. Now we have

$$3.7x + (\mathbf{-4.1x}) = 4.1x + (\mathbf{-4.1x}) + 13$$
$$\Leftrightarrow \quad -0.4x = 13$$

All additions and subtractions have been removed; only a multiplication by -0.4 remains. Dividing both sides by this number produces $x = -32.5$.
The solution set is $\{-32.5\}$. ■

The preceding example is done in detail to establish the idea of undoing operations. Often, some of the steps in solving an equation can be done in your head. If you can carry out steps without writing them down, fine. Also, you should not write additional steps whenever you can recognize the solution set. Nothing is gained by changing $4x = 8$ to $x = 2$ on paper.

Here are two last comments on this method of solving equations. If there are operations that can be carried out on one side of an equation, it may be a good idea to do such tidying up first. Secondly, be flexible in applying the guidelines just given.

EXAMPLE Solve $4 - 2(3x + 5x) + x - 7 = x - 3 + (9x - x)(-2)$.

$$
\begin{aligned}
& 4 - 2(3x + 5x) + x - 7 = x - 3 + (9x - x)(-2) && \text{Combine like terms.} \\
\Leftrightarrow \quad & 4 - 2(8x) + x - 7 = x - 3 + (8x)(-2) && \text{Multiply.} \\
\Leftrightarrow \quad & 4 - 16x + x - 7 = x - 3 - 16x && \text{Combine like terms.} \\
\Leftrightarrow \quad & -3 - 15x = -15x - 3
\end{aligned}
$$

But the left-hand side is the same expression as the right-hand side, so any real number is a solution.

The solution set is $\mathbb{R}$. We could write

$$4 - 2(3x + 5x) + x - 7 \equiv x - 3 + (9x - x)(-2)$$ ■

EXAMPLE Solve $\sqrt{29}(x - 13) = 58$.

$$
\begin{aligned}
& \sqrt{29}(x - 13) = 58 \\
\Leftrightarrow \quad & \frac{1}{\sqrt{29}} \cdot \sqrt{29}(x - 13) = \frac{1}{\sqrt{29}} \cdot 58 \\
\Leftrightarrow \quad & x - 13 = \frac{58}{\sqrt{29}}
\end{aligned}
$$

The solution set is $\left\{13 + \frac{58}{\sqrt{29}}\right\}$.

Note: Undoing the last operation performed, multiplication by $\sqrt{29}$, is more efficient than carrying out this multiplication and then undoing the remaining operations. ■

EXAMPLE Solve $\frac{2x - 3}{4} = \frac{2x - 3}{4} + 5$.

Don't forget to be flexible and, most importantly, to look at the statement. No real number is equal to the sum of itself and 5.
The solution set is { }, the empty set. ■

Some equations containing rational expressions are equivalent or almost equivalent to linear equations. We must use special care applying Rule 2. If the multiplier we choose is not a constant, we may unwittingly multiply by zero. For this reason, be sure to list any restrictions necessary to make the multiplier nonzero.

EXAMPLE Solve $\frac{5}{x - 3} = \frac{2}{x + 1}$.

$$\frac{5}{x - 3} = \frac{2}{x + 1}$$

Multiply both sides by $(x - 3)(x + 1)$ to undo the divisions.

$$\Leftrightarrow \quad 5(x + 1) = 2(x - 3), \qquad x \neq 3 \text{ and } x \neq -1$$

Note: Replacing x with either 3 or -1 would make the multiplier zero.

$$\Leftrightarrow \quad 5x + 5 = 2x - 6, \qquad x \neq 3 \text{ and } x \neq -1$$

$$\Leftrightarrow \quad 3x = -11, \qquad x \neq 3 \text{ and } x \neq -1$$

So, $\frac{-11}{3}$ is a candidate for a solution, and it satisfies all the conditions on x.
The solution set is $\{-\frac{11}{3}\}$. ■

EXAMPLE Solve $\dfrac{2x}{x - 2} = \dfrac{4}{x - 2}$.

$$\frac{2x}{x - 2} = \frac{4}{x - 2}$$

Multiply both sides by $x - 2$ to undo the divisions.

$$\Leftrightarrow \quad 2x = 4, \qquad x \neq 2$$

$$\Leftrightarrow \quad x = 2, \qquad x \neq 2$$

No replacement for x can satisfy both of these requirements.
The solution set is $\{\ \}$. ■

Let us look at another way of rewriting an equation that preserves its solution set. Consider $|2x - 5| = 12$. Adding 5 to both sides will not undo the subtraction because of the order of operations. In a sense, we cannot get our hands on the operations protected by the absolute value operation. Is there an operation that reverses taking the absolute value as division reverses multiplication?

We rewrite by using the definition of **absolute value.** How can the absolute value of a number be 12? This happens only if the number is -12 or 12. Thus the equation $|2x - 5| = 12$ is equivalent to the compound statement $2x - 5 = -12$ or $2x - 5 = 12$. Each part of the compound statement is solved to find the complete solution set.

$$|2x - 5| = 12$$

$$\Leftrightarrow \quad 2x - 5 = -12 \quad \text{or} \quad 2x - 5 = 12$$

$$\Leftrightarrow \quad 2x = -7 \quad \text{or} \quad 2x = 17$$

The solution set is $\{-3.5, 8.5\}$.

EXAMPLE **a.** Solve $\sqrt{3}|\sqrt{6} - 2x| = \sqrt{18}$.

b. Approximate the solutions of $\sqrt{3}|\sqrt{6} - 2x| = \sqrt{18}$ to three decimal places.

a. $$\sqrt{3}|\sqrt{6} - 2x| = \sqrt{18}$$

Divide by $\sqrt{3}$ to undo the multiplication.

$$\Leftrightarrow \quad |\sqrt{6} - 2x| = \sqrt{6}$$

$$\Leftrightarrow \quad \sqrt{6} - 2x = \sqrt{6} \quad \text{or} \quad \sqrt{6} - 2x = -\sqrt{6}$$

$$\Leftrightarrow \quad -2x = 0 \quad \text{or} \quad -2x = -2\sqrt{6}$$

The solution set is $\{0, \sqrt{6}\}$.

b. **The solutions are 0 and the approximation of $\sqrt{6}$ to three decimal places 2.449.** ■

EXAMPLE Solve $|7x - 19| = -3$.

Use your common sense. How can the absolute value of a number be -3? There is no replacement for x that will do this.

The solution set is { }. ■

Equations are used to model or describe real-world situations. As an example here is an ancient riddle about Diophantus, a Greek mathematician.

He was a boy for one-sixth of his life, finished school after another twelfth of his life, married a seventh of his life later, and after five years had a son. This son died after living half of his father's (full) age. Diophantus lived in grief for four more years and then died. How old was Diophantus when he died?

We begin by choosing the variable A to represent the unknown age. Next, each English phrase is translated into an algebraic equivalent.

$$\frac{1}{6}A + \frac{1}{12}A + \frac{1}{7}A + 5 + \frac{1}{2}A + 4 = A$$

Now we have an equation that represents the information given by the riddle, and we solve it to find that Diophantus' age at his death was 84 years.

Translation into algebraic expressions may be the most difficult part of handling these applications of algebra. Most phrases have straightforward equivalents.

Illustration

English expression	*Algebraic expression*
Five more than the number required	$5 + n$
Six years older than Jack	$6 + J$
Twice as many pounds	$2z$
The sum of 7 and a number	$(7 + n)$ □

There are several phrases that often present problems. You should make special note of them.

EXAMPLE Translate the following English phrase into an algebraic expression.

12 less than an unknown number $\quad$ $\boldsymbol{x - 12}$

Note the changed order in the algebraic expression. ■

EXAMPLE Translate the following English sentence into an algebraic equation.

There are five times as many quarters as dimes.

If we let d represent the number of dimes and q represent the number of quarters, q represents a larger number than d because there are more quarters than dimes. "Five times as many" refers to a multiplication, so we have $\boldsymbol{q = 5d}$. The smaller number is multiplied by 5 to produce the larger number. ■

EXAMPLE Translate the English sentence into an algebraic equation.

Twice the sum of 7 and a number is 3 less than the number.

Letting n represent the number, **the translation is $2(7 + n) = n - 3$.** ■

Sometimes translations are easier if several variables are used. In the following examples we explore how to modify our methods of solving one-variable equations to handle this situation. With practice you may not need a variable for each unknown quantity, but variables are cheap, so don't worry about using as many as *you* need.

EXAMPLE In checking the receipts for the evening at the movie theater, the manager notes that \$2310.75 were taken in, but the tally of the number of adult and children's tickets has been lost. (It's been one of those nights—the popcorn machine broke, too.) He remembers correctly there were a total of 575 tickets and he knows that the price of an adult ticket is \$4.50 and the price of a child's ticket is \$2.25. Can he determine how many of each were sold and complete his report?

Let A represent the number of adult tickets sold
C represent the number of children's tickets sold

Then $A + C = 575$ is a translation of the statement about the total number of tickets sold, and $4.50A + 2.25C = 2310.75$ is a translation of the statement about the amount of receipts. The problem with this translation of English sentences into mathematical statements is that there are two equations, not one. Rewrite $A + C = 575$ as $A = 575 - C$. Now we have a statement about A that we can substitute, or "patch," into the second equation.

$$
\begin{aligned}
& 4.50(575 - C) + 2.25C = 2310.75 \\
\Leftrightarrow\quad & 2587.50 - 4.50C + 2.25C = 2310.75 \\
\Leftrightarrow\quad & 2587.50 - 2.25C = 2310.75 \\
\Leftrightarrow\quad & -2.25C = -276.75 \\
\Leftrightarrow\quad & C = \frac{-276.75}{-2.25} \text{ or } 123
\end{aligned}
$$

Since $A = 575 - C$, $575 - 123$, or 452, adult tickets were sold. The final step is to give a complete statement that answers the original question.

Yes, the manager is able to determine that 452 adult tickets and 123 children's tickets were sold. ■

EXAMPLE Tommy's dad says that he will give Tommy 10¢ for each homework problem he does correctly and take 5¢ from Tommy if the problem is wrong. Tommy completes a set of 33 problems and neither one owes the other. How many problems did Tommy get right?

Let r be the number of problems right, and w be the number of problems wrong.

$$10r - 5w = 0 \qquad \text{The amount of money Tommy gets}$$

$$r + w = 33 \Leftrightarrow w = 33 - r \qquad \text{Number of problems}$$

Substitution from the equivalent form of the second equation into the first equation.

$$10r - 5(33 - r) = 0$$

$$\Leftrightarrow \quad 10r - 165 + 5r = 0$$

$$\Leftrightarrow \quad 15r = 165$$

Tommy got 11 problems right. ■

In many applications some information must be brought to the problem. For instance, in the next example we need to know the relationship between distance, rate, and time: $d = rt$. The toolkit contains any such outside information we need in this section.

EXAMPLE The racehorse Windswept heads into the home stretch of the track 16 yards ahead of Spencer's Beauty. Windswept maintains a constant speed of 20 miles per hour, and Spencer's Beauty increases her speed to 25 miles per hour. They cross the finish line at exactly the same time. Approximately how long is the home stretch to the nearest hundredth of a furlong?

We need to use $d = rt \Leftrightarrow t = d/r$. If r is measured in miles per hour, then d should be measured in miles. Since 1 mile ≡ 8 furlongs and 1 furlong ≡ 220 yards,

$$16 \text{ yards} \equiv 16 \text{ yards} \cdot \frac{1 \text{ furlong}}{220 \text{ yards}} \cdot \frac{1 \text{ mile}}{8 \text{ furlongs}} \text{ or } \frac{1}{110} \text{ mile}.$$

Let d be the length of the homestretch in miles that Windswept travels; then $d + \frac{1}{110}$ is the distance in miles Spencer's Beauty travels.

The time for Windswept to finish the race is the same as the time for Spencer's Beauty to finish the race.

$$\frac{d}{20} = \frac{d + \frac{1}{110}}{25}$$

$$\Leftrightarrow \quad 25d = 20d + \frac{2}{11}$$

$$\Leftrightarrow \quad 5d = \frac{2}{11}$$

$$\Leftrightarrow \quad d = \frac{2}{55}$$

We were asked to give the distance in furlongs.

$$\frac{2}{55}\text{ miles} = \frac{2}{55}\text{ miles} \cdot \frac{8\text{ furlongs}}{1\text{ mile}} = \frac{16}{55}\text{ furlongs} \approx 0.29\text{ furlong}$$

The length of the homestretch is approximately 0.29 furlong. ■

Here are some guidelines for working application problems.

1. Read the problem carefully. Be sure you understand what is discussed. Find any formula needed.
2. Choose a variable for each unknown quantity, if necessary. Identify each one; be sure they represent numbers.
3. Translate the English phrases into algebraic expressions. Remember that, in general, each clause or sentence will translate into an equation.
4. Each equation should represent a statement of facts given by the problem. Read your equations as English statements to be sure they're accurate and sensible translations.
5. Substitute if necessary and solve the equation.
6. Check to see if your answers are reasonable and that you have answered the question asked.

Some common errors in this sort of work are as follows.

1. Using a letter to represent something other than a number.

Illustration S = Sue
You probably needed something such as: S represents Sue's age now. □

2. Translating incorrectly.

Illustration There are half as many diamonds as rubies in her ring, translated incorrectly as $\frac{1}{2}d = r$. You can check this by considering which letter represents the smaller number. □

3. Mixing types of information.

Illustration A collection of silver and gold coins is worth \$5000, translated incorrectly as $s + g = 5000$. If s represents the number of silver coins and g represents the number of gold coins, this incorrect translation says the total number of coins is equal to an amount of money. □

4. Giving unreasonable answers.

Illustration Counting 13.5 chickens in the barnyard is probably incorrect. □

5. Answering only part of the question.

Illustration An answer of 13 minutes in response to a request for the time and distance required for a trip to the store is only a partial answer. □

EXAMPLE Farmer MacDonald has a total of 100 chickens, which are worth \$100 in all. If the hens are worth \$1 each, the roosters \$5 each, and the baby chickens 5¢, how many of each does she have?

One possibility is that Farmer Mac has 100 hens, each worth a dollar. Are there other possibilities? A direct translation using c for the number of chicks, h for the number of hens, and r for the number of roosters produces two equations, the first about the number of chickens and the second about the value of the chickens.

$$c + h + r = 100 \quad \Leftrightarrow \quad h = 100 - r - c$$
$$0.05c + 1h + 5r = 100$$

Making the substitution from the equivalent version of the first equation into the second gives

$$\begin{aligned} & 0.05c + 1(100 - r - c) + 5r = 100 \\ \Leftrightarrow \quad & -0.95c + 4r + 100 = 100 \\ \Leftrightarrow \quad & 4r - 0.95c = 0 \\ \Leftrightarrow \quad & 4r = .95c \end{aligned}$$

If we knew nothing else about the situation, we would be at a dead end, because this equation has two variables. However, we can rewrite the equation as

$$\begin{aligned} & 4r = \frac{95}{100}c \\ \Leftrightarrow \quad & r = \frac{95}{400}c \\ \Leftrightarrow \quad & r = \frac{19}{80}c \end{aligned}$$

Then we can read this equation as "The number of roosters is $\frac{19}{80}$ the number of chicks." Since the number of roosters must be a whole number, the number of chicks must be divisible by 80. There is only one number between 0 and 100 that is divisible by 80, and that is 80. Let's try it and see if it works as the number of chicks. If $c = 80$, $r = \frac{19}{80}(80)$ or 19, and $h = 100 - 80 - 19$, or 1.

Farmer MacDonald either has 80 baby chicks, 19 roosters, and 1 hen worth a total of \$100, or she has just 100 hens. ■

EXERCISES 1.1

***Note:* Exercises should do more than give practice in basic skills. Along with problems that give you experience with the required skills, each exercise set contains problems designed to extend the ideas of the section, to preview material, or to give your creativity a little exercise.**

In exercises 1 through 24, solve the equations.

1. $7x - 3(x - 2) = 8x + 5$

2. $y = 5 + 2(7y + 2)$

3. $3.2(a + 4) - 1.62 = 3.47a - 11$
(Also give the approximate solution correct to three decimal places.)

4. $3.2[a + 4] = 11$

5. $\dfrac{3c - 4}{c + 2} = 3 - \dfrac{10}{c + 2}$

6. $\dfrac{c - 3}{5} = \dfrac{c + 3}{4} + 1$
(Also give the approximate solution correct to three decimal places.)

7. $\dfrac{5x - 2.1}{3.7} = x$
(Also give the approximate solution correct to three decimal places.)
8. $4.3y = \dfrac{7 + 4y}{2.1} - \dfrac{y}{5}$
9. $\sqrt{2}x + 2\{x - \sqrt{7}\} = 1.3$
(Also give the approximate solution correct to three decimal places.)
10. $\sqrt[3]{7}w = \pi \cdot w - 8$
(Also give the approximate solution correct to three decimal places.)
11. $\dfrac{7}{x + 3} = \dfrac{4x}{x + 3}$
12. $\dfrac{7}{x + 3} = \dfrac{4.5}{x + 3}$
13. $\dfrac{5x}{x - 4} - 5 = \dfrac{4}{x + 1}$
14. $\dfrac{3}{2x - 1} + 2 = \dfrac{4x}{2x + 1}$
15. $|x - 5| = 17$
16. $|2x + \sqrt{3}| = 6$
(Also give the approximate solution correct to three decimal places.)
17. $|1.5x| = 0$
18. $|15x - 115| = 0$
19. $|3x + 1| = -1$
20. $-|3 - x| = 18$
21. $-|3 - x| = 0$
22. $-2|4 + 3x| = -8$
23. $|5x - 16| = |2x + 2|$
24. $|x| = \left|-0.6 + 3x - \dfrac{10x - 3}{5}\right|$

In exercises 25 through 28, solve and then write the solutions in standard form for complex numbers. Remember, $i^2 = -1$.

25. $(4 - 3i)x = 7 + i$
26. $(2 + 8i)y - (5 - 3i)y = -21$
27. $\dfrac{w}{6 - 3i} - 2 = \dfrac{5w}{6 - 3i}$
28. $\dfrac{3 + 5i}{x - 2} = \dfrac{4 - i}{x + 6}$

In exercises 29 through 32, state whether the given equations are equivalent.

29. $x = 3$ and $7x - 8 = 13$
30. $y = \sqrt{2}$ and $y = 1.41421$
31. $w = 1.3$ and $\dfrac{w}{20w - 26} = \dfrac{1.3}{20w - 26}$
32. $z = |8|$ and $z = 8$

In exercises 33 through 45, solve the given word problems.

33. In order to get the best interest rate possible for their \$10,000 car loan, Jane and Jim make several calls and narrow their choices to the following: 9.7% compounded monthly for 49 months, 10% compounded semiannually for 4 years, and 9.9% compounded quarterly for 4 years. They find that the compound interest formula is

$$A = P\left[1 + \frac{r}{m}\right]^n$$

where

A is the final amount
P is the principal
r is the interest rate in decimal form
m is the number of compounding periods per year
n is the total number of compounding periods

Use your calculator to determine which of the three plans they should choose in order to pay the least amount of interest.
34. Use the simple interest formula to give the simple interest rate that would produce the same amount of interest as each of the compound interest plans in exercise 33.
35. Luis likes to tease his sister Rosa, who is 5 years younger than he is. "When are you going to get married?" asks Rosa. "When five times my age is six times your age," replies Luis. When does Luis plan to marry?
36. A salesperson tries to convince Tom to purchase a TV with a price that has been reduced by 33%. The reduced price is \$294.75. The salesperson says that this is a great deal because the TV originally sold for more than \$500. Is the salesperson being truthful? If not, what was the price of the TV originally?
37. A Norman window is built using a rectangular window design surmounted with a semicircular window. The perimeter of the Norman window is 26.6 feet and the length of the rectangular section is $\frac{5}{3}$ the width of the window. What are the approximate dimensions of the Norman window correct to one decimal place?
38. An interior decorator plans to paint a 5-inch border around eight Norman windows like the ones described in exercise 37. She places an order for 2 quarts of paint. The contractor tells the decorator that she has not ordered enough paint. Has the decorator placed an incorrect order? (Assume a quart of paint will cover 50 square feet of surface area.)
39. Helena's great-aunt Mary has promised her \$3 for every A she makes on a test this term and a bonus of \$5 for each set of 10 A's she makes after the first five. There is a catch, though; Helena must pay Aunt Mary \$4 for any test grade below A. After taking 34 tests, Helena collects \$49. How many A's did she make?
40. Jackie is designing a ring for a client. It must have three times as many carats of rubies as of diamonds. The setting will cost about \$40; Jackie's average cost for rubies is \$147.20 per carat and \$215.30 per carat for dia-

monds. To the nearest tenth of a carat, how many carats will there be in the ring if her client can pay $500?

41. What relationship must hold between the radius and height of a can for the number of units in the volume to equal the number of units in the surface area? (See the exercises of Section A.1 for the formulas.)

42. Barbara and Jim plan a canoe trip for Saturday on a river they have not navigated before. The information provided by the parks service indicates that the entire trip should take $5\frac{1}{2}$ hours and cover 12.5 miles. The first $\frac{3}{4}$ of the trip is on rapidly moving water with few serious hazards. They are experienced canoeists and estimate they can average 3 miles per hour for this stretch. The last $\frac{1}{4}$ of the trip is over the backwaters of a lake with a headwind most of the time. How fast will Barbara and Jim have to paddle on the lake to be at the pick-up point in $5\frac{1}{2}$ hours? This is very important because Barbara's brother is an impatient sort and may not wait too long for them.

43. The chemistry lab this week requires 14 ounces of a 26% acid solution. Dot, the lab assistant, finds a gallon of 40% acid solution in the stock room and a gallon of 10% acid solution in the lab. (a) How much of each should she mix together to produce 14 ounces of the 26% solution? (b) The teacher comes in after Dot has made her calculations but before she actually mixes the solutions. She tells Dot that she has been saving the 10% solution for a demonstration and that is why she had removed it from the stock room. Dot then decides to mix water and the 40% solution to produce a 26% solution. How much water and how much of the 40% solution should she use?

44. Mr. Green runs a nursery and decides to introduce a large variety of rhododendron plants to his inventory. He buys 500 plants, 50 of each of 10 different types. He decides to advertise with a special offering of half-price on a group of 10 plants (one of each type). He limits this offer to the first seven purchasers. If Mr. Green paid $2769 for the shipment of 500 plants and wants to clear $800 in profit, what should he set as the full price of each plant? By the way, in the nursery business assume that $\frac{1}{10}$ of the plants will not be sold because of disease or drought.

45. Farmer MacDonald's chicks have grown up; she now has 81 hens. Each hen lays an average of $\frac{7}{9}$ egg per day. Every three days she ships all the eggs on hand, packed in 30-egg flats. How many flats does she ship at a time?

1.2 Quadratic Equations

Toolkit

For this section you may wish to review complex numbers, Section A.3; radicals, Section A.4; and rules for equations, Section 1.1. Some area formulas needed for the exercises are $A = lw$ for a rectangle and

$$A = \left[\frac{b_1 + b_2}{2}\right]h \text{ for a trapezoid.}$$

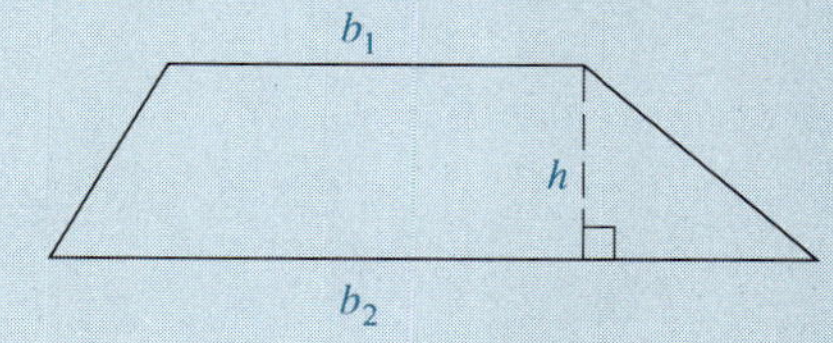

Suppose we are asked to find the dimensions of a room given the following information. The length of the room is 3 feet less than twice its width, and the area is 125 square feet. We follow the guidelines in Section 1.1.

1. We probably need the area formula, $A = lw$.
2. Let l represent the length and w represent the width, both measured in feet.
3. $l = 2w - 3$ The length is 3 less than twice the width.

4. $lw = 125$ The area is 125 square feet.
5. l can be replaced by $2w - 3$, so
$(2w - 3)w = 125$

Now we have a small difficulty; this equation is not linear, nor is it equivalent to a linear equation.

Definition An equation involving polynomial expressions in one variable of degree 2 but no higher is called a **quadratic equation.** Any quadratic equation can be written in the form $ax^2 + bx + c = 0$, $a \neq 0$.

This section presents techniques for solving quadratic equations. There are some quadratic equations that can be solved by inspection.

Illustration The solution set of $x^2 - 25 = 0$ is $\{-5,5\}$.
The solution set of $x^2 = 17$ is $\{-\sqrt{17}, \sqrt{17}\}$. □

As in Section 1.1 we need methods for transforming more complex equations into simpler ones. The technique of undoing operations can provide some help in solving these equations.

Illustration The equation $5x^2 = 20$, which is equivalent to $x^2 = 4$, has solution set $\{-2, 2\}$. □

Illustration The equation $12x^2 - 31 = 8$ is equivalent to $12x^2 = 39$, which is equivalent to $x^2 = \frac{13}{4}$; this equation has solution set

$$\left\{-\frac{\sqrt{13}}{2}, \frac{\sqrt{13}}{2}\right\}$$ □

Illustration

$$\begin{aligned} 7 - 3x^2 &= 34 \\ \Leftrightarrow \quad -3x^2 &= 27 \\ \Leftrightarrow \quad x^2 &= -9 \end{aligned}$$

The solution set is $\{-3i, 3i\}$. □

Illustration

$$\begin{aligned} -2(x - 1)^2 - 3 &= 15 \\ \Leftrightarrow \quad -2(x - 1)^2 &= 18 \\ \Leftrightarrow \quad (x - 1)^2 &= -9 \end{aligned}$$

$$\Leftrightarrow \quad x - 1 = 3i \quad \text{or} \quad x - 1 = -3i$$

The solution set is $\{1 + 3i, 1 - 3i\}$. □

In each of these examples we have finished by solving equations of the form $x^2 = A$ by inspection. Each of the examples began with an expression in the form $a[x - h]^2 + k = 0$. It appears that unravelling works if we have an equation of this form. Let's consider an equation not in this form, $x^2 + 5x = 6$. The third rule of equations in Section A.3 suggests another possible method.

Rule 3.
The product of numbers is zero if and only if one of the numbers is zero.

We can rewrite $x^2 + 5x = 6$ to obtain $x^2 + 5x - 6 = 0$ and factor the left-hand side, to have $(x + 6)(x - 1) = 0$. Rule 3 says $x + 6 = 0$ or $x - 1 = 0$, so the solution set for $x^2 + 5x = 6$ is $\{-6, 1\}$.

EXAMPLE Solve $11x^2 - 490x - 3000 = 0$.

$$11x^2 - 490x - 3000 = 0$$
$$\Leftrightarrow \quad (11x + 60)(x - 50) = 0$$

The solution set is $\left\{-\frac{60}{11}, 50\right\}$. ■

We still need a method for handling quadratic equations that cannot be factored easily or are not in the form $a(x - h)^2 + k = 0$. This technique for dealing with quadratic equations begins by transforming any quadratic equation, $ax^2 + bx + c = 0$, $a > 0$, into a form we know how to solve, $a(x - h)^2 + k = 0$. The heart of the matter is to produce a perfect square, $x^2 - 2hx + h^2$, or $(x - h)^2$.

EXAMPLE What needs to be added to the given expression to create a perfect square?

Expression	*Coefficient of x*	*Added term*	*Perfect square*
$x^2 + 16x$	16	**64**	$x^2 + 16x + 64$, or $(x + 8)^2$
$x^2 - 6x$	-6	**9**	$x^2 - 6x + 9$, or $(x - 3)^2$
$x^2 + 5x$	5	$\frac{25}{4}$	$x^2 + 5x + \frac{25}{4}$, or $\left[x + \frac{5}{2}\right]^2$
$x^2 - \frac{3}{2}x$	$-\frac{3}{2}$	$\frac{9}{16}$	$x^2 - \frac{3}{2}x + \frac{9}{16}$, or $\left[x - \frac{3}{4}\right]^2$
$x^2 + mx$	m	$\left[\frac{m}{2}\right]^2$	$x^2 + mx + \frac{m^2}{4}$, or $\left[x + \frac{m}{2}\right]^2$
$x^2 + \frac{m}{n}x$	$\frac{m}{n}$	$\left[\frac{m}{2n}\right]^2$	$x^2 + \frac{m}{n}x + \frac{m^2}{4n^2}$, or $\left[x + \frac{m}{2n}\right]^2$
$x^2 + 2hx$	$2h$	h^2	$x^2 + 2hx + h^2$, or $(x + h)^2$
$x^2 - 2hx$	$-2h$	h^2	$x^2 - 2hx + h^2$, or $[x - h]^2$

■

Let's write a sequence of equivalent equations beginning with $ax^2 + bx + c = 0$, $a > 0$, and ending with an equation of the form $a(a - h)^2 + k = 0$.

$$ax^2 + bx + c = 0$$

Our method of constructing a perfect square needs x^2 as a first term.

$$\Leftrightarrow \quad a\left[x^2 + \frac{b}{a}x \qquad \right] + c = 0$$

The left-hand side must be equivalent to the previous left-hand side.

$$\Leftrightarrow \quad a\left[x^2 + \frac{b}{a}x + \left(\frac{b}{2a}\right)^2\right] - a\left[\frac{b}{2a}\right]^2 + c = 0$$

Factor the perfect square.

$$\Leftrightarrow \quad a\left[x+\frac{b}{2a}\right]^2-\frac{b^2}{4a}+c=0$$

Combine the last two terms into a single term.

$$\Leftrightarrow \quad a\left[x+\frac{b}{2a}\right]^2-\left[\frac{b^2-4ac}{4a}\right]=0$$

We have the desired form; h is $-b/2a$ and k is $-(b^2-4ac)/4a$. Now we use our factoring technique to solve this equation.

$$\Leftrightarrow \quad a\left[x+\frac{b}{2a}\right]^2-\left[\frac{b^2-4ac}{4a}\right]=0$$

Divide each term by a.

$$\Leftrightarrow \quad \left[x+\frac{b}{2a}\right]^2-\left[\frac{b^2-4ac}{4a^2}\right]=0$$

Factor as the difference of squares.

$$\Leftrightarrow \quad \left[\left[x+\frac{b}{2a}\right]-\sqrt{\frac{b^2-4ac}{4a^2}}\right]\left[\left[x+\frac{b}{2a}\right]+\sqrt{\frac{b^2-4ac}{4a^2}}\right]=0$$

Either factor may be zero.

$$\Leftrightarrow \quad x+\frac{b}{2a}=\sqrt{\frac{b^2-4ac}{4a^2}} \quad \text{or} \quad x+\frac{b}{2a}=-\sqrt{\frac{b^2-4ac}{4a^2}}$$

$$\Leftrightarrow \quad x=-\frac{b}{2a}+\sqrt{\frac{b^2-4ac}{4a^2}} \quad \text{or} \quad x=-\frac{b}{2a}-\sqrt{\frac{b^2-4ac}{4a^2}}$$

These solutions can be simplified using our experience with radicals and rational expressions and the fact that a is a positive number.

$$-\frac{b}{2a}+\sqrt{\frac{b^2-4ac}{4a^2}}=-\frac{b}{2a}+\frac{\sqrt{b^2-4ac}}{2a} \quad \text{or} \quad \frac{-b+\sqrt{b^2-4ac}}{2a}$$

The other solution is $[-b-\sqrt{b^2-4ac}]/2a$. It is common to see the two solutions written as a single expression:

$$\frac{-b\pm\sqrt{b^2-4ac}}{2a}$$

Because we worked with a generic quadratic equation, we have produced a formula for the solutions of any quadratic equation.

The Quadratic Formula:
The solution set of $ax^2+bx+c=0$, $a\neq 0$, is

$$\left\{\frac{-b+\sqrt{b^2-4ac}}{2a}, \frac{-b-\sqrt{b^2-4ac}}{2a}\right\}$$

Note: the formula does not require $a>0$, although we used this fact in the discussion. (See exercises 41 and 42.)

Illustration Applying the *quadratic formula*, $6x^2+5x-21=0$ has solution set

$$\left\{\frac{-5+\sqrt{5^2-4(6)(-21)}}{2(6)}, \frac{-5-\sqrt{5^2-4(6)(-21)}}{2(6)}\right\}$$

which simplifies to

$$\left\{\frac{-5+\sqrt{529}}{12}, \frac{-5-\sqrt{529}}{12}\right\} = \left\{\frac{-5+23}{12}, \frac{-5-23}{12}\right\} \text{ or } \left\{\frac{3}{2}, \frac{-7}{3}\right\}.$$

Notice that we have rational solutions because $b^2 - 4ac$ is 529, a perfect square. □

Illustration Applying the quadratic formula, $6x^2 + 5x - 20 = 0$ has solution set

$$\left\{\frac{-5+\sqrt{5^2-4(6)(-20)}}{2(6)}, \frac{-5-\sqrt{5^2-4(6)(-20)}}{2(6)}\right\}$$

which simplifies to

$$\left\{\frac{-5+\sqrt{505}}{12}, \frac{-5-\sqrt{505}}{12}\right\}.$$

Notice that we have irrational solutions because $b^2 - 4ac$ is 505, a positive number that is not a perfect square. □

Illustration Applying the quadratic formula, $6x^2 + 5x + 20 = 0$ has solution set

$$\left\{\frac{-5+\sqrt{5^2-4(6)(20)}}{2(6)}, \frac{-5-\sqrt{5^2-4(6)(20)}}{2(6)}\right\}$$

which simplifies to

$$\left\{\frac{-5+\sqrt{455}i}{12}, \frac{-5-\sqrt{455}i}{12}\right\}.$$

Notice that we have complex nonreal solutions because $b^2 - 4ac$ is -455, a negative number. □

From the preceding illustrations, we see that part of the quadratic formula, $b^2 - 4ac$ determines the type of solutions of the quadratic equation $ax^2 + bx + c = 0$, a, b, c rational numbers. The expression $b^2 - 4ac$ is referred to as the *discriminant*.

Definition Given the quadratic equation $ax^2 + bx + c = 0$ with a, b, c rational numbers. If the discriminant, $b^2 - 4ac$, is

1. A perfect square, then the solutions of the equation are rational.
2. Positive and not a perfect square, then the solutions of the equation are irrational.
3. Negative, then the solutions of the equation are complex but not real.

You may have noticed that the first example, $6x^2 + 5x - 21 = 0$, can be solved by factoring. The equation $(2x - 3)(3x + 7) = 0$ has solution set $\{\frac{3}{2}, -\frac{7}{3}\}$. In practice, if you quickly recognize the factors, use the factoring method; otherwise, use the formula. Note that both techniques require that one side of the equation be zero. Until we deal with roots of complex numbers in Chapter 7, quadratic equations will have real coefficients. The following examples show several variations of quadratic equations.

EXAMPLE Solve $|8 - 9x + x^2| = 6$.

$$\begin{aligned} &|8 - 9x + x^2| = 6 \\ \Leftrightarrow\quad & 8 - 9x + x^2 = 6 \quad \text{or} \quad 8 - 9x + x^2 = -6 \\ \Leftrightarrow\quad & x^2 - 9x + 2 = 0 \quad \text{or} \quad x^2 - 9x + 14 = 0 \\ \Leftrightarrow\quad & x = \frac{9 \pm \sqrt{81 - 8}}{2} \quad \text{or} \quad (x - 7)(x - 2) = 0 \\ \Leftrightarrow\quad & x = \frac{9 \pm \sqrt{73}}{2} \quad \text{or} \quad x = 7 \quad \text{or} \quad x = 2 \end{aligned}$$

The solution set is $\left\{\frac{9 + \sqrt{73}}{2}, \frac{9 - \sqrt{73}}{2}, 2, 7\right\}$. ■

EXAMPLE Find the approximate solutions of

$$\frac{4}{x - 3} - \frac{7x}{x + 5} = 1$$

correct to four decimal places.

$$\begin{aligned} & \frac{4}{x - 3} - \frac{7x}{x + 5} = 1 \\ \Leftrightarrow\quad & 4[x + 5] - 7x[x - 3] = [x - 3][x + 5], && x \neq 3 \text{ and } x \neq -5 \\ \Leftrightarrow\quad & 4x + 20 - 7x^2 + 21x = x^2 + 2x - 15, && x \neq 3 \text{ and } x \neq -5 \\ \Leftrightarrow\quad & 0 = 8x^2 - 23x - 35, && x \neq 3 \text{ and } x \neq -5 \\ \Leftrightarrow\quad & x = \frac{23 \pm \sqrt{(-23)^2 - 4(8)(-35)}}{2(8)}, && x \neq 3 \text{ and } x \neq -5 \\ \Leftrightarrow\quad & x = \frac{23 \pm \sqrt{1649}}{16}, && x \neq 3 \text{ and } x \neq -5 \end{aligned}$$

The approximate solutions are -1.1005 and 3.9755. ■

At the beginning of this section we left an equation to be solved, $(2w - 3)w = 125$. This equation is equivalent to $2w^2 - 3w = 125$, or $2w^2 - 3w - 125 = 0$. The exact solutions are

$$\frac{3 + \sqrt{1009}}{4} \quad \text{and} \quad \frac{3 - \sqrt{1009}}{4}.$$

The last step for an application problem is to see if your answers are reasonable. The width of the room cannot be

$$\frac{3 - \sqrt{1009}}{4} \text{ feet,}$$

because this is a negative number of feet. However, the width of the room can be $(3 + \sqrt{1009})/4$ because this is approximately 17 feet. We were asked to find the length and width of a room given certain information. The width is

$$\frac{3 + \sqrt{1009}}{4} \text{ feet}$$

so the length is

$$\left[2\left(\frac{3 + \sqrt{1009}}{4}\right) - 3\right] \text{ feet,} \quad \text{or} \quad -\frac{3}{2} + \frac{\sqrt{1009}}{2} \text{ feet}$$

We do not give approximations for these numbers because we do not know what accuracy is needed.

In developing the quadratic formula, we found a way to create perfect squares. This turns out to be useful in Chapter 2.

EXAMPLE Create a perfect square whose first two terms are $x^2 + 7x$.

$$x^2 + 7x + \left[\frac{7}{2}\right]^2 \text{ or } \left[x + \frac{7}{2}\right]^2$$

■

EXAMPLE Rewrite $3x^2 + 8x$ in the form $a(x - h)^2 + k$.

Needed to make a perfect square
↓

$$3\left[x^2 + \frac{8}{3}x\right] = 3\left[x^2 + \frac{8}{3}x + \left(\frac{4}{3}\right)^2\right] - 3\left(\frac{4}{3}\right)^2$$

↑
Needed to give expression equal to original

$$= 3\left[x + \frac{4}{3}\right]^2 + \frac{-16}{3} = 3\left[x - \left[-\frac{4}{3}\right]\right]^2 + \frac{-16}{3}$$

■

If we know the factors of $ax^2 + bx + c$, we know the solutions of $ax^2 + bx + c = 0$: $a[x - r][x - s] = 0 \Leftrightarrow x = r$ or $x = s$, so the solution set is $\{r, s\}$. In fact, this works in the other direction, too. Knowing the solutions of $ax^2 + bx + c = 0$ gives the factors of $ax^2 + bx + c$. The quadractic formula can always be used to obtain the factors of a quadratic expression. If we solve the corresponding quadratic equation, we can find the factors of the expression.

EXAMPLE Factor $x^2 + 3x + 1$ into linear factors.

We first try to factor in the conventional trial-and-error fashion and get nowhere. Solving the corresponding quadratic equation, $x^2 + 3x + 1 = 0$, will help, because each solution can be used to create a factor.

$$x^2 + 3x + 1 = 0$$

$$\Leftrightarrow \quad x = \frac{-3 \pm \sqrt{9 - 4(1)(1)}}{2(1)}$$

$$\Leftrightarrow \quad x = \frac{-3 \pm \sqrt{5}}{2}$$

The linear factors of $x^2 + 3x + 1$ are

$$x - \frac{-3 + \sqrt{5}}{2} \quad \text{and} \quad x - \frac{-3 - \sqrt{5}}{2}$$

or

$$x + \frac{3 - \sqrt{5}}{2} \quad \text{and} \quad x + \frac{3 + \sqrt{5}}{2}$$

Thus

$$\mathbf{x^2 + 3x + 1 = \left(x + \frac{3 - \sqrt{5}}{2}\right)\left(x + \frac{3 + \sqrt{5}}{2}\right).}$$ ■

Here is an example to show what happens if the coefficient of x^2 is not 1.

EXAMPLE Factor $4x^2 - 3x + 1$ into linear factors.

$$\begin{aligned} 4x^2 - 3x + 1 &= 0 \\ \Leftrightarrow \qquad x &= \frac{3 \pm \sqrt{9 - 16}}{8} \\ \Leftrightarrow \qquad x &= \frac{3 \pm \sqrt{7}i}{8} \end{aligned}$$

So the solution set is $\left\{\frac{3 + \sqrt{7}i}{8}, \frac{3 - \sqrt{7}i}{8}\right\}$.

This means $4x^2 - 3x + 1$ has as factors

$$\left[x - \frac{3 + \sqrt{7}i}{8}\right] \quad \text{and} \quad \left[x - \frac{3 - \sqrt{7}i}{8}\right].$$

Multiplying these, we have

$$\left[x - \frac{3 + \sqrt{7}i}{8}\right]\left[x - \frac{3 - \sqrt{7}i}{8}\right] = \left[\left(x - \frac{3}{8}\right) - \frac{\sqrt{7}}{8}i\right]\left[\left(x - \frac{3}{8}\right) + \frac{\sqrt{7}}{8}i\right],$$

which gives

$$\left[\left(x - \frac{3}{8}\right)^2 - \left(\frac{\sqrt{7}}{8}i\right)^2\right] = \left[x^2 - \frac{3}{4}x + \frac{9}{64} + \frac{7}{64}\right], \text{ or } x^2 - \frac{3}{4}x + \frac{1}{4}.$$

We also need a constant factor of 4 to produce $4x^2 - 3x + 1$.

$$\mathbf{4x^2 - 3x + 1 = 4\left(x - \frac{3 + \sqrt{7}i}{8}\right)\left(x - \frac{3 - \sqrt{7}i}{8}\right)}$$ ■

EXERCISES 1.2

In exercises 1 through 20, solve the given equation.

1. $x^2 - 6x = -8$
2. $\frac{1}{2}x^2 - 3x + 4 = 0$
3. $6y - y^2 = 8$
4. $12x^2 - 5x - 2 = 0$
5. $\frac{1}{5}a^2 + 7a - \frac{2}{5} = 0$
   (Also give the approximate solutions correct to three decimal places.)
6. $3w^2 + 23w + 40 = 0$
   (Also give the approximate solutions correct to three decimal places.)
7. $3x^2 - 23x + 40 = 0$
8. $12x^2 + 168 = 158x$
9. $121x^2 - 88x + 16 = 0$
10. $12x^2 - 5x + 2 = 0$
11. $\frac{1}{13}x^2 + 5 = 0$
12. $5x^2 - 7.2x = 8.5x^2 - 9.1$
    (Also give the approximate solutions correct to two decimal places.)
13. $0.0001x^2 - 0.01x + 100 = 0$
14. $0.32w - 1.31w^2 = 5.7$
15. $\sqrt{5}b^2 - 2\sqrt{5}b + 5\sqrt{5} = 0$
16. $\pi a^2 - [\pi + 1]a = 8$
17. $|x^2 - 3| = 5$
18. $|x^2 - 4x + 3| = 2$
19. $|x^2 + 9| = -1$
20. $3\left|\frac{1}{3}w^2 + \frac{1}{5}w - \frac{1}{2}\right| = 18$

In exercises 21 through 24, write a quadratic equation that has the solution set given. (There is not a unique answer.)

21. $\left\{\frac{7}{3}, -\frac{1}{4}\right\}$
22. $\{2 + \sqrt{2}, 2 - \sqrt{2}\}$
23. $\left\{\frac{1 - 7\sqrt{13}}{8}, \frac{1 + 7\sqrt{13}}{8}\right\}$
24. $\left\{\frac{1 - 7\sqrt{13}i}{8}, \frac{1 + 7\sqrt{13}i}{8}\right\}$

In exercises 25 through 28, use the quadratic formula to factor the quadratic expression.

25. $x^2 - 5x + 1$
26. $x^2 + 5x - 1$
27. $3x^2 - 15x + 2$
28. $2x^2 + \frac{4}{3}x - 10$

In exercises 29 through 32, rewrite the quadratic expression in the form $a[x - h]^2 + k$.

29. $x^2 - 2x + 4$
30. $5x^2 + 2x - 3$
31. $0.7y^2 - 1.1y$
32. $\sqrt{2}a^2 + a + 5$

In exercises 33 through 40, use the discriminant to determine—without solving the equation—whether the solutions of the equation are rational, irrational, or complex and not real.

33. $3x^2 + 12x - 96 = 0$
34. $23.4x^2 - 13x + 11.8 = 0$
35. $\sqrt{2}a^2 + a + 5 = 0$
36. $\sqrt{2}a^2 + a - 5 = 0$
37. $\frac{1}{4} - \frac{2}{3}w - w^2 = 0$
38. $\frac{2}{5} = \frac{5}{3}w - \frac{1}{5}w^2$
39. $3.14y - 3 = 5y^2$
40. $3.14y = 5y^2$
41. Show that using $a = -3$, $b = 5$, $c = 2$ in the quadratic formula, you obtain the solution set for $-3x^2 + 5x + 2 = 0$; that is, check your results from the formula.
42. In developing the quadratic formula we used $a > 0$ to write

$$-\frac{b}{2a} + \sqrt{\frac{b^2 - 4ac}{4a^2}} \quad \text{as} \quad -\frac{b}{2a} + \frac{\sqrt{b^2 - 4ac}}{2a}$$

If we know only $a \neq 0$, then

$$-\frac{b}{2a} + \sqrt{\frac{b^2 - 4ac}{4a^2}} = -\frac{b}{2a} + \frac{\sqrt{b^2 - 4ac}}{2|a|}$$

and

$$-\frac{b}{2a} - \sqrt{\frac{b^2 - 4ac}{4a^2}} = -\frac{b}{2a} - \frac{\sqrt{b^2 - 4ac}}{2|a|}$$

Show that if $a < 0$, the numbers represented by the right-hand sides are

$$\frac{-b + \sqrt{b^2 - 4ac}}{2a} \quad \text{and} \quad \frac{-b - \sqrt{b^2 - 4ac}}{2a}$$

the same results obtained for $a > 0$.

In exercises 43 and 44, solve the equation for z.

43. $3z^2 - az - 2 = 0$
44. $(n + m)z - nz^2 = 4$

In exercises 45 through 50, solve the given word problems. Be sure to state your answers in complete sentences.

45. In a trapezoid, the upper base is twice as long as the lower base and the height is one-third of the lower base. Give the dimensions of the trapezoid if the area is 42 square meters.
46. José remembers that the area of his rectangular garden is 6557 square feet because this number happens to match the last four digits of his phone number. He also remembers that one side is 4 feet longer than the other. Can he determine the dimensions without having to measure again? If so, what are they?

47. The number of minutes Tracy runs is 5 less than the number of miles per hour she runs. What are her rate and time if she covers 5 miles while running? Approximate Tracy's speed to two decimal places. Does Tracy need to give up, speed up, or turn professional?

48. The members of a school club pay a total of \$720 to rent a bus for a field trip. At the last minute 10 members can't go on the trip, and the others are told they will each have to pay \$6 more. How many people actually went on the field trip?

49. Tim's house is 6 miles from the lighthouse as the crow flies. But to reach the lighthouse Tim must go east to the intersection and then turn left on Lighthouse Road. If he travels 3 miles more on Lighthouse Road than from his house to the intersection, approximately (to the nearest tenth) how many miles must Tim drive to visit his friend at the lighthouse? (Assume the roads are perpendicular at the intersection.) (If you need the Pythagorean theorem, see the toolkit for Section 2.1.)

50. Write instructions for a friend to use to create a perfect square whose first two terms are ax^2 and bx.

1.3 Radical Equations

Toolkit

Here we'll need the fact that $[\sqrt[n]{x}]^n = x$, n a natural number, and some facts about fractional exponents from Section A.4.

In the previous two sections we concentrated on polynomial equations. Here we look at one type of nonpolynomial equation. A **radical equation** is an equation that has a radical with a radicand containing a variable. Our idea of undoing operations can help solve such equations when only one type of root is taken in the statement.

Illustration The equations

$$\sqrt[3]{x} = 7$$
$$\sqrt{5z^2 - 8} = -3$$
$$\sqrt{y + 4} = 2y - 15\sqrt{y}$$
$$\sqrt[4]{x^3 - 3x^2 + x} = x + 2 \qquad \text{are all radical equations.}$$

The equation $\sqrt{2}y^2 - 3y = 8$ is not a radical equation. □

In the appendix we show that squaring a number and taking the square root of a number are not operations that undo each other exactly. This is also true for any even power and the same even root. Formally, the solution set of $v = w$ is a subset of the solution set of $v^n = w^n$ for any $n \in N$.

Illustration The equation $x = 2$ has solution set $\{2\}$.
The equation $x^2 = 4$ has solution set $\{-2, 2\}$.
We say $x = 2$ *implies* that $x^2 = 4$ and use the notation $x = 2 \Rightarrow x^2 = 4$. One statement implies another means if the first is true, the second is also true. This is not the same as saying the equations are equivalent. □

The basic technique for solving radical equations is to raise each side of the equation to the kth power if a kth root appears. We must test the real number solutions to this equation in the original radical equation only if k is an even number. If the new equation has a complex, nonreal solution, this number must always be tested in the original radical equation.

EXAMPLE Solve $\sqrt[3]{x} = 7$.

Raising each side to the third power,

$$\sqrt[3]{x} = 7 \Rightarrow [\sqrt[3]{x}]^3 = 7^3 \Rightarrow x = 7^3$$

The solution set is $\{7^3\}$, or $\{343\}$. ■

EXAMPLE Solve $\sqrt{x^2 - 9} = -4$.

Squaring each side,

$$\begin{aligned} & \sqrt{x^2 - 9} = -4 \\ \Rightarrow\quad & (\sqrt{x^2 - 9})^2 = (-4)^2 \\ \Leftrightarrow\quad & x^2 - 9 = 16 \\ \Leftrightarrow\quad & x^2 = 25 \end{aligned}$$

Possible solutions are -5 and 5, but neither makes the original statement true.
The solution set is $\{\ \}$. ■

Note: Being alert and reading the statement before squaring both sides of the equation, you could determine by inspection the solution set is $\{\ \}$ because square roots are never negative numbers.

Note: From Section A.2, $\{\ \}$ is a subset of every set, so the solution of the equation $\sqrt{x^2 - 9} = -4$ is contained in the solution set of $x^2 = 25$.

The following example shows that some difficulties can arise in carrying out the preceding method.

EXAMPLE Solve $3 + \sqrt{x - 2} = x - 5$.

Because a square root is used, we square both sides of the equation.

$$\begin{aligned} & 3 + \sqrt{x - 2} = x - 5 \\ \Rightarrow\quad & [3 + \sqrt{x - 2}]^2 = (x - 5)^2 \\ \Leftrightarrow\quad & 9 + 6\sqrt{x - 2} + (x - 2) = x^2 - 10x + 25 \end{aligned}$$

The result still contains a radical. If we isolate the radical first, a much simpler equation is produced when we square both sides.

$$\begin{aligned} & 3 + \sqrt{x - 2} = x - 5 \qquad \text{Original equation} \\ \Leftrightarrow\quad & \sqrt{x - 2} = x - 8 \\ \Rightarrow\quad & [\sqrt{x - 2}]^2 = (x - 8)^2 \\ \Leftrightarrow\quad & x - 2 = x^2 - 16x + 64 \\ \Leftrightarrow\quad & 0 = x^2 - 17x + 66 \\ \Leftrightarrow\quad & 0 = (x - 11)(x - 6) \end{aligned}$$

This equation has solutions 6 and 11, but because we were not able to produce equivalent equations at each step, we must check these numbers as possible solutions in the original equation.

Possible solution	6
Left-hand side : $3 + \sqrt{x - 2}$	$3 + \sqrt{6 - 2}$, or 5
Right-hand side : $x - 5$	$6 - 5$, or 1

So 6 is not a solution because the left and right sides are not equal.

Possible solution	11
Left-hand side : $3 + \sqrt{x - 2}$	$3 + \sqrt{11 - 2}$, or 6
Right-hand side : $x - 5$	$11 - 5$, or 6

So 11 is a solution.
The solution set is {11}. ■

As a general rule, isolate radicals as much as possible, one to a side of an equation. A particular equation may even then require raising to a power more than once.

EXAMPLE Solve $\sqrt[3]{x - 1} - 1.4 = 2.7$.

$$\begin{aligned} \sqrt[3]{x - 1} - 1.4 &= 2.7 \\ \Leftrightarrow \quad \sqrt[3]{x - 1} &= 4.1 \\ \Rightarrow \quad [\sqrt[3]{x - 1}]^3 &= [4.1]^3 \\ \Leftrightarrow \quad x - 1 &= 68.921 \end{aligned}$$

We do not need to test this real number in the original radical equation.
The solution set is {69.921}. ■

EXAMPLE Solve $\sqrt{2x + 1} - \sqrt{x} = 1$.

$$\begin{aligned} \sqrt{2x + 1} - \sqrt{x} &= 1 \\ \Leftrightarrow \quad \sqrt{2x + 1} &= \sqrt{x} + 1 \\ \Rightarrow \quad (\sqrt{2x + 1})^2 &= (\sqrt{x} + 1)^2 \\ \Leftrightarrow \quad 2x + 1 &= x + 2\sqrt{x} + 1 \\ \Leftrightarrow \quad 2x &= x + 2\sqrt{x} \\ \Leftrightarrow \quad x &= 2\sqrt{x} \\ \Rightarrow \quad x^2 &= 4x \\ \Leftrightarrow \quad x^2 - 4x &= 0 \\ \Leftrightarrow \quad x(x - 4) &= 0 \\ \Leftrightarrow \quad x = 0 \quad &\text{or} \quad x = 4 \end{aligned}$$

We must check these possible solutions.

Possible solution	0	4
LHS: $\sqrt{2x+1} - \sqrt{x}$	$\sqrt{0+1} - \sqrt{0}$ or 1	$\sqrt{8+1} - \sqrt{4}$ or 1
RHS: 1	1	1
Solution	Yes	Yes

The solution set is {0, 4}. ■

EXAMPLE Solve $\dfrac{3}{\sqrt[4]{x}} = 2$.

$$\frac{3}{\sqrt[4]{x}} = 2$$

$$\Rightarrow \quad \left(\frac{3}{\sqrt[4]{x}}\right)^4 = (2)^4$$

$$\Leftrightarrow \quad \frac{81}{x} = 16$$

$$\Leftrightarrow \quad \frac{81}{16} = x, \qquad x \neq 0$$

The possible solution is $\frac{81}{16}$. We must check this because we were not able to produce equivalent equations at each step.

Possible solution	$\frac{81}{16}$
LHS: $\dfrac{3}{\sqrt[4]{x}}$	$\dfrac{3}{\sqrt[4]{\frac{81}{16}}}$ or $\dfrac{3}{\frac{3}{2}}$ or 2
RHS: 2	2
Solution	Yes

The solution set is $\left\{\frac{81}{16}\right\}$. ■

An equation involving a fractional exponent with a variable base is equivalent to a radical equation and can be solved as we just did if only one fractional exponent is used, even if it appears more than once.

EXAMPLE Solve $(x - 2)^{1/5} = 3$.

$$\begin{aligned} & (x - 2)^{1/5} = 3 \\ \Rightarrow\quad & [(x - 2)^{1/5}]^5 = (3)^5 \\ \Leftrightarrow\quad & x - 2 = 243 \end{aligned}$$

The solution set is {243}. ■

EXAMPLE Solve $(x + 1)^{1/2} = 2 - (x + 3)^{1/2}$.

$$\begin{aligned} & (x + 1)^{1/2} = 2 - (x + 3)^{1/2} \\ \Rightarrow\quad & [(x + 1)^{1/2}]^2 = [2 - (x + 3)^{1/2}]^2 \\ \Leftrightarrow\quad & x + 1 = 4 - 4(x + 3)^{1/2} + (x + 3) \\ \Leftrightarrow\quad & -6 = -4(x + 3)^{1/2} \qquad \text{Isolate the radical.} \\ \Leftrightarrow\quad & \frac{3}{2} = (x + 3)^{1/2} \\ \Rightarrow\quad & \left(\frac{3}{2}\right)^2 = [(x + 3)^{1/2}]^2 \\ \Leftrightarrow\quad & \frac{9}{4} = x + 3 \end{aligned}$$

So, $-\frac{3}{4}$ is a possible solution.

Possible solution	$-\frac{3}{4}$
LHS: $[x + 1]^{1/2}$	$[-\frac{3}{4} + 1]^{1/2}$ or $[\frac{1}{4}]^{1/2}$ or $\frac{1}{2}$
RHS: $2 - [x + 3]^{1/2}$	$2 - [-\frac{3}{4} + 3]^{1/2}$ or $2 - \frac{3}{2}$ or $\frac{1}{2}$
Solution	Yes

The solution set is $\{-\frac{3}{4}\}$. ■

Some equations with more than one type of radical also can be solved by unravelling.

EXAMPLE Solve $\sqrt[3]{x - 1 + \sqrt{x}} = -1$.

$$\begin{aligned} & \sqrt[3]{x - 1 + \sqrt{x}} = -1 \\ \Rightarrow\quad & x - 1 + \sqrt{x} = (-1)^3 \\ \Leftrightarrow\quad & \sqrt{x} = -x \\ \Rightarrow\quad & x = x^2 \\ \Leftrightarrow\quad & 0 = x^2 - x \\ \Leftrightarrow\quad & 0 = x(x - 1) \end{aligned}$$

The numbers 0 and 1 have to be checked in the original radical equation.

Possible solution	0	1
LHS: $\sqrt[3]{x-1}+\sqrt{x}$	$\sqrt[3]{0-1}+\sqrt{0}$ or -1	$\sqrt[3]{1-1}+\sqrt{1}$ or 1
RHS: -1	-1	-1
Solution:	Yes	No

The solution set is $\{0\}$. ■

EXAMPLE Solve $\sqrt{2x-1}=\sqrt[4]{3x-2}$.

$$\begin{aligned}
&& \sqrt{2x-1} &= \sqrt[4]{3x-2}\\
&\Rightarrow & (\sqrt{2x-1})^4 &= (\sqrt[4]{3x-2})^4\\
&\Leftrightarrow & (2x-1)^2 &= 3x-2\\
&\Leftrightarrow & 4x^2-4x+1 &= 3x-2\\
&\Rightarrow & 4x^2-7x+3 &= 0\\
&\Leftrightarrow & (4x-3)(x-1) &= 0
\end{aligned}$$

The possible solutions are $\frac{3}{4}$ and 1.

We must check the possible solutions.

Possible solution	$\frac{3}{4}$	1
LHS: $\sqrt{2x-1}$	$\sqrt{2\left(\frac{3}{4}\right)-1}$ or $\sqrt{\frac{1}{2}}$	$\sqrt{2-1}$ or 1
RHS: $\sqrt[4]{3x-2}$	$\sqrt[4]{3\left(\frac{3}{4}\right)-2}$ or $\sqrt[4]{\frac{1}{4}}$ or $\sqrt{\frac{1}{2}}$	$\sqrt[4]{1}$ or 1
Solution	Yes	Yes

The solution set is $\left\{\frac{3}{4}, 1\right\}$. ■

You may have noticed that no complex numbers have been used as coefficients in this section's examples. This is not because the methods fail but because, in general, we do not have the tools with which to check the possible solutions. The next two examples show that we may or may not be able to check possible solutions.

EXAMPLE Solve $i\sqrt{x}=4$.

Squaring both sides gives $-x=16$, or $x=-16$. Thus a possible solution is -16, but we need to check it. Substituting in $i\sqrt{x}$, we have $i\sqrt{-16}$, or $i(4i)$. But $i(4i)$ is -4, not 4.

The solution set is $\{\ \}$. ■

EXAMPLE Solve $i\sqrt{x} = 4 + i$.

$$\begin{aligned} & i\sqrt{x} = 4 + i \\ \Rightarrow\quad & -x = 15 + 8i \\ \Leftrightarrow\quad & x = -15 - 8i \end{aligned}$$

We haven't defined $\sqrt{a + bi}$ yet, so we can't check the possible solution, $-15 - 8i$. This will be done in Section 7.5. ■

Even if the equation has real coefficients, there may be difficulties in checking possible solutions.

EXAMPLE Solve $x + \sqrt{x} = -1$.

$$\begin{aligned} & x + \sqrt{x} = -1 \\ \Leftrightarrow\quad & \sqrt{x} = -1 - x \\ \Rightarrow\quad & (\sqrt{x})^2 = (-1 - x)^2 \\ \Leftrightarrow\quad & x = 1 + 2x + x^2 \\ \Rightarrow\quad & 0 = 1 + x + x^2 \end{aligned}$$

Using the quadratic formula, possible solutions for the original equation are

$$-\frac{1}{2} + \frac{\sqrt{3}}{2}i \quad \text{and} \quad -\frac{1}{2} - \frac{\sqrt{3}}{2}i.$$

These need to be checked, but we do not yet know how to express

$$\sqrt{-\frac{1}{2} + \frac{\sqrt{3}}{2}i} \quad \text{or} \quad \sqrt{-\frac{1}{2} - \frac{\sqrt{3}}{2}i}$$

in standard form. We return to check these in Section 7.5. ■

EXERCISES 1.3

In exercises 1 through 39, solve the given equations.

1. $\sqrt{x - 3} = 5$
2. $\sqrt{2y + 7} = 13$
3. $\sqrt{7x - 8} = -1$
4. $\sqrt{4 + x} = -\frac{1}{2}$
5. $\sqrt[3]{w} = 107$
6. $\sqrt[3]{2x - 3} = 5$
7. $\sqrt[3]{y + 3} = -1$
8. $\sqrt[3]{7 - 2w} = 0.2$
9. $\sqrt[4]{3x + 2} = 8$
10. $\sqrt[4]{5x - 0.3} = 0.1$
11. $\sqrt[4]{3y + 2} = -8$
12. $\sqrt[4]{0.3 - 5x} = -0.1$
13. $\sqrt[135]{-x} = -1$
14. $\sqrt[134]{-u} = -1$
15. $\sqrt{x^2 - 5x + 1} = 1$
16. $\sqrt{x^2 - 5x + 1} = -1$
17. $\sqrt[3]{w^2 - 5w + 1} = 1$
18. $\sqrt[3]{x^2 - 5x + 1} = -1$
19. $\sqrt{2x^2 - 4x} = x + 1$
20. $\sqrt{x^3 + 1} = x + 1$
21. $\sqrt[3]{z^2 + 2z + 3} = z - 1$
22. $\sqrt{2v + 3} = \sqrt{v + 1}$
23. $\sqrt{2x - 3} = \sqrt{x + 1}$
24. $\sqrt{2x - 3} = 5 - \sqrt{3x}$
25. $\sqrt{2y} = \sqrt{y + 1} - 3$
26. $\sqrt{x - 1} = \sqrt[4]{x^2}$
27. $\sqrt[6]{3x^2} = \sqrt[3]{x + 2}$
28. $(2y + 3)^{1/2} = y$
29. $(a + 1)^{1/2} = 1 - a^{1/2}$
30. $x + \frac{1}{4} = x^{1/2}$
31. $(x + 1)^{1/4} = (x + 1)^{1/2}$
32. $(3w - 2)^{-1/3} = (w + 1)^{1/3}$
33. $(2x - 3)^{-1/4} = (2x - 5)^{-1/2}$
34. $\dfrac{1}{\sqrt{x^2 - 4x - 12}} = \dfrac{1}{\sqrt{x + 2}}$
35. $\dfrac{-1}{\sqrt[6]{y + 2}} = \dfrac{2}{\sqrt[3]{1 - y}}$

36. $\sqrt{12 + \sqrt{12 + \sqrt{12 + \cdots}}} = x$
(Refer to Section A.4, exercise 42.)

37. $\sqrt{2 + \sqrt{2 + \sqrt{2 + \cdots}}} = x$

38. $\sqrt{x + \sqrt{x + \sqrt{x + \cdots}}} = 2$

39. $\sqrt{y + \sqrt{y + \sqrt{y + \cdots}}} = 8$

40. Government regulations for handicapped ramps require that the ratio of x to r be no less than 0.995, where r is the length of the ramp and x is the distance from the base of the ramp to the structure. If the distance up to the base of a door is 2 feet, approximately how many feet from the structure should the ramp begin (correct to the nearest $\frac{1}{2}$ foot)?

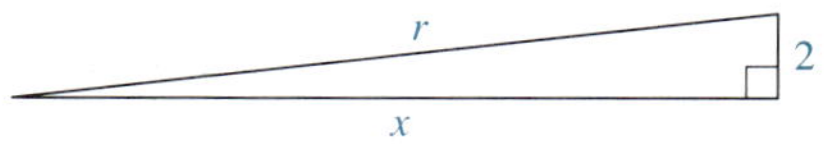

1.4 Miscellaneous Equations

Toolkit

This section uses the definition of quadratic in form from Section A.5.

Many types of algebraic expressions that do not fit in the categories in Sections 1.1 through 1.3 may be used in equations. The problem of solving equations with polynomials of degree greater than two was worked on very actively between 1500 and 1832. A cubic formula was developed for equations of third degree, and a formula was found for fourth-degree equations. Surprisingly, a French mathematician, Galois, was able to show no such formula was possible for equations of fifth or higher degree. Today the cubic formula is rarely used, but many third-degree and higher-degree equations—as well as a variety of other types—can be solved using ideas from Section 1.2. In this section we extend those techniques to handle nonquadratic equations.

First, we can apply the rule that a product is zero if and only if one of the factors is zero to many types of equations.

EXAMPLE Solve $x^3 - 7x^2 + 12x = 0$.

$$x^3 - 7x^2 + 12x = 0$$
$$\Leftrightarrow \quad x(x^2 - 7x + 12) = 0$$
$$\Leftrightarrow \quad x(x - 3)(x - 4) = 0$$
$$\Leftrightarrow \quad x = 0 \quad \text{or} \quad x = 3 \quad \text{or} \quad x = 4$$

Therefore, the solution set is {0, 3, 4}. ■

EXAMPLE Solve $x^3 = 1$.

$$x^3 - 1 = 0$$
$$\Leftrightarrow \quad (x - 1)(x^2 + x + 1) = 0$$
$$\Leftrightarrow \quad x - 1 = 0 \quad \text{or} \quad x^2 + x + 1 = 0$$

$$\Leftrightarrow \qquad x = 1 \quad \text{or} \quad x = \frac{-1 \pm \sqrt{1-4}}{2}$$ From the quadratic formula

The solution set is $\left\{1, \frac{1-\sqrt{3}i}{2}, \frac{1+\sqrt{3}i}{2}\right\}$.

Note: Undoing the cubing by taking the cube root of both sides produces only one of the three solutions. ■

EXAMPLE Solve $x^{2/3} - 5x^{1/3} - 6 = 0$.

The left-hand side is quadratic in form and factors as $(x^{1/3} - 6)(x^{1/3} + 1)$. If $(x^{1/3} - 6)(x^{1/3} + 1) = 0$, then $x^{1/3} - 6 = 0$ or $x^{1/3} + 1 = 0$. So $x^{1/3} = 6$ or $x^{1/3} = -1$. Cubing each side in both equations gives $x = 6^3$ or $x = (-1)^3$.

The solution set is $\{-1, 216\}$. ■

EXAMPLE Solve $x^{-4/3} - 4x^{-2/3} = 0$.

$$\begin{aligned}
& x^{-4/3} - 4x^{-2/3} = 0 \\
\Leftrightarrow \quad & x^{-4/3}(1 - 4x^{2/3}) = 0 \\
\Leftrightarrow \quad & x^{-4/3}(1 + 2x^{1/3})(1 - 2x^{1/3}) = 0 \\
\Leftrightarrow \quad & x^{-4/3} = 0 \quad \text{or} \quad 1 + 2x^{1/3} = 0 \quad \text{or} \quad 1 - 2x^{1/3} = 0 \\
\Leftrightarrow \quad & x^{-4/3} = 0 \quad \text{or} \quad x^{1/3} = -\tfrac{1}{2} \quad \text{or} \quad \tfrac{1}{2} = x^{1/3} \\
\Leftrightarrow \quad & x^{-4} = 0 \quad \text{or} \quad x = -\tfrac{1}{8} \quad \text{or} \quad \tfrac{1}{8} = x
\end{aligned}$$

The equation $x^{-4} = 0$ has no solutions, so **the solution set is** $\{-\frac{1}{8}, \frac{1}{8}\}$. ■

The quadratic formula may be used to solve any equation for which setting one side equal to zero gives a quadratic-in-form expression on the other side. In fact, this is the reason for the name quadratic in form. When used to solve $ax^2 + bx + c = 0$, $a \neq 0$, the formula gives replacements for x, the type of the middle term. Used for an equation that is quadratic in form, the formula will produce replacements for the type of the middle term.

EXAMPLE Approximate the real solutions of $x^6 - x^3 - 6 = 0$ correct to two decimal places.

The left side is quadratic in form and can be factored.

$$\begin{aligned}
& x^6 - x^3 - 6 = 0 \\
\Leftrightarrow \quad & (x^3 - 3)(x^3 + 2) = 0 \\
\Leftrightarrow \quad & (x - \sqrt[3]{3})(x^2 + \sqrt[3]{3}x + \sqrt[3]{3^2})(x + \sqrt[3]{2})(x^2 - \sqrt[3]{2}x + \sqrt[3]{2^2}) = 0
\end{aligned}$$

The second and fourth factors produce complex solutions that are nonreal.

The approximate real solutions are 1.44 and −1.26. ■

EXAMPLE Solve $\sqrt[3]{x^2} - 2\sqrt[3]{x} = 4$.

Rewriting the equation as $\sqrt[3]{x^2} - 2\sqrt[3]{x} - 4 = 0$, the left-hand side is quadratic in form, but its factors are not obvious. The quadratic formula will give us replacements for $\sqrt[3]{x}$, the type of the middle term.

$$\sqrt[3]{x} = \frac{2 + \sqrt{4 + 16}}{2} \quad \text{or} \quad \sqrt[3]{x} = \frac{2 - \sqrt{4 + 16}}{2}$$

$$\Leftrightarrow \quad \sqrt[3]{x} = 1 + \sqrt{5} \quad \text{or} \quad \sqrt[3]{x} = 1 - \sqrt{5}$$

$$\Rightarrow \quad x = (1 + \sqrt{5})^3 \quad \text{or} \quad x = [1 - \sqrt{5}]^3$$

The solution set is $\{(1 + \sqrt{5})^3, (1 - \sqrt{5})^3\}$. ■

EXAMPLE Solve $3\left[\frac{1}{x} + 1\right]^2 - 4\left[\frac{1}{x} + 1\right] - 5 = 0.$

Because the equation is quadratic in form, we can use the quadratic formula. The type of the middle term is $\left[\frac{1}{x} + 1\right]$.

$$\Leftrightarrow \quad \frac{1}{x} + 1 = \frac{4 \pm \sqrt{16 - 4(3)(-5)}}{2(3)}$$

$$\Leftrightarrow \quad \frac{1}{x} + 1 = \frac{4 \pm \sqrt{76}}{6}$$

$$\Leftrightarrow \quad \frac{1}{x} = \frac{2 \pm \sqrt{19}}{3} - 1$$

$$\Leftrightarrow \quad \frac{1}{x} = \frac{-1 \pm \sqrt{19}}{3}$$

$$x = \frac{3}{-1 \pm \sqrt{19}}$$

The solution set is $\left\{\frac{3}{-1 - \sqrt{19}}, \frac{3}{-1 + \sqrt{19}}\right\}$. ■

EXERCISES 1.4

In exercises 1 through 36, solve the equation.

1. $x^3 - 3x^2 - 10x = 0$

2. $w^3 - 3w^2 - w = 0$

3. $x^4 - 4x^2 + 4 = 0$

4. $2x^4 + 3x^2 = 1$

5. $x^6 + 7x^3 - 8 = 0$

6. $2z^7 + 7z^6 = 5z^5$

7. $5x^6 + 1 - 6x^3 = 0$

8. $4x^6 + \frac{17}{3}x^3 = \frac{5}{3}, \quad x \in \mathbb{R}$

9. $x^{-2} + 3x^{-3} = 0$

10. $x^{-6} - 9x^{-8} = 0$

11. $x^{-4} + 5x^{-5} - 14x^{-6} = 0$

12. $x^{-1} - 22x^{-2} + 121x^{-3} = 0$

13. $3x^{-5} - 5x^{-6} = x^{-7}$

14. $3x^{-5} + x^{-7} = 2x^{-6}$

15. $0 = 2x^{2/3} - x^{1/3} - 1$

16. $2x^{2/5} - x^{1/5} - 1 = 0$

17. $1 = 2(u - 1)^{2/5} - (u - 1)^{1/5}$

18. $0 = x^{4/3} + x^{-2/3} - 2x^{1/3}$

19. $x - 2\sqrt{x} + 1 = 0$

20. $\frac{2}{9}y^{-4/3} - \frac{4}{3}y^{-5/3} = 0$

21. $x^{2/3} = 2x^{1/3}$

22. $0 = \frac{2}{3}p^{-1/3} - \frac{4}{3}p^{-2/3}$

23. $-\frac{2}{3}x^{-4/3} + \frac{4}{9}x^{-5/3} = 0$

24. $x = 3x^{4/3}$

25. $2(w - 1)^{-3/4} - 3(w - 1)^{1/4} = 0$

26. $0 = 2 - (x - 3)^{4/3}, \quad x \in \mathbb{R}$

27. $\frac{4}{3}x^{1/3} = -\frac{4}{9}x^{-2/3}$

28. $\frac{2}{3}x^{1/5} + \frac{4}{9}x^{-4/5} = 0$

29. $0 = z^{-1} + 3z^{2/3}$

30. $x^{-1/2} + x^{3/2} = 0$

31. $-x^{1/4} = x^{1/2}, x \in \mathbb{R}$

32. $0 = \frac{(x - 1)^{1/2}(x + 2)}{x - 2}$

33. $\left[\dfrac{3}{x+7}\right]^2 + \dfrac{9}{x+7} = 8$

34. $5\left[\dfrac{1}{2y-3}\right]^2 - \dfrac{4}{2y-3} + 3 = 0$

35. $(x^2 + x + 1)(x^2 - 5x - 6) = 0$

36. $4 = \dfrac{w}{\sqrt{w-2}}$

37. Look back at the polynomial equations you have solved in this set of exercises. Make a prediction about the number of solutions that a polynomial equation of degree n will have.

In exercises 38 and 39, use the following information to solve each cubic equation. There is a formula similar to the quadratic formula that can be used to solve cubic equations. It is called the cubic formula and is given next.

The cubic equation $y^3 + py^2 + qy + r = 0$ has solution set

$$\left\{A + B, -\frac{A+B}{2} + \frac{A-B}{2}\sqrt{-3}, -\frac{A+B}{2} - \frac{A-B}{2}\sqrt{-3}\right\},$$

with

$$A = \sqrt[3]{-\frac{b}{2} + \sqrt{\frac{b^2}{4} + \frac{a^3}{27}}}, \qquad B = \sqrt[3]{-\frac{b}{2} - \sqrt{\frac{b^2}{4} + \frac{a^3}{27}}},$$

$$a = \frac{1}{3}(3q - p^2), \qquad b = \frac{1}{27}(2p^3 - 9pq + 27r).$$

38. $z^3 = 1$

39. $x^3 + 8 = 0$

Chapter 1: Summary

In this chapter you have seen a variety of equations involving one variable. The most important techniques we use for solving these equations are common sense, undoing operations, and the quadratic formula. Some suggestions for using algebraic equations to answer questions about real-world situations were given in Section 1.1

Key Concepts

equation	
solution	
to solve	
equivalent equations	$\Leftrightarrow$
linear equations	
absolute value equations	
quadratic equations	*quadratic formula, discriminant*
radical equations	*check possible solutions*

Skills

You should be able to:

1. Solve a linear equation.
2. Translate English phrases to algebraic expressions.
3. Solve quadratic and other equations by factoring.
4. Solve quadratic and other equations by the quadratic formula.
5. Determine the type of roots of a quadratic equation with rational coefficients.
6. Construct a perfect square given two terms.

SUMMARY EXERCISES

The starred exercises may serve as a practice chapter test. Exercise answers are in the answer section at the back of the book.

In exercises 1 through 18, solve the equation.

1. $-3(4x - 9) + 17x = 43 - 5x$
2. $7.2m + 1.4(3.2 - m) = 0$
* 3. $\frac{2}{3}t + 4 = 2t - \frac{5}{3}$
4. $2\sqrt{17}y + 11\sqrt{15} = \sqrt{6}[3y - 5]$
Also give the approximate solution correct to three decimal places.
5. $(6 - 9i)x + (4 + 8i) = (2 + 3i)(-4 + i)x$, $i = \sqrt{-1}$
* 6. $(1.3 - 2.7i) + (5.8 + 1.4i)x = 2.9 + (-4.6 + 1.9i)x$, $i = \sqrt{-1}$
7. $\frac{1.4}{x - 4} + \frac{2.6}{4 - x} = 3.5$
8. $\frac{x}{x + 3} - \frac{7}{x + 2} = \frac{1}{x + 3}$
* 9. $1 + \frac{25}{x - 5} = \frac{2x + 15}{x - 5} - x$
10. $4|3x - 7| - 24 = 0$
11. $-2|x^2 - 4x + 6| = -12$
* 12. $|5 - \pi y| = |3\pi y - 4|$
13. $5x^2 - 7x + 3 = -1$
14. $\sqrt{3x^2 + 4x} = 2$
* 15. $88x^2 - 6336 = -528x$
16. $x^2 + 113 = 0$
17. $-4(2x - 9)^2 = 12$
* 18. $18x - 3x^2 = 21$
19. Find k so that $6x^2 - kx + 19 = 0$ has exactly one solution.
20. Use the discriminant to determine what kind of roots $-2x^2 + 3x - 5 = 0$ has.
* 21. Write two equations of different degree, each of which has solution set $\{9\}$.
22. To cook spaghetti properly you need 0.42 liters of water plus 0.13 additional liters of water for each ounce of spaghetti. If Aman has 1.68 liters of water in the pot, how much spaghetti is he planning to cook (to the nearest half ounce)?
23. When $7x^2 - 28x - 5$ is rewritten in the form $a[x - h]^2 + k$, a is ______, h is ______, and k is ______.
* 24. Since ancient times rectangles for which the ratio of width to length is equal to the ratio of length to the sum of width and length have been considered pleasing by Western artists. Such a rectangle is called a *golden rectangle* and the ratio of width to length, the *golden ratio*. Find the width in terms of the length for a golden rectangle and calculate the golden ratio.

In exercises 25 through 30, solve the equation.

25. $\sqrt{3x - 5} = \sqrt{2x - 4} + 1$
26. $\sqrt{9 - 40x} - 1 = 2\sqrt{3 - 13x}$
* 27. $\sqrt[6]{x + 3} = \sqrt[3]{2x}$
28. $\frac{1}{\sqrt[5]{2 - x}} + \frac{2}{\sqrt[10]{x}} = 0, x \in \mathbb{R}$
29. $(x - 4)^{2/3} = 4^{1/3}$
* 30. $(x + 10)^{1/2} + 2 = x$
* 31. The West End Kids Club just had its annual picnic. Each sixth-grader ate 5 miniburgers and 2 hotdogs; each fourth-grader had 2 miniburgers and 1 hotdog. If 389 miniburgers and 168 hotdogs were devoured, how many sixth-graders attended the picnic?

In exercises 32 through 40, solve the equation.

32. $x^{1/2} - 4x^{1/4} = 12$
* 33. $2(x + 3)^{-2} = (x + 3)^{-1} + 1$
34. $\frac{3}{(2x - 1)^4} + \frac{2}{(2x - 1)^2} - 5 = 0, x \in \mathbb{R}$
35. $x^3 - x = 4x^2$
* 36. $2x^{8/5} - 5x^{3/5} - 3x^{-2/5} = 0$
37. $x^4 = 1$
38. $|\sqrt[3]{x} - 5| = 7$
* 39. $|x^4 - x^3 - 5x^2| = 3x^2$
40. $\frac{(x + 4)^{1/2}(3x - 1)}{x^3(x + 2)^{1/3}} = 0$
41. Harlan has a job interview in Tunica at 10 A.M. He must go home to Jasper, which is 10 miles, and from Jasper to Tunica, which is 18 miles. Usually he can drive 7 miles per hour faster on the first part than on the second. Yesterday a trial trip took him 30 minutes. Approximately how long (to the nearest tenth of a minute) did it take to drive from Jasper to Tunica?
* 42. The national flag of Zwab is a purple rectangle surrounded by a red border of uniform width. Every flag must have length equal to one unit more than twice its width. The width of the border is always one-fourth of the total width. If the Zwabian flag you bought yesterday had an area of 36 square kweebs, what is the exact area of the red part?

Analytic Geometry and Equations of Two Variables

- 2.1 Introduction to Analytic Geometry
- 2.2 Linear Graphs
- 2.3 Parabolas
- 2.4 The Conic Sections
- 2.5 Translations, Expansions, and Reflections

2.1 Introduction to Analytic Geometry

Toolkit

You will need the Pythagorean theorem: The square of the length of the hypotenuse of a right triangle is equal to the sum of squares of the lengths of the sides.

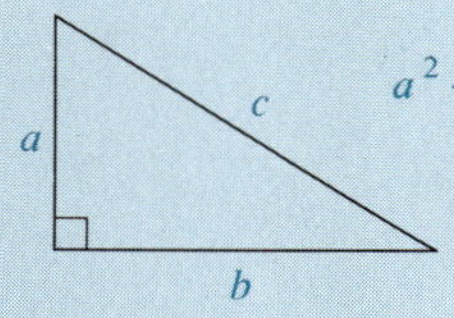

$$a^2 + b^2 = c^2$$

Also, you will need some properties of absolute value from Section A.3, including $|a - b| = |b - a|$.

Before we discuss solving two-variable equations, we need a way to describe their solutions. René Descartes (1596–1650) developed the notion of a **real-number plane** to represent pairs of real numbers just as we use the number line to represent single real numbers. Place one number line perpendicular to another so that they intersect at the zero on each line. This intersection is called the *origin*. The convention is to have one line vertical with positive numbers up and the other horizontal with positive numbers to the right. Each of the lines is called an *axis*. Each point in the plane containing these lines is assigned an ordered pair of numbers, called its *coordinates*. The first coordinate gives its position right or left of the vertical axis. The second coordinate tells its position above or below the horizontal axis.

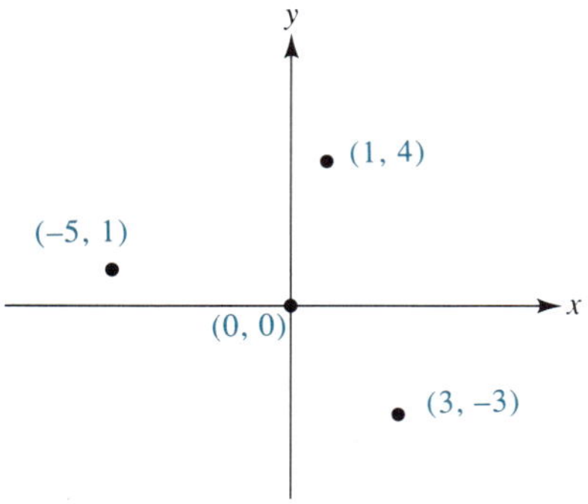

You may have noticed that no scale has been marked on our axes. No matter how carefully we plotted a point's location the physical representation would not be completely accurate. We handle this problem by labeling any point we wish to identify. When you label a dot (4.1, 7.3), then its coordinates are 4.1 and 7.3. As we proceed, we will talk more about suggested labeling.

The most common naming scheme is to denote the first coordinate by x and the second coordinate by y. If you wish to use other variables, label the axes and use the first coordinate to measure horizontal distance. To identify a particular point we may also assign a letter, as A: (1, 4). This means that point A is located at (1, 4). The four regions of the plane created are called *quadrants* and are numbered for reference, counterclockwise from upper right. The points in a quadrant share some characteristics. The axes themselves do not belong to any quadrant. You can see that any point on the x-axis must have y-coordinate 0; A is on the x-axis if and only if A is located at $(x, 0)$. Similarly, B is on the y-axis if and only if B is located at $(0, y)$. All the points on a particular horizontal line must share the same second coordinate. On any vertical line, all points must have the same first coordinate.

This can be described in set-builder notation as $\{(x, y)|x = a\}$ for a vertical line and $\{(x, y)|y = b\}$ for a horizontal line.

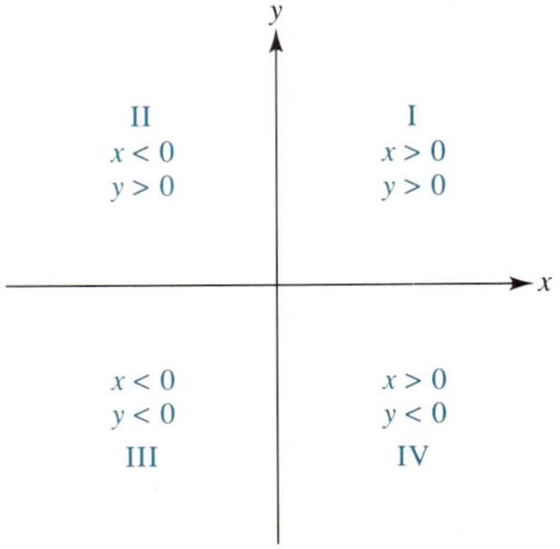

Descartes actually developed the coordinate plane to study geometric problems in algebraic terms. One fundamental idea in geometry is **distance in the plane**: How far is it from one point to another? We already have a way to calculate distance if the two points are on the same number line. The absolute value of a number is its distance from zero; -3 is $|-3|$, or 3, units from zero. This could also be written $|-3 - 0|$, or 3.

Definition In general, $|a - b|$ units is the *distance from a to b* on a number line.

Illustration The distance from 4 to -7 is $|4 - (-7)|$, or 11, units.
The distance from -7 to 4 is $|-7 - 4|$, or 11, units.
The distance from -2.8 to -3.7 is $|-2.8 - (-3.7)|$, or 0.9, units.
The distance from a to b is $|a - b|$ units, which is the same as the distance from b to a, $|b - a|$ units. □

If two points are described by giving ordered pairs, how do we find the distance between them? When they have the same first coordinate, the points determine a vertical line. The distance from A: (x, y_1) to B: (x, y_2) is $|y_1 - y_2|$ units because this vertical line serves as a number line, where location is given by the second coordinates. In a similar fashion, the distance between two points with the same second coordinate, A: (x_1, y) and B: (x_2, y), is $|x_1 - x_2|$ units.

Illustration

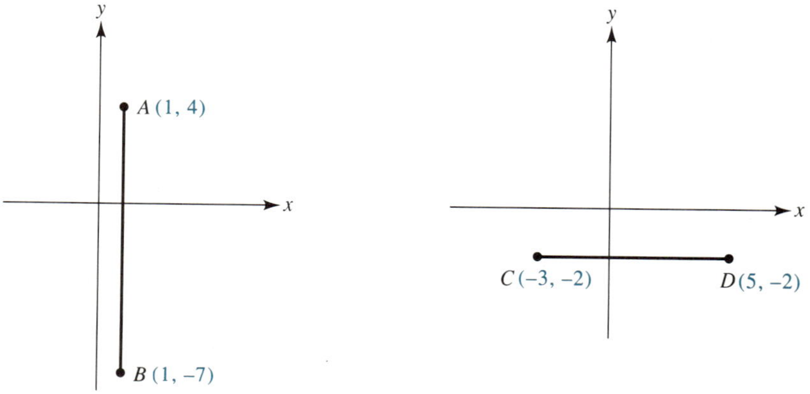

The distance from A to B is $|4 - (-7)|$, or 11, units.
The distance from C to D is $|-3 - 5|$, or 8, units. □

To find the distance between two points, such as A: (1, 4) and D: (5, −2), which do not have any coordinates in common, we apply the Pythagorean theorem. Consider Figure 2.1a. The vertical line through A contains points with coordinates $(1, y)$. The horizontal line through D contains points $(x, -2)$. The point of intersection for these two lines is E: (1, −2). We know how to find the distance from E to A and from E to D. The length of $\overline{AE}$ (segment AE) is 6 units; the length of $\overline{DE}$ is 4 units.

FIGURE 2.1a

There is a right angle at E, so the length of $\overline{AD}$, denoted by AD, can be found by solving $AD^2 = AE^2 + DE^2$.

$$AD^2 = 6^2 + 4^2 \Leftrightarrow AD^2 = 52 \Rightarrow AD = \sqrt{52}$$

If we analyze our calculations, we can describe a general formula for finding AB given the coordinates of A and B. For A: (x_1, y_1) and B: (x_2, y_2),

$$AB = \sqrt{|x_1 - x_2|^2 + |y_1 - y_2|^2}, \quad \text{or} \quad AB = \sqrt{(x_1 - x_2)^2 + (y_1 - y_2)^2}.$$

Because squaring z or $-z$ produces the same result, we do not need to take the absolute values before squaring. It makes no difference which point you identify as (x_1, y_1).

EXAMPLE Find the distance from A: (4.2, −3.7) to B: (−2.1, −11.9).

$$\begin{aligned} AB &= \sqrt{[4.2 - (-2.1)]^2 + [-3.7 - (-11.9)]^2} \\ &= \sqrt{(6.3)^2 + (8.2)^2} \\ &= \sqrt{39.69 + 67.24} \\ &= \sqrt{106.93} \end{aligned}$$

The distance from A: (4.2, −3.7) to B: (−2.1, −11.9) is $\sqrt{106.93}$ units. ■

This new distance formula is a generalization of the earlier cases. If A and B are on the same vertical line, $x_1 = x_2$ and

$$\begin{aligned} & AB = \sqrt{(x_1 - x_2)^2 + (y_1 - y_2)^2} \\ \Leftrightarrow\quad & AB = \sqrt{(y_1 - y_2)^2} \\ \Leftrightarrow\quad & AB = |y_1 - y_2|. \end{aligned}$$

Section A.4

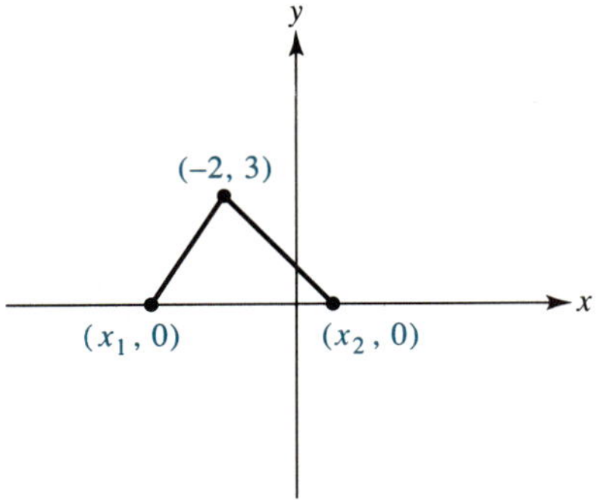

EXAMPLE Describe all points on the x-axis that are 4 units from $(-2, 3)$.

A sketch helps us to see there are probably two such points. We can write an algebraic description of what is required. Any point on the x-axis is of the form $(x, 0)$, so

$$
\begin{aligned}
& 4 = \sqrt{(-2 - x)^2 + (3 - 0)^2} \\
\Leftrightarrow\quad & 4 = \sqrt{(x + 2)^2 + 9} \\
\Leftrightarrow\quad & 16 = (x + 2)^2 + 9 \qquad \text{Because each side represents a nonnegative number} \\
\Leftrightarrow\quad & 7 = (x + 2)^2 \\
\Leftrightarrow\quad & -\sqrt{7} = x + 2 \quad \text{or} \quad \sqrt{7} = x + 2 \\
\Leftrightarrow\quad & x = -2 - \sqrt{7} \quad \text{or} \quad x = -2 + \sqrt{7}
\end{aligned}
$$

The points $(-2 - \sqrt{7}, 0)$ and $(-2 + \sqrt{7}, 0)$ are on the x-axis and are 4 units from $(-2, 3)$. ■

Considering geometric problems in terms of the coordinates of points and algebraic statements is the domain of analytic geometry. Let's look at another geometric idea before moving on to solving two-variable equations.

Definition The **midpoint** of a line segment is halfway between the endpoints of the segment.

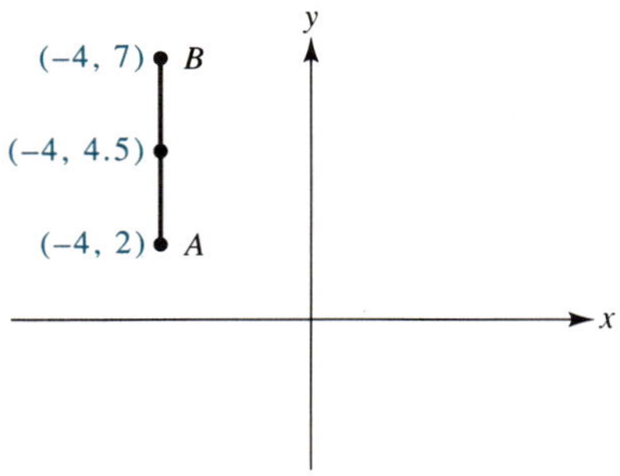

If you know the coordinates of the endpoints, can you give the coordinates of the midpoint? As before, the horizontal and vertical line cases are easy. The point halfway between (5, 3) and (7, 3) is certainly (6, 3). The midpoint of $\overline{AB}$, with A: $(-4, 2)$ and B: $(-4, 7)$, must have first coordinate -4. Its second coordinate is halfway between 2 and 7; this number is $(2 + 7)/2$, or 4.5.

What about the general case, the midpoint of $\overline{AB}$, A: (x_1, y_1), B: (x_2, y_2)? A good guess is to average the first coordinates and to average the second coordinates because that worked in the previous two cases. Figure 2.1b also suggests this is a reasonable idea. Our sketch shows only one possible arrangement of A and B; we really should prove our guess. One idea for a proof is in Exercise 15; here we do another to practice using the distance formula.

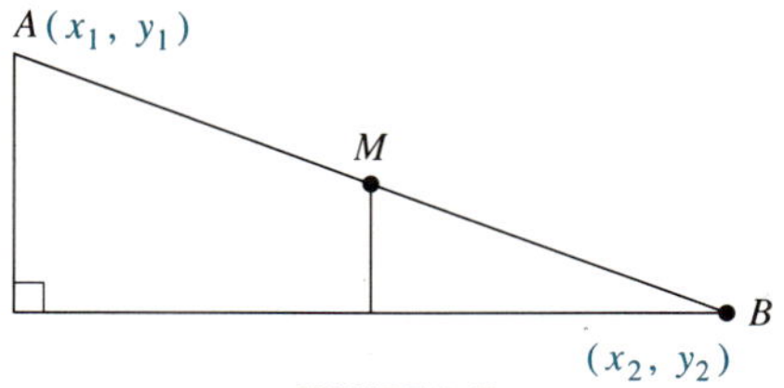

FIGURE 2.1b

For A, B, and M in Figure 2.1b, we want to prove $AM = MB$. We know

$$AM = \sqrt{\left[x_1 - \frac{x_1 + x_2}{2}\right]^2 + \left[y_1 - \frac{y_1 + y_2}{2}\right]^2}$$

and

$$MB = \sqrt{\left[\frac{x_1 + x_2}{2} - x_2\right]^2 + \left[\frac{y_1 + y_2}{2} - y_2\right]^2}$$

$$AM = \sqrt{\left[x_1 - \frac{x_1 + x_2}{2}\right]^2 + \left[y_1 - \frac{y_1 + y_2}{2}\right]^2}$$

$$\Leftrightarrow \quad AM = \sqrt{\left[\frac{2x_1 - x_1 - x_2}{2}\right]^2 + \left[\frac{2y_1 - y_1 - y_2}{2}\right]^2}$$

$$\Leftrightarrow \quad AM = \sqrt{\left[\frac{x_1 - x_2}{2}\right]^2 + \left[\frac{y_1 - y_2}{2}\right]^2}$$

$$\Leftrightarrow \quad AM = \sqrt{\left[\frac{x_1 + x_2 - 2x_2}{2}\right]^2 + \left[\frac{y_1 + y_2 - 2y_2}{2}\right]^2}$$

We want to produce something that looks like *MB*.

$$\Leftrightarrow \quad AM = \sqrt{\left[\frac{x_1 + x_2}{2} - x_2\right]^2 + \left[\frac{y_1 + y_2}{2} - y_2\right]^2}$$

We have exactly *MB*.

$$AM = MB$$

We do more analytic geometry in Section 2.4. Analytic geometry takes advantage of algebraic techniques to handle geometric problems. You may also use geometric methods in dealing with algebraic ideas. An idea that helps do this is that of a graph.

Definition A two-dimensional **graph** is the geometric representation of a set of ordered pairs.

Enough points should be labeled to distinguish the graph.

Illustration The graph of $\{(x, y) | y = 3\}$ is

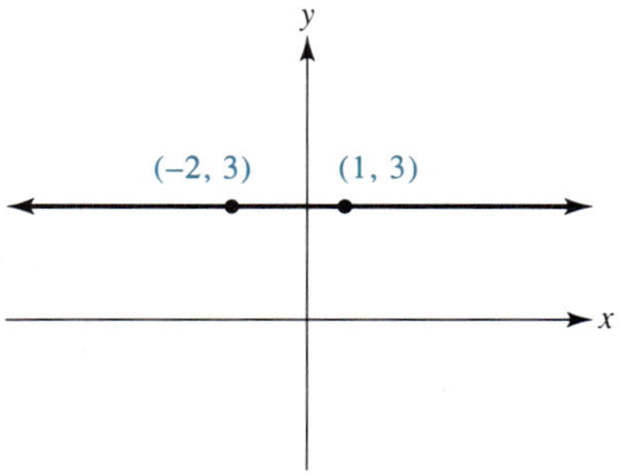

The graph of $\{(s, t) | s \geq 4 \text{ and } t \geq 2\}$ is

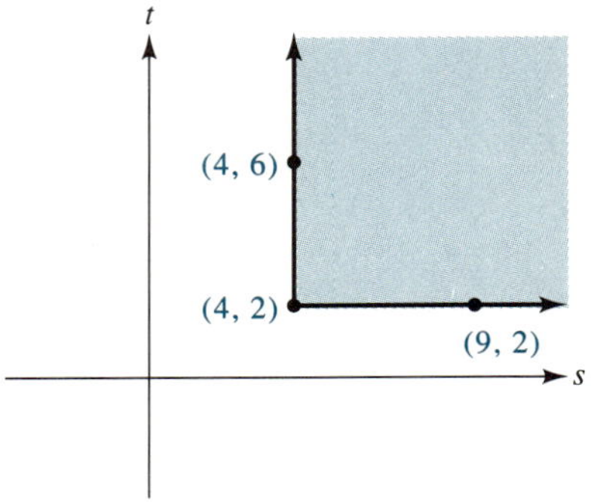

The graph of $\{(2, -1), (\pi, 7), (-1.6, \sqrt{3})\}$ is

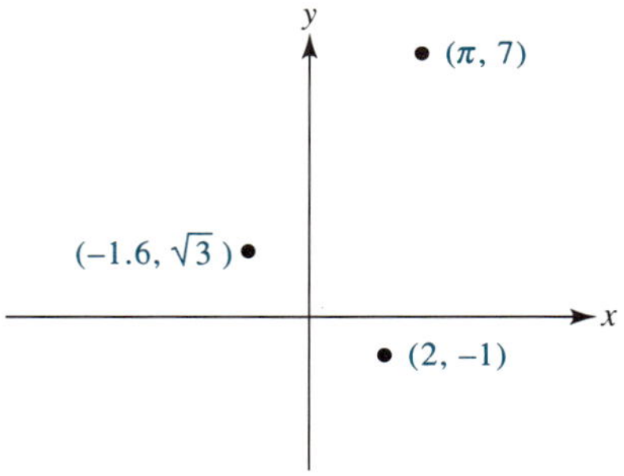

The graph of $\{(x, y | x = -4\}$ is

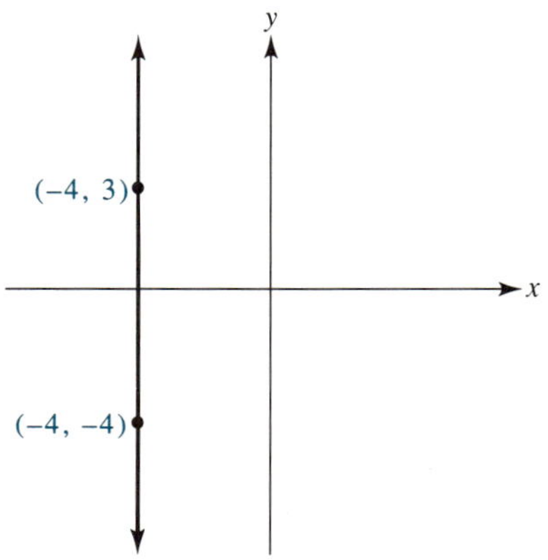

□

EXERCISES 2.1

In exercises 1 through 4, locate each point in the same set of axes. You may want to use your calculator to approximate the position of the point. (Remember that this is only an approximation and is not correct for labeling.)

1. $(3.4, -2)$
2. $(-\sqrt{12}, -7.33)$
3. $(-\pi, 2\pi)$
4. $(\sqrt[3]{17}, \sqrt{\pi})$

In exercises 5 through 14, find the distance between the given points. In exercises 11 through 14, also give the approximate distance correct to 4 decimal places.

5. $(7, 0), (-13.4, 0)$
6. $(5, -4), (5, -146)$
7. $(c, 12), (c, -2)$
8. $(\pi, a), (e, a)$
9. $(4, -3), (0, 0)$
10. $(-7, 10), (-2, -2)$
11. $(4.1, 3.5), (-3.9, 11.6)$
12. $(-5, -8), (13.5, -2.4)$
13. $(7, -\pi), (-3, 2)$
14. $(\sqrt[3]{12}, 4.33), (172, \sqrt{17})$
15. Develop the midpoint formula for the coordinates of M using the fact that ΔABC is similar to ΔMBD and that the ratios of corresponding parts of similar triangles are equal.

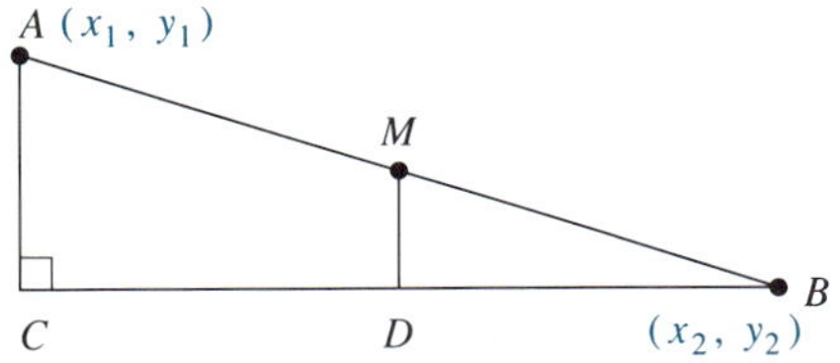

16. Develop the midpoint formula using the fact that the point on the real number line halfway between a and b, $a < b$, is $a + (b - a)/2$.

In exercises 17 through 22, find the indicated point.

17. The point halfway from $(27, -3)$ to $(-16, -9)$
18. The point halfway from $(-\sqrt{8}, 4.1)$ to $(3.7, \sqrt[3]{5})$
19. The point one-third of the way from $(0, 0)$ to $(5, 6)$
20. The point one-third of the way from $(27, -3)$ to $(-16, -9)$
21. The point one-third of the way from $(-1, 4)$ to $(-3, -10)$
22. The point one-third of the way from (a, b) to (c, d)

In exercises 23 through 26, give the points described.

23. The points on the y-axis 4 units from $(2, 2)$
24. The points on the x-axis 7 units from $(5, -1)$
25. The points on the x-axis 6.2 units from $(e, 0)$
26. The points on the y-axis a units from $(-1, 3)$, $a > 1$

In exercises 27 through 34, draw the indicated graph.

27. The graph of $\{(x, y) | x = -17.1\}$
28. The graph of $\{(x, y) | y = \sqrt{7}\}$
29. The graph of $\{(x, y) | 2 \le y \le 18\}$
30. The graph of $\{(x, y) | x \ge -\pi\}$
31. The graph of $\{(x, y) | x \ge 4 \text{ and } y \le 6\}$
32. The graph of $\{(x, y) | -3.1 \le x \le 5 \text{ and } -4 \le y \le 19\}$
33. The graph of $\{(x, y) | x \ge 7 \text{ or } y \le -2\}$
34. The graph of $\{(x, y) | -2 \le x \le -0.39 \text{ or } -4 \le y \le 0\}$
35. A figure can be verified to be a square or a rectangle if the lengths of the diagonals are the same and the diagonals bisect each other. Is the four-sided figure with corner points $(2, 0)$, $(4, 4)$, $(1, 7)$, and $(-1, 3)$ a rectangle that is not a square, a square, or neither?
36. Graph all the points in a plane three units from $(2, -8)$ and describe the graph in words.
37. In the city of Acia, the streets run north and south, the avenues run east and west, and all blocks are the same size. The fire stations have marked each intersection with a pair of integers according to the following plan. The intersection $(0, 0)$ is at the center of town; intersection $(-4, 6)$ is four blocks west and 6 blocks north of the city's center.

What is the shortest distance, in number of blocks, you can travel to drive from

a. $(0, 0)$ to $(-4, 6)$?
b. $(-4, 6)$ to $(2, 5)$?

38. Write a formula for Acians to use to compute how far, in city blocks, it is from A: (m_1, n_1) to B: (m_2, n_2) when traveling on the roads.
39. The citizens of Acia have decided to build a subway. Underground, they won't be restricted to following the streets and avenues. In block units, what is the minimum subway tunnel length they need to have stations at the intersections labeled $(0, 0)$, $(2, 5)$, $(-3, 3)$ and $(1, -2)$?

2.2 Linear Graphs

We saw in the last section that the vertical line through (-4.1) is the graph of $\{(x,y)|x = -4\}$. We can interpret the graph as a picture of all the ordered pairs (x,y) that satisfy the equation $x = -4$. If we consider an equation of first degree involving two variables—for example, $2x - y = 6$—we know there is an infinite number of pairs (x,y) that are solutions for this equation. Among these solutions are $(3, 0)$, $(0, -6)$, $(2, -2)$, and $[\sqrt{3}, -6 + 2\sqrt{3}]$. One way to represent this set of solutions is to use set-builder notation. The solution set for $2x - y = 6$ is $\{(x, y)|2x - y = 6\}$. Somehow this doesn't seem satisfying. Another option is to try a graph of the solution set. Here we are very fortunate.

Definition Any equation of the form $ax + by = c$, a and b not both zero, has a solution set whose graph is a straight line. For this reason such equations are called **linear equations.**

You will investigate this further in the exercises.

EXAMPLE Graph the solution set of $2x - y = 6$.
Only one line can pass through two distinct points, so if we find two solutions of the equation, the graph is determined. We found before that $(3, 0)$ and $(0, -6)$ are solutions.

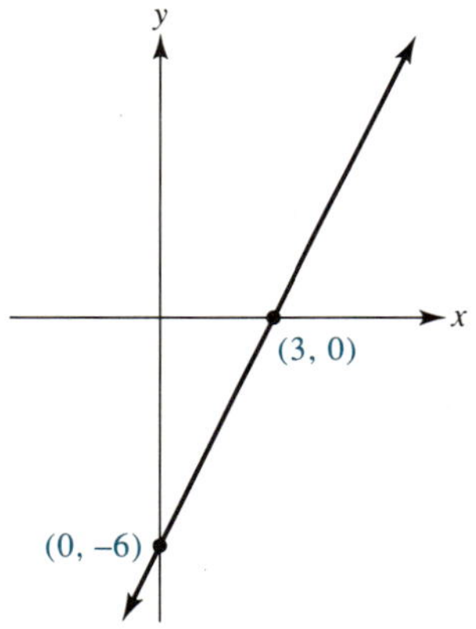

At least two points must be labeled to identify the line. ■

The solutions we decided to use in the last example have several important features. It is usually easier to determine a value for x or y so that $(x, 0)$ or $(0, y)$ is a solution than it is to find other solutions. These points are quickly located on the axes, and their relative positions are clear. Any two solutions can be used, but $(x, 0)$ and $(0,y)$ do have advantages. They also have geometric interpretations: They are the points of intersection with the x- and y-axes.

Definition The *x-intercept* of $ax + by = c$ is $(c/a, 0)$, and the *y-intercept* is $(0, c/b)$, provided that both ordered pairs are solutions of the equation.

Not every linear equation has both an x- and a y-intercept. Think about what happens if $a = 0$ or $b = 0$.

Illustration The graph of $\{(x, y) | y = 1.5\}$ has y-intercept $(0, 1.5)$ and no x-intercept.

The graph of $\{(x, y) | x = -\sqrt{3}\}$ has no y-intercept. Its x-intercept is $(-\sqrt{3}, 0)$.

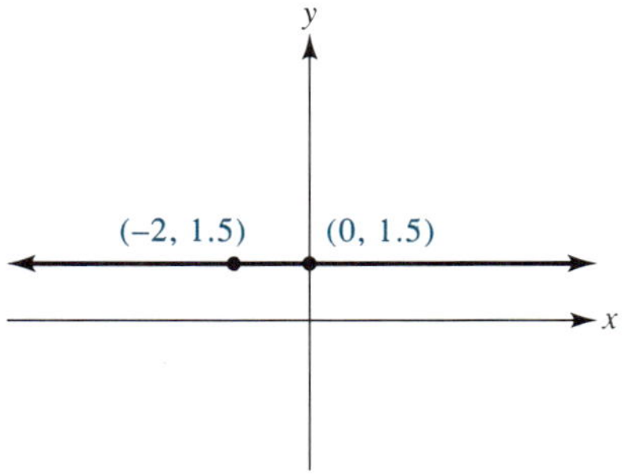

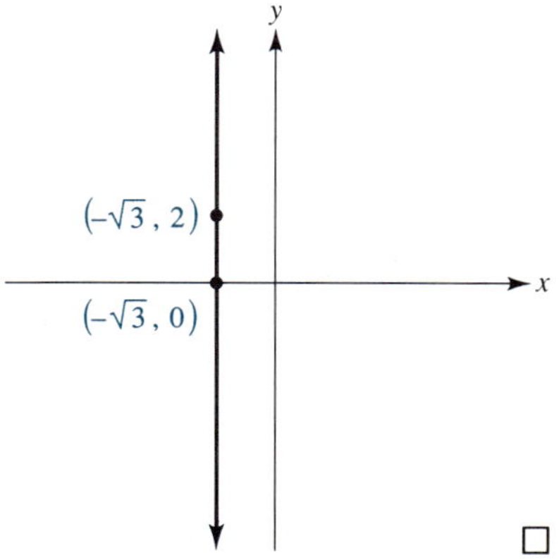

□

Note: For the remainder of this chapter we will use $x = r$ to mean $\{(x, y) \mid x = r\}$ and $y = s$ for $\{(x, y) \mid y = s\}$. This is done for brevity, and you should be aware of the context when interpreting $x = r$ or $y = s$.

Using the coordinate plane, we can also identify a line by identifying a single point on the line and describing the slant of the line relative to the horizontal axis.

Illustration The three lines through $(3, 1)$ have different slants with respect to the x-axis. See Figure 2.2a.

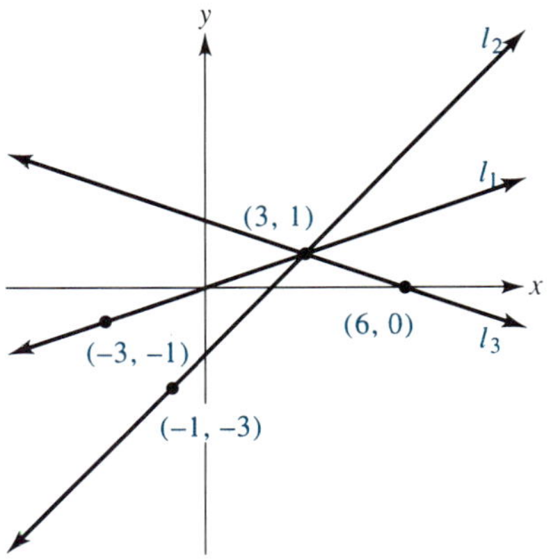

FIGURE 2.2a

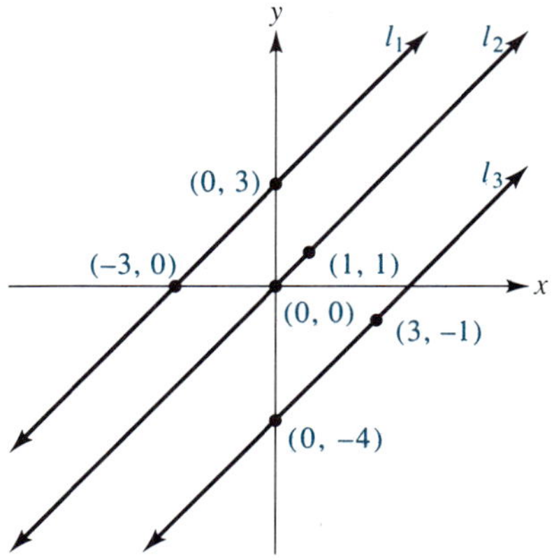

FIGURE 2.2b

The three lines in Figure 2.2b have the same slant but different y-intercepts. □

One way to describe this slant is to assign a number, called the *slope,* to each nonvertical line by calculating the ratio of the change in second coordinates to the corresponding change in first coordinates between two points.

Definition The slope, m, of a line through (x_1, y_1) and (x_2, y_2) is $(y_1 - y_2)/(x_1 - x_2)$.

EXAMPLE Find the slope of each line in Figure 2.2a.

The line l_1 passes through $(-3, -1)$ and $(3, 1)$. **Its slope m_1 is**

$$\frac{-1 - 1}{-3 - 3}, \quad \text{or} \quad \mathbf{\frac{1}{3}}.$$

The slope m_2 of l_2 is

$$\frac{-3 - 1}{-1 - 3}, \quad \text{or} \quad \mathbf{1}.$$

Line l_3 has slope

$$\frac{0 - 1}{6 - 3}, \quad \text{or} \quad \mathbf{\frac{-1}{3}}.$$

Note: You may use either point as (x_1, y_1), but you must be consistent in using the definition. For example, for line l_1 the slope can also be calculated as follows:

$$m_1 = \frac{1 - (-1)}{3 - (-3)} = \frac{1}{3}.$$

■

The slope of a line is the same no matter which two points on the line you use for the calculation of m. One of the exercises outlines a proof of this.

In Figure 2.2a, l_2 and l_1 go uphill as we travel from left to right; this produces their positive slopes. The negative slope of l_3 indicates that the graph goes downhill from left to right. What would a slope of 0 say about the graph? The line l_2 goes uphill more steeply than l_1. This is reflected in the relative sizes of their slopes. A line with larger positive slope goes uphill more rapidly than one with a smaller positive slope. This is because we are comparing how y changes as we change x in moving from left to right. For lines with negative slopes, a larger absolute value of the slope means a steeper slant downhill.

Vertical lines were left out of our definition of slope. Any vertical line is the graph (of the solution set) of $x = c$. All points on the line are of the form (c, y) for some value of y. If you attempt to use $m = (y_1 - y_2)/(x_1 - x_2)$, the denominator will be zero. We assign no slope to a vertical line. Because x does not change value, it makes no sense to ask how y changes when x changes. A line with no slope is not the same as a line with a slope of 0.

EXAMPLE Does the line that is the graph of $3y - 2x = 12$ go uphill or downhill from left to right?

To use the definition of slope, we need two points on the line. The intercepts are quickly found, $(0, 4)$ and $(-6, 0)$.

$$m = \frac{4 - 0}{0 - (-6)} = \frac{2}{3}$$

The line goes uphill because its slope is positive. ■

The lines l_1, l_2, l_3 in Figure 2.2b should have the same slope.

$$m_1 = \frac{0 - 3}{-3 - 0} \text{ or } 1, \qquad m_2 = \frac{1 - 0}{1 - 0} \text{ or } 1, \qquad m_3 = \frac{-4 - (-1)}{0 - 3} \text{ or } 1$$

We see that these lines are parallel. If two lines are parallel, then the lines have the same slope or both have no slope.

Lines may also be perpendicular. A line perpendicular to a line with positive slope (uphill) must have negative slope (downhill). Also, if a line rises (or falls) steeply, one perpendicular to it must fall (or rise) slowly.

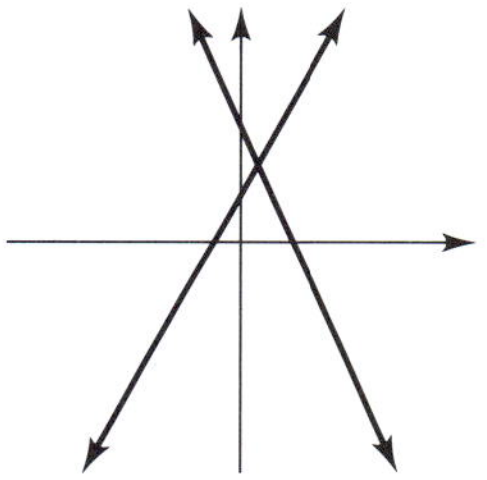

Both steep;
not perpendicular

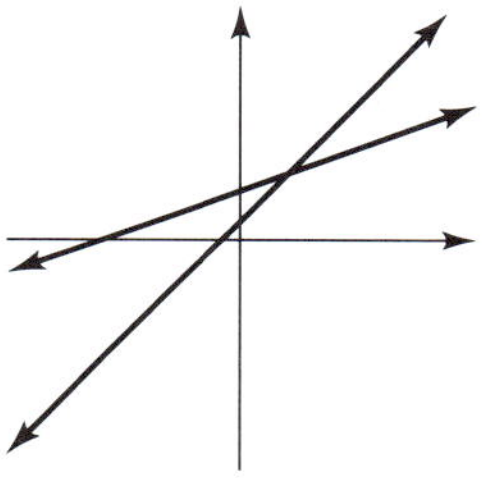

Both uphill;
not perpendicular

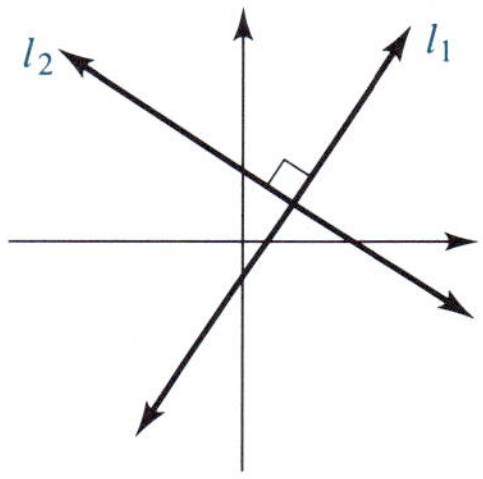

l_1 uphill rapidly;
l_2 down slowly;
$l_1 \perp l_2$

If l_1 with slope m_1 and l_2 with slope m_2, with each slope nonzero, are perpendicular, then $m_1 = -1/m_2$. A vertical line is perpendicular to a horizontal line, although we do not have an algebraic slope relationship for such lines. (See Exercise 39.)

The slope of a line tells how y changes for a unit change in x. For this reason rewriting $3y - 2x = 12$ to show how y depends on x should show us the slope.

$$3y - 2x = 12 \quad \Leftrightarrow \quad 3y = 2x + 12 \quad \Leftrightarrow \quad y = \frac{2}{3}x + 4$$

Each change of 1 unit in x produces a change of $\frac{2}{3}$ unit in y. As we saw earlier, the graph of this equation does have slope $\frac{2}{3}$. You may also recognize the 4 as the second coordinate of the y-intercept. If you substitute 0 for x in $y = \frac{2}{3}x + 4$, y will have value 4. For this reason $y = mx + b$ is called the *slope-intercept form* for a linear equation; m is the slope and $(0, b)$ is the y-intercept. This form gives us another way to find the slope.

EXAMPLE What is the slope of any line perpendicular to the graph of $-2x + 5y = 7$?

We find the slope of the graph of $-2x + 5y = 7$ by rewriting the equation in slope-intercept form.

$$-2x + 5y = 7 \quad \Leftrightarrow \quad 5y = 2x + 7 \quad \Leftrightarrow \quad y = \frac{2}{5}x + \frac{7}{5}$$

The slope is $\frac{2}{5}$.
The slope of any line perpendicular to the graph of $-2x + 5y = 7$ is $-\frac{5}{2}$. ■

This second method of finding a line's slope is essentially algebraic. In order to use it, we need an equation for which the line is a graph. Given sufficient information about a line, how do you find an equation that describes it? Obviously, if you somehow knew the slope and the y-intercept, it would be a matter of substituting values into the slope-intercept form.

Illustration The line with slope -0.4 and y-intercept $(0, 1.3)$ is the graph of $y = -0.4x + 1.3$. □

Let's consider the case of two known points $(3, -2)$ and $(-5, 1)$. We know there is only one line through these two points; what is its equation? We can find its slope,

$$\frac{-2 - 1}{3 - (-5)} = -\frac{3}{8}, \text{ and}$$

begin by writing $y = -\frac{3}{8}x + b$. Because each point represents a solution of the equation, substituting either pair of values for x and y would determine b, and we would have the desired equation. One solution of $y = -\frac{3}{8}x + b$ is $(-5, 1)$.

$$1 = -\frac{3}{8}(-5) + b \quad \Leftrightarrow \quad b = -\frac{7}{8}$$

The equation is $y = -\frac{3}{8}x - \frac{7}{8}$.

There is a slightly more efficient way to handle our original problem by using the first definition of slope. Any point (x, y) on the line through $(3, -2)$ and $(-5, 1)$ may be used to calculate the slope. In fact,

$$\frac{-2 - 1}{3 - (-5)} = \frac{y - 1}{x - (-5)}.$$

We can rewrite this as follows:

$$\frac{-2 - 1}{3 - (-5)} = \frac{y - 1}{x - (-5)}$$

$$\Leftrightarrow \quad \frac{-3}{8} = \frac{y - 1}{x + 5}$$

$$\Leftrightarrow \quad y - 1 = -\frac{3}{8}(x + 5), \qquad \text{if } x \neq -5$$

But we know $(-5, 1)$ is on the line and $y - 1 = -\frac{3}{8}(x + 5)$ is satisfied by $(-5, 1)$, so $y - 1 = -\frac{3}{8}(x + 5)$ is one equation of the line. This is an example of point-slope form; the slope $-\frac{3}{8}$ is visible, as is one of the given points, $(-5, 1)$. A linear equation written as $y - k = m(x - h)$ is in *point-slope form; m* is the slope of the graph of the equation and (h, k) is a point on the graph.

EXAMPLE **a.** Write an equation of the line through $(3, -5)$ with slope 0.3.

a. Using the point-slope form of a linear equation, with point $(3, -5)$ and slope

0.3, **an equation of the line is $y - (-5) = 0.3(x - 3)$.** An equivalent equation is $y = 0.3x - 5.9$, which is in slope-intercept form.

b. Write an equation of the line parallel to the line $y = -\frac{2}{3}x + 11$ and through the point (4, 4).

b. Our line must have slope $-\frac{2}{3}$ to be parallel to the line $y = -\frac{2}{3}x + 11$. We use this and the point (4, 4) to write an equation: $\mathbf{y - 4 = -\frac{2}{3}(x - 4)}$.

c. Write an equation of the line containing $(-2, -4)$ and $(3, -1)$.

c. We use the two points to compute the slope of the line:

$$\frac{-4 - (-1)}{-2 - 3}, \quad \text{or} \quad \frac{3}{5}$$

Now we can use either of the points given to write the equation in point slope form:

$$\mathbf{y - (-1) = \frac{3}{5}(x - 3)}$$ ■

All equations $ax + by = c$, a and b not both zero, have graphs that are straight lines. With the exception of those whose graphs are vertical lines, a linear equation can be rewritten in point-slope form as $y - k = m(x - h)$. There is not a unique way to do this; different pairs h and k may be used. In a sense the graphs of all linear equations belong to the same family, the family of lines. We will find it useful as we proceed to have a representative for this family, sort of a typical member. We call such a family representative a **standard graph.**

Standard Graph 1: $y = x$

The graph is a line through (0, 0) with slope 1. Suggested labeling is two points on the line.

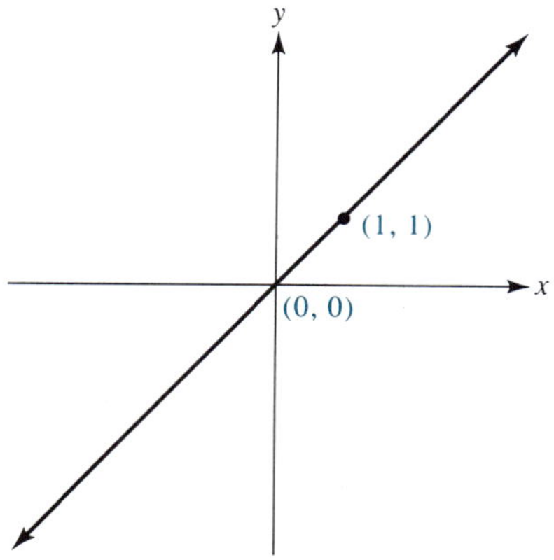

We have chosen the simplest linear equation using both x and y. In Section 2.5 we will look more closely at how the other members of the family are variations of this standard graph.

After we discussed first-degree, one-variable equations, we looked at absolute value equations involving first-degree expressions. Now is a good time to examine equations involving the absolute value of two-variable, first-degree expressions. Consider

$$|5x - 3y| = 8.$$

There is an infinite number of pairs (x, y) that are solutions. Again, perhaps the best we can do is a graph of the solution set. This solution set is the collection of all the pairs satisfying $5x - 3y = 8$ and all the pairs satisfying $5x - 3y = -8$, because

$$|5x - 3y| = 8 \quad \Leftrightarrow \quad 5x - 3y = 8 \quad \text{or} \quad 5x - 3y = -8$$

The graph of each of these equations is a line; the graph of $|5x - 3y| = 8$ is both lines. Formally, the solution set is the union of the solution sets of $5x - 3y = 8$ and $5x - 3y = -8$.

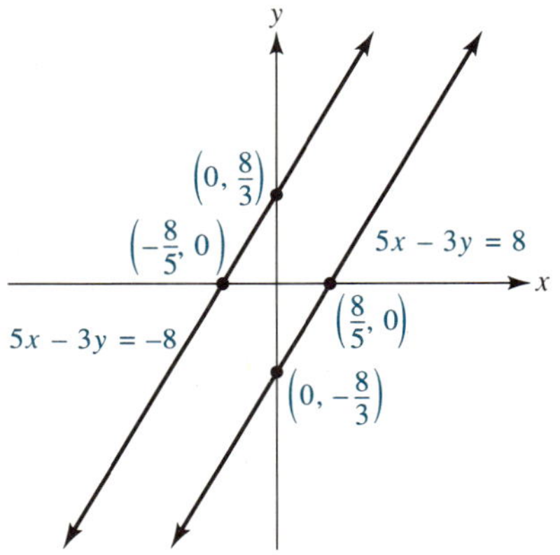

Sometimes we may want to graph only part of a line.

EXAMPLE Graph the solution set of $y = 2x - 1, y > -2$.

Only pairs (x, y) with $y > -2$ can be solutions. For $y = -2$, $y = 2x - 1$ yields $x = -0.5$. Any points below $(-0.5, -2)$ must be deleted. The point $(-0.5, -2)$ is not in the solution set but serves as a boundary point for those above it. We mark it with an open circle.

The graph of $y = 2x - 1, y > -2$ is

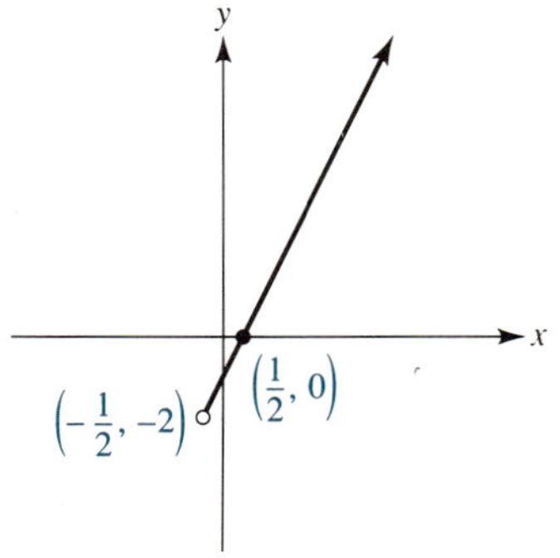

■

EXERCISES 2.2

In exercises 1 through 6, assume x and y are the variables and classify the equations as linear or nonlinear.

1. $4.2x - 3.55y = 398$
2. $a(x - 3) + c = \frac{b}{d}y$, with $b, d \neq 0$
3. $4y^2 - 8x = 2y$
4. $13 = \sqrt{1.4x + 3.3y}$
5. $\frac{2x - 5}{-3} = y$
6. $\frac{2x - 5}{x} = y$

In exercises 7 through 16, sketch the graph of the equation and state the x- and y-intercepts and the slope, if they exist.

7. $4x + 2y = 18$
8. $3x - 4y = 12$
9. $y = \frac{1}{4}x + 2$
10. $y = -\frac{1}{4}x + 2$
11. $y = -3.1x - 4$
12. $y = -4.1x - 4$
13. $x = -10$
14. $y = \sqrt{19}$
15. $y = \sqrt{80x}$
16. $y = \sqrt{80x} - \pi$

In exercises 17 through 26, state the slope of the line that:

17. Contains $(-3, 5)$ and $(13, 4)$.
18. Contains $(2.4, 5.1)$ and $(-3.9, 9.0)$.
19. Contains $(2a, -5b)$ and $(7a, b)$.
20. Contains $[-2.1\pi, \sqrt{3.6}]$ and $(-2.9\pi, -1)$. Also give the approximate slope correct to one decimal place.
21. Is parallel to the line $3x + 11y = 9$.
22. Is parallel to the line $a_1x - a_2y = a_3$; $a_1, a_2 \neq 0$.
23. Is parallel to the line $y = 18$.
24. Is parallel to the line $x = -7$.
25. Is perpendicular to the line $y = 45.3x - 249$.
26. Is perpendicular to the line $3x - 6y = \sqrt{3}$.

In exercises 27 through 36, write the equation of the line that:

27. Contains $(4, 8)$ and $(-3, 0)$.
28. Contains $(-7.3, 9.9)$ and $(-3, -4.5)$.
29. Contains $[\sqrt{7}, 4]$ and $[\sqrt{7}, -3.5]$.
30. Contains (a, c) and (b, c), $a \neq b$.
31. Is parallel to the line $3x + 5y = 1$ and contains $(0, -1)$.
32. Has slope $-\frac{3}{4}$ and contains $(-3, 5)$.
33. Is parallel to the line $x = -3.33$ and contains $(-1, -1)$.
34. Is parallel to the line $y = \sqrt{5}$ and contains $(3, 9)$.
35. Is perpendicular to the line $y = 5x - 3$ and contains $(\pi, 0)$.
36. Is perpendicular to the line $5x + 4y = 20$ and has y-intercept $(0, 5)$.
37. The shortest distance between two points is the length of the line segment between them. We can use this to check whether three points are on the same line. The points E, F, and G are on the line if $EF + FG = EG$. Check that the three solutions E: $(0, \frac{2}{3})$, F: $(\frac{2}{5}, 0)$, G: $(1, -1)$ of $5x + 3y - 2 = 0$ lie on the same line by computing EF, FG, and EG.
38. Verify that the slope of a straight line is the same no matter which two points are selected. Use the following figure and what you know about similar triangles. (See Exercise 15, Section 2.1.)

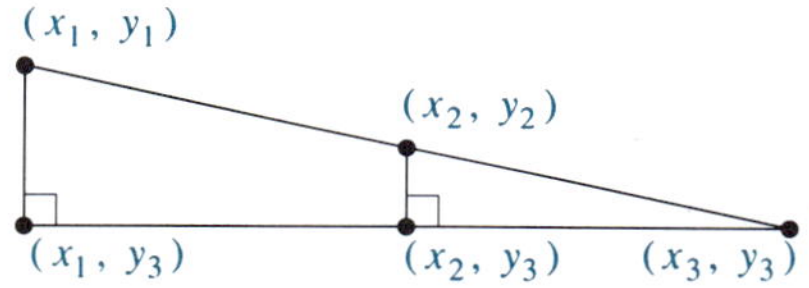

39. Prove that if two lines are perpendicular and neither is vertical or horizontal, they have slopes that are negative reciprocals of each other. Use the following figure and the fact that perpendicular lines form a right angle. The triangle with vertices $(0, 0)$, (a, b), and (c, d) is therefore a right triangle; $a, b, c, d \neq 0$.

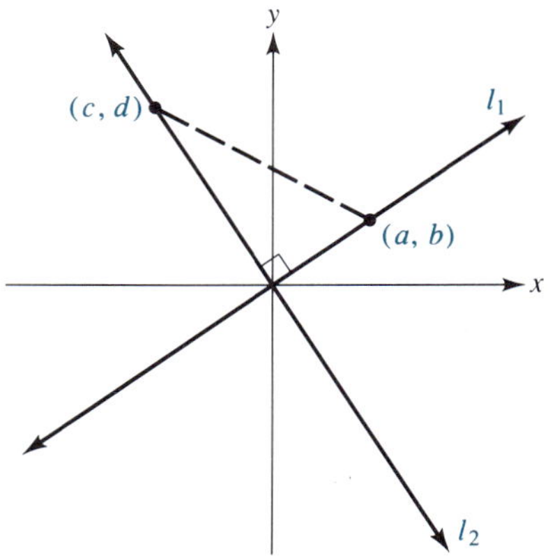

In exercises 40 through 49, sketch the graph of the equation.

40. $|2x + y| = 3$
41. $|-9x + y| = 8$
42. $|2x + y| = -3$
43. $|-9x + y| = 0$
44. $y = 5x - 13, x < 0$
45. $x = -2y + 8, x \geq -1$
46. $x + y = 8.33, -2 \leq x \leq 3$
47. $2y - 4x = 16, 1 < x < 10$

48. $y = 5x - 13, y > 3$
49. $y = \sqrt{6}, y < -2$

(Graphing calculator) In exercises 50 through 53, enter the equation of the straight line, (a) position the cursor to find the approximate *x*-intercept, and (b) use a solving routine to find the approximate *x*-intercept. (c) Compare your result to the exact *x*-intercept.

50. $y = 3x - 2.4$
51. $y = -\sqrt{3}x - 2.3$
52. $4.5x + 2.4y = \sqrt{7.99}$
53. $\sqrt{3}x - 5.891y = 3\sqrt{7.43}$

2.3 Parabolas

Two-variable equations in which one of the variables is squared have a family of graphs just as those for linear equations do. We discuss a geometric definition for members of this family in Section 2.4. Here we begin by examining the simplest such equations:

$$y = x^2, \qquad x = y^2.$$

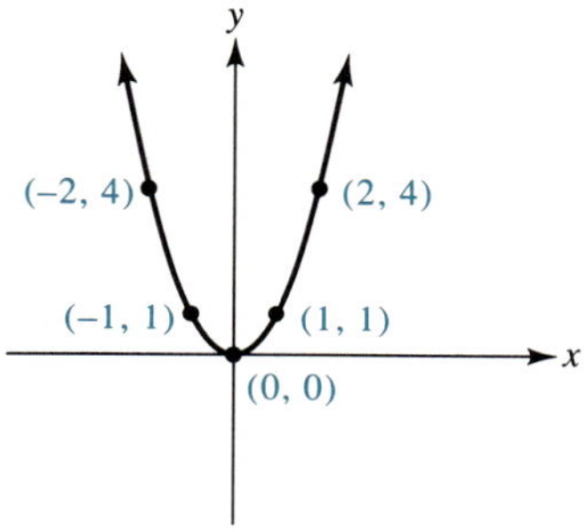

The graph of $y = x^2$ must have only nonnegative y-values. The origin belongs to the graph, and opposite values of x produce the same y-value. There are no restrictions on x-values, and y-values grow larger with no limit on them. In connecting the points we have located, we do so with the simplest curve possible. Only linear equations or absolute value equations with linear expressions have graphs that are composed of lines or half-lines. This is discussed in more detail later in Section 4.5.

Standard graph 2: $y = x^2$

The graph lies in the first and second quadrants. The smallest value of y is 0. If (a, b) belongs to the graph, so does $(-a, b)$. Suggested labeling is the turning point (0, 0) and a point to the right and to the left to establish the scale.

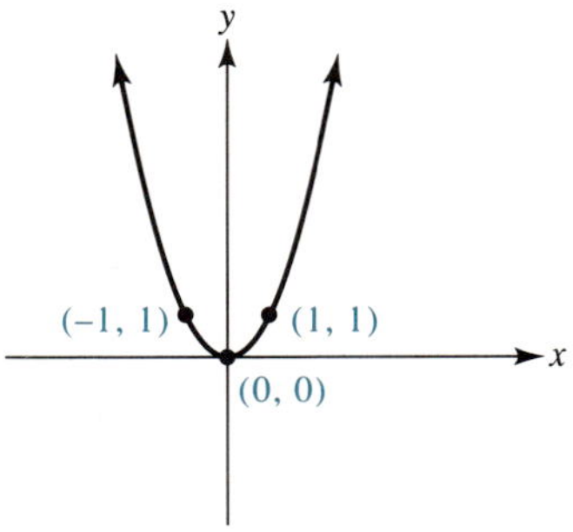

Note: The arrowheads at the ends of the graph indicate it continues.

Standard graph 2 is an example of a parabola. The turning point (0, 0) is its *vertex*. In the remainder of this section we discuss the other types of equations whose graphs are parabolas. The suggested labeling for any parabola is similar to that of the standard graph.

The other simple example, $x = y^2$, interchanges the roles of x and y. Its graph must do the same thing. Instead of being symmetric around the positive y-axis, it is symmetric around the positive x-axis.

The graph of $x = y^2$ is

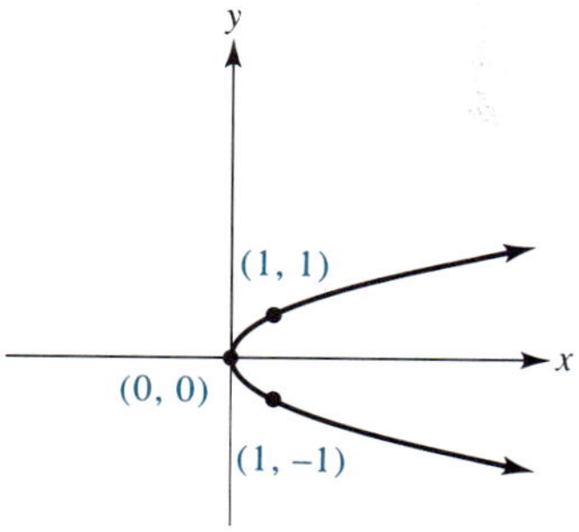

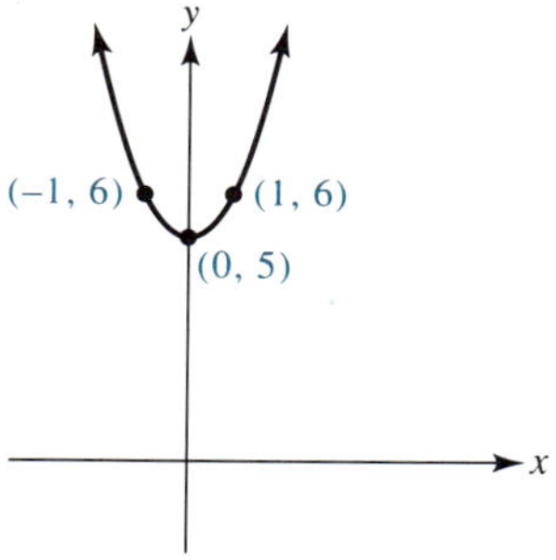

This is the graph of $y = x^2$ with the names of the axes exchanged. We can limit our discussion to equations where x is the squared variable because similar equations with y squared will have graphs that look the same except thay will be "horizontal," as is the graph of $x = y^2$.

Suppose you add a constant to the x^2-term. What happens to the graph? For example, what is the graph of $y = x^2 + 5$? Each y-value is *increased* by 5, so each point on the standard graph is moved *up* 5 units. The vertex is now (0, 5). Adding a negative constant (or subtracting a positive constant) will pull the standard parabola down that many units.

Illustration The graph of $y = x^2 - 3$ has vertex $(0, -3)$.

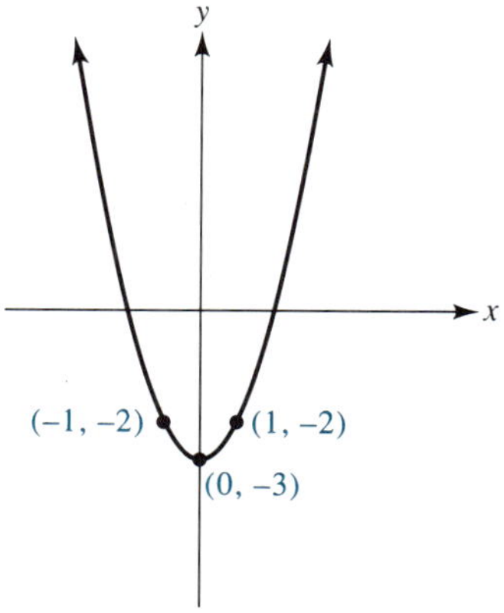

□

The graph of $y = x^2 + k$ is a parabola that opens upward with vertex $(0, k)$. Its shape is exactly that of the standard, $y = x^2$.

If we can move the standard parabola up or down, what about right or left? Adding a constant to x^2 adds to the y value. To move right or left, we need to affect the x-value. The way to do this is to add a constant to x before the squaring takes place, as in $y = (x + 2)^2$. The vertex of the graph of $y = (x + 2)^2$ is the point with smallest y-value. The value of $(x + 2)^2$ is always nonnegative; it is 0 when $x = -2$. The vertex is $(-2, 0)$. Each point on the parabola will be two units to the left of a point on $y = x^2$.

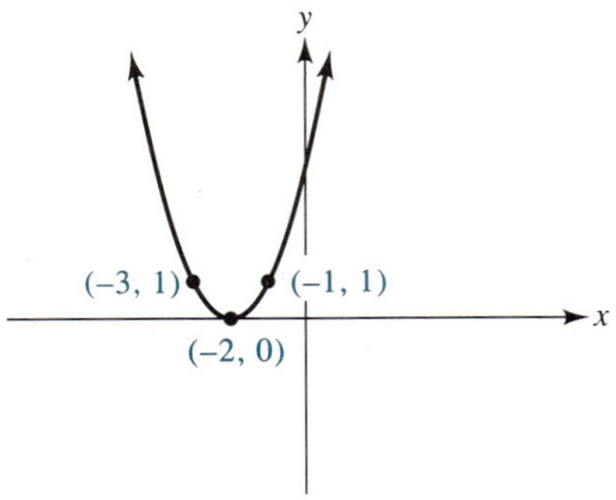

We can make the description of this variation a bit easier by writing $y = [x - (-2)]^2$ or, in general, $y = (x - h)^2$. The graph of $y = (x - h)^2$ is a parabola that opens upward with vertex $(h, 0)$. Its shape is exactly that of the standard $y = x^2$.

EXAMPLE Sketch the graph of $y = (x - 1)^2 - 3$.
Comparing $y = (x - 1)^2 - 3$ to $y = x^2$, we see the vertex has been shifted to $(1, -3)$. (Both kinds of shifts have taken place.) The graph still opens upward with the shape of the standard.
The graph of $y = (x - 1)^2 - 3$ is

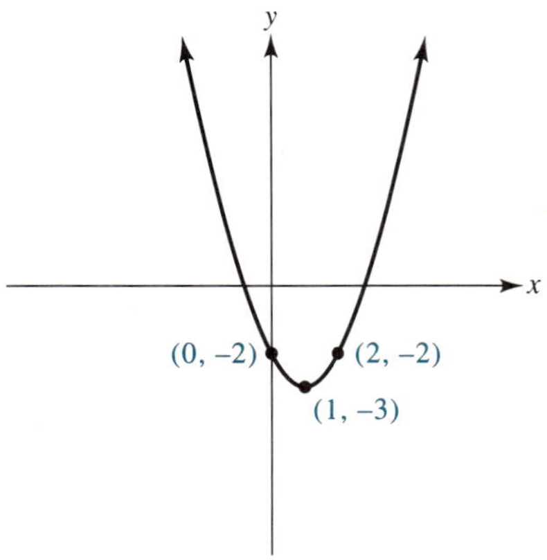

■

How do we change the shape of the standard parabola while keeping the characteristic form of a parabola? (We're fudging a bit, of course, because the characteristic form is not formally described until the next section.) The only possibilities are to have the parabola open more rapidly or slowly.

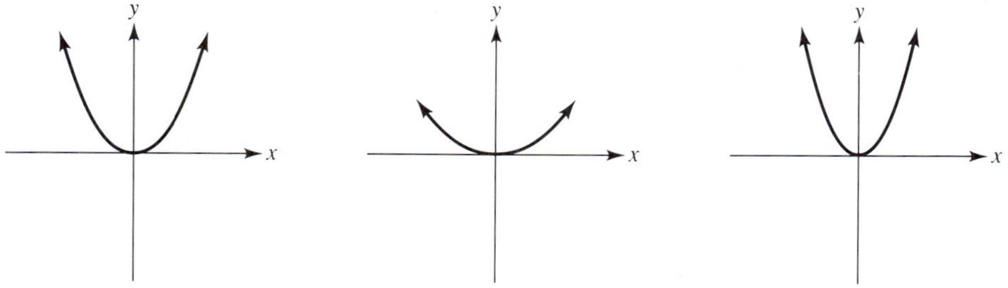

Standard Graph (2) Parabola *A* Parabola *B*

Comparing parabola *A* to the standard parabola, we see *A* has smaller *y*-values for corresponding *x*-values. Parabola *B* has larger *y*-values for the same *x*'s. By multiplying the x^2-term by a constant, we can produce this effect. Here are some examples.

Equation	$x = 0$	$x = 1$	$x = -1$	$x = 2$
$y = x^2$	(0, 0)	(1, 1)	(−1, 1)	(2, 4)
$y = 5x^2$	(0, 0)	(1, 5)	(−1, 5)	(2, 20)
$y = 0.5x^2$	(0, 0)	(1, 0.5)	(−1, 0.5)	(2, 2)

Notice that the vertex has not been moved. One possibility is not covered in these examples, that of a negative multiplier. The solutions of the equation $y = -2x^2$ have only nonpositive *y*-values; its graph opens downward. The absolute value of -2 determines the spread of the parabola.

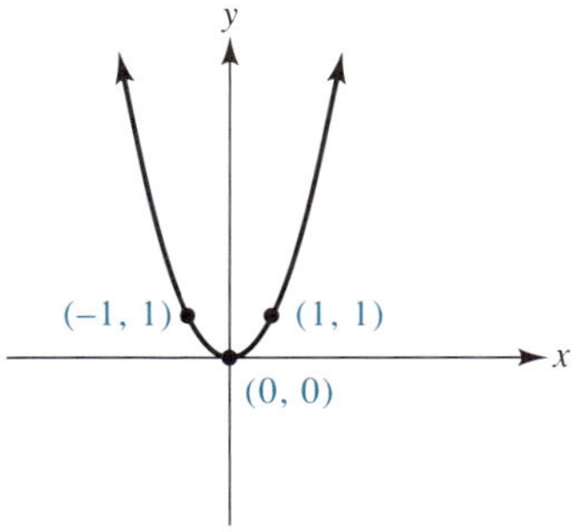

$y = x^2$

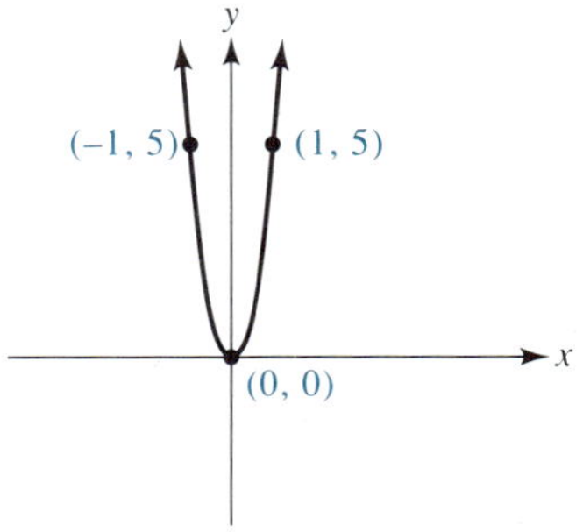

$y = 5x^2$

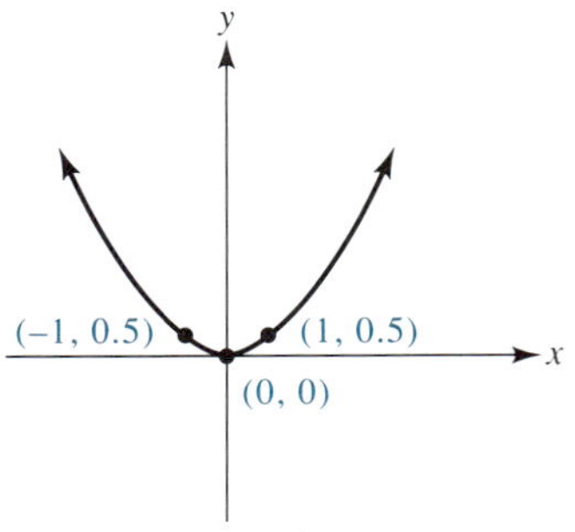

$y = 0.5x^2$

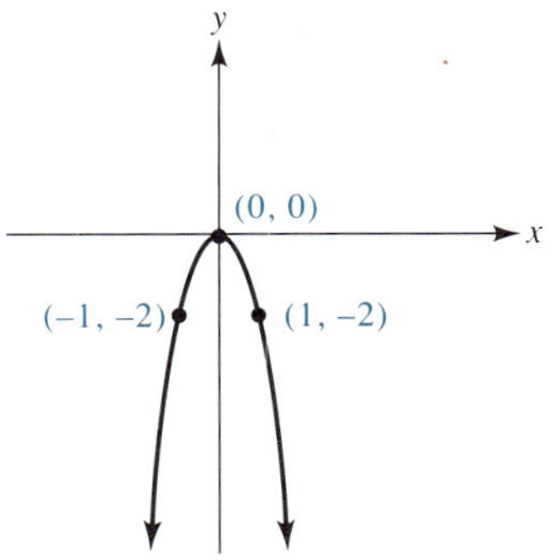

$y = -2x^2$

We can now pull together all the types of changes we've seen. The graph of $y = a(x - h)^2 + k$, $a \neq 0$, is a parabola with vertex (h, k). It opens upward if $a > 0$, downward if $a < 0$. The absolute value of a determines the rate of opening. Absolute values greater than 1 produce "skinnier" parabolas than the standard; absolute values less than 1 give parabolas "fatter" than the standard.

EXAMPLE Sketch the graph of $y = -1.5(x + 2)^2 + 1$.
The vertex is $(-2, 1)$ and the parabola opens downward. It is skinnier than the standard parabola because $|-1.5| > |1|$. Locate the vertex and sketch a parabola opening downward.

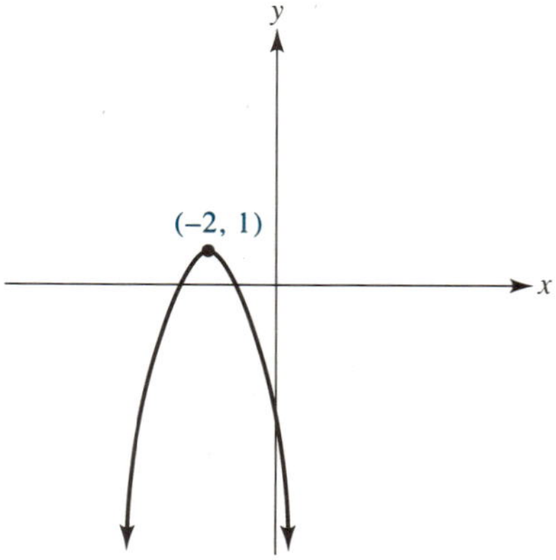

Label a point to the right and one to the left of the vertex to establish the scale.

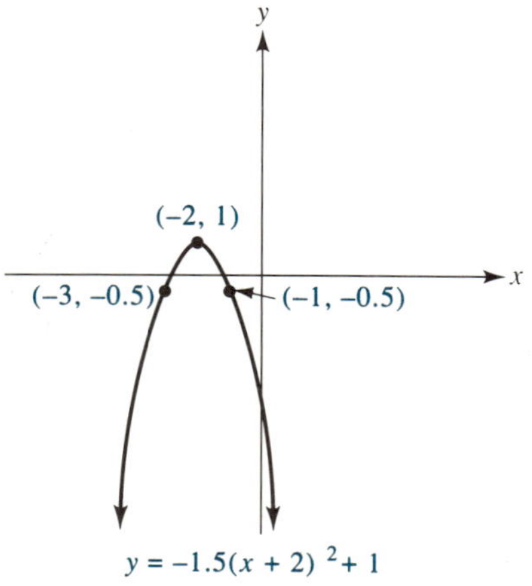

■

EXAMPLE Sketch the graph of $x = 2(y - 1)^2 + 4$.
Remember this graph will be a horizontal parabola like $x = y^2$. The 4 increases the x-values and the 1 affects y-values. The vertex is $(4, 1)$. The 2 makes the parabola skinnier than $x = y^2$. It opens to the right, the positive direction, because $2 > 0$.

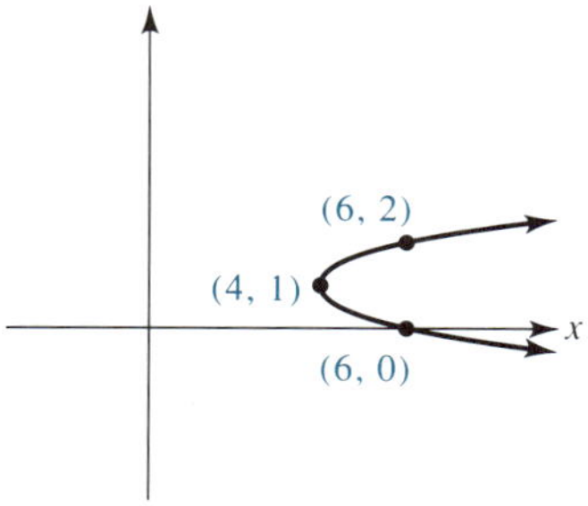

■

Now we're ready to tackle the general form of a second-degree two-variable equation with only one of the variables squared. The basic idea is to rewrite such equations in a form we have already handled,

$$y = a(x - h)^2 + k \qquad (\text{or } x = a(y - k)^2 + h)$$

The general form is

$$y = ax^2 + bx + c, \quad a \neq 0 \qquad (\text{or } x = ay^2 + by + c, \quad a \neq 0).$$

You may remember doing problems somewhat like this in Section 1.2. The technique we use is completing the square.

To rewrite $y = 4x^2 - 3x + 7$ in the form $y = a(x - h)^2 + k$, we begin by factoring the a, 4, out of the first two terms.

$$\begin{aligned} & y = 4x^2 - 3x + 7 \\ \Leftrightarrow \quad & y = 4\left[x^2 - \frac{3}{4}x \quad\right] + 7 \end{aligned}$$

Next we determine the necessary constant term to make $x^2 - \frac{3}{4}x + c$ a perfect square. The square of half of $-\frac{3}{4}$ is $\left(-\frac{3}{8}\right)^2$, or $\frac{9}{64}$. Adding $\frac{9}{64}$ inside the parentheses is the equivalent of adding $4\left(\frac{9}{64}\right)$ to the right-hand side; we must subtract this number from the right-hand side to have an equivalent equation.

$$\begin{aligned} & y = 4\left[x^2 - \frac{3}{4}x \quad\right] + 7 \\ \Leftrightarrow \quad & y = 4\left[x^2 - \frac{3}{4}x + \frac{9}{64}\right] + 7 - \frac{36}{64} \\ \Leftrightarrow \quad & y = 4\left[x - \frac{3}{8}\right]^2 + \frac{103}{16} \end{aligned}$$

Thus, the graph of $y = 4x^2 - 3x + 7$ opens upward $(4 > 0)$ and is skinny $(|4| > 1)$, with vertex $\left(\frac{3}{8}, \frac{103}{16}\right)$.

EXAMPLE Sketch the graph (of the solution set) of

$$y = -\frac{1}{3}x^2 - \frac{1}{6}x + \frac{1}{12}.$$

We begin by rewriting $y = -\frac{1}{3}x^2 - \frac{1}{6}x + \frac{1}{12}$ in the form $y = a(x - h)^2 + k$.

$$
\begin{aligned}
& y = -\frac{1}{3}x^2 - \frac{1}{6}x + \frac{1}{12} \\
\Leftrightarrow \quad & y = -\frac{1}{3}\left[x^2 + \frac{1}{2}x \quad \right] + \frac{1}{12} \\
\Leftrightarrow \quad & y = -\frac{1}{3}\left[x^2 + \frac{1}{2}x + \left(\frac{1}{4}\right)^2\right] + \frac{1}{12} - \left(-\frac{1}{3}\right)\left(\frac{1}{4}\right)^2 \\
\Leftrightarrow \quad & y = -\frac{1}{3}\left[x + \frac{1}{4}\right]^2 + \frac{5}{48} \\
\Leftrightarrow \quad & y = -\frac{1}{3}\left[x - \left(-\frac{1}{4}\right)\right]^2 + \frac{5}{48}
\end{aligned}
$$

The graph is a downward opening parabola that is fatter than the standard $(|-\frac{1}{3}| < 1)$; it has vertex $(-\frac{1}{4}, \frac{5}{48})$.

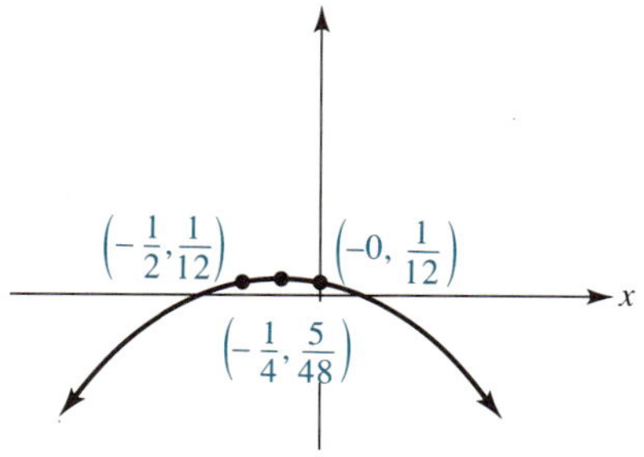

In some ways the vertex is the most important point on a parabola. For vertical parabolas it is the high or low point, and its y-coordinate is the maximum or minimum value of y. We now have an algebraic way to find this maximum or minimum value without sketching the graph.

EXAMPLE Jane Jones has been hired to analyze the production costs of various products for the Standard Company. She finds that the cost of producing each bookcase (style 239a) varies according to the number produced each day. It is expensive to open the plant and produce only a few bookcases each day. On the other hand, if too many orders are taken, overtime pay and other costs drive up the unit cost for a day. She collects data over a period of time and varied conditions and develops an equation to model the situation. Her equation is $C = 0.2x^2 - 7.8x + 178$ for producing from 0 to 35 bookcases per day, where C is the total cost of production per day and x represents the number of bookcases. What is the minimum cost per day and how many bookcases produce this cost?

The graph of $C = 0.2x^2 - 7.8x + 178$ is an upward-opening parabola, so its vertex is a low point. Completing the square will help us identify this point.

$$
\begin{aligned}
& C = 0.2x^2 - 7.8x + 178 \\
\Leftrightarrow \quad & C = 0.2(x^2 - 39x \quad) + 178 \\
\Leftrightarrow \quad & C = 0.2(x - 19.5)^2 + 101.95
\end{aligned}
$$

The vertex is (19.5, 101.95). Hence, **the minimum cost is \$101.95 to produce 19.5 bookcases per day.**

Note: Jane must interpret this result for her model; it may mean that one bookcase is half finished one day and completed the next day. ■

EXERCISES 2.3

In exercises 1 through 20, sketch the graph of the equation. Label the vertex and one other point.

1. $y = x^2 - 5$
2. $y = x^2 + 3.4$
3. $x = y^2 + \pi$
4. $x = y^2 - \sqrt{2}$
5. $y = -2x^2$
6. $y = 0.25x^2$
7. $x = [1 - \sqrt{13}]y^2$
8. $x - 9y^2 = 0$
9. $y = (x - 5)^2$
10. $y = (x + 3.4)^2$
11. $x = (y + \pi)^2$
12. $x = [y - \sqrt{2}]^2$
13. $y = [\sqrt{2} - x]^2$
14. $y = -[\sqrt{2} - x]^2$
15. $y = 2(x + 3)^2 - 1$
16. $y = -4(x - 1)^2 - 3$
17. $x = -(y - 8)^2 + 2$
18. $x = 2.1(y + 3.3)^2 + 5.9$
19. $y = -\pi(2\pi - x)^2$
20. $x = -(38.9 - y)^2 - 3$

In exercises 21 through 28, use the method of completing the square in order to find the vertex. Determine whether the graph opens up or down or right or left.

21. $y = x^2 - 8x + 17$
22. $y = x^2 + 7x - 3$
23. $x = y^2 + 5y + 3$
24. $x = y^2 - 4.2y - 1.3$
25. $y = 2x^2 - 8x + 13$
26. $y = -2x^2 + 8x - 13$
27. $y = -3x^2 - x + 2$
28. $x = \frac{1}{4}y^2 + \frac{1}{2}y + 2$
29. What is the vertex of the graph of $y = 2x^2 - ax - 3$?
30. What is the vertex of the graph of $x = -ay^2 + aby + c$?

In exercises 31 through 34, what is the minimum or maximum value of *y*?

31. $y = 2x^2 - 8x + 13$
32. $y = 2x^2 - 8x + 3$
33. $y = -2x^2 + 8x - 13$
34. $y = -3x^2 - x + 5$
35. What is the maximum value of the product of a number and 5 minus the number?
36. Sketch the graph of $y = x^2 - 5x - 6$ by first factoring the right-hand side to determine the x-intercepts. Next complete the square on the right-hand side to determine the vertex. What is the relationship between the x-intercepts and the vertex?
37. Use the quadratic formula to find the x-intercepts of the graph of $y = ax^2 + bx + c$, $a \neq 0$. Complete the square on the right-hand side to find the vertex of the graph of the equation. What is the relationship between the x-intercepts and the vertex?
38. For what values of x is the graph of $y = -x^2 + 4x + 12$ above the x-axis?

In exercises 39 through 44, sketch the graph of the equation.

39. $y = 2x^2 - 7, x \geq 0$
40. $y = (x - 7)^2 + 1, x > 7$
41. $x = (y + 3)^2, y < -3$
42. $x = -y^2 + 5, y \geq 0$
43. $y = 5x^2 - 2x + 1, x < \frac{1}{5}$
44. $x = 0.2y^2 - 8.2, y \leq 0$
45. State the restriction necessary to produce the graph of the right half of the parabola $y = x^2 - 7x - 4$.
46. State the restriction necessary to produce the bottom half of the parabola $x = -y^2 - 4y - 4$.
47. The vertical position p measured in feet of a projectile at time t seconds is given by $p = -16t^2 + 64t + 80$. What is the maximum height the projectile will attain? At what time will this occur?
48. Allie and Mary decide to fence part of the back yard as a play area for their children. A fence constructed by the county already exists across the back of the yard. Allie's father gave them 300 feet of fencing. They decide to construct a rectangular play area using the existing back fence for one side. What dimensions should the area be in order to maximize the space for play? What factors other than maximum area might Allie and Mary consider before arriving at a decision?
49. George is a very scientific gardener. He wants his new apricot orchard to produce as many pounds of fruit as possible, but he knows that the more trees planted per acre, the less fruit each produces. After careful reading, he finds that if 25 trees are planted per acre, each can be expected to produce 43 pounds of apricots. Each additional tree per acre will decrease the yield by 1 pound per tree. Based on this information, how many trees should George plant on each of the 4 acres? How many pounds of apricots would this produce?
50. After collecting data from her cattle-farming operation, Irene models the food cost per head per day with the

equation $C = 0.02n^2 - 8.16n + 835.57$, where C is the food cost per head per day in dollars and n is the number of head of cattle. How many head of cattle should Irene have in order to minimize the cost of food per head per day? What is the cost associated with this number of cattle?

(Graphing calculator) In exercises 51 through 54, enter the equation of the parabola, (a) position the cursor to find the approximate x-intercepts, (b) use a solving routine to find the approximate x-intercepts, and (c) position the cursor to find the approximate vertex point.

51. $y = \sqrt{3}(x - 3.5)^2 + \pi$
52. $y = -9.57(x + \sqrt{2})^2 - 7.33$
53. $y = 8.4x^2 - 5.2x - 3.1$
54. $y = -6x^2 + \sqrt[3]{7}\,x + 1.1$

2.4 The Conic Sections

Toolkit

You'll need the distance formula from Section 2.1. Also, the distance from a point to a line is the distance from the point to the nearest point on the line.

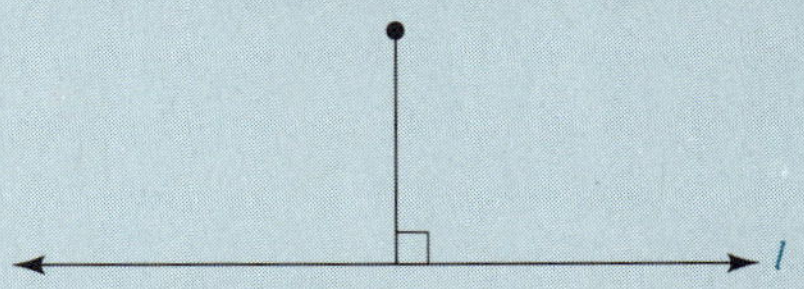

In Section 1.3 we used the fact that $u = v \Rightarrow u^2 = v^2$. If we know u and v are nonnegative, then $u = v \Leftrightarrow u^2 = v^2$.

In this section we discuss the other families of curves that are graphs of equations involving two-variable expressions of at most second-degree. We do this first from a geometric point of view. In Section 2.3 we mentioned that any parabola satisfies a geometric definition. Applying analytic geometry techniques to this definition, we will see that the equations we identified in the last section are the only ones that are associated with vertical or horizontal parabolas.

Definition A *parabola* is the set of all points in a plane equidistant from a given point, the focus, and a given line, the directrix.

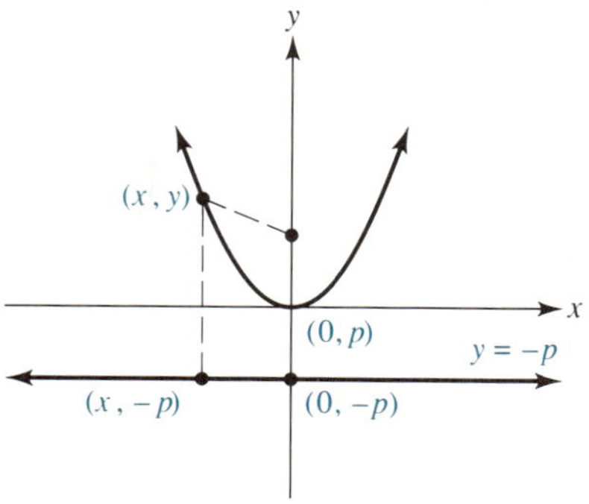

To use analytic geometry to describe a set of points satisfying this definition, we are allowed to place the axes wherever they will make our calculations simpler. One choice for this situation is to place the axes so that the focus is $(0, p)$ and the directrix is $y = -p$. We draw the picture for $p > 0$. Notice that the vertex, $(0, p)$, is a point equidistant from the focus and the directrix. If (x, y) represents any point on the parabola, then we have

$$\sqrt{(x - x)^2 + (y + p)^2} = \sqrt{(x - 0)^2 + (y - p)^2}$$

Because the left-hand side and the right-hand side are known to be nonnegative

$$\Leftrightarrow \quad (y + p)^2 = x^2 + (y - p)^2$$

$$\Leftrightarrow \quad y^2 + 2py + p^2 = x^2 + y^2 - 2py + p^2$$

$$\Leftrightarrow \quad 4py = x^2$$

$$\Leftrightarrow \quad y = \frac{x^2}{4p}$$

We recognize this as the type of equation whose graph was said to be a vertical parabola with vertex (0, 0). If $p > 0$, the parabola opens upward, as in our picture. Choosing axes so that the focus is on the x-axis and the vertex is at (0, 0) will produce an equation of the form $x = ay^2$. Other placements of the axes, focus, and directrix can be found to produce the other forms of equations associated with parabolas in Section 2.3. The important point here is that we have shown the connection between the geometric definition of a parabola and the algebraic definition given previously.

Illustration Parabolas are extremely useful for many applications. One of the most widely known is the parabolic surface or mirror. This surface is created by rotating a parabola about its axis. In a telescope the light waves from a distant object are nearly parallel to the axis of the parabola and reflect off the surface to the focus of the parabola. In certain headlights, if the light source is at the focus, the light rays are reflected off the parabolic surface in rays that are parallel to the axis.

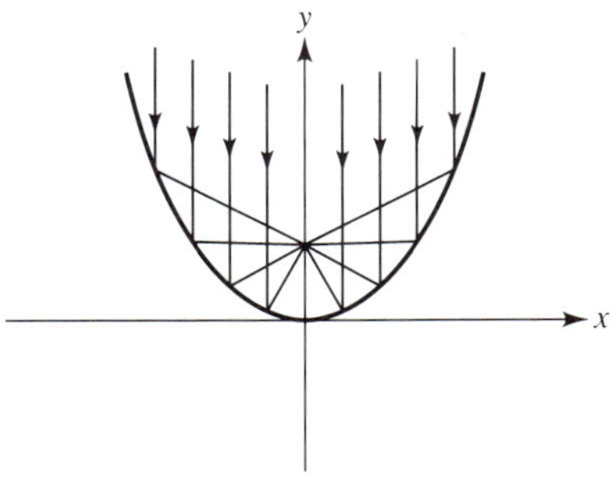

Cross section of telescope mirror

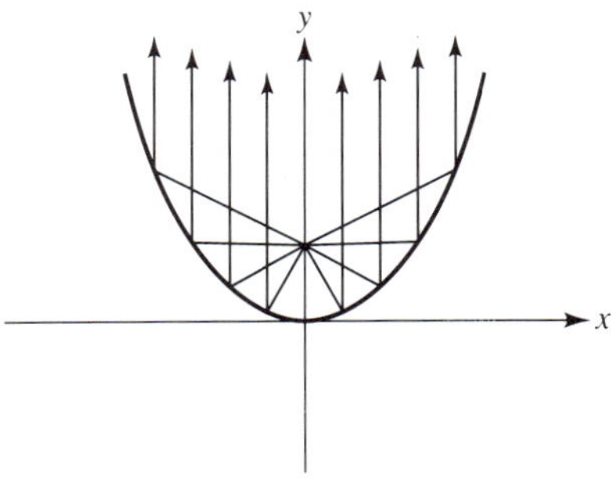

Cross section of headlight

□

Another common geometric figure, the circle, has a simpler geometric description.

Definition A *circle* is the set of all points in a plane at a given distance, the radius, from a given point, the center.

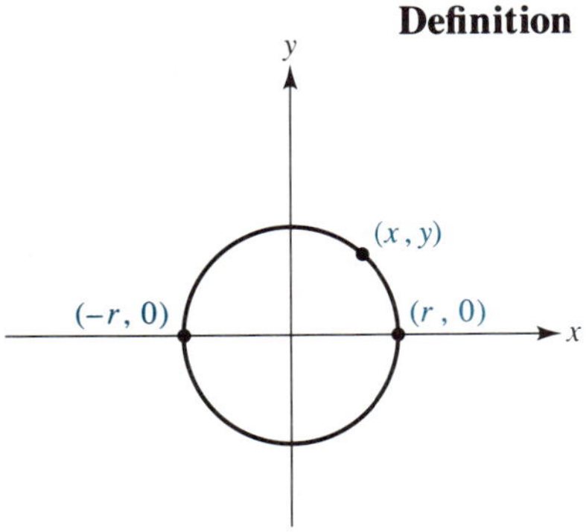

The most straightforward choice for the placement of the axes to obtain an algebraic description of a circle is to place the center of the circle at the origin. If (x, y) is a point on the circle and r is the given distance, then

$$\sqrt{(x - 0)^2 + (y - 0)^2} = r$$

$$\Leftrightarrow \quad x^2 + y^2 = r^2 \qquad \text{Because } r \geq 0$$

Any circle with center (0, 0) and radius r is the graph of $x^2 + y^2 = r^2$. We investigate how to describe a circle whose center is (h, k) in the next section, but if you think back to your work with parabolas (Section 2.3), you may be able to see how to do this.

The circle with center (0, 0) and radius 1 is called the *unit circle* and will serve as the standard graph for this family of curves.

Standard Graph 3: $x^2 + y^2 = 1$

The graph is a circle with center (0, 0) and radius 1. Suggested labels are the x- and y-intercepts.

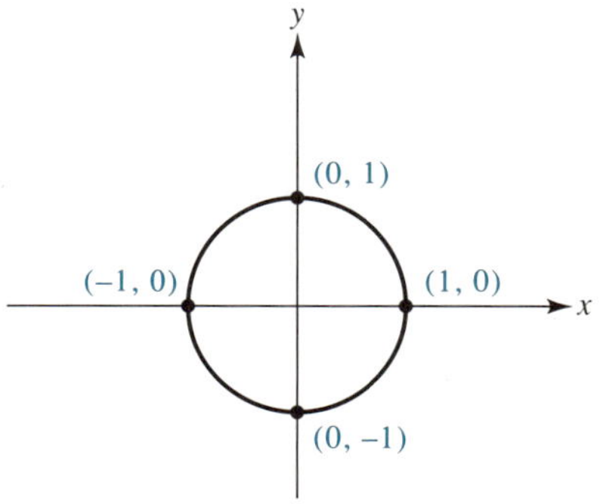

EXAMPLE Sketch the graph of $x^2 + y^2 = 24$.

The graph is a circle with center (0, 0) and radius $\sqrt{24}$. We can estimate that $\sqrt{24}$ is slightly less than $\sqrt{25}$, which is 5, if we are concerned with scale. Labeling the intercepts is sufficient.

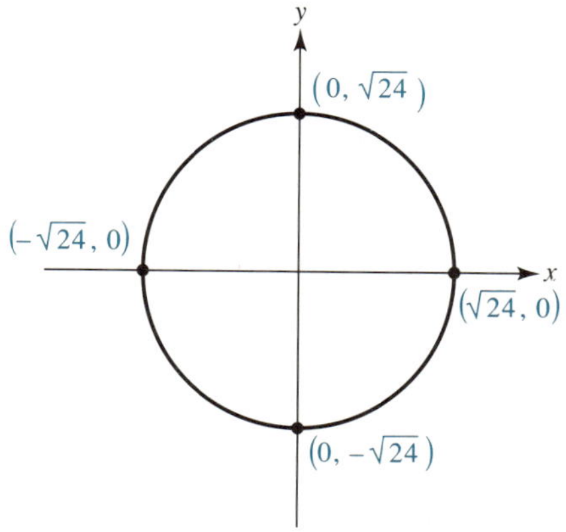

■

EXAMPLE Give an equation for the following graph.

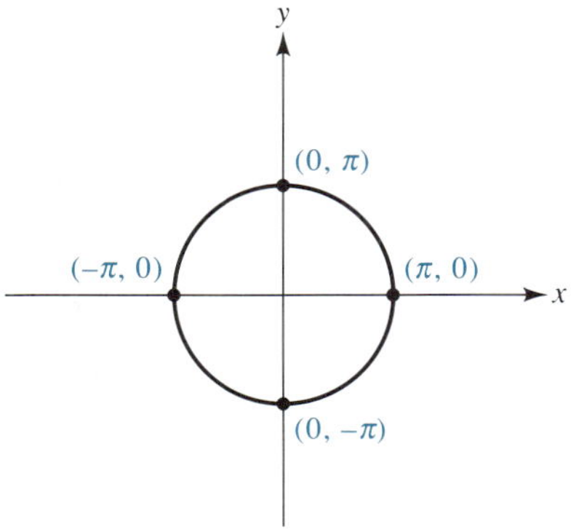

The graph is a circle with center (0, 0) and radius π.
An equation for the graph is $x^2 + y^2 = \pi^2$.

Geometers consider a circle to be a special case of an ellipse.

Definition Geometrically an *ellipse* is the set of points in a plane such that the sum of the distances from a point in the set to two given points, the foci, is a constant.

Imagine a string pinned down at A and B and a pencil holding the string taut at C.

Slowly move the pencil to the left around A, keeping the string taut. Continue around until the pencil returns to the starting point. The figure traced out is an ellipse with foci A and B; $AC + CB$ represents the length of the string. To find an algebraic description, we place the axes so that the center of the ellipse is (0, 0) and the foci are $(f, 0)$ and $(-f, 0)$. The center of the ellipse is the midpoint of the line segment connecting the foci. For convenience we name the constant sum of the distances $2a$.

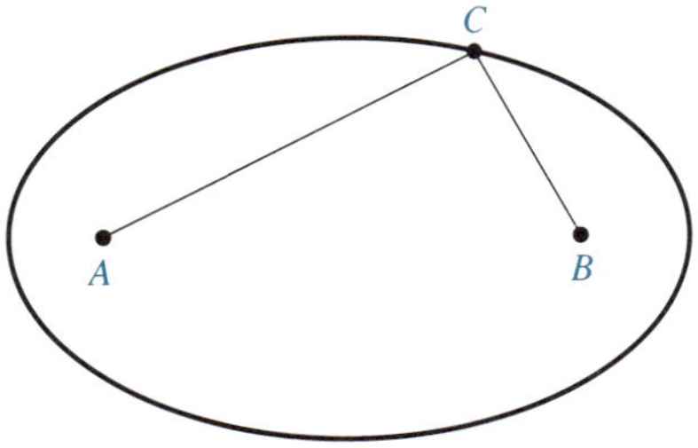

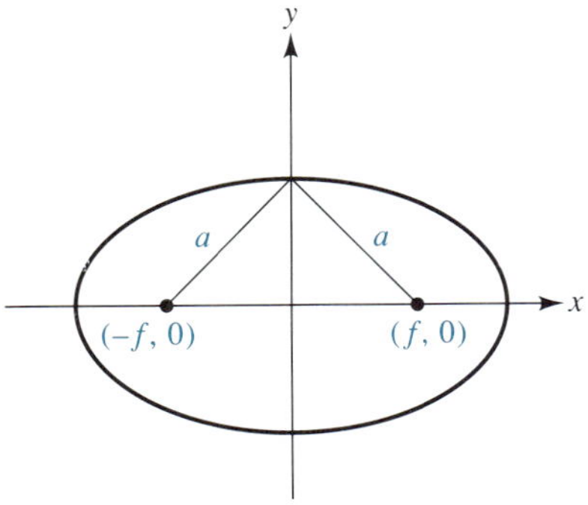

Note: The distance from a focus to a y-intercept is a.

If $a = f$ or $a = 0$, a degenerate ellipse is produced. Let (x, y) represent a point on the ellipse. We must have

$$\sqrt{(x - f)^2 + (y - 0)^2} + \sqrt{(x + f)^2 + (y - 0)^2} = 2a$$

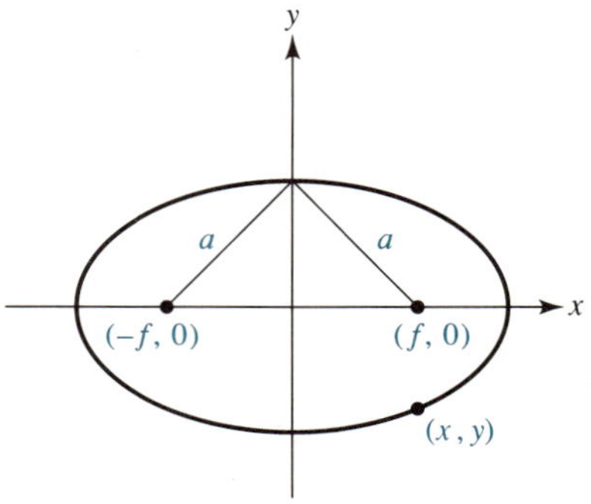

An equation without radicals might be easier to use.

$$\sqrt{(x-f)^2+(y-0)^2}+\sqrt{(x+f)^2+(y-0)^2}=2a$$

$$\Leftrightarrow \quad \sqrt{(x-f)^2+y^2}=2a-\sqrt{(x+f)^2+y^2}$$

Because each side of the equation represents a nonnegative number

$$\Leftrightarrow \quad (x-f)^2+y^2=4a^2-4a\sqrt{(x+f)^2+y^2}+(x+f)^2+y^2$$

$$\Leftrightarrow \quad x^2-2xf+f^2+y^2=4a^2-4a\sqrt{(x+f)^2+y^2}+x^2+2xf+f^2+y^2$$

$$\Leftrightarrow \quad -4xf-4a^2=-4a\sqrt{(x+f)^2+y^2}$$

$$\Leftrightarrow \quad xf+a^2=a\sqrt{(x+f)^2+y^2}$$

$$\Leftrightarrow \quad x^2f^2+2xfa^2+a^4=a^2x^2+2xfa^2+f^2a^2+a^2y^2$$

$$\Leftrightarrow \quad a^4-f^2a^2=(a^2-f^2)x^2+a^2y^2$$

$$\Leftrightarrow \quad a^2(a^2-f^2)=(a^2-f^2)x^2+a^2y^2$$

In the nondegenerate case, $a \neq f$ and $a \neq 0$, so we can divide both sides by $a^2(a^2-f^2)$. Then

$$a^2(a^2-f^2)=(a^2-f^2)x^2+a^2y^2, \qquad a \neq f, a \neq 0$$

$$\Leftrightarrow \quad 1=\frac{x^2}{a^2}+\frac{y^2}{a^2-f^2}$$

Let's examine our picture once more and label the y-intercept with $y > 0$, $(0, b)$. By the Pythagorean theorem, $b^2+f^2=a^2$, or $b^2=a^2-f^2$. We can rewrite our description of the ellipse as $1=x^2/a^2+y^2/b^2$. The ellipse has y-intercepts $(0, b)$ and $(0, -b)$. To find its x-intercepts, substitute 0 for y. The x-intercepts are $(a, 0)$ and $(-a, 0)$.

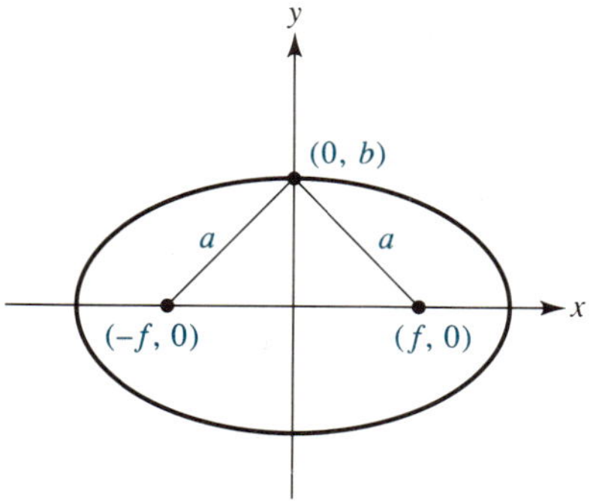

The graph of

$$\frac{x^2}{a^2}+\frac{y^2}{b^2}=1$$

is an ellipse with center $(0, 0)$, x-intercepts $(a, 0)$ and $(-a, 0)$, and y-intercepts $(0, b)$ and $(0, -b)$. Its foci are $(-f, 0)$ and $(f, 0)$ if $f^2=a^2-b^2$; the foci are $(0, -f)$ and $(0, f)$ if $f^2=b^2-a^2$. You can see now that a circle is a special case of an ellipse. If $a=b$, then $x^2/a^2+y^2/b^2=1$ is equivalent to $x^2/a^2+y^2/a^2=1$, or $x^2+y^2=a^2$.

There is no distinguished member of the family of noncircular ellipses to choose as a representative. This is an important geometric shape, though, so we wish to have a standard graph. We use the general form itself as a representative.

Standard Graph 4: $\dfrac{x^2}{a^2} + \dfrac{y^2}{b^2} = 1,\ a \neq b$

The graph is an ellipse with center at (0, 0), x-intercepts $(a, 0)$ and $(-a, 0)$, and y-intercepts $(0, b)$ and $(0, -b)$. If $a^2 > b^2$, its foci are $(f, 0)$ and $(-f, 0)$, where $f = \sqrt{a^2 - b^2}$. If $b^2 > a^2$, the foci are $(0, f)$ and $(0, -f)$, where $f = \sqrt{b^2 - a^2}$. Suggested labels are the x- and y-intercepts.

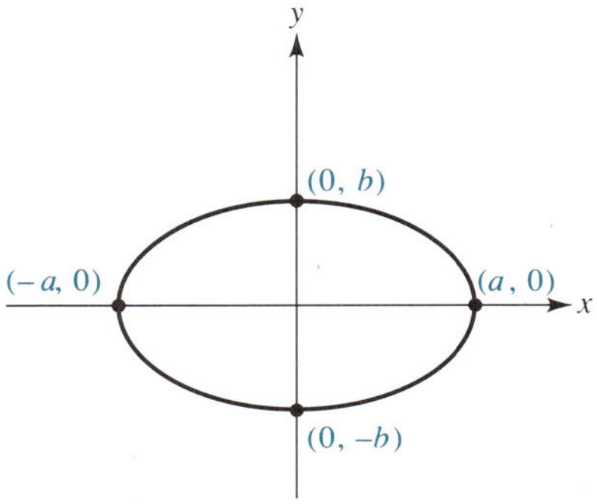

EXAMPLE Sketch the solution set of $\dfrac{x^2}{16} + \dfrac{y^2}{5} = 1$.

The graph is an ellipse with center (0, 0), x-intercepts (4, 0) and (−4, 0), and y-intercepts $(0, \sqrt{5})$ and $(0, -\sqrt{5})$.

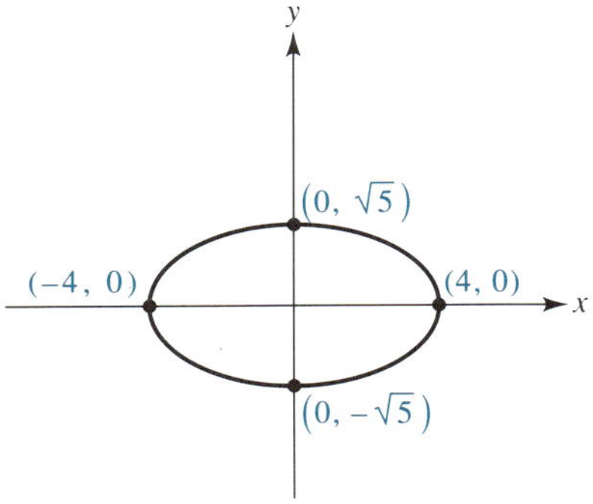

EXAMPLE Sketch the graph of $\dfrac{x^2}{4} + \dfrac{y^2}{25} = 1$ and $y \geq 0$.

The graph of $x^2/4 + y^2/25 = 1$ is an ellipse with intercepts (2, 0), (−2, 0), (0, 5), (0, −5). Because $y \geq 0$, we have only the top half of this ellipse.

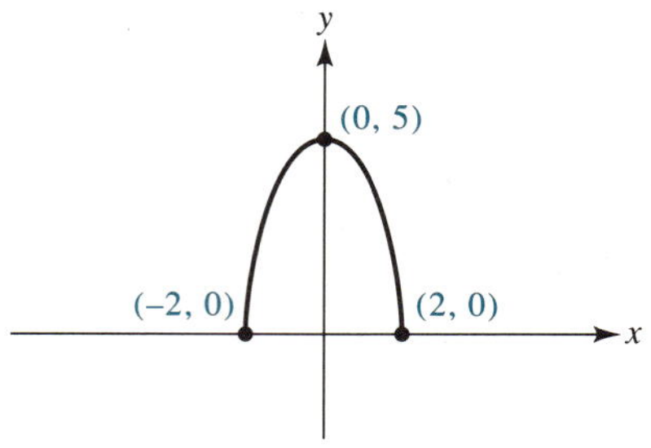

Note: This is the same graph you would sketch for $y = 5\sqrt{1 - x^2/4}$ because

$$\frac{x^2}{4} + \frac{y^2}{25} = 1$$

$$\Leftrightarrow \qquad y^2 = 25\left(1 - \frac{x^2}{4}\right)$$

$$\Leftrightarrow \qquad y = 5\sqrt{1 - \frac{x^2}{4}} \quad \text{or} \quad y = -5\sqrt{1 - \frac{x^2}{4}}$$

Thus

$$\frac{x^2}{4} + \frac{y^2}{25} = 1 \quad \text{and} \quad y \geq 0 \quad \Leftrightarrow \quad y = 5\sqrt{1 - \frac{x^2}{4}}$$

$$\frac{x^2}{4} + \frac{y^2}{25} = 1 \quad \text{and} \quad y \leq 0 \quad \Leftrightarrow \quad y = -5\sqrt{1 - \frac{x^2}{4}}$$

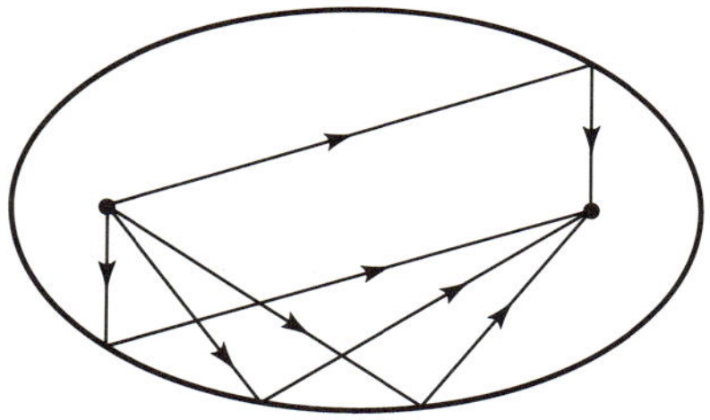

An ellipse has the property that light or sound generated at one focus will be reflected to the other focus. This fact has been used recently in employing ultrasound to break up kidney stones.

What sort of figure results from making the difference between the distances from a point to two distinct points, A and B, a constant?

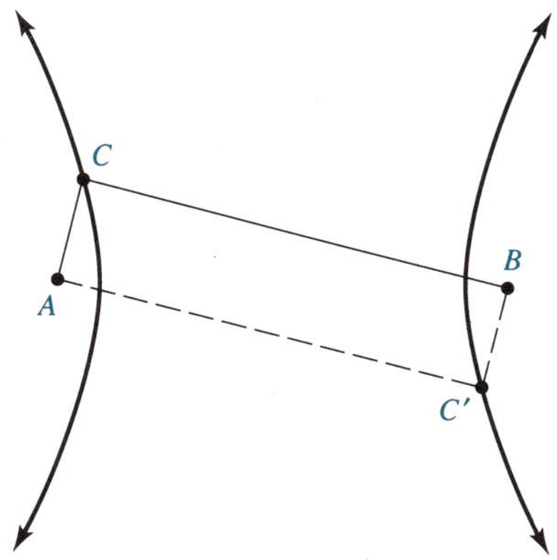

$CB - CA = K$
$C'A - C'B = K$

This means if D is any point on the figure, $|DA - DB| = K$

Definition A *hyperbola* is the set of points in a plane such that the difference of the distances from a point in the set to two given points, the foci, is a positive constant.

Again we place the axes so that the foci are $(-f, 0)$ and $(f, 0)$. It also simplifies the algebraic computation to use $2a$ to represent the constant difference.

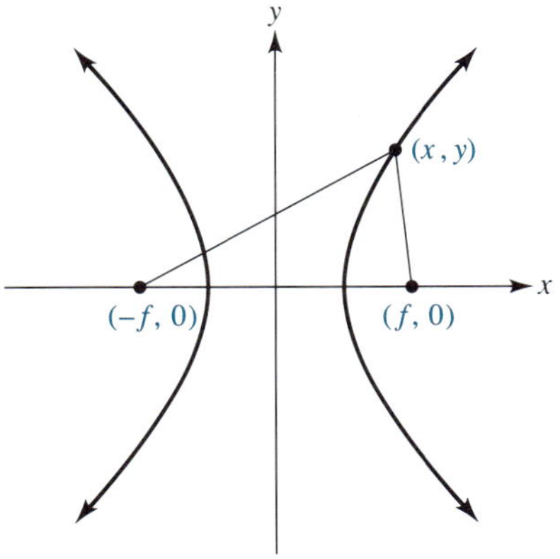

We can see what a measures in our picture by considering the right-hand x-intercept, $(z, 0)$. Its distance from $(-f, 0)$ is $f + z$; it is $f - z$ units from $(f, 0)$. The difference $(f + z) - (f - z)$ is $2a$. We have $2z = 2a$, or $z = a$. The x-intercepts are $(a, 0)$ and $(-a, 0)$. For a point (x, y) on the hyperbola in the first quadrant, we have

$$\sqrt{(x-(-f))^2+(y-0)^2} - \sqrt{(x-f)^2+(y-0)^2} = 2a$$

$$\Leftrightarrow \quad \sqrt{(x+f)^2+y^2} = 2a + \sqrt{(x-f)^2+y^2}$$

$$\Leftrightarrow \quad (x+f)^2+y^2 = 4a^2 + 4a\sqrt{(x-f)^2+y^2} + (x-f)^2+y^2$$

$$\Leftrightarrow \quad 4xf - 4a^2 = 4a\sqrt{(x-f)^2+y^2}$$

$$\Leftrightarrow \quad xf - a^2 = a\sqrt{(x-f)^2+y^2}$$

$$\Leftrightarrow \quad x^2f^2 - 2xfa^2 + a^4 = a^2(x-f)^2 + a^2y^2$$

$$\Leftrightarrow \quad x^2f^2 - 2xfa^2 + a^4 = a^2x^2 - 2xfa^2 + f^2a^2 + a^2y^2$$

$$\Leftrightarrow \quad x^2f^2 - x^2a^2 - a^2y^2 = f^2a^2 - a^4$$

$$\Leftrightarrow \quad (f^2 - a^2)x^2 - a^2y^2 = (f^2 - a^2)$$

The x-intercepts must lie between the foci, so $f > a$. (Because the foci are distinct points, $f \neq a$.) We can rewrite our equation by dividing by $a^2(f^2 - a^2)$.

$$(f^2 - a^2)x^2 - a^2y^2 = a^2(f^2 - a^2), \qquad a \neq 0, a \neq f$$

$$\Leftrightarrow \quad \frac{x^2}{a^2} - \frac{y^2}{f^2 - a^2} = 1$$

To have a neater equation we define b^2 to be $f^2 - a^2$. We will see the geometric significance of b after stating the algebraic definition. This definition has two versions.

The graph of

$$\frac{x^2}{a^2} - \frac{y^2}{b^2} = 1$$

is a hyperbola whose branches open to the left and right and which has x-intercepts $(-a, 0)$ and $(a, 0)$ and foci $(f, 0)$ and $(-f, 0)$, where $f = \sqrt{a^2 + b^2}$.

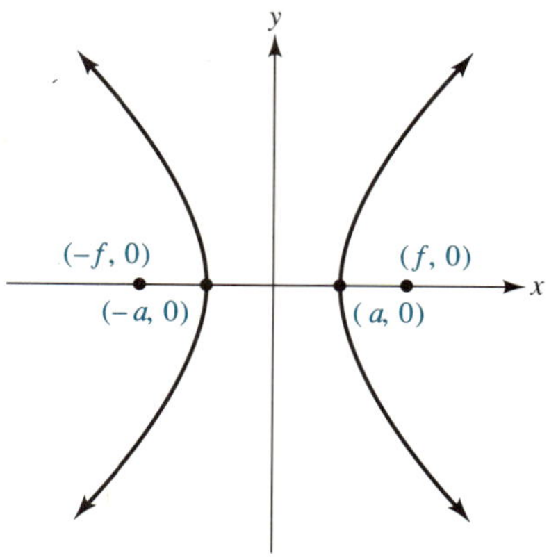

The graph of

$$\frac{y^2}{b^2} - \frac{x^2}{a^2} = 1$$

is a hyperbola whose branches open upward and downward and which has y-intercepts $(0, -b)$ and $(0, b)$ and foci $(0, f)$ and $(0, -f)$, where $f = \sqrt{a^2 + b^2}$.

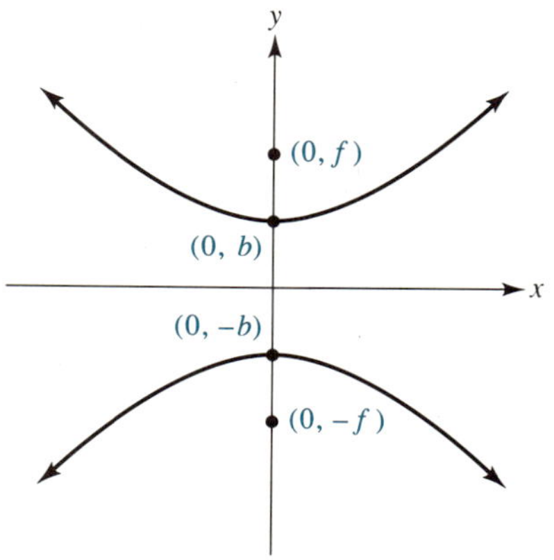

Rewriting $x^2/a^2 - y^2/b^2 = 1$ to have y in terms of x produces

$$y = \frac{b}{a}\sqrt{x^2 - a^2} \quad \text{or} \quad y = -\frac{b}{a}\sqrt{x^2 - a^2}.$$

Remember a is a constant, so for very large or very small values of x, there is not much difference between values of x^2 and values of $x^2 - a^2$. For example, if a is 2, when x is 100,000, $x^2 = 10{,}000{,}000{,}000$ and $x^2 - a^2 = 9{,}999{,}999{,}996$. When x is $-100{,}000$, $x^2 = 10{,}000{,}000{,}000$ and $x^2 - a^2 = 9{,}999{,}999{,}996$. For x's with very large absolute values,

$$y \approx \frac{b}{a}x \quad \text{or} \quad y \approx -\frac{b}{a}x.$$

The equations

$$y = \frac{b}{a}, \quad \text{and} \quad y = -\frac{b}{a}x$$

have linear graphs; these lines approximate the behavior of the hyperbola for extreme values of x. The geometric meaning of the constant b is that the points (a, b) and $(-a, -b)$ are on $y = (b/a)x$; $(a, -b)$ and $(-a, b)$ lie on $y = -(b/a)x$. These points are sufficient to determine these approximating lines and can aid us in sketching $x^2/a^2 - y^2/b^2 = 1$.

EXAMPLE Sketch the graph of $\dfrac{x^2}{2.25} - \dfrac{y^2}{1.69} = 1$.

Matching the equation to our patterns, we see the graph is a hyperbola opening left and right with $a = \sqrt{2.25}$, or 1.5, and $b = \sqrt{1.69}$, or 1.3. Locate (1.5, 1.3), (1.5, -1.3), (-1.5, 1.3), and (-1.5, -1.3) and mark the graphs of $y = (1.3/1.5)x$ and $y = -(1.3/1.5)x$ with dashes.

The branches of the hyperbola have x-intercepts (1.5, 0) and (-1.5, 0). As they spread, they approach more closely the dashed lines.

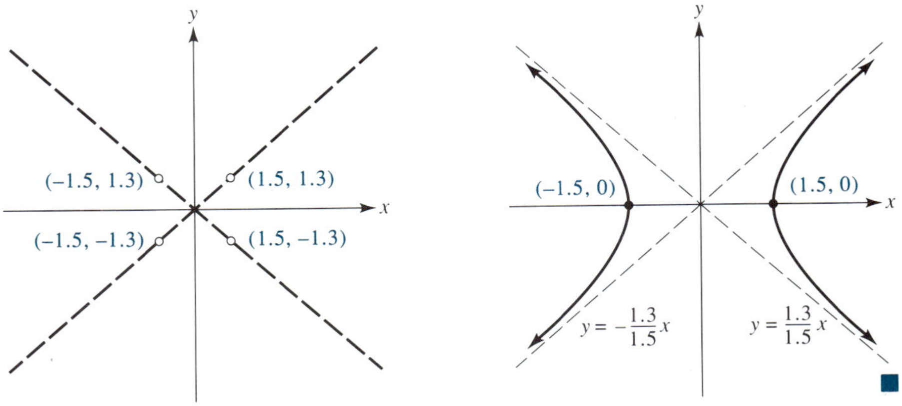

Approximating lines such as $y = (b/a)x$ and $y = -(b/a)x$ for $x^2/a^2 - y^2/b^2 = 1$, are called *asymptotes*. We will see them again later with other types of graphs. One exercise for this section shows that $y = (b/a)x$ and $y = -(b/a)x$ are also asymptotes for $y^2/b^2 - x^2/a^2 = 1$. The asymptotes are not part of the graph but are included to identify it as a hyperbola.

EXAMPLE Sketch the graph of $\dfrac{y^2}{5} - \dfrac{x^2}{3} = 1$.

We recognize this as a hyperbola that opens upward and downward with y-intercepts $(0, \sqrt{5})$ and $(0, -\sqrt{5})$. Locate these points and mark $[\sqrt{3}, \sqrt{5}]$, $(-\sqrt{3}, \sqrt{5})$, $(\sqrt{3}, -\sqrt{5})$, and $(-\sqrt{3}, -\sqrt{5})$ for the asymptotes.

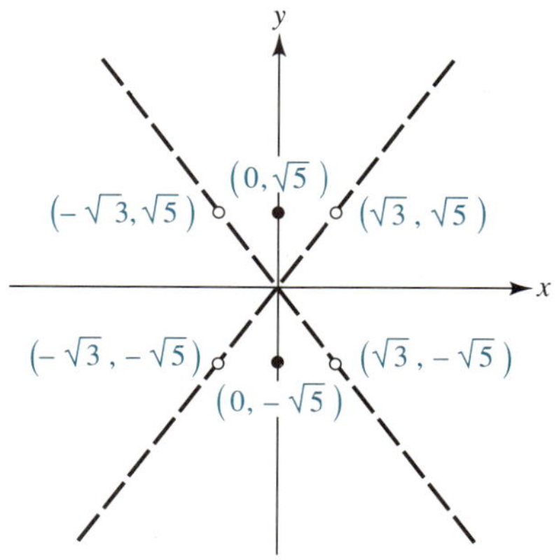

Draw the asymptotes as dashed lines. Label them with the associated equations to identify them as asymptotes. Sketch the branches.

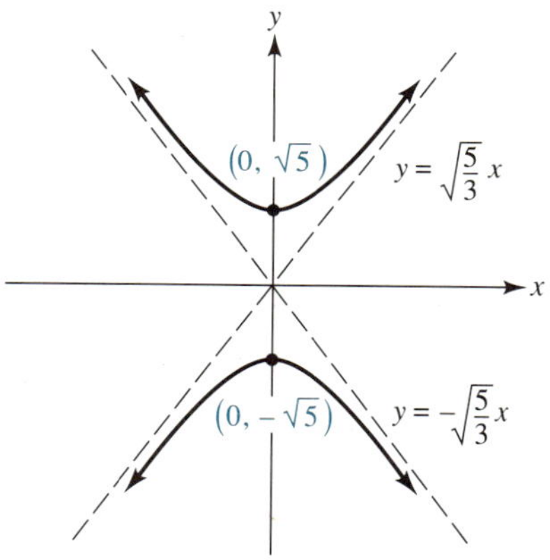

■

Hyperbolas occur in nature and have been used to define architectural forms. One arm of a hyperbola can be seen as the lower edge of the shadow cast on a wall by a lamp shade. Also, some comets appear to have orbits that are a single arm of a hyperbola. Other comets travel in parabolic or elliptical orbits. In our solar system, the planets trace out paths that are ellipses.

We do not choose a standard hyperbola to represent this family. A close relative appears as a standard graph in Section 2.5. When you sketch a hyperbola, you should label its asymptotes and a point on each branch. You may wonder if a branch of a hyperbola is a parabola. It is not. The geometric definitions of a parabola and a hyperbola produce equations that are not variations of each other. Having only one of the variables squared in the defining equation is fundamentally different from having both variables squared.

Except for being described as sets of points that satisfy some requirement, it may not seem that parabolas, circles, ellipses, and hyperbolas are related. In fact, the early Greek geometers studied these figures as different results of one action. A right circular cone is a three-dimensional object of the type illustrated here.

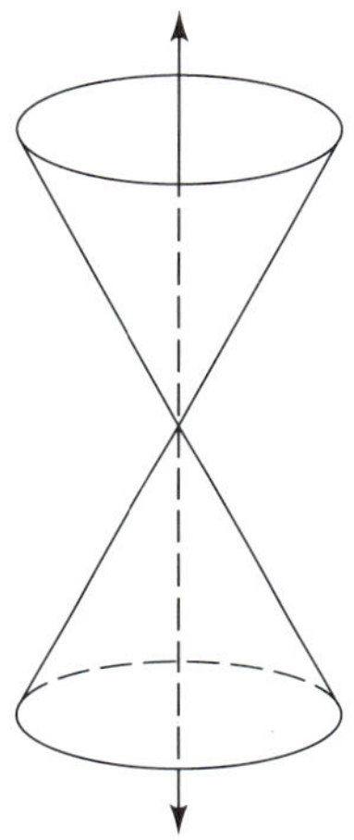

Consider the intersection of a plane with this cone. Some possible results are illustrated.

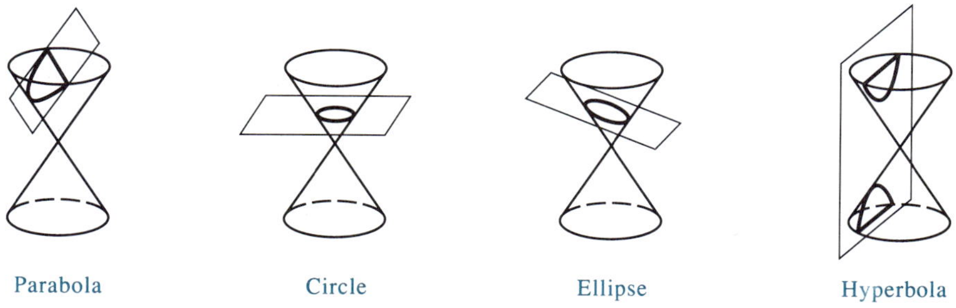

It is this view of parabolas, circles, ellipses, and hyperbolas that give them the name conic sections, or conics.

Algebraically the conic sections share the characteristic that each is the graph of a two-variable quadratic equation. The other side of the coin is whether any two-variable quadratic equation has a solution set whose graph is a conic section. The graph of $ax^2 + bx + cy^2 + dy + e = 0$, a more general two-variable quadratic, is investigated in the next section. Here is a catalog of our results thus far.

<table>
<tr><th>Form</th><th>Typical Graph</th><th>Special Features</th></tr>
<tr><td>$y = a(x - h)^2 + k$,
$a \neq 0$</td><td>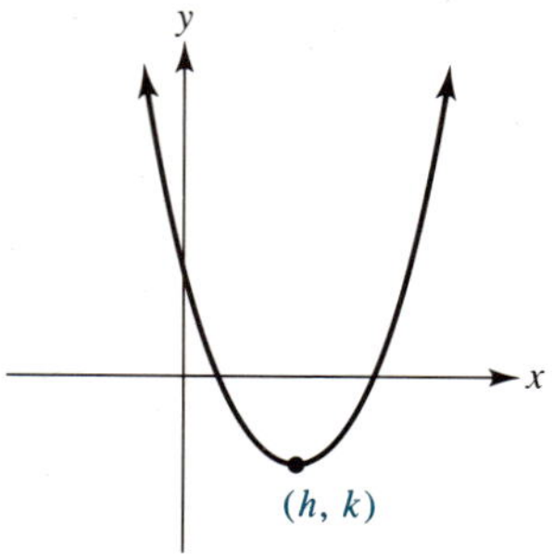
</td><td>parabola;
opening upward if $a > 0$;
vertex (h, k)</td></tr>
<tr><td>$x = b(y - k)^2 + h$,
$b \neq 0$</td><td>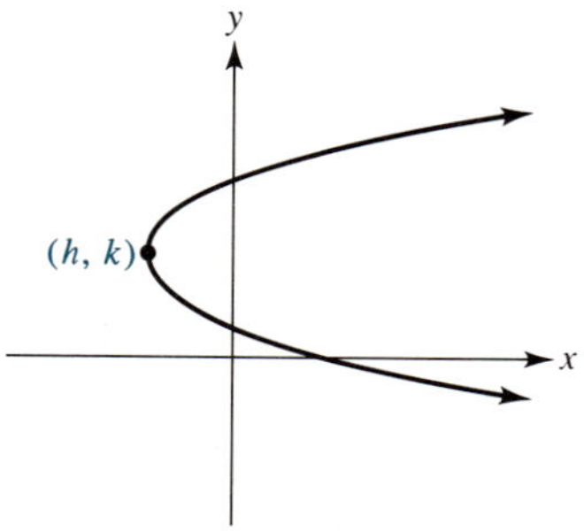
</td><td>parabola;
opening right if $b > 0$;
vertex (h, k)</td></tr>
<tr><td>$x^2 + y^2 = r^2$</td><td>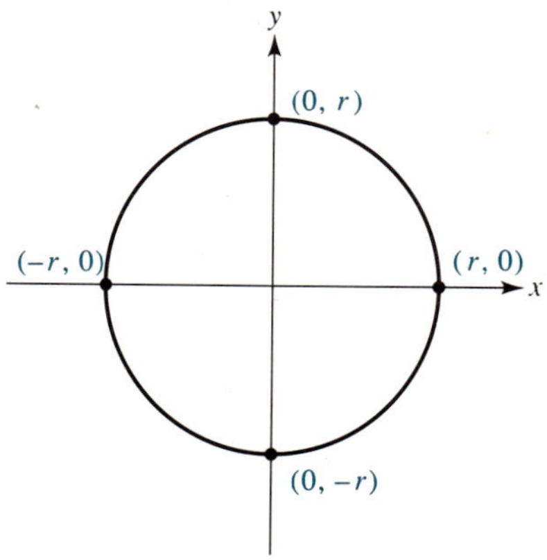
</td><td>circle;
center $(0, 0)$;
radius r</td></tr>
</table>

<table>
<tr><th>Form</th><th>Typical Graph</th><th>Special Features</th></tr>
<tr><td>$\frac{x^2}{a^2} + \frac{y^2}{b^2} = 1$</td><td>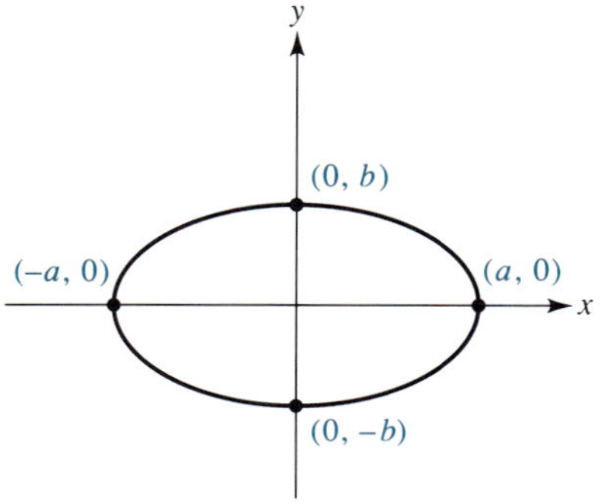
</td><td>ellipse;
intercepts
$(a, 0)$, $(-a, 0)$,
$(0, b)$, $(0, -b)$</td></tr>
<tr><td>$\frac{x^2}{a^2} - \frac{y^2}{b^2} = 1$</td><td>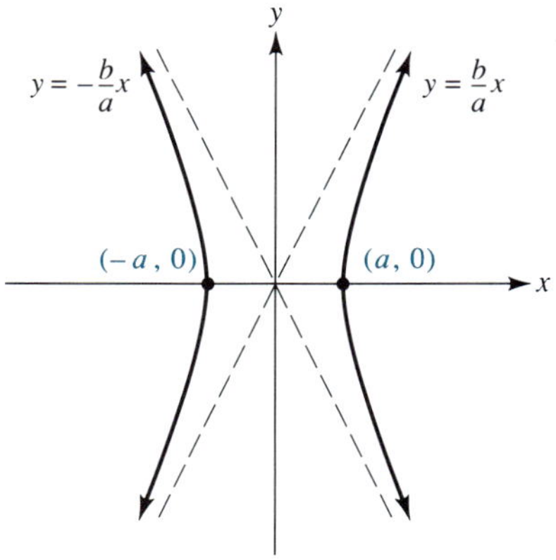
</td><td>hyperbola;
opening right and left;
x-intercepts
$(a, 0)$, $(-a, 0)$</td></tr>
<tr><td>$\frac{y^2}{b^2} - \frac{x^2}{a^2} = 1$</td><td>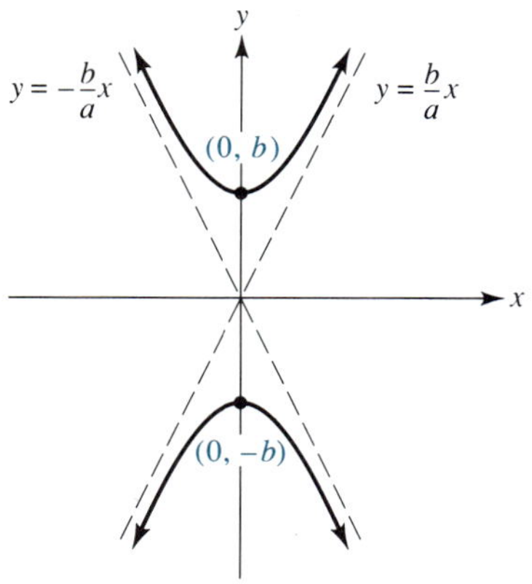
</td><td>hyperbola;
opening upward and downward;
y-intercepts
$(0, b)$, $(0, -b)$</td></tr>
</table>

EXERCISES 2.4

In exercises 1 through 22, sketch the graph of the equation in the *xy*-plane.

1. $x^2 + y^2 = 16$

2. $x^2 + y^2 = 17$

3. $x^2 + y^2 = 0$

4. $x^2 + y^2 = -4$

5. $y = 4x^2$

6. $1 = x^2$

7. $0 = x^2$

8. $x = -3y^2$

9. $y^2 = -3$

10. $y^2 = 3$

11. $\frac{x^2}{25} + \frac{y^2}{4} = 1$

12. $\frac{x^2}{13} + \frac{y^2}{35} = 1$

13. $\frac{x^2}{25} + \frac{y^2}{16} = 0$

14. $\frac{y^2}{-2} - \frac{x^2}{1} = -3$

15. $\frac{y^2}{-2} - \frac{x^2}{1} = 3$

16. $\frac{x^2}{9} - \frac{y^2}{25} = 1$

17. $4x^2 - 4y^2 = 1$

18. $4x^2 - 4y^2 = 25$

19. $\frac{y^2}{6} - \frac{x^2}{32} = 1$

20. $\frac{y^2}{6} - \frac{x^2}{32} = -1$

21. $4x^2 - 4y^2 = 0$

22. $y^2 = x^2 - \frac{1}{3}$

23. The graph of an equation of the form

$$ax^2 + by^2 + cx + dy + e = 0$$

is called a *degenerate conic* if it is a line, a pair of lines, a single point, or the empty set. Look at the graphs you drew in exercises 1 through 22. Which do you think might be classified as degenerate conics?

24. Compare the graph of

$$\frac{(x-2)^2}{16} + \frac{(y+1)^2}{25} = 1$$

with the graph of

$$\frac{x^2}{16} + \frac{y^2}{25} = 1.$$

25. Derive the form of the equation of a parabola if the focus is $(0, r)$ and the x-axis is the directrix.

26. Derive the form of the equation of a parabola if the focus is $(a, 0)$ and the directrix is $x = -a$.

In exercises 27 through 30, write an equation for each graph.

27.

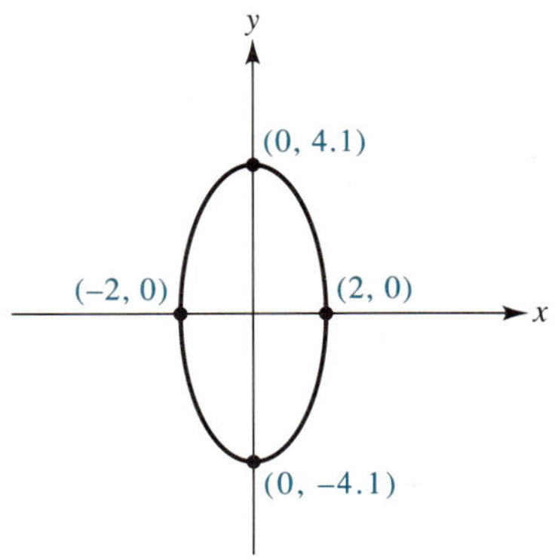

28.

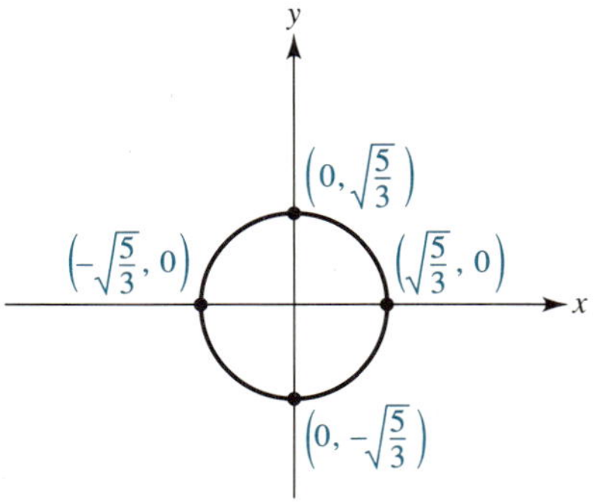

29.

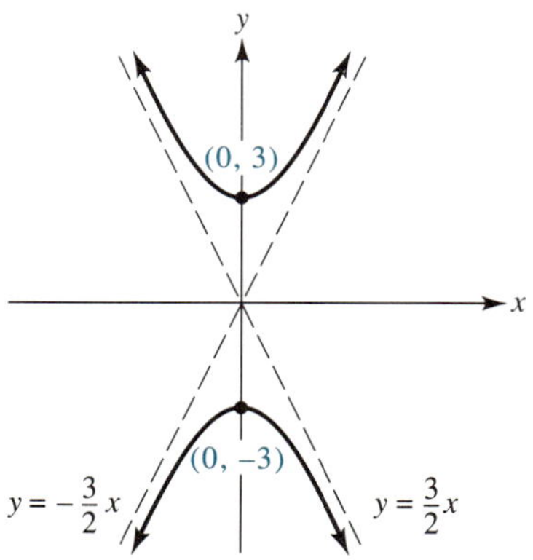

30.

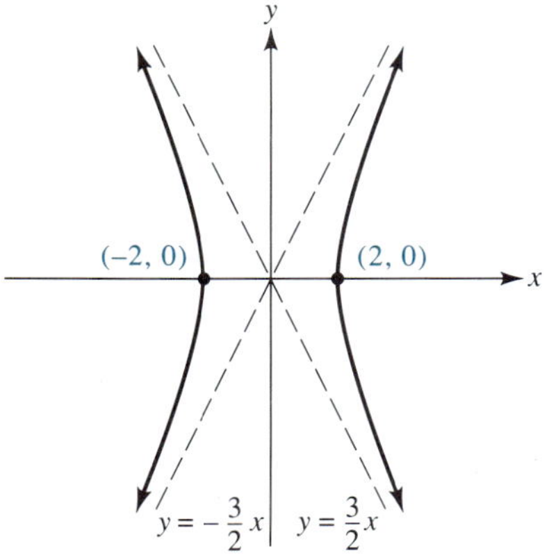

In exercises 31 through 36, if you have access to software that graphs conics, use it to test your conclusions.

31. We saw that a circle is a special case of an ellipse by examining $x^2/a^2 + y^2/b^2 = 1$ algebraically. In geometric terms, what must happen for an ellipse to be a circle?

32. Using the geometric definition of a parabola at the beginning of this section, we derived $y = x^2/(4p)$ as an algebraic description. Different values of p will produce different graphs. What happens to these graphs as p becomes very large?

33. What happens to the graph of $x^2/a^2 + y^2/b^2 = 1$ if $|a|$ is much, much larger than $|b|$?

34. What happens to the graph of $x^2/a^2 + y^2/b^2 = 1$ if $|a|$ and $|b|$ are very close to zero?

35. What happens to the graph of $x^2/a^2 - y^2/b^2 = 1$ if $|a|$ is much, much larger than $|b|$?

36. What happens to the graph of $x^2/a^2 - y^2/b^2 = 1$ if $|a|$ and $|b|$ are very close to zero?

37. Solve $y^2/b^2 - x^2/a^2 = 1$ for y and determine an equation for each asymptote.

In exercises 38 through 40, describe the three-dimensional graph of each equation based upon what we've done in this section. Also, name a familiar object with that shape.

38. $x^2 + y^2 + z^2 = 4$

39. $x^2 + \frac{y^2}{4} + z^2 = 1$

40. $y = x^2 + z^2$

In exercises 41 through 50, sketch the graph of the solution set for each statement.

41. **a.** $x^2 + y^2 = 25$ and $y \le 0$

b. $y = -\sqrt{25 - x^2}$

42. **a.** $4x^2 + 4y^2 = 1$ and $x \ge 0$

b. $x = \sqrt{\frac{1}{4} - y^2}$

43. **a.** $\frac{x^2}{16} + \frac{y^2}{100} = 1$ and $x \le 0$

b. $x = -4\sqrt{1 - \frac{y^2}{100}}$

44. **a.** $\frac{x^2}{36} + \frac{y^2}{17} = 1$ and $y \ge 0$

b. $y = \sqrt{17\left(1 - \frac{x^2}{36}\right)}$

45. **a.** $\frac{x^2}{9} - \frac{y^2}{40} = 1$ and $x \ge 0$

b. $x = 3\sqrt{1 + \frac{y^2}{40}}$

46. **a.** $\frac{x^2}{14} - \frac{y^2}{3} = -1$ and $y \le 0$

b. $y = -\sqrt{3\left(\frac{x^2}{14} + 1\right)}$

47. $y = \sqrt{25 + x^2}$

48. $x = -\sqrt{1 - 0.5y^2}$

49. $y = -\sqrt{1 + \frac{x^2}{16}}$

50. $x = \sqrt{16 + y^2}$

51. (Graphing calculator) A graphing calculator will not sketch the graph of every equation that you enter. Experiment with the equations in exercises 41 through 50 to determine which equations can be graphed directly as entered using your calculator. Can you give a condition necessary for the equation to be graphed? Is there a way to produce the graph of the complete equation by combining several steps or routines?

2.5 Translations, Expansions, and Reflections

In Section 2.3 we investigated how to write the equation of a parabola as $y = a(x - h)^2 + k$ [or $x = b(y - k)^2 + h$] and how to sketch its graph quickly by viewing it as a **variation** of the standard graph.

Let's consider $y = x^2$ again. Subtracting a constant, h, from x before squaring shifts the standard graph h units to the right if $h > 0$ and $|h|$ units left if $h < 0$.

Adding a constant k after the squaring shifts the standard graph k units up if $k > 0$ and $|k|$ units down if $k < 0$. These shifts are commonly called translations, horizontal or vertical. Vertical translations can also be produced by subtracting a constant from y; $y - k = x^2 \Leftrightarrow y = x^2 + k$. Both kinds of translations are the result of subtracting a constant from a variable before any other operation is applied. Translations may be applied to any graph.

EXAMPLE Write an equation for a circle of radius 1 with center (3, 2).
This circle is the unit circle $x^2 + y^2 = 1$ shifted 3 units right and 2 units up. To make a horizontal shift of 3 units, we subtract 3 from x before squaring. Similarly, the vertical shift of 2 units up is produced by subtracting 2 from y before squaring.

The graph of $(x - 3)^2 + (y - 2)^2 = 1$ is a circle of radius 1 with center (3, 2). ■

EXAMPLE Sketch the graph of $\dfrac{(x + 1)^2}{4} + \dfrac{(y - 1)^2}{9} = 1$.

The equation matches most closely that of a standard ellipse. The graph is shifted left 1 unit and up 1 unit. Without the shifts, its intercepts would be (2, 0), (−2, 0), (0, 3), and (0, −3). These extreme points become (1, 1), (−3, 1), (−1, 4) and (−1, −2).

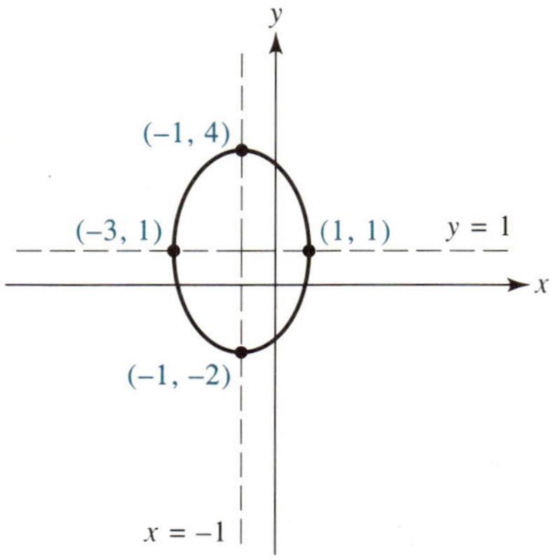

Notice the ellipse is still a maximum of 4 units wide and 6 units high; the shape has not been changed. ■

EXAMPLE Sketch the graph of $x^2 - 6x - y^2 = 7$.
In this form the equation does not look like that of any familiar graph. However, if we can use $x^2 - 6x$ to produce something of the form $(x - h)^2$, it may.

$$x^2 - 6x - y^2 = 7$$

$$\Leftrightarrow \quad x^2 - 6x + (3)^2 - y^2 = 7 + (3)^2 \qquad \text{Complete the square.}$$

$$\Leftrightarrow \quad (x - 3)^2 - y^2 = 16$$

$$\frac{(x - 3)^2}{16} - \frac{y^2}{16} = 1$$

This is an equation of a hyperbola that opens right and left and is translated 3 units right, because x-values have been affected.

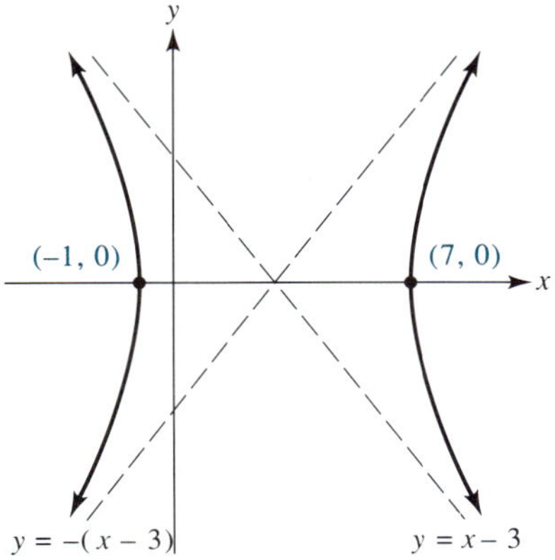

The asymptotes are also shifted 3 units right. For $x^2/16 - y^2/16 = 1$ the asymptotes are $y = \frac{4}{4}x$ and $y = -\frac{4}{4}x$; for $(x - 3)^2/16 - y^2/16 = 1$, the asymptotes are $y = 1(x - 3)$ and $y = -1(x - 3)$. The change to x must be made before any other operation. ■

In the equation $y = a(x - h)^2 + k$, the constant a has two possible effects on the graph as compared to the standard parabola. If $a \neq 1$, the rate of growth of the parabola is changed, and it is fatter or skinnier than the standard. Also, if $a < 0$, the graph opens downward. Let's look first at rate-of-growth changes for other graphs. A positive constant factor a applied after the characteristic operation for that family of graphs is called an expansion-contraction factor. The constant a is an expansion factor if it speeds up the rate of growth of the second coordinates. For a vertical parabola, this means a skinnier, or narrower, graph. Expansion-contraction factors can be applied either to expressions involving x or to expressions involving y.

What happens to the graph of $x^2 + y^2 = 1$ if we multiply the x^2-term by 1.44? Since

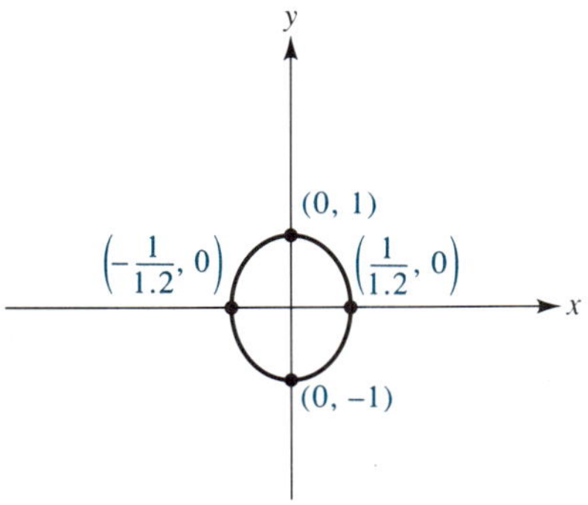

$$1.44x^2 + y^2 = 1 \quad \Leftrightarrow \quad \frac{x^2}{\frac{1}{1.44}} + \frac{y^2}{1} = 1,$$

the graph is an ellipse with center (0, 0) and intercepts (1/1.2, 0), (−1/1.2, 0), (0, 1) and (0, −1). We have distorted the unit circle in the horizontal dimension. Multiplying the y^2-term by a positive number will change the graph in the vertical dimension. Using the same positive factor for both terms produces a new circle.

Illustration The graph of $2.5x^2 + 2.5y^2 = 1$ is a circle with center (0, 0) and radius $\sqrt{1/2.5}$, or $\sqrt{2/5}$, because $2.5x^2 + 2.5y^2 = 1 \Leftrightarrow x^2 + y^2 = 1/2.5 \Leftrightarrow x^2 + y^2 = \frac{2}{5}$. □

The changes produced using a positive factor on one or both variable terms of $x^2/a^2 + y^2/b^2 = 1$ are similar to those produced for circles.

For hyperbolas a positive factor applied to the first term of $x^2/a^2 - y^2/b^2 = 1$ (or $y^2/b^2 - x^2/a^2 = 1$) changes the x-intercepts (or y-intercepts) and affects the shape of the branches by changing the asymptotes.

EXAMPLE Sketch the graphs of

$$\frac{y^2}{1} - \frac{x^2}{9} = 1 \quad \text{and} \quad \frac{1}{25} \cdot \left(\frac{y^2}{1}\right) - \frac{x^2}{9} = 1.$$

The solution set of $y^2/1 - x^2/9 = 1$ forms a hyperbola with y-intercepts (0, 1) and (0, −1) and asymptotes $y = \frac{1}{3}x$ and $y = -\frac{1}{3}x$.

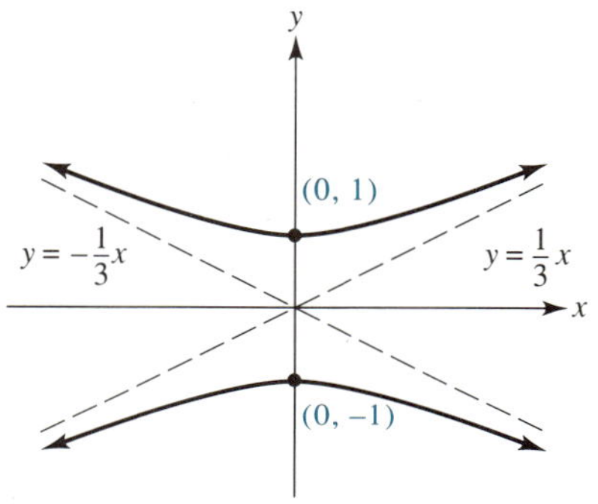

The graph of $(\frac{1}{25})(y^2/1) - x^2/9 = 1$ has y-intercepts (0, 5) and (0, −5) and asymptotes $y = \frac{5}{3}x$ and $y = -\frac{5}{3}x$.

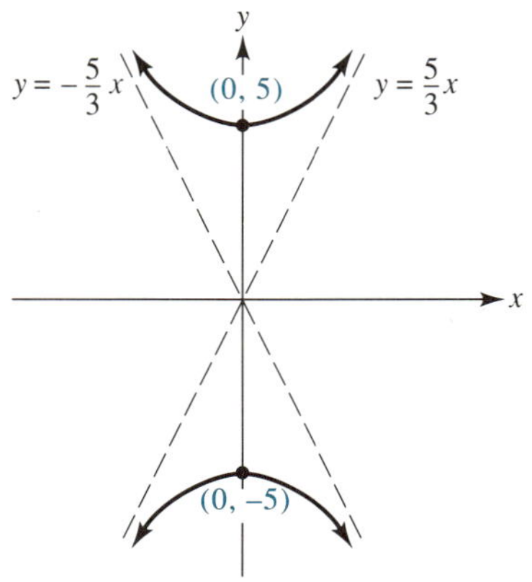

Multiplying the second term of $x^2/a^2 - y^2/b^2 = 1$ (or $y^2/b^2 - x^2/a^2 = 1$) by a positive constant changes the asymptotes and thus the spread of the branches. In this case the intercepts are unaffected.

One way to view the effect of a positive constant factor is that it changes the scale on the horizontal axis if it is applied to the term with x and changes the vertical scale if it is applied to the term with y.

Illustration

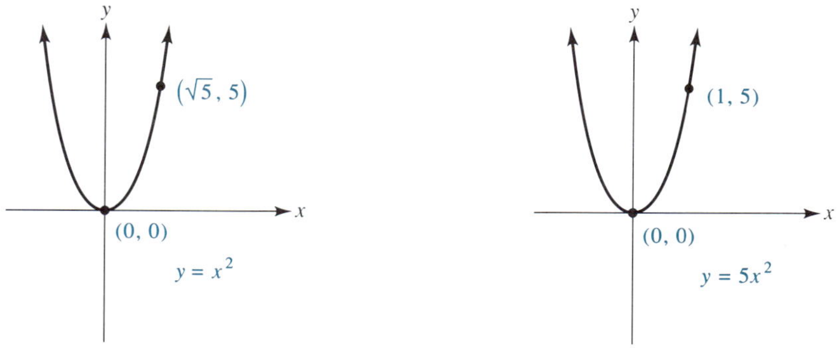

The graphs can be drawn to look exactly the same by changing the size of a unit on the x-axis. This is why labels are needed to identify the graph and why you must read graphs carefully. □

Using a negative multiplier produces results that are not so easily described. Some of these are investigated in the exercises. The effect of one other sort of change is easy to describe, however. Let's compare the graphs of $y = x$ and $y = -x$ and the graphs of $x = y^2$ and $-x = y^2$.

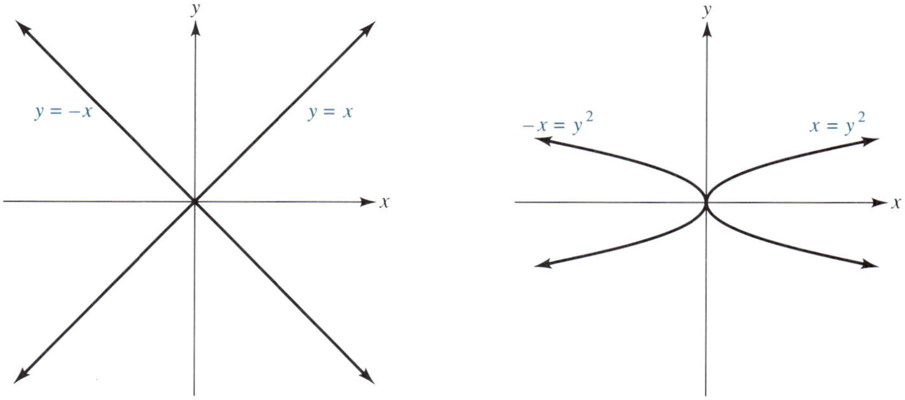

This is not much to go on, but we notice that if (a, b) belongs to the original graph, $(-a, b)$ must belong to the one obtained by replacing x with $-x$. The new graph is a mirror image of the original, with the y-axis playing the role of the mirror. We can produce a similar result for any variation of $y = x$ or $x = y^2$ that is a horizontal shift of the original graph. Such variations have replaced x with $x - h$. Using $-(x - h)$ instead of x reflects the original in the y-axis and shifts it h units horizontally.

EXAMPLE Sketch the graphs of $-(x - 5) = y^2$ and $x = y^2$ on the same axes.
The graph of $-(x - 5) = y^2$ is the graph of $x = y^2$ reflected in the y-axis and then shifted 5 units to the right.

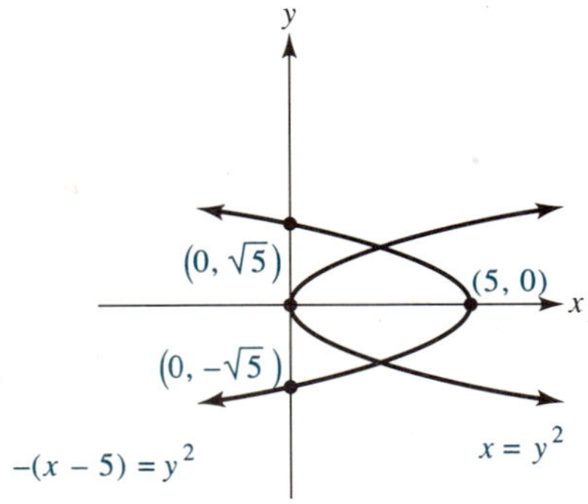

Vertical shifts may be combined with horizontal shifts, reflections and expansions.

EXAMPLE Sketch the graph of $(y - 1)^2 = -2(x + 4)$.

$$(y - 1)^2 = -2(x + 4) \quad \Leftrightarrow \quad (y - 1)^2 = 2[-(x + 4)]$$

so the graph is the standard horizontal parabola reflected in the y-axis, shifted 4 units left and 1 unit up, with an expansion factor of 2.

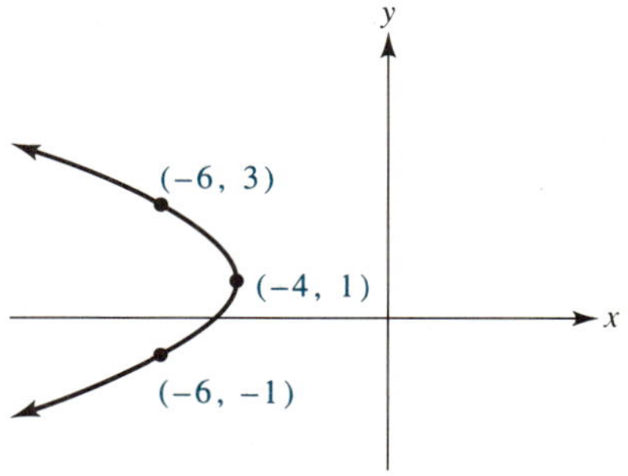

As you may have guessed, replacing y with $-(y - k)$ in the equations $y = x$ and $y = x^2$ produces graphs that are copies of the originals reflected in the x-axis and then shifted k units vertically. Changing $(x - h)$ to $-(x - h)$ or $(y - k)$ to $-(y - k)$ in equations for circles, ellipses, or hyperbolas does not change the graphs. Algebraically, whether a number or its opposite is squared, the result is the same. Geometrically, reflection in the x- or y-axis gives the original graph in these cases.

Our examples for translations, expansions, and reflections have been lines and conics, but the same ideas can be applied to any graph and its defining equation. Take care of reflection in the x- or y-axis first; then note the results of any translations. Expansions (or contractions) are usually dealt with by labeling some points to establish the scale.

At this point we have shown that any equation of the form $Ax^2 + Bx + Cy^2 + Dy + E = 0$ describes a conic section, possibly a degenerate one. Completing the square for each of $Ax^2 + Bx$ and $Cy^2 + Dy$ and rewriting as $a(x - h)^2 + b(y - k)^2 = c$ will display the information needed to classify the conic and to identify the important features of the graph. In the exercises you will practice going

Illustration

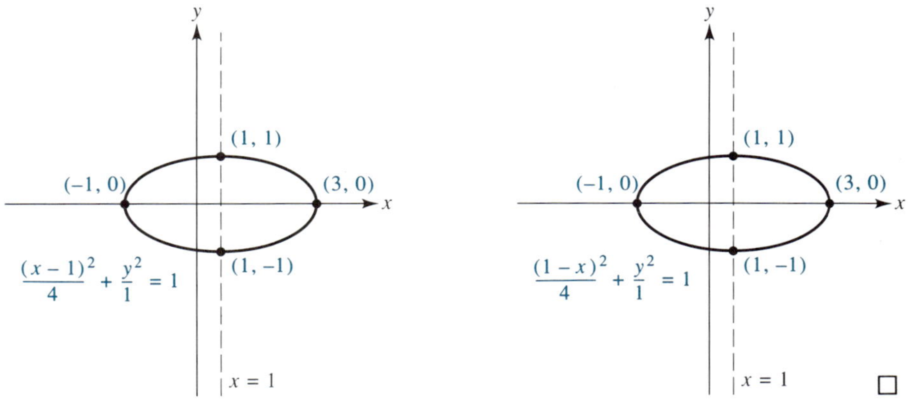

□

in the other direction, too; given the graph of a conic, describe it algebraically as a second-degree two-variable polynomial equation. We have not quite finished with this topic, because a second-degree two-variable polynomial equation might have an xy-term. Our rewriting techniques will not transform $Ax^2 + Bxy + Cy^2 + Dx + Ey + F = 0$ into $a(x - h)^2 + b(y - k)^2 = c$. Do such equations also define conics? We'll return to this in Section 9.4.

We promised a close relative of the hyperbolas discussed as a standard graph. The graph of $y = 1/x$ is a hyperbola—but one that is not a shifted, reflected, or expanded version of those we saw earlier. It is a variation of another kind discussed in Section 9.4. When x is positive, so is y. Negative values of x give negative values of y. Very large values of x result in values of y very close to 0; (1000, 0.001) and (100,000, 0.00001) are solutions of $y = 1/x$. In the other direction, very small values of x also require values of y very close to 0; (−1000, −0.001) and (−100,000, −0.00001) are solutions. There are no points $(0, c)$ on this graph. When x is positive and very close to 0, y is a very large number; (0.001, 1000) and (0.0001, 10,000) are solutions. For negative values of x close to 0, y is a very small number; (−0.01, −100) and (−0.00001, −100,000) are solutions.

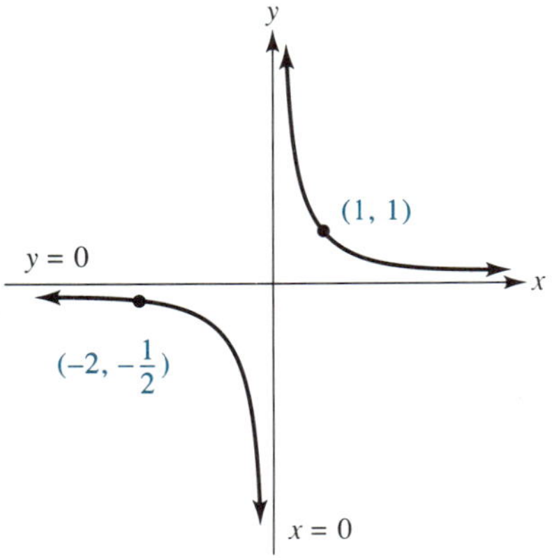

Standard Graph 5: $y = \frac{1}{x}$

The solutions fall into two categories: (a, b), $a > 0$, $b > 0$, and (c, d), $c < 0$, $d < 0$. There are no x-intercepts or y-intercepts. The x-axis is an asymptote for the graph. The y-axis is also an asymptote for the graph. Suggested labels are the asymptotes and a point on each branch to establish the scale. The ideas of reflection, translation, and expansion can be used with this new standard graph.

EXAMPLE Sketch the graph of $y = \frac{3}{x - 2}$.

Subtracting 2 from x before taking the reciprocal shifts this standard graph to the right 2 units. This makes $x = 2$ an asymptote. The factor 3 changes the scale.

The graph of $y = \frac{3}{x - 2}$ is

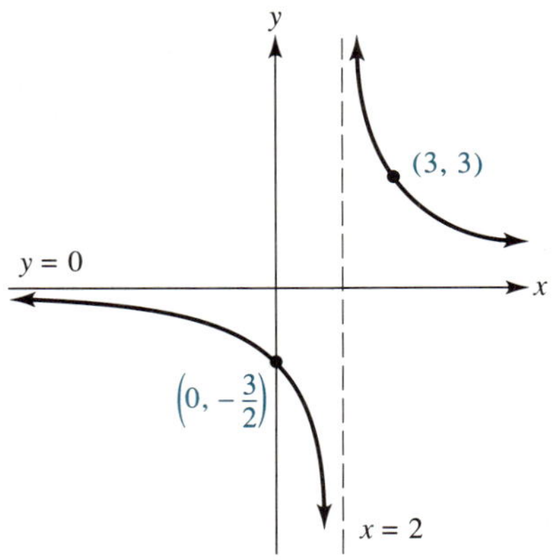

EXAMPLE Sketch the graph of $y = \dfrac{1}{1.5 - x} + 4$.

This graph is also a variation of $y = 1/x$. Rewriting $y = 1/(1.5 - x) + 4$ as $y = 1/[-(x - 1.5)] + 4$, we see the graph of $y = 1/x$ has been reflected in the x-axis and shifted 1.5 units right and 4 units up.

The graph of $y = \dfrac{1}{1.5 - x} + 4$ is

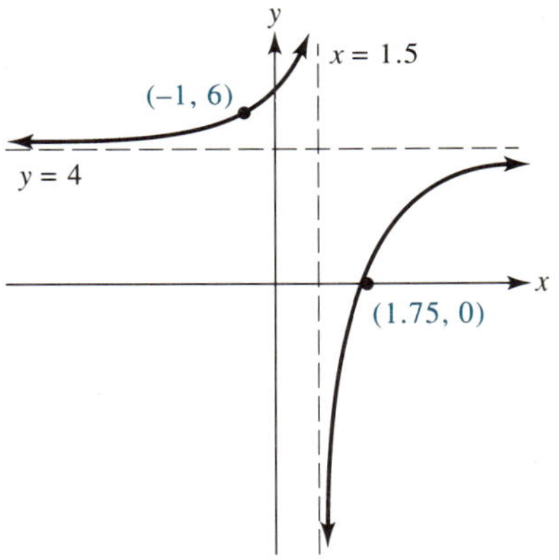

Note the effect of the shifts on the asymptotes.

Using translations, expansions, and reflections allows us to sketch efficiently the variations of standard graphs as well as others that are familiar. In Chapter 4 we look at some other graphing techniques. It is always a good idea to double check a few points even if you choose not to label them on the graph.

EXERCISES 2.5

In exercises 1 through 26, sketch the graph of the equation.

1. $(x - 4)^2 + (y - 1)^2 = 25$
2. $(x + 3.2)^2 + (y - 3)^2 = 1$
3. $4(x + 5)^2 + 4(y + 9)^2 = 1$
4. $7(x - 3.2)^2 + 7(y - 4.4)^2 = 90$
5. $\dfrac{(x + 1)^2}{16} + \dfrac{(y - 3)^2}{25} = 1$
6. $\dfrac{(x - 6.7)^2}{9} + \dfrac{(y - 2.1)^2}{4} = 1$
7. $4(y + 3)^2 + 6(x - 9)^2 = 36$
8. $4(x - 0.5)^2 + 6(y - 1.2)^2 = 1$
9. $\dfrac{(x + 1)^2}{16} - \dfrac{(y - 3)^2}{25} = 1$
10. $(x + 3.2)^2 - (y - 3)^2 = 1$
11. $\dfrac{(y - 2.1)^2}{4} - \dfrac{(x - 6.7)^2}{9} = 1$
12. $4(x + 5)^2 - 4(y + 9)^2 = -1$
13. $(x + 3.2)^2 - 9(y - 3)^2 = 16$
14. $4(x + 5)^2 - 4(y + 9)^2 = -25$
15. $y = \dfrac{2}{x - 4} + 1$
16. $y = \dfrac{-1}{2x + 1} - 5$
17. $y = \dfrac{3}{4 - x} - 3.2$
18. $x = \dfrac{3}{4 - y} + \pi$
19. $x^2 - 6x + y^2 + 4y = 3$
20. $x^2 - 4x - 4y^2 - 8y = 4$
21. $x^2 + 4y^2 - 6x + 16y + 25 = 0$
22. $x^2 - 4x + 4y^2 + 8y + 4 = 0$
23. $1 = xy - 3y$
24. $xy + 4y - 3x = 14$ (Try using long division after solving for y.)
25. $25x^2 - 50x + 9y^2 + 250 = 0$
26. $4x^2 + 32x - 5y^2 + 20y + 44 = 0$

In exercises 27 through 32, give an equation of the graph with the same shape as the graph of

27. $y = \dfrac{1}{x}$, shifted 7.3 units left and 150 units up.
28. $x^2 - 3y^2 = 9$, shifted 1 unit right.
29. $y = x$, shifted 6 units left and $\sqrt{2}$ units down, with expansion factor of -0.3.
30. $\dfrac{x^2}{16} + \dfrac{y^2}{9} = 1$, shifted a units up and b units left, $a \geq 0$ and $b \geq 0$.
31. The unit circle shifted $\sqrt[3]{7}$ units left and $-\frac{3}{4}$ unit vertically.
32. $y^2 - x^2 = 1$ shifted 0.00003 unit right.

In exercises 33 through 36, give an algebraic description for each graph.

33.

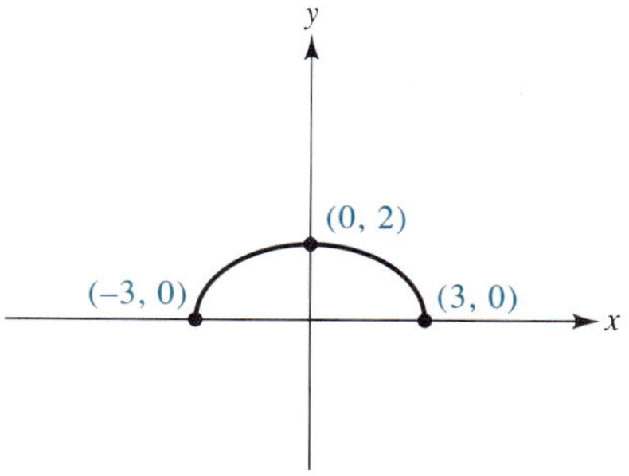

34.

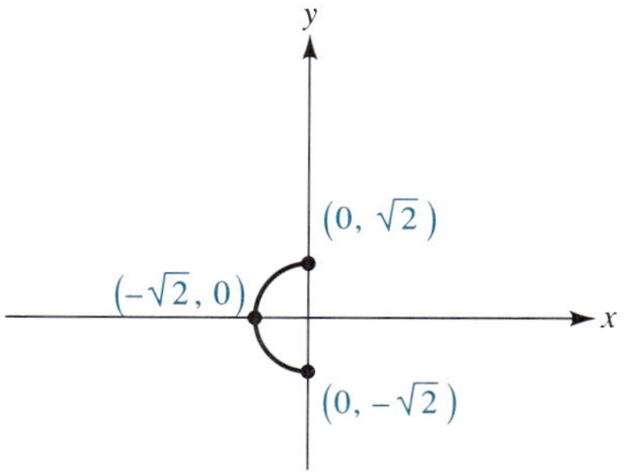

35. A vertical parabola with vertex (1, 2) and passing through (4, 29).

36.

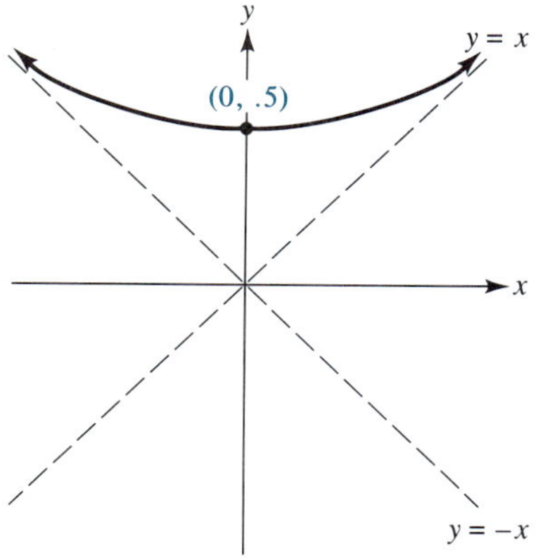

37. Describe the graph that results from multiplying the x^2 term of $x^2 + y^2 = 1$ by a number larger than 1.
38. Describe the graph that results from multiplying the x^2 term of $x^2 + y^2 = 1$ by a number between 0 and 1.

39. Describe the graph that results from multiplying the x^2 term of $x^2 + y^2 = 1$ by a number less than -1.

40. Describe the graph that results from multiplying the x^2 term of $x^2 + y^2 = 1$ by a number between 0 and -1.

In exercises 41 through 44, describe the three-dimensional graph of each equation based upon what we have done in this section.

41. $(x - 2)^2 + y^2 + (z + 5)^2 = 4$

42. $(x + 5)^2 + \dfrac{(y - 4)^2}{9} + (z - 8)^2 = 1$

43. $y = x^2 + z^2 - \pi$

44. $y = (x + 1)^2 + (z - 1.3)^2$

45. (Graphing calculator) Use your calculator to sketch the graph of each of the equations $y + x^2$, $y = \frac{1}{100}x^2$, and $y = x^2 + 100$. Does your display present the graph you expected? If not, try changing the parameters (width and height) of the display screen until you have a representative display.

46. (Graphing calculator) Use your calculator to sketch the graph of each of the equations $y = (x - 12.4)^2$, $y = x^2 - 12.4$, and $y = (2x + 3.5)^2 + 100$. Does your display present the graph you expected? If not, try changing the parameters (width and height) of the display screen until you have a representative display.

47. (Graphing calculator) Describe how the ideas of the section could be used to choose parameters for the display screen before graphing.

Chapter 2: Summary

Equations with two variables often have infinite solution sets of ordered pairs; we are more likely to sketch a graph of the solution set than give it in set notation. Consequently, most of this chapter dealt with two-dimensional graphs. The conics (parabola, ellipse, circle, and hyperbola) were defined both geometrically and algebraically. We saw that any quadratic equation of the form $Ax^2 + Bx + Cy^2 + Dy + E = 0$ defines a conic, perhaps degenerate.

Five standard graphs were introduced; they will be used in the rest of the book. The ideas of reflection, translation, and expansion will also appear numerous times.

Key Concepts

real-number plane	*origin, axis*
distance in the plane	
midpoint	
graph	
linear equations	*x-intercept, y-intercept, slope; slope-intercept form, point-slope form, general form*
standard graphs	(1) $y = x$ (2) $y = x^2$ (3) $x^2 + y^2 = 1$ (4) $\dfrac{x^2}{a^2} + \dfrac{y^2}{b^2} = 1$ (5) $y = \dfrac{1}{x}$
conic sections	*parabolas, circles, ellipses, hyperbolas*
variations	*translations (shifts), expansions, reflections*

Skills

You should be able to:

1. Locate points in the plane.
2. Compute the distance between two points in the plane and compute the midpoint of the line segment connecting two points.
3. Find the slope of a straight line given two points on the line or given the equation of the line.
4. Write the equation of a straight line given: (a) two points on the line, (b) the slope and a point on the line, (c) a point and the equation of a line perpendicular or parallel to the given line.
5. Sketch the standard graphs and their variations.
6. Write the equation of a standard graph or of a variation of a standard graph.

SUMMARY EXERCISES

The starred exercises can serve as a practice chapter test. The answers are in the answer section at the back of the book.

1. Locate $(\pi - 4, -\sqrt{2})$, $\left(-\frac{17}{3}, 3.1\right)$, and $(4 + \sqrt{3}, \sqrt[3]{9})$ in the xy-plane.

In exercises 2 through 5, use A: (4.9, −3.2) and B: (−1.5, −5.99). Find:

2. The point two-thirds of the way from B to A.
3. The midpoint of the segment connecting A and B.
4. The distance between A and B.
5. Four-thirteenths of the distance between A and B approximated to two decimal places.

In exercises 6 through 11, sketch the graph of:

6. $\{(x, y) | x \geq \pi \text{ or } y \leq -4\}$
7. $\{(x, y) | -1.1 \leq x \leq 2 \text{ and } 3 \leq y \leq \sqrt{18}\}$
8. $\{(x, y) | -1.1 \leq x \leq 2 \text{ or } 3 \leq y \leq \sqrt{18}\}$
9. $\{(x, y) | y = -\frac{2}{3}x + 4\}$
10. $\{(x, y) | x = -\sqrt[3]{5}\}$
11. $\{(x, y) | 4x - 3y = 15 \text{ and } x > -2\}$
12. Give the slope, x-intercept, and y-intercept of the graph of $ax + by = c$. Be sure to state any restrictions on the constants.

In exercises 13 through 17, write an equation of the line.

13. Containing (4, −3) and (7, 4).
14. Containing (a, b) and (c, d).
15. With slope $-\frac{7}{3}$ and containing (4, 0).
16. Parallel to the line $2x - \sqrt{3}y = 4$ and containing (0, 1.3).
17. Perpendicular to the line $y = \pi - 2$ and containing (−3.45, −5.44).

* 18. Describe any two points on the same vertical line.

19. Find the points on the line perpendicular to $y = -x + 3$ that are two units from the intersection of the two lines at (−1, 4).

In exercises 20 through 23, sketch the graph of:

20. $y = -2(x - 4)^2 + 1$

* 21. $x = 3y^2 - 8y - 3$

22. $x = -y^2 + 3.2y, y < 10$
23. $y = 5x^2 - 20x + 24, -5 \leq x \leq 5$

* 24. In how many points could the graphs of $y = ax^2 + bx + c, a \neq 0$, and $x = dy^2 + ey + f, d \neq 0$, intersect?

25. What is the maximum value of y if $y = -2x^2 - 6x + 3$?
26. What is the minimum value of y if $y = -2x^2 - 6x + 3$?

In exercises 27 through 32, sketch the graph of:

* 27. $\dfrac{x^2}{13} + \dfrac{y^2}{25} = 1$

28. $\dfrac{x^2}{13} - \dfrac{y^2}{25} = 1$
29. $\dfrac{x^2}{13} + \dfrac{y^2}{13} = 1$

* 30. $\dfrac{x^2}{13} + \dfrac{y^2}{25} = -1$

31. $\dfrac{x^2}{13} - \dfrac{y^2}{25} = -1$
32. The hyperbola with vertices (−3, 0) and (3, 0) and asymptotes $y = -\frac{5}{2}x$ and $y = \frac{5}{2}x$.

* **33.** In discussing the ellipse in Section 2.4, we said that if $a = f$ or $a = 0$, a degenerate ellipse is produced. Describe the graph produced in each case.

34. A degenerate conic can be produced by having the plane that intersects the cone contain the vertical axis of the cone. A pair of straight lines is produced. Give an equation of this graph.

In exercises 35 through 42, sketch the graph of:

35. $\dfrac{(x-4)^2}{13} + \dfrac{(y+2)^2}{25} = 1$

* **36.** $\dfrac{(x+3.1)^2}{13} - \dfrac{(y+1)^2}{25} = 1$

37. $\dfrac{x^2}{13} + \dfrac{(y-18)^2}{13} = 1$

38. $\dfrac{(x-\sqrt{2})^2}{13} + \dfrac{(y+3.3)^2}{25} = -1$

* **39.** $\dfrac{(x-\sqrt{2})^2}{13} - \dfrac{(y+3.3)^2}{25} = -1$

40. The hyperbola with vertices $(-6, 2)$ and $(0, 2)$ and asymptotes $y = -\frac{5}{2}(x+3) + 2$ and $y = \frac{5}{2}(x+3) + 2$

41. $2x^2 - 3y^2 - 8x + 6y - 13 = 0$

* **42.** $\sqrt{5}\,x^2 + \sqrt{5}\,y^2 + 5x + 10y = 0$

43. Sketch the graphs of $y = \sqrt{1 - x^2}$ and $x^2 + y^2 = 1$, $y \geq 0$.

44. Give two statements similar in form to those of exercise 43 whose graph is the bottom half of the unit circle.

* **45.** Write equations for two different circles with center $(-3, 7)$.

CHAPTER THREE

Inequalities

- 3.1 One-variable Inequalities I
- 3.2 One-variable Inequalities II
- 3.3 Two-variable Inequalities
- 3.4 Compound Inequalities

Mathematicians often handle a new problem by turning it into one they've seen before. Transforming a problem in this way can be a very effective problem-solving technique. In this chapter we solve inequalities by using the methods of Chapters 1 and 2 for solving equations. All solutions are restricted to real numbers. Remember that the complex numbers are not ordered (Section A.3), and so we will not discuss inequalities for them.

3.1 One-variable Inequalities I

Toolkit

The ideas of sets, interval notation, and number-line graphs from Section A.2 are used in this section. Each set of exercises in this chapter contains several preview problems. Doing these as you come to them will help in later sections.

If we examine a very simple **inequality,** $x < 5$, the solutions are apparent, all real numbers less than 5. Using set-builder notation, the solution set is $\{x \mid x < 5\}$. A graph of these solutions is

Notice that 5 is a **boundary number** that separates solutions from nonsolutions. Let's extend this idea in a meatier example.

In Section 1.1 we saw that first-degree, one-variable equations often have a single solution. We begin our study of inequalities with the same type of inequality. If we wish to solve

$$5 - 2x \leq 7 + 8x,$$

we first find any solutions to

$$5 - 2x = 7 + 8x,$$

since these numbers are boundary numbers that separate regions of the number line where the expression $5 - 2x$ has value less than that of $7 + 8x$ from those regions where $5 - 2x$ has value greater than that of $7 + 8x$. The equation $5 - 2x = 7 + 8x$ has one solution, $-\frac{1}{5}$, which is also a solution of the original inequality. This number, $-\frac{1}{5}$, divides the real number line into two regions.

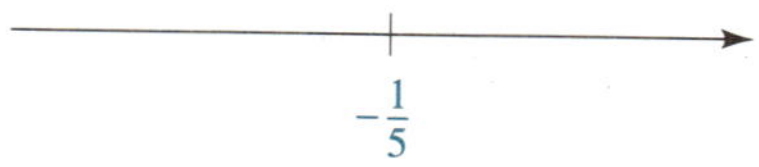

Choose any number larger than $-\frac{1}{5}$ and test to determine if it is a solution of the original inequality, $5 - 2x \le 7 + 8x$. Let's try 0.

Left-hand side: $5 - 2(0)$, or 5

Right-hand side: $7 + 8(0)$, or 7

Since $5 \le 7$, 0 is a solution. Try any other number you like that is greater than $-\frac{1}{5}$; it will also be a solution. What happens if we use a number less than $-\frac{1}{5}$, such as -1?

Left-hand side: $5 - 2(-1)$, or 7

Right-hand side: $7 + 8(-1)$, or -1

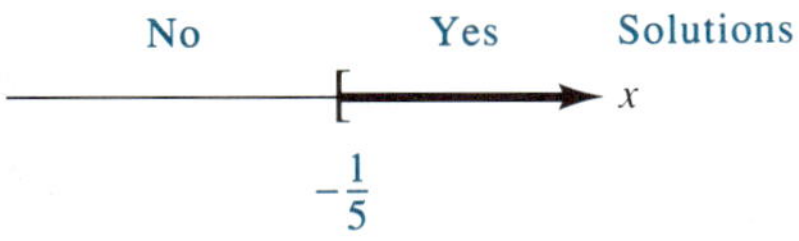

The left-hand side is larger than the right-hand side with this replacement, so -1 is not a solution. In fact, any number you choose that is less than $-\frac{1}{5}$ will fail to satisfy the inequality.
The solution set is

$$\left\{x \,\middle|\, -\frac{1}{5} \le x\right\}.$$

A key point is that solutions "stick together"; if one number in a region is a solution, then all numbers in that region are solutions. The nonsolutions behave the same way.

Our method for solving these inequalities (and in fact all one-variable inequalities) is a two-step process.

1. Find boundary numbers by solving the corresponding equation and determine whether they are solutions of the original inequality.
2. Determine which regions represent solutions by testing.

EXAMPLE Solve $3.7x - 8 > 4.1x + 5$.
First, we find all solutions of the corresponding equation.

$$3.7x - 8 = 4.1x + 5$$
$$\Leftrightarrow \quad -13 = 0.4x$$
$$\Leftrightarrow \quad -32.5 = x$$

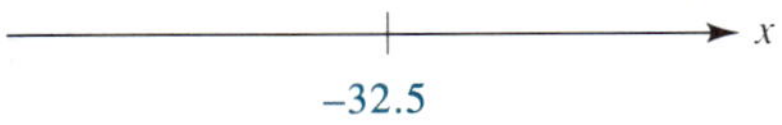

There is only one boundary number, -32.5. It is not included in the solution set.
Second, we test the two regions left and right of -32.5.
Zero is a good choice because the calculations are simplified. Zero does not satisfy the original inequality; $3.7(0) - 8$ is not greater than $4.1(0) + 5$.

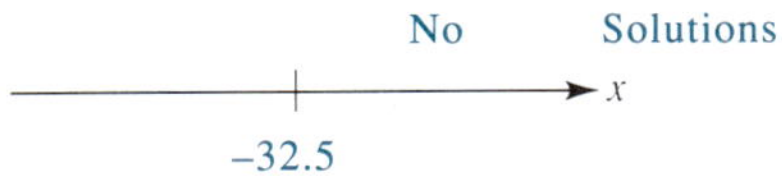

We record our result.

Now we test some number on the left; say -100.

LHS: $3.7x - 8$: $3.7(-100) - 8$, or -378

RHS: $4.1x + 5$: $4.1(-100) + 5$, or -405

We see -100 is a solution, so each number to the left of -32.5 is a solution.

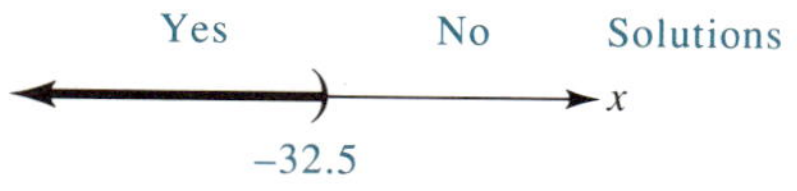

The solution set is $\{x|x < -32.5\}$.
Note: This could also be given as $(-\infty, -32.5)$. ■

A nice feature of this process is that it works, with only small modifications, for all inequalities.

EXAMPLE Solve $|2x - 5| > 12$.
First solve the equation $|2x - 5| = 12$. This is equivalent to $2x - 5 = 12$ or $2x - 5 = -12$, and solutions are $\frac{17}{2}$ and $-\frac{7}{2}$. We have two boundary numbers and three regions. The boundary numbers are not solutions of the inequality.

Test number	-4	0	10
LHS: $\lvert 2x - 5\rvert$	13	5	15
RHS: 12	12	12	12
Solution	Yes	No	Yes

The graph of the solution set is

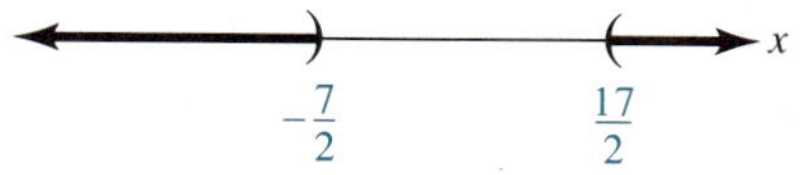

The solution set is $\{x|x < -\frac{7}{2} \text{ or } \frac{17}{2} < x\}$, or $(-\infty, -\frac{7}{2}) \cup (\frac{17}{2}, +\infty)$. ■

Remember, in testing a region you may use any number in that region; choose numbers that are easy to work with.

EXAMPLE Solve $-5|8 + 4x| > 10$.
Because the left-hand side will be nonpositive for any choice of x, the solution set is $\{\ \ \}$. To demonstrate how to handle the case of no boundary numbers, we show the boundary-number method here also.

We solve $-5|8 + 4x| = 10$ first.

$$-5|8 + 4x| = 10 \quad \Leftrightarrow \quad |8 + 4x| = -2$$

and this equation has no solutions. Since there are no boundary numbers, we have only one region.

We test 0 and find

Test number	0
LHS: $-5\|8 + 4x\|$	-40
RHS: 10	10
Solution	No

The left-hand side is not greater than the right-hand side; hence, 0 and all numbers in the same region are not solutions.

The solution set is { }. ■

EXAMPLE Find all numbers within 3 units of 7.

To be within 3 units of 7 means the distance from the number n to 7 is less than or equal to 3 units, $|n - 7| \leq 3$. To solve this inequality we first find the boundary number(s) by solving the corresponding equation.

$$|n - 7| = 3$$
$$\Leftrightarrow \quad n - 7 = 3 \quad \text{or} \quad n - 7 = -3$$

The boundary numbers are 10 and 4, and they both satisfy the inequality.

Second, we test the regions determined by the boundary numbers for solutions by choosing a number within each region.

Test number	0	7	100
LHS: $\|n - 7\|$	7	0	93
RHS: 3	3	3	3
Solution of $\|n - 7\| \leq 3$	No	Yes	No

The graph of the solution set follows.

The numbers 4 through 10 are within 3 units of 7.
Note: We did not give the solution set because the original instructions were not to solve an inequality. ■

In the appendix we discuss properties of inequalities similar to those for equations. You may have expected to use ideas like those to solve inequalities. In fact it is possible to do so, but the method of this chapter is more general and, for most problems, more efficient. However, there are times when what is required is to rewrite an inequality, not to solve it, and then these properties are a good tool.

EXAMPLE If $4 - 2x < 1.5$, how does $2x + 3$ compare to 0.3?

$$\begin{aligned} & 4 - 2x < 1.5 \\ \Leftrightarrow\quad & -2x < -2.5 \\ \Leftrightarrow\quad & 2x > 2.5 \\ \Leftrightarrow\quad & 2x + 3 > 5.5 \end{aligned}$$

Since $2x + 3$ is greater than 5.5, $2x + 3$ is certainly greater than 0.3; $2x + 3 > 0.3$. ■

EXERCISES 3.1

In exercises 1 through 24, solve the given inequality. Either set-builder or interval notation may be used.

1. $2(x - 3) \geq 7x - 14$
2. $5x + 72 < -3(x + 1) - 4$
3. $2.3 - y < 13.21$
4. $\pi > \frac{5}{2}a$
5. $-(w - 6 + 3w) \leq \sqrt{28}$
6. $\sqrt[3]{7}z + 84 > 7z$
7. $(w - 4)^2 + 3 > 16 + w^2,\ w \in W$
8. $2a^3 - 6a^2 \leq 2(a - 1)^3$
9. $(w - 4)^2 + 3 \geq 16 - 8w + w^2$
10. $(w - 4)^2 + 3 \geq 19 - 8w + w^2$
11. $|3x - 4| > 2$
12. $|7 + a| \leq 13$
13. $5|3x - 4| \geq \sqrt{13}$
14. $26|7 + a| > 13$
15. $|3x - 4| > -2$
16. $|3x - 4| \leq -2$
17. $30 < 6|14 - 3a|,\ a \in I$
18. $|2x - 1| \geq |5 + 7x|$
19. $|x + 3| + 7 \leq |x|$
20. $|7 - 8x| + |x + 1| < -3$
21. $\dfrac{c - 3}{5} \geq \dfrac{c + 3}{4}$
22. $\dfrac{c - 3}{5} \geq \dfrac{c + 3}{4} + 1,\ c \in I$
23. $\dfrac{5x - 2.1}{3.5} < x$
24. $4.3y \leq \dfrac{7 + 4y}{2.1} - \dfrac{y}{5}$

In exercises 25 through 30, write a statement of inequality to describe

25. All numbers within 100 units from 7.
26. All numbers more than 100 units from 7.
27. All numbers less than π units from -7.
28. All numbers not less than 13.285 units from a.
29. The distance between $a - 3$ and 5 is not more than $\dfrac{132}{5}$ units.
30. The distance between a and b is more than $\dfrac{3 + \pi}{8}$ units.

In exercises 31 and 32, apply the properties of equations and inequalities to the first statement to determine the correct inequality or equality symbol for the second statement.

31. **a.** $3x - 5 < -8x + 7;\ 11x\ ?\ 12$

b. $5 + 6(2 - x) < 20$; $x - 2 \quad ? \quad -\frac{5}{2}$

c. $15x^2 \le 12$; $\frac{3}{x^2} \quad ? \quad 3.75$

32. a. $x = \pi$; $x^3 \ ? \ \pi^2$

b. $a > \frac{2}{3}$; $5a - 13 \quad ? \quad -10$

c. $a > \frac{2}{3}$; $13 - 5a \ ? \ 10$

33. A track star has grades of 75, 68, 58, and 65 on unit tests. The average of these four grades comprises two-thirds of the final grade and the final exam grade, the other one-third. What can the star make on the final exam to earn a grade of C or better in the course and maintain eligibility? (A grade of 70 to 79 is a C.) Would you bet your money on the star's making the grade?

34. Deborah's goal is to be an honor graduate at the end of the year. This means she must have a grade-point average (GPA) of at least 3.5 earned over 180 hours. She currently has a GPA of 3.4 earned over 140 hours. If an A earns 4 points per hour, a B, 3 points, a C, 2 points, a D, 1 point, and an F, 0 points, what is the minimum number of points that Deborah must earn over the next 40 hours of coursework to be an honor graduate?

35. Todd has purchased a new car, and with taxes, insurance, car payments, and his regular living expenses he must earn at least \$1530 per month to make ends meet. He works 22 days a month, 8 hours a day, for \$6.75 per hour. He also earns a commission of 7.5% on whatever he sells. What should his total sales be each month in order to meet his budget?

36. Social security restrictions require that Mrs. Jones earn no more than \$388 per month. She works part time at McConald's, which pays \$3.35 per hour. Assuming that she works 22 days per month, how many hours per day should she work in order to retain her full social security benefits?

37. To qualify as a lightweight for a title bout, Murray the Mauler must weigh no more than 150 pounds the morning of the fight. His current weight is 163 pounds. He figures he can lose 0.03 pounds for each 90 calories below his normal 2500-calorie intake per day. How many calories can he consume per day from now till the fight day, which is 60 days away, and qualify as a lightweight?

38. Jennifer has been hired to work at data entry for a bank. Rodney is training her, and the bank needs them both to work on a project as soon as possible. Rodney can do the entire project alone in 8 days. What must Jennifer's rate be in order for them both to be able to finish the project in 5 days or less?

39. (Preview) Graph $|x| + |y| = 1$. We will call this graph the *unit box*.

40. (Preview) Consider traveling counterclockwise on the unit box. How far is it from (1, 0) to

a. (0, 1)? **b.** (−1, 0)? **c.** (0, −1)?

d. (1, 0) traveling all around the box?

41. (Preview) **a.** Give all the trip lengths that will take you from (1, 0) to (0, 1). (*Hint:* There are infinitely many.)

b. Give the lengths of all trips from (1, 0) to (−1, 0).

c. Give the lengths of all trips from (1, 0) to (0, −1).

3.2 One-variable Inequalities II

Let's see how the process described in the last section works for inequalities that use one variable but are not first-degree.

EXAMPLE Solve $y^2 < 4$.

You may be able to puzzle out that any number between −2 and 2 is a solution, and the solution set is $\{y \mid -2 < y < 2\}$. If not, try the technique of Section 3.1.

The solution set of $y^2 = 4$ is $\{-2, 2\}$. There are two boundary numbers and three regions. The boundary numbers are not solutions.

Test each region. The number −3 is not a solution; 0 is a solution; 3 is not a solution.

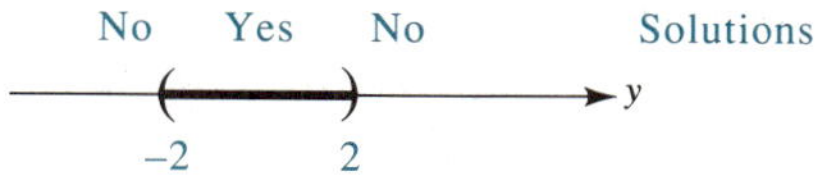

Hence the solution set is $\{y \mid -2 < y < 2\}$. ■

If the quadratic formula is used to solve an equation for the boundary numbers, you may find it useful to approximate them with a calculator in order to choose appropriate test values.

EXAMPLE Solve $(2w - 3)w \geq 125$.

The boundary numbers are the solutions of $(2w - 3)w = 125$. Rewriting this as $2w^2 - 3w - 125 = 00$ and using the quadratic formula, we find solutions

$$\frac{3 + \sqrt{1009}}{2} \quad \text{and} \quad \frac{3 - \sqrt{1009}}{2}.$$

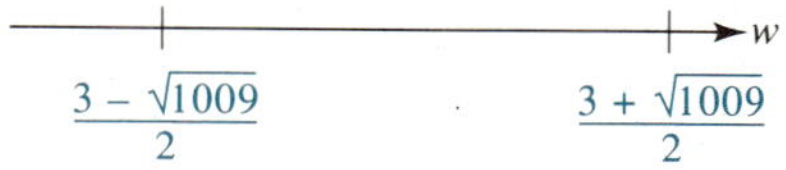

Estimating mentally or approximate with a calculator,

$$\frac{3 - \sqrt{1009}}{2} \approx -14 \quad \text{and} \quad \frac{3 + \sqrt{1009}}{2} \approx 17$$

so -20, 0, and 20 are reasonable test numbers. The graph of the solution set is

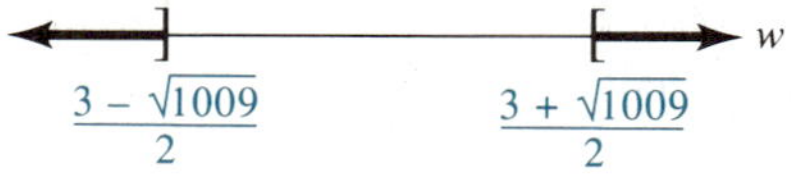

The solution set is

$$\left\{w \,\middle|\, w \leq \frac{3 - \sqrt{1009}}{2} \quad \text{or} \quad \frac{3 + \sqrt{1009}}{2} \leq w\right\}.$$ ■

EXAMPLE Solve $x^2 + x < -1$.

Solving $x^2 + x = -1$ gives $(-1 \pm \sqrt{3}i)/2$ as solutions. These complex numbers cannot be boundary numbers on the real-number line. We have no boundary numbers and one region. Zero as a test number does not satisfy the inequality.

Hence, the solution set is { }.

Note: The solution set for $x^2 + x = -1$ is $\{-1 \pm \sqrt{3}i/2\}$ or { } if only real-number solutions are required. ■

Our techniques for solving equations of degree greater than two also depend on factoring and the quadratic formula. In any case we begin by writing an equivalent equation with one side equal to 0. This makes possible a refinement of our testing procedure for inequality solutions.

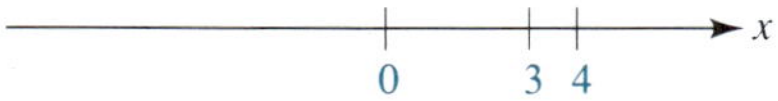

Let's look at the process when applied to solving $x^3 - 7x^2 \geq -12x$. First, we solve $x^3 - 7x^2 = -12x$ by rewriting it as $x^3 - 7x^2 + 12x = 0$ and factoring, to produce $x(x - 3)(x - 4) = 0$ and solutions 0, 3, and 4. We know that $x^3 - 7x^2 \geq -12x$ is equivalent to $x(x - 3)(x - 4) \geq 0$, and we can test using this latter inequality. Why take the trouble to rewrite in factored form with one side zero? We need only to check the sign of the left-hand side in this form. We want values of x that make the left side positive or 0. The boundary numbers occur where the left side is zero. The left-hand side will be positive or negative depending on the signs of the factors.

Choose any number less than zero. The first factor x is negative for this value; $x - 3$ is also negative; $x - 4$ is negative. The product of three negative numbers is negative, so no number less than 0 can satisfy the inequality. Notice we did no calculations, just noted the sign of each factor. Choose any number between 0 and 3. The factor x is positive for this number, $x - 3$ is negative, $x - 4$ is negative, and the product is positive. This is a region of solutions.

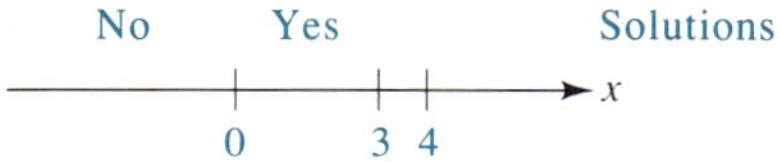

Between 3 and 4, x is positive and $x - 3$ is also positive, but $x - 4$ is negative; hence, the product is negative. This is not what we want.

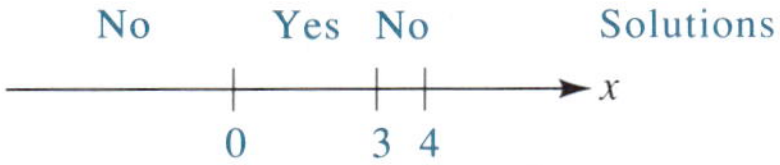

If you choose any number greater than 4, all three factors are positive, and their product is positive.

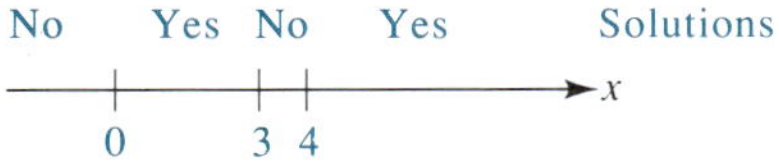

We can summarize this in a chart.

Test region	$(-\infty, 0)$	$(0, 3)$	$(3, 4)$	$(4, +\infty)$
LHS: $x(x - 3)(x - 4)$	Negative	Positive	Negative	Positive
RHS: 0	0	0	0	0
Solution of $x(x - 3)(x - 4) \geq 0$	No	Yes	No	Yes

The boundary numbers are included in the solution set, so the graph is

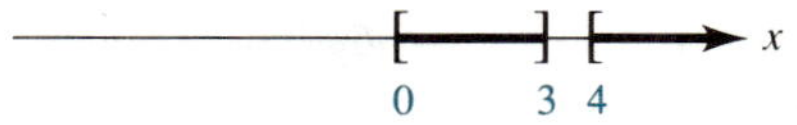

The solution set is $\{x \mid 0 \leq x \leq 3 \quad \text{or} \quad 4 \leq x\}$.

EXAMPLE Solve $(2 - x)(x - 1)^2 \leq 0$.
The solutions of $(2 - x)(x - 1)^2 = 0$ are 2 and 1, so we have three regions to test. Both boundary numbers are solutions of the inequality.

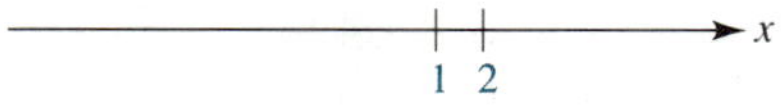

Test region	$(-\infty, 1)$	$(1, 2)$	$(2, +\infty)$
LHS: $(2 - x)(x - 1)^2$	Positive	Positive	Negative
RHS: 0	0	0	0
Solution of $(2 - x)(x - 1)^2 \leq 0$	No	No	Yes

Note: The factor $2 - x$ should be examined carefully. It is the opposite of $x - 2$.

The graph of the solution set is

The solution set is $\{x \mid x = 1 \quad \text{or} \quad x \geq 2\}$. Did you expect the solution regions to alternate with the nonsolution regions?

Note: You may still test by carrying out calculations for a particular test number, but that is more work. ■

The next examples show how the boundary-number method works for some other types of inequalities and introduce a small modification for restrictions.

EXAMPLE Graph the solution set of $\dfrac{x - 3}{x + 2} \geq 6$.

To solve

$$\frac{x - 3}{x + 2} = 6,$$

we multiply both sides by $x + 2$ and note the restriction on the original equation.

$$\frac{x - 3}{x + 2} = 6$$

$$\Leftrightarrow \quad x - 3 = 6(x + 2) \quad \text{and} \quad x \neq -2$$
$$\Leftrightarrow \quad x - 3 = 6x + 12 \quad \text{and} \quad x \neq -2$$
$$\Leftrightarrow \quad -5x = 15 \quad \text{and} \quad x \neq -2$$
$$\Leftrightarrow \quad x = -3 \quad \text{and} \quad x \neq -2$$

We see -3 is a boundary number because it is a solution for $(x - 3)/(x + 2) = 6$. *The extension of our process is that any forbidden values, such as -2 in this case, are also boundary numbers*. We now have two boundary numbers and three regions. The boundary number -3 is a solution of the inequality. For test numbers, we use -4, -2.5, and 0. We could have rewritten the original equation so that one side was 0 and tested by the sign method. Exercise 33 asks you to do the problem this way to compare the two methods.

Test number	-4	-2.5	0
LHS: $\dfrac{x-3}{x+2}$	$\dfrac{-7}{-2}$	$\dfrac{-5.5}{-0.5}$	$\dfrac{-3}{2}$
RHS: 6	6	6	6
Solution of $\dfrac{x-3}{x+2} \geq 6$	No	Yes	No

The graph is

Note: The number -2 does not belong to the solution set. We marked it in a special way on the first number line so that we'd remember it is a restricted value. ■

EXAMPLE Solve $0 \geq \dfrac{(x + 1.5)(2x + 1)^2}{x^3}$.

The boundary numbers are 0, -1.5, -0.5, obtained from solving

$$0 = \frac{(x + 1.5)(2x + 1)^2}{x^3}$$

and noting the restriction $x \neq 0$. The boundary numbers -1.5 and -0.5 are solutions of the inequality.

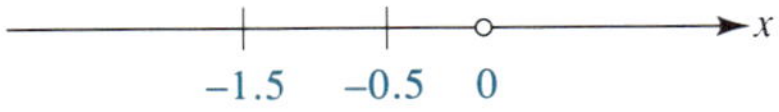

Because the right-hand side of the inequality is in factored form and the left-hand side is zero, we can test each region by simply deciding whether the right-hand side is positive or negative for a number within that region.

Test region	$(-\infty, -1.5)$	$(-1.5, -0.5)$	$(-0.5, 0)$	$(0, +\infty)$
LHS: 0	0	0	0	0
RHS: $\dfrac{(x + 1.5)(2x + 1)^2}{x^3}$	Positive	Negative	Negative	Positive
Solution	No	Yes	Yes	No

The graph of the solution set is

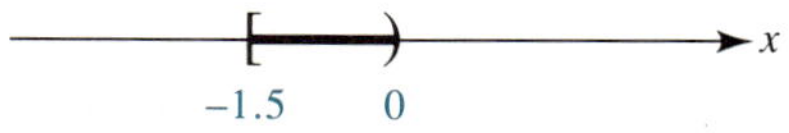

The solution set is $\{x \mid -1.5 \le x < 0\}$. ■

EXAMPLE Solve $3 + \sqrt{x - 2} > x - 5$.
First solve the corresponding equation.

$$\begin{aligned} 3 + \sqrt{x - 2} &= x - 5 \\ \Leftrightarrow \quad \sqrt{x - 2} &= x - 8 \qquad (*) \\ \Rightarrow \quad (\sqrt{x - 2})^2 &= (x - 8)^2 \\ \Leftrightarrow \quad x - 2 &= x^2 - 16x + 64 \\ \Leftrightarrow \quad 0 &= x^2 - 17x + 66 \\ \Leftrightarrow \quad 0 &= (x - 11)(x - 6) \end{aligned}$$

Because equivalent equations were not produced at each stage (*), the two possible solutions 6 and 11 must be checked in the original equation. Only 11 is a solution. Both the original inequality and equation have an implicit restriction that x cannot be less than 2 because we have restricted ourselves to real-number solutions. The test line is:

There are only two regions to test. Checking these we find the following.

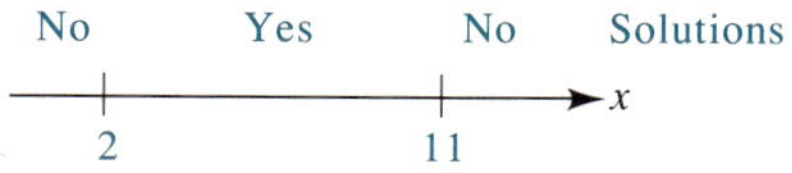

The solution set is $\{x \mid 2 \le x < 11\}$. ■

Now we can give a complete description of the process for solving a one-variable inequality:

1. Solve the corresponding equation. Mark as boundary numbers on a real-number line all solutions of the equation, all isolated restricted values, and the endpoints of any restricted region for the inequality. Test the boundary numbers to determine if they are solutions. It is a good idea to mark the restricted values in a special way. A hollow dot is appropriate for an isolated restricted value.
2. Test the regions marked on this number line by determining if a number in the region is a resolution of the original inequality. If one side of the inequality is 0, you may test regions by checking the sign of the other side.

The number line graphs and the tables we have used are just one way that you might organize your work. Use them when you need to do so, but remember they are neither necessary nor part of the answer.

EXERCISES 3.2

In exercises 1 through 32, solve the inequality.

1. $x^2 > 25$
2. $3a^2 \geq 75$
3. $-x^2 \leq 25$
4. $3a^2 < -75$
5. $5x^2 \geq 85$
6. $2b^2 < 13$
7. $12x^2 - 31 \leq 8$
8. $12x^2 > 5(x + 2)$
9. $88x - 16 < 121x^2$
10. $x(3x - 23) \geq -40$
11. $0.32w - 1.31w^2 \leq 5.7$
12. $0.4x^2 - 3x - 2 > 0$
13. $\frac{2}{3}c - \frac{4}{7}c^2 \leq \frac{1}{4}$
14. $\frac{1}{1000}a^2 < a$
15. $|x^2 - 3| \geq 5$
16. $|x^2 - 4x + 3| \leq 2$
17. $|x^2 + 9| < -1$
18. $-|5x^2 - 7x + 3| > -1$
19. $-|2x^2 - 3| \leq 5$
20. $|5x^2 - 3x + 1| \leq |x^2 + x|$
21. $\frac{x + 3}{x - 5} \geq 5x$
22. $\frac{x + 3}{x - 5} < 5x$
23. $\frac{1}{x} - \frac{1}{x - 8} \leq 3.2$
24. $a - \frac{1}{a} \geq 3a$
25. $x^3 - 3x^2 - 4x + 12 > 0$
26. $x^3 - 3x^2 - 4x < 0$
27. $\frac{x^3 - 3x^2 - 4x + 12}{x - 3} \leq 0$
28. $\frac{x^3 - 3x^2 - 4x + 12}{(x - 3)(x + 2)} \leq 0$
29. $\frac{(5x - 3)^2(x + 1)^3}{(x - \pi)(2x - 4)^4} \geq 0$
30. $(y - 4)^3(4 - y)^2(2y - 3)(3 - 2y)^3 < 0$
31. $w - 5 < \sqrt{2 - w}$
32. $1 + \sqrt{x + 3} \geq x$
33. Solve $\frac{x - 3}{x + 2} \geq 6$ by first rewriting it with the right-hand side 0. Which procedure do you prefer: this one or that used in the example of this section? Why?

In exercises 34 through 44, give all values of the variable for which the expression is (a) positive, (b) negative, (c) nonnegative, and (d) nonpositive. Express each of the four answers in interval notation.

34. $(x - 2)^2$
35. $(x - 2)^3$
36. $(2 - x)^2$
37. $(2 - x)^3$
38. $(x - 2)(x - 3)$
39. $(3a + 1)(a - 5)$
40. $(x - 2)^2(x - 3)$
41. $(3a + 1)(a - 5)^2$
42. $(x - 2)(x - 3)^3$
43. $(3a + 1)^3(a - 5)$
44. $(x - 2)^2(x - 3)^2$
45. $(3a + 1)^3(a - 5)^2$
46. The area of a room in the new hotel must be no larger than 300 square feet. If the length must be twice the width, what can the maximum length of the room be? What can the width of the room be if the specifications required measurements in increments of 0.5 feet only?
47. The distance between twice a number and its reciprocal is greater than the distance between the number of its opposite. Give the set of all such numbers.
48. The product of three consecutive integers is negative. Give all such triples of integers.
49. The product of three consecutive integers is not positive. Give all such triples of integers.
50. (Preview) For the unit box (Exercise 39, Section 3.1), suppose a trip of negative length means to go clockwise.
a. Where does a trip of length $-\sqrt{2}$ units from (1, 0) take you?

b. A trip of length $-3\sqrt{2}$ units from (1, 0) takes you to what point?

c. A trip of length $-5\sqrt{2}$ units from (1, 0) on the unit box takes you to what point?

51. (Preview) Give two positive and two negative trip lengths that will take you on the unit box from (1, 0) to each point.
a. (−1, 0) **b.** (0, 1) **c.** (0, 1) **d.** (1, 0)

52. (Preview) For each trip length given, tell where such a trip would take you from (1, 0) on the unit box.
a. $-6\sqrt{2}$ units **b.** $19\sqrt{2}$ units **c.** $218\sqrt{2}$ units
d. $-45\sqrt{2}$ units **e.** $-13\sqrt{2}$ units

3.3 Two-Variable Inequalities

Two-variable equations require two-dimensional graphs to represent their solution sets. To solve inequalities with two variables, we find boundary lines or curves rather than boundary numbers and test regions in the plane rather than on the number line. The plan of attack is the same as before: Solve the corresponding equation to find the boundaries and then practice **region testing.**

On the number line we have shaded intervals that belong to solution sets and used the symbols [,], and (,) to distinguish between boundary points that belong or do not belong to the solution set. In the two-dimensional case we also shade the sections of the plane that belong to the solution set. We use solid **boundary lines** or **curves** if those points belong to the solution set and dashed lines or curves if those points are not solutions.

EXAMPLE Graph the solution set of $5x - 2y \leq 8$.

The graph of $5x - 2y = 8$ is a line that is the boundary between solutions and nonsolutions.

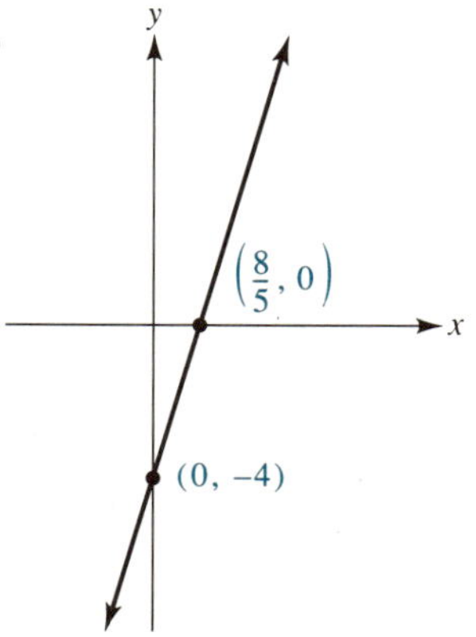

Because the points on this line are solutions of the inequality, we draw a solid line. The plane is divided into two regions. The point (0, 0) is easy to use as a test point, and because 5(0) − 2(0) is less than or equal to 8, all points in the region containing (0, 0) represent solutions. The point (2, 0) belongs to the other region and is not a solution. (Check it.) The graph of the solution set is

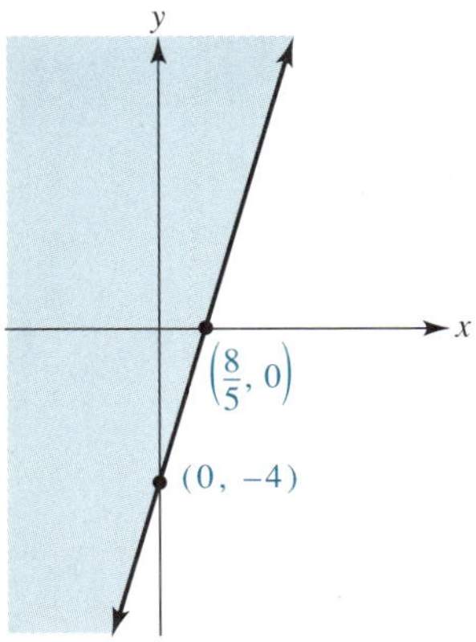

If you can solve the corresponding equation, you can solve the inequality. Here are some examples with a variety of boundary curves.

EXAMPLE Sketch the graph of the solution set of $4x^2 - y^2 > 16$.

The equation $4x^2 - y^2 = 16$ defines a hyperbola that intersects the x-axis at (2, 0) and $(-2, 0)$. The solutions of the equation are not solutions of this inequality, so we draw a dashed hyperbola.

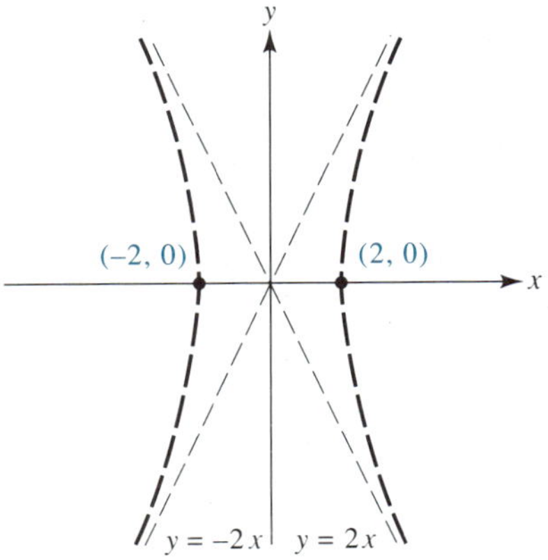

There are three regions in the plane to be tested. Be sure to verify the region in which a chosen test point lies. Here $(-3, 0)$, (0, 0), and (3, 0) are clearly in the three different regions.

Test point	$(-3, 0)$	(0, 0)	(3, 0)
LHS: $4x^2 - y^2$	36	0	36
RHS: 16	16	16	16
Solution	Yes	No	Yes

The graph of the solution is

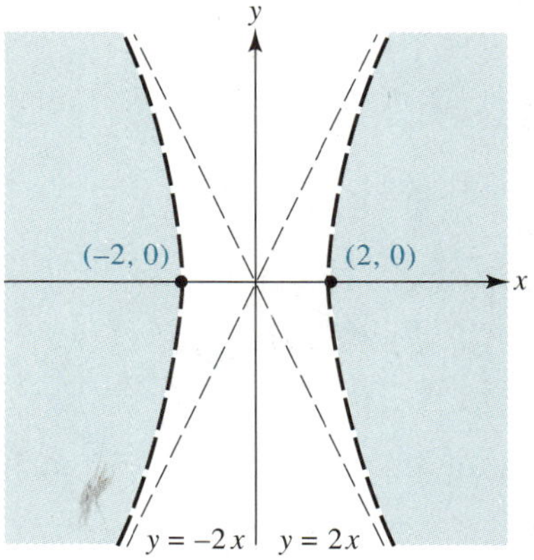

EXAMPLE Find all points within 4 units of the point $(-2, 5)$.

To be within 4 units of $(-2, 5)$ means the distance from the point (x, y) to $(-2, 5)$ is less than or equal to 4 units.

$$\sqrt{(x - (-2))^2 + (y - 5)^2} \le 4$$

To find the boundary curve, we solve the corresponding equation.

$$\sqrt{[x - (-2)]^2 + [y - 5]^2} = 4$$

$$\Leftrightarrow \quad (x + 2)^2 + (y - 5)^2 = 16 \qquad \text{Because each side is nonnegative}$$

The solution set for this equation is a circle with center $(-2, 5)$ and radius 4.

Now we test $(-2, 5)$ in the original inequality to determine if the points interior to the circle are solutions of the inequality. Since $\sqrt{0^2 + 0^2} \le 4$ is a true statement, the interior points are solutions. Testing $(0, 0)$, we see that $\sqrt{2^2 + (-5)^2} \le 4$ is not true, so the exterior points are not solutions.

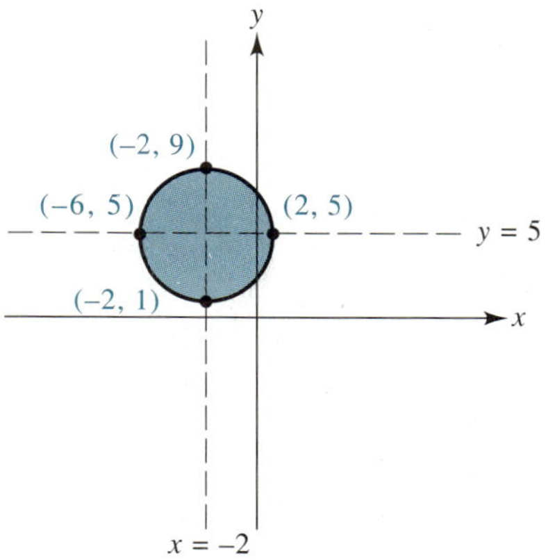

All the points on the circle defined by $(x + 2)^2 + (y - 5)^2 = 16$, together with the points interior to this circle, are within 4 units of the point $(-2, 5)$.

EXAMPLE Graph the solution set of $|x + y| \leq 1$.

The equation $|x + y| = 1$ is equivalent to $x + y = 1$ or $x + y = -1$. Each of these equations defines a straight line. The boundaries belong to the solution set.

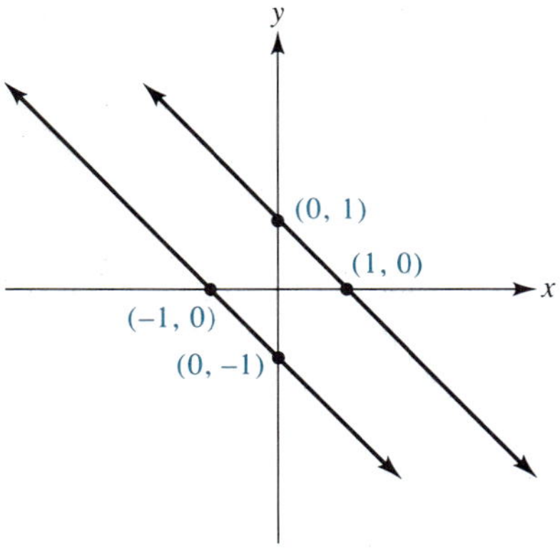

These two lines define three regions. Using $(-2, 0)$, $(0, 0)$, and $(2, 0)$ as test points, we obtain the following.

Test point	$(-2, 0)$	$(0, 0)$	$(2, 0)$
LHS: $\lvert x + y\rvert$	2	0	2
RHS: 1	1	1	1
Solution	No	Yes	No

The graph of the solution set of $|x + y| \leq 1$ is

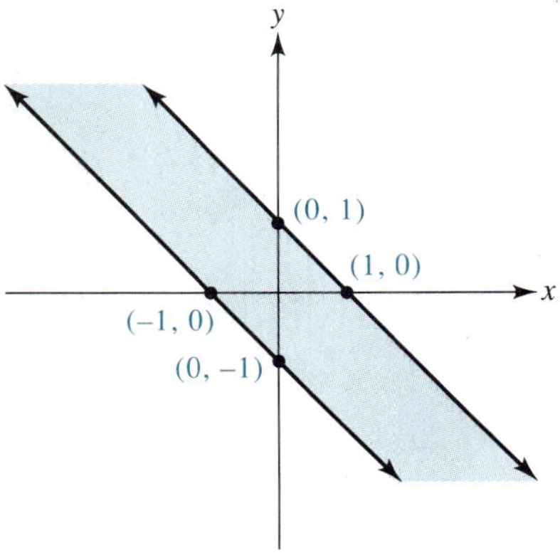

■

EXERCISES 3.3

In exercises 1 through 24, sketch the graph of the solution set.

1. $2x + y \leq 7$
2. $7x - 3 > 18y$
3. $5y \geq 2x + 4$
4. $4.1x - 3.2y \leq 6.3$
5. $x^2 + y^2 < 25$
6. $x^2 + y^2 \geq 27.04$
7. $2x^2 + 8y^2 > 32$
8. $5y^2 < 240 - 6x^2$
9. $x^2 - y^2 \geq 25$
10. $y^2 - x^2 < 27.04$
11. $2x^2 - 8y^2 < 32$
12. $5y^2 < 6x^2 + 240$
13. $y > x^2 - 7$
14. $y \leq 2(x - 3)^2 + 1$
15. $x - y^2 + 2y - 1 \geq 0$
16. $3y^2 - 18y < 4x + 1$
17. $|x - y| > 2$
18. $|2x + 3y| < 6$
19. $|3.1x - 4.2y| \leq -6.2$
20. $|3.1x - 4.2y| > 0$
21. $|x| + y \leq 1$
22. $x + |y| \geq 1$
23. $|x| + |y| > 1$
24. $|x + 1| > y$

In exercises 25 through 30, write the inequality whose graph is the set of all points

25. Within 4 units of the origin.
26. Above the horizontal line through the point $(2, -7)$.
27. Exterior to the ellipse centered at $\left(-5, \frac{1}{2}\right)$ with maximum vertical height 4 units and maximum horizontal width 3 units.
28. Within $\frac{2}{3}$ unit of $(7, 3.1)$
29. To the right of the line through the points $(-1, 4)$ and $(3, 0)$.
30. On or below the parabola with vertex $(-3, 6)$, opening upward, with expansion factor of 2.

In exercises 31 through 34, write an inequality whose graph is the shaded region and its boundaries.

31.

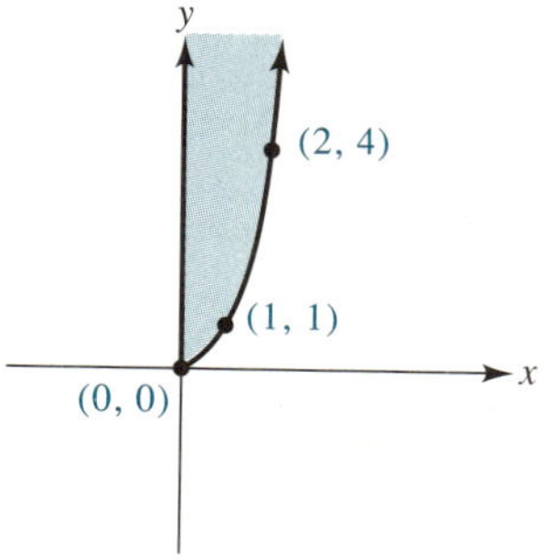

32.

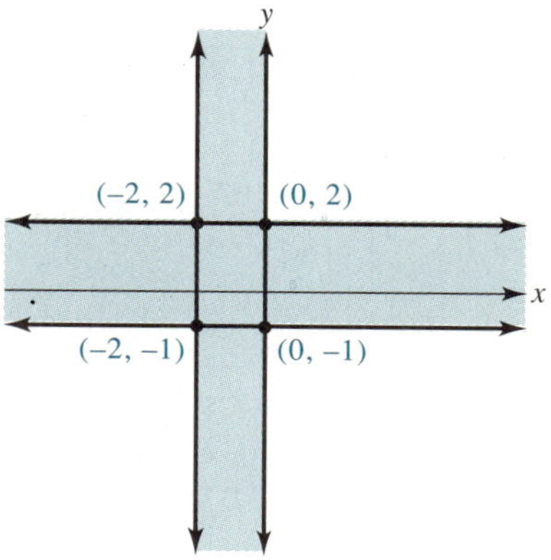

33.

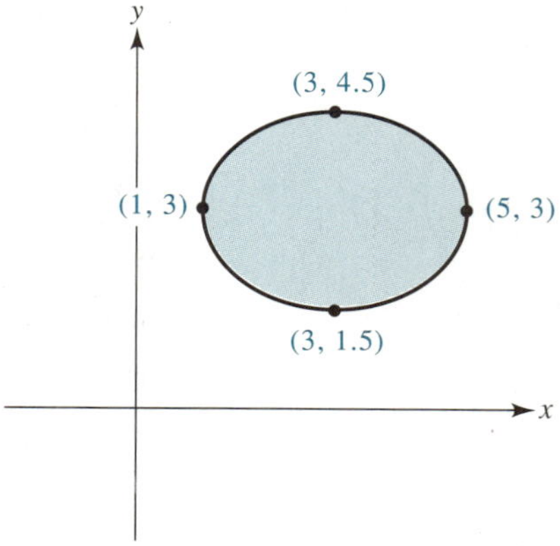

34.

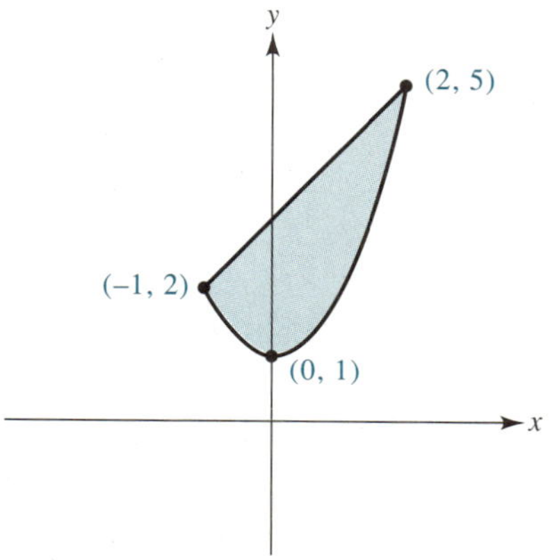

35. (Preview) Each side of the unit box has a midpoint. Give the coordinates for each of these four points.

36. (Preview) What is the shortest positive trip length from (1, 0) to each point?
a. $(\frac{1}{2}, \frac{1}{2})$ **b.** $(-\frac{1}{2}, \frac{1}{2})$ **c.** $(-\frac{1}{2}, -\frac{1}{2})$
d. $(\frac{1}{2}, -\frac{1}{2})$

37. (Preview) Give the coordinates of the point reached by traveling from (1, 0).
a. $\frac{15}{2}\sqrt{2}$ units **b.** $\frac{3}{2}\sqrt{2}$ units
c. $-\frac{7}{2}\sqrt{2}$ units **d.** $-\frac{9}{2}\sqrt{2}$ units

3.4 Compound Inequalities

Toolkit

For the compound statement *P* and *Q* to be true, both *P* and *Q* must be true. The compound statement *P* or *Q* is true if either *P* or *Q* is true.

Occasionally the inequalities we need to solve are compound statements. These **compound inequalities** may be constructed using *and* or *or*. For example, suppose the expression $3x^2 - 7$ must produce values between 2 and 8; we have $2 < 3x^2 - 7$ and $3x^2 - 7 < 8$. Another situation is if either of two regions is acceptable; we may write $3x + 2y > 6$ or $x^2 + 2y^2 \leq 4$.

EXAMPLE Solve $2 < 3x^2 - 7 < 8$.

The inequality to be solved can be read $2 < 3x^2 - 7$ *and* $3x^2 - 7 < 8$, so we solve each simple inequality first. The boundary numbers are obtained from the corresponding equations.

$$\begin{aligned} & 2 = 3x^2 - 7 & & 3x^2 - 7 = 8 \\ \Leftrightarrow\ & 0 = 3x^2 - 9 & \Leftrightarrow\ & 3x^2 - 15 = 0 \\ \Leftrightarrow\ & 0 = 3(x^2 - 3) & \Leftrightarrow\ & 3(x^2 - 5) = 0 \end{aligned}$$

We have four boundary numbers, $-\sqrt{3}$, $\sqrt{3}$, $-\sqrt{5}$, and $\sqrt{5}$, and five regions.

Now we test the regions.

Test number	−10	−2	0	2	10
Solution of $2 < 3x^2 - 7$	Yes	Yes	No	Yes	Yes
Solution of $3x^2 - 7 < 8$	No	Yes	Yes	Yes	No
Solution of $2 < 3x^2 - 7 < 8$	No	Yes	No	Yes	No

The graph of the solution set is

The solution set is

$$\{x \mid -\sqrt{5} < x < -\sqrt{3} \text{ or } \sqrt{3} < x \sqrt{5}\}.$$

Note: This solution set is the intersection of the solution sets for the simple inequalities: $\{x|x < -\sqrt{3} \text{ or } x > \sqrt{3}\} \cap \{x| -\sqrt{5} < x < \sqrt{5}\}$. ■

EXAMPLE Graph the solution set of $x > y$ or $x^2 + 2y^2 \leq 4$.

The equations corresponding to the two inequalities produce a boundary line and a boundary curve, an ellipse. The points on the line do not satisfy the first inequality, but those on the ellipse satisfy the second inequality and belong to the solution set. We sketch a dashed line and a solid ellipse.

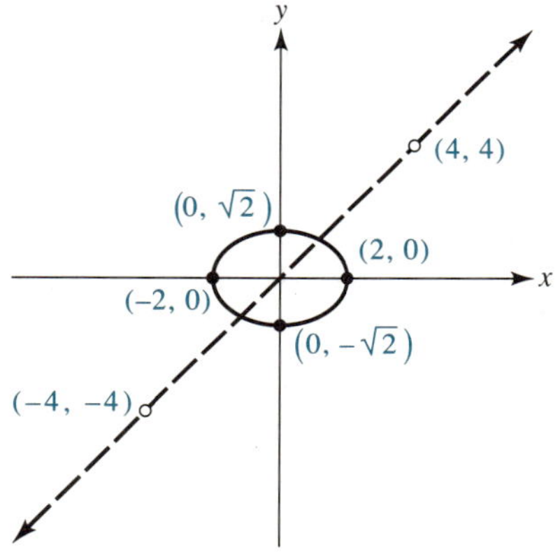

These two boundaries define four regions. We test the regions for solutions and nonsolutions using the points $(-3, 0)$, $(-1, 0)$, $(1, 0)$, and $(3, 0)$.

Test point	$(-3, 0)$	$(-1, 0)$	$(1, 0)$	$(3, 0)$
Solution of $x > y$	No	No	Yes	Yes
Solution of $x^2 + 2y^2 \leq 4$	No	Yes	Yes	No
Solution of $x > y$ or $x^2 + 2y^2 \leq 4$	No	Yes	Yes	Yes

The graph of the solution set is

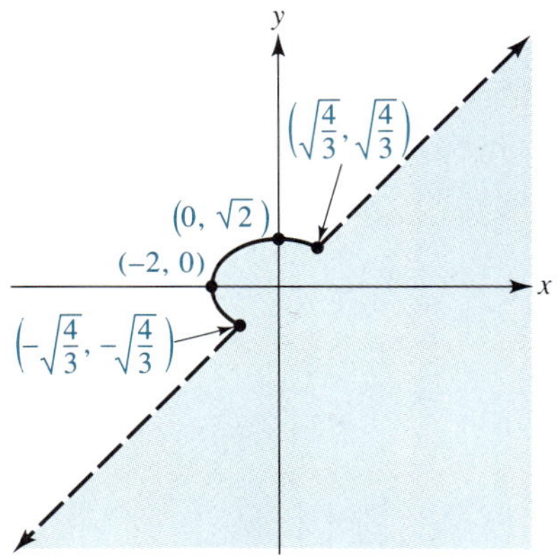

Note: This solution is the union of the solution sets of the simple inequalities. ■

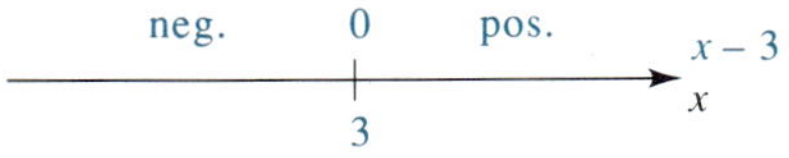

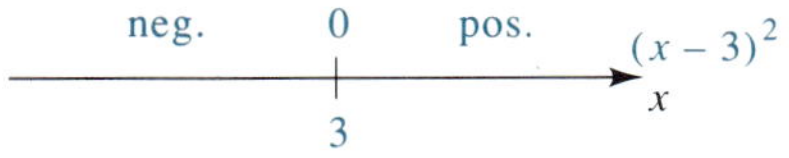

As a change of pace, let's return to the one-variable inequalities and pull together some ideas that will make our method more efficient.

You have probably noticed that for many of the inequalities you have solved, the regions of solutions and nonsolutions alternate. If we can determine when this will happen, we can streamline region testing considerably. Consider the expression $x - 3$. When x is replaced by 3, $x - 3$ is 0. If a number greater than 3 is substituted for x, $x - 3$ is positive. A number less than 3 produces a negative value. We see that 3 is the boundary between negative and positive values for $x - 3$. However, if the expression is $(x - 3)^2$, 3 does not separate negative from positive values. This means if you are solving an inequality by testing the signs of expressions in each region, you may ignore any linear factors with even powers because you know the expression will have a positive value there.

EXAMPLE Solve $\dfrac{(x - 2)(x + 4)^2}{(1 - x)^3(x + 5)^4} \le 0.$

Boundary numbers for

$$\frac{(x - 2)(x + 4)^2}{(x - 1)^3(x + 5)^4} \le 0$$

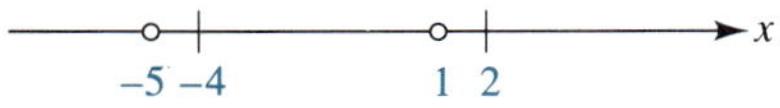

are -5, -4, 1, and 2.

We have five regions to test, but notice that $(x + 4)^2$ and $(x + 5)^4$ will be positive except at -4 and -5, respectively, and cannot affect the sign of the result. Thus we really have only three regions to test: left of 1, between 1 and 2, and right of 2. We know the sign of $1 - x$ will change as we pass 1, and the sign of $(x - 2)^3$ will change as we pass 2. When x is 100, the left-hand side is negative, so we have

The solution set is $\{x \mid x < -5 \text{ or } -5 < x < 1 \text{ or } 2 \le x\}$. ■

EXAMPLE Solve $3x^{-4} - 7x^{-5} + 4x^{-6} \le 0$.

Remember that our method of solving inequalities works for any case where you know how to solve the corresponding equation. We solve $3x^{-4} - 7x^{-5} + 4x^{-6} = 0$ to find the boundary numbers.

$$\begin{aligned} & 3x^{-4} - 7x^{-5} + 4x^{-6} = 0 \\ \Leftrightarrow \quad & x^{-6}(3x^2 - 7x + 4) = 0 \\ \Leftrightarrow \quad & x^{-6}(3x - 4)(x - 1) = 0 \end{aligned}$$

The boundary numbers are $\frac{4}{3}$, 1, and 0. Of these, $\frac{4}{3}$ and 1 are solutions of the inequality. We have to test only three regions because x^{-6} will not change sign as we pass 0.

Test region	$(0, 1)$	$(1, \frac{4}{3})$	$(\frac{4}{3}, +\infty)$
LHS	Positive	Negative	Positive
RHS	0	0	0
Solution	No	Yes	No

The graph of the solution set is

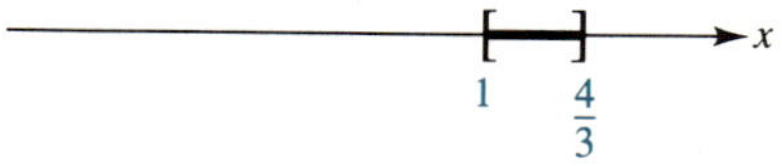

The solution set is $\{x \mid 1 \le x \le \frac{4}{3}\}$. ■

EXAMPLE Graph the solution set of $\frac{7}{5}x^{2/5} - \frac{2}{5}x^{-3/5} \ge 0$.

The equation

$$\frac{7}{5}x^{2/5} - \frac{2}{5}x^{-3/5} = 0$$

is equivalent to

$$\frac{1}{5}x^{-3/5}(7x - 2) = 0.$$

The only solution is $\frac{2}{7}$, and 0 is a restricted value because $x^{-3/5}$ is undefined for 0.

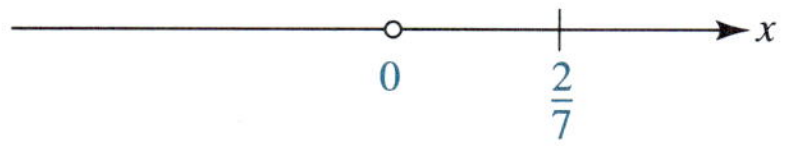

Test region	$(-\infty, 0)$	$(0, \frac{2}{7})$	
Factor: $x^{-3/5}$	Negative	Positive	
Factor: $7x - 2$	Negative		
LHS: $x^{-3/5}(7x - 2)$	Positive		
RHS: 0	0		
Solution	Yes		

The graph of the solution set is

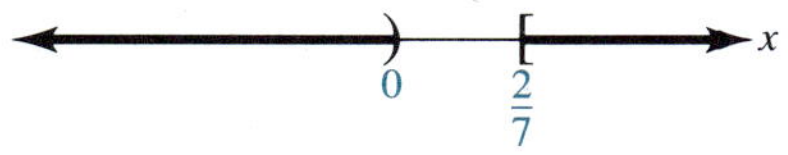

EXERCISES 3.4

In exercises 1 through 8, sketch the one-dimensional graph of the solution set. Verify a section of your graph by evaluating for a given number or point.

1. $0 \le x^2 - 9 \le 7$ **2.** $3 < y^2 + 2 < 83$

3. $-1 \le x^2 - 9 \le 13$ **4.** $1 \le y^2 + 2 < 90$

5. $0 \le x^2 + 3x - 4 \le 14$

6. $-2 < 3a^2 + 5a - 1 < 8$

7. $5 < 4w^2 - 3w < -16$ **8.** $1 < a^3 < 8$

In exercises 9 through 20, sketch the two-dimensional graph of the solution set. Verify a section of your graph by evaluating for a given ordered pair or point.

9. $x > y$ or $2x + 3y < 7$

10. $x < y$ and $2x + 3y \ge 7$

11. $x \ge -1$ and $x + 5y^2 < 7$

12. $5 < y$ and $2x^2 + 4y^2 \ge 16$

13. $x^2 + y^2 \le 4$ and $y < 2x^2$

14. $x^2 + y^2 \le 4$ or $y < 2x^2$

15. $x \ge 6$ and $x^2 - y^2 \ge 1$

16. $x^2 + y^2 < 3$ and $x^2 + y^2 \ge 7$

17. $x^2 + y^2 < 3$ or $x^2 + y^2 \ge 7$

18. $x^2 + y^2 > 3$ and $x^2 + y^2 \le 7$

19. $x^2 + y^2 > 3$ or $x^2 + y^2 \le 7$

20. $x^2 + y^2 < 3$ and $x^2 + y^2 \le 7$

In exercises 21 through 38, sketch the one-dimensional graph of the solution set.

21. $x(x + 3)^2(x - 2)^3 > 0$

22. $(3x - 1)^4(2 - x)(x + 5)^3 \le 0$

23. $(x - 7.2)^5(7.1 - x)^2(x - 7.15)^8(7.3 - x)^{17} > 0$

24. $(y + \sqrt{7})^7(y + \sqrt{8})^8(y + \sqrt{9})^9(y + \sqrt{10})^{10} \ge 0$

25. $\dfrac{(2x + 3)^2(x - 1)}{x(x - 5)^3(x + 3)^2} \ge 0$

26. $\dfrac{(y - 8)^8(y + 5)^6(3 - y)^5}{(y + 15)^{10}} < 0$

27. $\dfrac{-12a^8(3 - a)^6(a + \pi)^7}{a^3 - 5a^2 - 6a} > 0$

28. $\dfrac{x^5 - 6x^4 - 5x^3 + 30x^2}{(7x - 1)^3(x - 3)^2(x - \sqrt{3})^4} \le 0$

29. $x^{-6} + 3x^{-5} \le 0$

30. $y^{-7} - 5y^{-8} > 6y^{-9}$

31. $2a^{-11} - 18a^{-13} < 0$

32. $w^{-11} + 4x^{-13} + 4w^{-15} \ge 0$

33. $x^{-1/3} - x^{-4/3} \le 0$

34. $a^{-5/3} + a^{-8/3} > 0$

35. $y^{-2/3} > 2y^{-5/3} - y^{1/3}$

36. $w^{-8/3} - w^{-1/3} \le 0$

37. $2x^{1/4} - x^{-3/4} - 18x^{-7/4} \geq 0$

38. $\frac{2}{3}x^{-1/4} - \frac{1}{3}x^{-5/4} - 6x^{-9/4} \leq 0$

In exercises 39 and 40, sketch the two-dimensional graph of the solution set.

39. $x^{5/2} + 11x^{3/2} + 11x^{1/2} < 0$

40. $\frac{1}{5}y^{-5/2} - \frac{11}{5}y^{-7/2} + \frac{11}{5}y^{-9/2} \geq 0$

In exercises 41 through 46, give the values of the variable for which the expression is (a) positive, (b) negative, (c) nonnegative, and (d) nonpositive.

41. $(x - 2)^{3/2}$ **42.** $(x - 2)^{2/3}$

43. $(x - 2)^{5/3}$ **44.** $(2 - x)^{3/2}$

45. $(2 - x)^{2/3}$ **46.** $(2 - x)^{5/3}$

In exercises 47 through 51, write a compound inequality that will have the given graph.

47.

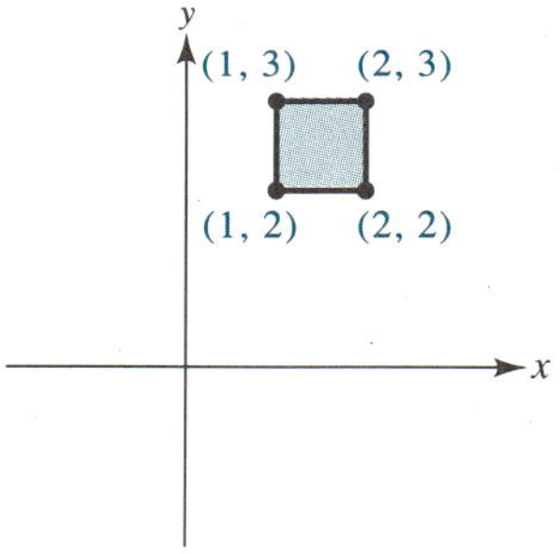

48.

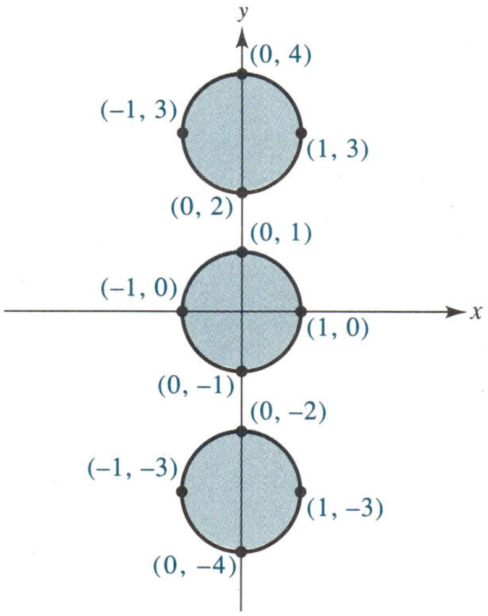

49.

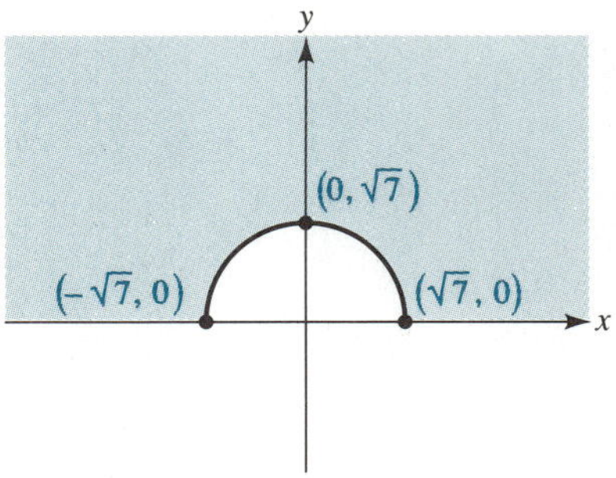

50.

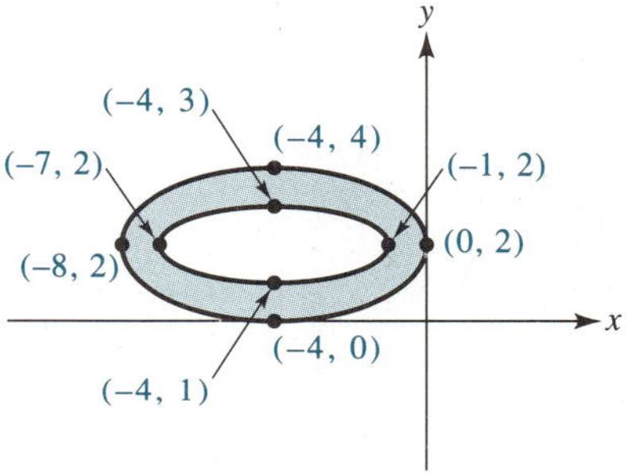

51.

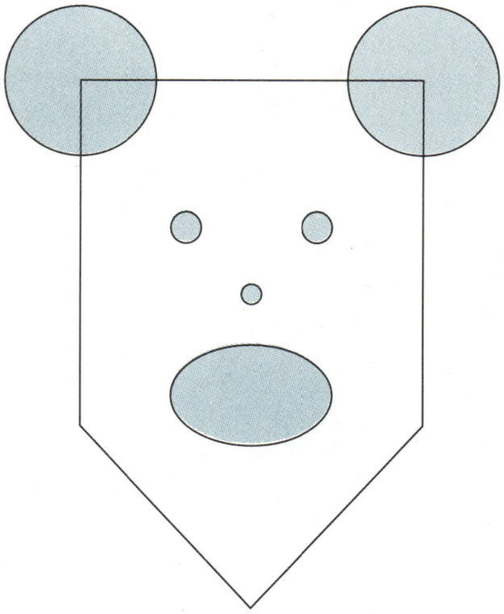

(You may use any scale you wish. The answer in the back of the book represents one of infinitely many choices.)

52. (Preview) Give a pattern for all trip lengths on the unit box from (1, 0) to each point.
a. (0, 1) **b.** (−1, 0) **c.** (0, −1) **d.** (1, 0)
Be sure to include positive and negative trip lengths. (*Hint:* It may help to make a list of some of the trip lengths from (1, 0) to (0, 1). Check that the pattern you give will produce these lengths.)

53. (Preview) **a.** Give a pattern for all trip lengths on the unit box from (1, 0) to $\left(-\frac{1}{2}, \frac{1}{2}\right)$.
b. Give a pattern for all trip lengths on the unit box from (1, 0) to $\left(-\frac{1}{2}, -\frac{1}{2}\right)$.

54. (Preview) On the unit box, where does a trip of length 1 unit take you from (1, 0)? If you travel -1 unit from (1, 0) on the unit box, what are the coordinates of your destination?

Chapter 3: Summary

A general method for solving inequalities is: (1) Solve the corresponding equation. The solutions give boundary numbers or curves. Any isolated restricted values as well as endpoints or edges of restricted regions are also boundary numbers or boundary curves, respectively. (2) Determine if boundaries belong to the solution. (3) Then test the regions determined by the boundaries. This may be done with particular values or, in some cases, by checking the signs of factors.

Key Concepts

inequality	
boundary number	*solution of equation, restricted value*
region testing	
boundary line or curve	*arithmetical calculation, sign checks*
compound inequalities	

Skills

You should be able to:

1. Find boundary numbers for one-variable inequalities.
2. Find boundary curves for two-variable inequalities.
3. Test regions to determine location of solution sets.
4. Write solution sets for one-variable inequalities.
5. Graph solution sets for two-variable inequalities.

SUMMARY EXERCISES

The starred exercises can serve as a practice chapter test. The answers for these are at the end of the answer section at the back of the book.

In exercises 1 through 22, solve the inequality.

1. $4x - 15 \le \frac{7}{2}[2 - 10x]$

2. $ay + 4y - 2 > 13.3 - 3y$, where a is a constant greater than 5

* **3.** $0 \ge -4 + \sqrt{8w} - 3w$

4. $x^2 - 5x + 6 < 0$

5. $(2u - 1)^2 - u^3 + 4 > 5 - u^3$

* **6.** $2x^2 - 5x + 6 \ge 0$

7. $2x^2 - 5x + 6 \le 0$

8. $|5x + 7| > 9$

* 9. $3|2 - x| - 13 \leq 0$

10. $|5x + 7| > -3$

11. $3|2 - x| + 13 < 0$

* 12. $|x^2 - 5| < 4$

13. $|w^2 + 3w| \geq |5 - w|$

14. $\dfrac{2x - 4}{5 - x} \leq 0$

* 15. $\dfrac{2x - 4}{5 - x} - 3 \leq 6$

16. $\dfrac{(3x - 1)(x + 7)}{4x - 9} > 0$

17. $\dfrac{x(3x - 1)^2(x + 7)}{4x - 9} > 0$

* 18. $\dfrac{x^2(6 - x)(x + 5)^5}{x^3 - 3x^2 - 5x + 15} \leq 0$

19. $\sqrt{x + 8} < 42$

20. $\sqrt{x + 8} > 42$

* 21. $\sqrt{x + 8} < -42$

22. $\sqrt{x + 8} > -42$

23. For what numbers is it true that the quotient of the number and the number increased by 7 is negative?

* 24. A plot of ground has dimensions 18 feet by 30 feet. A border x feet wide is to be planted in red zinnias with the center planted in yellow zinnias. How wide can the border be in order that the area covered in red is larger than the area covered in yellow?

25. The product of two consecutive odd numbers is divided by 3 more than the first odd number. For what numbers is this quotient nonnegative?

In exercises 26 through 32, solve the inequality.

26. $x^{-4} + x^{-5} - 12x^{-6} \leq 0$

* 27. $x^{-11} + x^{-12} - 12x^{-13} \leq 0$

28. $x^{8/3} + x^{5/3} - 12x^{2/3} \leq 0$

29. $x^{-7/3} + x^{-4/3} - 12x^{-1/3} \leq 0$

* 30. $x^{3/2} + x^{1/2} - 12x^{-1/2} \leq 0$

31. $x^{2/3} + x^{-1/3} - 12x^{-4/3} \leq 0$

32. $y^{-7/8}(3 - y)^{14}(2y + 5)^3y^5 > 0$

In exercises 33 through 42, sketch the graph of the solution set in the *xy*-plane.

* 33. $x^2 + (y - 4)^2 > 25$

34. $y \leq -(x + 3)^2 - 4$

35. $0 < x^2 + 3 \leq 19$

* 36. $x - 3y < 4$ and $x \leq y$

37. $y^2 \leq 4 - x^2$ and $x \leq y$

38. $|x - y| < 1$

* 39. $\dfrac{x^2}{16} + \dfrac{(y - 3)^2}{4} > 1$

40. $\dfrac{x^2}{16} + \dfrac{(y - 3)^2}{4} > 0$

41. $x^2 + y^2 \leq 1$ or $x^2 + y^2 \geq 5$

* 42. $\dfrac{x^2}{16} - \dfrac{(y - 3)^2}{4} > 1$

In exercises 43 through 50, write a mathematical statement that describes all points in the *xy*-plane.

43. In the third quadrant.

44. Inside the unit circle.

* 45. Above and including the standard parabola.

46. Outside the rectangle with corner points $(-5, 9)$, $(-5, 2)$, $(-1, 9)$, and $(-1, 2)$.

47. Below the line containing $(9, \sqrt{2})$ and $(8, \sqrt{3})$.

* 48. In the first quadrant and inside the parabola $x = y^2$.

49. To the right of the line $x = -\pi$ or to the left of the line $x = -490$.

50. Above the top section of the hyperbola $\dfrac{y^2}{4} - \dfrac{(x - 1)^2}{1} = 1$.

CHAPTER FOUR

Functions

- 4.1 Definitions and Notation
- 4.2 Graphing Techniques I
- 4.3 Asymptotes
- 4.4 Properties of Graphs and Functions
- 4.5 Graphing Techniques II

4.1 Definitions and Notation

Toolkit

For this chapter you'll need factoring methods, exponent rules, ways of writing sets, the definition of absolute value from the appendix, and the graphing techniques from Chapter 2.

Suppose we have a graph, a set of points, which are ordered pairs of real numbers. Might we be able to describe the collection with set-builder notation?

Illustration **a.**

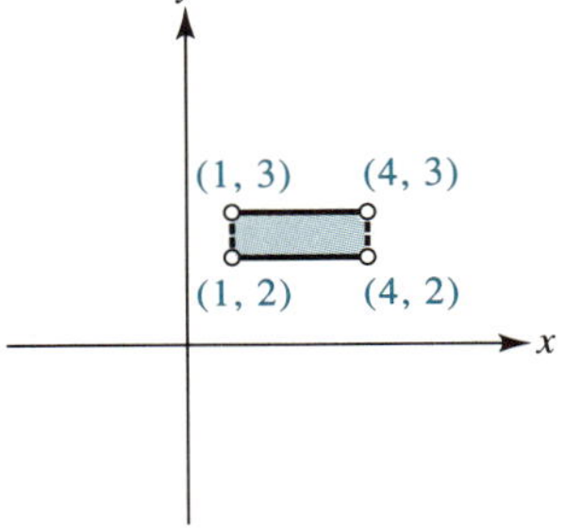

The shaded region is $\{(x, y) \mid 1 < x < 4 \text{ and } 2 \leq y \leq 3\}$.

b.

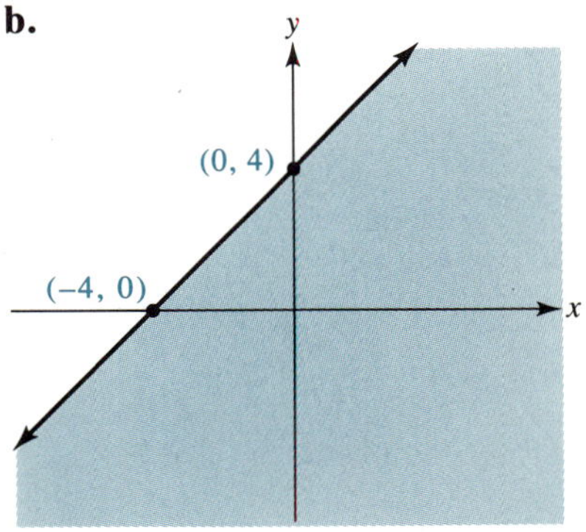

The shaded region is $\{(x, y) \mid x \geq y - 4\}$.

c.

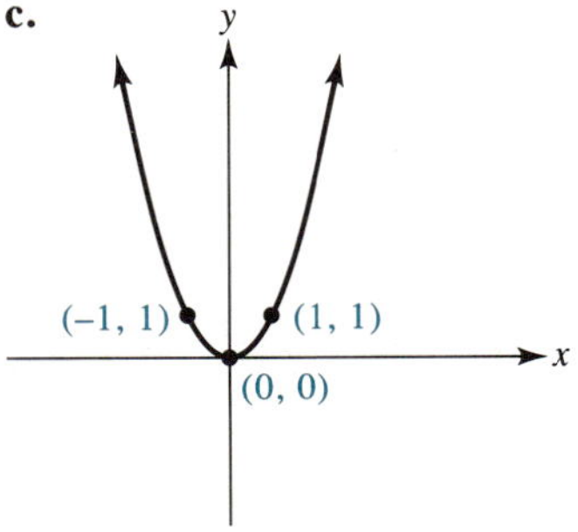

$\{(x, y) \mid y = x^2\}$ □

Let's extend this idea a bit.

Definition A **relation** is a set of ordered pairs.

Notice that this is a good deal more general than a graph because this definition does not specify that the entries are numbers.

Illustration $\{(x, y) \mid x$ is a person's name and y is that person's social security number$\}$ is a relation. □

For most of our work here, we deal with ordered pairs of real numbers. In fact, we're concerned mainly with relations that have the following special property.

Definition A **function** is a relation for which the choice of a first coordinate uniquely determines the second coordinate.

In the previous illustrations, (c) is a function; if a value for x is chosen, there is only one value of y that together with the choice of x will satisfy the equation $y = x^2$. The fact that two different choices for x, say 2 and -2, may yield the same value for y doesn't contradict the definition.

Definition The variable denoting the first coordinate is the *independent variable;* that for the second coordinate is the *dependent variable*.

Sometimes it's helpful to think of a function in a different way. One idea is that of a number-manipulating machine where numbers, the first coordinates, go in and numbers, the second coordinates, come out.

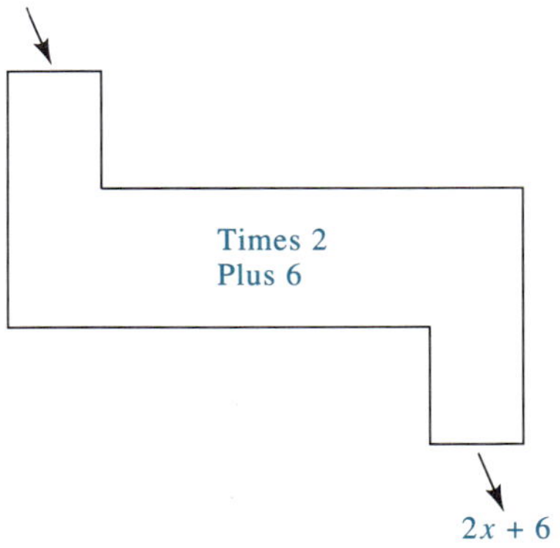

Represents
$\{(x, y) \mid y = 2x + 6\}$

Another way of looking at a function is as a mapping, a way to get from one set of numbers to another.

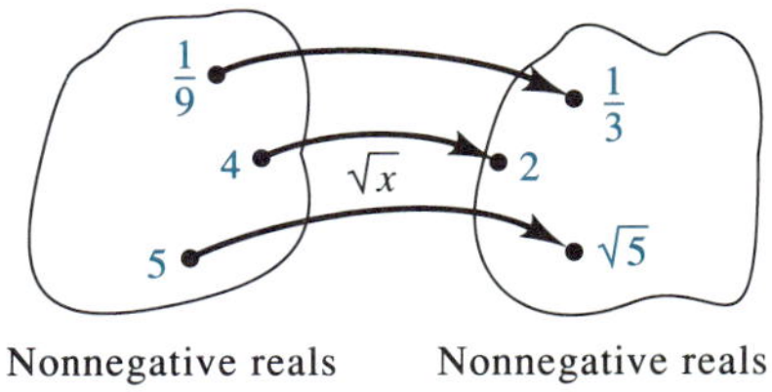

Represents
$\{(x, y) \mid y = \sqrt{x}\}$

Definition Since we often talk about the set of possible values for the independent variable, we give this set a name, the *domain* of the function. The *range* of the function is the set of values for the dependent variable.

The function is a mapping from the domain to the range. For a function it's sufficient to describe the domain along with the function's rule. Often the domain won't be stated explicitly, and any number for which we can carry out the calculations in the rule is allowable.

EXAMPLE What is the domain of $f = \left\{(x, y) \mid y = \sqrt{2.3 - x}\right\}$?
In this section we are using ordered pairs of real numbers, so x must be chosen so that $2.3 - x \geq 0$.
The domain is $\{x \mid 2.3 \geq x\}$. ■

Illustration $\{(x, y) \mid x + 2y = 7, x \in I\} \neq \{(x, y) \mid x + 2y = 7\}$
The domain for the function on the left side is the set of integers, whereas that for the function on the right side is the set of real numbers. Therefore, the sets are not equal.

$$\left\{(x, y) \,\middle|\, \frac{x^2 - 4}{x - 2} = y\right\} \neq \{(x, y \mid x + 2 = y\}$$

The domain for the left-hand side is the set of reals except 2, whereas that for the right-hand side is the set of all real numbers.

$$\left\{(x, y) \,\middle|\, \frac{x^2 - 9}{x + 3} = y\right\} = \{(x, y) \mid x - 3 = y, x \neq -3\}$$

The domain for each is the set of all real numbers except -3. □

Instead of using set-builder notation, we can give a function by enumeration or, as we've seen, by a graph. More common is an abbreviated form of set-builder notation consisting of just the rule.

Illustration The function $\{(x, y) \mid y = 2x - 4\}$ can be written $y = 2x - 4$. □

However it is described, we must be able to confirm that the relation is a function. Here's a case where it's easier to say what we don't want. To be a function the relation should not contain two pairs of the form (a, b), (a, c) with $b \neq c$. If the set is enumerated, we must inspect the pairs listed. If it's given as a graph, two points (a, b) and (a, c), $b \neq c$, would lie on the same vertical line. Thus if some vertical line intersects a graph in more than one point, the graph does not represent a function. When a rule is given, we must consider if a choice of the independent variable can lead to two different values of the dependent variable.

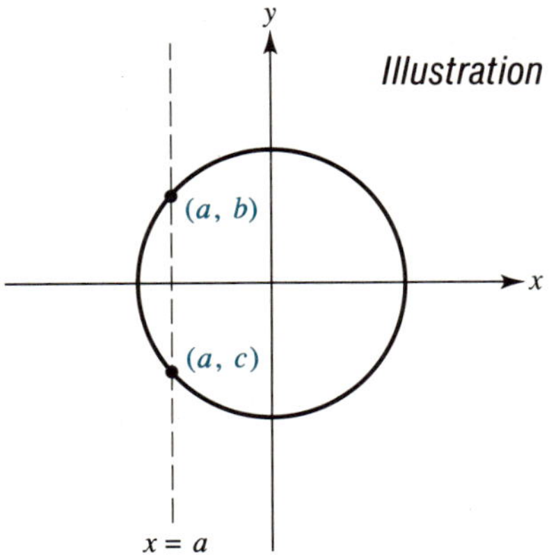

Illustration The relation $\{(2, 1), (3, 1), (4, 1)\}$ is a function.
The relation $\{(2, 1), (2, 5), (3, 1), (4, 1)\}$ is not a function because 2 is paired with both 1 and 5.
This is not the graph of a function; it fails the vertical line test.
The relation $\{(x, y) \mid y^2 - x = 25\}$ is not a function; it is a relation containing both $(0, 5)$ and $(0, -5)$. A function is described by $y = \sqrt{7 - 2x}$. □

A modification of naming sets is very useful when the sets are functions. Instead of abbreviating $f = \{(x, y) \mid y = 7x^2 - 3\}$ as $y = 7x^2 - 3$, we write $f(x) = 7x^2 - 3$, where $f(x)$ is read as "f of x." This notation makes it easy to refer to two functions, such as $g(x) = 6x - 4$ and $f(x) = 7x^2 - 3$, in the same discussion. It also emphasizes which is the independent variable and gives a convenient way to state this question: What is the value of the dependent variable in the function g when x is replaced by 7? We write $g(7) = ?$.

Definition Using the preceding notation the value of the independent variable is called the *argument* of the function.

In the expression $f(a + 2) + 12$, the argument of f is $a + 2$.

EXAMPLE For $g(x) = 6x - 4$ and $f(x) = 7x^2 - 3$, evaluate $g(7), f(4), g(a)$, and $f(a^2)$.

$$g(7) = 6(7) - 4, \text{ or } \mathbf{38}$$
$$f(4) = 7(4)^2 - 3, \text{ or } 112 - 3, \text{ or } \mathbf{109}$$
$$g(a) = \mathbf{6a - 4}$$
$$f(a^2) = 7(a^2)^2 - 3, \text{ or } \mathbf{7a^4 - 3}$$

■

We need to see where a function whose domain is a set of numbers fits into the order-of-operations list of Section A.1, because such a function operates on a number in much the same way the square root and absolute value operations do. We recapitulate and extend the list in Section A.1.

1. *Symbols of grouping are a signal to carry out the operations within them first.* Symbols of grouping include parentheses, brackets, braces, the vinculum (which groups both the numerator and denominator), absolute value bars, and the radical sign. These symbols may be doing double duty by also indicating an operation—e.g., the radical sign. The parentheses used to set off a function's argument are symbols of grouping. The function value must be found next.
2. *Exponents are used next.*
3. *Multiplications and divisions follow.* These are done in "reading order" from left to right.
4. *The operation of negation, taking the opposite of a number, is done next.*
5. *Additions and subtractions are now done in "reading order" from left to right.*

EXAMPLE With $g(x) = 6x - 4$ and $ssq(x) = 7x^2 - 3$, evaluate $g(3) + 6$, $\text{ssq}(a + 3)$, $4g(2)$, $\dfrac{\text{ssq}(a)}{6}$, and $g\left(\dfrac{-4}{3}\right)$.

$$g(3) + 6 = (6(3) - 4) + 6, \text{ or } \mathbf{20}$$

$$\text{ssq}(a + 3) = 7(a + 3)^2 - 3, \text{ or } \mathbf{7a^2 + 42a + 60}$$

$$4g(2) - 5 = 4(6 \cdot 2 - 4) - 5, \text{ or } \mathbf{27}$$

$$\frac{\text{ssq}(a)}{6} = \mathbf{\frac{7a^2 - 3}{6}}$$

$$g\left(-\frac{4}{3}\right) = 6\left(-\frac{4}{3}\right) - 4, \text{ or } \mathbf{-12}$$

■

In Section A.5 we defined a *term* to be an expression containing only the operations of exponentiation, multiplication and division. We now extend this definition to allow the use of functions also.

Definition An expression that contains exponentiation, multiplication, division, the absolute value operation, and function values is called a **generalized term.**

Some of the most commonly used functions have two- or three-letter names. It should be clear from the context whether abs, for example, represents the product of variables or the name of a function. Later we'll discuss numerous functions that have generally recognized names. When working with these functions—for example, sin for the sine function—it's not uncommon to omit the parentheses around the argument. When this is done, the argument is assumed to be only the next generalized term.

Illustration Many computer languages use abs to name the function that produces the absolute value of a number; abs(x) means $|x|$. With this definition,

$$\text{abs}(-4 + (-3)) = |-4 + (-3)|, \text{ or } 7;$$
$$\text{abs } 4 - 3 = |4| - 3, \text{ or } 1;$$
$$\text{abs}\frac{\sqrt{2x}}{7} - 9 = \left|\frac{\sqrt{2x}}{7}\right| - 9.$$ □

We haven't exhausted the ways in which a function can be described. A function may behave very differently on different pieces of its domain; then a piecewise definition is appropriate. In this situation care must be taken at any boundary number in the domain to be sure a function is defined.

Illustration

$$f(x) = \begin{cases} x^2, & x \leq 1 \\ 3x - 6, & x > 1 \end{cases}$$

Note: If both parts of the domain included 1, f would not be a function.

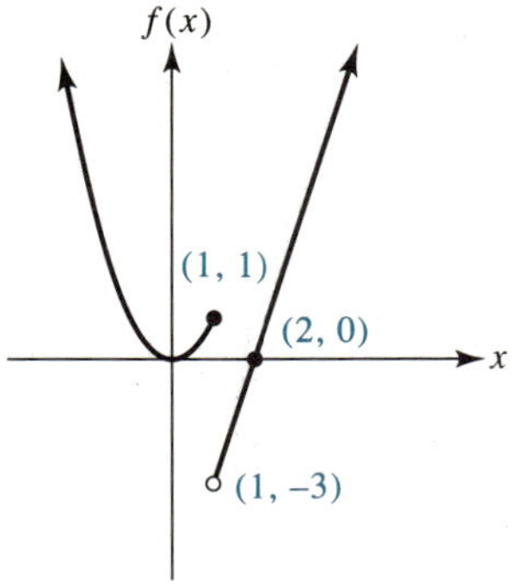

□

Some functions can't be described by an algebraic pattern, even a piecewise one. Some very common functions are of this sort. Function values for these are often approximated and stored in tables or can be approximated using special keys on a calculator. If a particular function isn't popular enough to have a standard calculator key, we may still have to deal with a table. If the domain is large, only a partial table of approximate function values may be possible.

Illustration The function f is defined for x in [0, 1]. Some approximate function values are given in Table 4.1a.

$$f(0.68) \approx 5.9248$$
$$f(0) \approx 5.0000$$
$$f(0.81) \approx 6.3122$$

□

TABLE 4.1a
Function values for f approximated to four decimal places

	Second decimal place in the value of x									
x	**0.00**	**0.01**	**0.02**	**0.03**	**0.04**	**0.05**	**0.06**	**0.07**	**0.08**	**0.09**
0.0	**5.0000**	5.0002	5.0008	5.0018	5.0032	5.0050	5.0072	5.0098	5.0128	5.0162
0.1	5.0200	5.0242	5.0288	5.0338	5.0392	5.0450	5.0512	5.0578	5.0648	5.0722
0.2	5.0800	5.0882	5.0968	5.1058	5.1152	5.1250	5.1352	5.1458	5.1568	5.1682
0.3	5.1800	5.1922	5.2048	5.2178	5.2312	5.2450	5.2592	5.2738	5.2888	5.3042
0.4	5.3200	5.3362	5.3528	5.3698	5.3872	5.4050	5.4232	5.4418	5.4608	5.4802
0.5	5.5000	5.5202	5.5408	5.5618	5.5832	5.6050	5.6272	5.6498	5.6728	5.6962
0.6	5.7200	5.7442	5.7688	5.7938	5.8192	5.8450	5.8712	5.8978	**5.9248**	5.9522
0.7	5.9800	6.0082	6.0368	6.0658	6.0952	6.1250	6.1552	6.1858	6.2168	6.2482
0.8	6.2800	**6.3122**	6.3448	6.3778	6.4112	6.4450	6.4792	6.5138	6.5488	6.5842
0.9	6.6200	6.6562	6.6928	6.7298	6.7672	6.8050	6.8432	6.8818	6.9208	6.9602

We can approximate some of the missing values for a function such as f in the example above by a technique called *linear interpolation*. Suppose we want to approximate $f(0.273)$. The idea is to locate 0.273 between two given values of x and then to find a number in the same position between corresponding function values.

Illustration

0.003 units

x: 0.27 0.273 0.28

0.01 units

0.273 is 0.003/0.01, or 0.3 of the way from 0.27 to 0.28.

d units

$f(x)$: 5.1458 y 5.1568

0.0110 units

y is approximately 0.3 of the way from 5.1458 to 5.1568.

$$d = 0.3(5.1568 - 5.1458) = 0.0033$$

So

$$y \approx 5.1458 + 0.0033$$
$$f(0.273) = y \approx 5.1491$$

As a check on our work we note that our approximation of $f(0.273)$ is between $f(0.27)$ and $f(0.28)$. We always want to be sure that any result obtained is *reasonable*. □

In the preceding illustration, function values become larger as values of x increase. Should we change our procedure if function values become smaller as x increases? Where will the fact that function values behave this way show up in the calculations? Some of the exercises explore these questions.

Linear interpolation gives a good approximation if the graph of the function is smooth and rising or falling gently between the two values of x used. This is because we are approximating that piece of the graph with a straight line. (Problem 45 explains this further.)

Let's look at one more family of functions that will serve as illustrations later.

Definition For any positive integer p, we define a **function mod_p** whose domain is the set of whole numbers, as follows: $\text{mod}_p(n)$ is the remainder when n is divided by p.

Illustration

$$\text{mod}_3(2) = 2 \quad \text{since} \quad 2 = 0 \cdot 3 + \mathbf{2}$$
$$\text{mod}_3(17) = 2 \quad \text{since} \quad 17 = 5 \cdot 3 + \mathbf{2}$$
$$\text{mod}_7(12) = 5 \quad \text{since} \quad 12 = 1 \cdot 7 + \mathbf{5}$$
$$\text{mod}_{12}(0) = 0 \quad \text{since} \quad 0 = 0 \cdot 12 + 0$$
$$\text{mod}_5(16) - 2 = -1 \quad \text{since} \quad 16 = 3 \cdot 5 + \mathbf{1} \text{ and } 1 - 2 = -1$$ □

We have defined a collection of mod functions, one for each positive integer. Here is the graph of one of them, $g(x) = \text{mod}_3(x)$.

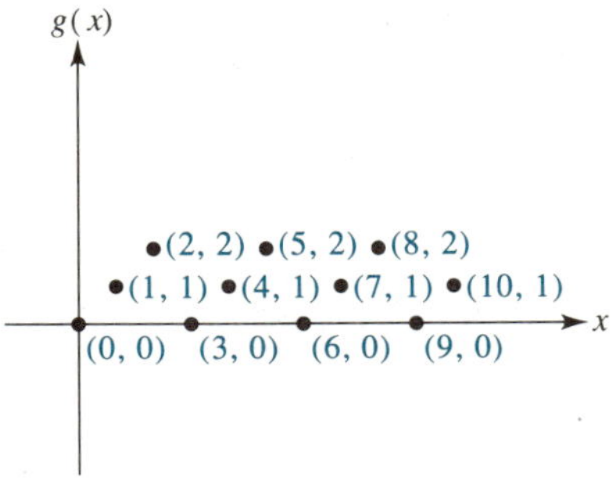

The graph continues to the right.

EXAMPLE What is the range of $f(n) = \text{mod}_4(n)$?

The function can produce only numbers that are remainders from division by 4.

The range of f is {0, 1, 2, 3}. ■

What is the range of $g(n) = \text{mod}_p(n)$ for any positive integer p?

EXERCISES 4.1

In exercises 1 through 4, indicate if the given relation is a function.

1. $\{(3, 1), (2, 1), (1, 1), (0, 1)\}$
2. $\{(6, 3), (4, 2), (10, 5), (0, 0)\}$
3. $\{(2, 1), (1, 2), (3, 6), (2, -1)\}$
4. $\{(\sqrt{5}, 5), (\sqrt{2}, 2), (-\sqrt{5}, 5), (-\sqrt{2}, 2), (\sqrt{3}, 3)\}$

In exercises 5 through 14, indicate if the given relation is a function of *x* or not.

5.

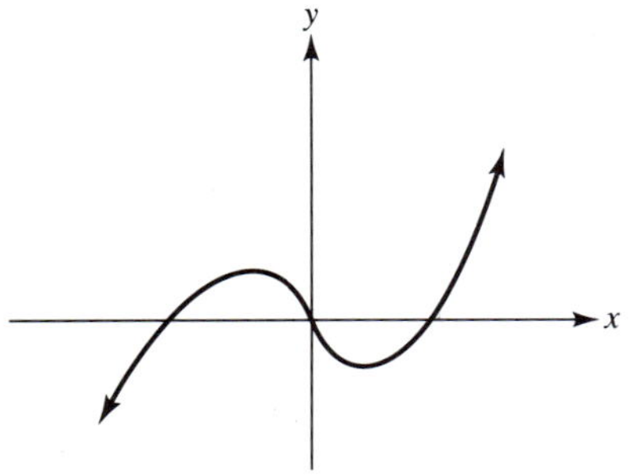

6.

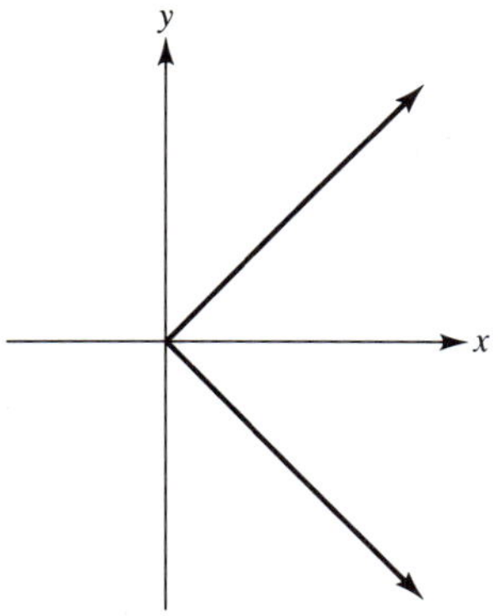

7.

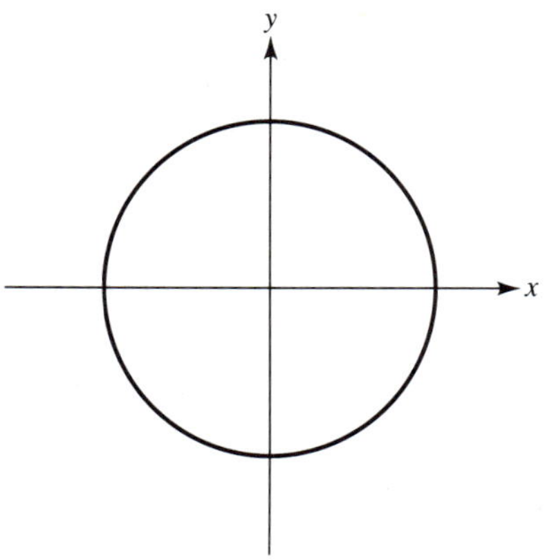

8.

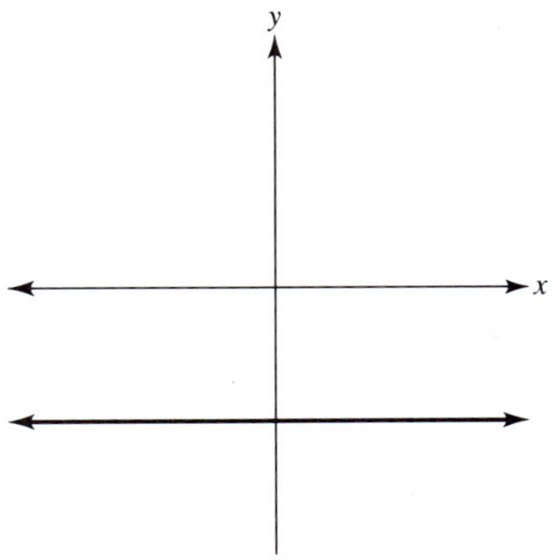

9. $f = \{(x, y) \mid y \leq 2x\}$
10. $g = \{(x, y) \mid x + y^2 = 4\}$
11. $h = \{(x, y) \mid 2x + 3y = 7\}$
12. $k = \{(x, y) \mid y = |x + 4|\}$
13. $f(x) = \begin{cases} 3x - 1, & x < 1 \\ x^2, & x \geq 1 \end{cases}$
14. $g(x) = \begin{cases} (x - 2)^2, & x \leq 2 \\ 5 - 4x, & x \geq 2 \end{cases}$

In exercises 15 through 20, give the domain and range of each function. Each should be a subset of $\mathbb{R}$.

15. $m(x) = 6 - 19x$
16. $h(x) = (x - 3)^2 + 4$
17. $g(x) = \text{mod}_7(x)$
18. $H(x) = \dfrac{23}{x + 7}$
19. $q(x) = \sqrt{x^2 - 8}$
20. $p(x) = |x - 5|$

In exercises 21 through 24, evaluate and state any necessary restrictions on *r*, *s*, and *t* so that the expression describes a real number.

21. $f(x) = 2x^2 - 5$
    a. $f(1)$ **b.** $f(2)$ **c.** $f(r)$
    d. $f(2s)$ **e.** $f(r + t)$
22. $g(x) = \sqrt{4 - 3x}$
    a. $g(-1)$ **b.** $g(0)$ **c.** $g(-r)$
    d. $g\left(\frac{1}{3}s\right)$ **e.** $g(r - t)$
23. $h(x) = \dfrac{6}{x - 3}$
    a. $h(0)$ **b.** $h(3)$ **c.** $h(r^2)$
    d. $h(5s)$ **e.** $h(2r + t)$
24. $k(x) = |4.7 - 2.5x|$
    a. $k(1)$ **b.** $k(6)$ **c.** $k\left(\dfrac{1}{r}\right)$
    d. $k(0.5s)$ **e.** $k(r - 2t)$

In exercises 25 through 30, sketch the graph.

25. $m(n) = \text{mod}_5(n)$

26. $d(n) = \text{div}_4(n)$, where $\text{div}_p(n)$ is the largest integer k such that $kp \le n$. Roughly speaking, $\text{div}_p(n)$ in the integer part of $\frac{n}{p}$. The domain of d is the whole numbers and p is any positive integer. For instance, $d(6) = 1$ because $1 \cdot 4 \le 6$ but $2 \cdot 4 > 6$.

27. $G(x) = [\![x]\!]$, where $[\![x]\!]$ represents the greatest integer less than or equal to x. For example, $G(3.85) = 3$ and $G(-3.7) = -4$.

28. $F(N) = \text{div}_3(n) + 2$ (See exercise 26.)

29. $g(n) = \text{mod}_4(n + 2)$

30. $h(x) = 2[\![x]\!]$

31. Use Table 4.1b and linear interpolation to approximate the requested values.
a. $g(3.71)$ **b.** $g(4.25)$ **c.** $g(2.76)$ **d.** $g(3.53)$

32. Use Table 4.1c and linear interpolation to approximate the following values.
a. $h(1.233)$ **b.** $h(0.946)$ **c.** $h(1.479)$ **d.** $h(0.332)$

TABLE 4.1b

x	$g(x)$	x	$g(x)$
2.5	11.50	3.5	9.50
2.6	11.48	3.6	9.08
2.7	11.42	3.7	8.62
2.8	11.32	3.8	8.12
2.9	11.18	3.9	7.58
3.0	11.00	4.0	7.00
3.1	10.78	4.1	6.38
3.2	10.52	4.2	5.72
3.3	10.22	4.3	5.02
3.4	9.88	4.4	4.28
		4.5	3.50

TABLE 4.1c
Function values for h approximated to four decimal places

	Second decimal place in the value of x									
x	**0.00**	**0.01**	**0.02**	**0.03**	**0.04**	**0.05**	**0.06**	**0.07**	**0.08**	**0.09**
0.0	1.0000	1.0000	1.0000	1.0000	1.0000	1.0001	1.0001	1.0002	1.0003	1.0004
0.1	1.0005	1.0007	1.0009	1.0011	1.0014	1.0017	1.0020	1.0025	1.0029	1.0034
0.2	1.0040	1.0046	1.0053	1.0061	1.0069	1.0078	1.0088	1.0098	1.0110	1.0122
0.3	1.0135	1.0149	1.0164	1.0180	1.0197	1.0214	1.0233	1.0253	1.0274	1.0297
0.4	1.0320	1.0345	1.0370	1.0398	1.0426	1.0456	1.0487	1.0519	1.0553	1.0588
0.5	1.0625	1.0663	1.0703	1.0744	1.0787	1.0832	1.0878	1.0926	1.0976	1.1027
0.6	1.1080	1.1135	1.1192	1.1250	1.1311	1.1373	1.1437	1.1504	1.1572	1.1643
0.7	1.1715	1.1790	1.1866	1.1945	1.2026	1.2109	1.2195	1.2283	1.2373	1.2465
0.8	1.2560	1.2657	1.2757	1.2859	1.2964	1.3071	1.3180	1.3293	1.3407	1.3525
0.9	1.3645	1.3768	1.3893	1.4022	1.4153	1.4287	1.4424	1.4563	1.4706	1.4851
1.0	1.5000	1.5152	1.5306	1.5464	1.5788	1.5955	1.6125	1.6125	1.6299	1.6475
1.1	1.6655	1.6838	1.7025	1.7214	1.7408	1.7604	1.7804	1.8008	1.8215	1.8426
1.2	1.8640	1.8858	1.9079	1.9304	1.9533	1.9766	2.0002	2.0242	2.0486	2.0733
1.3	2.0985	2.1240	2.1500	2.1763	2.2031	2.2302	2.2577	2.2857	2.3140	2.3428
1.4	2.3720	2.4016	2.4316	2.4621	2.4930	2.5243	2.5561	2.5883	2.6209	2.6540
1.5	2.6875	2.7215	2.7559	2.7908	2.8261	2.8619	2.8982	2.9349	2.9722	3.0098
1.6	3.0480	3.0866	3.1258	3.1654	3.2055	3.2461	3.2871	3.3287	3.3708	3.4134
1.7	3.4565	3.5001	3.5442	3.5889	3.6340	3.6797	3.7259	3.7726	3.8199	3.8677
1.8	3.9160	3.9649	4.0143	4.0642	4.1148	4.1658	4.2174	4.2696	4.3223	4.3756
1.9	4.4295	4.4839	4.5389	4.5945	4.6507	4.7074	4.7648	4.8227	4.8812	4.9403

In exercises 33 through 36, give a domain for the rule that will define a function with the requested range.

33. $g(t) = 6t - 4$; range $= [-8, 8]$
34. $f(x) = (x - 4)^3 + 2$; range $= [0, +\infty)$
35. $h(x) = \sqrt{9 - x^2}$; range $= (1, 2)$
36. $k(t) = |2t - 5|$; range $= [0, 4]$

Functions needn't be restricted to mapping a subset of the real numbers to a subset of the real numbers. Consider each of the functions in exercises 37 through 40 and give the domain and requested function values.

37. $h(x) = \left(\frac{x}{7}, \pi x\right)$
a. Domain of h **b.** $h(\frac{4}{3})$
c. $h(-6)$ **d.** $h(\sqrt{6})$

38. $f(x) = (\text{div}_3 x, \text{mod}_3 x)$ (See Exercise 26.)
a. Domain of f **b.** $f(5)$
c. $f(12)$ **d.** $f(128)$

39. $q(x, y) = \begin{cases} n, & \text{if } (x, y) \text{ lies in quadrant } n \\ 0, & \text{if } (x, y) \text{ is not in any quadrant} \end{cases}$
a. Domain of q **b.** $h(-3, 7)$
c. $h(0, 3)$ **d.** $h(\pi, -\pi)$

40. $g(x) =$ (height in centimeters of x, weight in kilograms of x), where x represents a living person
a. Domain of g **b.** $g(\text{you})$
c. $g(\text{your father})$ **d.** $g(\text{Michael Jordan})$

In exercises 41 through 44, give a functional notation description of the function given by enumeration.

41. $\{(1, 8), (2, 11), (3, 14), (4, 17), \ldots\}$
42. $\{(1, 1), (2, 3), (3, 5), (4, 7), \ldots\}$
43. $\{(1, 1), (2, 8), (3, 27), (4, 64), \ldots\}$
44. $\{(1, 2), (2, 8), (3, 18), (4, 32), (5, 50), \ldots\}$

45. Here's a small piece of the function q with domain $[0, 10]$. Suppose we need to approximate $q(6.3)$. Linear interpolation estimates $q(6.3)$ as y, the second coordinate of the point $(6.3, y)$ on the line through $(6, 2.457)$ and $(7, 3.228)$.

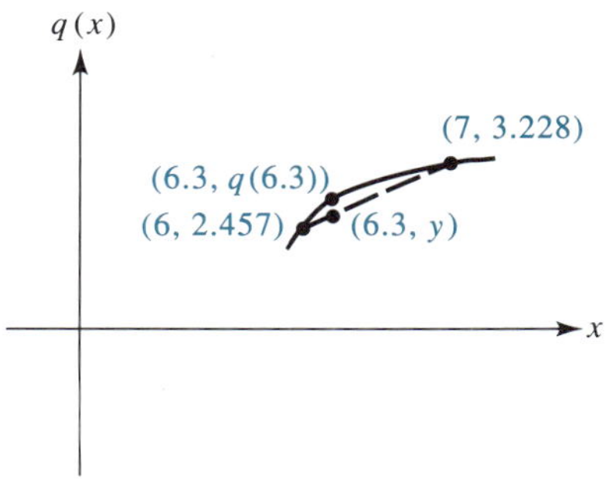

a. What is the slope of this line? (Slope is defined in Section 2.2.)
b. Write the slope of the line through $(6, 2.457)$ and $(6.3, y)$.
c. Use the fact that the slope in parts a and b must be equal to solve for y.

The steps described in the text for linear interpolation are equivalent to what you have just done to calculate y as an approximation for $q(6.3)$.

46. Define $f(n) = 1 + 2 + \cdots + n$, $n \in N$. Experiment with $f(1)$, $f(2)$, $f(3)$, $f(4)$, and others to find a pattern that would allow you to calculate $f(n)$ without adding all these terms.

4.2 Graphing Techniques I

Toolkit

You may need to review the properties of absolute value in Section A.3 and the graphing techniques discussed in Sections 2.4 and 2.5. Beginning with the exercises of this section, we continue with the preview ideas begun in Chapter 3.

In Chapter 2 we worked with some **standard graphs** on which we used the general graphing techniques of translation, reflection, and expansion. Of these graphs the line, $y = x$, the parabola, $y = x^2$, and the hyperbola, $y = 1/x$, are functions.

Illustration

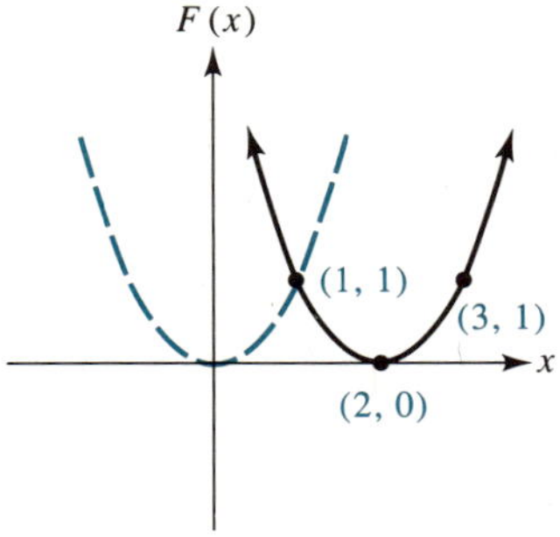

$F(x) = (x - 2)^2$
The standard graph
$f(x) = x^2$ is shown as a dashed line.

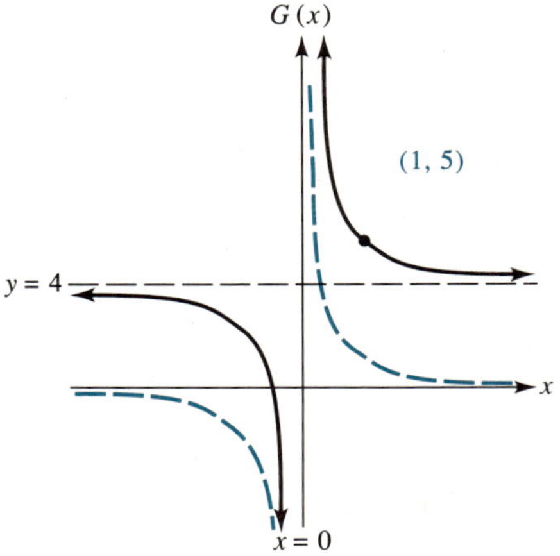

$G(x) = 1/x + 4$
The standard graph
$g(x) = 1/x$ is shown as a dashed line.

□

Standard Graph 6: $f(x) = x^3$

The function f has value zero when x is 0, is positively valued when x is positive, and is negatively valued when x is negative. Verification of the curving of the graph must be done using calculus. The only distinguished point is (0, 0); it is where the curving changes. Suggested labels are the distinguished point, a point to the left, and one to the right to establish the scale.

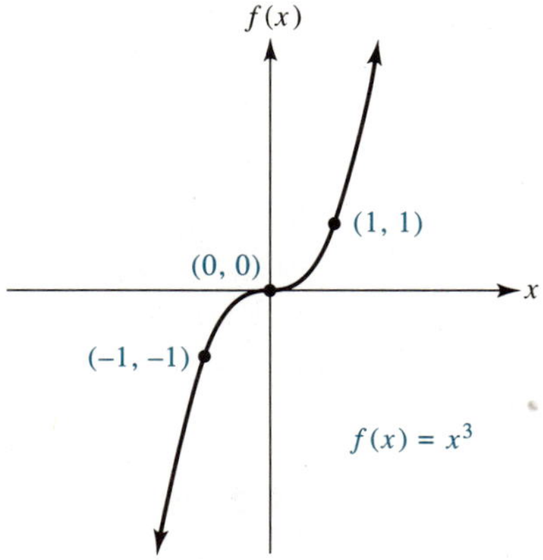

Standard Graph 7: $g(x) = |x|$

The function g has value zero when x is 0 and is positively valued when x is not 0. If $x \geq 0$, $g(x) = x$; for $x < 0$, $g(x) = -x$, so the graph is the union of two rays. Suggested labels are the vertex (the turning point) and a point to the right and to the left.

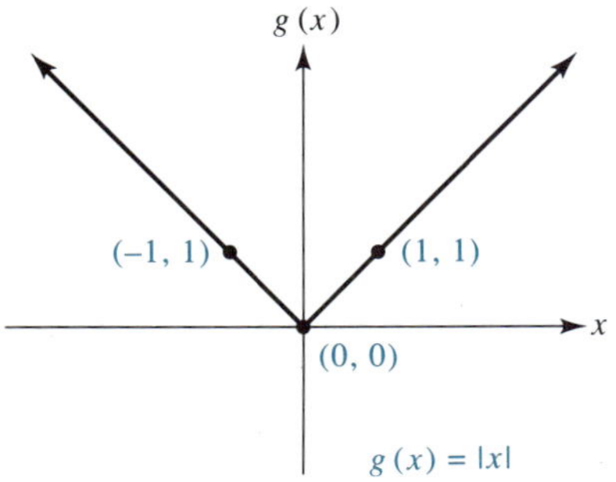

Standard Graph 8: $h(x) = \sqrt{x}$

The domain of h is $\{x \mid x \geq 0\}$, and the range is $\{y \mid y \geq 0\}$, so the graph lies entirely in the first quadrant, except for the only distinguished point, the "beginning" point (0, 0). Suggested labels are the beginning point and one more point.

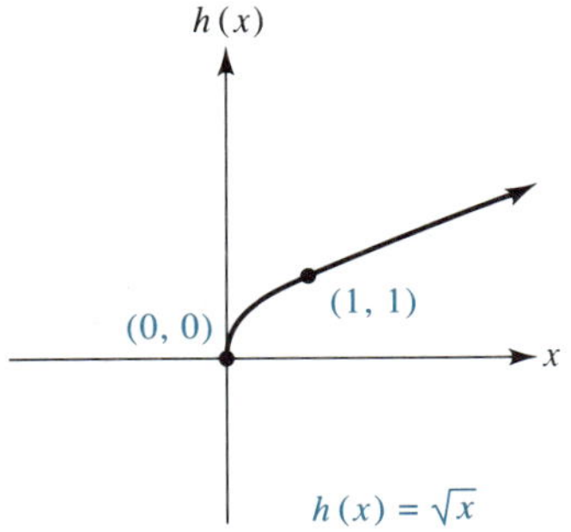

Notice that we can produce an $h(x)$ larger than any number we choose by picking a large enough x. For example, if we choose 10,000 for $h(x)$, then using 100,000,003 for x gives an $h(x)$ larger than 10,000. Therefore, the curve must continue to rise without bound as x grows larger.

Let's see how our graphing techniques apply to these new standard graphs. A horizontal shift of a standard graph is the result of adding a constant to the independent variable before the characteristic operation is applied. A vertical shift of a standard graph is the result of adding a constant after the characteristic operation is applied.

EXAMPLE Sketch the graph of each of the following:

$$f(x) = \sqrt{x-3}, \qquad g(x) = |x| + 5, \qquad h(x) = (x-1)^3 - 2$$

Sketch the graph of the related standard graph on the same coordinate system using dashed curves.

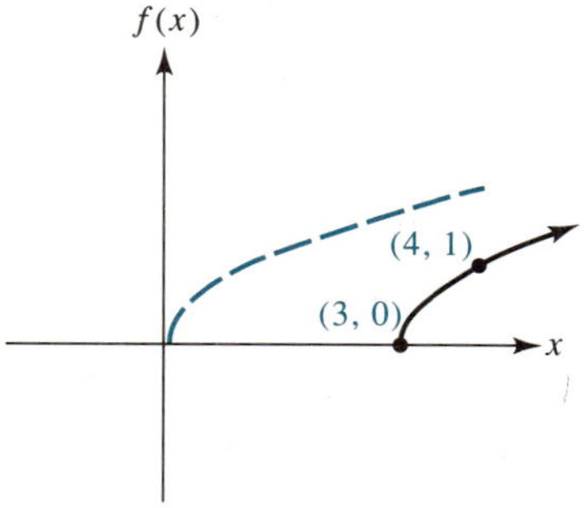

$f(x) = \sqrt{x-3}$

Horizontal shift 3 units right

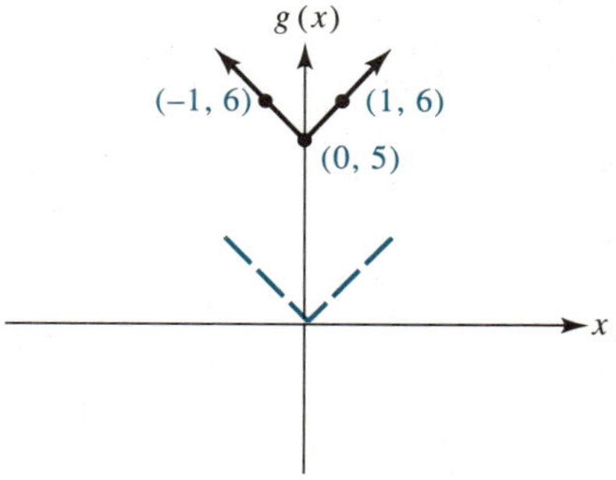

$g(x) = |x| + 5$

Vertical shift 5 units up

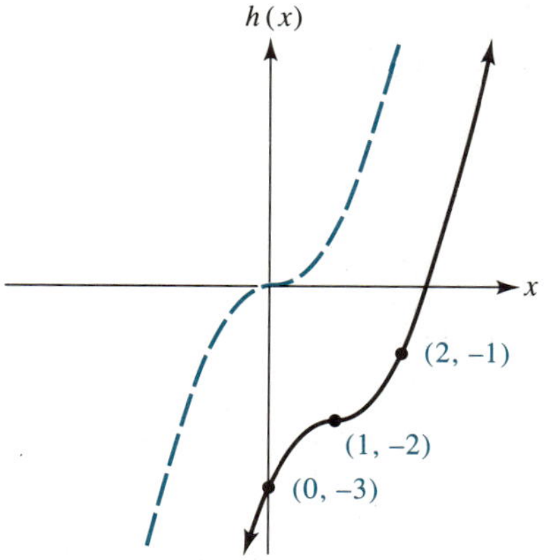

$h(x) = (x-1)^3 - 2$

Horizontal shift 1 unit right,
and vertical shift 2 units down ■

Standard graphs 1–8 will be reflected in the y-axis if we replace x by its opposite. A reflection in the x-axis is produced by taking the opposite of the result of the characteristic operation; this is the equivalent of replacing y by its opposite. In Section 4.4 we consider reflection in the line $y = x$.

EXAMPLE Sketch the graph of each of the following:

$$j(x) = \sqrt{-x}, \qquad k(x) = -|x|, \qquad m(x) = (-x)^3$$

Sketch the graph of the related standard graph on the same coordinate system using dashed curves.

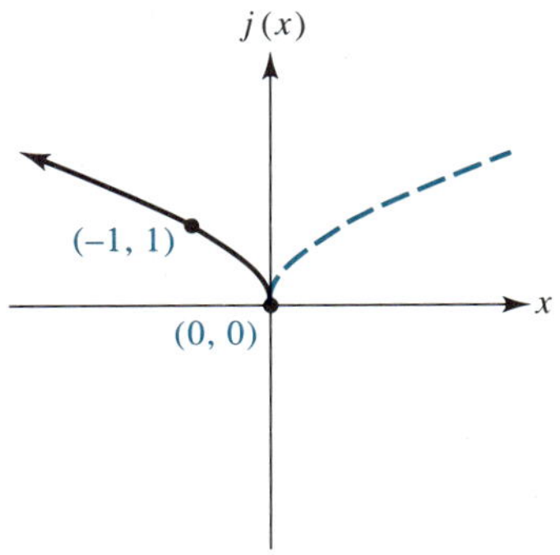

$j(x) = \sqrt{-x}$

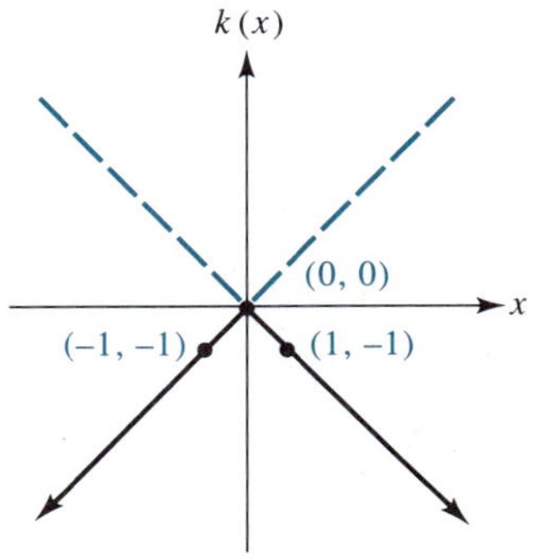

$k(x) = -|x|$

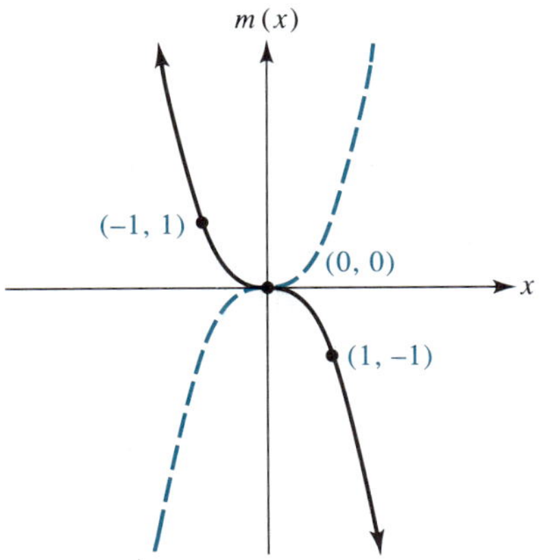

$m(x) = (-x)^3$

Note: $m(x) = (-x)^3$ is equivalent to $m(x) = -x^3$, so the graph may be viewed as a reflection in the x- or y-axis. ■

The vertical scale, or rate at which a standard graph changes, is affected by the constant factor in the generalized term with the characteristic operation.

EXAMPLE Sketch the graph of each of the following:

$$r(x) = 3|x|, \qquad s(x) = \tfrac{1}{3}\sqrt{x}, \qquad t(x) = 2.5x^3$$

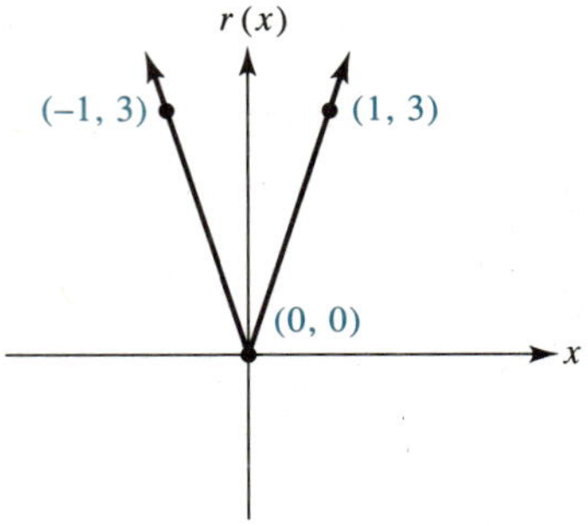

$r(x) = 3|x|$

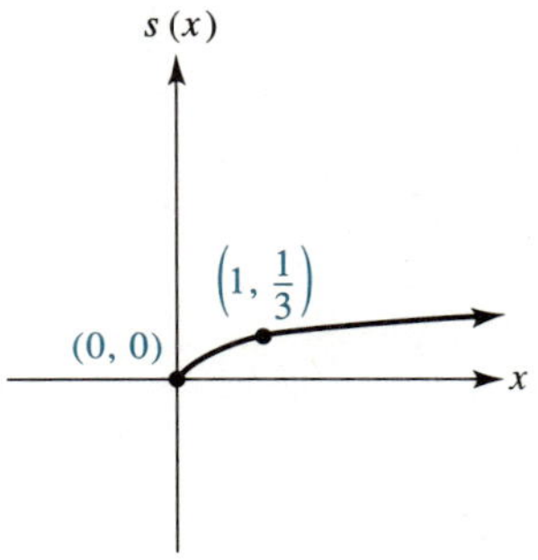

$s(x) = \frac{1}{3}\sqrt{x}$

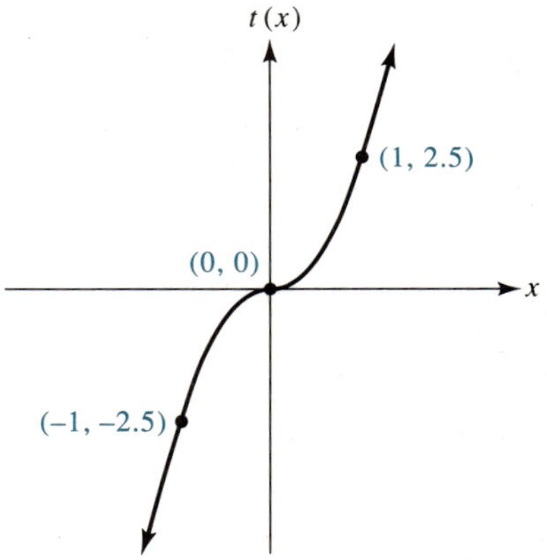

$t(x) = 2.5x^3$

$r(x) = 3|x|$ changes three times as fast as the standard, $y = |x|$.
$s(x) = \frac{1}{3}\sqrt{x}$ grows one-third as fast as the standard, $y = \sqrt{x}$.
$t(x) = 2.5x^3$ grows 2.5 times as fast as the standard, $y = x^3$. ■

EXAMPLE Sketch the graph of $f(x) = -\frac{1}{2}\sqrt{x + 3}$.
The standard graph, reflected in the x-axis, is shifted 3 units left, and scaled by a factor of $\frac{1}{2}$.

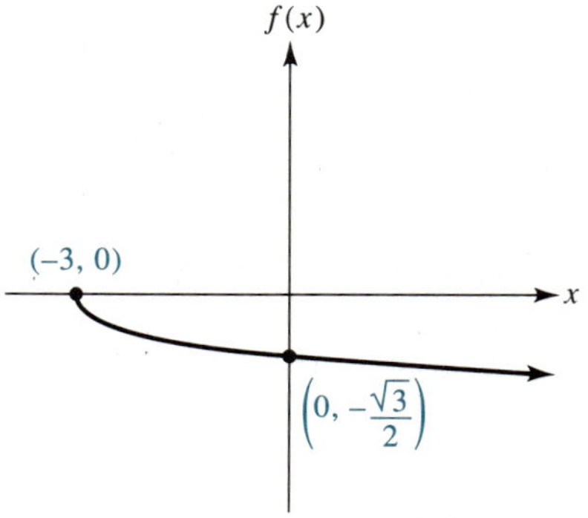

■

Not all graphs are variations of our standard graphs, but these techniques do work in general and can greatly simplify graph-sketching tasks. We can sometimes rewrite the definition of a function to use these techniques in cases where it doesn't appear possible to do so at first glance.

EXAMPLE Sketch the graph of $f(x) = \sqrt{2x - 5}$.

We can rewrite $f(x)$ by using factoring and an exponent rule to obtain

$$f(x) = \sqrt{2\left(x - \frac{5}{2}\right)} = \sqrt{2}\sqrt{x - \frac{5}{2}}.$$

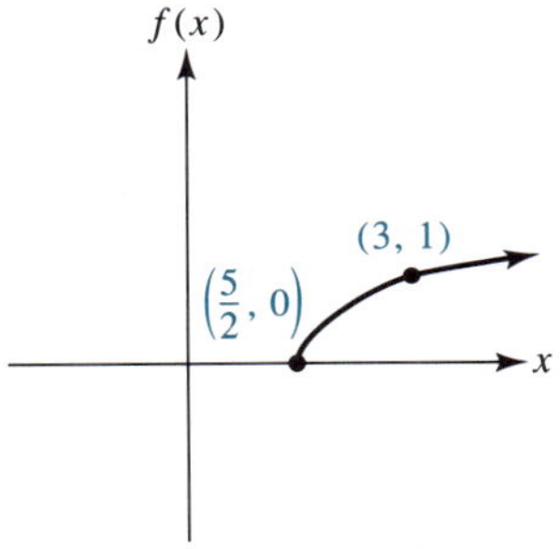

The graph of f is the standard graph $y = \sqrt{x}$ shifted $\frac{5}{2}$ units right with a vertical expansion factor of $\sqrt{2}$. ■

EXERCISES 4.2

In exercises 1 through 10, sketch the related standard graph as a dashed curve and the graph of the given function as a solid curve.

1. $f(x) = (x - 2)^3$
2. $g(x) = 3|x| - 5.6$
3. $h(x) = \sqrt{x + 2}$
4. $m(x) = 2.3x^3 + 1.7$
5. $n(x) = -6\sqrt{x - 5}$
6. $p(x) = |x - 2.5|$
7. $q(x) = -(x - 1)^3 + 7$
8. $r(x) = 2\sqrt{x - 4} - 3$
9. $s(x) = 5 - |x + 10|$
10. $t(x) = \dfrac{\sqrt{4 - x}}{3} + \dfrac{2}{3}$

In exercises 11 through 20, sketch the graph of the function.

11. $f(x) = \text{mod}_5(x + 1) - 3$
12. $g(x) = [\![x + 2]\!] - 1$
13. $h(x) = 2\ \text{div}_3(x + 6)$
14. $m(x) = -\text{mod}_4(2x)$
15. $n(x) = 3(x + 7)^2 - 5$
16. $p(x) = -\frac{1}{3}\left(x + \frac{1}{2}\right) - \frac{1}{4}$
17. $r(x) = 6\left(x - \frac{1}{2}\right) + 4$
18. $q(x) = -(x - 2.5)^2 + 3.1$
19. $s(x) = \dfrac{4.5}{x - 0.1}$
20. $t(x) = -\dfrac{2}{x} + 3$

In exercises 21 through 24, write a function whose graph is the graph of

21. The standard cubic function shifted 2 units left and 4 units up.
22. The standard square root function reflected in the y-axis and shifted down 3.5 units.
23. Mod_4 reflected in the x-axis and shifted 3 units right.
24. The standard absolute value function with vertex at $(-3, 7)$.

In exercises 25 through 28, tell how the graph of the function compares to the related standard graph.

25. $g(x) = -5(x + 8)^2 + 7$
26. $h(x) = 10|x - 15|$
27. $m(x) = 23 - \sqrt{x - 12}$
28. $n(x) = \frac{1}{2}(x - 4)^3 - 6$

In exercises 29 through 36, rewrite the function to find the vertical expansion factor, translations, and reflections. Describe what effect these have on the related standard graph.

29. $F(x) = |4x + 7|$
30. $P(x) = |5 - 3x|$
31. $r(x) = (-2x + 3)^3$
32. $s(x) = (1 - 5x)^3$
33. $t(x) = (0.7x + 2.1)^2$
34. $R(x) = (\sqrt{3} - \sqrt{6x})^2$
35. $S(x) = \sqrt{10 - 5x}$
36. $T(x) = [\![3x - 6]\!]$

In exercises 37 through 40, use the graph of *f* to sketch the graph of the equation.

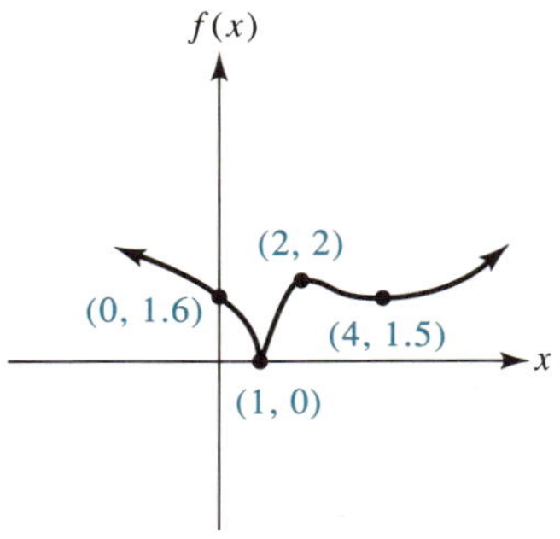

37. $y = f(x - 2)$
38. $y = f(x) + 3$
39. $y = -2f(x)$
40. $y = 0.5f(x + 1) - \sqrt{3}$

In exercises 41 through 44, sketch the graph of the function.

41. $f(x) = \begin{cases} 2x + 13, & x \leq -5 \\ \sqrt{25 - x^2}, & -5 < x \end{cases}$

42. $g(x) = \begin{cases} \sqrt{x^2 - 1}, & x < 0 \\ 2x^2 + 1, & 0 < x \end{cases}$

43. $h(x) = \begin{cases} -\sqrt{-2 - x}, & x < -2 \\ 3\sqrt{1 - x^2/4}, & -2 \leq x \leq 2.5 \\ 2, & 2.5 < x \end{cases}$

44. $j(x) = \begin{cases} 0.5(x + 3)^2, & x < -1 \\ (2 - x)^3, & -1 < x < 1 \\ |8 - x|, & 1 \leq x \end{cases}$

45. (Graphing calculator) Sketch standard graphs 1 through 8 using a graphing calculator. Which of the features of the standard graphs are clearly shown on the calculator graph and which are not?

Exercises 46 through 48 are preview exercises. Let pc(*t*) be the first coordinate of the point reached on the unit box by traveling *t* units from (1, 0).

46. Find each of the following.
a. $pc(\sqrt{2})$ **b.** $pc(2\sqrt{2})$ **c.** $pc(0)$
d. $pc(-\sqrt{2})$ **e.** $pc(-3\sqrt{2})$

47. **a.** What is the domain of pc?
b. What is the range of pc?

48. Find each of the following.
a. $pc(\sqrt{1/2})$ **b.** $pc(-3/2\sqrt{2})$
c. $pc(17/2\sqrt{2})$ **d.** $pc(1)$

4.3 Asymptotes

Toolkit

This discussion uses the exponent definition and rules of exponents from Section A.4 and the boundary-number concept from Chapter 3.

The standard graph $f(x) = 1/x$ and its variations share two unusual features. The function values for every member of the family become larger or smaller without bound as x approaches the value for which the function is undefined. We emphasize this by sketching a dashed vertical line through the excluded value of x. The function values for each member of the family approach a fixed number as x grows very large or very small. We show this with a dashed horizontal line.

EXAMPLE Sketch the graph of $m(x) = \dfrac{1}{x + 3} + 2$.

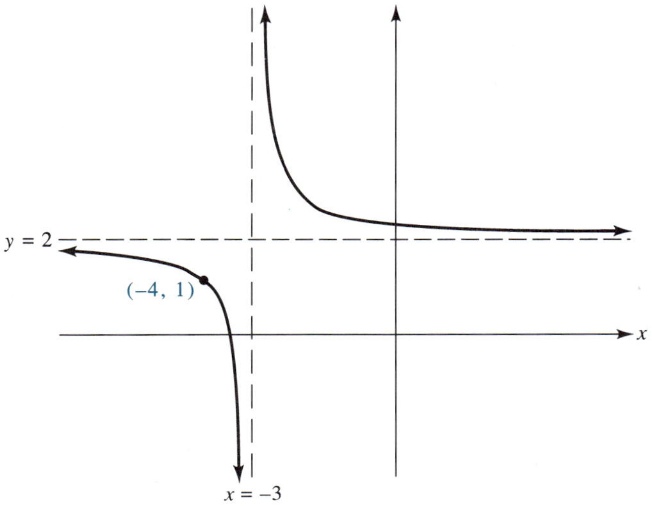

The graph of m is the standard graph $f(x) = 1/x$ shifted 2 units up and 3 units to the left. We call the boundary line $x = -3$ a *vertical asymptote* and the boundary line $y = 2$ a *horizontal asymptote*. ■

In this section we'll investigate rational functions, which may have vertical or horizontal asymptotes.

Definition A **rational function** is a function whose rule is a rational expression involving one variable.

That is, a rational function can be written as $f(x) = P(x)/Q(x)$ where $P(x)$ and $Q(x)$ are each polynomials in x. Any number that makes the denominator zero must be excluded from the domain of the function. Surely something of interest happens at such numbers when we graph the function.

EXAMPLE Sketch the graph of $f(x) = \dfrac{4}{x - 3}$.

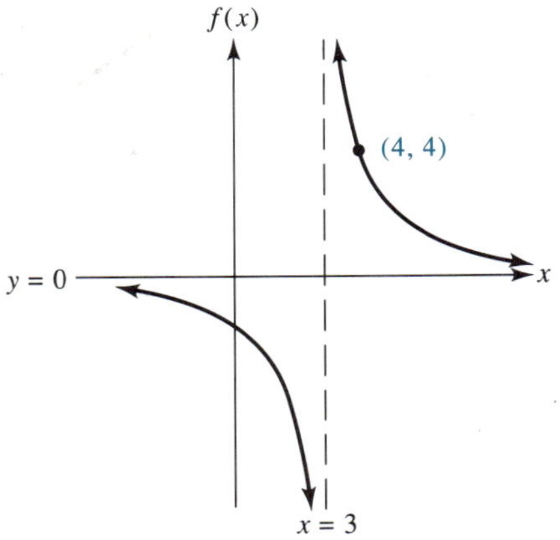

The function f is rational because 4 is a polynomial of degree zero and $x - 3$ is a first-degree polynomial. The graph of f can also be thought of as the graph of the standard function $y = 1/x$ shifted 3 units to the right with a scaling factor of 4. ■

EXAMPLE Sketch the graph of $g(x) = \dfrac{x^2 - x - 6}{x - 3}$.

Since $\dfrac{(x - 3)(x + 2)}{x - 3} = x + 2, x \neq 3$, the graph is a straight line with a hole at (3, 5). ■

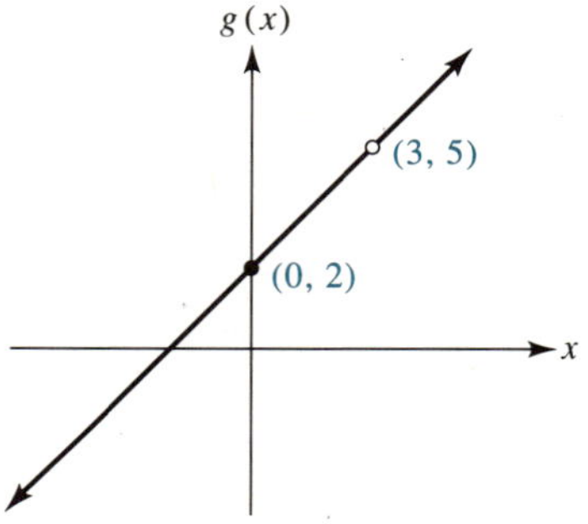

■

When will the graph of a rational function have a vertical asymptote and when will it have a hole? This question is easy to answer if we rewrite the function's rule in lowest terms, noting any restrictions no longer implicit. In this form any noted (explicit) restriction creates a hole and any remaining implicit restriction creates a vertical asymptote. Remember that an implicit restriction is created by a factor of the denominator that can have the value 0 after the rule has been rewritten in lowest terms.

EXAMPLE Sketch the graph of $h(x) = \dfrac{x^2 - 5x - 6}{x^3 - x}$.

$$h(x) = \frac{(x - 6)(x + 1)}{x(x - 1)(x + 1)}, \frac{x - 6}{x(x - 1)}, x \neq -1$$

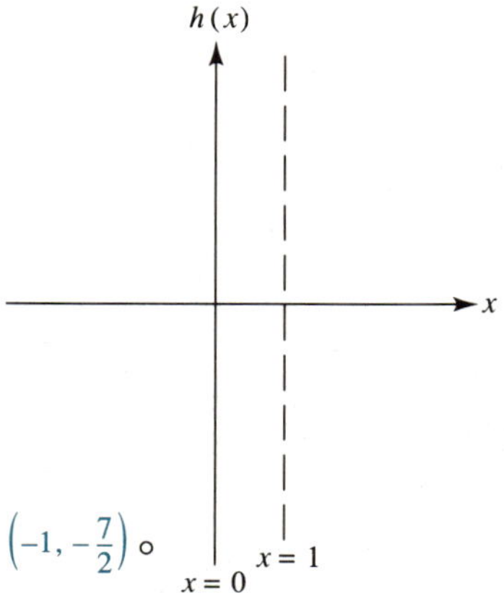

An explicit restriction is $x \neq -1$, and implicit restrictions are $x \neq 1$ and $x \neq 0$. The graph of h has a hole at $(-1, -\frac{7}{2})$ and vertical asymptotes of $x = 1$ and $x = 0$.

Note: $\dfrac{x - 6}{x(x - 1)} = -\dfrac{7}{2}$ when $x = -1$.

We are now faced with a dilemma. This rational function is not related to any of our standard graphs. The best we can do here is plot some points, examine whether the function values are positive, zero, or negative, and connect points between asymptotes using the simplest possible smooth curve. We can use the concept of boundary numbers to help make these decisions. The value of $h(x)$ is 0 when x is 6 and is undefined when x is 0 or 1. We mark these boundary numbers on the number line and determine the sign of the function value within the four regions.

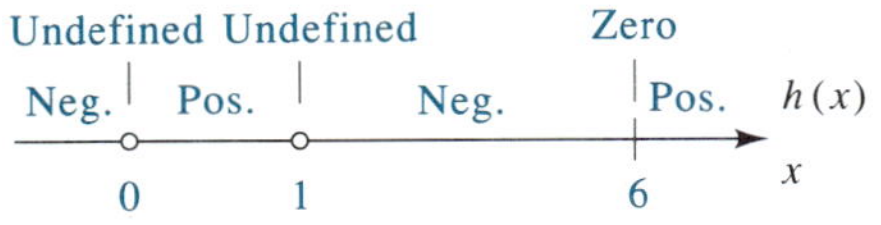

Now we can complete the graph.

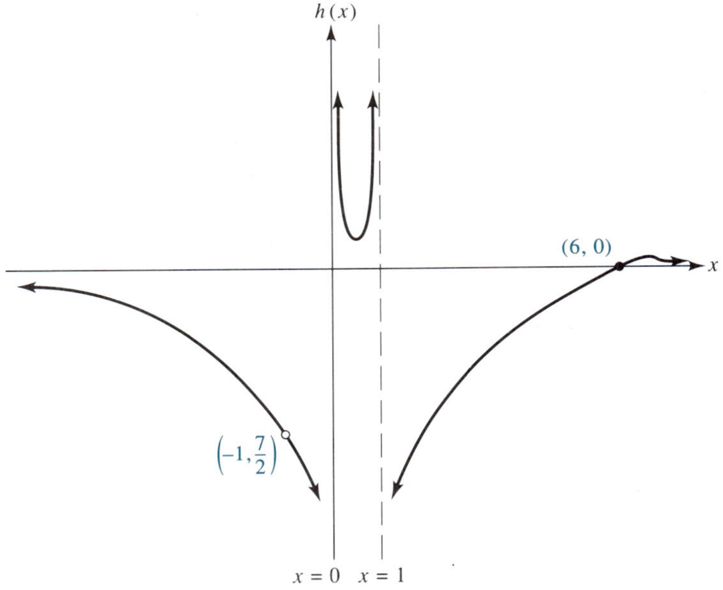

This is the best graph we can sketch with the tools available. But something is troubling about it. Where are the high and low points within a region? Is the x-axis really an asymptote? If so, we have the graph crossing a horizontal asymptote. Remember, a horizontal asymptote helps us approximate the behavior of the function as x grows very large or very small. It says nothing about what happens in between. The next example will help us decide more about horizontal asymptotes. ■

A convenient way to discover any horizontal asymptotes of a function is to rely on our experience in simple cases. As x increases or decreases without bound, $1/x$ approaches 0. In fact as x increases (or decreases) without bound, for any counting number n, $1/x^n$ approaches 0. To use these facts, we divide both the numerator and denominator of the function's rule by the highest power of x that appears in the denominator. This creates some terms that are multiples of $1/x^n$.

EXAMPLE Sketch the graph of $f(x) = \dfrac{2x^2 - 3x - 5}{6 - x^2}$.

If $x \neq 0$, then dividing numerator and denominator by x^2,

$$f(x) = \frac{2 - \dfrac{3}{x} - \dfrac{5}{x^2}}{\dfrac{6}{x^2} - 1}$$

As x increases (or decreases) without bound, $f(x)$ approaches $2/-1$, or -2. The graph of f has horizontal asymptote $y = -2$ on both the left and the right. If we solve

$$\frac{2x^2 - 3x + 5}{6 - x^2} = -2,$$

we find a solution $\frac{17}{3}$. This shows that $f(\frac{17}{3}) = -2$, and the graph of f crosses the line $y = -2$. Writing the function rule in factored form will help us gather the other information we need to sketch the graph.

$$f(x) = \frac{(2x - 5)(x + 1)}{(\sqrt{6} + x)(\sqrt{6} - x)}$$

The boundary numbers are $-\sqrt{6}$, -1, $\frac{5}{2}$, and $\sqrt{6}$. Between these numbers the function values are either positive or negative. We'll again use the number line to present this information.

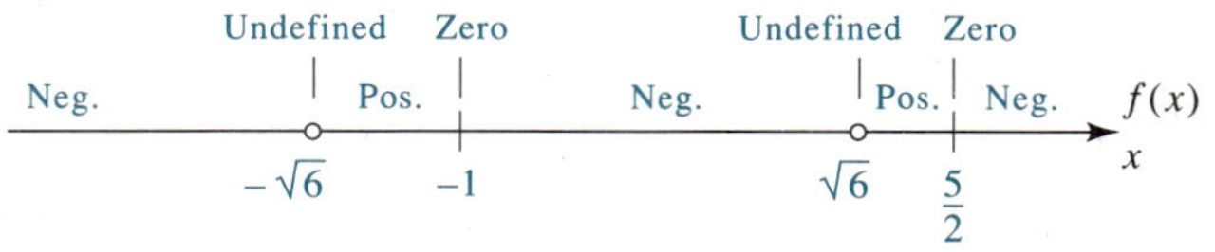

Combining all our information and using simple smooth curves, we produce the graph.

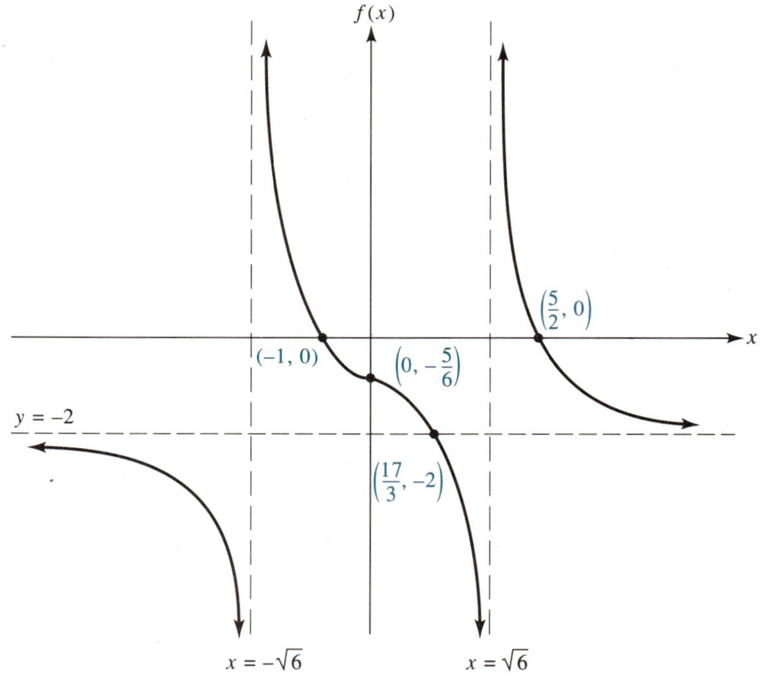

■

EXAMPLE Sketch the graph of $g(x) = \dfrac{x^2}{x - 7}$.

If $x \neq 0$ and we divide numerator and denominator by x, as we did in the previous example,

$$g(x) = \frac{x}{1 - \dfrac{7}{x}}$$

It does not appear that the graph has a horizontal asymptote. As x increases (decreases) without bound, $g(x)$ increases (decrease) without bound. The only boundary numbers are 0 and 7. Let's use the information displayed on the number line and plot additional points to confirm what the graph does as x becomes very large or very small.

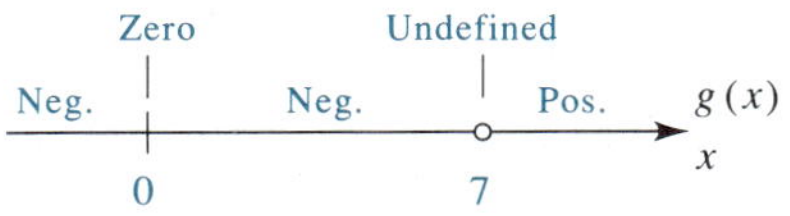

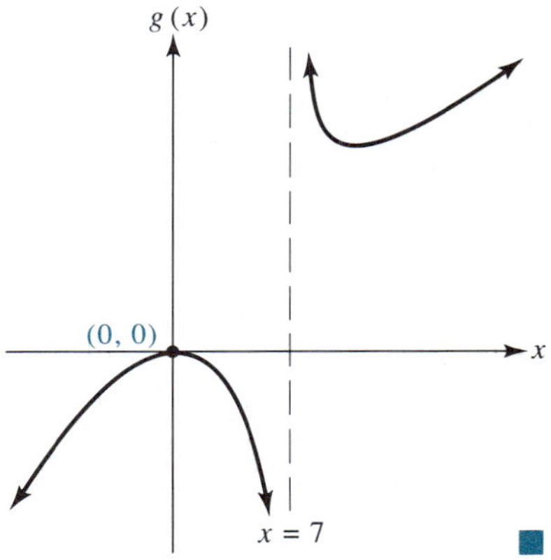

The last three examples illustrate a general property of the graphs of rational functions. If the degree of the numerator is less than or equal to the degree of the denominator, then there is a horizontal asymptote. If the degree of the numerator is greater than the degree of the denominator, there is no horizontal asymptote. (Exercise 39 explores this further.)

Everything we have done in this section with rational functions can also be used in sketching the graphs of functions whose rules are fractional expressions involving one variable.

Rational and fractional functions are not the only functions whose graphs have asymptotes. We'll give an example of one such function. For this chapter it will be just a quick introduction; we'll return to this family in the next chapter. A little background is necessary. In Section A.4, we defined a^r for any rational number r with suitable restrictions on a. If we choose a positive base and let the exponent vary, $f(x) = a^x$ is a function whose domain is the rational numbers. If we sketch the graph of $f(x) = 2^x$, the graph seems solid, although it has infinitely many holes—one for each irrational number.

EXAMPLE Sketch the graph of $f(x) = 2^x$, x a rational number.

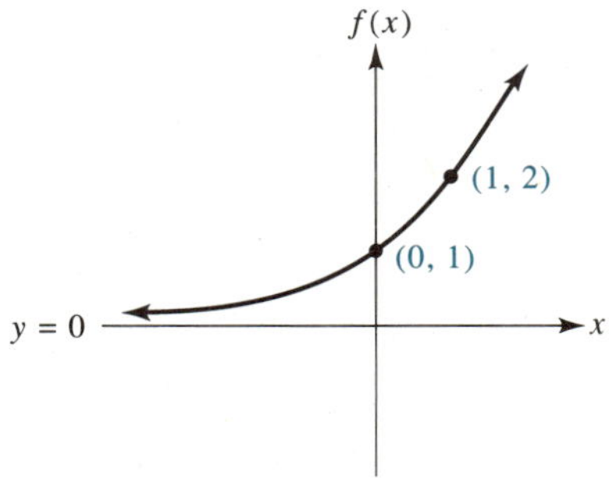

As x gets larger without bound, so does 2^x. As x gets smaller without bound, 2^x approaches 0, so f has a horizontal asymptote, $y = 0$. ■

By filling in the holes for which x is irrational, so that the curve is smooth, we can define 2^x for irrational values of x. The y-coordinate of the point on the graph with x-coordinate p is 2^p. Of course, this definition doesn't give us an efficient way to calculate such numbers. In fact there is no algebraic way to do this. The values must be approximated by using a table or a calculator. The exponent rules in Section A.4 are assumed to hold for all real-number exponents.

Any positive number except 1 can be used as a base for a member of this family, $f(x) = a^x$, with domain the real numbers. The number 1 is excluded as a base because $y = 1^x$ is not typical of the family. We wish to choose a base for the standard graph of this family. The choice may seem a bit unusual, because you may not have seen the constant e before. The number e is a naturally occurring irrational constant like π. This constant is used in mathematical models for many natural phenomena and for calculating continuously compounded interest. We investigate some of these situations in Chapter 5. For the moment we need only an approximation for e; 2.7182818 is accurate to seven decimal places.

Standard Graph 9: $g(x) = e^x$

Since $e > 2$, the graph of g will climb faster than that of $f(x) = 2^x$, but it will still cross the y-axis as (0, 1). A horizontal asymptote is $y = 0$. Suggested labels for this family are the horizontal asymptote, the y-intercept and one other point for setting the scale.

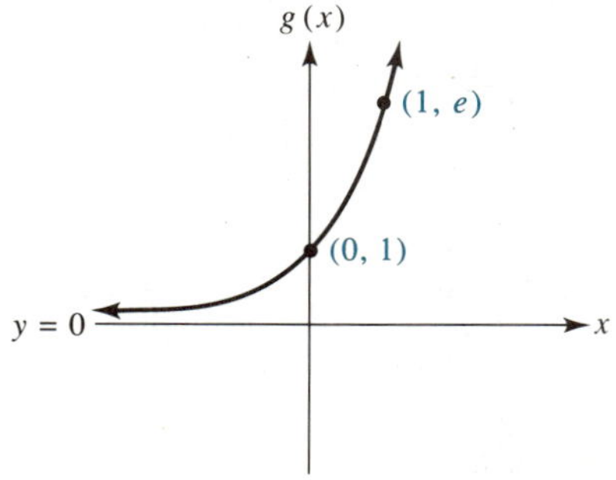

Another name for this function is exp, and you may find EXP as a label on a calculator key or as a built-in function in some computer programming languages.

As we look as more functions we'll find it useful to have more than an intuitive definition of an asymptote. A precise definition requires the concept of a limit, but we can give a working definition. We write $t \Rightarrow \infty$ to represent "t increases without bound," $t \Rightarrow -\infty$ for "t decreases without bound," $t \Rightarrow a$ for "t gets closer and closer to a," $t \Rightarrow a^+$ for "t gets closer and closer to a and $t > a$," and $t \Rightarrow a^-$ for "t gets closer and closer to a and $t < a$." Then we can make the following definition.

Definition The graph of f has a *vertical asymptote*, $x = c$, if as $x \Rightarrow c^+$, $f(x) \Rightarrow +\infty$ or $f(x) \Rightarrow -\infty$, or if as $x \Rightarrow c^-$, $f(x) \Rightarrow +\infty$ or $f(x) \Rightarrow -\infty$. The graph of f has a *horizontal asymptote, y = a*, if as $x \Rightarrow +\infty$ or $x \Rightarrow -\infty$, $f(x) \Rightarrow a$.

Illustration

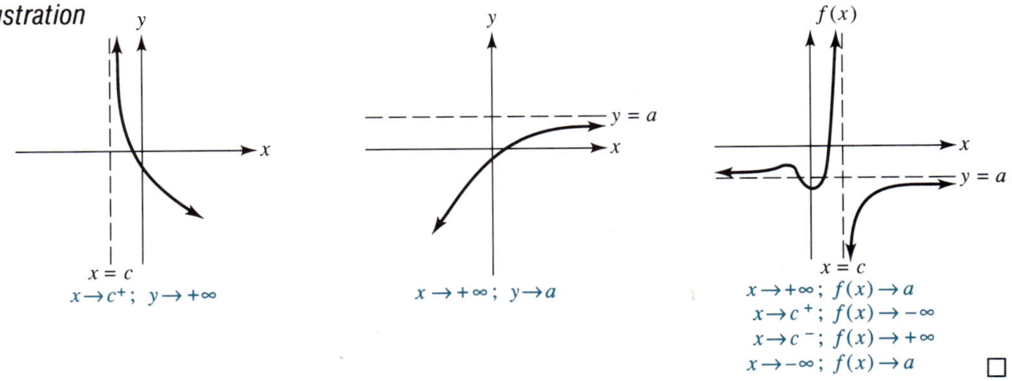

□

EXERCISES 4.3

In exercises 1 through 20, sketch the graph of each of the following. Label all vertical and horizontal asymptotes, holes, and intercepts. (If you have a graphing calculator, use it for exercises 1 through 16.)

1. $f(x) = \dfrac{x - 3}{x + 2}$

2. $s(x) = \dfrac{x + 5}{3x - 4}$

3. $h(x) = \dfrac{x^2 - 8x + 15}{x - 3}$

4. $q(x) = \dfrac{x^2 + 7x + 12}{x + 4}$

5. $g(x) = \dfrac{9x^2 - 25}{5 - 3x}$

6. $r(x) = \dfrac{4x^2 - 49}{7 - 2x}$

7. $m(x) = \dfrac{x + 4}{16 - x^2}$

8. $p(x) = \dfrac{5x - 1}{1 - 25x^2}$

9. $k(x) = \dfrac{3x^2 + 19x - 14}{2x^2 + 11x - 21}$

10. $j(x) = \dfrac{x^2 - 10x + 25}{3x^2 - 4x - 15}$

11. $t(x) = \dfrac{x^2 - 5x - 6}{3x^3 - 2x^2 - 5x}$

12. $F(x) = \dfrac{(x - 1)(x + 2)(2x - 5)}{(x - 1)^2(x + 2)}$

13. $t(x) = \dfrac{x^4 - x}{x^2 - x}$

14. $S(x) = \dfrac{(x - 1)(x + 2)(2x - 5)}{(x - 1)^2(x + 2)^3}$

15. $G(x) = \dfrac{(x + 1)(x + 2)(5 - 2x)}{(x - 1)^2(x - 2)^3}$

16. $M(x) = \dfrac{(x + 1)^2(x + 3)(2x - 5)}{(1 - x)(x + 2)^2}$

17. $H(x) = \begin{cases} \dfrac{2}{x - 3}, & x < 3 \\ 7 - x, & x > 3 \end{cases}$

18. $J(x) = \begin{cases} \dfrac{2}{x + 1}, & x < -1 \\ 8 + 3x, & x > -1 \end{cases}$

19. $K(x) = \begin{cases} x^2 - 2, & x \le 0 \\ \dfrac{5x - 1}{x}, & x > 0 \end{cases}$

20. $R(x) = \begin{cases} \dfrac{x^2 + 4}{x}, & x < 0 \\ x^2 + 4, & x \ge 0 \end{cases}$

In exercises 21 through 24, approximate the value of $f(a)$ using the horizontal asymptote of the graph of f.

21. $f(x) = \dfrac{2x^2 - 4x + 3}{5x^2 + x}$, $a = 100{,}000$

22. $f(x) = \dfrac{3x^3 + 4x^2 + 3}{7x^5 - 10x^4 + 3x^3 + x^2 - 8x - 1}$, $a = 945{,}688{,}234$

23. $f(x) = \dfrac{13x^2}{4x^2 - 3x - 5}$, $a = 4{,}000{,}000{,}000$

24. $f(x) = \dfrac{-13x^2}{4x^2 - 3x - 5}$, $a = -4{,}000{,}000{,}000$

In exercises 25 through 30, use the rules of exponents and perform the indicated operations.

25. $(2^{\sqrt{2}})^2$

26. $\dfrac{3^{\pi}}{2^{\pi/2}}$

27. $(e^{0 \cdot 5t})(e^{1 \cdot 3t})$

28. $(6^{\sqrt{5}} \cdot 3^{\sqrt{7}})^{\sqrt{3}}$

29. $\dfrac{a^{2-\sqrt{3}}}{a^{5-2\sqrt{3}}}$, $a \in N$

30. $(2x^{\sqrt{3}} - 3^{\sqrt{3}})^2$, $x \in N$

In exercises 31 through 38, sketch the graph of the given function.

31. $f(x) = (\frac{1}{2})^x$

32. $g(x) = 6^x$

33. $h(x) = e^{-x} + 2$

34. $m(x) = -e^x - 3$

35. $n(x) = e^{2x}$

36. $p(x) = 2e^x$

37. $q(x) = \frac{1}{2}e^{x-1}$

38. $r(x) = 0.6e^{3-x}$

39. Examine the rational functions you graphed in Exercises 1 through 20 and experiment with

$$f(x) = \frac{x^2}{x + 1}, \qquad g(x) = \frac{2x^2}{x + 1}, \qquad h(x) = \frac{3x^2}{x + 1}$$

and others if necessary to confirm the description of horizontal asymptotes for a rational function in terms of the degrees of the numerator and denominator that is given in this section. Can anything special be said about the behavior of the graphs of these three functions as x increases or decreases without bound?

40. Use the fact that for $r \in \mathbb{R}$ and $r > 0$, $1/x^r \Rightarrow 0$ when $x \Rightarrow +\infty$ to sketch the graph of each of the following.

a. $f(x) = \dfrac{x^{12/5} - 8x^{7/5} + 12x^{2/5}}{x^{14/5} - 4x^{9/5} - 12x^{4/5}}$

b. $g(x) = \dfrac{x^{8/3} - 4x^{5/3} - 5x^{2/3}}{x^{7/3} - 6x^{4/3} + 5x^{1/3}}$

41. (Preview) Sketch the graph of the function pc defined in Exercise 46, Section 4.2, for $0 \le t \le 4\sqrt{2}$.

42. (Preview) Sketch the graph of the function pc.

4.4 Properties of Graphs and Functions

We've now seen a variety of graphs and so have a background for a discussion of graphs in general. There are many features of graphs of sufficient interest that special terms are defined for them. We deal mainly with graphs of functions in this section.

For example, as we scan a graph from left to right, does it go uphill or downhill? Does it behave this way for the entire graph or does it behave differently on different intervals? Do different values of x produce different values of y? Or are values of y repeated in some regular pattern? We'll give both algebraic and geometric definitions as answers to these questions.

Definition A function is *increasing* on the interval $[a, b]$ if for any numbers c and d in $[a, b]$ with $c < d$. $f(c) < f(d)$.
The graph of f is uphill from a to b.

Roughly speaking, greater values of the independent variable produce greater values of the dependent variable.

Definition If $f(c) > f(d)$ for any numbers c and d in $[a, b]$ with $c < d$, then f is *decreasing* on $[a, b]$, and its graph is downhill from a to b.

Illustration

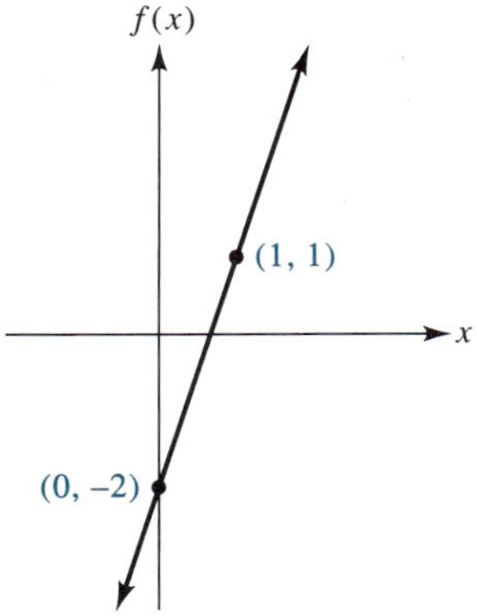

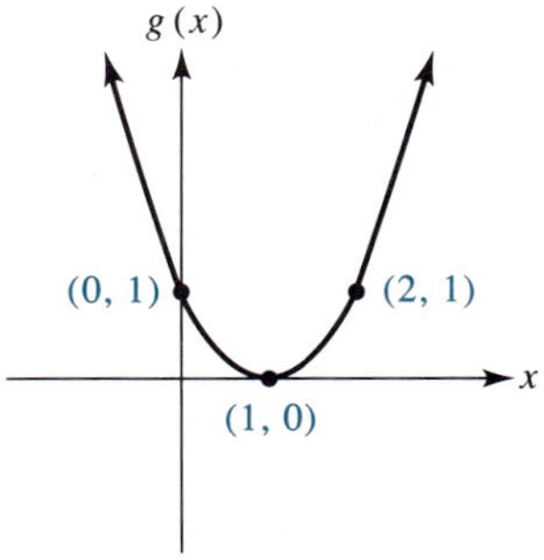

$f(x) = 3x - 2$ is increasing on $(-\infty, +\infty)$; $g(x) = (x - 1)^2$ is increasing on $[1, +\infty)$ and decreasing on $(-\infty, 1]$.
Note: g is not increasing or decreasing at 1; the definitions refer to an interval, not individual points. □

Definition A function is *one-to-one,* written 1-1, if for any two different numbers in the domain of f, the function values are different.

That is, for any x_1 and x_2 in the domain of f, if $x_1 \neq x_2$, then $f(x_1) \neq f(x_2)$. Is there a way to look at a graph and tell if it is the graph of a 1-1 function? Consider what we don't want: different x's producing the same y. This would correspond to points (x_1, y), (x_2, y), $x_1 \neq x_2$, on the graph. These could be detected by a horizontal line test. If a horizontal line intersects the graph in more than one point, the function is not 1-1.

Illustration

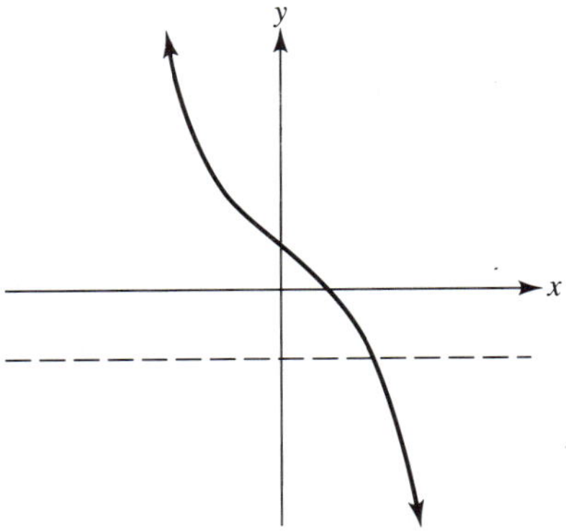

1-1

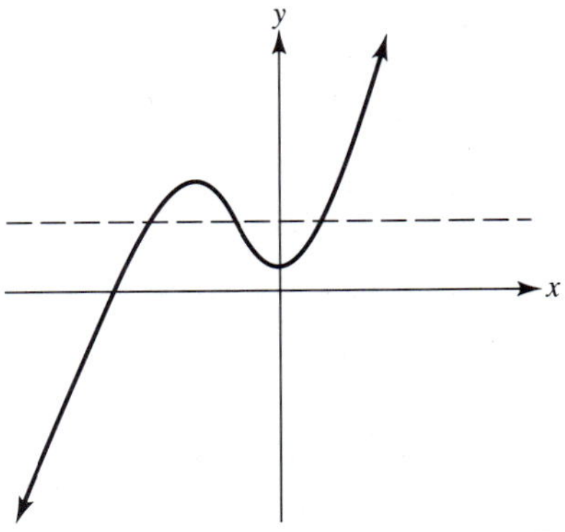

Not 1-1 □

Another way to describe a 1-1 function is to note that f is 1-1 only if for any c in the range of f, there is a unique b such that $f(b) = c$.

Definition Some graphs are *symmetric with respect to a line;* that is, some line divides the graph into two halves that are mirror images of each other.

The standard graphs $y = x^2$ and $y = |x|$ exhibit this property for the line $x = 0$. Of particular interest for the graphs of functions are *symmetry with respect to the y-axis* and a related idea, *symmetry with respect to the origin*. The graph of $f(x) = x^3$ is an example of symmetry with respect to the origin. If this graph is cut at the origin and one of the halves is reflected in the x-axis, the new figure is symmetric with respect to the y-axis. If we could test algebraically for such symmetries, we could sketch half a graph and use symmetry to draw the other half.

Illustration

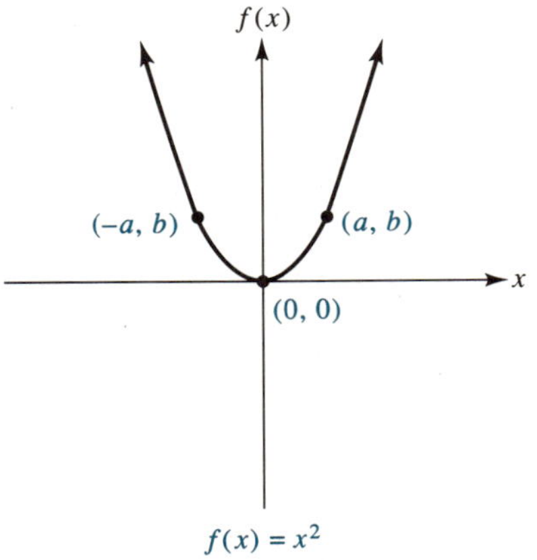

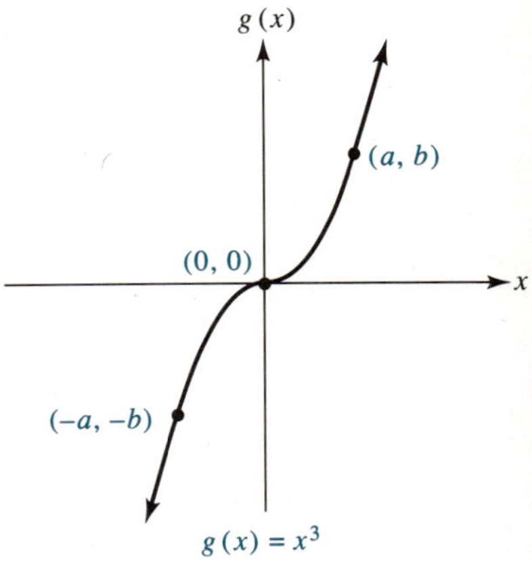

The graph of $f(x) = x^2$ is symmetric with respect to the y-axis, and the graph of $g(x) = x^3$ is symmetric with respect to the origin. □

To be symmetric with respect to the y-axis, the graph of a relation must contain $(-a, b)$ whenever it contains (a, b), since these points are mirror images in the y-axis. This is easy to test for functions.

Definition Is $f(x) = f(-x)$ for every x in the domain of f? If so, the graph of the function is symmetric with respect to the y-axis, and the function is called an **even function.**

To be symmetric with respect to the origin, the graph of a relation must contain $(-a, -b)$ whenever it contains (a, b), since reflection in the x-axis carries (a, b) to $(a, -b)$, and $(a, -b)$ and $(-a, -b)$ are mirror images in the y-axis. Again we can easily test for this property for functions.

Definition If $f(-x) = -f(x)$ for every x in the domain of f, the graph of the function is symmetric with respect to the origin, and the function is called an **odd function.**

Even and odd functions take their names from the models $f(x) = x^2$ and $g(x) = x^4$ (even) and $h(x) = x^3$ and $k(x) = x^5$ (odd).

Of course a graph may be symmetric with respect to the x-axis, but such a graph couldn't represent a function. (Why?)

Illustration

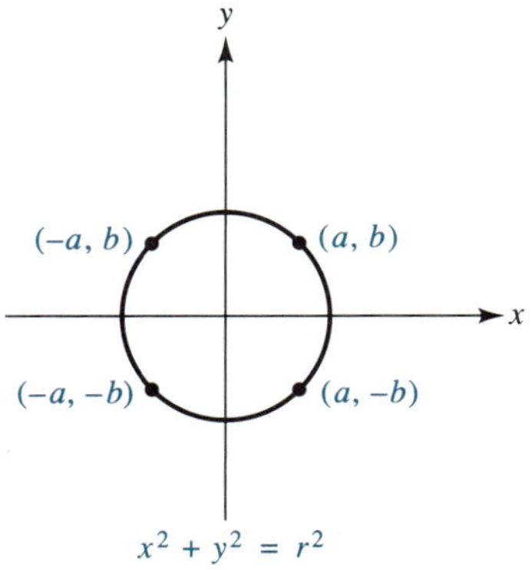

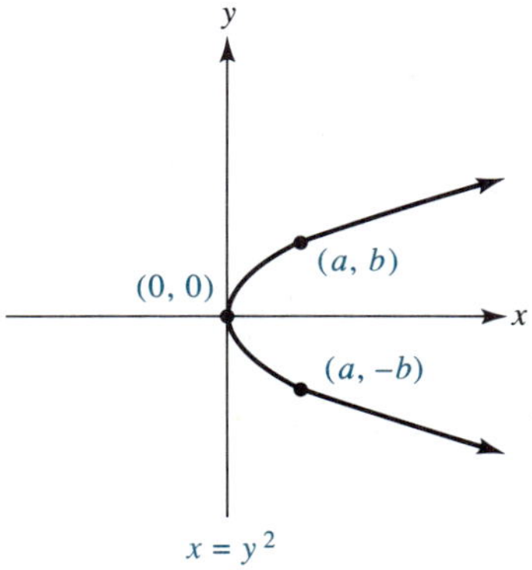

The graph of the relation $x^2 + y^2 = r^2$ is symmetric with respect to the x-axis, the y-axis, and the origin. The graph of the relation $x = y^2$ is symmetric with respect to the x-axis. Notice that neither of these relations is a function. □

EXAMPLE Is $f(x) = x^{2/5}$ even, odd, or neither?

We note $f(-32) = (-32)^{2/5} = (-2)^2 = 4$ and $f(32) = 32^{2/5} = 2^2 = 4$. Testing one case does not prove that f is even but might give us a feel for what is happening. We can prove f is even by checking $f(-r)$. $f(-r) = (-r)^{2/5} = [(-r)^2]^{1/5} = (r^2)^{1/5} = r^{2/5} = f(r)$ for any $r \in \mathbb{R}$.

Therefore, f is even.

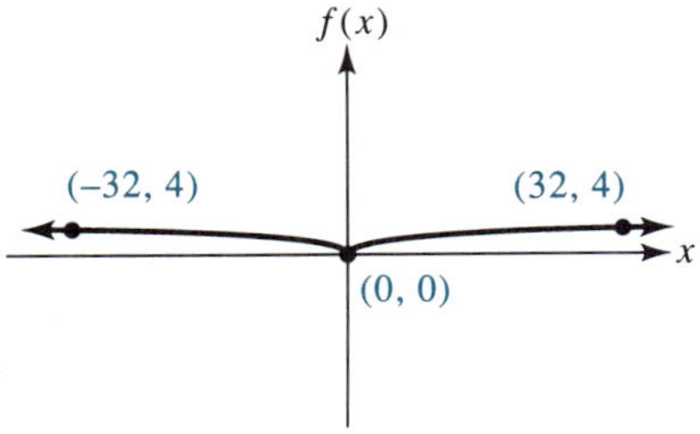

Note: The graph of f is not smooth; it has a spike (cusp) at (0, 0). ■

Illustration Since $g(-x) = -g(x)$ for all possible values of x, $g(x) = x^{1/3}$ is an odd function.

$$g(-x) = (-x)^{1/3} = -x^{1/3} = -g(x)$$

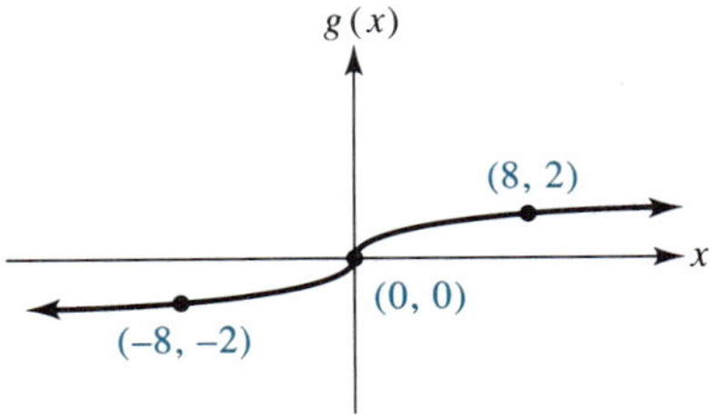

□

When a translation is applied to a graph that is symmetric with respect to the y-axis or the origin, the line or point of symmetry is also shifted.

EXAMPLE Sketch the graph of $F(x) = (x - 3)^{2/5}$.

The graph of F is the graph of $f(x) = x^{2/5}$ shifted 3 units right. It is symmetric about the line $x = 3$ with a cusp at (3, 0).

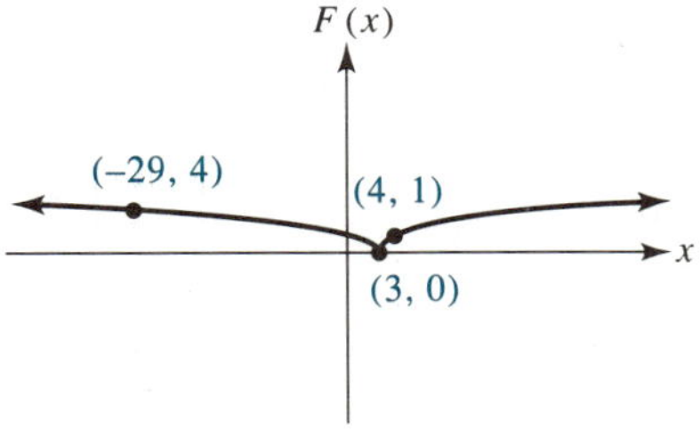

■

Another case of symmetry with respect to a line comes in handy for some graphs we consider in Chapters 5 and 6.

Definition A graph is *symmetric with respect to the line* $y = x$ if whenever (a, b) belongs to the graph so does (b, a).

Such graphs do not always represent functions.

Illustration

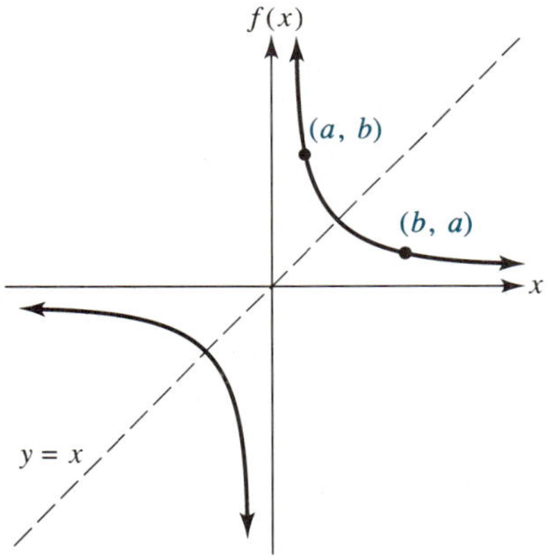

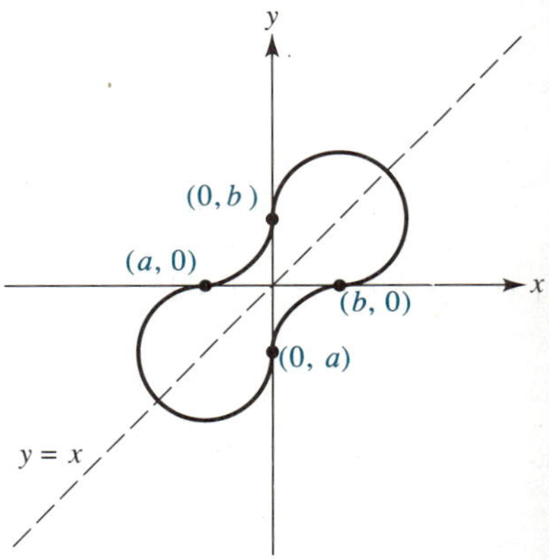

Notice $y = x$ cuts each graph into halves that are reflections of each other. □

Many functions repeat function values over and over in a regular pattern in much the same way the amount of air in your lungs or water's depth in a tidal pool varies with time. These are **periodic functions.**

Definition Formally, f is *periodic* if $f(x) = f(x + p)$ for some fixed p. The smallest positive p for which this is true is the *period* of f.

Illustration Let $f(n)$ be the digit in the nth decimal place in the decimal expansion of $\frac{11}{37}$. Since $\frac{11}{37} = 0.297297\ldots$, $f(2) = f(5) = f(8) = 9$, and we see that $f(n) = f(n + 3)$ for $n \in N$. It is also true that $f(n) = f(n + 6)$ for $n \in N$, but the period of f is 3, since 3 is the smallest positive number p for which $f(n) = f(n + p)$. □

If f has period p, then $f(x) = f(x + kp)$ for each integer k that gives a value of $x + kp$ in the domain of f. In the last example, only nonnegative integers k produce values of $n + 3k$ in the domain of f. (See Exercises 43–48.)

EXAMPLE Find the domain, range, and period of the function g represented by the following graph.

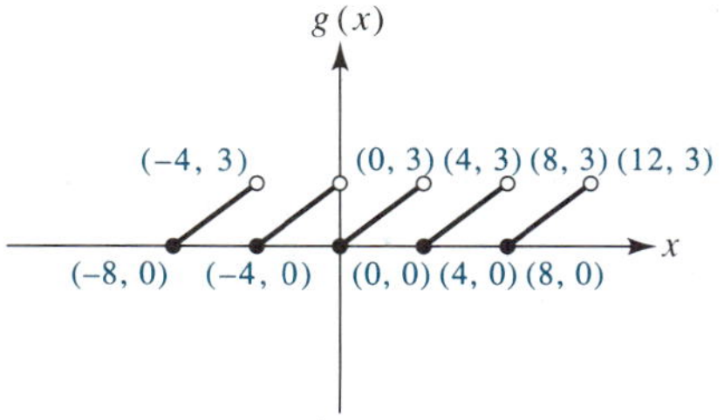

The domain of g is $[-8, 12)$, its range is $[0, 3)$, and the period is 4. ■

Definition Each repeated piece of the graph of a periodic function is called a *cycle*.

In the example above, the piece from (0, 0) to (4, 3) is a cycle of the graph of g.

EXERCISES 4.4

In exercises 1 through 12, sketch the graph and describe any symmetries.

1. $f(x) = x^2 - 2$
2. $g(x) = (x - 2)^2$
3. $h(x) = -(2 - x)^2$
4. $k(x) = |x + \frac{4}{3}|$
5. $m(x) = \frac{1}{2}|x - 3| - 1$
6. $t(x) = (-x)^3 + 1$
7. $F(x) = -x^3 + 1$
8. $G(x) = (x + 1)^3$
9. $H(x) = \sqrt{x - 1}$
10. $K(x) = \sqrt{1 - x}$
11. $M(x) = -\sqrt{x - 1}$
12. $T(x) = -\sqrt{1 - x}$

13.–24. Use interval notation to describe where the functions in 1 through 12 are increasing.

In exercises 25 through 30, restrict the domain for each function to create a 1-1 function with the same range as that of the original.

25. $f(x) = x^2 - 2$
26. $g(x) = (x - 2)^2$
27. $h(x) = -(2 - x)^2$
28. $k(x) = |x + \frac{4}{3}|$
29. $m(x) = \frac{1}{2}|x - 3| - 1$
30. $t(x) = (-x)^4 + 1$

In exercises 31 through 36, without graphing decide if the function is even, odd or neither.

31. $f(x) = 12x^3 - 3x + 4$
32. $g(x) = 158x^4 + 47x^2 - 29$
33. $h(x) = x^{1/3} + 8x^{5/3}$
34. $k(x) = x^{2/3} - 7x^{4/3} + 9x^{6/5}$
35. $t(x) = -38|x| - 7$
36. $m(x) = 17|x - 2|$
37. Select the equations whose graphs would be symmetric with respect to the line $y = x$.

a. $y = \frac{1}{x}$
b. $y = x^2$
c. $x^2 + y^2 = 4$
d. $y + x = 0$

38. Describe any symmetries that might apply.

a.

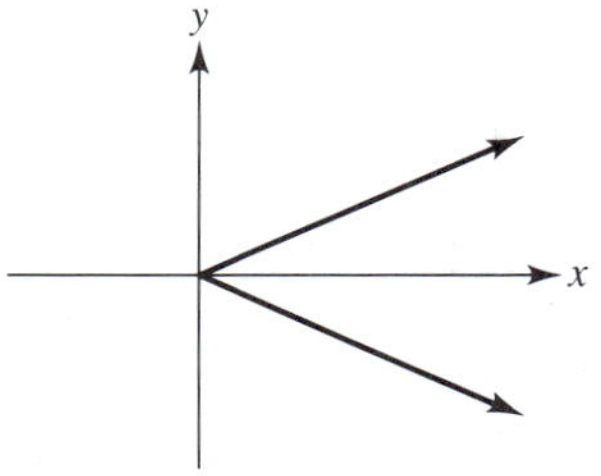

b.

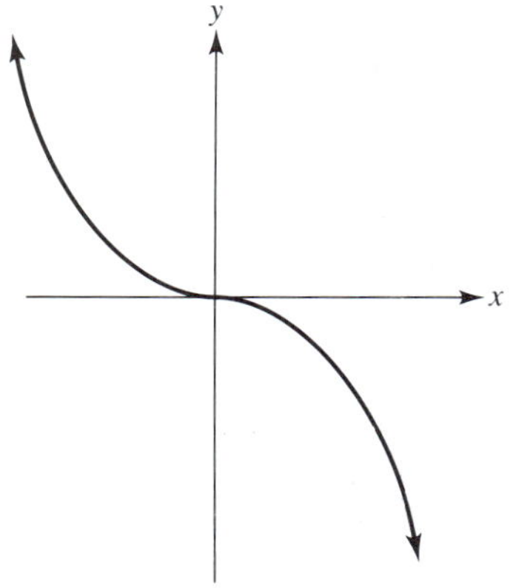

c.

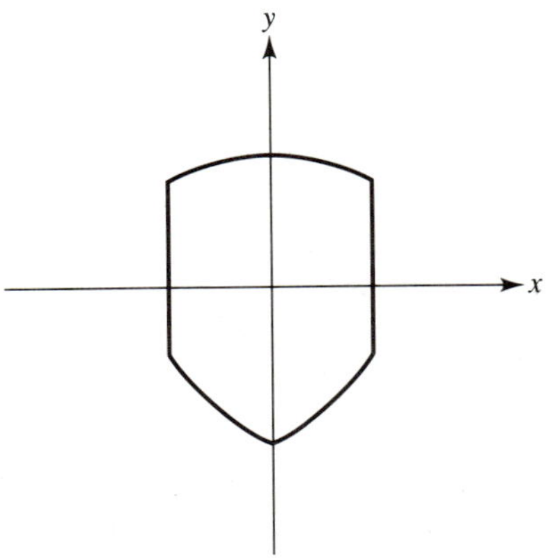

d.

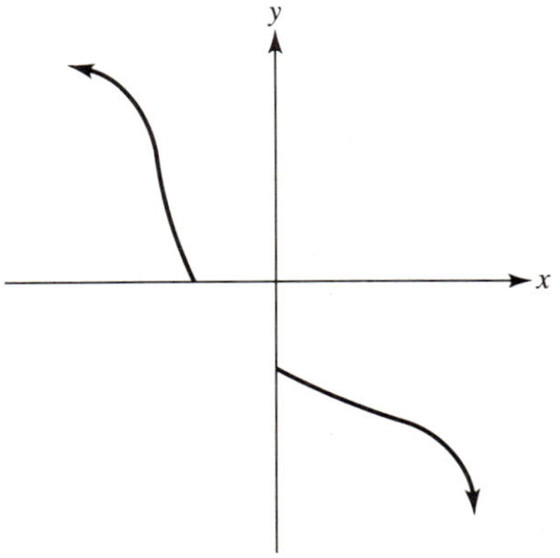

39. Give a 1-1 function with domain the natural numbers and range the positive even integers.
40. Give a 1-1 function with domain the natural numbers and range the positive odd integers.
41. Give a 1-1 function with domain the natural numbers and range the nonnegative multiples of 5.
42. Give a 1-1 function with domain the natural numbers and range the negative multiples of 7.

In exercises 43 through 47, find the period of the given function.

43. $f(n) = \text{mod}_5(n)$ (See the end of Section 4.1.)
44. $h(n) = \text{mod}_{17}(n)$
45. $k(n)$ is the $(n + 1)^{\text{st}}$ digit in the decimal expansion of $\frac{12}{13}$.
46. *g(n) is the* n^{th} digit in the decimal expansion of $\frac{3}{7}$.
47. $F(n) = i^n$ for $n \in N$ and $i = \sqrt{-1}$.
48. What is the period of $f(n) = \text{mod}_k(n)$, $k \in N$?
49. Give both geometric and algebraic definitions for a graph to be symmetric with respect to the x-axis. (This is done in the text for symmetry with respect to the y-axis.)
50. Would the graph of each of the following equations be symmetric with respect to the x-axis or another horizontal line? Describe any such line of symmetry.
a. $x = 2y^2$ **b.** $x = 2(2 - y)^2 + 3$
c. $x = -3\,|y - 1|$ **d.** $x = -(y + 1)^3$
51. Suppose the graph of f is symmetric with respect to the line $x = r$. Give both geometric and algebraic definitions for this.
52. What's the connection between increasing, decreasing, and one-to-oneness?
53. Show that for $f(x) = 1/x$, it is true that $f(p) \cdot f(q) = f(pq)$ for all nonzero p and q. Is this true for other functions?
54. Let $f(x) = ax + b$. Give conditions so that $f(p + q) = f(p) + f(q)$, for all values of p and q.

In exercises 55 through 60, state whether g is increasing, decreasing, or neither on the stated interval.

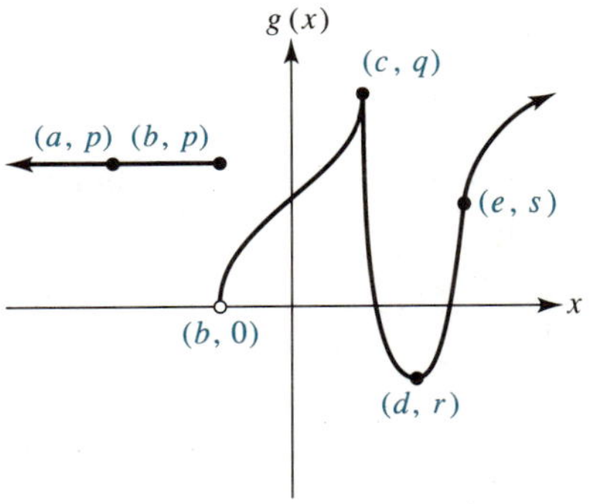

55. $[a, b)$ 56. $(b, c]$ 57. $(c\ e]$
58. $[d, e]$ 59. $[c, d]$ 60. $[a, c)$
61. Suppose g is a periodic function with period 4 and dom_g is the natural numbers. Using $g(1) = 1.7$, $g(2) = 2.6$, $g(3) = -3.1$, and $g(4) = -4.5$, find each value.
a. $g(5)$ **b.** $g(15)$ **c.** $g(42)$
62. The function h has period 2.5 and domain the integers. You are told $h(1) = 16.1$, $h(3) = 25.3$, and $h(7) = 49.7$. Find each value.
a. $h(5.5)$ **b.** $h(17)$ **c.** $h(21)$
63. If $\text{dom}_k = \mathbb{R}$ and k has period 5, give a pattern that describes all $x \in \mathbb{R}$ for which $k(x) = k(-2)$.
64. For the function m with period 1.3 and domain $\mathbb{R}$, give a pattern for all $x \in \mathbb{R}$ for which $m(x) = m(4.7)$.
65. (Preview) For the function pc (Exercise 46, Section 4.2), what is the period?
66. (Preview) Is pc even, odd, or neither?

4.5 Graphing Techniques II

We certainly haven't found out all there is to know about graphing functions, nor can we hope to do so in this book. There are a few other elementary techniques we should add to our package though.

Let's begin with functions defined by polynomials. Not all polynomial functions are variations of our standard graphs, $f(x) = x$, $g(x) = x^2$, and $h(x) = x^3$. In general in this course we can produce only rough sketches for polynomial functions, and we can do this only by making a large assumption.

Before looking at these more general polynomial functions, let us take care of a very simple case.

Definition A function defined by $f(x) = c$ is a **constant function.**

Each constant function has a horizontal line as its graph.

EXAMPLE Sketch the graph of $f(x) = 3$.

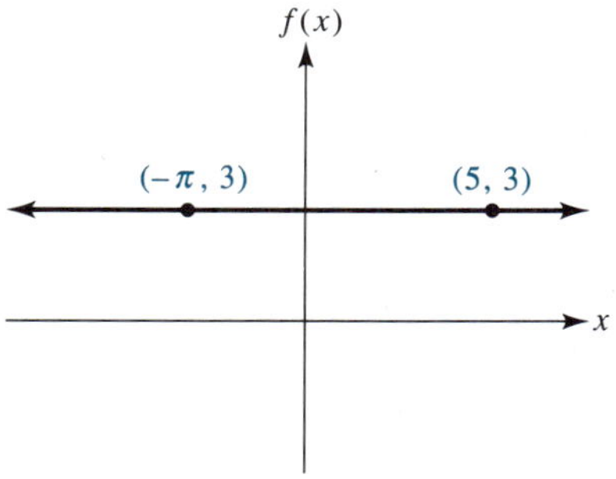

Remember, a line is determined by two points, so you should label at least two points. ■

Our method of sketching the graph of a polynomial function is to find as many x-intercepts of the graph as possible, determine whether the function values are positive or negative between the x-intercepts, and then use this information along with the polynomial principle to sketch the graph.

Definition **Polynomial principle** for graphing: The graph of a polynomial function is smooth and continuous. It has no vertical asymptotes, no holes, and no cusps (spikes).

Definition The x-coordinate of an x-intercept of the graph of a polynomial function is called a **zero of the function.**

EXAMPLE Sketch the graph of $f(x) = (x - 7)(x + 3)(2x + 5)$.
The function f has zeros 7, -3, and $-\frac{5}{2}$. A number line chart helps us keep track of whether the function values are positive or negative between these zeros.

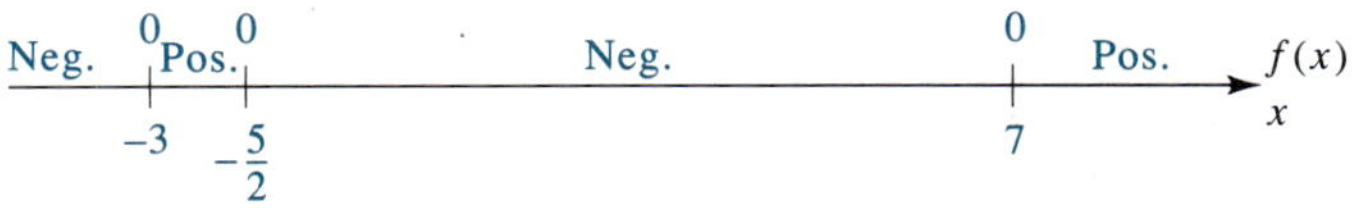

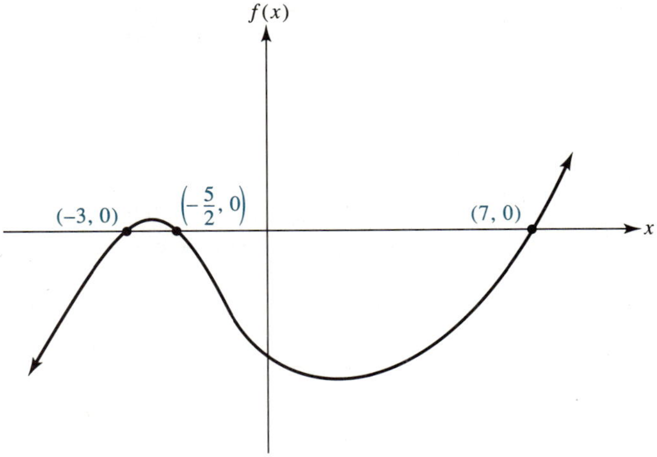

Notice there is not a vertical scale marked because there is no way at present to determine how high or low the loops go. Also, there is no evidence to suggest that these local highs and lows occur halfway between the zeros. ■

EXAMPLE Graph $g(x) = x(x + 1)^2(x - 2)$.
The zeros of g are -1, 0, and 2.

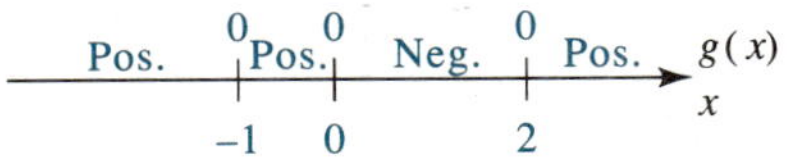

We need to be careful at points such as $(-1, 0)$ not to create a cusp, because g is a polynomial function.

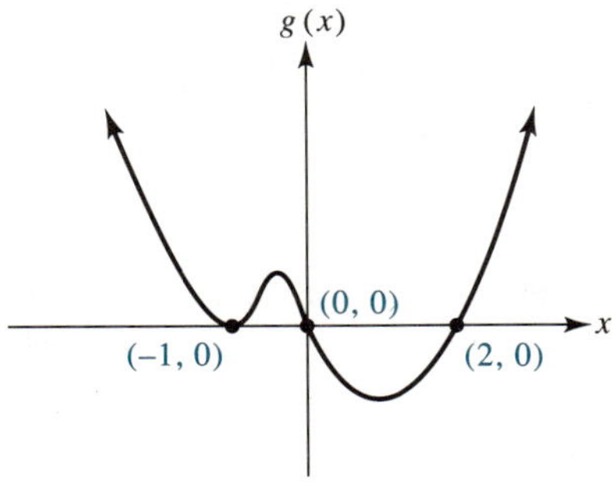

■

EXAMPLE Graph $h(x) = 23x - x^3$.
Factoring the rule for h gives $h(x) = x(\sqrt{23} + x)(\sqrt{23} - x)$. Therefore, h has zeros $-\sqrt{23}$, 0, and $\sqrt{23}$. Notice that for this application of factoring, the only restriction on coefficients is that they be real numbers.

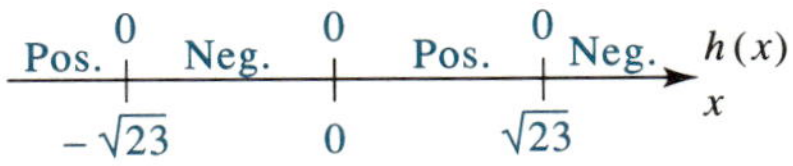

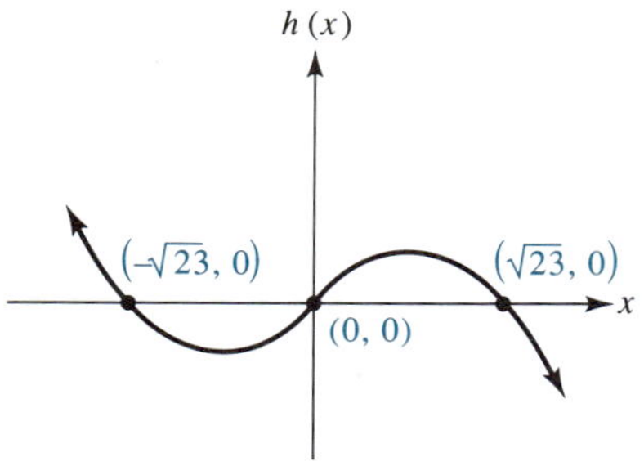

In these examples, seeing the function in factored form was a great advantage. Our previous methods of factoring may not suffice for us to obtain such a neat form for all polynomial functions. In Chapter 8, we develop one more tool to help with this problem.

Let's review our graphing techniques. We begin by finding the domain of the function. Thus far the most important restrictions on domains are numbers that would make a denominator zero or produce a negative radicand for an even root; the mod and div functions have the natural numbers as their domain.

We decide if the graph is a variation of one of the standard graphs. If so, we apply the appropriate reflection, translation, and expansion to the corresponding standard graph.

If the function is defined by a polynomial, a rational expression or a fractional expression, we find the numbers that identify zeros, holes, and asymptotes, if any.

Dividing numerator and denominator of a rational or fractional function by the highest power of the variable that appears in the denominator helps us decide if the function has a horizontal asymptote. (Keeping track of this information and the sign of the function values on a number-line chart may be useful.) If the function is a polynomial function, we use the polynomial principle to complete the graph.

If the function doesn't fall into any of these categories, we can check for symmetries, periodicity, intercepts, and changes in the sign of the function values. Using a calculator is a good way to produce estimates of second coordinates. We then complete our sketch using all available information to make the simplest connections possible. Remember, line segments are relatively rare in our graphs; curves are the usual connections. We can test if a line segment should connect two points by calculating the y-values for three equally spaced x-values. If the y-values are also equally spaced, then we use a line segment. (This is also discussed in Section 5.4.)

EXERCISES 4.5

In exercises 1 through 16, sketch the graph of each of the functions. (If you have a graphing calculator, use it for these exercises.)

1. $f(x) = 2x^2 - 3x + 1$
2. $g(x) = -2x^2 + 3x - 1$
3. $h(x) = x^3 - 5x^2 - 4x + 20$
4. $j(x) = x^4 - 5x^3 - 4x^2 + 20x$
5. $f(x) = x(x - 3)^2(2 + x)$
6. $g(x) = x(x + 2)^2(3 - x)$
7. $h(x) = x(x + 2)^3(3 - x)$
8. $m(x) = (x - 3)^3(x + 2)^2$
9. $f(x) = -x(x - 4)(x + 1)(x^2 - 24)$

10. $g(x) = -x(x - 4)(x + 1)(24 - x^2)$
11. $h(x) = -x(x - 4)(x + 1)^2(24 - x^2)$
12. $n(x) = -x(x - 4)(x + 1)^3(24 - x^2)$
13. $F(x) = |x^2 - 5|$
14. $G(x) = |x^3 - 7x^2 + 6x|$
15. $H(x) = |x(x + 2)^3(3 - x)|$
16. $N(x) = |x^4 + 3x^2 - 10|$

In exercises 17 through 28, sketch the graph of each of the given functions.

17. $g(x) = x^{2/3}$
18. $f(x) = x^{1/5}$
19. $h(x) = \dfrac{1}{x^2}$
20. $m(x) = \dfrac{1}{x^3}$
21. $n(x) = \left[\!\left[x - \dfrac{1}{2}\right]\!\right]$ (See Exercise 27, Section 4.1.)
22. $p(x) = [\![x]\!]^2$
23. $q(x) = -3[x + 1]^{2/3}$
24. $r(x) = \dfrac{1}{x^2 - 3}$
25. $s(x) = \dfrac{1}{x^3 - 3}$
26. $t(x) = [\![x^2]\!]$
27. $u(x) = \text{mod}_5(x^2)$
28. $v(x) = (\text{mod}_5 x)^2$
29. **a.** Sketch $f(x) = x^3 - 3x^2 + x$.
    b. How does the graph of $g(x) = (x^3 - 3x^2 + x) - 1$ compare to the graph of f?

In exercises 30 through 33, use the technique suggested by exercise 29 to sketch the graph of each of the given functions.

30. $f(x) = 2x^3 + x^2 - 3x + 5$
31. $g(x) = 3x^4 + 7x^3 + 2x^2 - 4$
32. $h(x) = x^4 - 2x^3 - 3x^2 - 3$
33. $m(x) = x^4 - 3x^3 - 6x^2 + 18x + 2$

In exercises 34 through 37, sketch the graph of each of the given functions.

34. $f(x) = \sqrt[4]{x - 1}$
35. $g(x) = \sqrt[5]{3 - x}$
36. $h(x) = -4\sqrt[7]{12 - 5x}$
37. $k(x) = -1.5\sqrt[6]{2.3 - x}$

In exercises 38 through 40, find a number p halfway between two adjacent zeros. Compute $f(p)$, $f(p - 0.01)$, and $f(p + 0.01)$. Could $(p, f(p))$ be the highest or lowest point between the zeros you chose?

38. $f(x) = 2x^3 + x^2 - 3x$
39. $g(x) = 3x^4 + 7x^3 + 2x^2$
40. $h(x) = x^4 - 2x^3 - 3x^2$
41. (Preview) We want to define a new function related to the function pc. The new function ps, defined by ps(t), is the second coordinate of the point reached on the unit box by traveling t units from (1, 0).
    a. What is the domain of ps? What is its range?
    b. Find $\text{ps}(-2\sqrt{2})$, $\text{ps}(\frac{3}{2}\sqrt{2})$, and $\text{ps}(-\frac{7}{2}\sqrt{2})$.
42. (Preview) **a.** What is the period of ps?
    b. Sketch the graph of ps.

Chapter 4: Summary

The idea of a function is a very powerful one in mathematics. A variety of ways to represent a function were presented here. In this chapter some general concepts for functions and graphs were also defined: increasing, decreasing, 1-1, symmetric, periodic, even, and odd. Four more standard graphs were introduced: the cubic, absolute value, square root, and natural exponential.

Key Concepts

relation	
function	*domain, range, independent variable, dependent variable, argument*
generalized term	
function mod$_p$	
standard graphs	$y = x^3, y = |x|, y = \sqrt{x}, y = e^x$
rational function	
asymptote	*horizontal, vertical*

properties of graphs	*symmetry with respect to x-axis, y-axis, origin, line $x = y$*
properties of functions	*increasing, decreasing, 1-1*
constant function	
polynomial principle	
zero of the function	
even and odd functions	
periodic functions	*period, cycle*

Skills

You should be able to:

1. Recognize a relation that is a function.
2. Perform linear interpolation for a 1-1 function.
3. Apply reflection, translation, and expansion/contraction to the graph of a function.
4. Find horizontal and vertical asymptotes.
5. Sketch the graph of a polynomial function, a rational function, and any standard graph or a variation of one.
6. Describe where a function is increasing; describe where a function is decreasing.
7. Tell whether a graph is symmetric with respect to the x-axis, y-axis, origin, or line $y = x$.
8. Tell whether a function is 1-1, even, odd, or periodic.

SUMMARY EXERCISES

The starred exercises can serve as a practice chapter test. The answers are in the answer section at the back of the book.

In exercises 1 through 3, identify the relation as a function or not a function.

1. $\{(x, y) \mid x^2 + y^2 = 25\}$
2. $\{(x, y) \mid x^2 + y = 25\}$
3. $\{(x, y) \mid x + y^2 = 25\}$

In exercises 4 through 6, use $f(x) = \sqrt{x - 3}$ and $g(x) = 2x + 5$ to find the requested numbers. State any restrictions on a and h needed to make the result a real number.

4. $f(a^2 + 2)$
5. $g\left[\dfrac{1}{h + 2}\right]$
6. $f[g(a)]$

In exercises 7 through 9, use $S(x) = \begin{cases} -1, & x < 0 \\ 0, & x = 0. \\ 1, & x > 0 \end{cases}$

* 7. Sketch the graph of S and tell whether the function is even, odd, or neither.

8. Sketch the graph of $f(x) = 3S(x) + 2$.
9. Compare the graphs of S and $g(x) = S(x - 1)$.

In exercises 10 through 12, use linear interpolation to find approximate function values for the given values of x using the indicated table within this section.

10. Table 4.Sa
    a. 0.64
    b. 1.362
    c. 0.99

11. Table 4.Sb
    a. 0.146
    b. 0.802
    c. 0.573

* 12. Table 4.Sc
    a. −2.358
    b. −1.504
    c. −1.937

TABLE 4.Sa
Function values for f approximated to four decimal places.

x	$f(x)$	x	$f(x)$
0.0	5.0000	1.0	9.0000
0.1	5.3001	1.1	9.7641
0.2	5.6016	1.2	10.6736
0.3	5.9081	1.3	11.7561
0.4	6.2256	1.4	13.0416
0.5	6.5625	1.5	14.5625
0.6	6.9296	1.6	16.3536
0.7	7.3401	1.7	18.4521
0.8	7.8096	1.8	20.8976
0.9	8.3561	1.9	23.7321

TABLE 4.Sb
Function values for g approximated to four decimal places

	Second decimal place in the value of x									
x	**0.00**	**0.01**	**0.02**	**0.03**	**0.04**	**0.05**	**0.06**	**0.07**	**0.08**	**0.09**
0.0	1.0000	1.0000	1.0000	1.0000	1.0000	1.0001	1.0001	1.0002	1.0003	1.0004
0.1	1.0005	1.0007	1.0009	1.0011	1.0014	1.0017	1.0020	1.0025	1.0029	1.0034
0.2	1.0040	1.0046	1.0053	1.0061	1.0069	1.0078	1.0088	1.0098	1.0110	1.0122
0.3	1.0135	1.0149	1.0164	1.0180	1.0197	1.0214	1.0233	1.0253	1.0274	1.0297
0.4	1.0320	1.0345	1.0370	1.0398	1.0426	1.0456	1.0487	1.0519	1.0553	1.0588
0.5	1.0625	1.0663	1.0703	1.0744	1.0787	1.0832	1.0878	1.0926	1.0976	1.1027
0.6	1.1080	1.1135	1.1192	1.1250	1.1311	1.1373	1.1437	1.1504	1.1572	1.1643
0.7	1.1715	1.1790	1.1866	1.1945	1.2026	1.2109	1.2195	1.2283	1.2373	1.2465
0.8	1.2560	1.2657	1.2757	1.2859	1.2964	1.3071	1.3180	1.3293	1.3407	1.3525
0.9	1.3645	1.3768	1.3893	1.4022	1.4153	1.4287	1.4424	1.4563	1.4706	1.4851

TABLE 4.Sc
Function values for h approximated to four decimal places

	Second decimal place in the value of x									
x	**0.00**	**0.01**	**0.02**	**0.03**	**0.04**	**0.05**	**0.06**	**0.07**	**0.08**	**0.09**
−3.0	14.500	14.365	14.232	14.099	13.967	13.836	13.706	13.577	13.449	13.321
−2.9	13.194	13.069	12.944	12.820	12.697	12.575	12.453	12.333	12.213	12.094
−2.8	11.976	11.859	11.742	11.627	11.512	11.398	11.285	11.173	11.062	10.951
−2.7	10.842	10.733	10.624	10.517	10.411	10.305	10.200	10.096	9.992	9.890
−2.6	9.788	9.687	9.587	9.487	9.389	9.291	9.194	9.097	9.002	8.907
−2.5	8.813	8.719	8.626	8.535	8.443	8.353	8.263	8.174	8.086	7.999
−2.4	7.912	7.826	7.741	7.656	7.572	7.489	7.406	7.325	7.244	7.163
−2.3	7.084	7.004	6.926	6.849	6.772	6.695	6.620	6.545	6.471	6.397
−2.2	6.324	6.252	6.180	6.109	6.039	5.969	5.900	5.832	5.764	5.697

	Second decimal place in the value of x									
x	0.00	0.01	0.02	0.03	0.04	0.05	0.06	0.07	0.08	0.09
−2.1	5.631	5.565	5.499	5.435	5.371	5.308	5.245	5.183	5.121	5.060
−2.0	5.000	4.940	4.881	4.823	4.765	4.707	4.651	4.595	4.539	4.484
−1.9	4.430	4.376	4.322	4.270	4.217	4.166	4.115	4.064	4.014	3.965
−1.8	3.916	3.868	3.820	3.773	3.726	3.680	3.634	3.589	3.544	3.500
−1.7	3.457	3.413	3.371	3.329	3.287	3.246	3.205	3.165	3.126	3.087
−1.6	3.048	3.010	2.972	2.935	2.898	2.862	2.826	2.791	2.756	2.721
−1.5	2.688	2.654	2.621	2.588	2.556	2.524	2.493	2.462	2.432	2.402
−1.4	2.372	2.343	2.314	2.286	2.258	2.230	2.203	2.176	2.150	2.124
−1.3	2.099	2.073	2.049	2.024	2.000	1.977	1.953	1.930	1.908	1.886
−1.2	1.864	1.843	1.822	1.801	1.780	1.760	1.741	1.721	1.702	1.684
−1.1	1.666	1.648	1.630	1.613	1.596	1.579	1.562	1.546	1.531	1.515
−1.0	1.500									

In exercises 13 through 30, sketch the graph of the given function.

13. $f(x) = \dfrac{3x - 4}{2x + 1}$
14. $h(x) = \begin{cases} \dfrac{1}{x + 1}, & x \leq 0 \\ \dfrac{1}{x}, & x > 0 \end{cases}$
* 15. $g(x) = \dfrac{x^3 - 8}{x^4 - 4}$
16. $p(x) = x^3 - 6x^2 + 9x$
17. $m(x) = (x - 1)^2(x + 2)(x^2 - 3)$
* 18. $q(x) = \left(x + \dfrac{1}{2}\right)^3 - \dfrac{3}{2}$
19. $F(x) = \begin{cases} x^2 - 2x + 3, & x < 2 \\ -2x + 7, & x \geq 2 \end{cases}$
20. $G(x) = \sqrt{3 - 2x}$
* 21. $H(x) = x\sqrt{x - 2}$
22. $M(x) = |5.7x - 3.2| + 6.1$
23. $J(x) = 3x^{1/5}$
* 24. $S(x) = -2x^{2/5}$
25. $f(x) = |x| \cdot |x + 1|$
26. $g(x) = -2e^x$
* 27. $h(x) = 1.5e^{x+2}$
28. $R(x) = -\text{mod}_4(x + 2)$
29. $r(x) = -2[\![x]\!] + 1$
* 30. $n(x) = \left|\dfrac{1}{x}\right|$

In exercises 31 through 33, use the function *f* described by this graph to sketch the graph of each of the following functions.

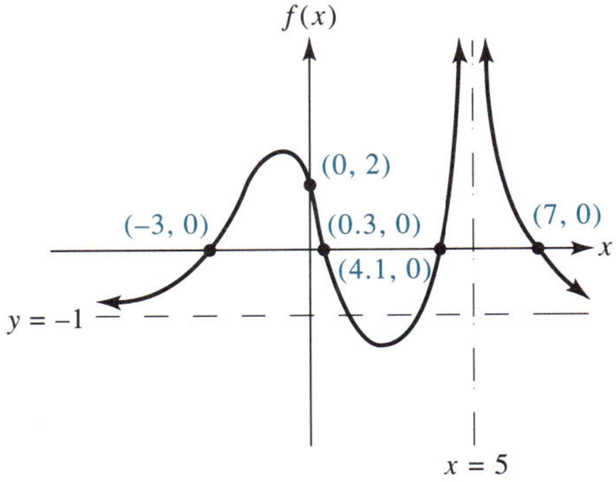

31. $g(x) = f(x - 3)$
32. $h(x) = f(x) + 4$
* 33. $j(x) = -f(x + 1)$
34. Give a variation of $f(x) = x^3$ named F whose graph is the graph of f with a vertical expansion factor of 1.5 and shifted 2.5 units left.
35. Suppose all you know about a function is its name, g, and its domain, the reals. Give a variation of g named G whose graph is the graph of g that is translated 4 units up and 5 units left and is reflected in the x-axis.
* 36. Give a variation of $h(x) = x^{2/3}$ named H whose graph is the graph of h translated 3 units right and 2 units down.

37. Let $f(n)$ = n^{th} digit in the decimal expansion of $\frac{1}{17}$. What is the period of f?

38. What is the period of $g(n) = \text{mod}_6(2n)$?

* **39.** Define a function with period 7.

40. If $f(x) = x/(x + 2)$, find b so that $f(b) = 6$. Is there only one b?

41. If $g(x) = x^2 - 3.2$, find b so that $g(b) = 5.7$. Is there only one b?

* **42.** If $h(x) = \sqrt[3]{x - 4}$, find b so that $h(b) = c$. Is there only one b?

43. Define a new function with the same rule as $g(x) = 3x^2 - 5$ so that the new one is decreasing on its domain and has the same range as g.

44. Define a new function with the same rule as $f(x) = -5x^2 + 6$ so that the new one is 1-1 and has the same range as f.

45. Define a new function with the same rule as $f(x) = x^2 - 4x + 7$ so that the new one is 1-1 and has the same range as g.

46. Define a new function with the same rule as $h(x) = |3 - \pi x|$ so that the new one is 1-1 and has the same range as h.

* **47.** Define a new function with the same rule as $h(x) = -|x + 1| - 4$ so that the new one is increasing on its domain and has range $[-10, -5]$. Is your function 1-1?

CHAPTER FIVE

Algebra of Functions

- 5.1 New Functions from Old
- 5.2 Composition and Inverse Functions
- 5.3 Exponential and Logarithmic Functions
- 5.4 New Functions from Conditions

5.1 New Functions from Old

Toolkit

This chapter uses the ideas of Chapter 4 extensively. You may also find it helpful to review the properties of real numbers in Section A.3, particularly the verbal definitions. Sets and set operations are covered in Section A.2.

Patterns and similarities are very important in mathematics. Many of the exercises have explored patterns. For example, in Section A.2, some exercises demonstrate the similarities between the operations of union and intersection on sets and the operations of addition and multiplication on real numbers. We'll see such similarities in this section.

Unless specifically stated, any function in this chapter has domain and range that are subsets of the real numbers.

There are many ways to give the definition of a specific function—for example, as a graph, as a set, as a table, as a rule plus a domain. Another possibility is to define it using existing functions.

Since our discussions will involve several functions and their domains, it's helpful to have a notation to identify a domain. We use dom_f to represent the domain of function f.

If we have two functions, say f and g, that have been defined, we can create a new function $f + g$ as follows: $(f + g)(x)$ is defined to be $f(x) + g(x)$. A function described by a rule, as $f + g$ is, must also have a domain specified. When will $f(x) + g(x)$ make sense? Only when $x \in \text{dom}_f$ and $x \in \text{dom}_g$ does $f(x) + g(x)$ represent a real number. Note that if $x \in \text{dom}_f$ and $x \in \text{dom}_g$, then $x \in \text{dom}_f \cap \text{dom}_g$. The new function can only be defined where both summand functions are defined.

Definition Given two functions f and g, $\boldsymbol{f + g}$ is defined by $(f + g)(x) = f(x) + g(x)$ for every $x \in \text{dom}_f \cap \text{dom}_g$.

EXAMPLE Give the domain of $f + g$ if $f(x) = \sqrt{x - 4}$ and $g(x) = e^x$.

The domain of f is $\{x|x \geq 4\}$ and the domain of g is the set of reals.
The domain of $f + g$ is $\{x|x \geq 4\}$ because this is $\text{dom}_f \cap \text{dom}_g$. ■

You may have guessed that if we can build a new function as the sum of two existing functions, we can probably do the same for differences, products, and quotients. We certainly can, but there's something going on here that may be obscured by using names (sum, difference, product, quotient) and symbols $(+, -, \cdot, /)$ that we already use for real number arithmetic. We've actually defined new operations. When we write $f(x) + g(x)$, we're representing with the plus sign the addition of two real numbers, $f(x)$ and $g(x)$. When we write $f + g$, we're using the $+$ to represent the addition of two functions, f and g. It's convenient—although a bit confusing—to use addition and $+$ in both cases.

Here are the definitions for the difference, product, and quotient of two functions.

Definition Given two functions f and g, $\boldsymbol{f - g}$ is defined by $(f - g)(x) = f(x) - g(x)$ for every $x \in \text{dom}_f \cap \text{dom}_g$.

EXAMPLE Given $f(x) = 3x - 2$ and $g(x) = \sqrt{4 - x}$, write $(f - g)(x)$ and give its domain.

$$(f - g)(x) = f(x) - g(x) = \mathbf{3x - 2 - \sqrt{4 - x}}$$

$\text{dom}_{f-g} = \text{dom}_f \cap \text{dom}_g$, so $\mathbf{dom}_{f-g}$ **is** $\{x|x \le 4\}$, which is the intersection of the set of real numbers and $\{x|x \le 4\}$. ■

Definition Given two functions f and g, $\boldsymbol{f \cdot g}$ is defined by $(f \cdot g)(x) = f(x) \cdot g(x)$, with $\text{dom}_{fg} = \text{dom}_f \cap \text{dom}_g$.

EXAMPLE Given $f(x) = 1/x$ and $g(x) = x^2 + 3$, write $(f \cdot g)(x)$ and give its domain.

$$(f \cdot g)(x) = f(x) \cdot g(x) = \left(\frac{1}{x}\right) \cdot (x^2 + 3)$$

$\text{dom}_{fg} = \text{dom}_f \cap \text{dom}_g$, so dom_{fg} is the intersection of the set of real numbers except 0 and the set $\mathbb{R}$; the intersection is the set of real numbers except 0.

$\mathbf{(f \cdot g)(x) = \dfrac{x^2 + 3}{x}}$; $\mathbf{dom}_{fg}$ **is the set of all reals except 0.** ■

A concern of a different sort arises in defining f/g, the quotient of two functions. Suppose $f(x) = 3x - 2$ and $g(x) = 4 - x$. If $(f/g)(x)$ is to be $(3x - 2)/(4 - x)$, as you'd expect, what is the domain of f/g? It can't be $\text{dom}_f \cap \text{dom}_g$ because $4 \in \text{dom}_f \cap \text{dom}_g$, $f(4) = 10$, $g(4) = 0$, and $(f/g)(4)$ is undefined. We need to eliminate any values from the domain for which the denominator function has value 0.

Definition Given two functions f and g, f/g is defined by $(f/g)(x) = f(x)/g(x)$, with $\text{dom}_{f/g} = \{x|x \in \text{dom}_f \cap \text{dom}_g \text{ and } g(x) \ne 0\}$.

EXAMPLE Describe $(f/g)(x)$ and its domain if $f(x) = \sqrt{x}$ and $g(x) = x^2 - 5x + 6$.

$$\left(\frac{f}{g}\right)(x) = \frac{f(x)}{g(x)} = \frac{\sqrt{x}}{x^2 - 5x + 6}, \text{ or } \frac{\sqrt{x}}{(x - 2)(x - 3)}$$

$\text{dom}_{f/g} = \{x|x \in \text{dom}_f \cap \text{dom}_g \text{ and } g(x) \ne 0\}$. Here dom_f is the nonnegative reals and dom_g is the reals, so $\text{dom}_f \cap \text{dom}_g$ is the nonnegative reals. We exclude 2 and 3 because $g(2) = g(3) = 0$.

Therefore, $\mathbf{\left(\dfrac{f}{g}\right)(x) = \dfrac{\sqrt{x}}{x^2 - 5x + 6}}$ **and** $\mathbf{dom}_{f/g} = \{x|x \ge 0 \text{ and } x \ne 2 \text{ and } x \ne 3\}$. ■

Now that we've defined these four operations on functions, we wish to investigate how they behave. The way in which the basic operations on real numbers behave is discussed in Section A.3. These properties are the fundamental assumptions we make about the arithmetic of the real numbers. In some exercises in

Section A.2, we explore informally some of these properties for union and intersection of sets. Here we can do a more formal investigation of these properties with regard to our new operations on functions.

Is the addition of functions associative? An operation is associative if whenever it is to be used twice, the same result is obtained regardless of the order in which the two applications are done. Therefore, we must show $(f + g) + h$ is the same function as $f + (g + h)$.

The simplest proofs begin with the relevant definitions. Two functions are the same if they have the same domain and produce the same result for each element in that domain.

In this case, $((f + g) + h)(x)$ is $(f + g)(x) + h(x)$ if $x \in \text{dom}_{f+g} \cap \text{dom}_h$. We can rewrite this, since $(f + g)(x) + h(x)$ is $[f(x) + g(x)] + h(x)$ if $x \in (\text{dom}_f \cap \text{dom}_g) \cap \text{dom}_h$. Doing the same for $f + (g + h)$, we have $(f + (g + h))(x)$ is $f(x) + [g(x) + h(x)]$ if $x \in \text{dom}_f \cap (\text{dom}_g \cap \text{dom}_h)$. The two numbers $[f(x) + g(x)] + h(x)$ and $f(x) + [g(x) + h(x)]$ are equal because addition is associative for the real numbers. In this text we assume $(\text{dom}_f \cap \text{dom}_g) \cap \text{dom}_h$ is the same as $\text{dom}_f \cap (\text{dom}_g \cap \text{dom}_h)$. This is an application of the associative property for intersection. (You may convince yourself that this is reasonable in Exercise 42, Section A.2.)

Yes, this is very detailed work, and you may not be fond of proofs, but it's time to try a few simple ones and begin to build your proof-writing muscles. Even if you never have to write a proof, except in exercises, you will have to read some. Whatever craft or sport you've done yourself, you can better understand and appreciate someone else's efforts in the same field.

The distributive property of real numbers is used in many places. Is there a similar property for functions?

EXAMPLE Prove $f \cdot [g + h] = f \cdot g + f \cdot h$ for any functions f, g, and h.

Remember we must show the two sides have the same domain and produce the same result for each number in this set. The left-hand side (LHS) has domain $\text{dom}_f \cap \text{dom}_{g+h} = \text{dom}_f \cap (\text{dom}_g \cap \text{dom}_h)$. But since intersection is associative, the parentheses are unnecessary, so dom_{LHS} is $\text{dom}_f \cap \text{dom}_g \cap \text{dom}_h$. For the right-hand side (RHS), $\text{dom}_{\text{RHS}} = \text{dom}_{fg} \cap \text{dom}_{fh} = (\text{dom}_f \cap \text{dom}_g) \cap (\text{dom}_f \cap \text{dom}_h) = \text{dom}_f \cap \text{dom}_g \cap \text{dom}_f \cap \text{dom}_h = \text{dom}_f \cap \text{dom}_g \cap \text{dom}_h$, and the domains are the same. We are just writing formally the fact that each side is defined only for those numbers on which all three functions f, g, and h operate.

Now for any $x \in \text{dom}_f \cap \text{dom}_g \cap \text{dom}_h$, we must show each side produces the same result. We do this by working with one side of the equation, not both.

	$(f \cdot [g + h])(x) = f(x) \cdot [g + h](x)$	Definition of function product
$\Leftrightarrow$	$(f \cdot [g + h])(x) = f(x) \cdot [g(x) + h(x)]$	Definition of function sum
$\Leftrightarrow$	$(f \cdot [g + h])(x) = f(x) \cdot g(x) + f(x) \cdot h(x)$	Distributive property of reals
$\Leftrightarrow$	$(f \cdot [g + h])(x) = (f \cdot g)(x) + (f \cdot h)(x)$	Definition of function multiplication
$\Leftrightarrow$	$(f \cdot [g + h])(x) = (f \cdot g + f \cdot h)(x)$	Definition of function addition

We have produced the right-hand side. ■

The style of proof here is a direct proof; we begin with definitions and known facts and try to proceed directly to the conclusion. Let's look at another property. Can we define an identity function for function multiplication and, perhaps, reciprocal functions? A multiplicative identity function should behave as follows: if i is this function, $f \cdot i$ should be the same as f for each function f. We begin as before with the relevant definition. $(f \cdot i)(x)$ is $f(x) \cdot i(x)$ for every $x \in \text{dom}_f \cap \text{dom}_i$. We want $f(x) \cdot i(x) = f(x)$ for all $x \in \text{dom}_f$. This tells us two things: we'd better not lose any of dom_f when we intersect it with dom_i, and since there's only one real number (the number 1) such that $r \cdot 1 = r$ for every $r \in \mathbb{R}$, $i(x)$ must equal 1 for every $x \in \text{dom}_i$. If i is to be the multiplicative identity function for every f, dom_i must be $\mathbb{R}$ or $\text{dom}_f \cap \text{dom}_i$ may not equal dom_f for every f.

Definition The multiplicative identity function, i, is defined as $i(x) = 1$, with domain all real numbers.

Does each function f have a reciprocal function? We can try the obvious, a function denoted $1/f$ defined by $(1/f)(x) = 1/(f(x))$. We must also give its domain and check to see if it behaves as a reciprocal should. Clearly, we must exclude from the domain any numbers for which $f(x)$ is 0, but then $f \cdot (1/f)$ may not be i, the identity function for all real numbers. Our conclusion is that not all functions have reciprocal functions. Remember a function is defined by both its rule and its domain.

Illustration

$$f(x) = 3x \qquad \text{dom}_f = \mathbb{R}$$
$$g(x) = \frac{1}{3x} \qquad \text{dom}_g = \mathbb{R} \text{ except } 0$$
$$(f \cdot g)(x) = 1, \quad x \neq 0 \qquad \text{dom}_{f \cdot g} = \mathbb{R} \text{ except } 0$$

So, f and g are not reciprocals of each other.

$$h(x) = x^2 + 1 \qquad \text{dom}_h = \mathbb{R}$$
$$H(x) = \frac{1}{x^2 + 1} \qquad \text{dom}_H = \mathbb{R}$$
$$(h \cdot H)(x) = 1 \qquad \text{dom}_{h \cdot H} = \mathbb{R}$$

So, h and H are a pair of reciprocals. □

We finish with an example of one way to disprove a statement.

EXAMPLE Prove or disprove this statement: division of functions is commutative.

The statement is not true. An operation is commutative if the order of operands doesn't affect the result. A counterexample for function division is if $f(x) = 2x$ and $g(x) = x + 1$, $(f/g)(x) = 2x/(x + 1)$, but $(g/f)(x) = (x + 1)/2x$. These are not the same function. ■

EXERCISES 5.1

In exercises 1 through 10, given f and g, write $f + g$ and $f \cdot g$; describe the domain in interval notation, where possible.

1. $f(x) = \dfrac{x+3}{x-2}$; $g(x) = x - 2$
2. $f(x) = x^3 - 4x + 5$; $g(x) = 6x^2 - 3x$
3. $f = \{(1, 4), (3, 7), (2, 9), (5, 8)\}$;
   $g = \{(1, -7), (2, \frac{1}{3}), (3, 9), (5, 8)\}$
4. $f = \{(1, 0), (2, 1), (4, 2), (8, 3), (16, 4), \ldots\}$;
   $g = \{(1, \frac{1}{2}), (2, 1), (4, 2), (6, 3), (8, 4), \ldots\}$
5. $f(x) = x + 4$; $g(x) = \dfrac{\sqrt{x} - 2}{\sqrt{x} + 2}$
6. $f(x) = \left|\dfrac{x-3}{1-x}\right|$; $g(x) = |x - 1|$
7. $f(n) = \text{mod}_5(n)$; $g(n) = n/5$, $n \in W$
8. $f(x) = 12x - 7.2$; $g(x) = 4.3x^2 + 9.1x - 6.8$
9. $f(x) = \sqrt[3]{x+3}$; $g(x) = \sqrt[3]{4 - 2x^2}$
10. $f(x) = \sqrt[4]{16 - x^4}$; $g(x) = \sqrt[6]{32 - x^5}$

In exercises 11 through 20, given f and g, write $f - g$ and f/g; describe the domain in interval notation, where possible.

11. $f(x) = \dfrac{x^2 + 7x + 12}{x^2 + 8x + 15}$; $g(x) = \dfrac{x^2 + 7x + 10}{x^2 + 6x + 8}$
12. $f(x) = e^{2x}$; $g(x) = 3x$
13. $f(x) = (1.01)^{0.5x}$; $g(x) = (1.01)^x$
14. $f(x) = |x^2 + 2x|$; $g(x) = |x + 2|$
15. $f(n) = \text{div}_3(n)$; $g(n) = 6n$
16. $f(x) = ax^2 + bx + c$; $g(x) = mx + d$
17. $f(x) = \sqrt{2x}$; $g(x) = \sqrt[3]{16x}$
18. $f(x) = \dfrac{1.21 - x^2}{2x^2 + 4.2x + 2.2}$; $g(x) = x - 1.1$
19. $f(x) = \sqrt{x - 4}$; $g(x) = \sqrt{x - 4}$
20. $f(n) = \text{mod}_5(n)$; $g(n) = 10n$, $n \in N$

In exercises 21 through 24, determine whether each of the functions has a reciprocal function. State the domain of each.

21. $f(x) = e^x$
22. $g(x) = 3x - 4$
23. $h(x) = \sqrt{1 - 5x}$
24. $j(x) = x^2 + 2x + 4$

(Preview) Each natural number can be written as the product of prime numbers in exactly one way, if you disregard the order of the factors. For example, $6 = 2 \cdot 3$, or $3 \cdot 2$; $225 = 3 \cdot 3 \cdot 5 \cdot 5$. We define a function ct that counts the number of prime factors for each natural number:

$\text{ct}(6) = 2 \quad \text{ct}(225) = 4 \quad \text{ct}(12) = 3$

Exercises 25 through 28 refer to the function ct.

25. Find ct(7), ct(81), ct(210), and ct(512).
26. Show that for any natural numbers m and n, $\text{ct}(mn) = \text{ct}(m) + \text{ct}(n)$.
27. Show that $\text{ct}(m/n) = \text{ct}(m) - \text{ct}(n)$ if m is divisible by n.
28. Show that $\text{ct}(m^n) = n\,\text{ct}(m)$ if $n \in N$.

In exercises 29 through 32, assume all functions have the real numbers as domain and range. Prove that each of the following properties holds for these functions.

29. Associative for function multiplication
30. Identity for function addition
31. Commutative for function multiplication
32. Commutative for function addition

In exercises 33 and 34, find a counterexample to show the following properties do not hold.

33. Commutative for function subtraction
34. Associative for function subtraction

In exercises 35 through 37, suppose b @ c is $(b + c)/2$, with b and c integers. Determine whether the operation @ has the following properties.

35. Commutative
36. Associative
37. Identity

In exercises 38 through 41, suppose $a * b$ is $|a - b|$, with a and b integers. Determine whether the operation * has the following properties.

38. Commutative
39. Associative
40. Identity
41. Inverse
42. Suppose that f and g are both even functions. Prove that $f + g$ and $f - g$ are both even functions. (Even and odd functions are defined in Section 4.4.)
43. If f and g are both odd functions, what can be said about $f + g$ and $f - g$? Prove that your claim is true.

5.2 Composition and Inverse Functions

Toolkit

You'll need the idea of a 1–1 function from Section 4.4.

In some ways all that we did in Section 5.1 was old news. The proofs of the properties of the four operations defined on functions relied on the corresponding properties for real numbers. If this were all we could do with functions, it would hardly seem worth the trouble. Section 5.1 was mostly to start you thinking about combining functions and about properties that might apply for such operations.

Another way to produce a new function from two existing functions f and g is composition. It's defined as if we linked the two function machines. The notation for "f composed with g" is $f \circ g$, and this function is defined as $(f \circ g)(x)$ is $f(g(x))$ with suitable domain. We'll come back to the domain question after we're sure what $f(g(x))$ says. First we evaluate $g(x)$ and then find f of that number. This gives us some clues about the domain. It must be a subset of dom_g, since it is possible for $g(x)$ to be defined but lie outside dom_f. We want only those elements x of dom_g for which $g(x) \in \text{dom}_f$.

Definition For two functions f and g, f composed with g is defined by $(f \circ g)(x) = f(g(x))$ and $\text{dom}_{f \circ g} = \{x \mid x \in \text{dom}_g \text{ and } g(x) \in \text{dom}_f\}$.

EXAMPLE Find $(f \circ g)(x)$ if $f(x) = 7(x + 2)$ and $g(x) = \sqrt{x}$.

$$(f \circ g)(x) = f(g(x)) = f(\sqrt{x}) = \mathbf{7(\sqrt{x} + 2)}$$

It may help to think of the action of the second function in words: "f adds 2 and then multiplies by 7." Here's how the function machines would be connected.

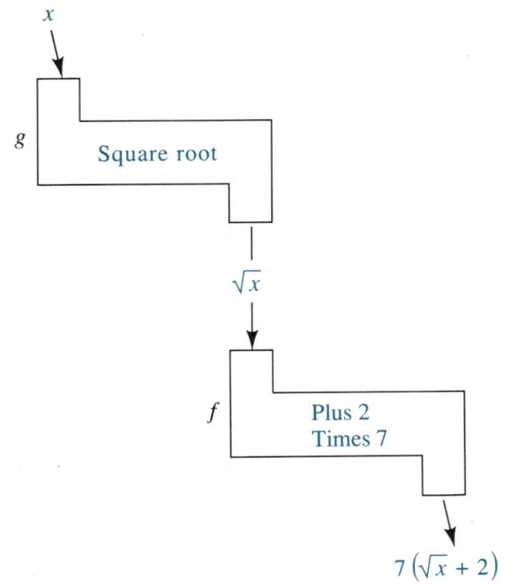

The domain of f composed with g is the nonnegative real numbers. ■

EXAMPLE Find $(g \circ f)(x)$ if $f(x) = 7(x + 2)$ and $g(x) = \sqrt{x}$.

$(g \circ f)(x) = g(f(x)) = g(7(x + 2)) = \sqrt{7(x + 2)}$.
Notice that **the domain of g composed with f is $\{x \mid x \geq -2\}$** because f produces nonnegative function values for these numbers, and the domain of g is the set of nonnegative real numbers.

Note: The function $t(x) = \sqrt{7(x + 2)}$ could also have been formed as the composition $h \circ k$, where $h(x) = \sqrt{7x}$ and $k(x) = x + 2$. ■

We've described how to compute the composition of two functions algebraically. What about a geometric approach? Let's begin with the graphs of h and k on the same set of axes. It's helpful also to have $y = x$ shown as a dashed line. We'll use the functions from the last example. To sketch the graph of $h \circ k$, we repeat the following steps to establish the shape.

1. Choose a value for x in the domain of k; locate $(x, k(x))$.

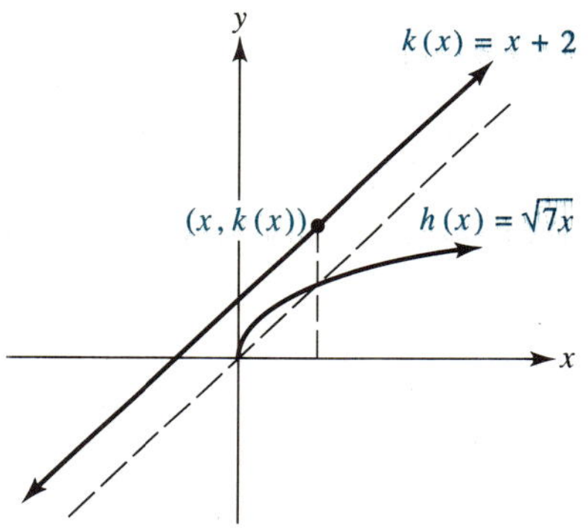

FIGURE 5.2a

2. Since we need $k(x)$ to be the argument of h, we'll measure $k(x)$ horizontally by moving right or left from $(x, k(x))$ to reach $(k(x), k(x))$ on the line $y = x$.

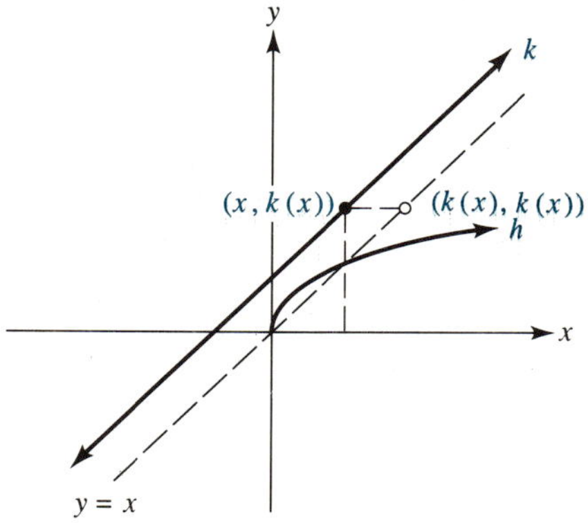

FIGURE 5.2b

3. Move vertically to find $(k(x), h(k(x))$ on the graph of h. If this isn't possible, x is not in the domain of $h \circ k$; begin with another value for x.

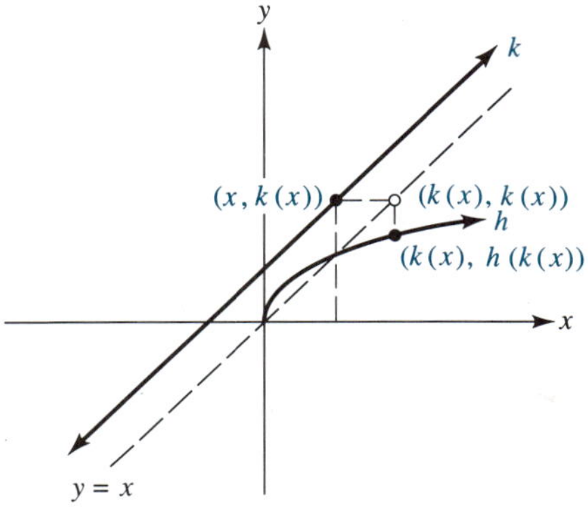

FIGURE 5.2c

4. The second coordinate $h(k(x))$ should be paired with x; move horizontally to $(x, h(k(x)))$.

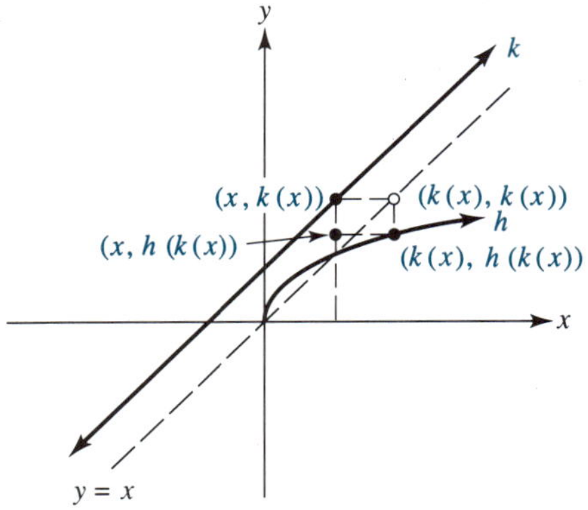

FIGURE 5.2d

Mark this point for the new graph. You'll find that with a bit of practice, you'll be able to move a pencil lightly over the graphs of h and k to find points for the graph of $h \circ k$.

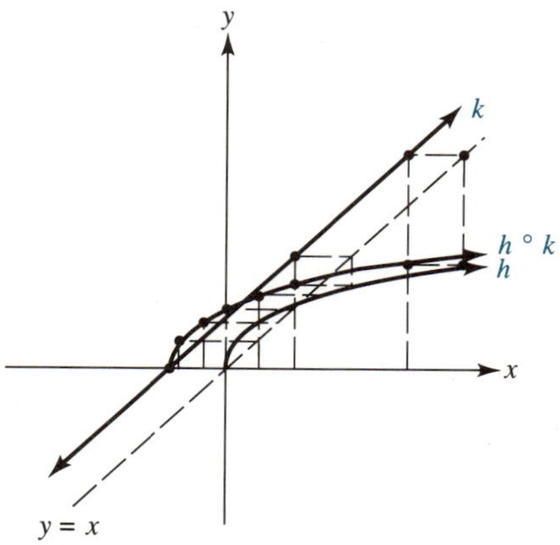

FIGURE 5.2e

For the graph of $k \circ h$, locate $(x, h(x))$, $(h(x), h(x))$, $(h(x), k(h(x)))$ and, finally, $(x, k(h(x)))$, move vertically, horizontally, vertically, and horizontally again.

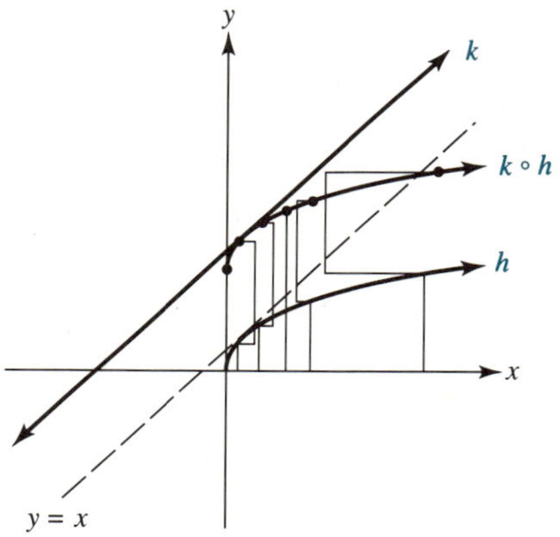

FIGURE 5.2f

Composition is an operation that we perform on functions, not on numbers. Unlike addition, subtraction, multiplication, and division of functions, there is no corresponding operation on real numbers. Now we do have something to explore. How does composition behave? The two earlier examples show that it is not a commutative operation; $f \circ g$ is not always the same function as $g \circ f$. An investigation of the associative property is done in the exercises.

Is there an identity function for composition? If there is, can we define inverse functions with respect to composition? As before, we look at the basic definition. A composition identity would have to leave any function unchanged by composition. If we call such an identity function I, we need $f \circ I$ to be the same as f; for example, $(f \circ I)(x) = f(x)$ for every x in dom_f. Since $(f \circ I)(x)$ is $f(I(x))$, there isn't much I can do to any number x. For $f(I(x))$ to be $f(x)$ for every function f, $I(x)$ must be x for every real number x.

Composition isn't commutative, so we check to see if $I \circ f$ is f, given the description of I. For any function f in this chapter, its range is a subset of $\mathbb{R}$ so $(I \circ f)(x) = I(f(x)) = f(x)$ for every x in dom_f, and $I \circ f$ is the same as f.

Definition The identity function for composition, I, is defined as $I(x) = x$, with domain the real numbers.

Do functions have composition inverses? For each function f, can we find a function inv_f so that $\text{inv}_f \circ f$ is the same as I? This is an important question, so we'll look at it in several ways. Formally, we want $(\text{inv}_f \circ f)(x) = \text{inv}_f(f(x)) = I(x) = x$ for all x in $\mathbb{R}$. But we have several ways of describing functions, as mappings, for instance. As a mapping, inv_f must return us to where we began with f.

Illustration

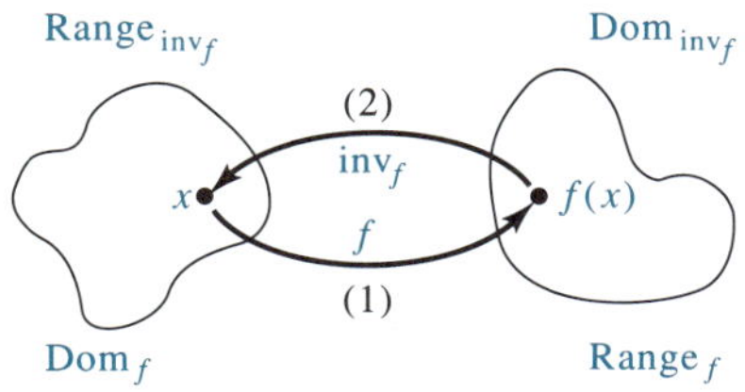

□

If we think of a function as a set of ordered pairs, then the ordered pairs of inv_f must be those of f reversed.

Illustration For $f = \{(2,3), (4, 1), (5, 7)\}$, we need $\text{inv}_f = \{(3, 2), (1, 4), (7, 5)\}$, so that

$$(\text{inv}_f \circ f)(2) = \text{inv}_f(f(2)) = \text{inv}_f(3) = 2$$

$$(\text{inv}_f \circ f)(4) = \text{inv}_f(f(4)) = \text{inv}_f(1) = 4$$

$$(\text{inv}_f \circ f)(5) = \text{inv}_f(f(5)) = \text{inv}_f(7) = 5$$

Note: We could also think of the pairs in f as the reverse of those in inv_f. □

Our last illustration suggests two problems, one obvious, one perhaps not so obvious. In the illustration $(\text{inv}_f \circ f)(x) = x$ for $x \in \{2, 4, 5\}$, which is not all the reals. Many useful functions are not defined for every real number, so we will relax the condition that $\text{inv}_f \circ f$ is the same as I by looking for inverse functions such that $(\text{inv}_f \circ f)(x) = x$ for every $x \in \text{dom}_f$. The hidden problem is apparent in a slightly different illustration.

Illustration If $g = \{(2, 3), (4, 3), (5, 7)\}$, looking for an inverse we might try $G = \{(3, 2), (3, 4), (7, 5)\}$. But G is not a function, so there is no inv_g. □

We have two tasks now: identify functions that do have composition inverses and tell how to describe the inverse function if it exists. The difficulty with g in the previous example is that it's not 1-1. For any 1-1 function f, the set of ordered pairs formed by reversing the ordering in f's pairs is also a function. We have a geometric test for one-to-oneness: the horizontal line test. If some horizontal line crosses the graph of the function more than once, the function is not 1-1. This gives a convenient way to test any function given by a graph or whose graph we can visualize. If it is 1-1, a function has a composition inverse; for example there is a function inv_f such that $(\text{inv}_f \circ f)(x) = x, x \in \text{dom}_f$.

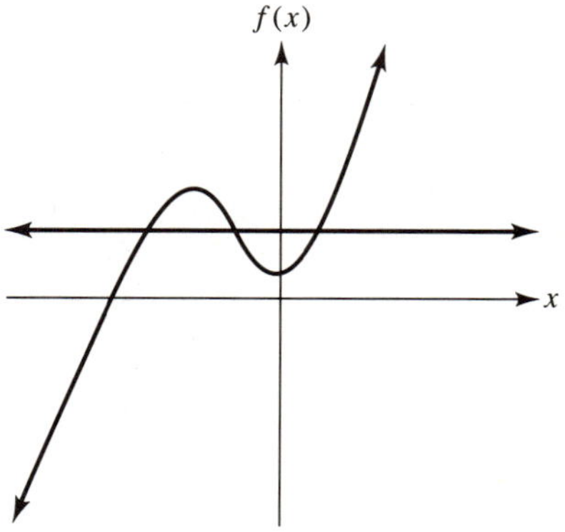

f is not 1-1
The graph fails the horizontal line test.

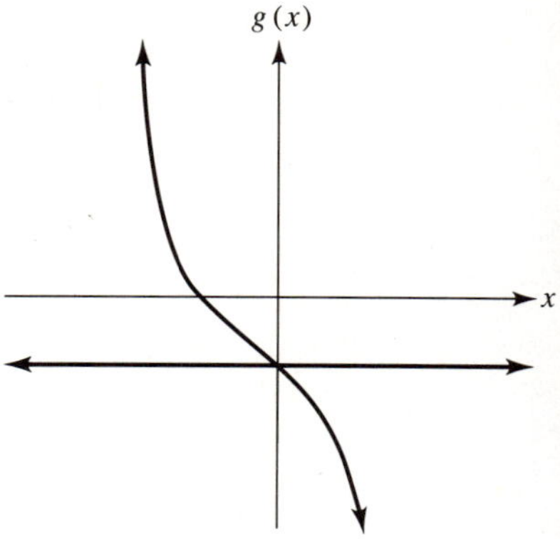

g is 1-1
The graph passes the horizontal line test.

If a 1-1 function is given by a set or a table, we can easily describe its inverse by reversing the paired values. This makes it easy to see that the roles of domain and range are also interchanged. The domain of *f* is the range of the inverse of *f*, and vice versa.

Illustration If *f* is defined by the table on the left, we can build the table for inv_f.

x	$f(x)$
-3	18
-1.4	6.37
0	$-\pi$
134	$\sqrt{5}$

x	$\text{inv}_f(x)$
18	-3
6.37	-1.4
$-\pi$	0
$\sqrt{5}$	134

□

If we have the graph of a 1-1 function, we can sketch the graph of its inverse function by including (b, a) whenever (a, b) belongs to the original graph.

Illustration

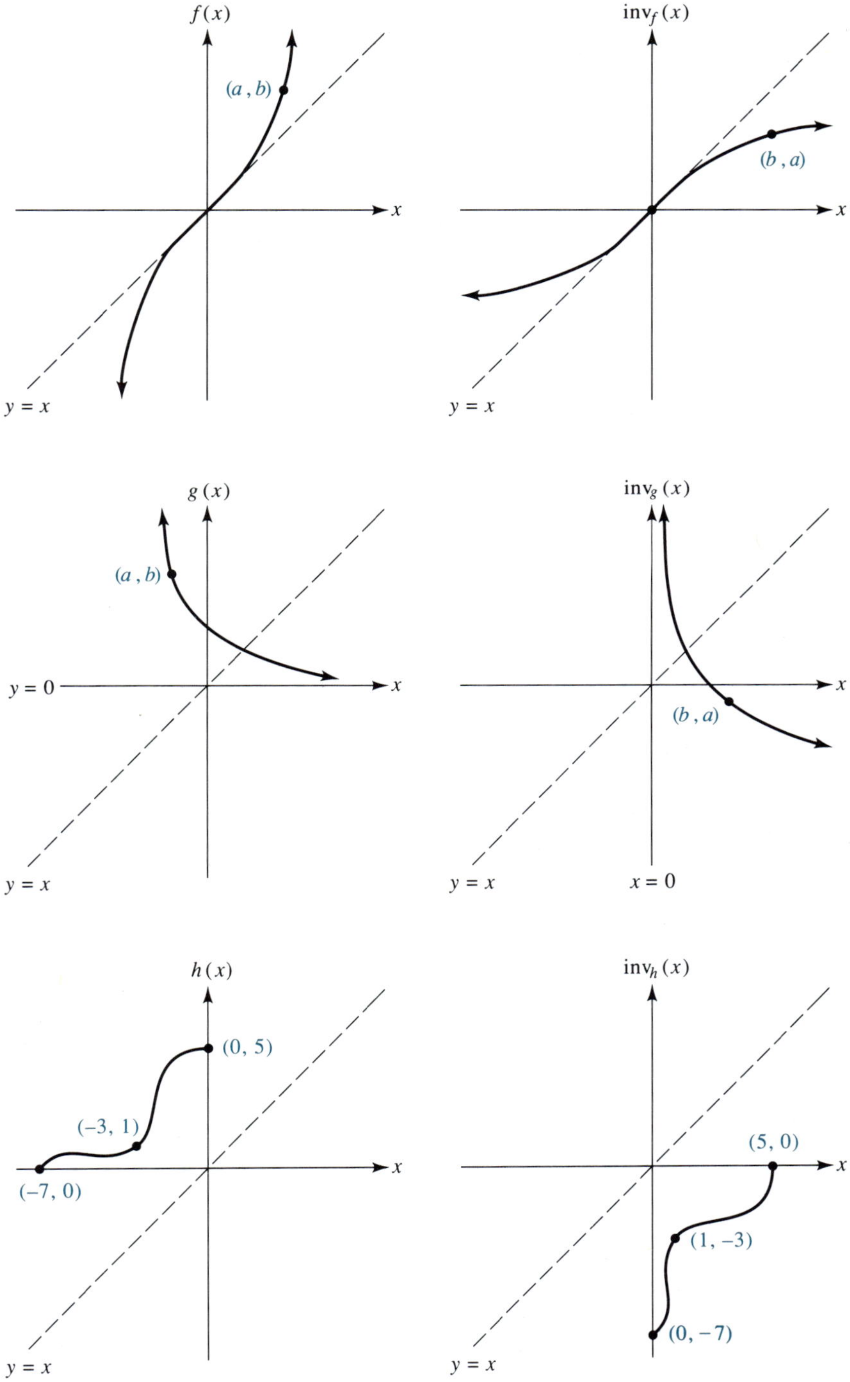

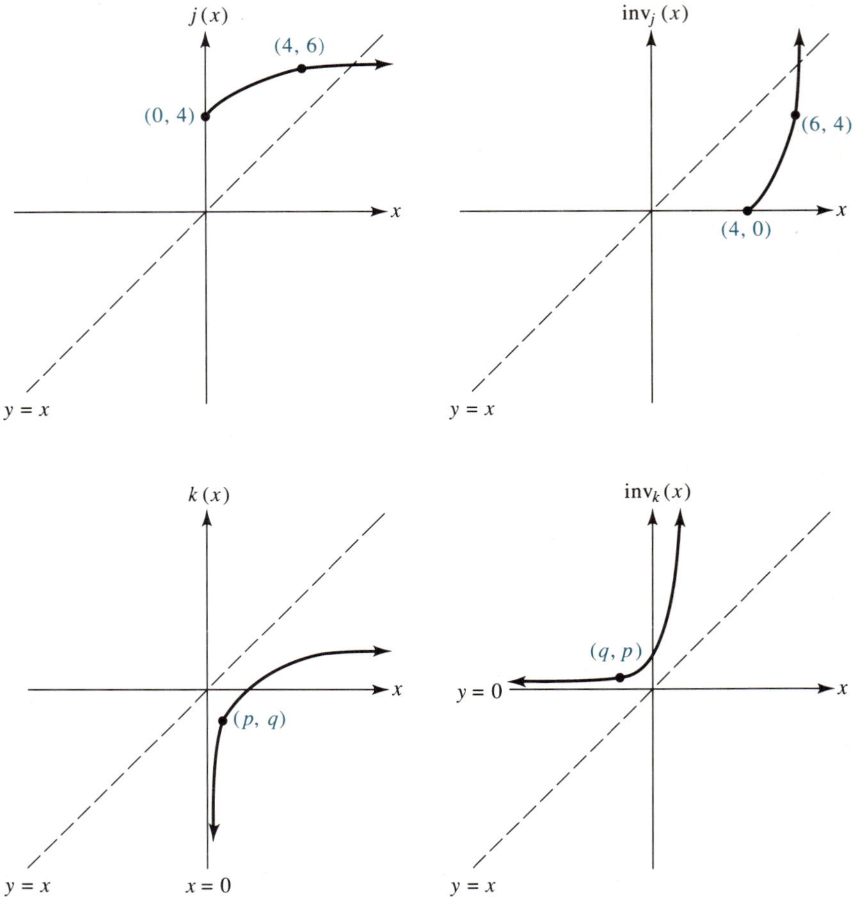

We can draw several conclusions about inv_f from this graphing procedure. If f increases, so does inv_f. If f is decreasing, inv_f is also. A geometric description of the graph of inv_f is the reflection of the graph of f in the line $y = x$. Notice that the line $y = x$ is the graph of I, the composition identity. □

From our last two examples we see that if f is 1-1, so is inv_f, and so inv_f must have an inverse. Surely $\text{inv}_{\text{inv}_f}$ should be f. Let's check. We see $(f \circ \text{inv}_f)(x) = f(\text{inv}_f(x))$, but if $(x, \text{inv}_f(x))$ belongs to inv_f, $(\text{inv}_f(x), x)$ belongs to f, and so $f(\text{inv}_f(x)) = x$. The functions f and inv_f are a pair of composition inverses. Although composition is not commutative in general, it is nearly so for inverses; $(f \circ \text{inv}_f)(x) = x$ for $x \in \text{dom}_{\text{inv}_f}$ and $(\text{inv}_f \circ f)(x) = x$ for $x \in \text{dom}_f$. If $\text{dom}_{\text{inv}_f} = \text{dom}_f$, then $f \circ \text{inv}_f = \text{inv}_f \circ f$.

Often functions are described by giving a rule and a domain. In this case we first try to visualize the graph or examine the rule to see if the function is 1-1. If this is difficult to decide, we usually proceed optimistically to look for an inverse function and carefully test what we find. The technique when a rule is given is to "unravel" the rule—to try to give the independent variable in terms of the dependent one.

EXAMPLE Find the inverse of $f(x) = 4x - 15$.

We know f is 1-1 because its graph is a nonhorizontal line. We write $y = 4x - 15$ and solve for x, obtaining $(y + 15)/4 = x$. Since our bias is to use x rather than y as the name of the independent variable, we can describe this function as **$\text{inv}_f(x) = (x + 15)/4$.** Remember it's the rule—not the name of the variable—that is important. The following chart summarizes what we just did.

f	$\mathbf{y = 4x - 15}$ or $x = (y + 15)/4$
inv_f	$x = 4y - 15$ or $\mathbf{y = (x + 15)/4}$

Notice that the roles of x and y are reversed. We should check that $(f \circ \text{inv}_f)(x) = x$ and $(\text{inv}_f \circ f)(x) = x$.

$$(f \circ \text{inv}_f)(x) = f\left(\frac{x + 15}{4}\right) = 4\left(\frac{x + 15}{4}\right) - 15 = x, \qquad \mathbb{R} = \text{dom}_{\text{inv}_f}$$

$$(\text{inv}_f \circ f)(x) = \text{inv}_f(4x - 15) = \frac{(4x - 15) + 15}{4} = x, \qquad \mathbb{R} = \text{dom}_{\text{inv}_f}$$

■

EXAMPLE Find the inverse of $g(x) = 12x^2$.

We know g is not 1-1 and therefore will not have an inverse, but it can serve as an example of how we recognize algebraically that there is no inverse function. Writing $y = 12x^2$ and solving for x, we first obtain $y/12 = x^2$. The next step would give us two possible values for x in terms of y, $\sqrt{y/12}$, and $-\sqrt{y/12}$, but this would violate the definition of a function. A step that gives more than one possible value is our algebraic signal that the original function was not 1-1 and does not have an inverse.

Therefore, **g has no inverse.**

Note: By restricting the domain to nonnegative real numbers, the new function would have the same range as g but would be 1-1 and, therefore, have an inverse.

■

We haven't said much about domains in our examples; we've been relying on our old principle of whatever works in the rule is okay. There are cases with inverses where we have to use more care.

EXAMPLE Find the inverse of $h(x) = \sqrt{\dfrac{-x}{3}}$, $[-75, -3]$.

The function h is 1-1 and so has an inverse. To find its rule, we write $y = \sqrt{-x/3}$ and solve for x, obtaining $x = -3y^2$, or $\text{inv}_h(x) = -3x^2$. But the range of h is the interval $[1, 5]$, so we must be careful to define **$\text{inv}_h(x) = -3x^2$ with domain the real numbers 1 through 5.** Again let's summarize with a chart.

	Rule	Domain	Range
h	$y = \sqrt{\dfrac{-x}{3}}$, or $x = -3y^2$	$[-75, -3]$	$[1, 5]$
inv_h	$x = \sqrt{\dfrac{-y}{3}}$, or $\mathbf{y = -3x^2}$	$[1, 5]$	$[-75, -3]$

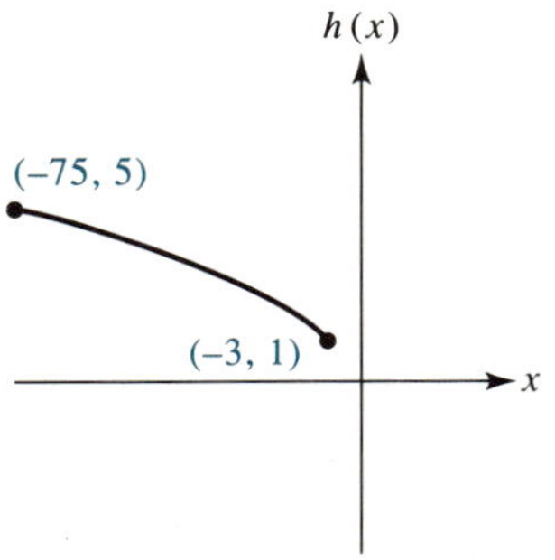

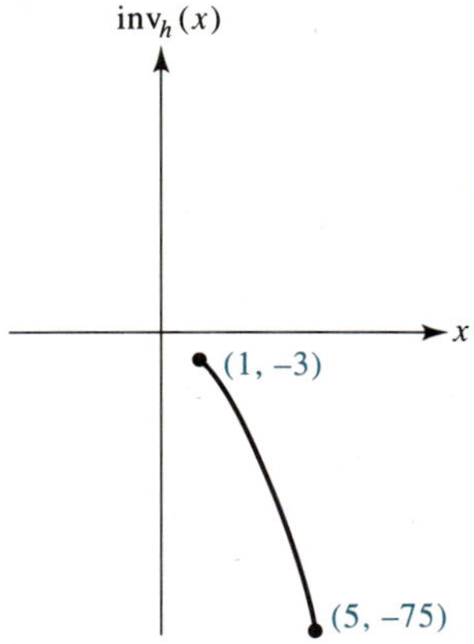

In a sense the inverse function undoes the action of the original function, and we'll often use it for exactly that purpose. Using the inverse of a function to unravel an equation guarantees that we will produce an equivalent equation.

EXAMPLE Solve $x^{1/3} = 4.1$.
Cubing both sides of the equation produces an equivalent equation, $x = (4.1)^3$, or 68.921, because $f(x) = x^3$ and $g(x) = x^{1/3}$ are a pair of inverses.

$$(f \circ g)(x) = f(g(x)) = f(x^{1/3}) = (x^{1/3})^3 = x, \qquad x \in \mathbb{R}, \text{ with } \mathbb{R} \text{ being dom}_g.$$

Therefore, **the solution set is {68.921}.**

EXAMPLE Solve $\sqrt{x} = -1.5$.
The solution set is { }; there is no number whose square root is -1.5. What would happen if we squared both sides of the equation? Squaring both sides of the equation does not produce an equivalent equation.

$$\sqrt{x} = -1.5 \text{ has solution set } \{\ \}$$
$$x = 2.25 \text{ has solution set } \{2.25\}$$

This happens because $f(x) = \sqrt{x}$ and $g(x) = x^2$ are not a pair of inverses. For example, $(f \circ g)(-4) = f(g(-4)) = f((-4)^2) = f(16) = 4$, not -4. The basic problem is that g is not 1-1. Look at the graphs.

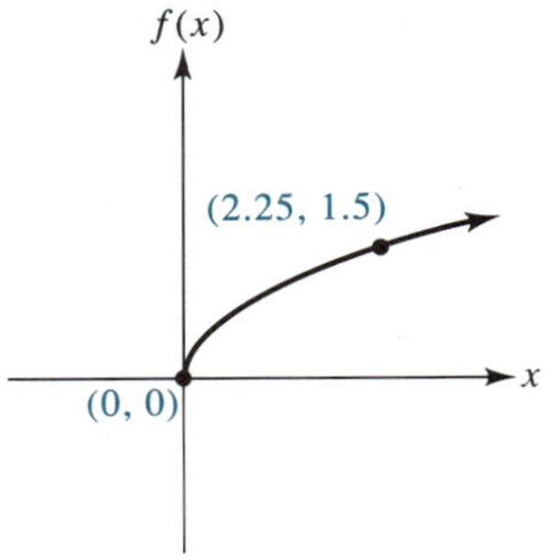

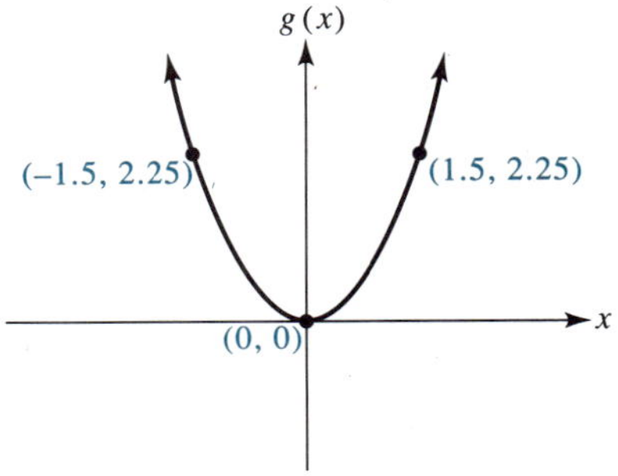

One last comment about inverse functions is needed. The function inv_f is often denoted by f^{-1}. We have $(f^{-1} \circ f)(x) = x$ and $(f \circ f^{-1})(x) = x$ with appropriate domain descriptions. The f^{-1} notation sometimes causes confusion because we use a similar notation to denote reciprocals of real numbers—for instance, $7^{-1} = \frac{1}{7}$. However, we use $1/f$ for reciprocal functions and in general $1/f$ is not the same as f^{-1}. After this section we will use the f^{-1} notation for the function inv_f. On some calculators you may see the inverse function key labeled INV. The key labeled $1/x$ will compute reciprocals of real numbers.

EXERCISES 5.2

In exercises 1 through 10, find $(f \circ g)(x)$ and $(g \circ f)(x)$ and give the domain in interval notation, where possible.

1. $f(x) = \frac{2}{3}x - 4$; $g(x) = \frac{3}{2}x + 6$
2. $f(x) = x^2$; $g(x) = \sqrt{x}$
3. $f(x) = 5.3x^2 - 7.1x + 2.0$; $g(x) = 3 - x$
4. $f(x) = e^x$; $g(x) = x + 5$
5. $f(x) = \dfrac{x + 7}{x - 4}$; $g(x) = 5x + 8$
6. $f(x) = \sqrt[3]{8 - x^2}$; $g(x) = x^{2/3}$
7. $f(x) = \dfrac{x + 2}{3 - x}$; $g(x) = \dfrac{3x - 2}{x + 1}$
8. $f(x) = x + 7$; $g(x) = \text{div}_7(x)$
9. $f(x) = \dfrac{1}{x}$; $g(x) = \dfrac{1}{x}$
10. $f(n) = \text{mod}_5(n)$; $g(x) = [\![x]\!]$

In exercises 11 through 20, each $h(x)$ can be thought of as $(f \circ g)(x)$. Give a pair of possible functions f and g so that $h = (f \circ g)$. This is referred to as *function decomposition*.

11. $h(x) = \sqrt[3]{(x + 1)^2}$
12. $h(x) = (x - 3)^2 + 6$
13. $h(x) = |x^2 - 7x + 12|$
14. $h(x) = \dfrac{2x}{6x + 1}$
15. $h(x) = \sqrt{4 - x^2}$
16. $h(x) = \text{mod}_7(x + 1)$
17. $h(x) = \text{mod}_5(x) + 1$
18. $h(x) = |(x - \pi)^3|$
19. $h(x) = e^{4x^2}$
20. $h(x) = 6^{1.5x - 2}$

Given functions f, g, and h in exercises 21 and 22, determine whether $(f \circ (g \circ h)) = ((f \circ g) \circ h)$.

21. $f(x) = \dfrac{5}{x}$
$g(x) = \sqrt{3 - x}$
$h(x) = (x + 1)^2 - 2$

22. $f(x) = |2.7 - 3.8x|$
$g(x) = 5(x - 2)^3 + 1$
$h(x) = \dfrac{x + 7}{7 - x}$

In exercises 23 and 24, find the inverse function, f^{-1}, for the function f.

23. $f = \{(3, 5), (9, 2), (19, -9)\}$
24. $f = \{(0, 0), (1, 1), (2, 8), (3, 27), (4, 64), \ldots\}$

In exercises 25 through 28, sketch the graph of f^{-1}, given the graph of f.

25.

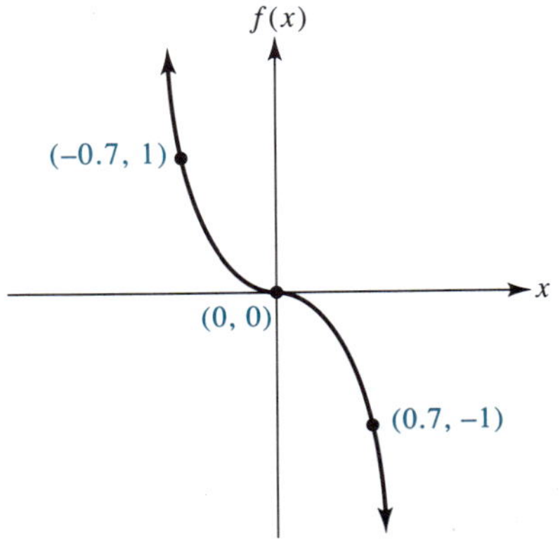

26.

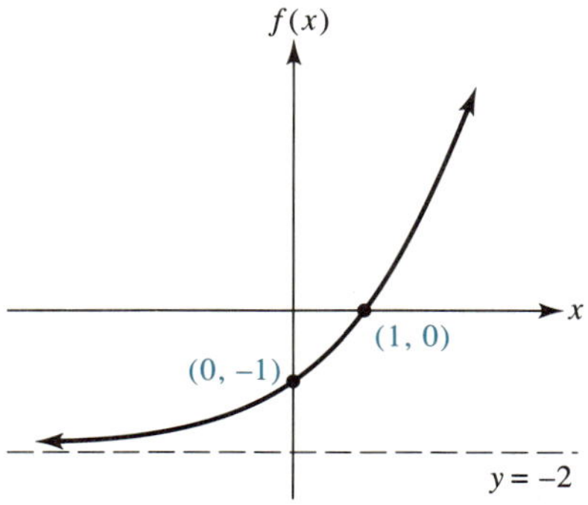

27.

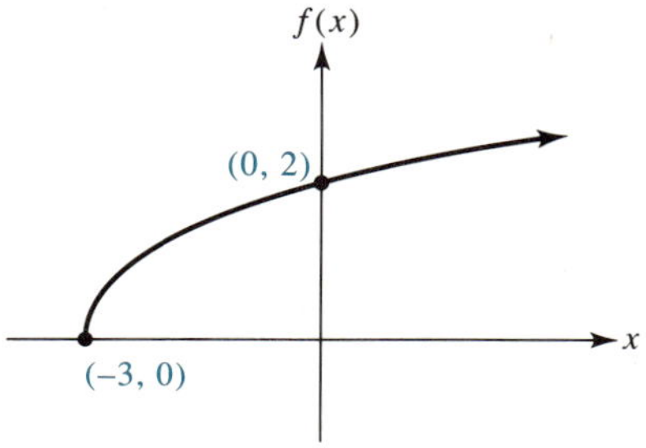

28.

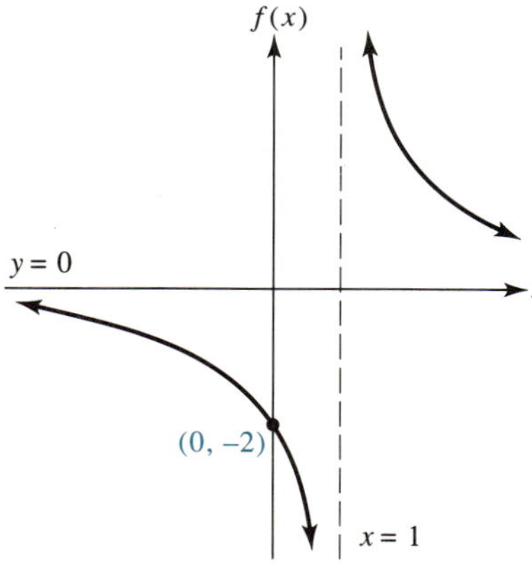

In exercises 29 through 38, find f^{-1} given f.

29. $f(x) = 3x + 5$

30. $f(x) = 6 - 0.5x$

31. $f(x) = x^5$

32. $f(x) = x^4, x \geq 0$

33. $f(x) = \dfrac{2x + 1}{x}$

34. $f(x) = \dfrac{3 - x}{2x + 5}$

35. $f(x) = \sqrt{x - 3}$

36. $f(x) = \sqrt[3]{x - 3}$

37. $f(x) = (x + 5)^2, x \geq -5$

38. $f(x) = (x - 5)^2, x < 5$

39. Prove that function composition is associative.

40. Find a function that is its own inverse.

41. (Graphing calculator) Use a graphing calculator to sketch the graphs of each of the pairs of functions f and g on the same coordinate system. Which of the pairs could not be inverses?

a. $f(x) = \ln(x - 1);\ g(x) = e^{x-1}$

b. $f(x) = \dfrac{4x + 3}{x - 2};\ g(x) = \dfrac{2x + 3}{x - 4}$

c. $f(x) = \sqrt[3]{4 - 2x};\ g(x) = 2 - 0.5x^3$

d. $f(x) = x^2 - 3x + 2;\ g(x) = \sqrt{x + 0.25} - 1.5$

42. Given the graph of h and the graph of H, sketch the graph of $h \circ H$.

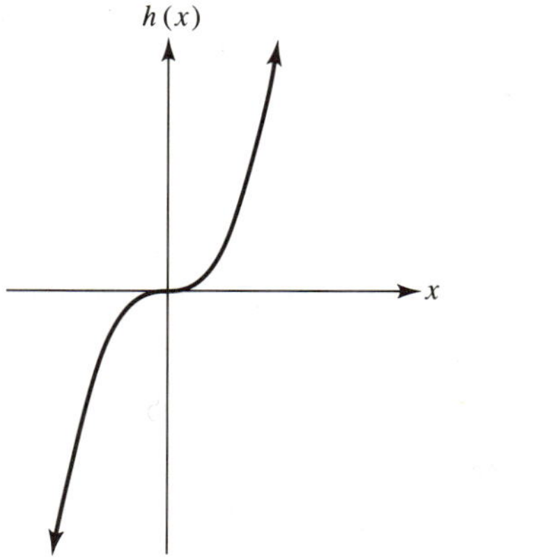

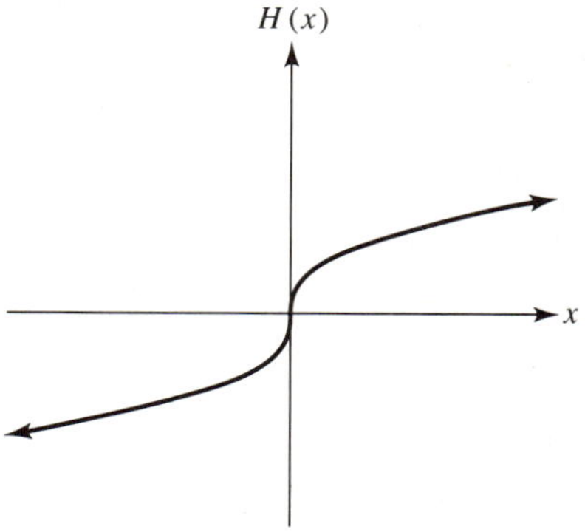

43. Given the graph of f and the graph of g, sketch the graph of $f \circ g$.

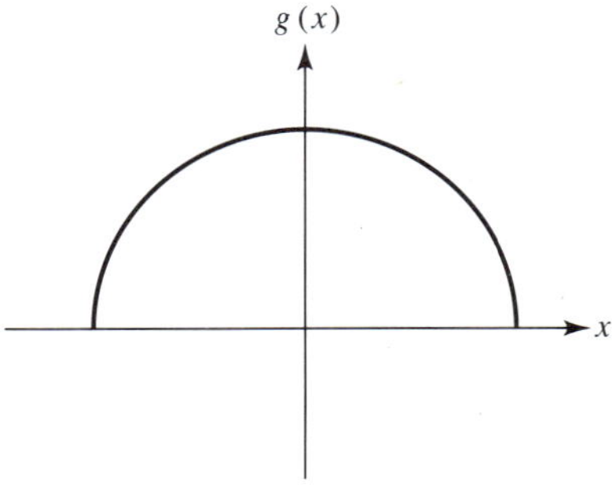

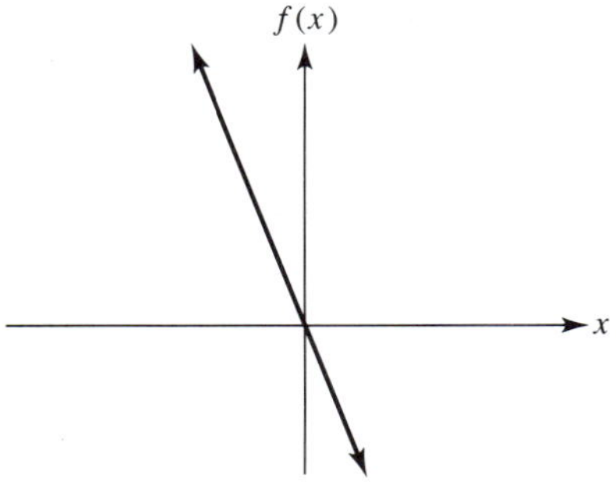

5.3 Exponential and Logarithmic Functions

Toolkit

The standard function $f(x) = e^x$ from Section 4.3 is used here.

We may not be able to carry out the algebraic process described in Section 5.2 for finding the inverse of a function given by a rule, even in a case where we know the function is 1-1.

Illustration Suppose we are interested in the inverse of $f(x) = x^5 + x^3$. We have no method for solving $y = x^5 + x^3$ for x, even though the function is increasing on its domain $\mathbb{R}$ and is therefore 1-1.

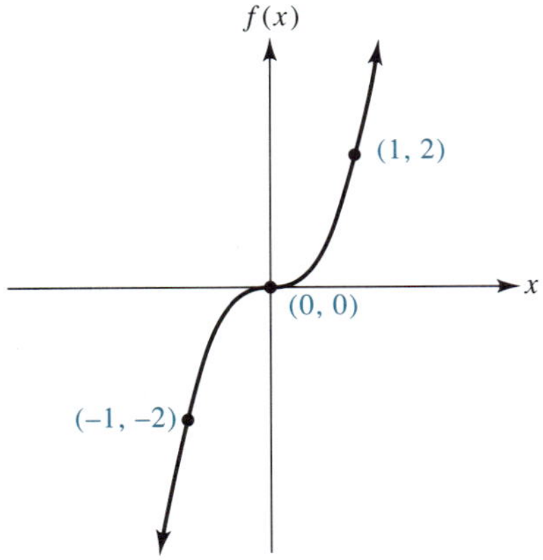

The inverse of f exists, but we have no algebraic method for expressing it. □

As another example we have the standard function $f(x) = e^x$ introduced in Section 4.3. Many natural growth and decay situations can be modeled using variations of this standard function.

Definition Members of the family $g_b(x) = b^x$, $b > 0$, $b \neq 1$, are called **exponential functions.**

Illustration The function $A(t) = Pe^{rt}$ describes the amount of money an investor has after investing a principal amount P at an annual rate r compounded continuously for t years. Thus $A(t) = 1000e^{0.08t}$ describes the amount of money in dollars accumulated at the end of t years when \$1000 is invested at 8% compounded continuously.

The function $N(t) = c2^t$ describes the number of bacteria in a culture at the end of t hours when the original number was c and the population doubles every hour. □

The graph of $f(x) = e^x$ shows that this function is 1-1 and so has an inverse function. We can sketch the graph of f^{-1} and investigate how it behaves. Some facts are apparent from the graph.

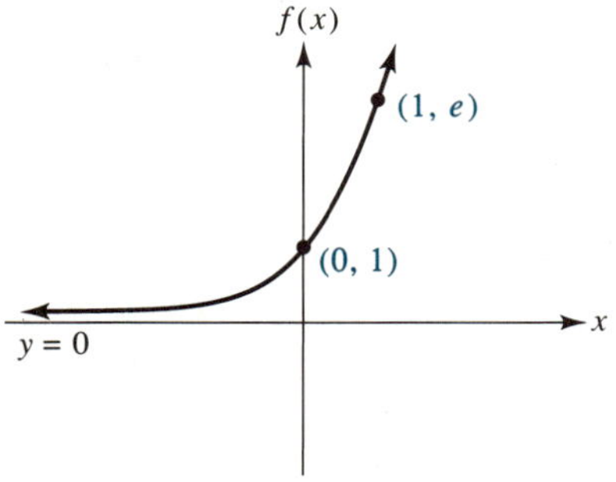

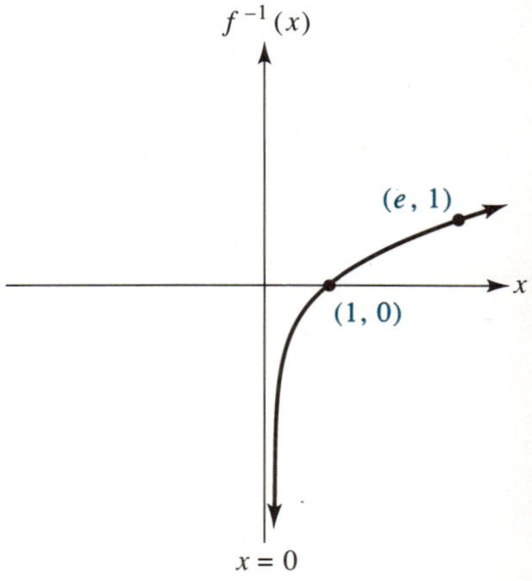

The domain of f^{-1} is the positive reals and the range is the real numbers. Also, f^{-1} is increasing and 1-1, the negative half of the y-axis is its vertical asymptote, and $f^{-1}(1)$ is zero.

We have no means at this point to solve $y = e^x$ for x, so let's see what else we can discover about f^{-1} just by using the fact that it is the inverse function for $f(x) = e^x$. Because $(f^{-1} \circ f)(x) = f^{-1}(f(x)) = f^{-1}(e^x) = x$ for all real numbers, for every number written as a power of e, f^{-1} "picks off" the exponent. Any positive real number has a representation as a power of e. We can see this by examining the graph of e^x; locate any positive real number r on the y-axis, and a horizontal line through that point will intersect the graph in (t, r), with $r = e^t$. We see from this that although $f^{-1}(r)$ is defined for each positive real r, we have to approximate this value unless r is given as a power of e.

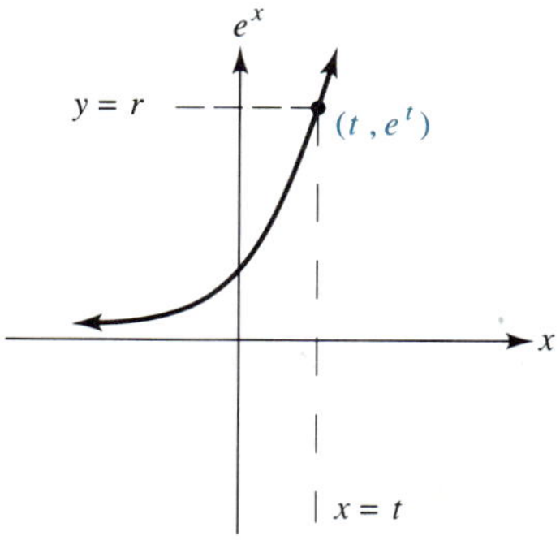

Illustration Using $f(x) = e^x$, we have $f^{-1}(e^5) = 5$, $f^{-1}(e^{7.2}) = 7.2$, and $f^{-1}(e^t) = t$. Since 8 is not an integral power of e, how can we approximate $f^{-1}(8)$? We know $2 < f^{-1}(8) < 2.1$ because $e^2 < 8 < e^{2.1}$. (Test this on your calculator.) We can do a closer approximation if required by considering $e^{2.01}$, $e^{2.02}$, and others, as necessary.

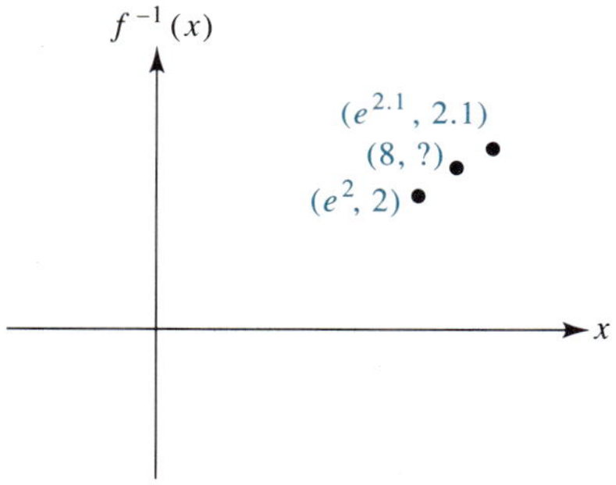

□

Usually we rely on tables or calculators to approximate values of f^{-1}, as we do for $f(x) = e^x$. The inverse of $f(x) = e^x$ is a very important function and has its own special symbol, ln. This is an abbreviation for "natural logarithm."

Standard Graph 10: $g(x) = \ln x$

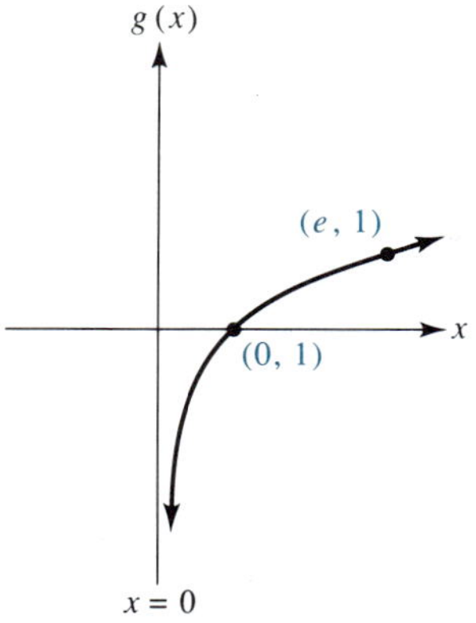

The function ln increases without bound and has the negative half of the y-axis as a vertical asymptote. Its domain is the positive reals (the range of $y = e^x$), and its range is the reals (the domain of $y = e^x$). Suggested labels for this function and its variations are the vertical asymptote, the x-intercept, and one other point for setting the scale.

Illustration The graphs of $f(x) = e^x$ and $f^{-1}(x) = \ln(x)$ are reflections of each other in the line $y = x$.

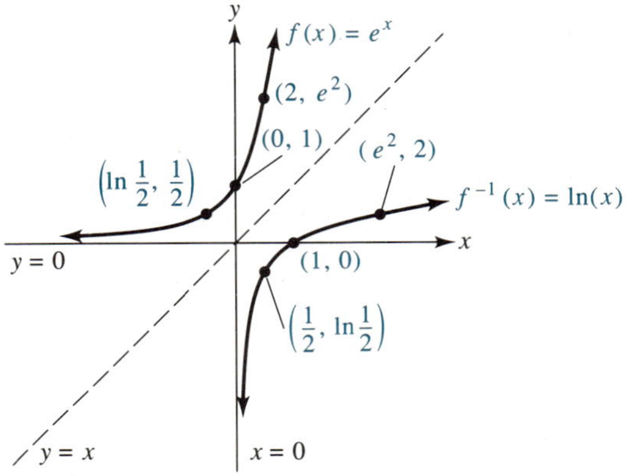

□

To emphasize that $g_b(x) = b^x$, $b > 0$ and $b \neq 1$, is an exponential function, we'll use $\exp_b$ as the function name instead of g_b. We read $\exp_b$ as "exponential base b." Each basic exponential function $\exp_b$, for a fixed $b > 0$ and $b \neq 1$, is 1-1 and has an inverse that—like ln—is called a **logarithmic function.** These inverse functions are named $\log_b$, read "logarithm, or log, base b."

Using this notation we have:

$$\exp_3(4) = 3^4, \text{ or } 81$$

$$\log_3(81) = 4$$

$$\exp_5(\log_5 t) = t,\ t > 0$$

$$\log_{1/2}(\exp_{1/2}12) = 12$$

Remember $\exp_b$ and $\log_b$ are inverse functions.

Sometimes we see mixed notation used.

$$6^{\log_6 8} = 8 \quad \text{since} \quad 6^{\log_6 8} = \exp_6(\log_6 8) = 8$$

$$\log_{1.2}(1.2)^5 = 5 \quad \text{since} \quad 1.2^5 = \exp_{1.2}5$$

In general, $\log_a(x) = y$ is equivalent to $a^y = x$, or $\exp_a(y) = x$. Using this notation, ln would be written $\log_e$. The ln function was named before function notation and the idea of a family of logarithmic functions was fully developed. We also have inherited log for $\log_{10}$; this is called the "common logarithmic function." We are beginning to see $\log_2$ shortened to lg because it is being used more frequently in computer science problems.

The functions $\exp_b$ and $\log_b$ are composition inverses of each other, and we must be careful in the use of that property when other operations or functions are involved. For example, it may not be possible to do anything with $\exp_a(\log_b(4))$ except find approximations if $a \neq b$. However, $\exp_b$ and $\log_b$ do have interesting properties with respect to some other operations. Let's develop these. For any exponential function $\exp_b$, the rules of exponents give us

$$\exp_b(p) \cdot \exp_b(q) = b^p \cdot b^q = b^{p+q} = \exp_b(p + q)$$

$$\frac{\exp_b(p)}{\exp_b(q)} = \frac{b^p}{b^q} = \exp_b(p - q)$$

$$[\exp_b(p)]^n = [b^p]^n = b^{pn} = \exp_b(pn)$$

Conversely, for the inverse function $\log_b$, we have:

If $p = b^k$ and $q = b^m$, $\log_b(pq) = \log_b(b^k \cdot b^m) = \log_b(b^{k+m}) = k + m = \log_b(p) + \log_b(q)$.
Similarly, $\log_b(p/q) = \log_b(p) - \log_b(q)$.
If $p = b^k$, $\log_b(p^n) = \log_b(b^{kn}) = kn = n \log_b p$.

Be careful not to create properties that do not exist. Consider $\log_3(72 + 9) = \log_3 81 = 4$. Is this a simple combination of $\log_3 72$ and $\log_3 9$? We know $3 < \log_3 72 < 4$ because $3^3 < 72 < 3^4$ and $\log_3$ increases because $\exp_3$ does; we also know $\log_3 9 = 2$. There seems to be no simple combination of this 2 and $\log_3 72$ that produces 4. In general, there is no rule for handling $\log_b(c + d)$ except to carry out the addition first.

Logarithms were first developed to handle multiplications, divisions, and powers required for navigation problems. The properties just developed allow us to express the logarithms of terms as sums, differences, and multiples of logarithms.

Given $b, p, q > 0$, $b \neq 1$,

Logarithm of a product:

$$\log_b(pq) = \log_b p + \log_b q$$

Logarithm of a quotient:

$$\log_b\left(\frac{p}{q}\right) = \log_b p - \log_b q$$

Logarithm of a power:

$$\log_b(p^n) = n \log_b p$$

Note: None of the properties applies to the log of a sum or difference. It is also important that you remember these rules apply to logs of *positive numbers*.

Today we're much more likely to evaluate or approximate a term using a calculator, but there are still situations where logarithms and their properties are useful.

EXAMPLE Given $\log_b 4 = 2.31$ and $\log_b 5 = 2.77$, find $\log_b 2$, $\log_b 20$, $\log_b 25$, and $\log_b 320$.

$\log_b 2 = \log_b 4^{1/2} = 0.5 \log_b 4 = 0.5(2.31)$, or **1.155**

$\log_b 20 = \log_b(4 \cdot 5) = \log_b 4 + \log_b 5 = 2.31 + 2.77$, or **5.08**

$\log_b 25 = \log_b 5^2 = 2 \log_b 5 = 2(2.77)$, or **5.54**

$\log_b 320 = \log_b(64 \cdot 5) = \log_b 4^3 + \log_b 5 = 3(2.31) + 2.77$, or **9.70** ■

EXAMPLE Rewrite $\ln \dfrac{5x^2z}{\sqrt{y}}$ using the properties of logarithms.

$$\ln \frac{5x^2z}{\sqrt{y}} = \ln(5x^2z) - \ln(\sqrt{y}) \qquad \text{ln of a quotient}$$

$$\Leftrightarrow \quad \ln \frac{5x^2z}{\sqrt{y}} = \ln 5 + \ln x^2 + \ln z - \ln y^{1/2} \qquad \text{ln of a product}$$

$$\Leftrightarrow \quad \ln \frac{5x^2z}{\sqrt{y}} = \mathbf{\ln 5 + 2 \ln x + \ln z - \frac{1}{2} \ln y} \qquad \text{ln of a power}$$

■

The properties of logarithms may also be used in solving equations. The primary importance of inverse functions is that one can be used to "undo" the other. This is precisely what we want to do in solving most equations.

EXAMPLE In how many years will an investment of \$1000 double in value if the rate of interest is 13% and interest is compounded continuously?

We begin with the formula for continuously compounded interest given at the beginning of this section.

$$A(t) = Pe^{rt}$$
$$2000 = 1000e^{0.13t}$$
$$\Leftrightarrow \quad 2 = e^{0.13t}$$

Now we can produce an equivalent equation by taking the natural logarithm of each side of the equation.

$$\Leftrightarrow \quad \ln 2 = \ln(e^{0.13t})$$
$$\Leftrightarrow \quad \ln 2 = 0.13t \qquad \text{Composition inverses}$$
$$\Leftrightarrow \quad \frac{\ln 2}{0.13} = t$$
$$5.33190 \approx t \qquad \text{We have obtained an approximation for } t.$$

The question was how many years it would take for the investment to double in value. Think about what our approximation for t represents. At some time between 5 and 6 years, the \$2000 will be attained. You might be tempted to choose 5 years by rounding down; that would be incorrect.

In 6 years the investment will have more than doubled in value. ■

EXAMPLE Solve $2^{3x-1} = 6$.

$$2^{3x-1} = 6$$
$$\Leftrightarrow \quad \ln(2^{3x-1}) = \ln 6$$
$$\Leftrightarrow \quad (3x - 1)\ln 2 = \ln 6 \qquad \text{ln of a power}$$
$$\Leftrightarrow \quad 3x - 1 = \frac{\ln 6}{\ln 2}$$
$$\Leftrightarrow \quad 3x = \frac{\ln 6}{\ln 2} + 1$$

$$\Leftrightarrow \qquad x = \frac{(\ln 6)/(\ln 2) + 1}{3}$$

$$\Leftrightarrow \qquad x = \frac{(\ln 6)/(\ln 2) + 1}{3} \cdot \frac{\ln 2}{\ln 2}$$

$$\Leftrightarrow \qquad x = \frac{\ln 6 + \ln 2}{3 \ln 2}$$

The solution set is $\left\{\dfrac{\ln 6 + \ln 2}{3 \ln 2}\right\}$. ■

EXAMPLE Solve $\log_4 x^3 - \log_4 2x = 1$.

Notice there is an implicit restriction on x: $x > 0$.

$$\log_4 x^3 - \log_4 2x = 1$$

$$\Leftrightarrow \qquad \log_4 \frac{x^3}{2x} = 1, \quad x > 0$$

Second property of logs; the restriction no longer is implied.

$$\Leftrightarrow \qquad \log_4 \frac{x^2}{2} = 1, \quad x > 0$$

$$\Leftrightarrow \qquad \exp_4\left(\log_4 \frac{x^2}{2}\right) = \exp_4 1, \quad x > 0$$

Inverse functions

$$\Leftrightarrow \qquad \frac{x^2}{2} = 4^1, \quad x > 0$$

$$\Leftrightarrow \qquad x^2 = 8, \quad x > 0$$

The solution set is $\{\sqrt{8}\}$. ■

There is one other property of logarithmic functions that is occasionally useful. Any problem involving logarithms can be handled using natural logarithms, but some formulas are written using log, and computer scientists frequently use $\log_2$. It is possible to change from an expression using one base to one using another base.

How is $\log_a(x)$ expressed using some other base b? We know $\log_a(x) = y$ is equivalent to $a^y = x$, so

$$\log_b(a^y) = \log_b(x)$$

$$\Leftrightarrow \qquad y \log_b(a) = \log_b(x)$$

$$\Leftrightarrow \qquad y = \frac{\log_b(x)}{\log_b(a)}, \quad \text{or} \quad \log_a(x) = \frac{\log_b(x)}{\log_b(a)}$$

Thus if we had a table or calculator button for $\log_b$, we could approximate values for $\log_a$.

EXAMPLE Approximate $\log_2 1000$ to four decimal places.

We have no $\log_2$ table or calculator key available, so we change to a base we can use.

$$\log_2 1000 = \frac{\ln 1000}{\ln 2} \approx \frac{6.90776}{0.69315} \approx 9.9658$$

The approximate value of $\log_2 1000$ is 9.9658, correct to four decimal places.

EXERCISES 5.3

In exercises 1 through 8, sketch the graph of each of the given functions.

1. $f(x) = 2^x$
2. $g(x) = (\frac{1}{3})^x$
3. $h(x) = 5^{-x}$
4. $j(x) = 4.1^x$
5. $F(x) = \log_2(x)$
6. $G(x) = \log_{1/3}(x)$
7. $H(x) = -\log_5(x)$
8. $J(x) = \log_{4.1}(x)$

In exercises 9 through 20, use the concepts of translation, expansion, and reflection to sketch the graph of each of the following functions.

9. $j(x) = 3^{x-2}$
10. $k(x) = -2^{x+5}$
11. $m(x) = 3^x - 2$
12. $n(x) = -2^x + 5$
13. $p(x) = 5^{2-x}$
14. $q(x) = \log_3(x - 2)$
15. $r(x) = \log_3(x) - 2$
16. $s(x) = \log_4(2x)$
17. $t(x) = 2\log_4(x)$
18. $u(x) = \log_4(3 - x)$
19. $v(x) = -\log(x + 1)$
20. $w(x) = -\log(x) + 1$

In exercises 21 and 22, determine whether the graphs of each pair of functions are the same. Explain why they are the same or why they are different.

21. $f(x) = 2\log_5(x)$; $F(x) = \log_5(x^2)$
22. $g(x) = 3\log_5(x)$; $G(x) = \log_5(x^3)$

In exercises 23 through 26, find the (composition) inverse for each of the functions given.

23. $f(x) = 7x$
24. $g(x) = b^{-x}$, $b > 1$
25. $h(x) = \log_{0.5}(x)$
26. $j(x) = \log_2(x - 5)$

In exercises 27 through 36, given $\log_b 2 \approx 0.3869$, $\log_b 3 \approx 0.6131$, and $\log_b 5 \approx 0.8982$, approximate each of the following correct to four decimal places.

27. $\log_b(4)$
28. $\log_b(30)$
29. $\log_b(\frac{9}{5})$
30. $\log_b(\frac{48}{25})$
31. $\log_5(2)$
32. $\dfrac{\log_b(2)}{\log_b(3)}$
33. $\log_b(10)$
34. $\log_b(\frac{1}{9})$
35. $\log_b\sqrt{6}$
36. $\log_b\sqrt{2.5}$

In exercises 37 through 40, rewrite each of the following expressions in terms of $\ln(x)$, $\ln(y)$, $\ln(z)$, and constants.

37. $\ln(6x^2\sqrt[3]{yz^2})$
38. $\ln\left(4\dfrac{x^3y}{z^{1/2}}\right)$
39. $\ln\left(\dfrac{\sqrt[3]{7x^2z^5}}{2\sqrt{y}}\right)$
40. $\ln\left(\dfrac{125y^3\sqrt[5]{3z}}{7x}\right)$

In exercises 41 through 44, rewrite each of the following expressions as the natural logarithm of a single term.

41. $\ln(x) - \ln(y) + \ln(z)$
42. $3\ln(x) + 0.5\ln(z)$
43. $\frac{1}{3}[\ln(7) + 2\ln(x) + 5\ln(z)] - \ln(2) - 0.5\ln(y)$
44. $3\ln(x) + \ln(y) - 0.5\ln(z) + \ln(4)$
45. Susan visits the local bank to gather information about investing her extra money. They inform her she can invest at 7.5% compounded monthly or at 8% compounded semiannually. Using the compound interest formula, $A(n) = P(1 + r/m)^{mn}$ (where $A(n)$ is the amount of the investment at the end of n years, P is the principal invested, r is the rate of interest as a decimal, and m is the number of compounding periods per year), she chooses to invest so that the return will be greatest after 3 years. Which plan did she choose?
46. The amount of radioactive substance present in a nonliving object decreases according to the formula $A(t) = A_0e^{-kt}$, where $A(t)$ represents the amount present at time t years, A_0 is $A(0)$, and k is the positive constant for the radioactive material measured in the object. If the original amount of radioactive substance is 24 grams and the constant for that substance is 0.00349, approximately how long will it take (correct to the nearest $\frac{1}{4}$year) for the amount of radioactive material to decrease to 12 grams? This is called the *half-life* of the radioactive material.
47. An advertisement for flea spray boasts that it will kill $N(t)$ fleas in time t minutes according to the formula $N(t) = 1.1^{0.05t}$. You have 4 hours before eight dinner guests will descend upon your 12-foot by 12-foot dining room, which now is crawling with approximately 2 fleas per square inch. Should you spend \$20 for the spray and plan to go ahead with the party, spray but call the guests and tell them to come later, or move the party to your neighbors? If you call, how much later should you tell the guests to come?
48. How long will it take to earn \$545 if you invest \$2998.99 at 10% interest compounded quarterly? (See exercise 45.)

49. Your lab experiment in chemistry this week requires that you determine whether a mystery solution is acidic or basic. Remembering that pH measures this characteristic, you look up the formula for computing the pH of a solution and find $\text{pH} = -\log[H^+]$, where $[H^+]$ is the hydrogen ion concentration in moles per liter. You run the experiment to compute $[H^+]$ three times, obtaining 0.00428, 0.00399, and 0.00403 moles per liter. What will you report as the pH of the mystery solution? If a solution is considered acidic if $\text{pH} < 7$ and basic if $\text{pH} > 7$, what can you predict about the solution if you know its $[H^+]$?

50. In 6 years a couple plans to retire and purchase a lake cottage. They estimate that they will need \$20,000 for a down payment. How much should they invest now at 8.25% compounded continuously?

51. On Saturdays a shopping center of x thousand square feet needs p parking places, where $\ln(p)$ is approximately $1.261 \ln(x) - 0.365$.

a. How many parking places should a 56,000-square-foot shopping center have?

b. If you need to answer questions like part (a) often, it would be convenient to have the relationship expressed as $p \approx f(x)$. Rewrite it in this form.

52. The number of parking places, p, needed for a sports club depends on x, the number of square feet occupied by the club. The relationship is approximately $\ln(p) = 0.72 \ln(x) + 2.35$. Restate this relationship in the form $p = f(x)$.

53. In a general-purpose office park, p parking places are needed each weekday for x thousand square feet of office space and $\ln(p) \approx 0.93 \ln(x) + 1.253$. If a nearby office park has 512 parking places, approximately how many square feet of office space are there? (Round to the nearest thousand.)

In exercises 54 through 65, solve the equation.

54. $7^{2x+1} = 8$

55. $3^{-x} = -1$

56. $2^{x^2-1} = 6$

57. $10^{x^2-3} = 5^x$

58. $\log_4(x^3) = \log_4(2x)$

59. $\log_5(3y^2) = \log_5(y) + 1$

60. $\log_{1/4}(2 - x) = 0.5 - \log_{1/4}(x + 1)$

61. $\ln(x^2 - 1) - \ln(x - 1) = 7$

62. $\log_2(\log_2 x^2) = 4$

63. $\log_{1/3}(6x) = \log_3(1.5)$

64. $5^{\log_5(2x)} = 15$

65. $9^{\log_3(x^2)} = 64$

66. An eccentric math teacher wills her nephew Steve $\log_{1.0002}(20{,}000{,}000)$ dollars. Steve would like to purchase a house worth \$2,000,000. Did his aunt leave him enough money? If not, should he settle for a \$20,000 car, a \$20 shirt, or a cup of coffee?

In exercises 67 through 73, identify the larger number, where possible.

67. $\ln(900)$ or $\log_{2.8} 783$

68. $5^{1/8}$ or $2^{1/3}$

69. $\frac{1}{8}\ln(5)$ or $\frac{1}{3}\ln(2)$

70. $\log_3 2^{90}$ or $56^{1.01}$

71. $\ln(15)^{-8}$ or $(-3)^3$

72. $\ln(15)^{-8}$ or $(-3)^{2.7}$

73. 3^{-1} or $\sqrt[9]{2^{-13}}$

74. (Graphing calculator) On the same coordinate system, sketch y = right-hand side and y = left-hand side of the given equation. Use the solving routine of your graphing calculator to approximate the intersection point(s). Give the approximate solution(s) correct to four decimal places.

a. $2^x = e^{1-x}$

b. $\log(x - 1) = \ln(2x - 3.5)$

5.4 New Functions from Conditions

Toolkit

We'll use the techniques from Section 2.2 for writing linear equations. Remember, the standard form for a quadratic equation is $y = a(x - h)^2 + k$, $a \neq 0$, and the general form for a quadratic equation is $y = ax^2 + bx + c$, $a \neq 0$.

Where do functions really appear? Outside of textbooks, functions are defined in attempts to model real-world situations. As we grow to adulthood, our height and weight increase. How are these measurements related to our age? Increasing the number of whistles your factory produces increases the costs; how? As an object falls its velocity increases. Can we make predictions about the velocity of this object, and what about other falling objects?

To develop a function that describes a situation in which we're interested, we begin by gathering information. In this text we will consider functions with one independent variable, so our information will have to be ordered pairs. This will mean using our best judgment about what the independent variable measures or what the most important independent variable is and concentrating on that. A rough graph of the ordered pairs collected may suggest to us one of the standard graphs. It's unlikely that the graph will look exactly like a standard one, so intuition and other knowledge of the situation is useful. There are some sophisticated techniques for developing a function from this stage, but we'll explore only some very simple cases in this text to get the flavor of such problems.

Actually, in Section 2.2 we did this sort of work. If there is reason to believe that the required function is linear, the techniques of Section 2.2 can be applied.

EXAMPLE The manager of a furniture factory has determined that the cost of producing tables depends only on x, the number of tables produced, and that this functional relationship is linear. Suppose the cost of producing 14 tables is \$300, whereas the cost of producing 20 tables is \$414. What is the linear function that describes this situation? What does this function predict for the cost to produce 32 tables?

The slope-intercept form of a linear equation is $y = mx + b$. We can let $C(x)$ represent the cost of producing x tables and use the production information provided to obtain two equations using two variables.

$$300 = m \cdot 14 + b, \quad \text{or} \quad 14m + b = 300$$
$$414 = m \cdot 20 + b, \quad \text{or} \quad 20m + b = 414$$

Solving for b in the first equation and substituting into the second, we obtain

$$20m + (300 - 14m) = 414 \iff 6m = 114, \quad \text{or} \quad m = 19$$

Substituting, we see that b is 34.

Therefore, **the function is $C(x) = 19x + 34$, where x is the number of tables.** Because $C(32) = 19(32) + 34$, **it would cost \$642 to produce 32 tables.** ■

Two ordered pairs supplied enough information for the manager to find a linear function. How many pieces of information (ordered pairs) will we need if we are going to use a quadratic function for our model? Suggested labels for a parabola are the vertex and one other point. If one measurement is known to represent the vertex, then one other pair will be sufficient.

EXAMPLE Give the function whose graph is a parabola with vertex (2, 7) passing through the point (4, 18).

We know the pattern for any variation of the standard parabola is

$$f(x) = a(x - h)^2 + k$$

$$f(x) = a(x - 2)^2 + 7 \qquad (h, k) \text{ represents the vertex.}$$

$$f(4) = a(4 - 2)^2 + 7 = 18 \qquad \text{The point (4, 18) is contained in the graph.}$$

Solving for a, we find a is $\frac{11}{4}$.
Therefore, **the function is $f(x) = \frac{11}{4}(x - 2)^2 + 7$.** ■

If we don't know the vertex, in either standard or general form there are three constants for which to find values, a, h, k or a, b, c. This requires three known points and the solution of a system of three equations. The solution could be obtained using substitution, as discussed in Section 1.1, but there are more efficient methods. We return to this problem when we have these other methods available in Section 9.1.

Illustration The furniture factory manager knows the cost of producing a certain model of chair depends on the number of chairs produced. He believes this is a quadratic relation. Again he collects data and finds the cost to produce 10 chairs is \$600; 30 chairs cost \$300; and 45 chairs cost \$265.50.
The general form of a quadratic equation is $f(x) = ax^2 + bx + c$. Using the three data points, we obtain:

$$\begin{aligned} f(10) &= a(10)^2 + b(10) + c = 600, &\text{or}\quad 100a + 10b + c &= 600 \\ f(30) &= a(30)^2 + b(30) + c = 300, &\text{or}\quad 900a + 30b + c &= 300 \\ f(45) &= a(45)^2 + b(45) + c = 265.50, &\text{or}\quad 2025a + 45b + c &= 265.50 \end{aligned}$$

The solution of this system is $(0.5, -40, 1050)$, so the quadratic function is $f(x) = 0.5x^2 - 40x + 1050$, where x is the number of chairs. □

A large number of natural phenomena can be described using linear, quadratic, or exponential functions. Let's see an example of an exponential function.

EXAMPLE A young married couple wishes to invest money now so that in 6 years they will have \$20,000 for a down payment on a house. The loan officer tells them to invest \$12,191.42 compounded continuously. What was the exponential function she was using?

The formula for continuously compounded interest is $A(t) = Pe^{rt}$. We must determine P and r.

$$\begin{aligned} 12{,}191.42 &= Pe^{r\cdot 0} &&A(0) = 12191.42 \\ 12{,}191.42 &= P &&\text{and } P = A(0). \\ 20{,}000 &= 12{,}191.42e^{6r} &&20{,}000 = A(6) \\ \Leftrightarrow \quad \frac{20{,}000}{12{,}191.42} &= e^{6r} \\ \Leftrightarrow \quad \ln\left(\frac{20{,}000}{12{,}191.42}\right) &= 6r \\ 0.0825 &\approx r \end{aligned}$$

Therefore, **the function was $A(t) \approx 12{,}191.42e^{0.0825t}$.** ■

As an aid in identifying the graph of a log function, special graph paper called log paper is available. Other similar aids exist, but, in general, it may be quite difficult to identify the type of a function from the graph of some data pairs. In addition to these geometric methods we also have algebraic methods. Let's examine one algebraic technique for determining the degree of a polynomial function. It is called the **method of finite differences** because it examines the differences between function values at equally spaced steps.

Consider the linear function $f(x) = 3.4x - 2.1$ and Table 5.4a. We know changing x by one unit changes $f(x)$ by 3.4 units, but uniform changes in $f(x)$ occur for any uniform changes in x.

TABLE 5.4a

x	$f(x)$
0.2	−1.42
0.4	−0.74
0.6	−0.06
0.8	0.62
1.0	1.30

We extend Table 5.4a by calculating the differences between successive values of $f(x)$.

TABLE 5.4a

x	$f(x)$	Differences in function values
0.2	−1.42	
		0.68
0.4	−0.74	
		0.68
0.6	−0.06	
		0.68
0.8	0.62	
		0.68
1.0	1.30	

The fact that these successive differences produce the same number is a characteristic of a linear function. Can you see why that difference is 0.68 here? By the way, this characteristic of linear functions is why our test in Section 4.5 for joining the points on a graph works.

Let's look at a quadratic function $g(x) = 1.5x^2 + 4.1$. We know changing x by one unit won't always change $g(x)$ by 1.5 units (see Table 5.4b).

TABLE 5.4b

x	$g(x)$	Differences in function values
1	5.6	
		4.5
2	10.1	
		7.5
3	17.6	
		10.5
4	28.1	
		13.5
5	41.6	

The entries in column 3 aren't identical, but there is a nice pattern. If we take differences between successive entries in this column, we always obtain 3. This is characteristic of second-degree functions.

If we have a set of data pairs in which values of the independent variable are equally spaced, we can test to see if a polynomial function is a good model for our data as follows. We construct a table of differences by finding differences between entries in the function value column and then between entries in the new column until a constant difference is obtained or we run out of values. See Tables 5.4c and 5.4d.

Illustration

TABLE 5.4c

x	$F(x)$	$F(x_i) - F(x_{i-1})$	Column (1) differences	Column (2) differences
1	6			
		42		
2	48		72	
		114		36
3	162		108	
		222		36
4	384		144	
		366		
5	750			
		(1)	(2)	(3)

TABLE 5.4d

x	$G(x)$	$G(x_i) - G(x_{i-1})$	Column (1) differences	Column (2) differences	Column (3) differences
1	6				
		18			
2	24		0		
		18		12	
3	42		12		−12
		30		0	
4	72		12		
		42			
5	114				
		(1)	(2)	(3)	(4)

□

The number of times we take differences before we obtain a constant difference gives the degree of the polynomial function.

We conclude that the function F in Table 5.4c is determined by a third-degree polynomial. In fact, you can check that we constructed the table using $F(x) = 6x^3$. We cannot draw such a conclusion about the function G given in Table 5.4d because we ran out of values for the subtraction before we found a constant difference.

After we have modeled a situation, the modeling function can be used either to interpolate values—approximate missing values between measured values—or to extrapolate—approximate values beyond measured values. Interpolation approximates the dependent variable for a value of the independent variable between those measured. You used a version of interpolation in Section 4.1. Extrapolation approximates the dependent variable for a value of the independent variable larger or smaller than any measured. Caution should be used in either application but especially in extrapolation, because the behavior of the real-world situation may change abruptly.

EXAMPLE It has been determined that a colony of marmots grows according to the function $P(t) = 1.243\,(2)^t$, where $P(t)$ is the population of the colony at time t days. Time 0 denotes noon yesterday. What is wrong with extrapolating from this to predict the number of marmots in the colony at noon 5 weeks from now?

From noon yesterday to noon 5 weeks from now, (7)(5) or 35 days will elapse, so

$$P(35) = 1.243(2)^{35}$$
$$\Leftrightarrow \quad P(35) \approx 4.2709155(10)^{10}$$

Since the world population is approximately 5 billion (or $5 \cdot 10^9$) people, this number of marmots is hard to visualize.

Just because the population can be predicted for a few days with the function *P*, we cannot expect it to work for extended periods. Too many other factors affect what happens. ■

Our natural language, as opposed to mathematical language, has ways of describing connections between variables that we need to be able to translate into functional descriptions.

We say pressure is proportional to the depth under water, meaning that as the depth increases, so does the pressure. Atmospheric pressure is said to be inversely proportional to the height above sea level; the atmospheric pressure decreases as we climb a mountain. What's missing in these statements is any description of whether these changes are fast or slow.

The mathematical translation of the statement about pressure under water is $p(d) = kd$, where $p(d)$ represents the pressure, d represents the depth, and k is a constant called the "constant of proportionality." The value of k depends on what units are used to measure $p(d)$ and d and the fact that the liquid in question is water. If we say y is *proportional* to x, the mathematical form is $y = kx$. The constant of proportionality must be determined by measurements. This may sound complicated, but all we have is a very simple linear function.

Illustration Every spring exerts a force proportional to the distance it has been stretched. If $F(x)$ represents force and x represents the distance stretched, $F(x) = kx$. Surely a thick, tightly coiled spring will exert more force for a given distance than will a thin, loosely coiled one, so we must have some information about the particular spring we are studying. We send Lisa to gather data, and she reports that 1 pound of force was required to stretch the spring 0.12 inches. Assuming that Lisa's measurements are correct, we conclude that $1 = k(0.12)$, or $k = \frac{25}{3}$, measured in pounds per inch. The functional model for this spring is $F(x) = \frac{25}{3}x$, with $F(x)$ measured in pounds and x measured in inches. □

If we say y is *inversely proportional* to x, the translation is $y = k/x$, with k a constant to be determined. Other English forms for these two cases are y varies as x, or $y = kx$, and y varies inversely as x or $y = k/x$.

Some other common expressions are y varies as the square of x, or $y = kx^2$, y varies inversely as the cube of x, or $y = k/x^3$, and similar combinations of these phrases. Occasionally the precise value of the constant is not important and we may write, for example, $y \propto x^3$ for y varies as the cube of x. One variable may depend on several other variables, and in such a case we might say y varies jointly as x and the square of z, or $y = kxz^2$, or y varies as the square of x and inversely as the square root of w, or $y = kx^2/\sqrt{w}$.

Based on our past experiences, we know that given a few measurements (the number required may vary), we can determine the constant of proportionality. The better the measurements, the better our estimate of k. This seems straightforward to us, but much of the basic scientific work done in past centuries was to determine constants of proportionality. This was in part because the concepts of function and function notation were not well developed. Describing natural phenomena with natural language only is very difficult. The phrases we've discussed here were first efforts at clarifying the situation.

EXERCISES 5.4

In exercises 1 through 4, assume a linear function is a good model for a given situation. Write the linear model for each pair of data points.

1. (0, 3), (−3, 18)
2. (0.2, 5000), (0.5, 6004)
3. (200,000, 18), (300,000, 19)
4. (0, 0), (456.9, 74.298)
5. Sue and Tom have an argument about the way to do exercises 1 through 4. Sue claims that the slope-intercept form of a linear equation is her choice as a model; Tom wants to use the general form $ax + by = c$. Which one is correct and why?

In exercises 6 through 9, assume a quadratic function is a good model for a given situation. Write the quadratic model for each set of data points.

6. (1, 5) and vertex (−2, 8)
7. (2, −4) and vertex (0, 0)
8. (2, 4), (5, 1), (0, 0)
9. (−1, 1), (0, 2), (2, 10)

In exercises 10 and 11, assume the exponential function $f(x) = a2^{kx}$ is a good model for a given situation. Write the function model for each pair of data points.

10. (0, 5), (−1, 1.25)
11. (0, −3), (10, −6)

In exercises 12 and 13, assume the natural logarithmic function $g(x) = a\ln(bx)$ is a good model for a given situation. Write the function model for each pair of data points.

12. (1, 0), (e, 18)
13. (0.5, 0), (e/4, −1)
14. a. It is sometimes possible to write a linear function f, an exponential function g, or a quadratic function h as a model from two data points. Write a model of each of these using the data points (0, 1.1), (2, 1.5). Use the exponential function $g(x) = ae^{bx}$ and write the coefficients obtained correct to three decimal places. For the quadratic model, use (0, 1.1) as the vertex point.
    b. Interpolate to approximate the function values for 1.3 and 1.4 from each of these models.
    c. Extrapolate to approximate the function values for 10 and 100 from each of these models.

In exercises 15 through 20, compute the constant of proportionality and write the function model for each.

15. y varies directly as x. When y is 204, x is −34.
16. $f(x)$ varies directly as x, and $f(-\pi) = 1.4$.
17. z varies directly as x and inversely as y. When x is 1 and y is 8, z is 102.
18. $g(x, y)$ varies directly as x and inversely as y. A data point is (34, −4, 16).

19. z varies jointly as the square of x and the cube root of y. A data point is (3, 8, 14).

20. $F(x, y)$ varies inversely as the square root of both x and y, and $F(24, 6)$ is -300.

In exercises 21 through 26, write a verbal description in terms of variation for each of the functions given.

21. $y = \dfrac{3}{x}$

22. $H(x) = 14.3x^3$

23. $f(x, y) = \pi\dfrac{\sqrt{x}}{y}$

24. $z = Cx^3y^2$

25. $w = \dfrac{x^2y^4}{z^3}$

26. $g(x, y\ z) = \dfrac{-213}{xyz}$

27. The pressure of an enclosed gas varies directly as the temperature and inversely as the volume. If a pressure of 100 pounds per square foot is exerted by 4 cubic feet of gas at 30 K (Kelvin), approximately what pressure (correct to three decimal places) will 150 cubic inches of gas exert at 50 K?

28. A manufacturer notes that it costs \$1200 to open the door of the plant each day, \$2500 to produce 150 items, and \$3600 to produce 220 items. Assuming the best model is quadratic, construct the system of equations from which the model function could be derived.

29. The amount of radioactive substance present at time t is given by $A(t) = A_0e^{-kt}$. Determine k if the half-life of the substance is 120 years. What is an approximation for k correct to five decimal places? (See exercise 46, Section 5.3.)

30. A spring has been stretched 4 inches beyond its normal length by a weight of 12 kilograms. Approximately what distance correct to four decimal places, will the spring be stretched by a weight of 12 pounds? (*Note:* 1 kilogram $\approx$ 2.2 pounds.)

In exercises 31 through 36, given the graph of the function, write an approximate function model.

31.

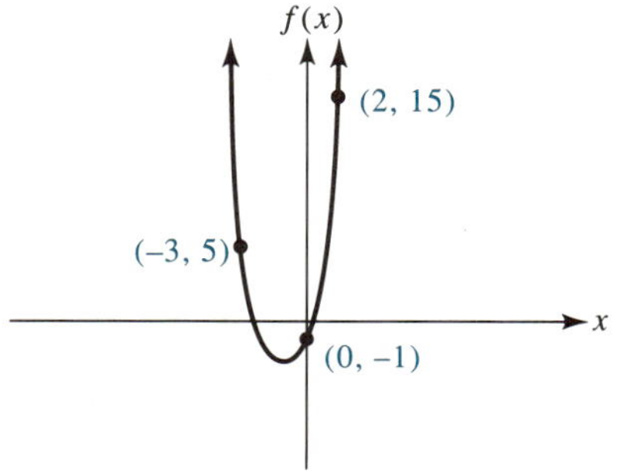

32.

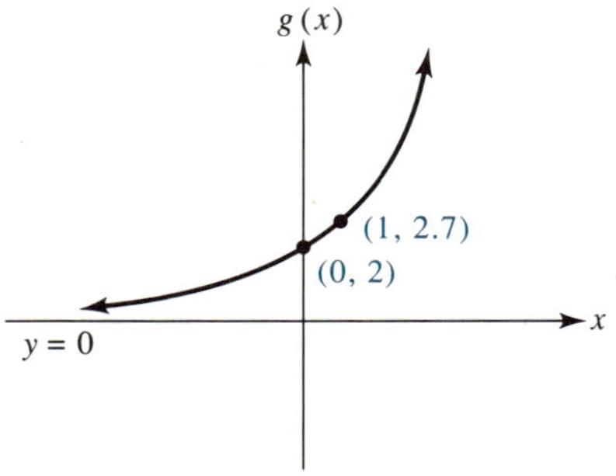

33.

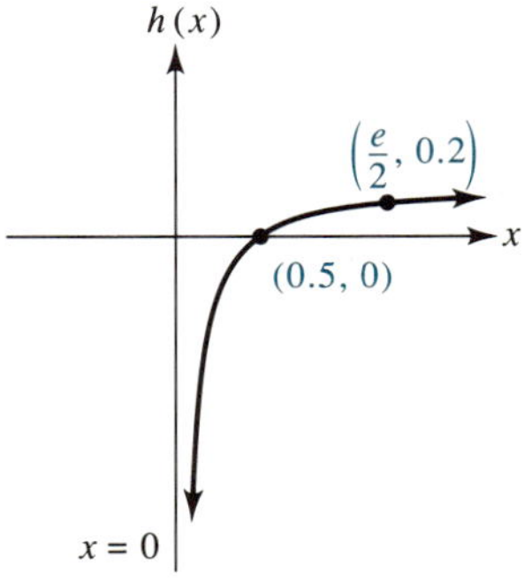

34.

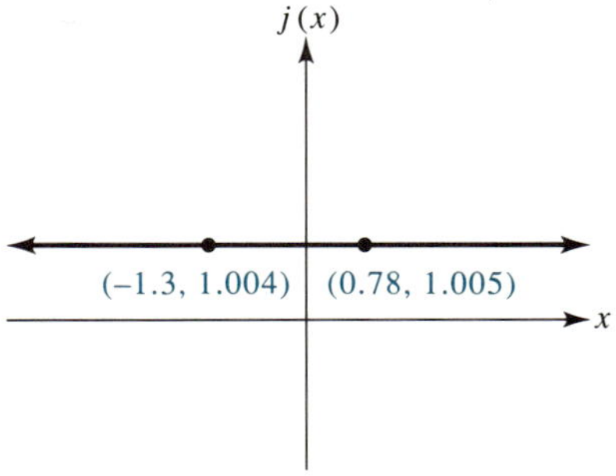

35.

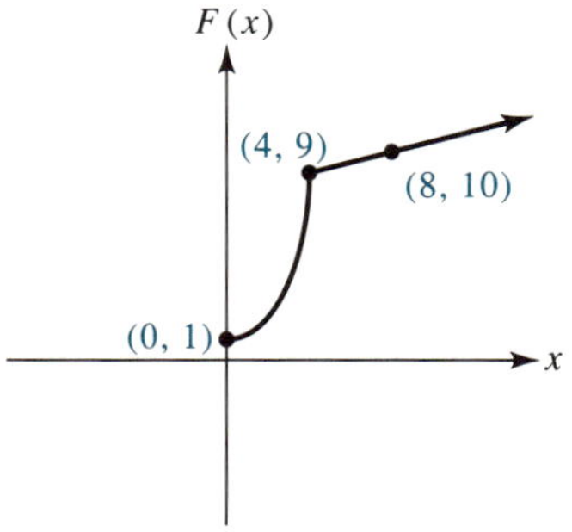

36.

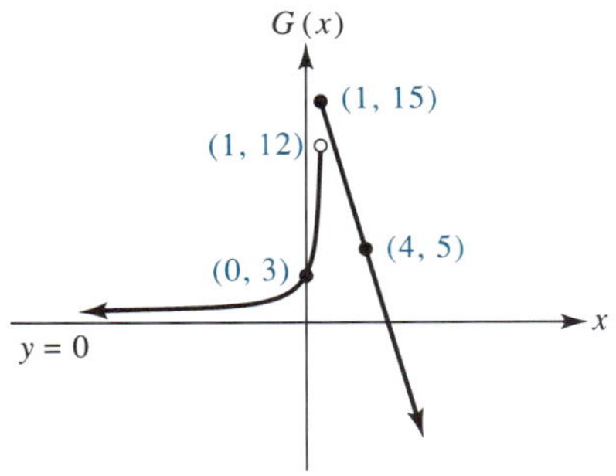

37. Suppose you need to write a model function of the following type: $f(x_1, x_2) = b_0 + b_1x_1 + b_2x_2 + b_{12}x_1x_2 + b_{11}x_1^2 + b_{22}x_2^2$. How many data points and what form of data point would you need?

38. Use the method of finite differences to determine whether the function given in Table 4.1b (page 120) is polynomial. If it is, find the rule for g.

39. Use the method of finite differences to determine whether the function given in Table 4.Sa (page 148) is polynomial. If it is, what is its degree?

40. (Graphing calculator) Use a graphing calculator to plot the graphs of the linear, exponential, and quadratic models you wrote for exercise 14.

41. You've been hired to survey the relationship between the number of seats in a movie theater and how many parking places are filled when a movie is showing. Based on your data, the number of parking places, p, seems to be a linear function of the number of seats, x.

a. Write the function if p is 328 for a 1300-seat theater and 178 for a 1000-seat theater.

b. For what size theaters do you think your function makes sense?

Chapter 5: Summary

The most important concepts developed in this chapter were function composition, inverse functions, and exponential and logarithmic functions. Applications of logarithmic and exponential functions are very important, and some were included here. Several ways to develop mathematical models were shown. Some work with proofs was done; more will be done in later chapters.

Key Concepts

arithmetic of functions	*addition, subtraction, multiplication, division*
algebraic properties of function operations	*commutative, associative, distributive, identity, inverse*
composition	*identity, function, domain*
inverse functions	$(f \circ f^{-1})(x) = x$, *domain*
exponential and logarithmic functions	*graphs, properties, half-life, doubling time*
direct proof	
method of finite differences	
variation	*direct, inverse, joint*

Skills

You should be able to:

1. Form the sum, difference, product, or quotient of given functions and describe the domain of the new function.
2. Do simple direct proofs.

3. Sketch the graph of f^{-1} given the graph of f.
4. Evaluate or approximate a value for a logarithmic or exponential function.
5. Sketch the graph of a logarithmic or exponential function.
6. Apply properties of exponential and logarithmic functions, especially in solving equations.
7. Define a linear, quadratic, exponential, or logarithmic function to satisfy certain conditions.
8. Translate variation statements from English to algebraic forms, and vice versa.
9. Define half-life and doubling time.

SUMMARY EXERCISES

The starred exercises can serve as a practice chapter test. The answers to these exercises are at the end of the answer section at the back of the book.

In exercises 1 through 4, where possible, write $f + g$, f/g, $f \circ g$, $g \circ f$, f^{-1}, and g^{-1} and state the domain for each.

1. $f(x) = \sqrt{2 - 3x}$; $g(x) = x^2 + 2$
2. $f(x) = 2x + 1$; $g(x) = |x|$
* 3. $f(x) = \ln x$; $g(x) = 2x$
4. $f(x) = \text{mod}_4 x$; $g(x) = 3x - 1$
5. If f is an odd function and g is an even function, what can be said about $f + g$ and $f - g$? Give evidence to support your claim.
* 6. If $h(x) = (f \circ g)(x)$ and $h(x) = \sqrt[4]{(2x - 1)^3}$, what are $f(x)$ and $g(x)$? Can you make a different choice for $f(x)$ and $g(x)$?

In exercises 7 through 10, sketch the graph of the given function.

7. $f(x) = 2^{x+5}$
8. $g(x) = \lg x^3$ (*Note:* lg denotes $\log_2$)
* 9. $h(x) = -\ln(x - 4) + 1$
10. $j(x) = -2e^{3x-1}$
11. Given $f(x) = x^n$, $n \in N$ and $g(x) = x^{1/n}$, $n \in N$, find $f \circ g$ and $g \circ f$. What can be said about f and g when n is an even natural number? What can be said about f and g when n is an odd natural number?

In exercises 12 through 14, rewrite each in terms of ln(a), ln(b), and ln(c) if possible.

* 12. $\ln\dfrac{7a^6}{\sqrt{2bc}}$
13. $\ln\sqrt[3]{4c^2(a + 3b)}$
14. $\log_b(ac)$

In exercises 15 and 16, rewrite as the natural logarithm of a single term if possible.

* 15. $\dfrac{1}{2}[3 \ln(x) - \ln(y)]$
16. $-\ln(x) + \ln[\ln(y)] + \ln(3)$

In exercises 17 through 22, solve the given equation.

17. $5^x = -1$
* 18. $\log_3(4x + 2) = 3$
19. $\dfrac{3^x}{3^2} = \pi$
20. $\ln(e^{x-1}) = 14$
* 21. $e^x \cdot e^2 = 3.2$
22. $\lg(x^2) - \ln(2) = \lg(x)$

In exercises 23 through 26, determine which number is larger, if possible. You may need to use logarithms in exercises 24 through 26.

23. $\log_5 1.467$ or $\log_6 1.549$
* 24. 2045^{100} or 1000^{204}
25. $(0.000008)^{1,000,000}$ or $(0.000006)^{1,050,000}$
26. $e^{1,000,000}$ or $\pi^{1,050,000}$
* 27. Write a description of the relationship between x, y, and z using variation if $y = 3x^4/\sqrt[3]{z}$.
28. The resistance of electricity through a wire varies directly as the length of the wire and inversely as the square of the diameter of the wire. Write the function model if a wire 0.4 meters long with a radius of 0.1 centimeters has resistance of 5 ohms.
29. The position of an object $s(t)$ projected from ground level is modeled by a quadratic function. What is the model function if the object is at 0 feet at time 0, 64 feet at 2 seconds, and 28 feet at $\frac{1}{2}$ second?
* 30. How long does it take an initial investment of \$4032 to triple in value if the rate is 10% and it is compounded quarterly? How long would it take under the same circumstances for \$1.34 to triple in value? What do you suspect from these results? Can you verify your suspicion? (The formula is found in exercise 45, Section 5.3.)

Trigonometric Functions

- 6.1 Preliminaries
- 6.2 The Trigonometric Functions
- 6.3 Graphs of the Trigonometric Functions
- 6.4 Graphing Variations of the Trigonometric Functions
- 6.5 The Inverse Trigonometric Functions
- 6.6 Equations Containing Trigonometric Expressions

Toolkit

In this chapter you'll need the Pythagorean theorem: for a right triangle, the sum of the squares of the lengths of the legs equals the square of the length of the hypotenuse.

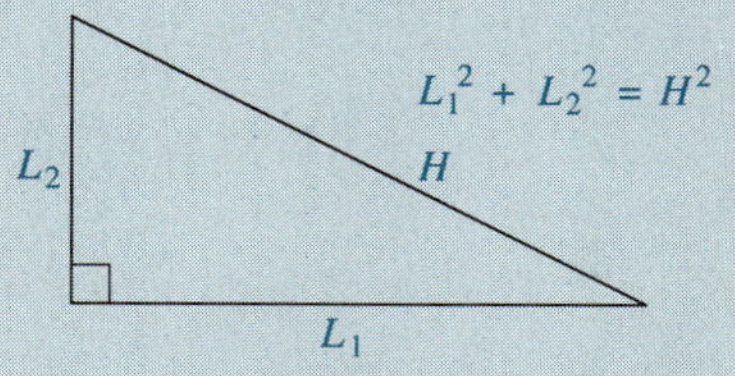

Also, recall that the graph of a circle with center (0, 0) is symmetric with respect to the *x*-axis, the *y*-axis, the line $y = x$, and the origin (Section 4.4). Periodic functions are also defined in Section 4.4.

Most of the functions we've studied so far are algebraic functions: the function values can be computed from a rule using arithmetic and algebraic operations. Other functions, such as logarithmic and exponential ones, do not have algebraic descriptions, and we often approximate function values for them. These functions are called "transcendental." In this chapter we'll develop a very important group of transcendental functions, the trigonometric functions. The trigonometric functions have many important applications. However, in this chapter we define the functions and explore their behavior before we attempt to use them in modeling real-world situations in Chapter 7.

Illustration

$f(x) = 2x^2 - 3x + 5$	Algebraic
$f(-3) = 2(-3)^2 - 3(-3) + 5$, or 32	Exact value
$g(x) = \sqrt{x^2 + 3} - \dfrac{1}{x}$	Algebraic
$g(0.5) = \sqrt{0.5^2 + 3} - \dfrac{1}{0.5} = \sqrt{3.25} - 2$	Exact value
$g(0.5) \approx -0.197$	Approximate value
$h(x) = 2\log_3 x + 5x^2$	Transcendental
$h(4) = 2\log_3(4) + 5 \cdot 4^2$	Exact value
$h(4) \approx 2(1.26186) + 80$, or 82.52372	Approximate value

□

6.1 Preliminaries

If you've dealt with trigonometric functions before, you may have used the restricted definitions given in terms of right triangles. Because we want to use the trig functions for a variety of applications, we begin with more general definitions. The connection between the general and restricted versions will be explained in Section 7.1.

We begin our discussion of the trigonometric functions by defining the wrapping function and using it as the basis of other definitions. The **wrapping function** W assigns to each real number a point. (If you've done the preview problems in Chapters 3 and 4, the following discussion will seem very familiar.) How is this assignment made? The circle with center (0, 0) and radius 1 has as its equation $x^2 + y^2 = 1$ (Section 2.4) and is called the unit circle. To find $W(t)$, we begin at (1, 0) and travel around the circle until we have gone a distance of $|t|$ units. The point at which we stop is $W(t)$. In which direction do we travel from (1, 0)? If t is positive, we travel in a counterclockwise direction; if t is negative, we travel clockwise.

$$W(t) = \begin{cases} \text{the point } t \text{ units from } (1, 0) \text{ measured counterclockwise along the unit circle if } t \geq 0. \\ \text{the point } |t| \text{ units from } (1, 0) \text{ measured clockwise along the unit circle if } t < 0. \end{cases}$$

EXAMPLE Locate $W(1)$ and $W(-1)$.

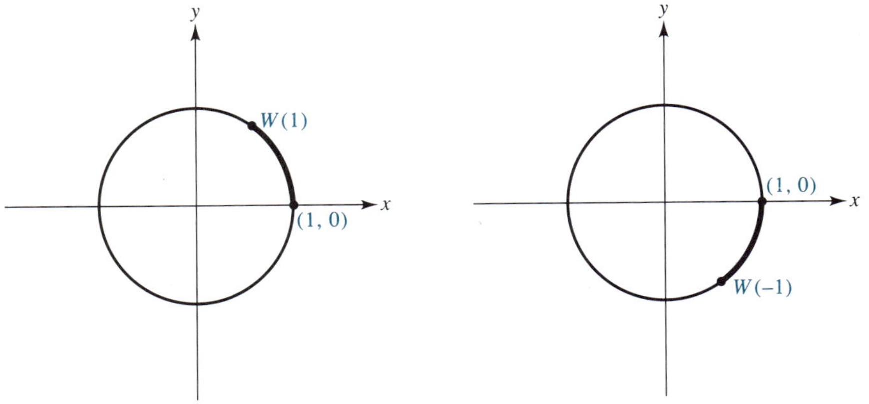

In each case the thicker arc of the circle should be the same distance as that from (0, 0) to (1, 0), which is 1 unit. ■

It is difficult to decide when we have gone exactly t units for most values of t. This means that as with the exponential functions, there are some values of W we can find easily and many for which we use approximations. We will use the symmetries of the circle and some elementary geometry to evaluate $W(t)$ for some simple cases.

The circumference of any circle is 2π times its radius, so the unit circle has a circumference of $2\pi \cdot 1$, or 2π. If we travel $\pi/2$, π, $3\pi/2$ and 2π units counterclockwise on the unit circle from (1, 0), we arrive at the points of intersection of the circle with the axes.

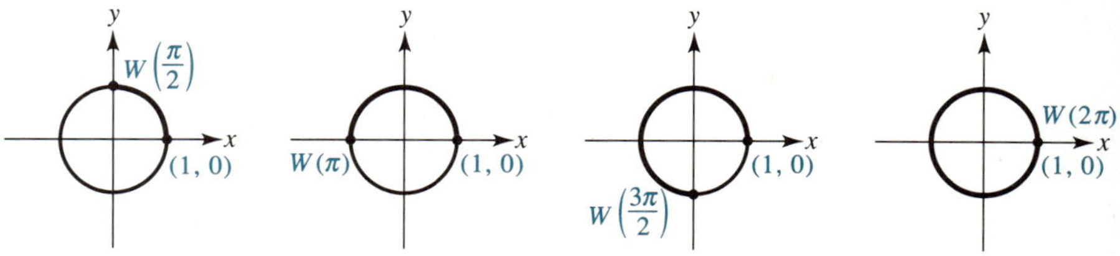

Our sketches show

$$W\left(\frac{\pi}{2}\right) = (0,\ 1), \qquad W(\pi) = (-1,\ 0),$$

$$W\left(\frac{3\pi}{2}\right) = (0,\ -1), \qquad W(2\pi) = (1,\ 0).$$

It's also easy to see that

$$W\left(-\frac{\pi}{2}\right) = (0,\ -1), \qquad W(-\pi) = (-1,\ 0),$$

$$W\left(-\frac{3\pi}{2}\right) = (0,\ 1), \qquad W(-2\pi) = (1,\ 0).$$

You may have noticed too that $W(0) = W(2\pi)$, or (1, 0). No restrictions are placed on t, so we may travel around the circle one or more times before arriving at $W(t)$, but any point on the circle can be reached by moving t units counterclockwise for t between 0 and 2π. This means W is a periodic function with period 2π, the length of a complete circuit.

Definition A number t for which $W(t)$ lies in the nth quadrant is sometimes called an nth quadrant number.

For example, $3\pi/4$ is a second-quadrant number.

EXAMPLE Evaluate $W(7\pi)$, $W(5\pi/2)$, $W(-28\pi)$, $W(-7\pi/2)$, and $W(155\pi/2)$.

$W(7\pi) = W(6\pi + \pi) = W(\pi) =$ **(−1, 0)**, since 6π represents 3 complete circuits counterclockwise.
$W(5\pi/2) = W(2\pi + \pi/2) = W(\pi/2) =$ **(0, 1).**
$W(-28\pi) = W(0) =$ **(1, 0)**, since -28π represents 14 complete circuits clockwise.
$W(-7\pi/2) = W(-2\pi - 3\pi/2) = W(-3\pi/2) =$ **(0, 1).**
$W(155\pi/2) = W(76\pi + 3\pi/2) =$ **(0, −1).**

Note: Since these points are on the x- or y-axis, the values that produce them are not referred to as quadrant numbers. ■

What other values of $W(t)$ can we find easily? $W(\pi/4)$ should be halfway from $W(0)$ to $W(\pi/2)$. This means $W(\pi/4)$ is on the line $y = x$ as well as on $x^2 + y^2 = 1$. Using (a, b) to represent $W(\pi/4)$, we can say $b = a$ and $a^2 + b^2 = 1$. Substitution gives us $a^2 + a^2 = 1$, or $a^2 = \frac{1}{2}$. The number a must be positive, since $W(\pi/4)$ is in the first quadrant, so $a = \sqrt{\frac{1}{2}}$, $b = \sqrt{\frac{1}{2}}$ and $W(\pi/4) = \left(\sqrt{\frac{1}{2}}, \sqrt{\frac{1}{2}}\right)$, or $(\sqrt{2}/2), \sqrt{2}/2)$.

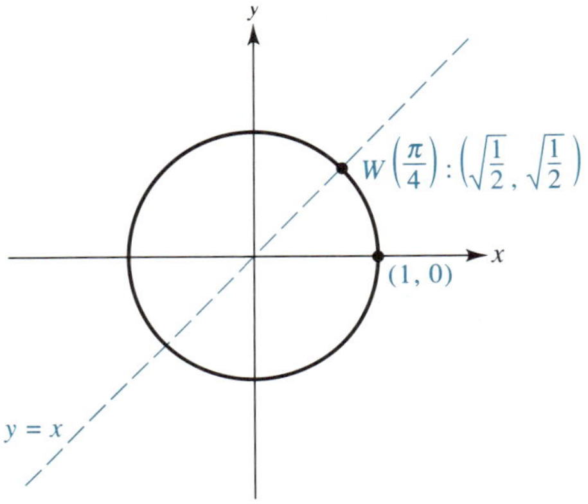

The point $W(3\pi/4)$ is halfway between $W(\pi/2)$ and $W(\pi)$, and a similar argument shows $W(3\pi/4) = \left(-\sqrt{\frac{1}{2}}, \sqrt{\frac{1}{2}}\right)$. Or, we could use the symmetry of $x^2 + y^2 = 1$ with respect to the y-axis to show that $W(3\pi/4)$ is $\left(-\sqrt{\frac{1}{2}}, \sqrt{\frac{1}{2}}\right)$. In fact the symmetries of the unit circle also yield $W(5\pi/4) = \left(-\sqrt{\frac{1}{2}}, -\sqrt{\frac{1}{2}}\right)$ and $W(7\pi/4) = \left(\sqrt{\frac{1}{2}}, -\sqrt{\frac{1}{2}}\right)$. More generally, these symmetries and the periodic nature of W allow us to concentrate on values of t such that $0 \leq t \leq \pi/2$.

EXAMPLE Evaluate $W(7\pi/4)$, $W(-3\pi/4)$, $W(17\pi/4)$, $W(-17\pi/4)$, and $W(28\pi/4)$.

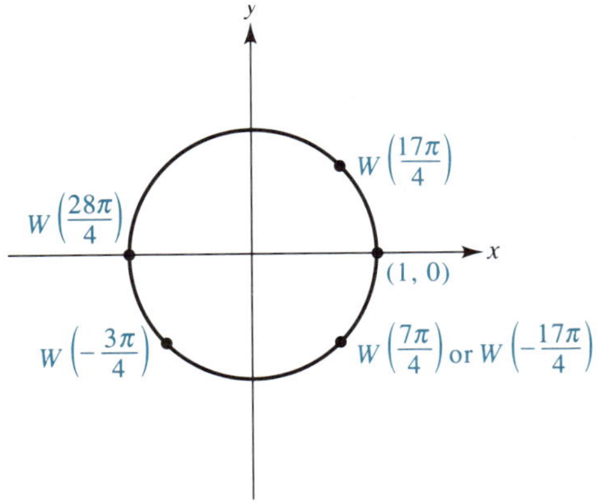

$W(7\pi/4) = \left(\sqrt{\frac{1}{2}}, -\sqrt{\frac{1}{2}}\right)$, since $\frac{7\pi}{4}$ is a fourth quadrant number.
$W(-3\pi/4) = \left(-\sqrt{\frac{1}{2}}, -\sqrt{\frac{1}{2}}\right)$, since $\frac{-3\pi}{4}$ is a third quadrant number.
$W(17\pi/4) = W(16\pi/4 + \pi/4) = W(\pi/4) = \left(\sqrt{\frac{1}{2}}, \sqrt{\frac{1}{2}}\right)$, since $16\pi/4$ represents 2 complete circuits counterclockwise.

$W(-17\pi/4) = W(-16\pi/4 - \pi/4) = \left(\sqrt{\tfrac{1}{2}}, -\sqrt{\tfrac{1}{2}}\right)$, since $-16\pi/4$, or -4π, represents 2 complete circuits clockwise.

$W(28\pi/4) = W(7\pi) = W(6\pi + \pi) = (-1, 0)$. ■

We use Figure 6.1a to calculate $W(\pi/3)$. Since $\pi/3 < \pi/2 < 2\pi/3$, we see $W(\pi/3)$ lies in quadrant I, and $W(2\pi/3)$ lies in quadrant II. These two points must divide the upper semicircle of length π into three equal pieces, from (1, 0)

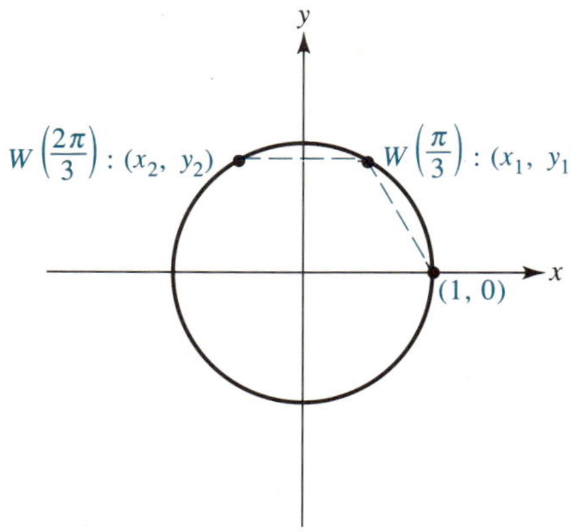

FIGURE 6.1a

to $W(\pi/3)$, $W(\pi/3)$ to $W(2\pi/3)$, and $W(2\pi/3)$ to $(-1, 0)$. This means that the line segments from (1, 0) to (x_1, y_1), or $W(\pi/3)$, and from (x_1, y_1) to (x_2, y_2), or $W(2\pi/3)$, are also equal. (Our intuition about this can be confirmed by checking this property in a high school geometry text.) Symmetry with respect to the y-axis means $x_2 = -x_1$, so the line segment from (x_2, y_2) to (x_1, y_1) is $2x_1$ units long. In Figure 6.1b we have a right triangle whose hypotenuse is of length $2x_1$, since it is the same length as the segment from $W(\pi/3)$ to $W(2\pi/3)$. From $(x_1, 0)$ to (1, 0) is $1 - x_1$ units; from $(x_1, 0)$ to (x_1, y_1) is y_1 units. The Pythagorean theorem assures us that

$$
\begin{aligned}
& (2x_1)^2 = (1 - x_1)^2 + y_1^2 \\
\Leftrightarrow \quad & 4x_1^2 = 1 - 2x_1 + x_1^2 + y_1^2 \\
\Leftrightarrow \quad & 4x_1^2 = 1 - 2x_1 + 1 \qquad \text{Because } x_1^2 + y_1^2 = 1 \\
\Leftrightarrow \quad & 0 = 2 - 2x_1 - 4x_1^2 \\
\Leftrightarrow \quad & 0 = 1 - x_1 - 2x_1^2 \\
\Leftrightarrow \quad & 0 = (1 + x_1)(1 - 2x_1)
\end{aligned}
$$

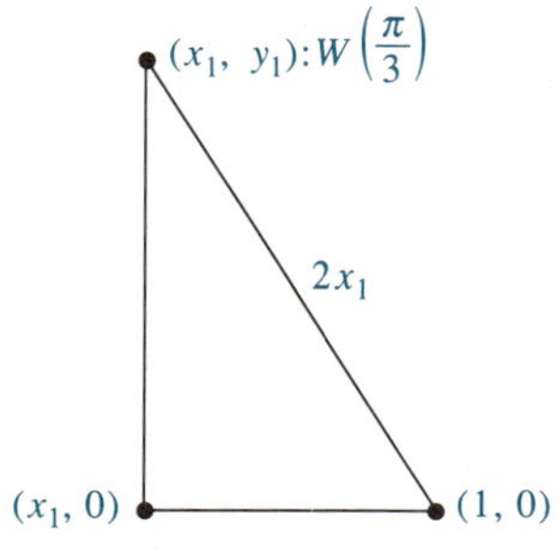

FIGURE 6.1b

and either $1 + x_1 = 0$ or $1 - 2x_1 = 0$. The point (x_1, y_1) is in quadrant I, so both x_1 and y_1 are positive, and $\frac{1}{2}$ is the correct value for x_1. To find y_1, we again use the fact that $x_1^2 + y_1^2 = 1$.

$$\left(\frac{1}{2}\right)^2 + y_1^2 = 1$$

$$\Leftrightarrow \qquad \frac{1}{4} + y_1^2 = 1$$

$$\Leftrightarrow \qquad y_1^2 = \frac{3}{4}$$

so

$$y_1 = \sqrt{\frac{3}{4}} \qquad \text{Because } y_1 > 0,\ y_1 \text{ is } \sqrt{\tfrac{3}{4}}.$$

Now we can say $W(\pi/3) = (x_1, y_1) = (\frac{1}{2}, \sqrt{\frac{3}{4}})$, or $(\frac{1}{2}, \sqrt{3}/2)$. Immediately, we also have $W(2\pi/3) = (-\frac{1}{2}, \sqrt{3}/2)$, $W(4\pi/3) = (-\frac{1}{2}, -\sqrt{3}/2)$, and $W(5\pi/3) = (\frac{1}{2}, -\sqrt{3}/2)$ using the symmetries of the circle.

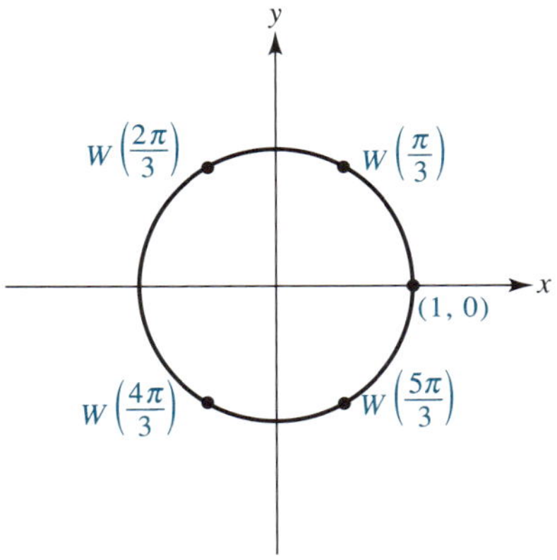

There is a symmetry of the unit circle we have not exploited yet. The unit circle is symmetric with respect to the line $y = x$ because if (a, b) belongs to the relation $x^2 + y^2 = 1$, so does (b, a). Writing $\pi/3$ as $2\pi/6$, we see that $W(\pi/6)$ and $W(\pi/3)$ divide the arc from $(1, 0)$ to $W(\pi/2)$, or $W(3\pi/6)$, into three equal arcs and that $W(\pi/6)$ and $W(\pi/3)$ are symmetrically placed with respect to $y = x$. Thus $W(\pi/6)$ must be $(\sqrt{3}/2, \frac{1}{2})$. See Figure 6.1c.

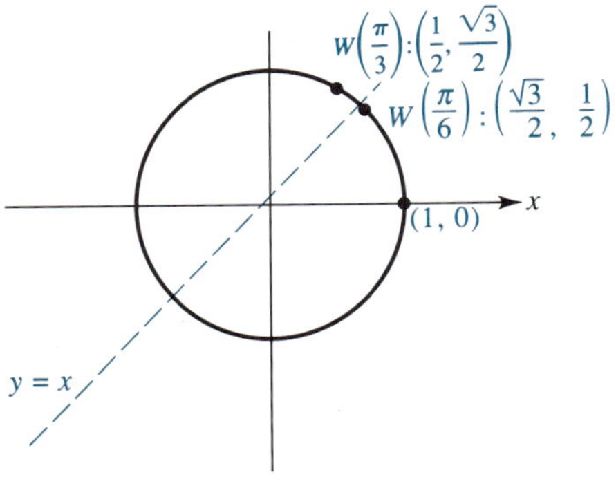

FIGURE 6.1c

Let's collect all our first-quadrant and axis points for future reference.

t	$W(t)$
0	(1, 0)
$\frac{\pi}{6}$	$\left(\frac{\sqrt{3}}{2}, \frac{1}{2}\right)$
$\frac{\pi}{4}$	$\left(\sqrt{\frac{1}{2}}, \sqrt{\frac{1}{2}}\right)$
$\frac{\pi}{3}$	$\left(\frac{1}{2}, \frac{\sqrt{3}}{2}\right)$
$\frac{\pi}{2}$	(0, 1)
π	(−1, 0)
$\frac{3\pi}{2}$	(0, −1)
2π	(1, 0)

Definition The values of t, $0 \leq t \leq \frac{\pi}{2}$, on this table are called reference numbers, since the values of W for these numbers, the symmetries of the unit circle, and the periodicity of W give us $W(t)$ for integral multiples of $\frac{\pi}{6}$, $\frac{\pi}{4}$, and $\frac{\pi}{3}$.

For this reason it's helpful to memorize values of W for the reference numbers.

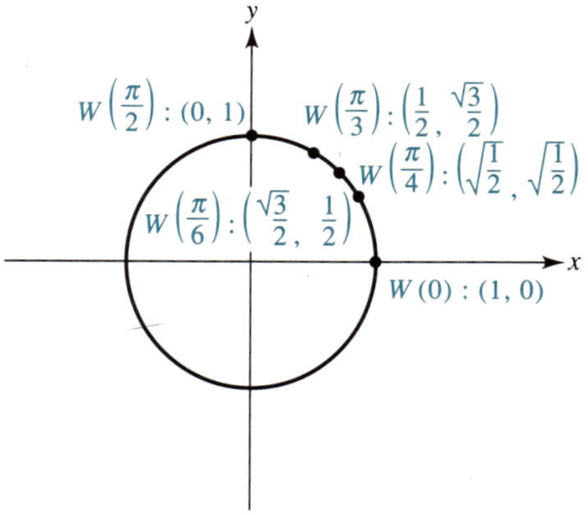

We can estimate the location of numbers that are not integral multiples of reference numbers by comparing them to integral multiples of reference numbers.

EXAMPLE Find an interval containing the number $\sqrt{5}$ whose endpoints are reference numbers or integral multiples of reference numbers.

An approximation for $\sqrt{5}$ is 2.2360679775, for $\frac{2\pi}{3}$ is 2.09439510239, and for $\frac{3\pi}{4}$ is 2.35619449019.

Therefore, $\sqrt{5} \in (2\pi/3, 3\pi/4)$.

Note: We have chosen the smallest such interval. ■

EXERCISES 6.1

In exercises 1 through 14, name the quadrant in which the point lies. Also find the coordinates of the point.

1. $W\left(\frac{5\pi}{6}\right)$
2. $W\left(\frac{7\pi}{6}\right)$
3. $W\left(-\frac{\pi}{3}\right)$
4. $W\left(-\frac{2\pi}{3}\right)$
5. $W\left(-\frac{5\pi}{6}\right)$
6. $W\left(-\frac{\pi}{6}\right)$
7. $W\left(\frac{3\pi}{4}\right)$
8. $W\left(\frac{5\pi}{4}\right)$
9. $W\left(\frac{13\pi}{6}\right)$
10. $W\left(\frac{25\pi}{3}\right)$
11. $W\left(-\frac{43\pi}{3}\right)$
12. $W\left(-\frac{101\pi}{4}\right)$
13. $W(147\pi)$
14. $W\left(\frac{83\pi}{2}\right)$

In exercises 15 through 24, identify the quadrant in which *W(t)* lies. Your calculator might be helpful for approximating numbers.

15. $W\left(\frac{\pi}{7}\right)$
16. $W\left(\frac{2\pi}{11}\right)$
17. $W(3.61)$
18. $W(-5)$
19. $W(-\sqrt{23})$
20. $W(6.251)$
21. $W(\pi + 4)$
22. $W(13 - 2\pi)$
23. $W(4e)$
24. $W(4e + \pi)$

In exercises 25 through 34, give a value of *t* that makes the statement true. *Note:* There are many possible values of *t*.

25. $W(t) = (-1, 0)$
26. $W(t) = \left(-\sqrt{\frac{1}{2}}, \sqrt{\frac{1}{2}}\right)$
27. $W(t) = \left(\frac{\sqrt{3}}{2}, -\frac{1}{2}\right)$
28. $W(t) = \left(-\sqrt{\frac{3}{4}}, -\frac{1}{2}\right)$
29. $W(t) = (0, 1)$
30. $W(t) = (0, -1)$
31. $W(t) = \left(-\frac{1}{2}, \sqrt{\frac{3}{4}}\right)$
32. $W(t) = \left(\frac{1}{2}, -\frac{\sqrt{3}}{2}\right)$
33. $W(t) = \left(\sqrt{\frac{1}{2}}, \sqrt{\frac{1}{2}}\right)$
34. $W(t) = \left(\sqrt{\frac{1}{2}}, -\sqrt{\frac{1}{2}}\right)$
35. Give a pattern that describes all the values of t that satisfy the statement in exercise 25.
36. Give a pattern that describes all the values of t that satisfy the statement in exercise 26.
37. Give a pattern that describes all the values of t that satisfy the statement in exercise 27.
38. Give a pattern that describes all the values of t that satisfy the statement in exercise 32.

In exercises 39 through 46, determine the smallest interval that (1) contains the given value of *t* and (2) has reference numbers or integral multiples of reference numbers as endpoints.

39. 4.2
40. -7.113
41. $\sqrt{2}$
42. $\sqrt{11}$
43. e
44. $3 - e$
45. $1 - \pi$
46. $\frac{18\pi}{7}$

6.2 The Trigonometric Functions

The wrapping function, W, is the basis for definitions of the six trigonometric functions. These functions were originally developed using right triangles; this is why they are called trigonometric (*trigon,* triangle: *metria,* measurement). Most modern applications require a more general definition, so we'll begin with this approach. In Section 7.1, we'll examine their restrictions to right triangles.

Definitions For any real number t, cosine(t) is the x-coordinate of $W(t)$. The common abbreviation for cosine is cos. The domain of cosine is $\mathbb{R}$.

For any real number t, sine(t) is the y-coordinate of $W(t)$. The common abbreviation for sine is sin. The domain of sine is $\mathbb{R}$.

EXAMPLE Use the results of Section 6.1 to find $\sin(\pi/4)$, $\cos(\pi/3)$, $\sin(-5\pi/6)$, $\sin(5\pi/4)$, $\cos(7\pi/6)$, and $\cos(5\pi/3)$.

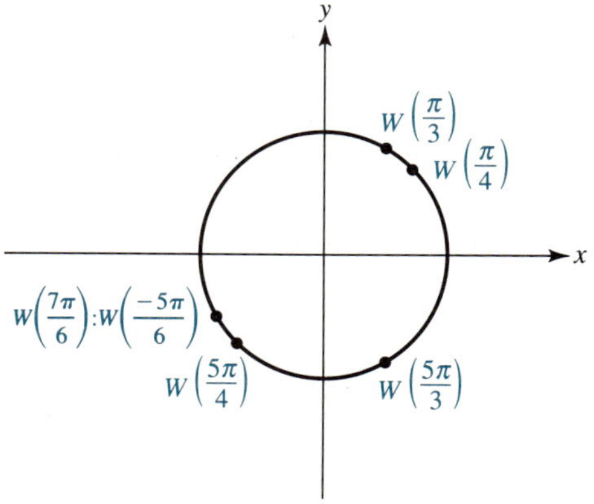

$\sin\left(\frac{\pi}{4}\right) = \sqrt{\frac{1}{2}}$ Since $W\left(\frac{\pi}{4}\right) = \left(\sqrt{\frac{1}{2}}, \sqrt{\frac{1}{2}}\right)$

$\cos\left(\frac{\pi}{3}\right) = \frac{1}{2}$ Since $W\left(\frac{\pi}{3}\right) = \left(\frac{1}{2}, \frac{\sqrt{3}}{2}\right)$

$\sin\left(-\frac{5\pi}{6}\right) = -\frac{1}{2}$ Because $-\frac{5\pi}{6}$ is a third quadrant number and $W\left(\frac{\pi}{6}\right) = \left(\frac{1}{2}, \frac{\sqrt{3}}{2}\right)$.

$\sin\left(\frac{5\pi}{4}\right) = -\sqrt{\frac{1}{2}}$

$\cos\left(\frac{7\pi}{6}\right) = -\frac{\sqrt{3}}{2}$

$\cos\left(\frac{5\pi}{3}\right) = \frac{1}{2}$ ■

It is acceptable to write $\sin t$ for $\sin(t)$ or $\cos \pi/4$ for $\cos(\pi/4)$. These functions have precedence over addition but not multiplication.

$\sin 2 \cdot \frac{\pi}{18}$, or $\sin \frac{\pi}{9} = \sin\left(\frac{\pi}{9}\right)$ $\sin 2 \cdot \frac{\pi}{18}$ is the same as $\sin\left(2 \cdot \frac{\pi}{18}\right)$.

$\cos \frac{\pi}{3} + \frac{\pi}{4}$, or $\cos\left(\frac{\pi}{3}\right) + \frac{\pi}{4} = \frac{1}{2} + \frac{\pi}{4}$ $\cos \frac{\pi}{3} + \frac{\pi}{4}$ is the same as $\left(\cos \frac{\pi}{3}\right) + \frac{\pi}{4}$.

□

The other trigonometric functions have restricted domains because they involve quotients. Here are their definitions.

Definition Let $(x_t, y_t) = W(t)$. Then

tangent(t), or $\tan(t)$, is $\frac{y_t}{x_t}$ for any t that is not an odd multiple of $\frac{\pi}{2}$— that is, $t \neq (2k + 1)\frac{\pi}{2}$, $k \in I$. (For each t that is an odd multiple of $\frac{\pi}{2}$, $W(t) = (0, y_t)$ and $\frac{y_t}{x_t}$ or $\tan(t)$, is undefined.)

secant(t), or $\sec(t)$, is $\frac{1}{x_t}$ for any $t \neq (2k + 1)\frac{\pi}{2}$, $k \in I$.

cosecant(t), or $\csc(t)$, is $\frac{1}{y_t}$ for any $t \neq k\pi$, $k \in I$.

cotangent(t), or $\cot(t)$, is $\frac{x_t}{y_t}$ for any $t \neq k\pi$, $k \in I$.

The sine, cosine, and tangent functions are used so frequently that many calculators have a key for each. To approximate values of the trigonometric functions as they are defined in this chapter, be sure your calculator is in radian mode. Evaluating the trigonometric functions for degree measurements is discussed in Section 7.1. The secant, cosecant, and cotangent functions are the reciprocals of the cosine, sine, and tangent functions, respectively, so values for them can be approximated by employing a reciprocal key. Remember, the reciprocal of a function value is generally not the same as the value of the (composition) inverse of that function. If your calculator has a $\cos^{-1}$ key, this will not produce values for the secant function; it will give values for the inverse of the cosine function.

You may have noticed the connections between the names of the six trigonometric functions: sine–cosine, secant–cosecant, and tangent–cotangent. The reason for these names is discussed in Section 7.4.

What properties do these functions have? The domains for each is listed next.

Function	**Domain**
cosine	The real numbers
sine	The real numbers
tangent	Any real number $t \neq (2k + 1)\pi/2$, k an integer
secant	Any real number $t \neq (2k + 1)\pi/2$, k an integer
cosecant	Any real number $t \neq k\pi$, k an integer
cotangent	Any real number $t \neq k\pi$, k an integer

What about the range of each trig function? Since $\cos(t)$ and $\sin(t)$ are each the coordinate of a point on the unit circle, the values of cosine and sine must lie between -1 and 1, inclusive. We also notice that $(\cos t)^2 + (\sin t)^2 = x_t^2 + y_t^2 = 1$ for all values of t. This fact is very useful, and we'll rely on it again and again. We often write $(\cos t)^2$ as $\cos^2 t$; we may use this style with any of the trigonometric functions.

Looking at the unit circle, we can see that there is no restriction on how close to zero $\cos(t)$ and $\sin(t)$ can come on either the positive or negative side. For instance, there must be t_1, t_2, t_3, and t_4, as labeled in Figure 6.2a. As a consequence, the reciprocal functions, secant and cosecant, have no bounds on how large or small their values become. However, because cosine and sine have range $[-1, 1]$, secant and cosecant cannot have values between -1 and 1 (exclusive). If sec(t) were $\frac{1}{2}$, then cos(t) would be 2, which is not possible.

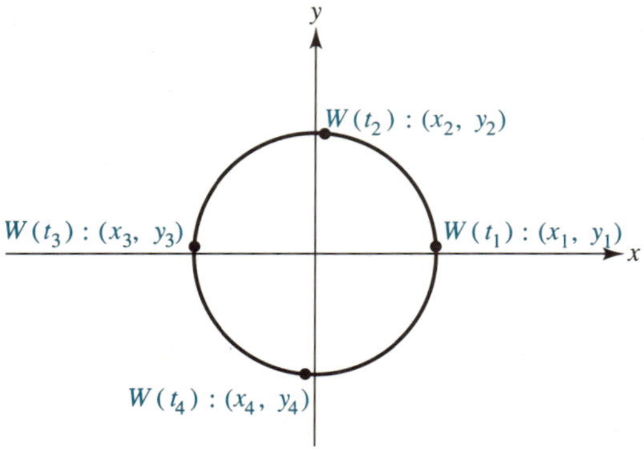

$0 < y_1 < 0.0001$ $0 < x_2 < 0.0003$ $-0.002 < y_3 < 0$ $-0.0005 < x_4 < 0$

FIGURE 6.2a

For a first-quadrant t, as sin(t) increases, cos(t) must decrease, and the quotient, tan t, increases. The values of tan t increase without bound. For a second-quadrant t, tan(t) is negative and increases (comes closer to 0) as t increases. There is no bound on how small tan(t) can become. The range of tangent and of its reciprocal, cotangent, is the set of real numbers.

Function	**Range**
cosine	$[-1, 1]$
sine	$[-1, 1]$
tangent	$(-\infty, +\infty)$
secant	$(-\infty, -1] \cup [1, +\infty)$
cosecant	$(-\infty, -1] \cup [1, +\infty)$
cotangent	$(-\infty, +\infty)$

It's a good idea to investigate the behavior of these functions using a calculator for approximations. See for yourself what happens to tan(x) as $x \to \pi/2$, for instance.

The fact that W, the wrapping function, is periodic suggests that the trigonometric functions may also be periodic. We explore this in the next section when we sketch the graphs of the trigonometric functions. It's easy to see, though, that because of W's periodicity, none of these functions is 1-1.

Do our new functions have any symmetries? Consider Figure 6.2b. No matter what quadrant we end in after traveling t units, cos(t) = cos($-t$) and sin(t) = $-$sin($-t$). This means cosine is an even function; its graph is symmetric with respect to the y-axis. The sine function is odd; its graph is symmetric with respect to the origin. The other four functions are tested for these properties in the exercises.

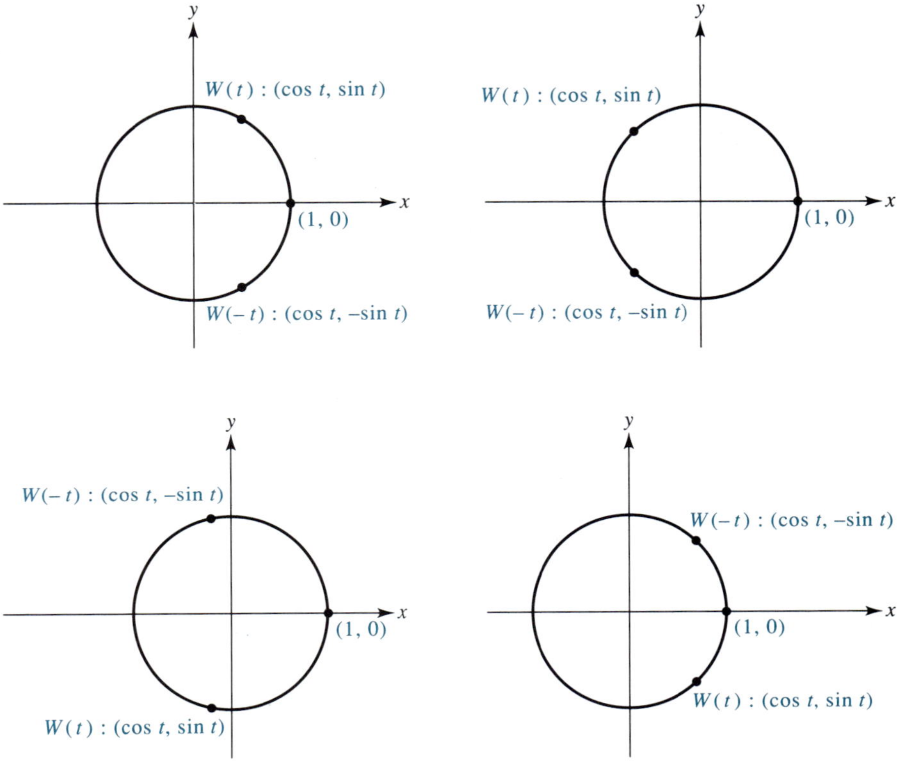

FIGURE 6.2b

The special values of W that we found in Section 6.1 give us a head start on finding values for the trigonometric functions. These are displayed in Table 6.2.

TABLE 6.2

t	$\sin t$	$\cos t$	$\tan t$	$\cot t$	$\sec t$	$\csc t$
0	0	1	0	Undefined	1	Undefined
$\frac{\pi}{6}$	$\frac{1}{2}$	$\sqrt{\frac{3}{4}}$	$\sqrt{\frac{1}{3}}$	$\sqrt{3}$	$\sqrt{\frac{4}{3}}$	2
$\frac{\pi}{4}$	$\sqrt{\frac{1}{2}}$	$\sqrt{\frac{1}{2}}$	1	1	$\sqrt{2}$	$\sqrt{2}$
$\frac{\pi}{3}$	$\sqrt{\frac{3}{4}}$	$\frac{1}{2}$	$\sqrt{3}$	$\sqrt{\frac{1}{3}}$	2	$\sqrt{\frac{4}{3}}$
$\frac{\pi}{2}$	1	0	Undefined	0	Undefined	1

These values are all for t's that are reference numbers, but using the symmetries of the circle as we did before, we can find function values for all integral multiples of these reference numbers.

EXAMPLE Use the properties just discussed to find the value of $\cos(-2\pi/3)$, $\sin(17\pi/4)$, $\tan(-5\pi/6)$, $\sec(5\pi/4)$, $\csc(7\pi/6)$, and $\cot(5\pi/3)$.

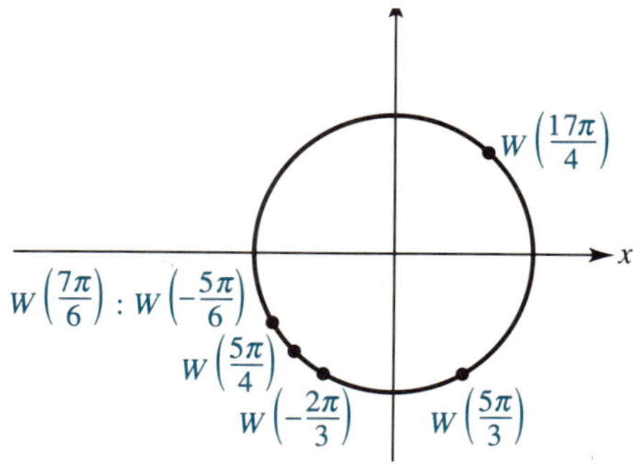

$$\cos\left(-\frac{2\pi}{3}\right) = \cos\left(\frac{2\pi}{3}\right), \text{ or } -\frac{1}{2},$$

Because cosine is an even number.

$$\sin\left(\frac{17\pi}{4}\right) = \sin\left(4\pi + \frac{\pi}{4}\right), \text{ or } \sin\left(\frac{\pi}{4}\right) = \sqrt{\frac{1}{2}}$$

$$\tan\left(-\frac{5\pi}{6}\right) = \tan\left(\frac{\pi}{6}\right), \text{ or } \sqrt{\frac{1}{3}}$$

Since $W\left(-\frac{5\pi}{6}\right)$ and $W\left(\frac{\pi}{6}\right)$ are symmetric points with respect to the origin

$$\sec\left(\frac{5\pi}{4}\right) = -\sqrt{2}$$

$$\csc\left(\frac{7\pi}{6}\right) = -2$$

$$\cot\left(\frac{5\pi}{3}\right) = -\sqrt{\frac{1}{3}}$$

■

The function values for the reference numbers are important enough to be memorized, but we can be very efficient about this. If we know $\sin(t)$ and $\cos(t)$, it is easy to evaluate $\sec(t)$, $\csc(t)$, $\tan(t)$, and $\cot(t)$, so perhaps only the first two columns of Table 6.2 need to be commited to memory. If you don't mind using the fact that $\cos^2 t + \sin^2 t = 1$, it's possible to get by with only sine or cosine values. Consider the following table; the simple pattern of the $\sin(t)$ values makes this list easy to recall.

t	**sin (t)**
0	$\sqrt{\frac{0}{4}}$, or $\frac{0}{2}$
$\frac{\pi}{6}$	$\sqrt{\frac{1}{4}}$, or $\frac{\sqrt{1}}{2}$
$\frac{\pi}{4}$	$\sqrt{\frac{2}{4}}$, or $\frac{\sqrt{2}}{2}$
$\frac{\pi}{3}$	$\sqrt{\frac{3}{4}}$, or $\frac{\sqrt{3}}{2}$
$\frac{\pi}{2}$	$\sqrt{\frac{4}{4}}$, or $\frac{\sqrt{4}}{2}$

In practice you'll probably find it easy to recall the common cosine-sine pairs because you will use them so often.

EXERCISES 6.2

In exercises 1 through 10, find the value of the trigonometric function. *Note:* This means find the *exact value*.

1. $\sin\left(\frac{5\pi}{6}\right)$ **2.** $\cos\left(-\frac{\pi}{3}\right)$

3. $\cos\left(-\frac{7\pi}{6}\right)$ **4.** $\sin\left(-\frac{2\pi}{3}\right)$

5. $\sin\left(-\frac{5\pi}{6}\right)$ **6.** $\cos\left(-\frac{\pi}{6}\right)$

7. $\cos\left(\frac{3\pi}{4}\right)$ **8.** $\sin\left(\frac{5\pi}{4}\right)$

9. $\sin\left(\frac{13\pi}{6}\right)$ **10.** $\cos\left(\frac{25\pi}{3}\right)$

11. If $\sin t = \frac{3}{5}$ and t is a first-quadrant number, then $\cos t$ = ______.

12. If $\cos t = \frac{4}{7}$ and t is a fourth-quadrant number, then $\sin t$ = ______.

13. If $\csc t = 2$ and t is a second-quadrant number, then $\tan t$ = ______.

14. If $\sec t = -\frac{13}{5}$ and t is a second-quadrant number, then $\cot t$ = ______.

In exercises 15 through 20, determine whether the function value is positive or negative.

15. $\sin\left(\frac{\pi}{7}\right)$ **16.** $\cos\left(\frac{2\pi}{11}\right)$

17. $\tan(3.61)$ **18.** $\cot(-5)$

19. $\sec(-\sqrt{23})$ **20.** $\csc(6.251)$

In exercises 21 through 32, give an approximate value of each function correct to five decimal places. Be sure your calculator is in radian mode.

21. $\sin\left(\frac{\pi}{7}\right)$ **22.** $\cos\left(\frac{2\pi}{11}\right)$

23. $\tan(3.61)$ **24.** $\cot(-5)$

25. $\sec(-\sqrt{23})$ **26.** $\csc(6.251)$

27. $\sin(\pi + 4)$ **28.** $\tan(13 - 2\pi)$

29. $\sec(4e)$ **30.** $\cos(4e + \pi)$

31. $\tan(1)$ **32.** $\cos\left(\frac{\sqrt{3}}{2}\right)$

In exercises 33 through 38, give a value of t that makes the statement true. (There are many values of t that will do.)

33. $\sin(t) = -1$ **34.** $\tan(t) = -1$

35. $\sec(t) = \sqrt{\frac{4}{3}}$ **36.** $\cos(t) = -\sqrt{\frac{3}{4}}$

37. $\csc(t) = 1$ **38.** $\sec(t) = 0$

In exercises 39 through 42, evaluate the given expressions.

39. $4\tan\left(\frac{3\pi}{4}\right) + 2\sin\left(-\frac{\pi}{3}\right)$

40. $\dfrac{\cos(5\pi/6)}{\tan(7\pi/4)}$

41. $-3 + \sec\left(-\dfrac{\pi}{2}\right) - \sin(\pi)$

42. $\cos\dfrac{3\pi}{4} + \dfrac{2\pi}{3}$

43. Give a pattern description for all values of t for which $\sin(t) = -\dfrac{\sqrt{3}}{2}$.

44. Give a pattern description for all values of t for which $\cos(t) = \sqrt{\dfrac{1}{2}}$.

45. Prove that secant is an even function and that cosecant, tangent, and cotangent are odd functions.

In exercises 46 and 47, use the following definition. A point that moves on a straight line so that the position $s(t)$ of the point at time t is given by either $s(t) = a\sin(bt)$ or $s(t) = a\cos(bt)$, a and b real numbers, is said to be in simple harmonic motion.

46. What is the position of a point in simple harmonic motion after 5 seconds if the position of the point $s(t)$ in feet at time t seconds is given by $s(t) = 12\sin(2t)$?

47. At what time is the position of a point in simple harmonic motion -0.5 foot if the position of the point $s(t)$ in feet at time t seconds is given by $s(t) = -3\cos(t)$?

48. Give two similarities between the cosine function and the function pc (Exercises 4.2).

49. Give two similarities between the sine function and the function ps (Exercises 4.5).

50. The sine and the cosine functions have the relationship $\sin^2 t + \cos^2 t = 1$. What is a similar relationship between ps and pc?

6.3 Graphs of the Trigonometric Functions

Toolkit

For this section you'll need techniques of graphing and properties of graphs from Sections 4.4 and 4.5 and the definition of cycle from Section 4.4

We must make some assumptions in order to sketch the graphs of the trigonometric functions. The graphs are connected except at those breaks implied by the domain restrictions (Table 6.2) and they are smooth, with no cusps or spikes. Experimenting with your calculator will help convince you of these properties, but some tools of calculus are required to prove that these assumptions are valid.

Let's begin with a sketch of cosine. We already have a collection of values for $\cos(t)$, $0 \le t \le \pi/2$ (Table 6.2), and we've seen that cosine is an even function. Figure 6.3a displays the results of using these facts.

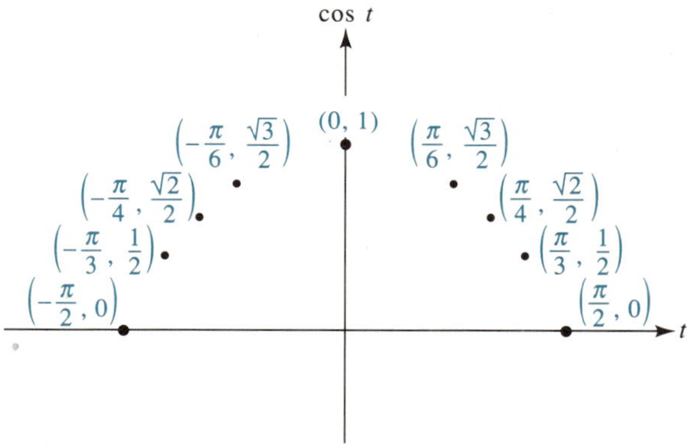

FIGURE 6.3a

Further, because cosine is even, we can concentrate on positive values of t and use reflection to obtain points to the left of the vertical axis. It also seems useful to mark divisions on the t-axis in terms of π rather than the usual integer or rational-number markings. Using our experience in Section 6.2, we can find values for some second- and third-quadrant t's (Table 6.3a). For future reference we have also included values of the sine function. (These may be found by using $\sin^2 t + \cos^2 t = 1$.)

TABLE 6.3a

t	$\frac{\pi}{2}$	$\frac{2\pi}{3}$	$\frac{3\pi}{4}$	$\frac{5\pi}{6}$	π	$\frac{7\pi}{6}$	$\frac{5\pi}{4}$	$\frac{4\pi}{3}$	$\frac{3\pi}{2}$
cos t	0	$-\frac{1}{2}$	$-\frac{\sqrt{2}}{2}$	$-\frac{\sqrt{3}}{2}$	-1	$-\frac{\sqrt{3}}{2}$	$-\frac{\sqrt{2}}{2}$	$-\frac{1}{2}$	0
sin t	1	$\frac{\sqrt{3}}{2}$	$\frac{\sqrt{2}}{2}$	$\frac{1}{2}$	0	$-\frac{1}{2}$	$-\frac{\sqrt{2}}{2}$	$-\frac{\sqrt{3}}{2}$	-1

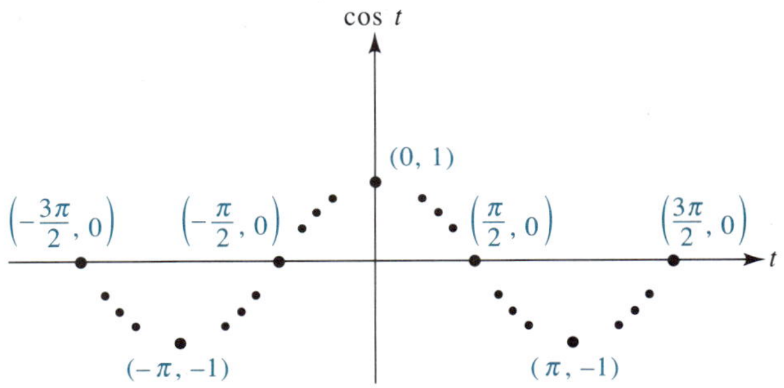

FIGURE 6.3b

These points are shown in Figure 6.3b. We now use the fact that the wrapping function W is periodic with period 2π. For any value of t, there is a t such that $0 \leq t < 2\pi$ and $t = t + 2k\pi$ for some integer k. For example, if $t = 27\pi/4$, $t = 3\pi/4 + 24\pi/4 = 3\pi/4 + 6\pi$, and if $t = -13\pi/6$, $t = 11\pi/6 - 4\pi$. This means $W(t) = W(t + 2k\pi) = W(t)$ and $\cos(t) = \cos(t)$ for t as described previously. This discussion shows our graph repeats the pattern established to both the right and left and *cosine is periodic with period 2π*. Connecting our points smoothly produces Figure 6.3c.

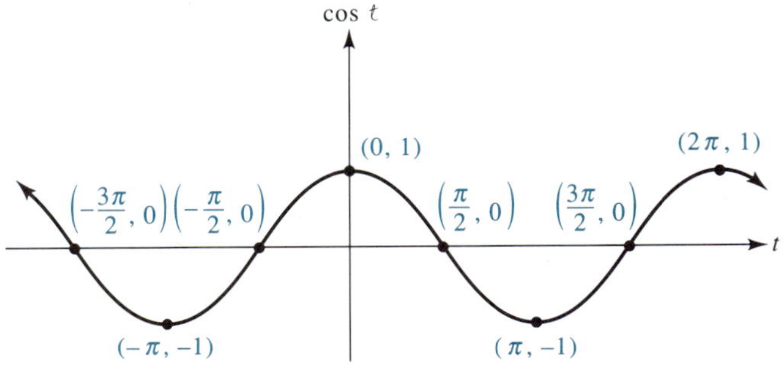

FIGURE 6.3c

Suggested labels for the graph of cosine or a variation of cosine are the high and low points and the points halfway between them for at least one complete cycle. For the graph of $y = \cos(x)$, these halfway points are the points of intersection with the x-axis, $((2k + 1)\pi/2, 0)$, $k \in I$. At least two complete cycles should be sketched.

In the process of developing the sketch of the graph of cosine, we've gathered most of the information needed to sketch the graph of sine. Using Tables 6.2 and 6.3a, we produce Figure 6.3d.

Since sine is an odd function, its graph is symmetric with respect to the origin, and we have Figure 6.3e. If we define t as we did in discussing the graph of cosine, $W(t) = W(t)$ implies $\sin(t) = \sin(t + 2k\pi) = \sin t$, and *sine is also periodic with period* 2π. Our completed sketch is Figure 6.3f.

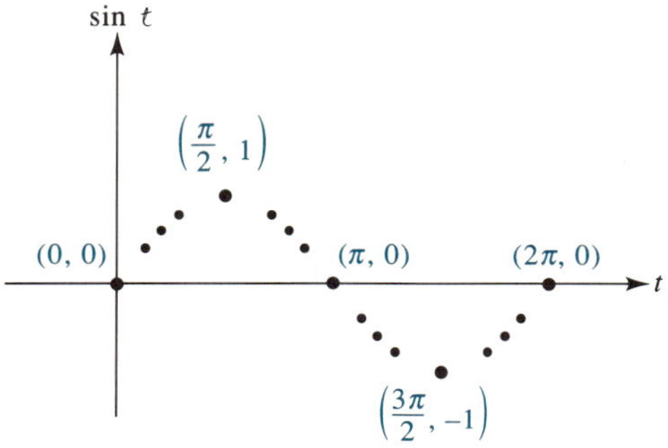

FIGURE 6.3d

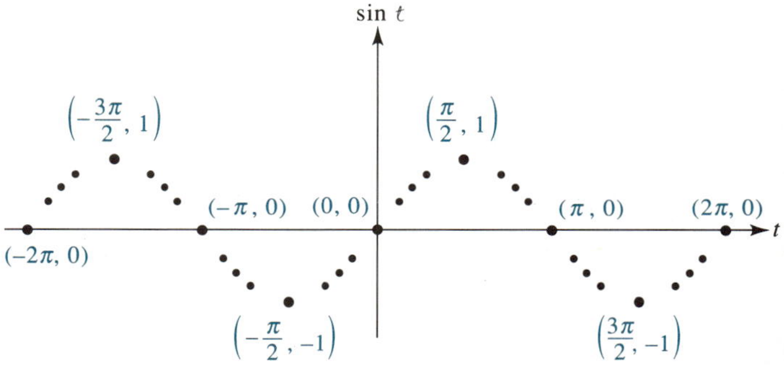

FIGURE 6.3e

Standard Graph 11: $q(t) = \sin t$

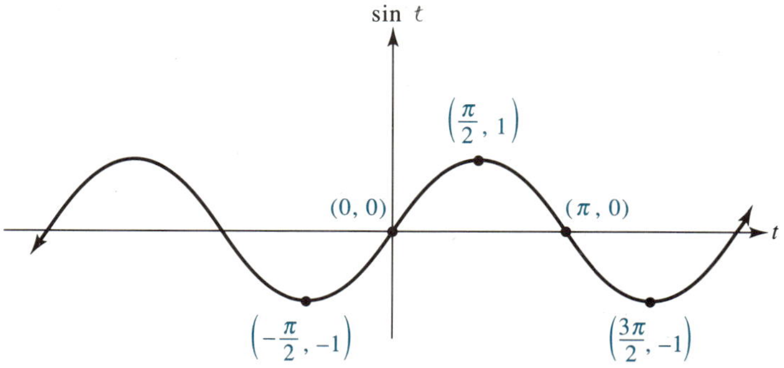

FIGURE 6.3f

Suggested labels for sine and its variations are the same as those for cosine and its variations. In fact, you've probably noticed that the graphs of sine and cosine are the same except for a horizontal shift of $\pi/2$. Generally, sine is taken as the standard and cosine is considered the variation:

$$\cos(t) = \sin\left(t + \frac{\pi}{2}\right).$$

(This fact is the origin of the names sine and cosine; we need a definition from Chapter 7 to give the full explanation, though.)

The graphs of the other four trig functions each have breaks. We'll work out the sketches of tangent and secant and leave cotangent and cosecant to be done in the exercises.

The tangent is undefined at each odd multiple of $\pi/2$. What is the behavior of the function near these values? Calculator-supported investigations such as those in Table 6.3b, suggest there are vertical asymptotes $t = (2k + 1)\pi/2$ with $\tan t \rightarrow -\infty$ as t approaches $(2k + 1)\pi/2$ from the right.

TABLE 6.3b

t	$\frac{\pi}{2} - 0.0001$	$\frac{\pi}{2}$	$\frac{\pi}{2} + 0.0001$	$\frac{3\pi}{2} - 0.0001$	$\frac{3\pi}{2}$	$\frac{3\pi}{2} + 0.0001$
$\approx$ **tan** t	10,000	Und.	−10,000	10,000	Und.	−10,000

Using the values from Tables 6.2 and 6.3a and symmetries, we can locate the points labeled in Figure 6.3g.

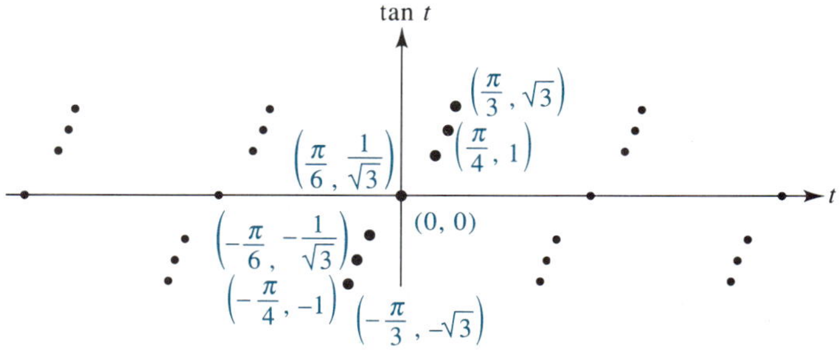

FIGURE 6.3g

Connecting the points smoothly and marking the asymptotes gives Figure 6.3h.

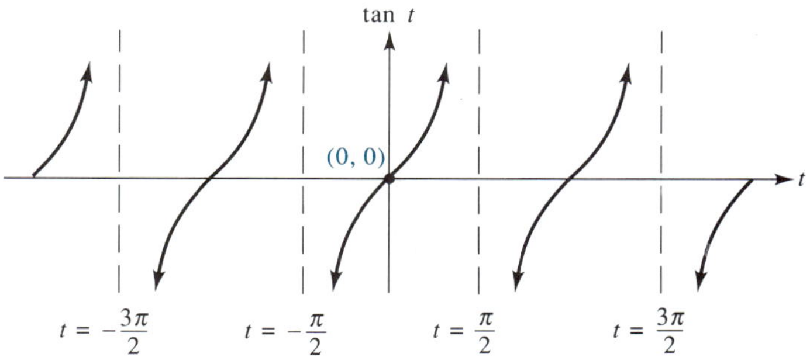

FIGURE 6.3h

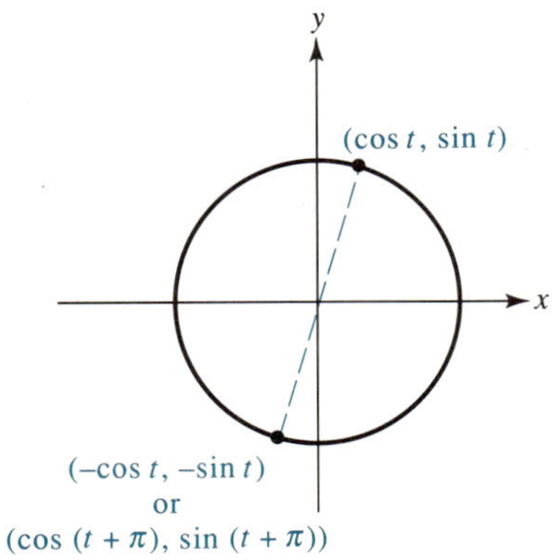

FIGURE 6.3i

This suggests tangent is periodic, as we suspected, but it appears to have a period of π, not 2π. Can we also confirm this algebraically?

$$\tan(t + \pi) = \frac{\sin(t + \pi)}{\cos(t + \pi)}$$

$$\Leftrightarrow \quad \tan(t + \pi) = \frac{-\sin t}{-\cos t} \qquad \text{Figure 6.3i}$$

$$\Leftrightarrow \quad \tan(t + \pi) = \tan t$$

Figure 6.3h shows there is no smaller positive number for which values of the tangent function repeat. Hence tangent has period π.

Standard Graph 12: $g(t) = \tan t$

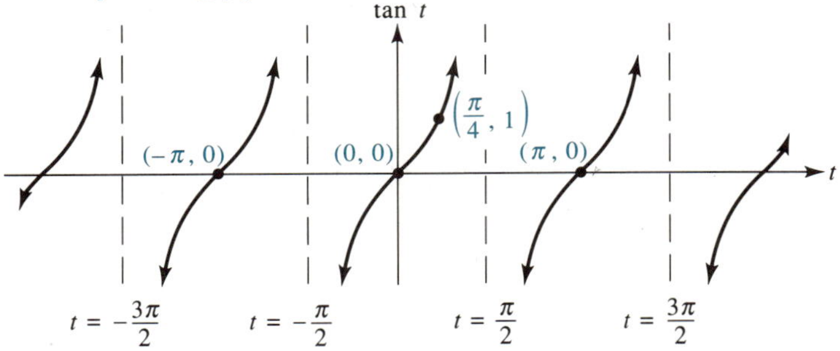

The graph continues left and right.

The tangent function is periodic with period π; it has vertical asymptotes $t = (2k + 1)\pi/2$, $k \in I$. Suggested labels for graphs of tangent and its variations are the asymptotes, the point where the curvature changes, and at least one other point to establish the scale in one cycle. At least two complete cycles should be sketched and some indication given that the pattern continues on both sides of these cycles if the domain is not artificially restricted. For $y = \tan(t)$, the curvature changes at the point of intersection with the t-axis. In general, these change points are halfway between the asymptotes.

The graph of secant also has vertical asymptotes, $t = (2k + 1)\pi/2$, $k \in I$. A sketch of $h(t) = \sec t$ can be quickly obtained by using $\sec t = 1/(\cos t)$ and noting that an increase (decrease) in the value of the cosine produces a decrease (increase) in the value of the secant.

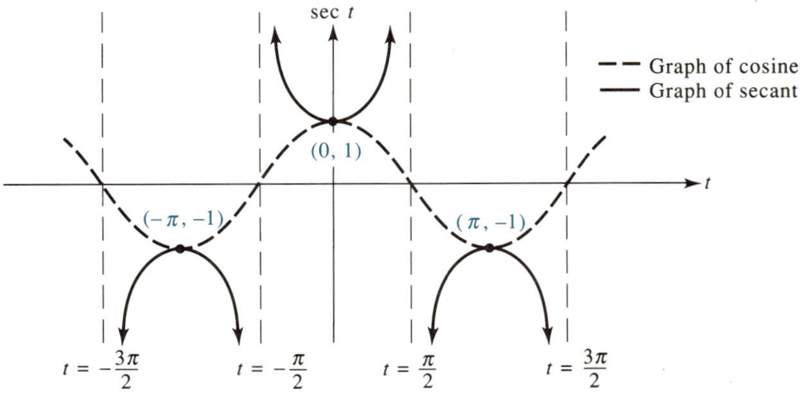

The periodic behavior of cosine means secant is also periodic. Our sketch shows *the period of secant is 2π.*

Standard Graph 13: $h(t) = \sec(t)$

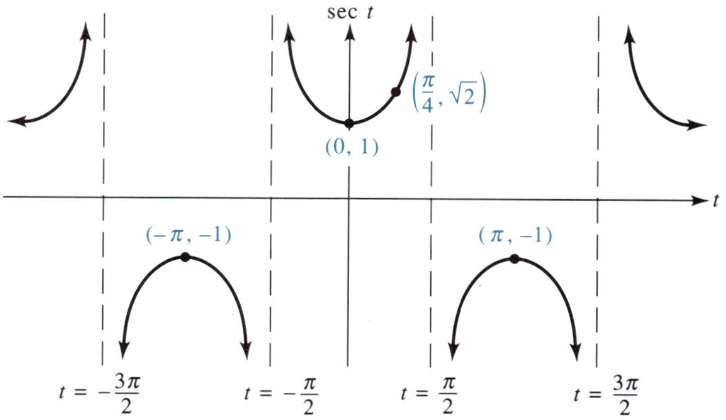

The graph continues left and right.

Suggested labels for the graph of secant or its variations are the asymptotes, high and low points, and at least one other point for scale in one cycle. At least two cycles should be sketched, and some indication given that the pattern continues, assuming the domain has not been artificially restricted.

EXERCISES 6.3

1. Sketch the graph of $g(t) = \cot t$ on the interval $[-\pi/2, \pi/2]$.
2. Sketch the graph of $h(t) = \csc t$ on the interval $[-\pi/2, \pi/2]$.
3. Sketch the graph of $G(t) = \cot t$.
4. Sketch the graph of $H(t) = \csc t$.

In exercises 5 through 10, tell where each of the trigonometric functions is increasing and where it is decreasing.

5. $f(t) = \sin t$
6. $g(t) = \cos t$
7. $h(t) = \tan t$
8. $F(t) = \cot t$
9. $G(t) = \sec t$
10. $H(t) = \csc t$

In exercises 11 through 16, state whether each of these properties holds for the stated function:

a. Has a graph symmetric with respect to the vertical axis.
b. Has a graph symmetric with respect to the horizontal axis.
c. Has a graph symmetric with respect to the origin.
d. Is an even function.
e. Is an odd function.
f. Is a 1-1 function.

11. $f(t) = \sin t$
12. $g(t) = \cos t$
13. $h(t) = \tan t$
14. $F(t) = \cot t$
15. $G(t) = \sec t$
16. $H(t) = \csc t$
17. Prove the period of secant is 2π.
18. Prove the period of cosecant is 2π.
19. Prove the period of cotangent is π.
20. Use the graphs of secant and cosecant to determine whether secant and cosecant have the property $f(t + \pi/2) = g(t)$. If so, which function plays the role of f?
21. Use the graphs of tangent and cotangent to determine whether tangent and cotangent have the property $f(t + \pi/2) = g(t)$. If so, which function plays the role of f?

(Preview and graphing calculator) In exercises 22 and 23, use a graphing calculator to sketch the graph of each function and give a verbal description of the graph.

22. $f(x) = \cos\left(x - \frac{\pi}{3}\right)$
23. $g(x) = -\sin\left(x + \frac{\pi}{6}\right) + 2$

6.4 Graphing Variations of the Trigonometric Functions

Toolkit

For this section you'll need the ideas of translation, expansion, and reflection of a graph from Section 2.5.

We have used the ideas of translation, expansion, and reflection to produce variations of known graphs. The extensive use of the trigonometric functions has resulted in a specialized vocabulary for some of these ideas.

Definition A horizontal shift of a trigonometric function is called a **phase shift.**

Remember that such shifts are produced by adding a constant to the independent variable of a function.

EXAMPLE Sketch the graph of $f(x) = \sin(x + \pi/4)$.

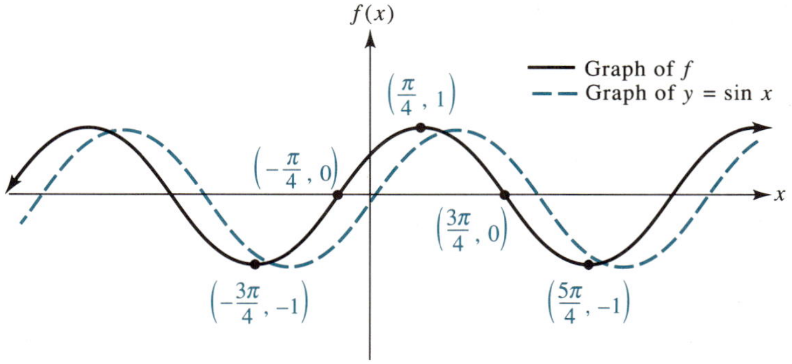

The graph of f is that of the sine function shifted $\pi/4$ units left. A vertical shift of the graph of a trigonometric function does not have a special designation. ■

EXAMPLE Sketch the graph of $g(x) = \cot(x) + 3$.

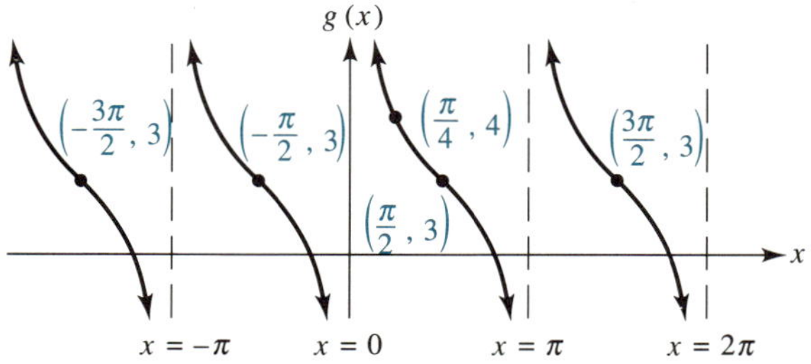

The graph continues left and right.

This is the graph of the cotangent function shifted 3 units up. *Note:* We label this as we would a variation of tangent: asymptotes, change point, and a point for scale.

Reflection in the x-axis is produced by taking the opposite of function values.

EXAMPLE Sketch the graph of $f(x) = -\sec(x)$.

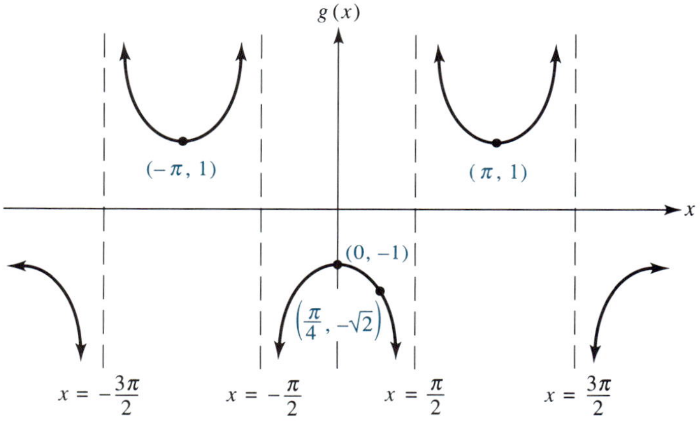

The graph continues left and right.

A vertical expansion factor is a constant multiplier applied to the generalized term with the characteristic operation or function. The function $m(x) = 3x^3$ is the result of applying an expansion factor of 3 to the standard function $f(x) = x^3$. The effect is to cause function values to change more rapidly. Such a factor may also affect the range of a function.

EXAMPLE Sketch the graph of $h(x) = 2\tan(x)$.

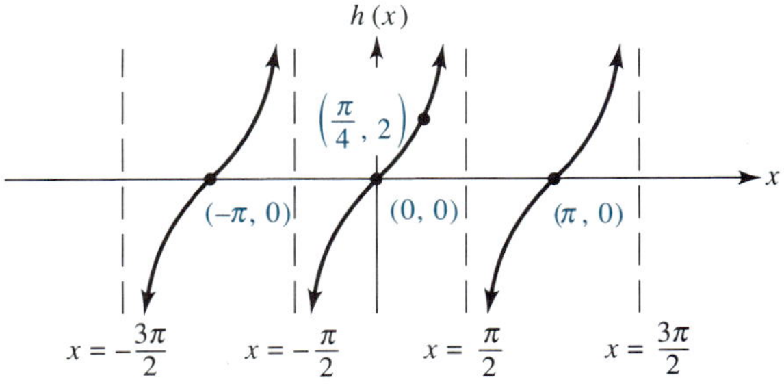

The graph continues left and right.

EXAMPLE Sketch the graph of $k(x) = 2\cos(x)$.

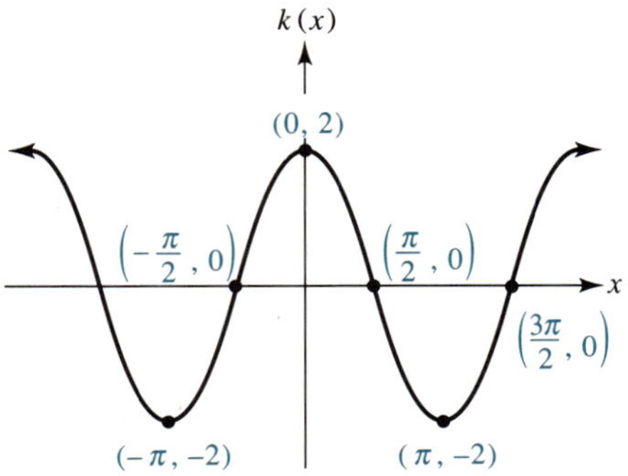

Note: The range of k is $[-2, 2]$.

Definition The **amplitude** of any variation of a sine or cosine function is defined to be one-half the difference between the maximum value and the minimum value of the function.

Illustration $k(x) = 2\cos(x)$ has amplitude 2, since—as we saw before—the range is $[-2, 2]$ and $[2 - (-2)]/2 = 4/2$, or 2.
$T(x) = 5\sin(x) - 7$ has amplitude 5, since its maximum is -2 and its minimum is -12; $[-2 - (-12)]/2 = 5$.
$g(x) = -3\cos(x + \pi/3)$ has amplitude 3. □

Amplitude is not defined for the other basic trigonometric functions because there are no bounds on their function values. Roughly speaking, amplitude describes the size of the range.

In our earlier discussion of ways to produce variations of a function, we did not examine in detail the effect of applying a multiplier to the independent variable. This is because the effect is very often similar to that produced by a vertical expansion factor.

Illustration The graph of $n(x) = |2x - 3|$ is the standard graph for $y = |x|$ with a vertical expansion factor of 2. It is shifted $\frac{3}{2}$ units right because $|2x - 3| = |2(x - \frac{3}{2})|$, or $2|x - \frac{3}{2}|$. □

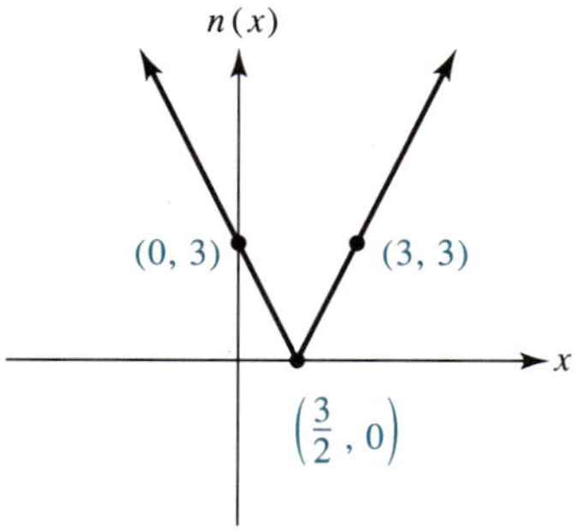

□

For periodic functions, more significant effects are produced by such a change.

Let's examine the behavior of $F(x) = \cos(\frac{1}{2}x)$. Using some values of x for which we can find exact values of F, we obtain Table 6.4.

TABLE 6.4

x	0	$\frac{\pi}{6}$	$\frac{\pi}{4}$	$\frac{\pi}{3}$	$\frac{\pi}{2}$	$\frac{2\pi}{3}$	$\frac{3\pi}{4}$	$\frac{5\pi}{6}$	π
$\cos x$	1	$\frac{\sqrt{3}}{2}$	$\frac{\sqrt{2}}{2}$	$\frac{1}{2}$	0	$-\frac{1}{2}$	$-\frac{\sqrt{2}}{2}$	$-\frac{\sqrt{3}}{2}$	-1
$\cos\left(\frac{1}{2}x\right)$	1			$\frac{\sqrt{3}}{2}$	$\frac{\sqrt{2}}{2}$	$\frac{1}{2}$			0

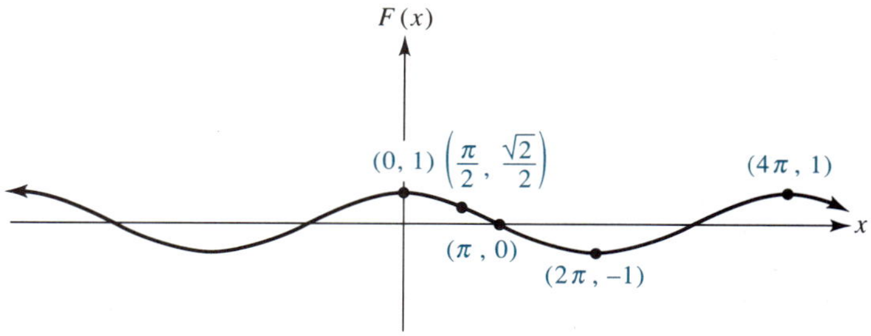

The factor $\frac{1}{2}$ slows the changes in $\cos(\frac{1}{2}x)$ as compared with $\cos(x)$; it takes twice as long for F to complete a cycle as it does for cosine. The period for F is 4π, as compared with 2π for the basic cosine function. A simple relationship between these numbers is $4\pi = 2\pi/(1/2)$. In general, if T is a trigonometric function, the period of $y = T(bx)$ is the period of T divided by $|b|$.

Illustration The graph of $k(x) = \tan(3x)$ is sketched next.

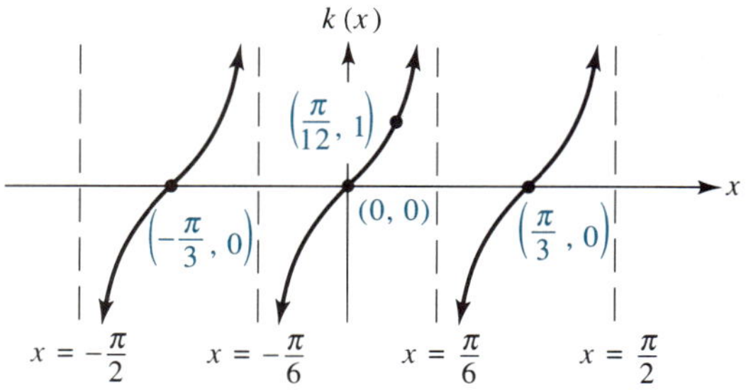

□

Note: The graph of k continues left and right and has period $\pi/3$.

We can now describe an efficient method for sketching the graph of any variation of a trig function T. We first outline the method and then demonstrate its use in several examples.

To sketch the graph of $AT(Bx - C) + D$, where T is a trig function, $A \neq 0$, and $B \neq 0$ (if A or B is zero, f is a constant function):

1. Decide if reflection in the x-axis is required by using the even or odd nature of

T to rewrite $f(x)$ as $aT(bx - c) + D$ with $b > 0$. If $a < 0$, the basic graph will be reflected in the x-axis.

2. Determine the period of f by dividing the period of T by b.
3. Determine the phase shift (horizontal translation) by rewriting $f(x)$ as $aT\left[b\left(x - \frac{c}{b}\right)\right] + D$. The phase shift is $\frac{c}{b}$.
4. Mark the numbers $\frac{c}{b}$ and $\left(\frac{c}{b} + \text{period of } f\right)$ on the horizontal scale. These are the endpoints of a cycle. Now locate the center of $\left[\frac{c}{b}, \frac{c}{b} + \text{period}\right]$ and the centers of the two intervals created. We saw earlier that the halfway points were generally distinguished in some way.
5. Sketch a copy of T, $0 \leq x \leq$ period of T, in the interval $\left[\frac{c}{b}, \frac{c}{b} + \text{period}\right]$, adjusting to match halfway points, reflecting if necessary, and translating D units vertically.
6. Extend the pattern for at least two complete cycles, indicating if it continues further.
7. Label the graph.

Basically, these steps are a summary of our general graphing techniques using some specific features of the trig functions.

EXAMPLE Determine if reflection in the x-axis will be required in sketching the graph of (a) $y = \tan(0.2x + 12)$, (b) $y = \cos(-5x + \pi)$, and (c) $y = \csc(-x - 4)$.

a. $\tan(0.2x + 12)$ fits the pattern already with $a = 1 > 0$; **no reflection is required.**

b. $\cos(-5x + \pi) = \cos(-(5x - \pi)) = \cos(5x - \pi)$, since cosine is an even function; **no reflection is needed.**

c. $\csc(-x - 4) = \csc(-(x + 4)) = -\csc(x + 4)$ because csc is an odd function; **a reflection is necessary.** ■

EXAMPLE Sketch the graph of $f(x) = 2 \sec(3x - \pi)$.

1. $f(x) = 2 \sec(3x - \pi)$ fits the pattern with $a = 2$ and $b = 3$. No reflection is required because $a > 0$.
2. The period of f is $\frac{2\pi}{3}$ because secant has period 2π.
3. The phase shift is $\frac{\pi}{3}$ since $f(x) = 2 \sec 3\left(x - \frac{\pi}{3}\right)$.
4. The horizontal scale should be marked at $\pi/3$ and $\pi/3 + 2\pi/3$, or π. The center of $[\pi/3, \pi]$ is $2\pi/3$. Mark the centers of $[\pi/3, 2\pi/3]$ and $[2\pi/3, \pi]$ as well.

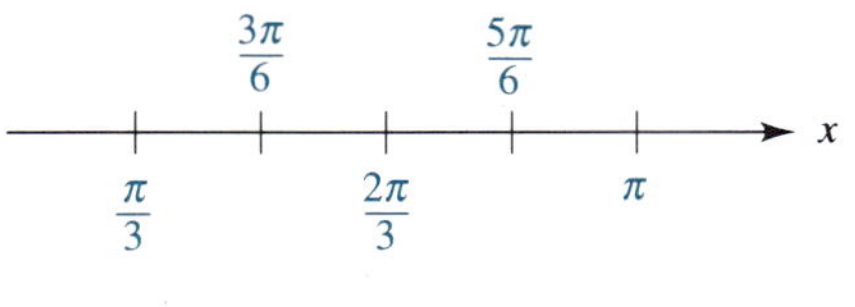

5. Adjust the graph of secant, $0 \le x \le 2\pi$, to fit on $[\pi/3, \pi]$.

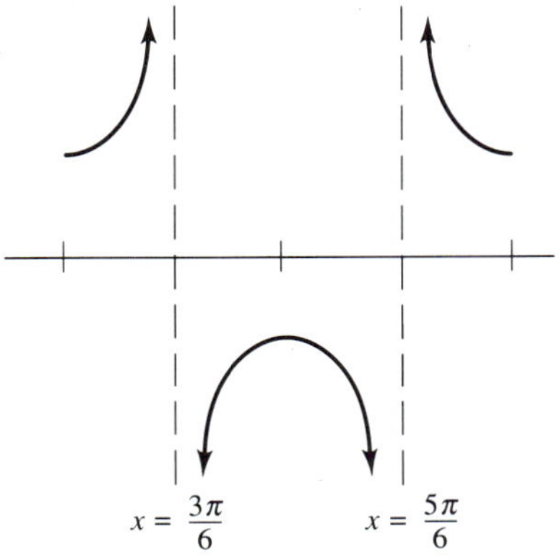

6.–7. Extend the pattern and label the graph.

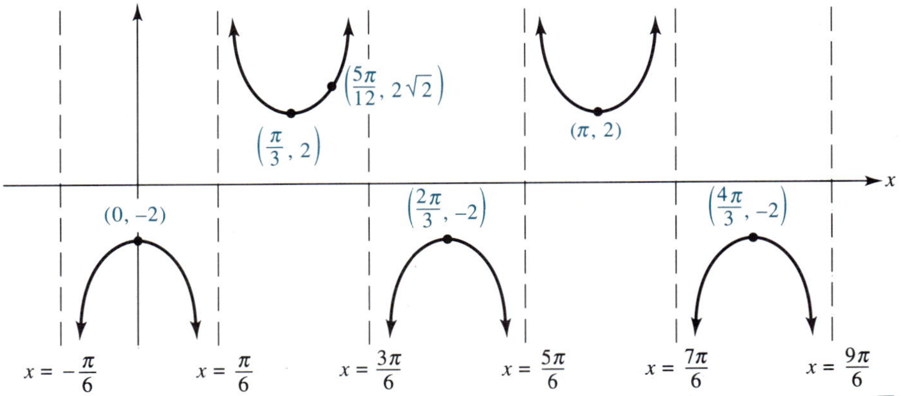

The graph continues left and right. ■

EXAMPLE Sketch the graph of $g(x) = \frac{1}{2}\sin(-2x + \pi)$.

1. $g(x) = \frac{1}{2}\sin(-2x + \pi) = \frac{1}{2}\sin[-(2x - \pi)] = -\frac{1}{2}\sin(2x - \pi)$ because sine is an odd function. Reflection in the line $y = 0$ is required.

2.–4. Writing $g(x)$ as $-\frac{1}{2}\sin 2(x - \pi/2)$, we see the period is $2\pi/2$, or π, and the phase shift is $\pi/2$. The numbers to be marked are $\pi/2$; $\pi/2 + \pi$, or $3\pi/2$; π; $3\pi/4$; and $5\pi/4$.

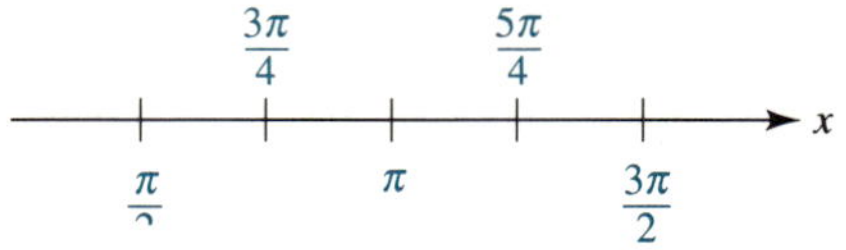

5.–7. We sketch a cycle of sine reflected in the x-axis, extend the pattern, and label the graph.

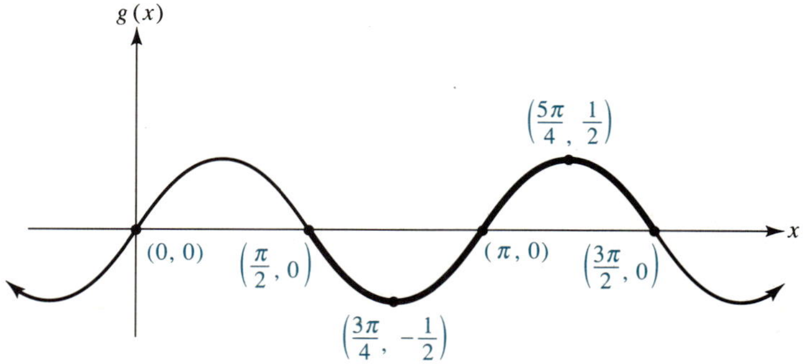

The heavy segment is the copy of $y = \sin(x)$ for $x \in [0, 2\pi]$ reflected in the x-axis and adjusted to fit in $[\pi/2, 3\pi/2]$. The amplitude of g is $\frac{1}{2}$. ■

EXAMPLE Sketch the graph of $h(x) = 3\tan(x + 4) + 2$.

1. No reflection is required, since $3 > 0$.

2.–4. The period is $\pi/1$, or π, and the phase shift is $-4/1$, or -4. We will mark -4 and $-4 + \pi$. Clearly, halfway between these values is $-4 + \pi/2$. What is the number halfway between -4 and $-4 + \pi/2$? What is the number between $-4 + \pi/2$ and $-4 + \pi$?

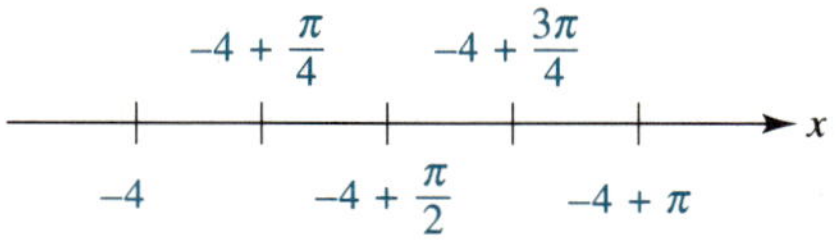

5.–7. The heavy section is the pattern cycle, and the entire graph has been shifted up 2 units.

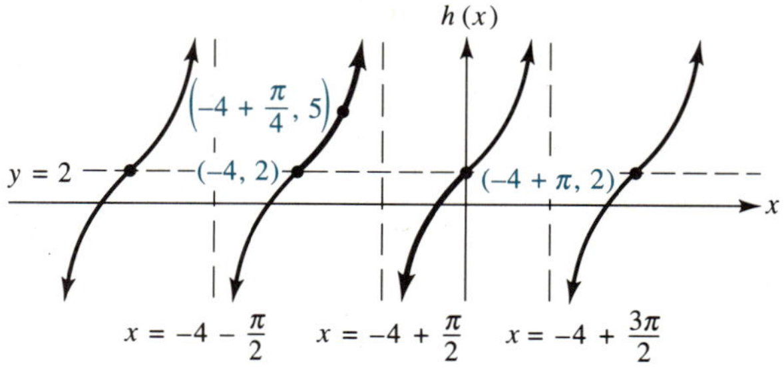

■

EXERCISES 6.4

In exercises 1 through 20, state the amplitude, period, and phase shift, and then sketch the graph.

1. $f(t) = -\sin(t)$

2. $g(t) = \sin(3t)$

3. $h(t) = \sin(0.5t)$

4. $F(t) = -\sin(2t)$

5. $G(t) = -\sin(4t)$

6. $H(t) = -\sin(0.5t)$

7. $j(t) = \sin(-t)$

8. $f(x) = \cos(-x)$

9. $g(t) = -\dfrac{1}{2}\sin(t - \pi)$

10. $h(t) = \sin\left(\dfrac{\pi}{2} - t\right)$

11. $g(x) = 3\sin(2\pi x)$

12. $f(t) = \sin\left(3t + \dfrac{\pi}{2}\right)$

13. $g(t) = -\sin(3t) + \dfrac{\pi}{2}$

14. $y = 4\cos(\pi - 0.2x)$

15. $y = 4\cos(2x - \pi) - 3$

16. $h(x) = a\cos[(bx) + c]$, where $a, b, c > 0$

17. $f(t) = \sec(t + \pi)$

18. $g(x) = -\dfrac{1}{2}\csc\left(\dfrac{x}{3}\right)$

19. $y = \tan(\pi - 2x)$

20. $h(t) - 1 = 2\tan(t)$

In exercises 21 through 26, sketch the graph of the function.

21. $m(x) = |\sin x|$

22. $M(x) = \sin|x|$

23. $N(x) = \cos|x|$

24. $n(x) = |\cos x|$

25. $p(x) = |\tan x|$

26. $P(x) = \tan|x|$

In exercises 27 through 32, write the sine function with the given period, phase shift, and amplitude.

	Period	Phase shift	Amplitude
27.	4π	π	3
28.	3π	4	π
29.	$\frac{1}{2}\pi$	5π	e
30.	4	$-\pi$	π
31.	3	$-e$	a
32.	$\frac{1}{2}$	-7	132

(Graphing calculator) In exercises 33 through 36, use a graphing calculator to sketch the graph of the function.

33. $f(x) = e^x\sin(x)$

34. $g(x) = \sin(\cos(x))$

35. $h(x) = \ln(\sin(x))$

36. $j(x) = \dfrac{1}{\tan(x - \pi/3)}$

6.5 The Inverse Trigonometric Functions

Toolkit

You'll need the definition of inverse functions and the relationship of the graphs of f and f^{-1} from Section 5.2.

Previously we have discussed inverse functions and methods for determining if they exist and how to find them if they do exist. An important use of inverse functions is in solving equations and inequalities. We've seen that the exponential and logarithmic functions are good examples of pairs of inverse functions, and we've used them to find solution sets.

As you might expect, the importance of the trigonometric functions means they are used frequently in equations and inequalities. Composition inverses for the trig functions would be very handy functions to have. Unfortunately, all the trigonometric functions are periodic and so are not 1-1, an essential requirement if a function is to have an inverse. The interest in having something like an inverse is so strong that we do the best we can by handling a related problem. The basic idea is to look at functions with the same rule as the trig functions but restrict their domains in such a way that the functions are 1-1. (We did this sort of thing in Exercises 4.4.)

Let's begin with $f(x) = \sin(x)$. We wish to define a new 1-1 function Sine with $\text{Sin}(x) = \sin(x)$ for some restricted domain. Is it possible to choose a section of the sine graph that covers the full range of sine but is only increasing or only decreasing?

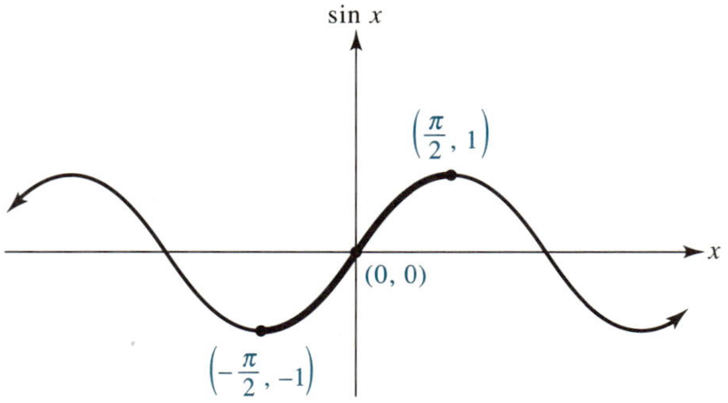

Such sections exist between each consecutive pair of odd multiples of $\pi/2$. We choose the one centered on (0, 0) for the convenience of small values of x.

Definition We define $\text{Sin}(x) = \sin(x)$, $x \in [-\pi/2, \pi/2]$.

The range of Sine is $[-1, 1]$, the same as for sine, so in a sense we have done the best we could, finding a 1-1 function as much like the sine function as possible. This new function, Sine, has an inverse, denoted by Sine^{-1} or Sin^{-1}.

$\text{Sin}^{-1}x = y$ if and only if $\text{Sin}(y) = x$.

Remember the interchange of domains and ranges for inverse functions; the domain of Sin^{-1} is $[-1, 1]$ and its range is $[-\pi/2, \pi/2]$.

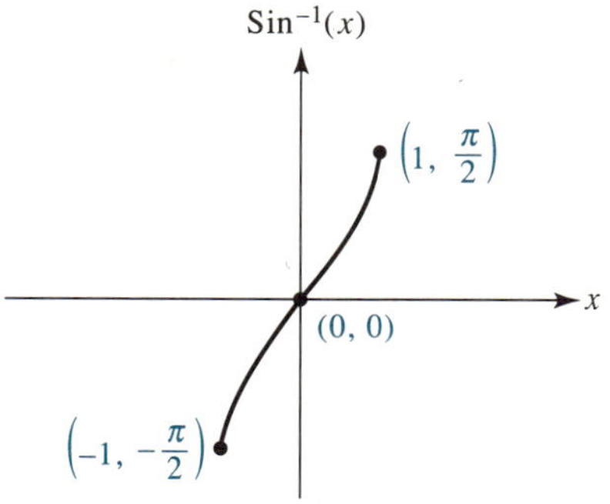

EXAMPLE Evaluate $\text{Sin}^{-1}(\sqrt{3}/2)$.
We know many numbers y for which $\sin(y) = \sqrt{3}/2$, but only $\pi/3$ lies in the domain of Sine.
$\text{Sin}^{-1}(\sqrt{3}/2) = \boldsymbol{\pi/3}$ because $\text{Sin}(\pi/3) = \sqrt{3}/2$ and $-\pi/2 < \pi/3 < \pi/2$. ■

EXAMPLE Evaluate $\text{Sin}^{-1}(-\frac{1}{2})$.
We must choose a number between $-\pi/2$ and $\pi/2$ for which $\text{Sin}(y) = -\frac{1}{2}$. Only $-\pi/6$ qualifies.
$\text{Sin}^{-1}(-\frac{1}{2}) = \boldsymbol{-\pi/6}$. ■

As is the case with the trigonometric functions, most values of Sin^{-1} have no exact expression other than $\mathrm{Sin}^{-1}(x)$.

EXAMPLE Approximate $\mathrm{Sin}^{-1}(0.3462)$ to five decimal places.
The number 0.3462 is in the domain of Sin^{-1} but is not equal to $\mathrm{Sin}(y)$, y an integer multiple of a reference number. Using a calculator we find: $\mathrm{Sin}^{-1}(0.3462) \approx$ **0.35352.**

Note: Because they produce the length of the arc (part of the unit circle) associated with the trig function value, the inverse trigonometric functions are sometimes named arcsine, arccosine, and so on. Your calculator may have an arcsin key rather than a $\sin^{-1}$ or inv sin key. ■

We next develop inverses for restrictions of cosine and tangent. Because the other three trig functions are the reciprocals of sine, cosine and tangent, it isn't as important to have (partial) inverses for them. These are left for you to develop in the exercises.

We wish to find a subset of $\mathrm{dom}_{\cos}$ that contains zero, that produces the range of cosine, and for which cosine is only increasing or only decreasing. Either $[-\pi, 0]$ or $[0, \pi]$ would do. The usual choice is $[0, \pi]$, as we generally prefer to work with positive numbers.

Definition We define $\mathrm{Cos}(x) = \cos(x)$, $x \in [0, \pi]$. The new function is 1-1 and has an inverse denoted by Cos^{-1}. $\mathrm{Cos}^{-1}x = y$ if and only if $\mathrm{Cos}(y) = x$.

The domain of Cos^{-1} is $[-1, 1]$ and its range is $[0, \pi]$.

EXAMPLE Solve $\cos(x) = -\frac{1}{2}$.
$\mathrm{Cos}^{-1}(-\frac{1}{2})$, or $2\pi/3$, is one solution, but cosine has the value $-\frac{1}{2}$ infinitely many times. Let's examine the unit circle again.

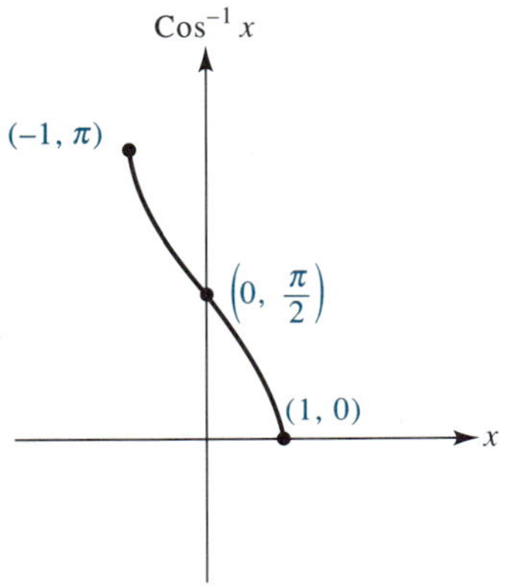

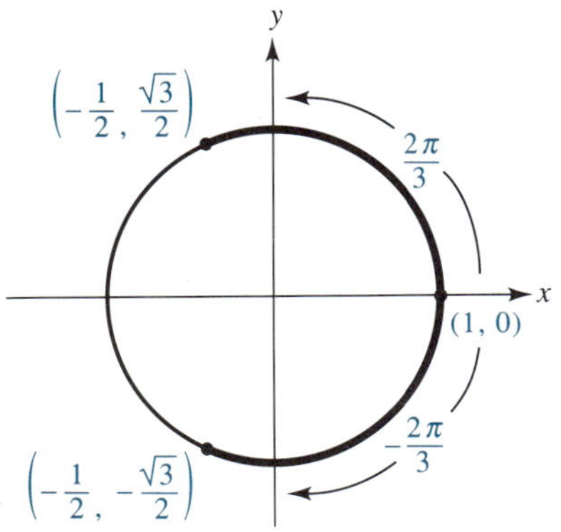

The third-quadrant point $W(-2\pi/3)$ also has first coordinate $-1/2$.

The complete solution is $\left\{ x \mid x = \frac{2\pi}{3} + 2k\pi \quad \text{or} \quad x = -\frac{2\pi}{3} + 2k\pi, k \in I \right\}$.

Note: Be sure you understand the difference between the following statements: Solve $\cos x = -\frac{1}{2}$. Solve $\text{Cos}^{-1}(-\frac{1}{2}) = x$. ■

The graph of the tangent function presents an obvious candidate for our restricted function, Tangent—namely the piece between $x = -\pi/2$ and $x = \pi/2$.

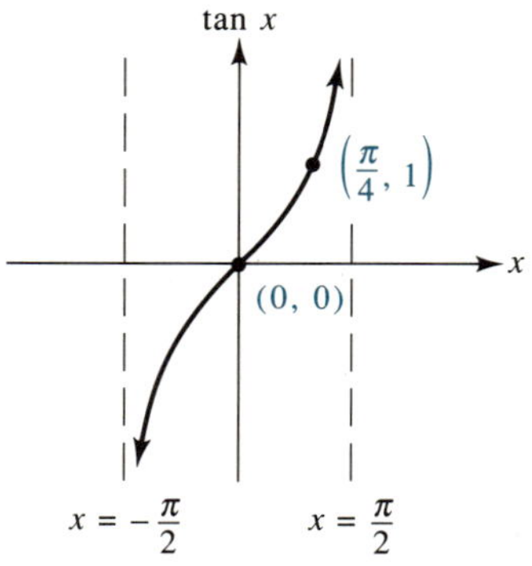

Definition We define $\text{Tan}(x) = \tan(x)$, $x \in (-\pi/2, \pi/2)$; $\text{Tan}^{-1}x = y$ if and only if $\text{Tan}(y) = x$.

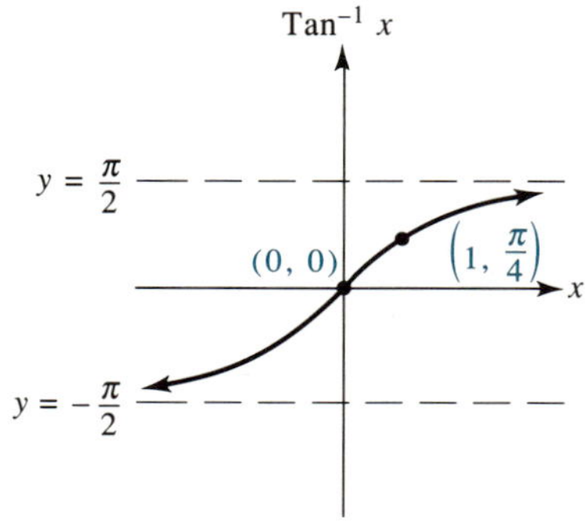

The domain of Tan^{-1} is $\mathbb{R}$ and its range is $(-\pi/2, \pi/2)$.

EXAMPLE Evaluate $\text{Tan}^{-1}(-1)$, $\text{Tan}^{-1}(-1.732)$, and $\text{Tan}^{-1}\left(\sqrt{1/3}\right)$.

$\text{Tan}^{-1}(-1) = $ **$-\pi/4$,** not $3\pi/4$, because $3\pi/4$ is not in the domain of Tangent.

The exact value for this expression is **$\text{Tan}^{-1}(-1.732)$.** Do not approximate unless you are asked to do so and are given the number of decimal places required.

$\text{Tan}^{-1}(\sqrt{1/3}) = $ **$\pi/6$** because $\text{Tan}(\pi/6) = \sqrt{1/3}$ and $\pi/6 \in (-\pi/2, \pi/2)$. ■

EXAMPLE Approximate the value of $2\,\text{Tan}^{-1}(-1.732)$ to four decimal places.

The approximate value is -2.0944. ■

EXERCISES 6.5

In exercises 1 through 10, find the function value.

1. $\mathrm{Sin}^{-1}\left(-\frac{1}{\sqrt{2}}\right)$
2. $\mathrm{Tan}^{-1}(\sqrt{3})$
3. $\mathrm{Cos}^{-1}(0.5)$
4. $\mathrm{Sin}^{-1}\left(\frac{\sqrt{3}}{2}\right)$
5. $\mathrm{Tan}^{-1}(-1)$
6. $\mathrm{Cos}^{-1}(-1)$
7. $\mathrm{Sin}^{-1}(-1)$
8. $\mathrm{Tan}^{-1}\left(\frac{1}{\sqrt{3}}\right)$
9. $\mathrm{Cos}^{-1}\left(\frac{\sqrt{3}}{2}\right)$
10. $\mathrm{Sin}^{-1}(-2^{-1})$

In exercises 11 through 24, find the approximate function value correct to five decimal places, where possible.

11. $\mathrm{Sin}^{-1}(0.3316)$
12. $\mathrm{Tan}^{-1}(73)$
13. $\mathrm{Cos}^{-1}(0.4)$
14. $\mathrm{Sin}^{-1}\left(\frac{2}{\sqrt{3}}\right)$
15. $\mathrm{Tan}^{-1}(-e)$
16. $\mathrm{Cos}^{-1}(-0.991)$
17. $\mathrm{Sin}^{-1}(\ln(0.22) + 3)$
18. $\mathrm{Tan}^{-1}(\sqrt{23})$
19. $\mathrm{Cos}^{-1}(\sqrt{3})$
20. $\mathrm{Sin}^{-1}(-\pi + 3)$
21. $\mathrm{Sin}^{-1}\left(\frac{\pi}{6}\right)$
22. $\mathrm{Tan}^{-1}\left(\frac{\pi}{4}\right)$
23. $\mathrm{Cos}^{-1}\left(\frac{2\pi}{3}\right)$
24. $\mathrm{Sin}^{-1}(3\pi)$

In exercises 25 through 34, find the value of the given expression, where possible.

25. $\mathrm{Sin}^{-1}\left(\cos\frac{\pi}{2} + \tan 4\pi\right)$
26. $\mathrm{Cos}^{-1}\left(\sin\frac{\pi}{3}\right)$
27. $\sin\frac{\pi}{6} + \mathrm{Sin}^{-1}1$
28. $\tan\left(\mathrm{Cos}^{-1}\left(\sin\frac{\pi}{3}\right)\right)$
29. $\mathrm{Cos}^{-1}\left(\tan\frac{\pi}{3}\right)$
30. $\mathrm{Tan}^{-1}\left(\tan\frac{\pi}{2}\right)$
31. $\tan\left(\mathrm{Tan}^{-1}\left(\frac{5}{2}\right)\right)$
32. $\mathrm{Tan}^{-1}\left(\tan\left(-\frac{2}{3}\right)\right)$
33. $\mathrm{Cos}^{-1}\left(\cos\frac{5\pi}{4}\right)$
34. $\sin\left(\mathrm{Sin}^{-1}\frac{\sqrt{3}}{2}\right)$

In exercises 35 through 40, find the approximate value of the given expression correct to three decimal places.

35. $\mathrm{Sin}^{-1}(\cos 7.2 + \tan 4)$
36. $\mathrm{Cos}^{-1}\left(\sin\left(-\frac{1}{3}\right)\right)$
37. $\sin\frac{\pi}{7} + \mathrm{Sin}^{-1}0.1$
38. $\tan(\mathrm{Cos}^{-1}(\sin 5.11))$
39. $\mathrm{Tan}^{-1}\left(\tan\left(\frac{5}{2}\right)\right)$
40. $\mathrm{Cos}^{-1}(\tan 1.05)$
41. Define Secant^{-1} in the style of the text and sketch the graph.
42. Define $\mathrm{Cosecant}^{-1}$ in the style of the text and sketch the graph.
43. Define $\mathrm{Cotangent}^{-1}$ in the style of the text and sketch the graph.

In exercises 44 through 47, graph the given function.

44. $y = 2\,\mathrm{Sin}^{-1}(x)$
45. $h(x) = -\frac{1}{3}\mathrm{Tan}^{-1}(x)$
46. $f(x) = \mathrm{Sin}^{-1}(2x)$
47. $g(x) = \mathrm{Tan}^{-1}(x - 1)$
48. Write a function that is a variation of Sin^{-1} that has the same graph as that of $y = \mathrm{Cos}^{-1}(x)$.

(Graphing calculator) In exercises 49 through 52, sketch the graph of the function. Describe how each is related to the corresponding simple inverse trig function.

49. $f(x) = 2\,\mathrm{Cos}^{-1}(x - 3)$
50. $g(x) = \mathrm{Sin}^{-1}(2x - 4) + 7$
51. $h(x) = 2\,\mathrm{Cos}^{-1}(3 - x)$
52. $j(x) = \mathrm{Sin}^{-1}(4 - 2x) + 7$

6.6 Equations Containing Trigonometric Expressions

Toolkit

You'll need techniques for solving equations from our earlier discussions in Chapter 1 and Chapter 5. You'll also use the inverse trig functions from the previous section and set union from Section A.2.

In the last section we solved a simple trigonometric equation with the aid of the inverse trig functions. Here we'll work with equations that may require some of our previous methods, such as factoring. The trigonometric functions have some special relationships; for instance, $\cos^2 t + \sin^2 t = 1$. These relationships give us some techniques for solving equations and inequalities involving trig functions that are not available for algebraic equations. We discuss these more fully in Chapter 7. This section will help consolidate your knowledge of the trigonometric functions and their (partial) inverses and review techniques of solving equations.

EXAMPLE Solve $\sin(x) + 1 = 3$.
$\sin(x) + 1 = 3$ is equivalent to $\sin(x) = 2$. The number 2 is not in the range of sine, so **the solution set is { }.** ■

EXAMPLE Solve $2 \cos(x) = 1.5$.
$2 \cos(x) = 1.5 \Leftrightarrow \cos(x) = 0.75$. Examining the graph of cosine, we see the solution set will be infinite.

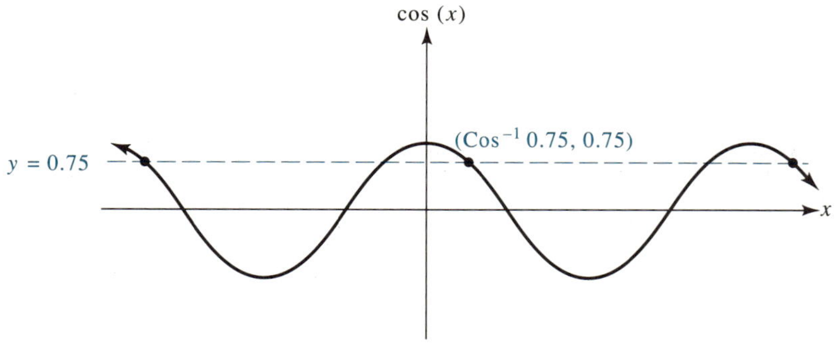

No reference number t satisfies $\cos(t) = 0.75$. We can represent the first-quadrant solution as $\text{Cos}^{-1}(0.75)$. The collection $\text{Cos}^{-1}(0.75) + 2k\pi$, k an integer, represents the x-coordinates of all the "right-hand" intersections of the cosine loops with $y = 0.75$. (These are marked with heavy dots.) We must also describe the "left-hand" intersections to obtain the complete solution set. If we can represent one of them, the others can be represented by adding multiples of 2π. We look more closely at the piece of the graph close to (0, 1).

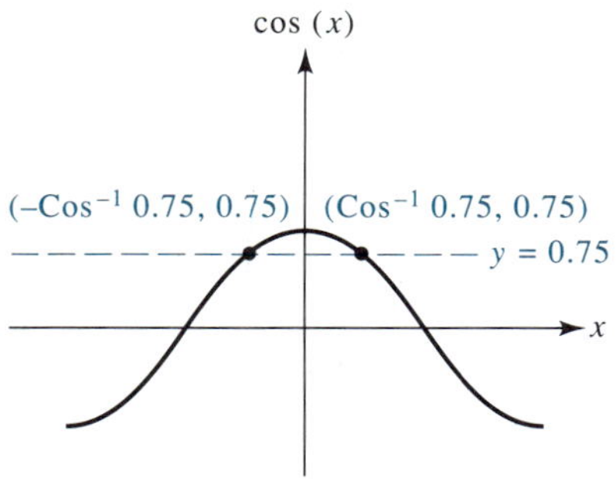

Now we see that one left-hand intersection must be $(-\text{Cos}^{-1}(0.75), 0.75)$. The solution set for $2\cos(x) = 1.5$ is

$$\{x|x = \text{Cos}^{-1}(0.75) + 2k\pi \quad \text{or} \quad x = -\text{Cos}^{-1}(0.75) + 2k\pi,\ k \in I\}.$$

This can also be written as

$$\{x|x = \text{Cos}^{-1}(0.75) + 2k\pi,\ k \in I\} \cup \{x|x = -\text{Cos}^{-1}(0.75) + 2k\pi,\ k \in I\}.$$

Note: $-\text{Cos}^{-1}(0.75) \neq \text{Cos}^{-1}(-0.75)$. ■

EXAMPLE Solve $\tan(3z) = \sqrt{3}$.
$\text{Tan}^{-1}\sqrt{3} = \pi/3$, so if $3z = \pi/3$, $\tan 3z = \sqrt{3}$. The period of tangent is π, so $3z = \pi/3 + k\pi$, $k \in I$, all satisfy $\tan 3z = \sqrt{3}$. Rewriting as $z = \pi/9 + k\pi/3$ gives a complete family of solutions.
The solution set is $\{z|z = \pi/9 + k\pi/3,\ k \in I\}$.

Note: The period of $\tan 3z$ is $\pi/3$, so if we solve for z by first obtaining $z = \pi/9$, we should add $k\pi/3$ to produce all solutions. It seems easier to add multiples of the period of the basic function first to avoid any errors in calculating the period of its variation. ■

EXAMPLE Solve $\sin^2(x) - 3\sin(x) - 4 = 0$.
This equation is quadratic in form, so we try factoring:

$$\sin^2(x) - 3\sin(x) - 4 = 0$$

$$\Leftrightarrow \quad (\sin(x) - 4)(\sin(x) + 1) = 0$$

Either $\sin(x) = 4$, an impossibility, or $\sin(x) = -1$, which occurs at $3\pi/2 + 2k\pi$, for every integer k.
The solution set is $\{3\pi/2 + 2k\pi,\ k \in I\}$. ■

EXAMPLE Solve $\dfrac{6}{\tan(t)} + \tan(t) = 5$.
Rewriting as we did in Section 1.4, we obtain

$$\frac{6}{\tan(t)} + \tan(t) - 5 = 0$$

$$\Leftrightarrow \quad \frac{\tan^2(t) - 5\tan(t) + 6}{\tan(t)} = 0$$

$$\Leftrightarrow \quad \tan^2(t) - 5\tan(t) + 6 = 0 \quad \text{and} \quad \tan(t) \neq 0.$$
$$\Leftrightarrow \quad (\tan(t) - 3)(\tan(t) - 2) = 0 \quad \text{and} \quad \tan(t) \neq 0.$$

The solution set is

$$\{\mathbf{Tan}^{-1}(3) + k\pi, k \in I\} \cup \{\mathbf{Tan}^{-1}(2) + k\pi, k \in I\}$$

because tangent has period π and no reference numbers produce $\tan(t) = 2$ or $\tan(t) = 3$. None of these values cause the denominator to be zero, so they are solutions of the original equation.

Note: If we choose to multiply both sides of the equation by $\tan(t)$, we are not guaranteed an equivalent equation. We must keep track of the restriction, which is no longer implicit. ■

EXAMPLE Solve $e^{\sin(2x)} = 2$.

We use ln, the inverse of this exponential function, to unravel the equation.

$$e^{\sin(2x)} = 2$$
$$\Leftrightarrow \quad \ln(e^{\sin(2x)}) = \ln 2$$
$$\Leftrightarrow \quad \sin(2x) = \ln 2$$

If $2x = \mathrm{Sin}^{-1}(\ln 2)$, the equation is satisfied. $\mathrm{Sin}^{-1}(\ln 2)$ is a first-quadrant number; there is also a second-quadrant solution, $\pi - \mathrm{Sin}^{-1}(\ln 2)$.

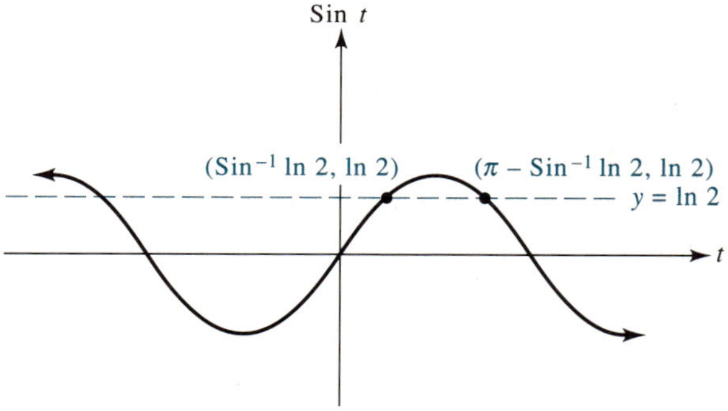

We solve $2x = \mathrm{Sin}^{-1}(\ln 2) + 2k\pi$ and $2x = (\pi - \mathrm{Sin}^{-1}\ln 2) + 2k\pi$ for x.

The solution set is

$$\left\{\frac{1}{2}\mathbf{Sin}^{-1}(\mathbf{ln}\ 2) + k\pi, k \in I\right\} \cup \left\{-\frac{1}{2}\mathbf{Sin}^{-1}(\mathbf{ln}\ 2) + \frac{\pi}{2} + k\pi, k \in I\right\}.$$ ■

Our general method of handling inequalities is applicable to inequalities involving trigonometric functions.

EXAMPLE Solve $\cos(x) < \frac{1}{2}$.

We begin as usual by solving the corresponding equation, $\cos(x) = \frac{1}{2}$. Since $\cos(\pi/3) = \frac{1}{2}$ and $\cos(-\pi/3) = \frac{1}{2}$, the set of boundary numbers is

$$\left\{\frac{\pi}{3} + 2k\pi, k \in I\right\} \cup \left\{-\frac{\pi}{3} + 2k\pi, k \in I\right\}.$$

These boundary numbers and the regions they define are partially displayed in the following chart, along with the results of testing the regions.

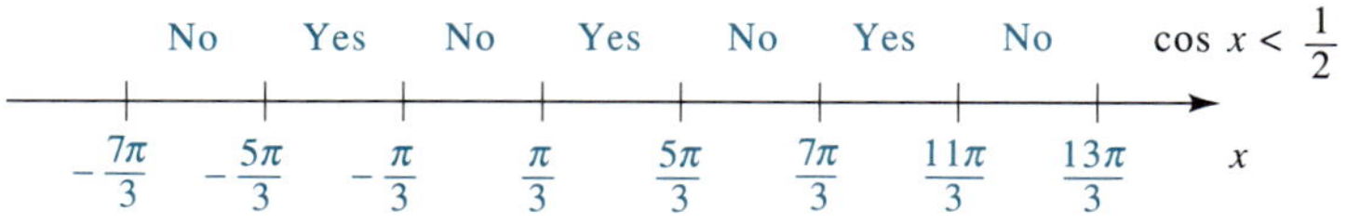

The inequality is true on intervals with left-hand endpoints $\pi/3 + 2k\pi$, $k \in I$ and length $4\pi/3$ units.

The solution set is the union of all intervals $(\pi/3 + 2k\pi, 5\pi/3 + 2k\pi)$, $k \in I$. ■

EXERCISES 6.6

In exercises 1 through 31, solve the given equation.

1. $\text{Cos}^{-1}(x) = \dfrac{\pi}{6}$
2. $\text{Tan}^{-1}(t) = -\dfrac{\pi}{4}$
3. $\text{Sin}^{-1}(t) = \dfrac{7\pi}{4}$
4. $\text{Cos}^{-1}(x) = -\dfrac{\pi}{3}$
5. $\tan(x) = \sqrt{3}$
6. $\sin(t) = -\sqrt{\dfrac{1}{2}}$
7. $\cos(t) = 0.3$
8. $\tan(x) = \dfrac{4}{7}$
9. $4 + \sin(x) = e$
10. $\sqrt{5} - \pi = \cot(x) + 2.4$
11. $\text{Cos}^{-1}(2x) = \dfrac{\pi}{6}$
12. $\text{Tan}^{-1}(0.5t) = -\dfrac{\pi}{4}$
13. $\text{Sin}^{-1}\left(\dfrac{1}{3}t\right) = \dfrac{7\pi}{4}$
14. $\text{Cos}^{-1}(3x) = -\dfrac{\pi}{3}$
15. $\tan(5x) = \sqrt{3}$
16. $\sin(0.4t) = -\sqrt{\dfrac{1}{2}}$
17. $\cos(t + 1) = 0.5$
18. $\tan(1 - x) = \dfrac{4}{7}$
19. $4 + \sin(3 - x) = e$
20. $\sqrt{5} - \pi = \cot 5(x + 2.4) + 3$
21. $2\sin^2(t) + 3\sin(t) + 1 = 0$
22. $4\cos^2(t) + 4\cos(t) + 1 = 0$
23. $4\sin^2(t) + 4\sin(t) + 1 = 0$
24. $2\sin^2(t) + \sin(t) + 1 = 0$
25. $\cos(t)\sin(t) = \sin(-t)$ *Hint:* The sine function is an odd function.
26. $\cos(t)\sin(t) = \cos(-t)$
27. $3\tan^2(t) + 5\tan(t) + 1 = 0$
28. $\tan(x) - \tan^3(x) = 0$
29. $\sin(t) = \dfrac{1 - \cos(t)}{\sin(t)}$
30. $\sin(t) - \sin^{1/2}(t) + \dfrac{1}{4} = 0$
31. $\cos(t) + \cos^{1/2}(t) + \dfrac{1}{4} = 0$

In exercises 32 through 36, solve the given inequality.

32. $\sin(x) \geq \dfrac{\sqrt{3}}{2}$
33. $\text{Cos}^{-1}(2x) \leq \dfrac{\pi}{6}$
34. $\text{Cos}^{-1}(3x) \leq \dfrac{\pi}{3}$
35. $\cos(t)\sin(t) > \sin(-t)$
36. $2\sin^2(t) + \sin(t) + 1 < 0$
37. (Graphing calculator) Find the approximate solutions to $2\cos x = 1.5$ in the interval $[-\pi, \pi]$ by graphing $f(x) = 2\cos x$ and $g(x) = 1.5$ in the same coordinate system and using the solving method of your calculator.

(Graphing calculator) In exercises 38 and 39, use the method of exercise 37 to find one approximate solution of the equation. (Although these equations don't turn up often, they do serve to illustrate this method of solution.)

38. $e^x\sin(x) = 4$
39. $e^x = \sin x$

Chapter 6: Summary

The purpose of this chapter was to develop the six trigonometric functions. The functions were defined in terms of the wrapping function in order to have them first in their most general form. Properties of these functions were investigated primarily in terms of their graphs. Inverse functions were defined for restrictions of the trigonometric functions. Three new standard graphs were designated: $y = \sin t$, $y = \tan t$, and $y = \sec t$.

Key Concepts

wrapping function	*values for reference numbers*
sin, cos, tan, cot, sec, csc	*graphs, values for reference numbers*
phase shift, period, amplitude	
inverse trigonometric functions	*domains, ranges, graphs*

Skills

You should be able to:

1. Find $W(t)$ for integral multiples of $\pi/6$, $\pi/4$, $\pi/3$, $\pi/2$, π.
2. Find values of the trigonometric functions for reference numbers and their integral multiples.
3. Find approximate values for the trigonometric functions.
4. Graph the trigonometric functions, including translations, expansions, and reflections.
5. Find amplitude, period, and phase shift.
6. Graph the inverse trigonometric functions.
7. Evaluate expressions containing the inverse trigonometric functions.
8. Solve equations and inequalities containing trigonometric expressions.

SUMMARY EXERCISES

The starred exercises can serve as a practice chapter test. The answers are at the end of the answer section at the back of the book.

In exercises 1 through 9, find the value of the given expression.

1. $W\left(-\frac{5\pi}{6}\right)$

2. $W\left(\frac{129\pi}{4}\right)$

* **3.** $W\left(-\frac{2\pi}{3} - 5\pi\right)$

4. $\sin\left(-\frac{\pi}{6} + 3\pi\right)$

5. $\cos(-318\pi)$

* **6.** $\tan\left(\frac{19\pi}{4}\right)$

7. $\sec\left(\pi - \frac{5\pi}{3}\right)$

8. $\csc\left(\frac{(4k+1)\pi}{2}\right), k \in I$

* **9.** $\cot\left(-\frac{7\pi}{6}\right)$

In exercises 10 through 15, find the approximate value correct to four decimal places.

10. $\sin(-3)$

11. $\tan\left(\frac{\pi}{12}\right)$

* 12. $\sec(2.8 - 3\pi)$

13. $\cos(2e)$

14. $\cot[6.1 + (2k + 1)\pi],\ k \in I$

* 15. $\csc\left(\sqrt{\frac{3}{2}}\right)$

In exercises 16 through 21, graph the given function.

16. $g(t) = \cos(4t)$

17. $g(t) = \cos\left(\frac{1}{4}t\right)$

* 18. $h(x) = 2\tan\left(x + \frac{\pi}{2}\right)$

19. $k(x) = -\csc\left(2x - \frac{\pi}{3}\right)$

20. $m(x) = 0.5\cot\left(3x + \frac{3\pi}{4}\right)$

* 21. $n(x) = 3\,\mathrm{Cos}^{-1}(x - 2)$

In exercises 22 through 33, complete the chart by answering yes or no and supplying the period.

		Period	1-1	Even	Odd
22.	sin				
23.	cos				
* 24.	tan				
25.	cot				
26.	sec				
* 27.	csc				
28.	Sin^{-1}				
29.	Cos^{-1}				
* 30.	Tan^{-1}				
31.	Cot^{-1}				
32.	Sec^{-1}				
* 33.	Csc^{-1}				

In exercises 34 through 39, give the amplitude (or expansion factor if amplitude is undefined), period, and phase shift for each of the given functions.

34. $f(x) = -2\csc(5x - 3)$

35. $h(t) = 3.5\sin\left(\frac{\pi}{8} - 2.3t\right)$

* 36. $m(x) = -\cos(-6.8x - 2.9)$

37. $g(t) = \frac{5}{6}\cot\left(-\frac{3}{8}t + \frac{\pi}{5}\right)$

38. $y = \frac{\sqrt{14}}{29}\tan(\sqrt{3}x - \sqrt{14})$

* 39. $y = \frac{2}{3}\sec\left(\frac{\pi}{7} - \frac{1}{5}x\right)$

In exercises 40 through 42, evaluate the given expression.

40. $\mathrm{Sin}^{-1}\left(-\frac{\sqrt{3}}{2}\right)$

41. $\mathrm{Cos}^{-1}\left(\cos\frac{\pi}{7}\right)$

* 42. $\mathrm{Tan}^{-1}\left(\sin\frac{7\pi}{6} + \cos\left(-\frac{\pi}{3}\right)\right)$

In exercises 43 through 48, solve the given equation or inequality. Be sure to give the complete solution set.

43. $\mathrm{Tan}^{-1}x = \frac{\pi}{3}$

44. $\cos(2t) \geq \frac{\sqrt{3}}{2}$

* 45. $\sin(t - 1) = \ln 2$

46. $6\sin^2 x - \sin x - 2 = 0$

47. $\cot(x)\cdot\cos(x) = 1 + \sin x$

* 48. $\tan x = \frac{3}{\tan x}$

Applications of the Trigonometric Functions

- 7.1 Right Triangles
- 7.2 Laws of Cosines and Sines
- 7.3 Basic Identities
- 7.4 Uses of the Basic Identities
- 7.5 Complex-Number Representation

7.1 Right Triangles

Toolkit

You'll need the fact that ratios of the lengths of corresponding sides of similar triangles are equal. Also, the sum of the measures of the angles in a triangle is 180°. The notation AB represents the measure of line segment $\overline{AB}$.

As our first application of the trigonometric functions, we return to their historical roots in right triangles. First we need some definitions. An *angle* is formed by two rays (half-lines) with a common starting point. For our purposes, it's convenient to modify this definition slightly. We identify one side of the angle as the *initial side* and the other as the *terminal side*. These names come from considering an angle to be formed by the motion of the terminal side away from the initial side; an angle is "opened up." You may wonder how to decide which side is which. This can be done only by placing the angle in standard position: one side is on the positive x-axis with starting point at the origin. This side is the initial side; the other is the terminal side.

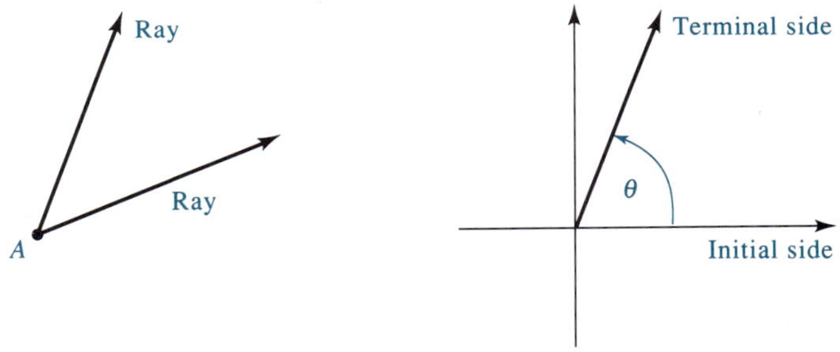

Standard position

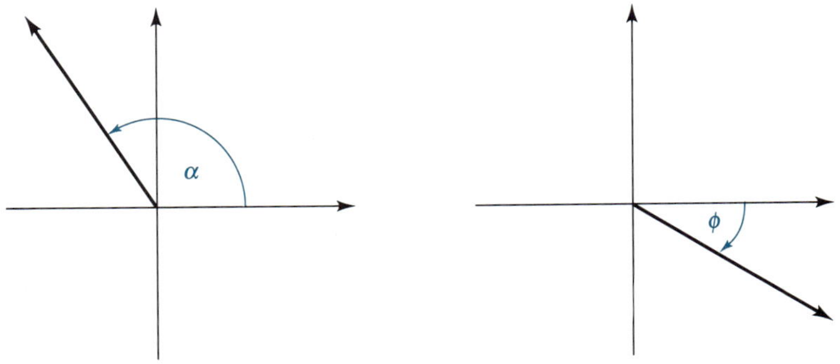

In the preceding pictures we have indicated the direction of motion of the terminal side away from the initial side. As in our discussion of the wrapping function, counterclockwise movement is positive; clockwise movement is negative.

Angles θ and α should be assigned positive measure and angle ϕ, negative measure. (It's common in trigonometry to name angles with Greek letters.)

How do we measure angles? One way is to use ideas from the last chapter. To measure an angle in standard position, draw a unit circle with center at the origin on the same set of axes. The point of intersection with the terminal side and the circle has coordinates (cos t, sin t).

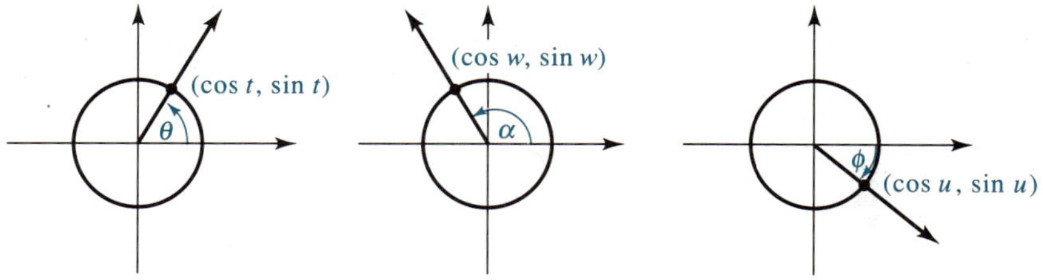

Here t is the length of the arc between the two sides of the angle θ. We define this to be the measure of the angle and may write $\theta = t$, $\alpha = w$, and $\phi = u$. Notice that t and w are positive, as we wanted, and u is negative, as required. We also use the name of the angle to represent its measurement. If we need to emphasize that t, w, and u represent measures, we denote the unit of measurement a **radian.**

θ is t radians; α is w radians; ϕ is u radians.

Definition Thus, 1 radian is the measure of an angle whose terminal side travels through 1 unit on the unit circle.

A radian is a dimensionless unit of measurement. This means if you multiply 6 radians by 4 feet, the result is 24 feet, not 24 radian-feet.

EXAMPLE Find the measure, in radians, of angle γ.

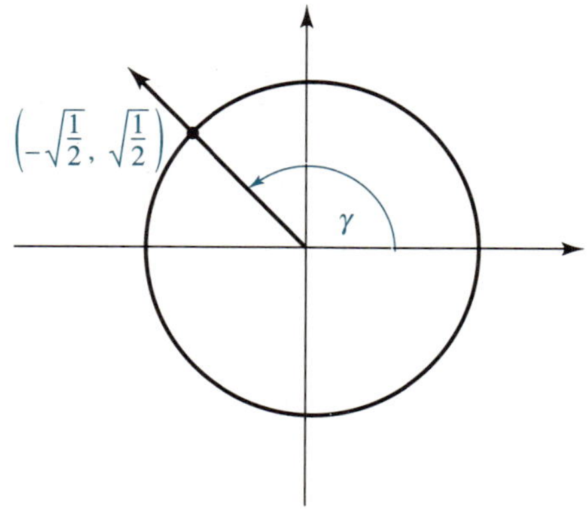

The point $\left(-\sqrt{\frac{1}{2}}, \sqrt{\frac{1}{2}}\right)$ is on the unit circle because $\left(-\sqrt{\frac{1}{2}}\right)^2 + \left(\sqrt{\frac{1}{2}}\right)^2 = \frac{1}{2} + \frac{1}{2}$, or 1. The measure of γ is t, where $\cos t = -\sqrt{\frac{1}{2}}$ and $\sin t = \sqrt{\frac{1}{2}}$. We recognize t as $3\pi/4$.

The measure of γ is $3\pi/4$ radians. ■

How important is the unit circle to this definition of an angle's measure? Could we use another circle? In fact we may measure the angle using any circle with center at the origin. Consider angle β in standard position and two circles centered at the origin with radii 1 and r (Figure 7.1a).

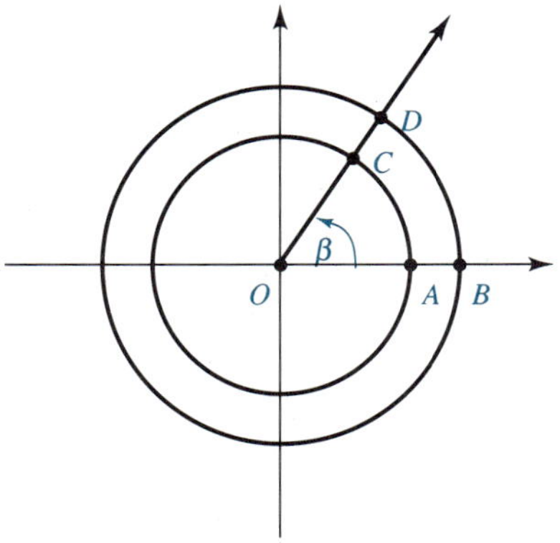

A: (1, 0)
B: (r, 0)
C: (cos t, sin t)
D: (x, y)

The length of the arc from A to C is t units.

β is t radians

FIGURE 7.1a

What are the coordinates of D in terms of some of the measures we know? Let's extract part of the picture and construct triangles EOC and FOD with perpendiculars at E and F. These are similar triangles, so the ratios of the lengths of corresponding sides are equal.

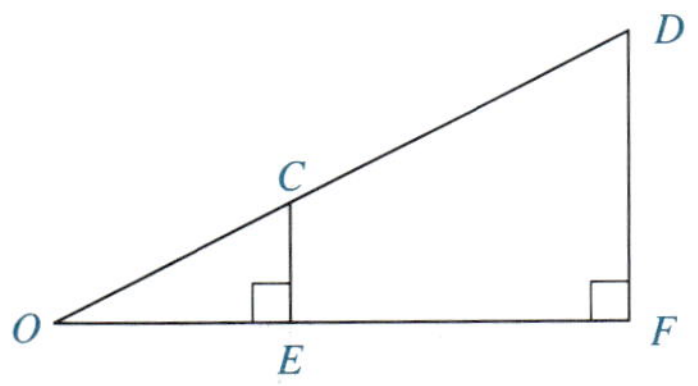

Using the information from Figure 7.1a, we have

$$\frac{1}{r} = \frac{\cos t}{x} = \frac{\sin t}{y}.$$

Thus, the coordinates of point D: (x, y) are $(r \cos t, r \sin t)$.

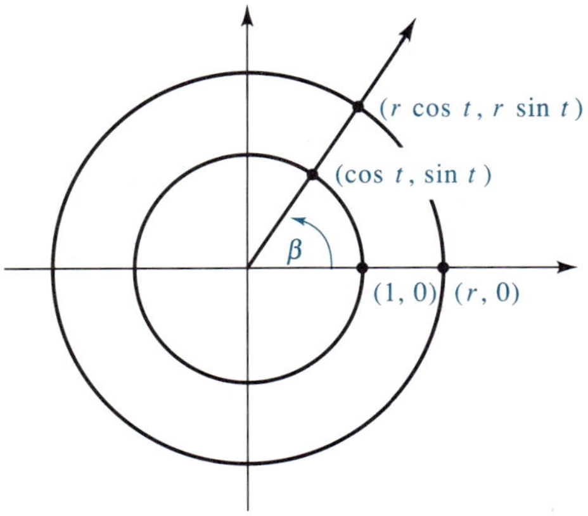

Note also that the arc from B to D has length rt because BOD and AOC are similar sectors. We can use a circle of radius r to calculate the measure of an angle as long as we take into account the radius r.

EXAMPLE Find the measure, in radians, of β if $(\sqrt{3}/4, -\frac{1}{4})$ is a point on the terminal side of β, as in Figure 7.1b.

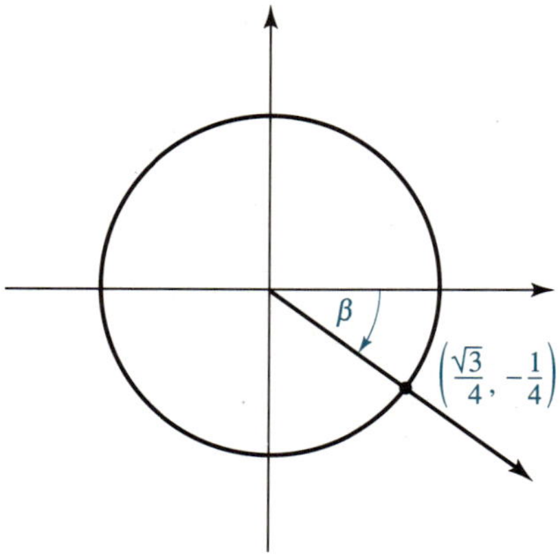

FIGURE 7.1b

Because $(\sqrt{3}/4)^2 + (-\frac{1}{4})^2 = \frac{3}{16} + \frac{1}{16}$, or $\frac{1}{4}$, this point lies on a circle of radius $\frac{1}{2}$ with center (0, 0). Now we can interpret $(\sqrt{3}/4, -\frac{1}{4})$ as $(\frac{1}{2}\cos t, \frac{1}{2}\sin t)$, and see that $\cos t = \sqrt{3}/2$, $\sin t = -\frac{1}{2}$. We recognize t as $-\pi/6$.
The measure of β is $-\pi/6$ radians. ■

This discussion shows that if we liked, we could give definitions of the trigonometric functions using any circle centered at the origin.

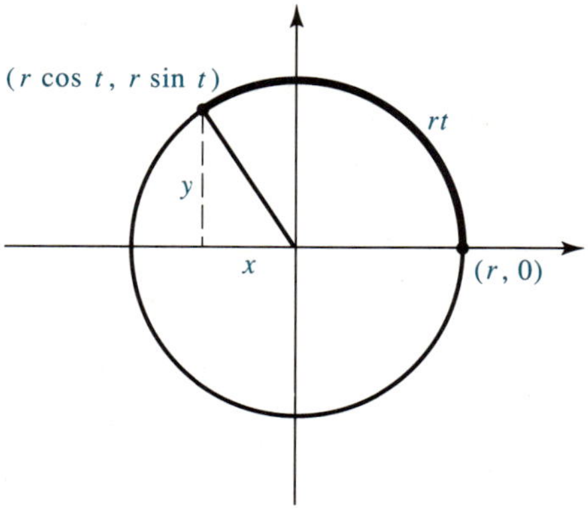

FIGURE 7.1c

Referring to Figure 7.1c,

$$x = r\cos t \quad \Rightarrow \quad \cos t = \frac{x}{r}$$

$$\sin t = \frac{y}{r}, \qquad \tan t = \frac{y}{x}, \qquad \cot t = \frac{x}{y},$$

$$\sec t = \frac{r}{x}, \qquad \csc t = \frac{r}{y}.$$

You can see the advantage of using a circle of radius 1 for our original definition.

This method of measuring angles is not very efficient because it's indirect. You probably know another way to measure angles directly with a protractor and another unit of measurement for angles, a degree. A degree is the measure of an angle that, in standard position, covers 1/360 of a complete circle. Angles may be subdivided into tenths and hundredths or into minutes and seconds by using the equivalences

$$1 \text{ degree} \equiv 60 \text{ minutes},$$
$$1 \text{ minute} \equiv 60 \text{ seconds}.$$

(The choices of 360 and 60 are part of our Babylonian heritage.) Degrees are denoted by °, minutes by ′, and seconds by ″. We read 13°4′15″ as 13 degrees 4 minutes 15 seconds. Your calculator can probably show degrees subdivided either in decimal form or as degrees, minutes, and seconds. A calculator will interpret a number you enter as radians or degrees according to the mode in which you set it.

What is the equivalence for degrees and radians? An angle θ that opens through a full circle has radian measure 2π and degree measure 360°. So, we have $2\pi \equiv 360°$ or, more simply, $\pi \equiv 180°$.

EXAMPLE Convert the reference angles $\pi/6$, $\pi/4$, $\pi/3$, and $\pi/2$ to degree measure.

$$\frac{\pi}{6} \equiv \frac{180°}{6}, \text{ or } \mathbf{30°} \qquad \frac{\pi}{4} \equiv \frac{180°}{4}, \text{ or } \mathbf{45°}$$

$$\frac{\pi}{3} \equiv \frac{180°}{3}, \text{ or } \mathbf{60°} \qquad \frac{\pi}{2} \equiv \frac{180°}{2}, \text{ or } \mathbf{90°}$$

■

The classical problem in trigonometry is to solve a triangle. This means given some information about a triangle, find, if possible, the measure of the remaining sides or angles. Before trying such problems, we will translate our definitions of the trigonometric functions into a restricted version in terms of right triangles.

We label our right triangles as follows: α, β, γ are the angles, γ is the right angle. The side facing α is a; the one facing β is b; and the one facing γ is c. If necessary, we label as A the vertex where α is; B where β is; and C where γ is. We also use the name of a side to represent its length.

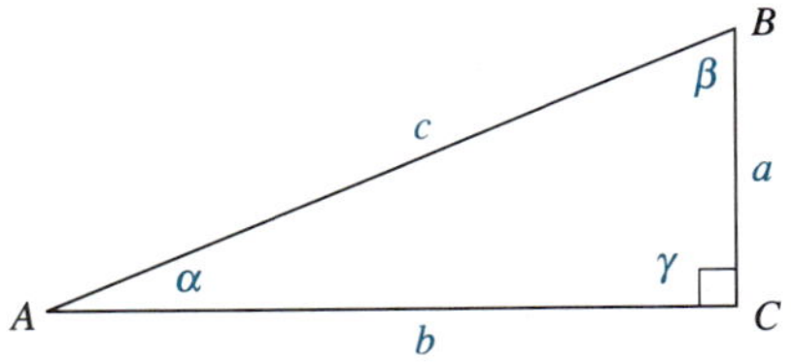

To define the trigonometric functions of α in terms of these measurements, place α in standard position and use a circle of radius c.

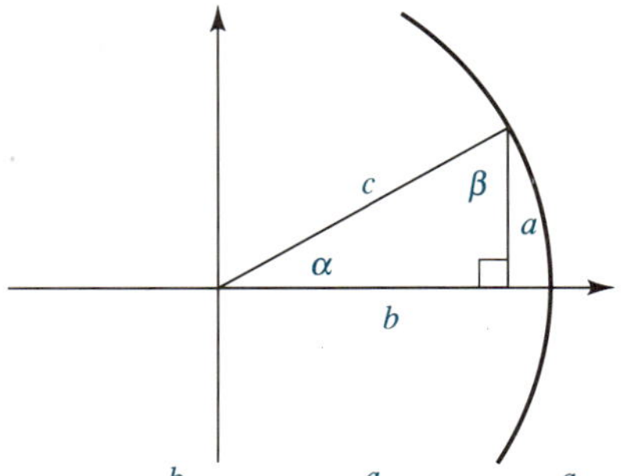

$$\cos\alpha = \frac{b}{c} \quad \sin\alpha = \frac{a}{c} \quad \tan\alpha = \frac{a}{b}$$

$$\sec\alpha = \frac{c}{b} \quad \csc\alpha = \frac{c}{a} \quad \cot\alpha = \frac{b}{a}$$

These can be interpreted as follows:

$$\cos\alpha = \frac{\text{length of adjacent side}}{\text{length of hypotenuse}}, \quad \text{abbreviated} \quad \cos\alpha = \frac{\text{adj}}{\text{hyp}},$$

$$\sin\alpha = \frac{\text{length of opposite side}}{\text{length of hypotenuse}}, \quad \text{abbreviated} \quad \sin\alpha = \frac{\text{opp}}{\text{hyp}},$$

$$\tan\alpha = \frac{\text{length of opposite side}}{\text{length of adjacent side}}, \quad \text{abbreviated} \quad \tan\alpha = \frac{\text{opp}}{\text{adj}}.$$

This allows us to calculate values of the trigonometric functions for either α or β when the right triangle is in any position.

Illustration

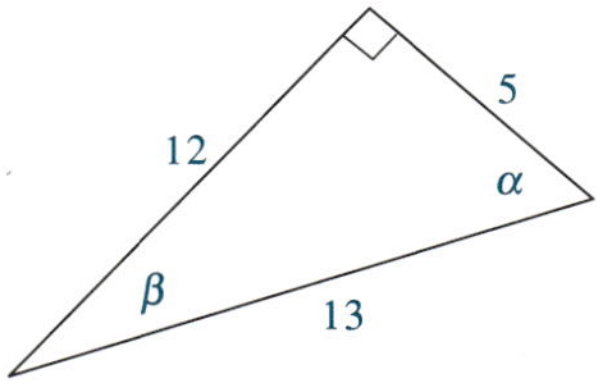

$$\cos\alpha = \frac{\text{adj}}{\text{hyp}}, \text{ or } \frac{5}{13}, \quad \sin\alpha = \frac{\text{opp}}{\text{hyp}}, \text{ or } \frac{12}{13}, \quad \tan\alpha = \frac{\text{opp}}{\text{adj}}, \text{ or } \frac{12}{5}$$

$$\cos\beta = \frac{\text{adj}}{\text{hyp}}, \text{ or } \frac{12}{13}, \quad \sin\beta = \frac{\text{opp}}{\text{hyp}}, \text{ or } \frac{5}{13}, \quad \tan\beta = \frac{\text{opp}}{\text{adj}}, \text{ or } \frac{5}{12}$$

Note: $\alpha + \beta = 90°$. Such angles are called *complementary angles*. The preceding values show that the value of any trigonometric function of α equals the corresponding *cofunction* of β, and vice versa. For example, sine α = cosine β and tangent β = cotangent α. The *co* in the cofunction names comes from *co*mplementary angles. □

Don't feel overwhelmed by all these definitions. You really have only two important new ones:

$$\text{sine } \alpha = \frac{\text{opp}}{\text{hyp}} \quad \text{and} \quad \text{cosine } \alpha = \frac{\text{adj}}{\text{hyp}}.$$

The other four functions can be described in terms of sine and cosine, as before.

Now let's solve some right triangles. One situation you already know well. If you are given the lengths of any two sides of a right triangle, the Pythagorean theorem provides a way to find the length of the third side. Our new definitions give a way to calculate the measures of angles of the triangle.

EXAMPLE Solve the right triangle *ABC* given $a = 4$ and $c = 5$.
A sketch of the triangle may help.

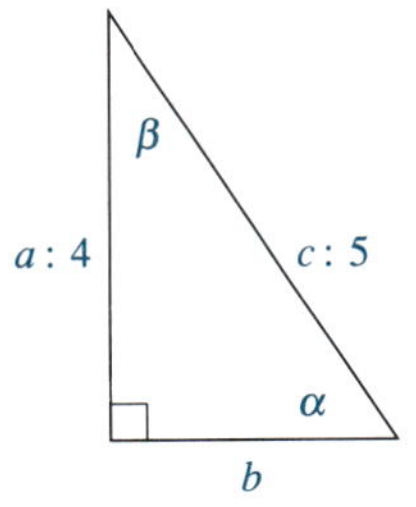

By the Pythagorean theorem, $4^2 + b^2 = 5^2$. $b^2 = 25 - 16 \Rightarrow b = 3$. Using one of the previous restricted definitions, $\sin \alpha = \frac{4}{5}$ and $\boldsymbol{\alpha = \text{Sin}^{-1}(\frac{4}{5}) \cdot 180°/\pi}$. Since $\alpha + \beta = 90°$, $\boldsymbol{\beta = 90° - \text{Sin}^{-1}(\frac{4}{5}) \cdot 180°/\pi}$.
We have found the missing parts of ΔABC, but you may not be satisfied with the answers. ■

In the rest of this section and in the next section, we will find approximate answers, if necessary, in solving triangles. There are many points of concern in deciding how such approximations should be done. Rather than attempting any discussion of these points, we make an arbitrary decision to report all approximate lengths to the nearest hundredth of the given unit and all approximate angles to the nearest hundredth of a degree or to the nearest second.

EXAMPLE Solve right triangle *ABC* (approximately), given $a = 4$ and $c = 5$.
We already know $b = 3$ (exactly) and $\alpha = \left(\text{Sin}^{-1}\frac{4}{5}\right) \cdot 180°/\pi$. To approximate α in degrees, put your calculator in degree mode and key in $\text{Sin}^{-1}\frac{4}{5}$. You will obtain approximately 53.13°. In degree mode the calculator automatically carries out the conversion. Hence, $\alpha \approx 53.13°$, or 53°7′48″ and $\beta \approx 90° - 53.13° = 36.87°$, or 36°50′12″.
For ΔABC, we have

$\boldsymbol{a = 4}$	$\boldsymbol{\alpha \approx 53.13°}$
$\boldsymbol{b = 3}$	$\boldsymbol{\beta \approx 36.87°}$
$\boldsymbol{c = 5}$	$\boldsymbol{\gamma = 90°}$

■

EXAMPLE Sixty feet from the base of a tree, a line from the ground to the top of the tree makes an angle of 43° with the ground. Assuming level ground, approximately how tall is the tree?

A sketch helps to organize our thoughts.

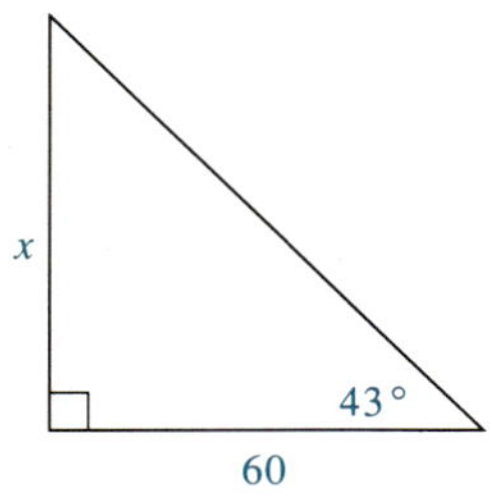

$$\tan 43^\circ = \frac{x}{60}$$ Why tangent? It uses the information we have.

$$\Leftrightarrow \quad x = 60 \tan 43^\circ$$

$$\Rightarrow \quad x \approx 55.95$$

The tree is approximately 55.95 feet tall. ■

It appears that if we are given two pieces of information about a right triangle, we can compute (at least approximately) the remaining parts of the triangle. In one case this isn't true. If you are given α and β, only the shape of the triangle can be determined, not its size. Consider the following illustration.

Illustration

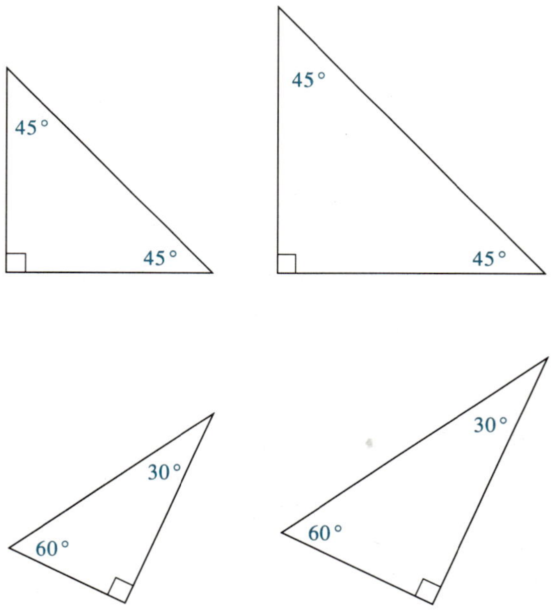

□

EXERCISES 7.1

In exercises 1 through 8, write the positive and negative measure of β in radians between -2π and 2π if the given point is on the terminal side of β, as in Figure 7.1b.

1. $(8, -8)$
2. $(-1, -\sqrt{3})$
3. $(4\sqrt{3}, 4)$
4. $(-3, 0)$
5. $\left(-\sqrt{\frac{1}{19}}, \sqrt{\frac{3}{19}}\right)$
6. $(108\sqrt{3}, -108)$
7. $(483.92, 0)$
8. $\left(\sqrt{\frac{1}{3.2}}, \sqrt{\frac{1}{3.2}}\right)$

In exercises 9 through 12, write the coordinates of the point on the circle with radius r that is on the terminal side of β.

9. $r = 2, \beta = 30°$ **10.** $r = 6, \beta = 120°$
11. $r = 12, \beta = -45°$ **12.** $r = \sqrt{13}, \beta = 210°$

In exercises 13 through 18, convert the angle measure from radians to degrees.

13. $-\frac{5\pi}{3}$ **14.** 4.2π
15. 3.4 **16.** 183
17. $\frac{19\pi}{6}$ **18.** $-\frac{487\pi}{4}$

In exercises 19 through 24, convert the angle measure from degrees to radians.

19. 240° **20.** 585°
21. 13° **22.** −400°
23. −150° **24.** 1170°

In exercises 25 through 30, find the approximate solution to right triangle ABC using the given information. Assume standard labeling.

25. $a = 13, b = 7$ **26.** $c = 432, \beta = 17°$
27. $\alpha = 64.2°, a = 32.1$ **28.** $\alpha = 45°, c = 100$
29. $\beta = 81.33°, a = 0.36$ **30.** $b = 5.24, c = 91.57$

In exercises 31 through 38, find the approximate measure of the indicated side or angle of the right triangle ABC using the given information. Assume standard labeling.

31. β, given $a = 34$ meters, $b = 16$ meters
32. c, given $\beta = 32°$, $b = 3.22$ inches
33. b, given $\alpha = 11.34°$, $c = 498.5$ feet
34. a, given $\cos\beta = 0.8793$, $c = 43.33$ yards
35. c, given $\tan\alpha = 3.69921$, $a = 700$ centimeters
36. α, given $a = 45.89$ zurks, $b = 392.33$ durks (1 durk = 0.3 zurks)
37. β, given $\tan\alpha = 0.0345$, $b = 4.30$ millimeters
38. γ, given $a = 4.993$ gads, $b = 3.209$ gads
39. Kaneesha and Mika are responsible for selecting a holiday tree for a major department store. The tree must be at least 50 feet tall but no more than 65 feet tall. To determine the approximate height of the tree in the forest, they use an instrument to determine the angle of elevation from the horizontal line of sight to the top of the tree. They find a tree that appears to be the correct height and take the following measurements: The angle of elevation is approximately 75.963757°, the distance from the instrument to the base of the tree is 10 feet, and the distance from the ground to the instrument is 5.5 feet. Assuming the ground is level, can they use this tree?
40. Kaneesha and Mika from exercise 39 are still looking for the perfect tree. They find two candidates and set the transit up between the two, taking the following approximate measurements: The angle of elevation to the first tree is 70.39° and to the second tree is 70.4231°, the distance from the transit to the base of the first tree is 20 feet 3 inches and to the second tree is 34 feet 2.5 inches, and the height of the transit is 5.5 feet. Can either of these trees be used?
41. We have discussed measuring angles in degrees and radians. A third method uses the measurement unit mil. A mil is approximately 1/1000 radian. What should the angle of elevation be, to the nearest hundredth of a mil, from a point 6 feet from the base of a vertical wall that is 11.34 feet tall to the top of the wall?
42. Jill and Tom are planning to elope. Tom has found a ladder that can be extended to a maximum of 30 feet. Jill is worried that it won't be long enough to reach the base of her window, which has an angle of elevation of 47° measured from the opposite side of the creek that runs at the side of the house. The distance from the opposite side of the creek to the base of the house is 22.78 feet. Will the ladder be long enough to reach or will young love be foiled because Tom failed trigonometry?

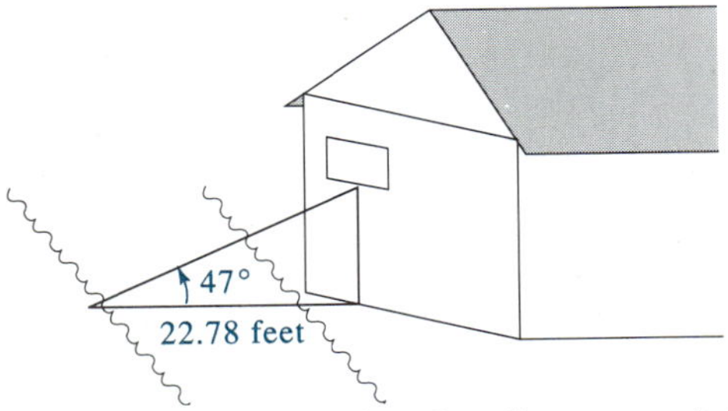

43. How many square feet of roofing tiles are needed for a shed roof that is 25 feet wide and makes an angle of 32.8° with the vertical? The horizontal distance from the bottom of the roof to the side of the house is 16.5 feet.

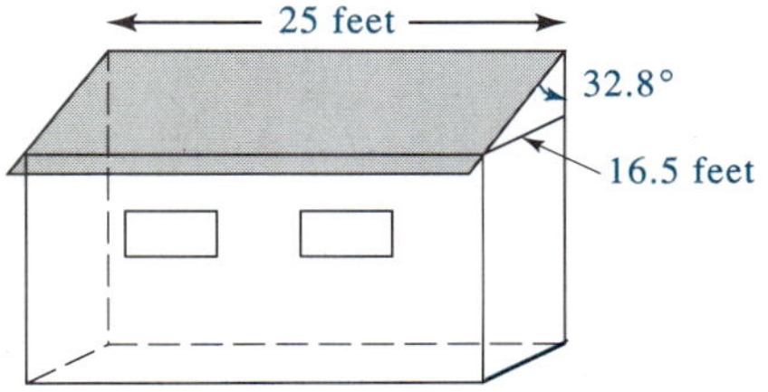

44. Sally knows that in her backyard the angle of elevation of the sun is 35° at 5:00 P.M. on October 13. How tall is a tree in her backyard if the length of its shadow is 300 feet?

45. The interstate highway system for a major metropolitan city consists of two perimeter highways and three others running through the central city. James wishes to travel from city A to city B. Which route is the shortest, traveling the outside perimeter, the inside perimeter, or the downtown system?

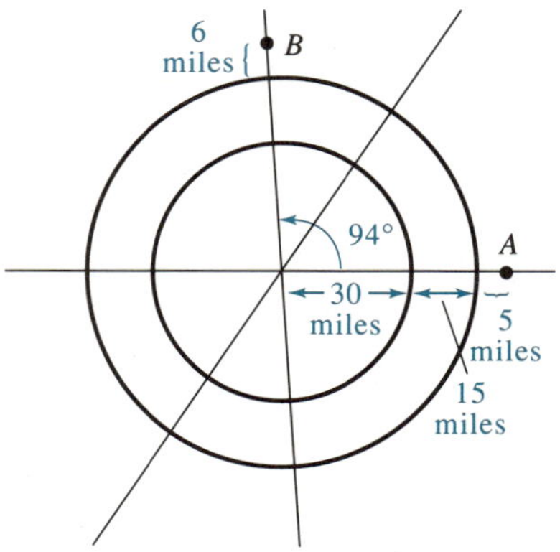

7.2 Laws of Cosines and Sines

Toolkit

You'll need to know the area of a triangle is computed by taking one-half the product of the base and the height.

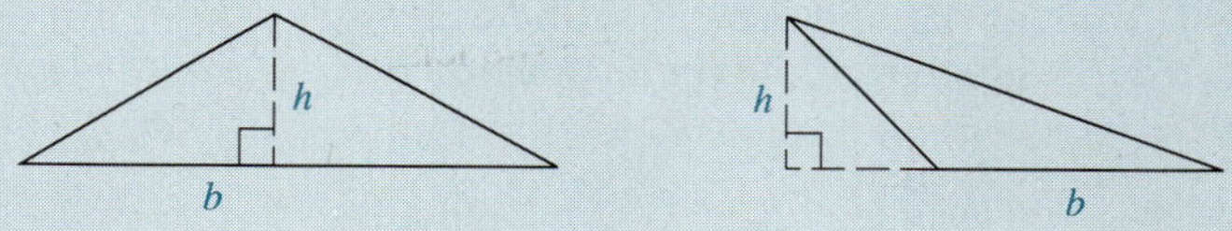

In this section we explore solving triangles that are not right triangles. The Pythagorean theorem does not apply to these triangles, and we first develop an extension of it that does apply. We continue the labeling convention of the last section.

Consider the triangle ABC with α in standard position. Here are sketches of two possibilities for α. The discussion will use Figure 7.2a; you will fill in the details for Figure 7.2b in Exercise 33.

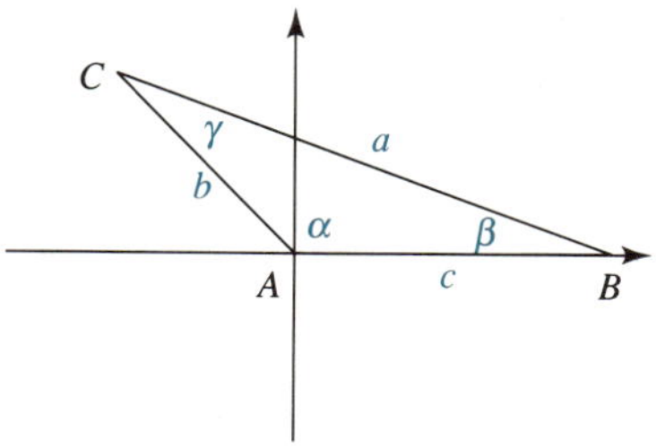

FIGURE 7.2a

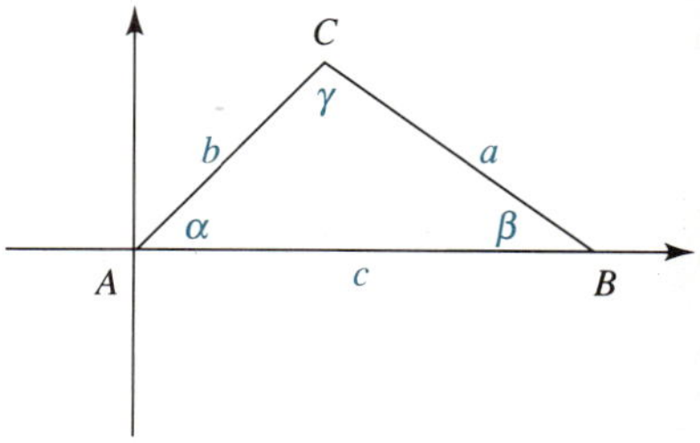

FIGURE 7.2b

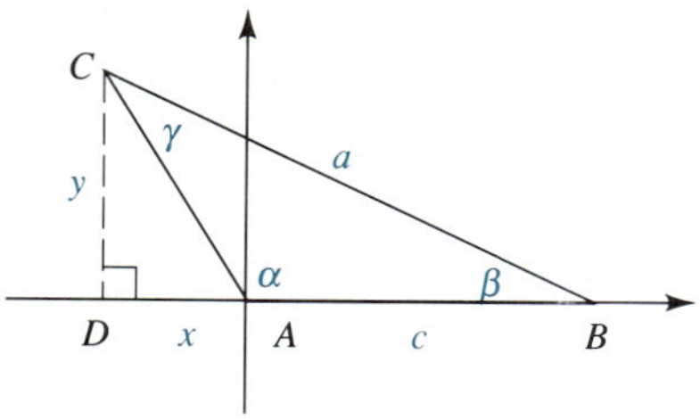

The coordinates of C are $(b \cos \alpha, b \sin \alpha)$, according to the discussion in Section 7.1. Draw a line from C perpendicular to the x-axis; call the intersection point D. Triangle BCD is a right triangle, so the Pythagorean theorem holds for it. We have labeled the two new line segments x and y, but $y = b \sin \alpha$ and $x = b \cos \alpha$. Note x is a negative number. The length of BD is $c + |x|$, or $c + (-x)$, because x is negative. It must be true that

$$
\begin{aligned}
& CD^2 + BD^2 = BC^2 \\
\Leftrightarrow \quad & y^2 + (c - x)^2 = a^2 \\
\Leftrightarrow \quad & (b \sin \alpha)^2 + (c - b \cos \alpha)^2 = a^2 \\
\Leftrightarrow \quad & b^2\sin^2\alpha + c^2 - 2bc \cos \alpha + b^2\cos^2\alpha = a^2 \\
\Leftrightarrow \quad & b^2(\sin^2\alpha + \cos^2\alpha) + c^2 - 2bc \cos \alpha = a^2 \\
\Leftrightarrow \quad & b^2 + c^2 - 2bc \cos \alpha = a^2.
\end{aligned}
$$

For triangle ABC we now have a formula close to the Pythagorean theorem; it has an additional term, $-2bc \cos \alpha$. We could describe this formula by saying the length of a side squared is the sum of the squares of the lengths of the other two sides minus twice the product of the lengths of the other two sides and the cosine of the angle between them. This is wordy, but it emphasizes that the labeling was arbitrary.

The Law of Cosines: For a triangle ABC with γ the angle opposite side c,

$$c^2 = a^2 + b^2 - 2ab \cos \gamma$$

The law of cosines is a true generalization of the Pythagorean theorem, since if c is the length of the hypotenuse of a right triangle, γ is 90° and $\cos \gamma$ is 0; thus $c^2 = a^2 + b^2 - 2ab \cos \gamma$ yields $c^2 = a^2 + b^2$.

EXAMPLE Two points at opposite ends of a lake are marked, and a third point that is 63 feet from one of these points and 72 feet from the other is selected. The lines determined by these points form an angle of 123° at the third point. How far is it across the lake between the selected points?

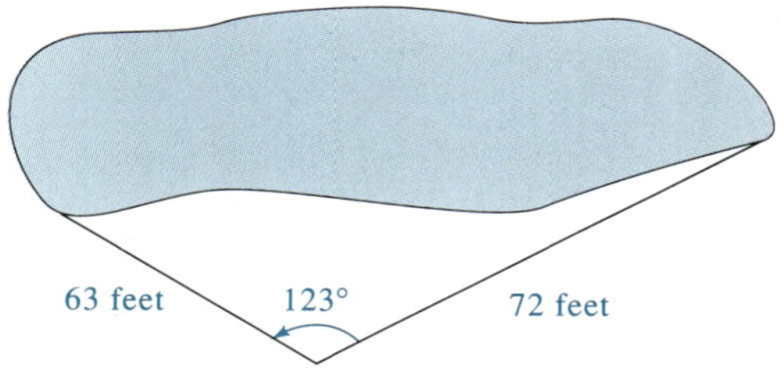

We wish to find the third side of a triangle for which we know two sides and the angle between them.

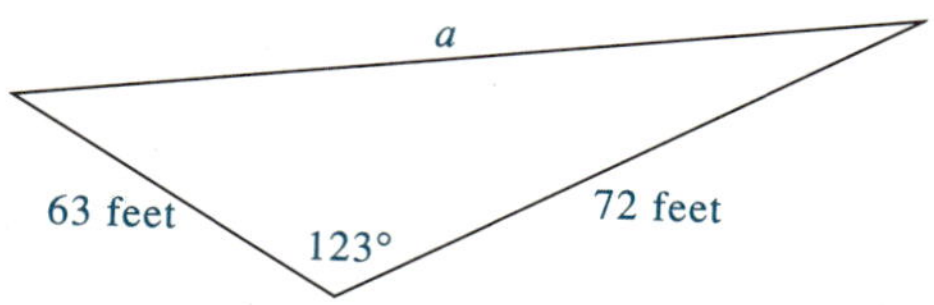

The law of cosines says

$$a^2 = 63^2 + 72^2 - 2(63)(72)\cos 123°$$
$$\Leftrightarrow \quad a^2 = 3969 + 5184 - 9072 \cos 123°$$
$$\Rightarrow \quad a^2 \approx 14{,}093.6533.$$

The length is approximately 118.72 feet.

Note: The side facing the largest angle in a triangle is the longest side. ■

EXAMPLE Solve ΔABC given $a = 6$, $b = 8.5$, and $c = 10$.

We need to find values for the angles α, β, and γ. It would be reasonable to use any of the versions of the law of cosines because we have values for all the variables but one.

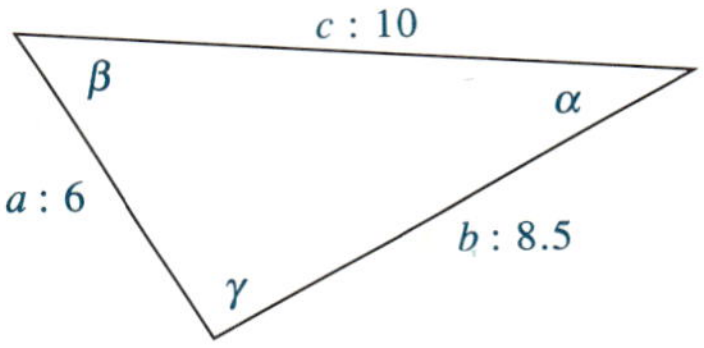

$$a^2 = b^2 + c^2 - 2ab \cos \alpha$$
$$6^2 = 8.5^2 + 10^2 - 2(8.5)(10)\cos \alpha$$
$$\Leftrightarrow \quad 36 = 172.25 - 170 \cos \alpha$$
$$\Leftrightarrow \quad \cos \alpha = \frac{136.25}{170}$$
$$\Rightarrow \quad \alpha = \text{Cos}^{-1}\left(\frac{136.25}{170}\right)$$
$$\Rightarrow \quad \alpha \approx 36.73°$$

Note: Any other solutions of $\cos \alpha = 136.25/170$ have degree measure larger than 180° or smaller than 0°. Similarly,

$$b^2 = a^2 + c^2 - 2ac \cos \beta$$
$$8.5^2 = 6^2 + 10^2 - 2(6)(10) \cos \beta$$
$$\Leftrightarrow \quad 72.25 = 136 - 120 \cos \beta$$
$$\Leftrightarrow \quad \cos \beta = \frac{63.75}{120}$$
$$\Rightarrow \quad \beta = \text{Cos}^{-1}\left(\frac{63.75}{120}\right)$$
$$\Rightarrow \quad \beta \approx 57.91°.$$

Finally, we have $\gamma \approx 180° - 36.73° - 57.91°$, so $\gamma \approx 85.36°$. You may also use the law of cosines a third time to approximate γ. This would serve as a check on your other work.
For triangle ABC,

$$\begin{aligned} a &= 6, & \alpha &\approx 36.73°, \\ b &= 8.5, & \beta &\approx 57.91°, \\ c &= 10, & \gamma &\approx 85.36°. \end{aligned}$$

■

We can use Figures 7.2a and 7.2b to develop area formulas for a triangle. They are repeated here as Figures 7.2c and 7.2d.

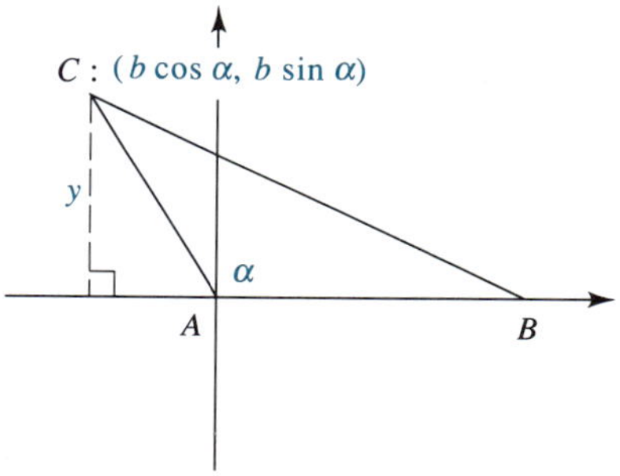

FIGURE 7.2c

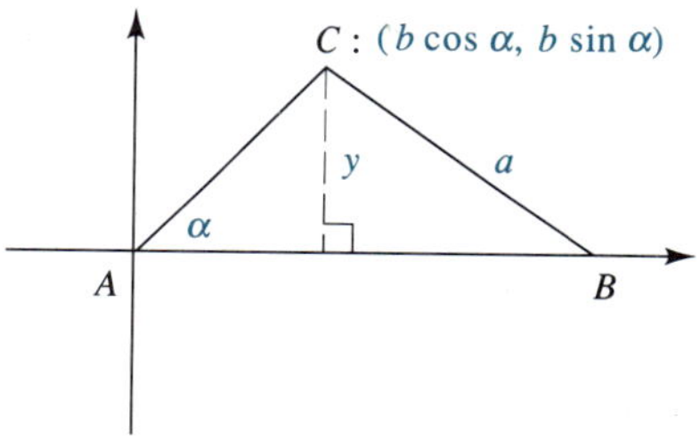

FIGURE 7.2d

The area of a triangle is $\frac{1}{2}$ · base · height. Using $\overline{AB}$ as the base, the height is y, or $b \sin \alpha$. The area of ΔABC is $\frac{1}{2} bc \sin \alpha$. Any side can serve as the base.

The *area of a triangle* is half the product of the length of two sides and the sine of the angle between those sides: $A = \frac{1}{2}bc \sin \alpha$.

EXAMPLE Compute the approximate area of ΔABC with α of approximate measure 102.71°, b of approximate length 13.1 meters, and c of approximate length 13.2 meters.

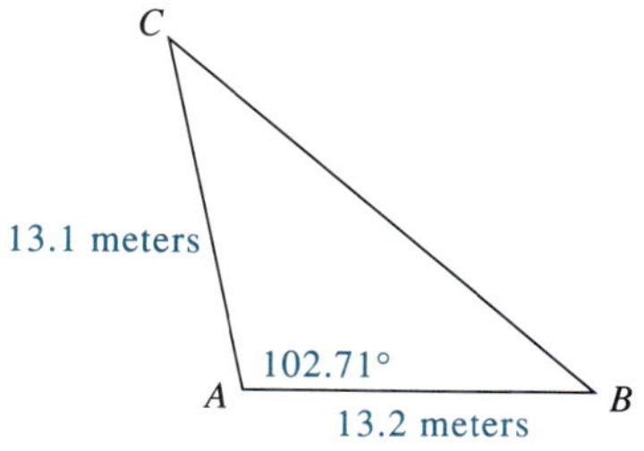

The area of ΔABC is $\frac{1}{2} bc \sin \alpha$.

$$\begin{aligned} \text{Area} &\approx \frac{1}{2}(13.2)(13.1)\sin(102.71°) \\ &\approx 86.46 \sin(102.71°) \end{aligned}$$

The area of triangle ABC is approximately 84.34 square meters. ■

Before we leave the topic of area, we note a formula for the area of a circular sector.

The area of circular sector AOB is $\theta \cdot r^2/2$, θ in radians. Why?

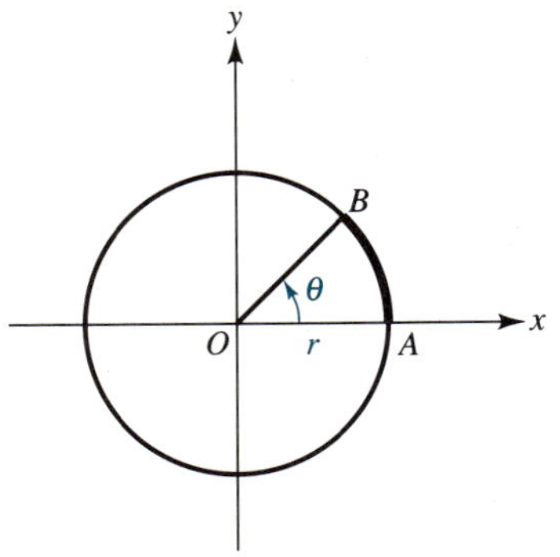

The area of a sector varies directly as the angle which it subtends, area $= k\theta$. A semicircle of radius r has area $\frac{1}{2}\pi r^2$, and any semicircle is a sector with angle π. Hence, $\frac{1}{2}\pi r^2 = k\pi$. Solving for k, we have k is $r^2/2$, and the area of AOB is $\theta r^2/2$.

EXAMPLE Compute the approximate area of the circular sector with radius 13.2 inches and subtending angle 61.2°.

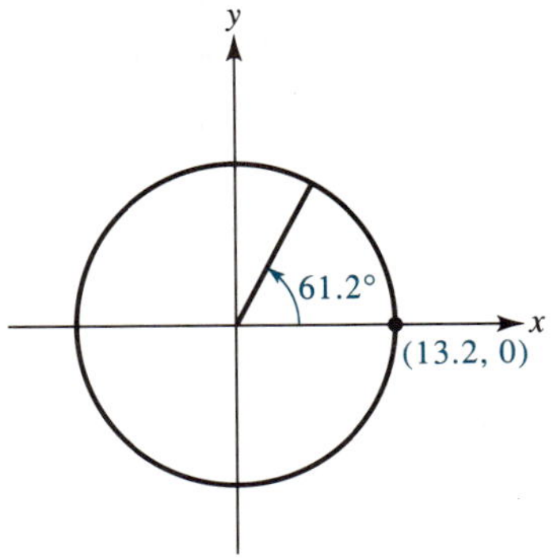

$$\text{area} = \frac{(61.2°)(\pi/180°)(13.2)^2}{2}$$

The area is approximately 93.06 square inches. ■

It may seem that the law of cosines has solved our problems in dealing with triangles that are not right triangles, but consider the following situation. Suppose $a = 6.2$, $b = 11.6$, and $\beta = 25°$ and you are asked to solve ΔABC. In two versions of the law of cosines, we lack values for two of the variables. For example,

$$a^2 = b^2 + c^2 - 2bc \cos \alpha$$
$$(6.2)^2 = (11.6)^2 + c^2 - 2(11.6)c \cos \alpha.$$

The third version produces a quadratic equation in c. There is another relationship between the sides and angles of any triangle to handle this case and others that the law of cosines will not. (*Note:* When you are given only the measure of two angles and one side, there is also not enough information to use the law of cosines.)

Consider once again Figures 7.2a and 7.2b, reproduced here as Figures 7.2e and 7.2f.

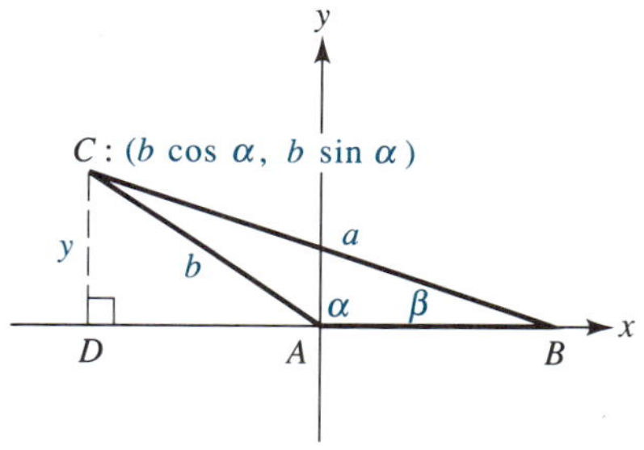

FIGURE 7.2e

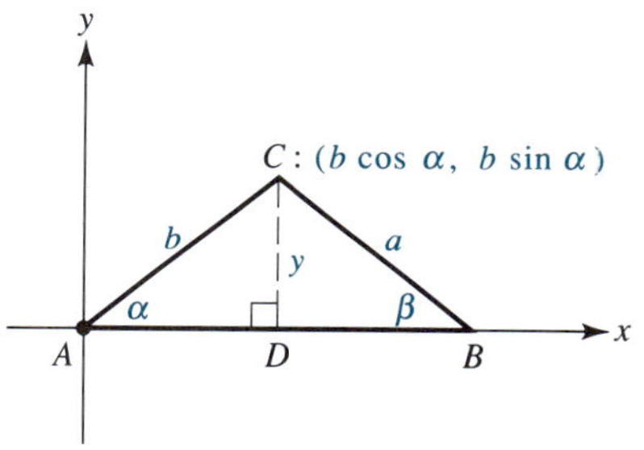

FIGURE 7.2f

As we've noted before, $y = b \sin \alpha$, but in ΔBCD, $\sin \beta = y/a$, or opp/hyp. Solving this last equation for y, $y = a \sin \beta$, and thus $b \sin \alpha = a \sin \beta$. This last statement is often written in its equivalent form

$$\frac{a}{\sin \alpha} = \frac{b}{\sin \beta}.$$

A similar argument shows $\dfrac{b}{\sin \beta} = \dfrac{c}{\sin \gamma}$.

The Law of Sines: $\dfrac{a}{\sin \alpha} = \dfrac{b}{\sin \beta} = \dfrac{c}{\sin \gamma}$; $\quad 0 < \alpha, \beta, \gamma < 180°$.

The ratios of the length of a side to the sine of the opposite angle are the same for each of the three sides of a triangle.

Sometimes an alternate version,

$$\frac{\sin \alpha}{a} = \frac{\sin \beta}{b} = \frac{\sin \gamma}{c},$$

is useful. Let's return to the earlier problem.

EXAMPLE Given $a = 6.2$, $b = 11.6$, and $\beta = 25°$, solve ΔABC.

We make a rough sketch of ΔABC.

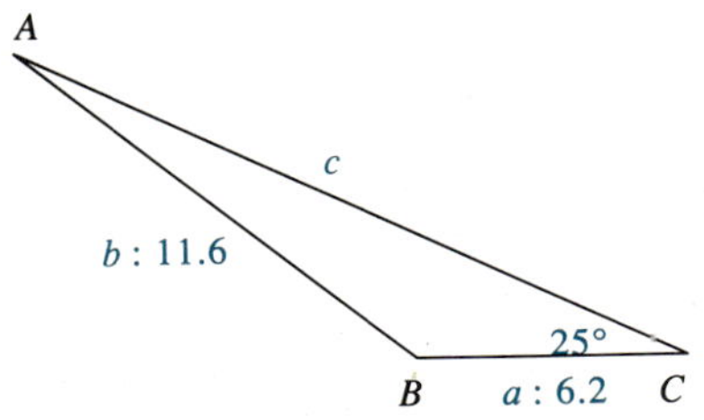

We already know only one version of the law of cosines gives an equation we can solve. Rather than solving this quadratic equation, we can try the law of sines.

$$\frac{11.6}{\sin 25°} = \frac{6.2}{\sin \alpha}$$

$$\Leftrightarrow \quad \sin \alpha = \frac{6.2}{11.6} \sin 25°$$

$$\Rightarrow \quad \alpha = \mathrm{Sin}^{-1}\left(\frac{6.2}{11.6} \sin 25°\right)$$

This value of α is a solution. We have $\alpha \approx 13.05°$, but we should check for all solutions between 0° and 180°. The angle $180° - 13.05°$, or 166.95°, also has sine approximately equal to (6.2/11.6) sin 25°. This equation has two solutions that are reasonable answers to our problem, 13.05° and 166.95°. If $\alpha \approx 13.05°$, then $\gamma = 180° - \alpha - \beta \approx 180° - 13.05° - 25°$, or 141.95°. If $\alpha \approx 166.95°$, then $\gamma \approx 180° - 166.95° - 25°$, or $-11.95°$, which is impossible.

Thus we have $\beta = 25°$, $\alpha \approx 13.05°$, and $\gamma \approx 141.95°$. The length c is still to be calculated.

$$\frac{c}{\sin\gamma} = \frac{11.6}{\sin 25°}$$

$$\Leftrightarrow \quad c = \frac{11.6}{\sin 25°} \cdot \sin \gamma$$

$$\Rightarrow \quad c \approx \frac{11.6}{\sin 25°} \cdot \sin 141.95°$$

$$\Rightarrow \quad c \approx 16.92$$

For ΔABC,

$\mathbf{a = 6.2,}$	$\boldsymbol{\alpha \approx 13.05°,}$
$\mathbf{b = 11.6,}$	$\boldsymbol{\beta = 25°,}$
$\mathbf{c \approx 16.92,}$	$\boldsymbol{\gamma \approx 141.95°.}$

■

Although only one of the possible solutions for α was a solution in the last example, it can happen that each candidate solution works.

EXAMPLE Solve ΔABC if $b = 2$, $c = 3$, and $\beta = 31°$.

Using the law of sines,

$$\frac{b}{\sin\beta} = \frac{c}{\sin \gamma}.$$

$$\frac{2}{\sin 31°} = \frac{3}{\sin \gamma}$$

$$\Leftrightarrow \quad \sin \gamma = \frac{3}{2} \sin 31°$$

$$\Rightarrow \qquad \gamma = \text{Sin}^{-1}(1.5 \sin 31°) \quad \text{or} \quad 180° - \text{Sin}^{-1}(1.5 \sin 31°)$$

$$\Rightarrow \qquad \gamma \approx 50.58° \quad \text{or} \quad 180° - 50.58°$$

$$\Rightarrow \qquad \gamma \approx 50.58° \quad \text{or} \quad 129.42°$$

(*Note:* Other solutions lie outside (0°, 180°).)
If $\gamma \approx 50.58°$, then $\alpha \approx 180° - 31° - 50.58°$, or 98.42°.
If $\gamma \approx 129.42°$, then $\alpha \approx 180° - 31° - 129.42°$, or 19.58°. There are two triangles that satisfy the original requirements. For each of these triangles we must calculate a.

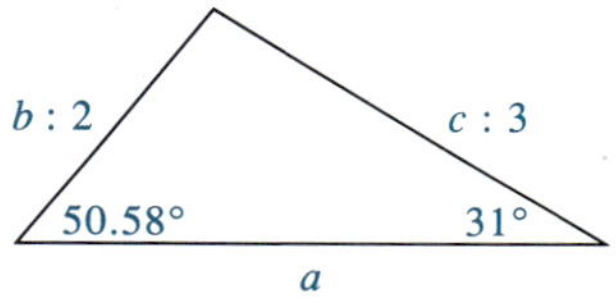

Triangle I

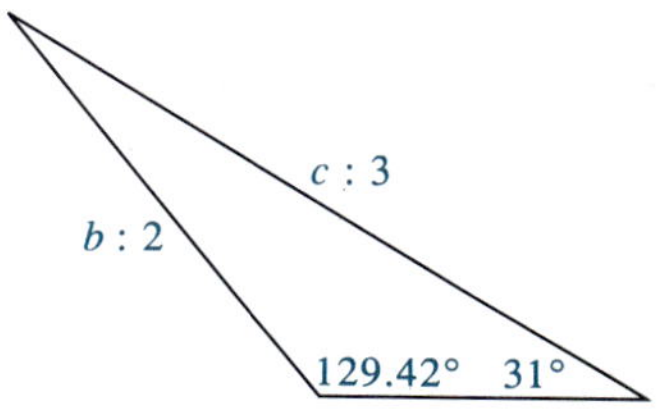

Triangle II

Triangle I

$$\frac{a}{\sin 98.42°} \approx \frac{2}{\sin 31°}$$

$$a \approx \frac{2}{\sin 31°} \cdot \sin 98.42°$$

$$a \approx 3.84$$

Triangle II

$$\frac{a}{\sin 19.58°} \approx \frac{2}{\sin 31°}$$

$$a \approx \frac{2}{\sin 31°} \cdot \sin 19.58°$$

$$a \approx 1.30$$

For ΔABC, either

$a \approx 3.84$	$\alpha \approx 98.42°$
$b = 2$	$\beta = 31°$
$c = 3$	$\gamma \approx 50.58°$

or

$a \approx 1.30$	$\alpha \approx 19.58°$
$b = 2$	$\beta = 31°$
$c = 3$	$\gamma \approx 129.42°$.

■

EXAMPLE Solve ΔABC if $a = 15$, $b = 8$, and $\beta = 34°$.
We have

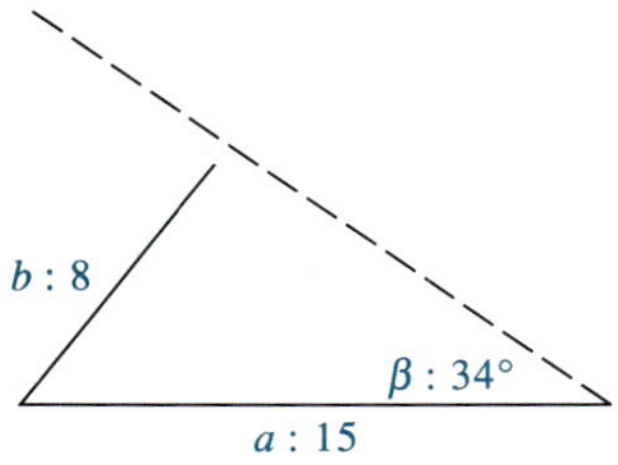

$$\frac{15}{\sin \alpha} = \frac{8}{\sin 34°}$$

$$\Leftrightarrow \qquad \sin \alpha = \frac{15}{8} \cdot \sin 34°$$

$$\Rightarrow \qquad \sin \alpha \approx 1.05,$$

but this is impossible. The range of the sine function is $[-1, 1]$.
No such triangle exists. ■

Using the law of sines to solve a triangle can lead to one of three cases: (1) a unique solution, one triangle that satisfies the original facts; (2) two triangles; or (3) no triangle. The second case is called the *ambiguous case* because the original data give no way to choose between the two solutions. It occurs because for each value r between 0 and 1, there are two angles with measure between 0° and 180° that have sine value equal to r. Both of these angles may fit with the original data.

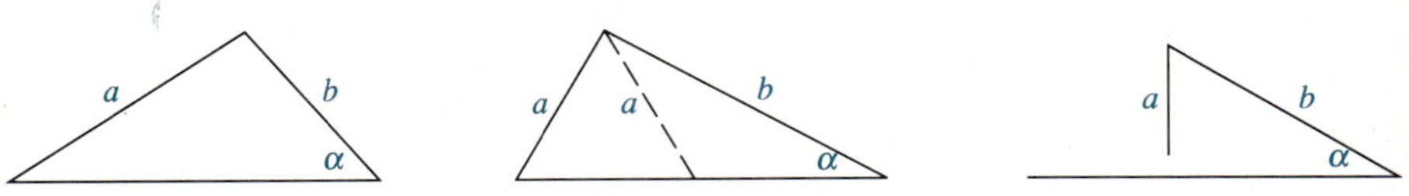

(1) Unique solution (2) Two triangles (3) No triangle

We are now in a situation similar to that at the end of Section 7.1. Given three pieces of information about a triangle and the laws of sines and cosines, we can determine the three missing parts *except* if the law of sines demonstrates the triangle is geometrically impossible or if the three facts are the measures of the angles. As we found for right triangles, knowing the measures of a triangle's angles determines only its shape, not the lengths of the sides.

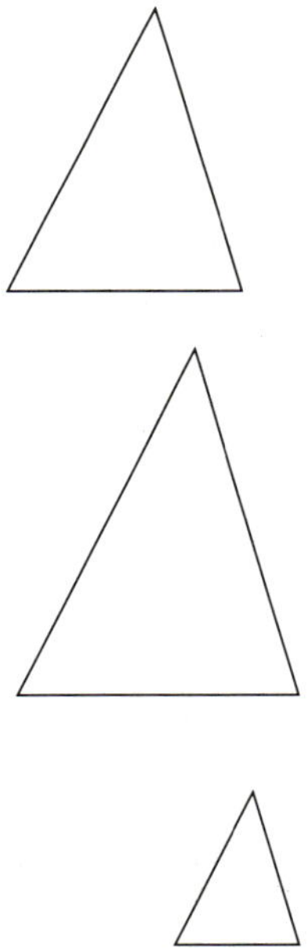

EXERCISES 7.2

In exercises 1 through 4, use the law of cosines to solve ΔABC approximately.

1. $a = 3, b = 7, c = 13$
2. $a = 4, b = 19.2, \gamma = 13.5°$
3. $b = 41.42, c = 14.66, \alpha = 3\pi/7$
4. $a = 17.3, b = 47.9, c = 23.5$

In exercises 5 through 8, use the law of sines to solve the triangle *ABC* approximately.

5. $\alpha = 29°, \beta = 93°, a = 14$
6. $\alpha = 124°, \gamma = 15°, c = 93.6$
7. $\beta = 100°, a = 16.3, b = 21.7$
8. $\alpha = 13.4°, a = 493.1, b = 942.3$

In exercises 9 through 14, solve ΔABC approximately.

9. $\beta = 54.3°, a = 103.5, c = 28.9$
10. $c = 132.64, b = 487.66, \alpha = 1.11$ radians
11. $\gamma = 79.2°, c = 1.78, a = 4.32$
12. $\gamma = 34°, c = 9.32, b = 7.41$
13. $\alpha = 27.52°, a = 0.7287, b = 1.4113$
14. $\beta = 54.3°, a = 92.8, b = 48.1$

In exercises 15 through 20, determine whether no triangle *ABC* exists, exactly one triangle *ABC* exists, or two triangles *ABC* exist.

15. $\alpha = 48.2°, a = 190, b = 246$
16. $\alpha = 48.2°, \beta = 93.2°, a = 3.14$
17. $b = 52.3, c = 41.2, \gamma = \pi/3$
18. $\beta = 64.32°, b = 10322, a = 10971$
19. $\beta = 151.3°, b = 514.21, c = 398.9$
20. $\gamma = 70.3°, c = 0.70321, b = 0.92143$

In exercises 21 through 24, compute (a) the area and (b) the approximate area, correct to one decimal place, of ΔABC.

21. $a = 4, b = 19.2, \gamma = 13.5°$
22. $b = 41.42, c = 14.66, \alpha = 134°$
23. $\beta = 54.3°, a = 103.5, c = 28.9$
24. $c = 132.64, b = 487.66, \alpha = 94.21°$

In exercises 25 through 30, compute (a) the area of the circular sector and (b) the approximate area of the circular sector correct to 2 decimal places.

25. $r = 4$ inches, $\theta = \dfrac{\pi}{3}$
26. $r = 13$ m, $\theta = \dfrac{7\pi}{11}$
27. $r = 16.2$ m, $\theta = 142°$
28. $r = 13{,}615$ yards, $\theta = 11.71°$
29. $r = \sqrt{172}$ centimeters, $\theta = 311.4°$
30. $r = 4.3182$ feet, $\theta = 1°$
31. Explain to a friend how to decide whether to use the law of cosines or the law of sines to solve a triangle.
32. Explain to a friend how you would know immediately from given measures when exactly one triangle exists.
33. Using Figure 7.2b, derive the law of cosines.
34. Thomas is surveying a piece of property and needs to find the distance between points *A* and *B*. The ground between the two is flat but covered with heavy undergrowth, including poison ivy. He also suspects that snakes abound there. He sets up a transit at point *C*, well away from the ivy infestation, and determines the measure of $\angle ACB$ to be 45.32°. He then measures the distance from *C* to *A*, which he finds to be 130 meters, and the distance from *C* to *B*, which he finds to be 210 meters. If Thomas makes the correct computation, what is the distance from *A* to *B*? What is the approximate distance correct to the nearest meter?
35. Maria and Hector purchase an old house and find lead pipes in the basement, one 27 feet long and the other 15 feet long. The pipes are joined with a fitting that creates an angle of 132°. What length pipe should they purchase to replace the lead pipes with one straight piece of plastic pipe?
36. Green Valley Landscape Company is to install a triangular flower bed at a new office complex. They are using very heavy treated timbers to form the sides of the bed. One timber is 20 feet long, another is 10 feet long, and the one remaining is 5 feet long. In order to minimize the labor involved, they will place the longest side first, then place the shortest by measuring the angle needed between these two sides, and finally place the timber that is 10 feet long. What angle, correct to the nearest $\frac{1}{2}°$, should they use?
37. Three rods of gold are to be joined to form a triangular pendant. If the lengths are $\frac{7}{8}$ inch, $\frac{1}{2}$ inch, and $\frac{5}{12}$ inch, what angles are formed between pairs of sides?
38. The specifications for a triangular brace in a piece of equipment require that the angles be 32°, 124°, and 24°. If the shortest side is to measure 56.7 cm, what are the lengths, correct to the nearest tenth of a centimeter, of the other two sides of the brace?
39. Maria and Hector decide to add a porch to the back of their house. The old roof has an angle of elevation of 46.2° and the porch roof is to have an angle of elevation of 30°. How far up the old roof will the new one join if the porch is to extend 14 feet from the house?
40. Susan is flying from point *A* to point *B*, 300 miles apart, to satisfy one of the requirements for obtaining her license. After flying for 23 minutes, her flight instructor tells her that she has been flying 3° off course since her departure. If her speed remains constant at 235 miles per hour, what angle turn must she make to arrive at *B*? How much time will she waste because of the error?
41. Bill sails N 40° E for 3 miles, turns through an angle β, sails 2.1 miles, turns through an angle α, and sails directly south back to his starting position. If he sailed no

more than 7 miles, approximately how far did he sail correct to three decimal places?

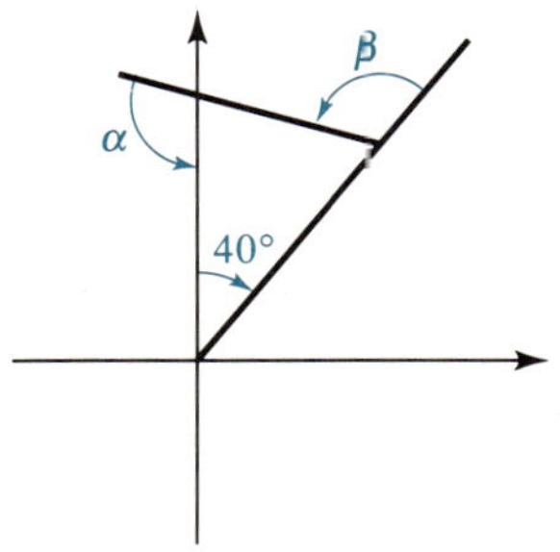

Use the following picture for exercises 42 and 43.

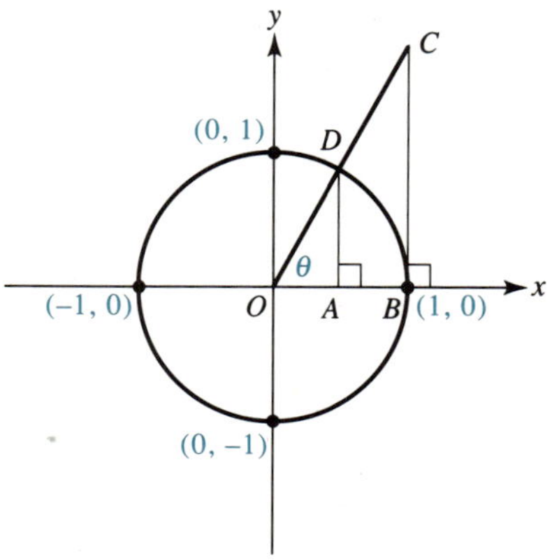

42. Describe BC, AD, and OA in terms of trigonometric function values.

43. Use trigonometric functions to describe the area of ΔOAD, of ΔOBC, and of sector OBD. Arrange the areas in order from least to greatest.

7.3 Basic Identities

Toolkit

In this section we will use the general trigonometric functions and facts about sine and cosine: Cosine is even because $\cos(-t) = \cos(t)$, and sine is odd because $\sin(-t) = -\sin(t)$.

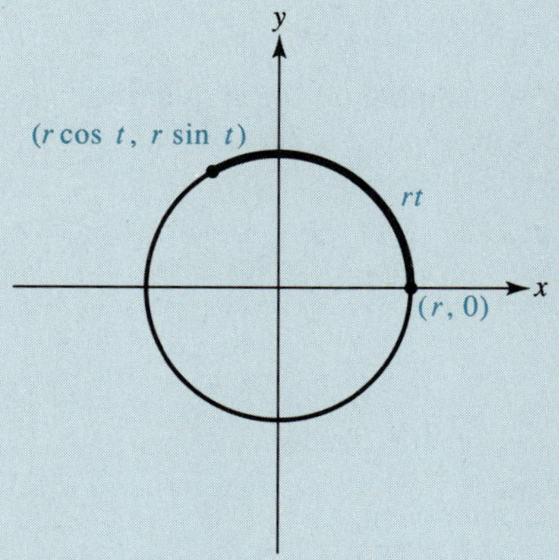

Also, remember not to approximate results unless you are asked to do so.

An **identity** is a mathematical statement that is true for all meaningful replacements of the variables. This means if the expression on one side has a restriction on the variable(s), the same restriction applies to the entire statement. We use $\equiv$ to indicate that a statement of equality is an identity; this is consistent with our earlier use of $\equiv$ to represent "is equivalent to."

Illustration $(x + 2)(x + 3) \equiv x^2 + 5x + 6$ is an identity; it is true for all replacements of x. $\dfrac{x^2 - 4}{x + 2} = x - 2$ is an identity. The left-hand side has an implicit restriction, $x \neq -2$; if we replace $(x^2 - 4)/(x + 2)$ with $x - 2$, we must note this restriction. □

Trigonometric identities are particularly important. They allow us to express trigonometric values in terms of known values and provide ways of solving equations. By now you know some of the most important trigonometric identities.

Identities

$\cos^2 t + \sin^2 t \equiv 1$	(1)
$\tan t \equiv \dfrac{\sin t}{\cos t}$	(2)
$\cot t \equiv \dfrac{\cos t}{\sin t}$	(3)
$\sec t \equiv \dfrac{1}{\cos t}$	(4)
$\csc t \equiv \dfrac{1}{\sin t}$	(5)

These identities can be easily manipulated to produce other useful identities. In practice, we often use $=$ rather than $\equiv$ in writing an identity, reserving $\equiv$ for situations in which we wish to emphasize the fact that the two sides are equivalent.

If you divide both sides of (1) by $\cos^2 t$, with the restriction that $\cos^2 t \neq 0$,

$$\cos^2 t + \sin^2 t = 1, \qquad \cos^2 t \neq 0$$

$$\Leftrightarrow \quad 1 + \frac{\sin^2 t}{\cos^2 t} = \frac{1}{\cos^2 t}, \qquad \cos^2 t \neq 0$$

$$\Leftrightarrow \quad 1 + \tan^2 t = \sec^2 t.$$

Replacing an expression with an equivalent one produces an equivalent equation.

Similarly, dividing (1) by $\sin^2 t$ yields

$$\frac{\cos^2 t}{\sin^2 t} + 1 = \frac{1}{\sin^2 t}, \qquad \sin^2 t \neq 0$$

$$\Leftrightarrow \quad \cot^2 t + 1 = \csc^2 t.$$

To produce the following two new identities,

$\mathbf{1 + \tan^2 t \equiv \sec^2 t}$ (6)

$\mathbf{\cot^2 t + 1 \equiv \csc^2 t}$ (7)

we used identities (2) through (5). Identities (2) through (5) allow us to rewrite any trigonometric expression using only the sine and cosine functions.

In Sections 6.1 and 6.2 we saw that some simple geometry gives trigonometric function values of $\pi/6$, $\pi/4$, $\pi/3$, $\pi/2$, and their integral multiples. An early problem in trigonometry was how to use these values to calculate values for other angles. For example, since $\pi/12 = \pi/3 - \pi/4$, can we use trig values of $\pi/3$ and $\pi/4$ to find $\cos \pi/12$? We begin by developing an identity for $\cos(t - u)$ in terms of $\cos(t)$ and $\cos(u)$. We will do one case in detail; the exercises explore the other possibilities.

An easy case to visualize is when $t > 0$, $u > 0$, and $t > u$.

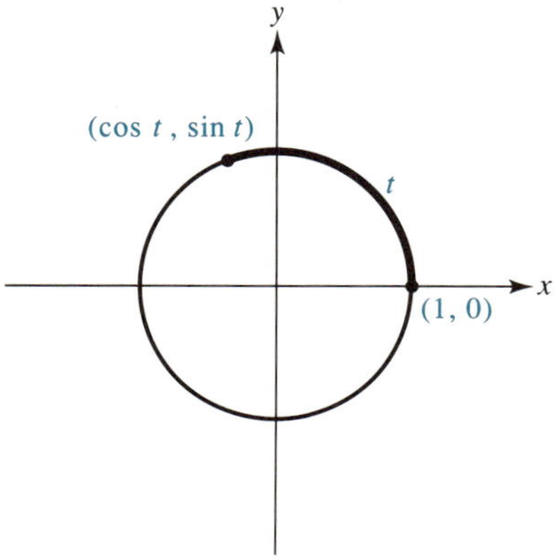

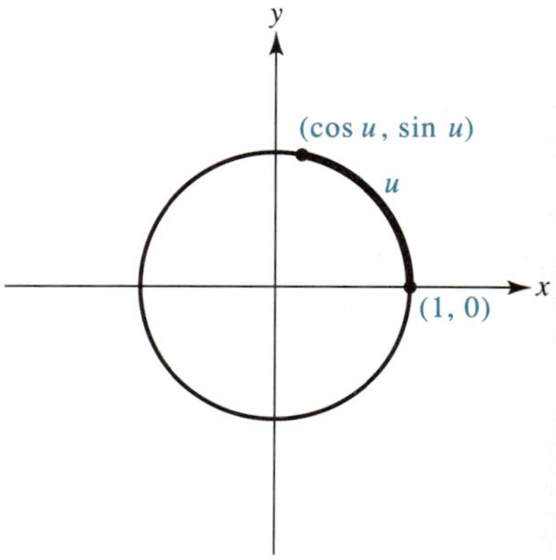

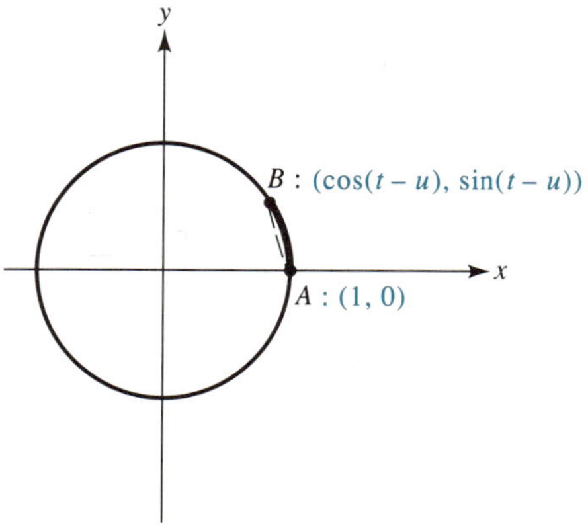

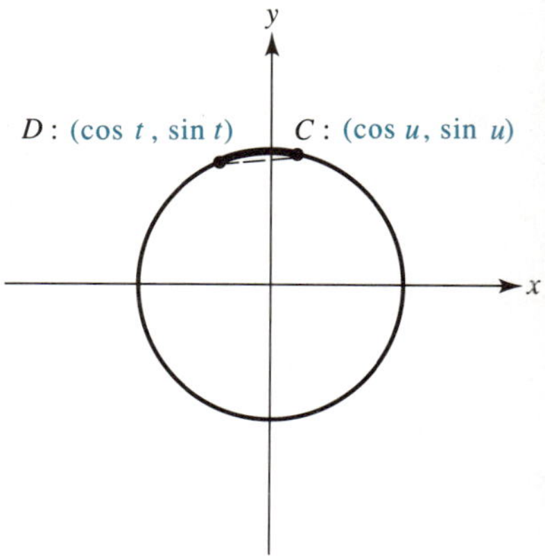

Both arcs $\widehat{AB}$ and $\widehat{CD}$ have length $t - u$, and hence the chords $\overline{AB}$ and $\overline{CD}$ also have equal lengths. We use the distance formula to compute these lengths.

$$AB = CD$$

$$\Leftrightarrow \quad \sqrt{(\cos(t-u)-1)^2 + (\sin(t-u))^2} = \sqrt{(\cos t - \cos u)^2 + (\sin t - \sin u)^2}$$

$$\Leftrightarrow \quad (\cos(t-u)-1)^2 + (\sin(t-u))^2 = (\cos t - \cos u)^2 + (\sin t - \sin u)^2$$

$$\begin{aligned}
\Leftrightarrow\quad \cos^2(t-u) - 2\cos(t-u) + 1 + \sin^2(t-u) &= \cos^2 t - 2\cos t\cos u + \cos^2 u + \sin^2 t - 2\sin t\sin u + \sin^2 u\\
\Leftrightarrow\quad 1 - 2\cos(t-u) + 1 &= 1 - 2\cos t\cos u + 1 - 2\sin t\sin u \qquad \text{Using identity (1)}\\
\Leftrightarrow\quad 2 - 2\cos(t-u) &= 2 - 2\cos t\cos u - 2\sin t\sin u\\
\Leftrightarrow\quad -2\cos(t-u) &= -2\cos t\cos u - 2\sin t\sin u\\
\Leftrightarrow\quad \cos(t-u) &= \cos t\cos u + \sin t\sin u\\
\mathbf{\cos(t-u)} &\equiv \mathbf{\cos t\cos u + \sin t\sin u} \qquad (8)
\end{aligned}$$

EXAMPLE Calculate exact values for cos $\pi/12$ and sin $\pi/12$.

$$\begin{aligned}
\cos\frac{\pi}{12} &= \cos\left(\frac{\pi}{3} - \frac{\pi}{4}\right)\\
\Leftrightarrow\quad \cos\frac{\pi}{12} &= \cos\frac{\pi}{3}\cos\frac{\pi}{4} + \sin\frac{\pi}{3}\sin\frac{\pi}{4}\\
\Leftrightarrow\quad \cos\frac{\pi}{12} &= \frac{1}{2}\cdot\sqrt{\frac{1}{2}} + \frac{\sqrt{3}}{2}\cdot\sqrt{\frac{1}{2}}\\
\Leftrightarrow\quad \cos\frac{\pi}{12} &= \frac{1}{2}\cdot\sqrt{\frac{1}{2}} + \frac{1}{2}\cdot\sqrt{\frac{3}{2}}\\
\Leftrightarrow\quad \cos\frac{\pi}{12} &= \frac{\sqrt{1}+\sqrt{3}}{2\sqrt{2}}, \quad \text{or} \quad \mathbf{\frac{\sqrt{2}+\sqrt{6}}{4}}
\end{aligned}$$

We can use the exact value for cos $\pi/12$ to calculate the exact value for sin $\pi/12$. Because $\cos^2 t + \sin^2 t \equiv 1$, $\cos^2(\pi/12) + \sin^2(\pi/12) = 1$ and

$$\left(\frac{\sqrt{2}+\sqrt{6}}{4}\right)^2 + \sin^2\frac{\pi}{12} = 1.$$

$$\begin{aligned}
\sin^2\frac{\pi}{12} &= 1 - \frac{8+2\sqrt{12}}{16}\\
\Leftrightarrow\quad \sin^2\frac{\pi}{12} &= 1 - \frac{8+4\sqrt{3}}{16}\\
\Leftrightarrow\quad \sin^2\frac{\pi}{12} &= 1 - \frac{2+\sqrt{3}}{4}\\
\Leftrightarrow\quad \sin^2\frac{\pi}{12} &= \frac{2-\sqrt{3}}{4}\\
\Rightarrow\quad \sin\frac{\pi}{12} &= \mathbf{\frac{\sqrt{2-\sqrt{3}}}{2}}
\end{aligned}$$

Note that only the positive solution to the quadratic equation is an appropriate answer because $\pi/12$ is a first-quadrant number. We look at another way of computing sin $\pi/12$ in exercise 27.

We could also calculate values of the other four trig functions using the cosine and sine values and identities (2) through (5). ■

Our example is an old-fashioned problem, and you're likely to make other uses of identity (8); before seeing some of these, let's tease some more identities out of (8).

We can write $\cos(t + u)$ as $\cos(t - (-u))$ and apply identity (8).

$$\cos(t + u) = \cos(t - (-u))$$

$$\Leftrightarrow \quad \cos(t + u) = \cos t \cos(-u) + \sin t \sin(-u)$$

$$\Leftrightarrow \quad \cos(t + u) = \cos t \cos u - \sin t \sin u \qquad \text{Because cosine is an even function and sine is an odd function}$$

$$\mathbf{\cos(t + u) \equiv \cos t \cos u - \sin t \sin u} \quad (9)$$

What about $\sin(t + u)$ and $\sin(t - u)$? First, we note

$$\cos\left(\frac{\pi}{2} - u\right) = \sin u \quad \text{and} \quad \sin\left(\frac{\pi}{2} - t\right) = \cos t.$$

$$\cos\left(\frac{\pi}{2} - u\right) = \cos\frac{\pi}{2} \cos u + \sin\frac{\pi}{2} \sin u \qquad \text{Identity (8)}$$

$$\Leftrightarrow \quad \cos\left(\frac{\pi}{2} - u\right) = 0 \cdot \cos u + 1 \cdot \sin u$$

$$\Leftrightarrow \quad \cos\left(\frac{\pi}{2} - u\right) = \sin u$$

$$\Leftrightarrow \quad \sin\left(\frac{\pi}{2} - t\right) = \cos\left(\frac{\pi}{2} - \left(\frac{\pi}{2} - t\right)\right) \qquad \text{Since } \sin u = \cos\left(\frac{\pi}{2} - u\right) \text{ for any number } u$$

$$\Leftrightarrow \quad \sin\left(\frac{\pi}{2} - t\right) = \cos\left(\frac{\pi}{2} - \frac{\pi}{2} + t\right)$$

$$\Leftrightarrow \quad \sin\left(\frac{\pi}{2} - t\right) = \cos t$$

These identities are the formal statements of the complementary angle relationship we noted in Section 7.1. To find an alternative expression for $\sin(t + u)$, we use $\sin(t + u) = \cos(\pi/2 - (t + u))$.

$$\sin(t + u) = \cos\left(\frac{\pi}{2} - (t + u)\right)$$

$$\Leftrightarrow \quad \sin(t + u) = \cos\left(\left(\frac{\pi}{2} - t\right) - u\right)$$

$$\Leftrightarrow \quad \sin(t + u) = \cos\left(\frac{\pi}{2} - t\right)\cos u + \sin\left(\frac{\pi}{2} - t\right)\sin u \qquad \text{Identity (8)}$$

$$\Leftrightarrow \quad \sin(t + u) = \sin t \cos u + \cos t \sin u$$

$$\mathbf{\sin(t + u) \equiv \sin t \cos u + \cos t \sin u} \quad (10)$$

Similarly,

$$\sin(t-u) = \cos\left(\frac{\pi}{2} - (t-u)\right)$$

$$\Leftrightarrow \quad \sin(t-u) = \cos\left(\left(\frac{\pi}{2} - t\right) + u\right)$$

$$\Leftrightarrow \quad \sin(t-u) = \cos\left(\frac{\pi}{2} - t\right)\cos u - \sin\left(\frac{\pi}{2} - t\right)\sin u$$

$$\Leftrightarrow \quad \sin(t-u) = \sin t \cos u - \cos t \sin u$$

$$\mathbf{\sin(t-u) \equiv \sin t \cos u - \cos t \sin u} \quad (11)$$

EXAMPLE Use identity (11) to verify identity (10):

$$\sin(u+v) \equiv \sin u \cos v + \cos u \sin v.$$

$$\sin(u+v) = \sin(u-(-v))$$

$$\Leftrightarrow \quad \sin(u+v) = \sin u \cos(-v) - \cos u \sin(-v) \qquad \text{Identity (11)}$$

$$\Leftrightarrow \quad \sin(u+v) = \sin u \cos v + \cos u \sin v \qquad \text{Because cosine is an even function and sine is an odd function}$$

$$\sin(u+v) \equiv \sin u \cos v + \cos u \sin v \qquad \text{Identity (10)}$$

■

Let's summarize the last group of identities, which are called the sum and difference formulas.

Sum and Difference Formulas

$\cos(u-v) \equiv \cos u \cos v + \sin u \sin v$	(8)
$\cos(u+v) \equiv \cos u \cos v - \sin u \sin v$	(9)
$\sin(u+v) \equiv \sin u \cos v + \cos u \sin v$	(10)
$\sin(u-v) \equiv \sin u \cos v - \cos u \sin v$	(11)

It may be helpful to state these identities in words. For example, identity (8) says: the cosine of the difference of two numbers is the cosine of the first times the cosine of the second plus the sine of the first times the sine of the second.

Two special cases of the sum and difference identities have applications in other areas. These are the double-angle formulas.

Consider $\cos(t+t)$. By identity (9),

$$\cos(t+t) = \cos t \cos t - \sin t \sin t$$

$$\Leftrightarrow \quad \cos(2t) = \cos^2 t - \sin^2 t$$

$$\mathbf{\cos(2t) \equiv \cos^2 t - \sin^2 t} \quad (12a)$$

If we wish, this last identity can be written using only cos t or only sin t.

$$\cos(2t) = \cos^2 t - \sin^2 t$$
$$\Leftrightarrow \quad \cos(2t) = \cos^2 t - (1 - \cos^2 t) \qquad \text{Identity (1)}$$
$$\Leftrightarrow \quad \cos(2t) = 2\cos^2 t - 1$$
$$\mathbf{\cos(2t) \equiv 2\cos^2 t - 1} \qquad (12b)$$

or

$$\cos(2t) = \cos^2 t - \sin^2 t$$
$$\Leftrightarrow \quad \cos(2t) = (1 - \sin^2 t) - \sin^2 t$$
$$\Leftrightarrow \quad \cos(2t) = 1 - 2\sin^2 t$$
$$\mathbf{\cos(2t) \equiv 1 - 2\sin^2 t} \qquad (12c)$$

If we examine sin($2t$), or sin($t + t$), identity (10) yields

$$\sin 2t = \sin t \cos t + \cos t \sin t$$
$$\Leftrightarrow \quad \sin 2t = 2 \sin t \cos t$$
$$\mathbf{\sin 2t \equiv 2 \sin t \cos t} \qquad (13)$$

EXAMPLE Find sin $2\pi/3$ using identity (13).

$$\sin\frac{2\pi}{3} = 2\sin\frac{\pi}{3}\cos\frac{\pi}{3}$$
$$\Leftrightarrow \quad \sin\frac{2\pi}{3} = 2\cdot\sqrt{\frac{3}{4}}\cdot\frac{1}{2}$$
$$\Leftrightarrow \quad \sin\frac{2\pi}{3} = \sqrt{\frac{3}{4}}$$

Note: This is not a very efficient way to find sin $2\pi/3$, but this example shows how the identity can be used. ■

You can see why the double-angle formulas were given their names, but if $2t$ is double t, then t is half of $2t$. We can interpret (12a), (12b), and (12c) as giving us information about t. For example,

$$\cos(2t) = 2\cos^2 t - 1$$
$$\Leftrightarrow \quad \cos^2 t = \frac{\cos(2t) + 1}{2}$$
$$\Leftrightarrow \quad \cos t = \sqrt{\frac{\cos 2t + 1}{2}} \quad \text{or} \quad \cos t = -\sqrt{\frac{\cos 2t + 1}{2}}.$$

Often this last statement is written

$$\mathbf{\cos\left(\frac{1}{2}u\right) = \sqrt{\frac{\cos u + 1}{2}} \quad or \quad \cos\left(\frac{1}{2}u\right) = -\sqrt{\frac{\cos u + 1}{2}}} \qquad (14)$$

and is called a **half-angle formula.** By rewriting (12c) in a similar way, you can obtain another half-angle formula,

$$\sin\left(\frac{1}{2}u\right) = \sqrt{\frac{1 - \cos u}{2}} \quad \text{or} \quad \sin\left(\frac{1}{2}u\right) = -\sqrt{\frac{1 - \cos u}{2}}. \qquad (15)$$

In applying either half-angle formula, we need more information about $\frac{1}{2}u$ to complete an evaluation.

EXAMPLE Find an exact value for $\cos(\pi/8)$.

We know $\pi/8 = (\frac{1}{2})(\pi/4)$. This is useful because $\cos \pi/4 = \sqrt{\frac{1}{2}}$. By identity (14)

$$\cos\frac{\pi}{8} = \sqrt{\frac{\cos \pi/4 + 1}{2}}$$

or its opposite, but since $\pi/8$ is a first-quadrant number, $\cos \pi/8$ must be positive.

$$\cos\frac{\pi}{8} = \sqrt{\frac{\cos \pi/4 + 1}{2}} = \sqrt{\frac{\sqrt{\frac{1}{2}} + 1}{2}}$$ ■

Half-Angle Formulas

$\cos\left(\frac{1}{2}u\right) \equiv \pm\sqrt{\frac{\cos u + 1}{2}}$	(14)
$\sin\left(\frac{1}{2}u\right) \equiv \pm\sqrt{\frac{1 - \cos u}{2}}$	(15)
$\tan\left(\frac{1}{2}u\right) \equiv \frac{1 - \cos u}{\sin u}$	(16)

We have included a half-angle formula for tangent in the table. You can explore its development in Exercise 22.

Identities can be combined to produce new identities. A virtually endless supply of identities can be produced, some useful, many not. By now you may feel there are already too many of them. Remember that most of the identities in this section are derived from $\cos^2 t + \sin^2 t = 1$ and $\cos(t + u) = \cos t \cos u - \sin t \sin u$. To help you remember the identities of this section and to give you some more practice with proofs, a block of exercises asks that you verify some identities. This means you must prove that the statement is true for all meaningful replacements of the variables. This is usually not done by producing the solution set but rather by showing that the given statement $P = Q$ is equivalent to the statement $P = P$, an obvious identity. Either side of the original equation may play the role of P.

Let's verify $\frac{1}{2}(1 - \cos 4t) + \cos^2 2t \equiv 1$. We begin by choosing the "worst" side to rewrite. Clearly the left-hand side is messier than the right.

$$\frac{1}{2}(1 - \cos 4t) + \cos^2 2t = 1$$

$$\Leftrightarrow \quad \frac{1}{2}[1 - (2\cos^2 2t - 1)] + \cos^2 2t = 1 \qquad \text{Identity (12b)}$$

$$\Leftrightarrow \quad \frac{1}{2}(2 - 2\cos^2 2t) + \cos^2 2t = 1$$

$$\Leftrightarrow \quad 1 - \cos^2 2t + \cos^2 2t = 1$$

$$\Leftrightarrow \quad 1 = 1$$

This last statement is in the form $P = P$, where P is the original right-hand side. We have verified the identity. This can be recorded by writing $\frac{1}{2}(1 - \cos 4t) + \cos^2 2t \equiv 1$, because we can replace the left-hand side by 1 for any value of t. In contrast, consider the next example.

EXAMPLE Is $2 = (1 + \cos 2x)(1 + \tan^2 x)$ an identity?

$$2 = (1 + \cos 2x)(1 + \tan^2 x)$$

$$\Leftrightarrow \quad 2 = (1 + 2\cos^2 x - 1)(1 + \tan^2 x) \qquad \text{Identity (12b)}$$

$$\Leftrightarrow \quad 2 = 2\cos^2 x \cdot \sec^2 x \qquad \text{Identity (6)}$$

$$\Leftrightarrow \quad 2 = 2\cos^2 x \cdot \frac{1}{\cos^2 x} \qquad \text{Identity (4)}$$

$$\Leftrightarrow \quad 2 = 2, \qquad \cos x \neq 0$$

The left-hand side is equal to the right-hand side as long as $\cos x \neq 0$. If we replace $(1 + \cos 2x)(1 + \tan^2 x)$ with 2, we must note the restriction $\cos x \neq 0$.
The original statement is an identity. ■

EXAMPLE Determine whether $\sec x - \cos x = \tan x \cdot \sin x$ is an identity.
Neither side is really more complicated than the other, so we could rewrite either.

$$\sec x - \cos x = \tan x \cdot \sin x$$

$$\Leftrightarrow \quad \frac{1}{\cos x} - \cos x = \tan x \cdot \sin x$$

$$\Leftrightarrow \quad \frac{1 - \cos^2 x}{\cos x} = \tan x \cdot \sin x \qquad \text{Identity (1)}$$

$$\Leftrightarrow \quad \frac{\sin^2 x}{\cos x} = \tan x \cdot \sin x$$

$$\Leftrightarrow \quad \frac{\sin x}{\cos x} \cdot \sin x = \tan x \cdot \sin x$$

$$\Leftrightarrow \quad \tan x \cdot \sin x = \tan x \cdot \sin x$$

The last statement is in the form $P = P$.
Therefore, **we have an identity;** $\sec x - \cos x \equiv \tan x \cdot \sin x$. ■

Something that may be worrying you about these examples is how some of the rewriting choices were made. To replace $\cos 4t$, should you use $\cos^2 2t - \sin^2 2t$, $2\cos^2 2t - 1$, or $1 - 2\sin^2 2t$? We'll confess that we've given you the tidy versions of these examples. You may have to try out several possibilities before you find the one that gives a neat result. After another example showing a common algebraic technique, we'll give you some suggestions for tackling these problems.

EXAMPLE Determine if $\dfrac{\sin\theta}{1 - \cos\theta} = \dfrac{1 + \cos\theta}{\sin\theta}$ is an identity.

There is an implicit restriction that $\sin\theta \neq 0$, so multiplying the right-hand side by $(\sin\theta)/(\sin\theta)$ will produce an equivalent fraction.

$$\frac{\sin\theta}{1 - \cos\theta} = \frac{1 + \cos\theta}{\sin\theta}$$

$$\Leftrightarrow \quad \frac{\sin\theta}{1 - \cos\theta} = \frac{\sin\theta}{\sin\theta} \cdot \frac{1 + \cos\theta}{\sin\theta}$$

$$\Leftrightarrow \quad \frac{\sin\theta}{1 - \cos\theta} = \frac{(\sin\theta)(1 + \cos\theta)}{\sin^2\theta}$$

$$\Leftrightarrow \quad \frac{\sin\theta}{1 - \cos\theta} = \frac{(\sin\theta)(1 + \cos\theta)}{1 - \cos^2\theta} \qquad \text{Identity (1)}$$

$$\Leftrightarrow \quad \frac{\sin\theta}{1 - \cos\theta} = \frac{(\sin\theta)(1 + \cos\theta)}{(1 - \cos\theta)(1 + \cos\theta)}$$

$$\Leftrightarrow \quad \frac{\sin\theta}{1 - \cos\theta} = \frac{\sin\theta}{1 - \cos\theta}, \; 1 + \cos\theta \neq 0$$

Since we already have the stronger restriction that $\sin\theta \neq 0$, the restriction $1 + \cos\theta = 0$ need not be stated.

$\dfrac{\sin\theta}{1 - \cos\theta} = \dfrac{1 + \cos\theta}{\sin\theta}$ **is an identity.** If $(1 + \cos\theta)/(\sin\theta)$ is replaced by $(\sin\theta)/(1 - \cos\theta)$, we must note that $\sin\theta \neq 0$. ■

A trigonometric equation $P = Q$ is an identity if you can show it is equivalent to $P = P$.

1. Begin by choosing a side to rewrite, usually the messier one. If there's not much difference, changing a sum or difference may be more efficient than rewriting a product.
2. To rewrite the chosen side, use algebraic methods such as factoring or replacing a fraction by an equivalent one as well as using basic trigonometric identities.
3. Keep an eye on where you're headed. What do you want the rewriting to produce?
4. If nothing else suggests itself, any trigonometric expression can be written using only sine and cosine functions.

Finally, you have to be able to recognize parts of the basic identities and be willing to experiment. You should find, as you did with factoring, that practice pays quick dividends.

EXAMPLE Is $1 + \cos t = \dfrac{\cos t}{\sec t - \tan t}$ an identity?

Rewriting the right-hand side in terms of sines and cosines,

$$1 + \cos t = \frac{\cos t}{\sec t - \tan t}$$

$$\Leftrightarrow \quad 1 + \cos t = \frac{\cos t}{\dfrac{1}{\cos t} - \dfrac{\sin t}{\cos t}}$$

$$\Leftrightarrow \quad 1 + \cos t = \frac{\cos t}{\dfrac{1 - \sin t}{\cos t}}$$

$$\Leftrightarrow \quad 1 + \cos t = \frac{\cos^2 t}{1 - \sin t}, \ \cos t \neq 0$$

$$\Leftrightarrow \quad 1 + \cos t = \frac{1 - \sin^2 t}{1 - \sin t}, \ \cos t \neq 0$$

$$\Leftrightarrow \quad 1 + \cos t = 1 + \sin t, \ \cos t \neq 0$$

The original statement is not an identity. ■

EXERCISES 7.3

In exercises 1 through 6, use identities (2) through (5) to write each expression in terms of sine and cosine only.

1. $\cot \theta \csc \theta$
2. $\sec^2 t + \csc^2 t$
3. $(\sin t)(\cot t + \csc t)$
4. $1 + \tan^2 t \cot^2 t + \cot^2 t$
5. $(1 + \tan^2\theta)(1 - \cos^2\theta)$
6. $\sec^{-2}\theta + \csc^{-2}\theta$

In exercises 7 through 16, use the half-angle formulas or sum and difference formulas to calculate the exact value for each expression.

7. $\sin\left(\dfrac{5\pi}{12}\right)$
8. $\cos\left(\dfrac{5\pi}{12}\right)$
9. $\sin\left(\dfrac{\pi}{8}\right)$
10. $\sin\left(-\dfrac{\pi}{8}\right)$
11. $\cos\left(\dfrac{7\pi}{12}\right)$
12. $\sin\left(\dfrac{7\pi}{12}\right)$
13. $\sin\left(\dfrac{11\pi}{12}\right)$
14. $\cos\left(\dfrac{11\pi}{12}\right)$
15. $\sin\left(\dfrac{3\pi}{8}\right)$
16. $\cos\left(\dfrac{3\pi}{8}\right)$
17. Compute the exact value of $\cos(3\pi/8)$ using

$$\cos\left(\frac{3\pi/4}{2}\right), \quad \cos\left(\frac{\pi}{8} + \frac{\pi}{4}\right), \quad \text{and} \quad \cos\left(\frac{\pi}{2} - \frac{\pi}{8}\right).$$

18. Use your calculator to find the approximate value of $\cos(3\pi/8)$. Compare this approximation to approximations obtained from each of the three answers to exercise 17.
19. Write an expression for $\tan(\alpha + \beta)$ in terms of $\tan \alpha$ and $\tan \beta$ only. (*Hint:* First write $\tan(\alpha + \beta)$ in terms of sine and cosine; then divide each term of the numerator and denominator by $\cos \alpha \cos \beta$.)
20. Rewrite $\tan(\alpha - \beta)$ in terms of $\tan \alpha$ and $\tan \beta$ only.
21. Write an expression for $\tan 2\theta$ in terms of $\tan \theta$ only.
22. **a.** Write an expression for $\tan \theta/2$ in terms of $\cos \theta$ only.

    b. Now supply the necessary steps to produce half-angle formula (16).
23. State identity (9) in words.
24. State identity (10) in words.
25. State identity (14) in words.
26. State the identity you developed in exercise 21 in words.
27. Use identity (10) to compute $\sin \pi/12$; compare the answer to that obtained in the example in this section (page 249). You might use your calculator to compare the answers. Because the calculator gives only approxi-

mate values, can you compare the two answers in some other way to determine if they are the same?

Exercises 28 through 31 contain answers that Jane obtained in doing her calculus homework and the corresponding answers in her textbook. Determine whether Jane's answer (a) is equivalent to that in the text (b).

28. **a.** $(\sin t - \cos t)(\sin t + \cos t)$
b. $2\sin^2 t - 1$

29. **a.** $(\cos\theta)(2\sin\theta + \cos\theta) - 1$
b. $\sin 2\theta - \sin^2\theta$

30. **a.** $\cos^4\theta - \sin^2 2\theta + \sin^4\theta$
b. $\cos^2 2\theta$

31. **a.** $\dfrac{2\tan\theta}{1 + \tan^2\theta}$
b. $\tan 2\theta$

32. Elisa says that since $1/(\sin t) \equiv \csc t$, she can replace $\sin t$ with $1/(\csc t)$. Is she correct? If not, give one value of t for which the left- and right-hand sides are not equal.

In exercises 33 through 45, determine whether the statements are identities or not. For each identity, state whether the two sides have the same restrictions.

33. $\dfrac{4 - 4\cos^2\theta + \sin^2 2\theta}{4\cos^4\theta} = \sec^4\theta - 1$

34. $(\sin\theta)(\cot\theta + \csc\theta) = 1 + \cos\theta$

35. $\dfrac{1}{\tan t + \cot t} = (\sin t)(\cos t)$

36. $\dfrac{1}{\sin^2 x + \sin x + 1} = \dfrac{\sin x - 1}{\sin^3 x - 1}$

37. $\sin\left(\dfrac{\pi}{2} - x\right) = \cos(2\pi - x)$

38. $\dfrac{1}{\cos(\pi/2 - x)} = \dfrac{1}{\sin(2\pi - x)}$

39. $\dfrac{2\sin x + 2\cos x}{\sin 2x} = \sec x + \csc x$

40. $\dfrac{\tan^2\theta}{\sec\theta + 1} = \dfrac{1 - \cos\theta}{\cos\theta}$

41. $(\cot\beta + \tan\beta)^2 = \csc^2\beta\sec^2\beta$

42. $(\tan x + \cot x)(\csc x - \sin x)(\sec x - \cos x) = 2$

43. $(\cos\gamma + \sin\gamma - 1)^2 = 2(1 - \sin\gamma)(1 - \cos\gamma)$

44. $\dfrac{\cot y - \cos y}{\cot y \cos y} = \dfrac{\cot y \cos y}{\cot y + \cos y}$

45. $(\sec t - \tan t)^7(\sec t + \tan t)^6 = \dfrac{1 + \sin t}{\cos t}$

46. Redevelop the identity $\cos(t - u) \equiv \cos t\cos u + \sin t \sin u$ when $t > 0$, $u > 0$, and $u > t$.

47. (Graphing calculator) Graph $y =$ left-hand side and $y =$ right-hand side of the equations of exercises 33, 40, and 45. Do the graphs clearly show the difference between an identity for which both sides have the same restrictions, an identity for which the sides have different restrictions, and a statement that is not an identity. Why are the graphs not reliable for distinguishing between the two classes of identities? You might want to sketch others.

7.4 Uses of the Basic Identities

Toolkit

You'll need ways of writing general solutions and the ideas of Section 6.6.

One way we use trigonometric identities is in rewriting expressions. For example, given a value for x, we could evaluate $\cos(2\,\mathrm{Sin}^{-1}x)$, but another form might be easier to handle.

Illustration $\cos(2\,\mathrm{Sin}^{-1}x) \equiv \cos^2(\mathrm{Sin}^{-1}x) - \sin^2(\mathrm{Sin}^{-1}x)$ or $1 - 2\sin^2(\mathrm{Sin}^{-1}x)$.
The second form has the advantage of fitting the definition of inverse functions exactly.

$$\sin^2(\mathrm{Sin}^{-1}x) = [\sin(\mathrm{Sin}^{-1}x)]^2 = x^2$$

We now have

$$\begin{aligned} &\cos(2\,\mathrm{Sin}^{-1}x) = 1 - 2\sin^2(\mathrm{Sin}^{-1}x) \\ \Leftrightarrow\quad &\cos(2\,\mathrm{Sin}^{-1}x) = 1 - 2x^2, \qquad x \in [-1, 1]. \end{aligned}$$

The right-hand side is easier to evaluate. The domain of Sin^{-1} is $[-1, 1]$, so we must write the restriction $x \in [-1, 1]$ explicitly on the right. Then, for example,

$$\cos\left(2\,\mathrm{Sin}^{-1}\frac{3}{4}\right) = 1 - 2\left(\frac{3}{4}\right)^2, \text{ or } -\frac{1}{8}. \qquad \square$$

We often need to apply trigonometric identities in order to rewrite equations in a form we can solve.

EXAMPLE Solve $\dfrac{1}{\csc\theta}\cdot\sec\theta = 0$.

Let's try rewriting in terms of sine and cosine functions. $1/(\csc\theta) \equiv \sin\theta$ is an identity because it's true when $\theta \neq k\pi$, $k \in I$.

$$\begin{aligned} &\frac{1}{\csc\theta}\cdot\sec\theta = 0 \\ \Leftrightarrow\quad &\sin\theta\sec\theta = 0, \qquad \theta \neq k\pi,\ k \in I \\ \Leftrightarrow\quad &\sin\theta\cdot\frac{1}{\cos\theta} = 0, \qquad \theta \neq k\pi,\ k \in I \\ \Leftrightarrow\quad &\tan\theta = 0, \qquad \theta \neq k\pi,\ k \in I \end{aligned}$$

There are no values of θ that make the compound statement true; therefore, **the solution set is { }.** ■

EXAMPLE Solve $\sin t + \cos 2t = 1$.

A good place to start is to make the arguments of the functions the same. We replace $\cos 2t$ with $1 - 2\sin^2 t$ because the equation already has a term $\sin t$.

$$\begin{aligned} &\sin t + \cos 2t = 1 \\ \Leftrightarrow\quad &\sin t + (1 - 2\sin^2 t) = 1 \\ \Leftrightarrow\quad &\sin t - 2\sin^2 t = 0 \\ \Leftrightarrow\quad &(\sin t)(1 - 2\sin t) = 0 \\ \Leftrightarrow\quad &\sin t = 0 \quad\text{or}\quad 1 - 2\sin t = 0 \end{aligned}$$

The solution set for $\sin t = 0$ is $\{k\pi, k \in I\}$.

$1 - 2\sin t = 0 \Leftrightarrow \sin t = \frac{1}{2}$. This has solution set $\{\pi/6 + 2j\pi, 5\pi/6 + 2j\pi, j \in I\}$.

The solution set for $\sin t + \cos 2t = 1$ is the union of these sets,

$$\left\{\frac{\pi}{6} + 2j\pi, \frac{5\pi}{6} + 2j\pi, k\pi; j, k \in I\right\}. \qquad ■$$

In the last example we used the fact that a product is zero only if at least one factor is zero. The technique of writing one side in factored form with the other side 0 is a very common method of solving equations. Here is a way to change a sum or difference of sines and cosines into a product. Consider $\cos(t + u) + \cos(t - u)$.

$$\begin{aligned} \cos(t + u) + \cos(t - u) &= (\cos t \cos u - \sin t \sin u) \\ &\quad + (\cos t \cos u + \sin t \sin u) \\ \Leftrightarrow \quad \cos(t + u) + \cos(t - u) &= 2 \cos t \cos u \end{aligned}$$

EXAMPLE Write $\cos 5x + \cos 3x$ as a product.

We need to determine t and u so that $t + u = 5x$ and $t - u = 3x$.

$$t - u = 3x \Leftrightarrow t = u + 3x,$$

so

$$\begin{aligned} t + u &= 5x \\ \Leftrightarrow \quad u + 3x + u &= 5x \\ \Leftrightarrow \quad 2u + 3x &= 5x \\ \Leftrightarrow \quad u &= x \end{aligned}$$

Therefore, t must be $4x$.

$$\begin{aligned} \cos 5x + \cos 3x &= \cos(4x + x) + \cos(4x - x) \\ \Leftrightarrow \quad \cos 5x + \cos 3x &= \mathbf{2 \cos 4x \cos x} \end{aligned}$$

What about $\cos(t + u) - \cos(t - u)$?

$$\begin{aligned} \cos(t + u) - \cos(t - u) &= (\cos t \cos u - \sin t \sin u) \\ &\quad - (\cos t \cos u + \sin t \sin u) \\ \Leftrightarrow \quad \cos(t + u) - \cos(t - u) &= -2 \sin t \sin u \end{aligned}$$

The sum and difference of $\sin(t + u)$ and $\sin(t - u)$ yield similar formulas.

Sum and Product Formulas

$\cos(t + u) + \cos(t - u) = 2 \cos t \cos u$		(a)
$\cos(t + u) - \cos(t - u) = -2 \sin t \sin u$		(b)
$\sin(t + u) + \sin(t - u) = 2 \sin t \cos u$		(c)
$\sin(t + u) - \sin(t - u) = 2 \cos t \sin u$		(d)

EXAMPLE Solve $\sin x = \sin 3x$.

Rewriting so that each side has x as its argument does not help. Instead, we use a sum and product formula.

$$\begin{aligned} \sin x &= \sin 3x \\ \Leftrightarrow \quad \sin 3x - \sin x &= 0 \\ \Leftrightarrow \quad 2 \cos 2x \sin x &= 0 \\ \Leftrightarrow \quad \cos 2x = 0 \quad &\text{or} \quad \sin x = 0 \end{aligned}$$

Use identity (d) with $t = 2x$ and $u = x$.

Solutions are $2x = \pi/2 + k\pi$, or $x = \pi/4 + k\pi/2$, $k \in I$, for the first equation and $x = 0 + j\pi, j \in I$ for the second.

The solution set is $\left\{\frac{\pi}{4} + \frac{k\pi}{2}, 0 + j\pi; k, j \in I\right\}$. ■

EXAMPLE Solve $\cos 3x - \cos x \geq 0$.

We need to obtain the solutions for

$$\cos 3x - \cos x = 0.$$

Using sum and product formula (*b*), the left-hand side can be rewritten to obtain $-2 \sin 2x \sin x = 0$

$$\Leftrightarrow \quad \sin 2x = 0 \quad \text{or} \quad \sin x = 0$$

$$\Leftrightarrow \quad 2x = 0 + j\pi, j \in I \quad \text{or} \quad x = 0 + k\pi, k \in I$$

$$\Leftrightarrow \quad x = 0 + \frac{j\pi}{2}, j \in I \quad \text{or} \quad x = 0 + k\pi, k \in I$$

The solutions of the equation are $j\pi/2$, $j \in I$. Using a number-line graph, we determine the regions in which the solutions of the inequality lie.

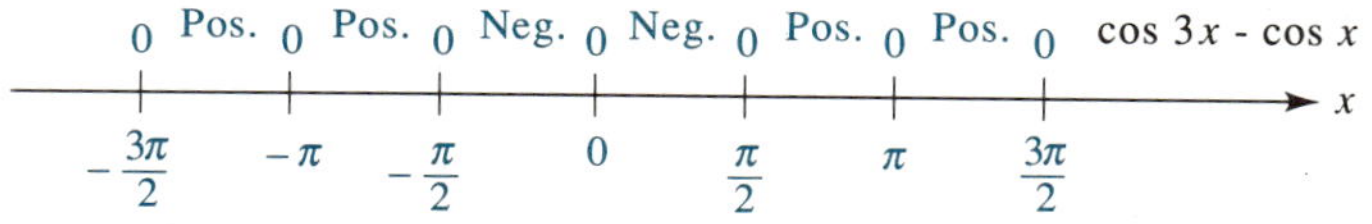

The solution is the union of all intervals

$$\left[\frac{\pi}{2} + 2j\pi, \frac{3\pi}{2} + 2j\pi\right], [0 + 2j\pi, 0 + 2j\pi], \qquad j \in I.$$

Note: The intervals $[0 + 2j\pi, 0 + 2j\pi]$ pick up the isolated zeros. ■

Expressions of the form $a \sin x + b \cos x$, $a > 0$, $b > 0$, often arise in describing the motion of vibrating or oscillating objects such as the end of a spring that has been compressed and then released. This form may be convenient for sketching a graph of $f(x) = a \sin x + b \cos x$, but an alternate form with a single term is also useful. The sum-product identities do not apply directly here. If it were possible to view a as the cosine of some angle, say t, and b as the sine of the same angle, $a \sin x + b \cos x$ would represent $\sin(x + t)$. We can't quite do this, but we can come close. Consider $t = \text{Tan}^{-1}(b/a)$; t must be a first-quadrant number, since $b/a > 0$.

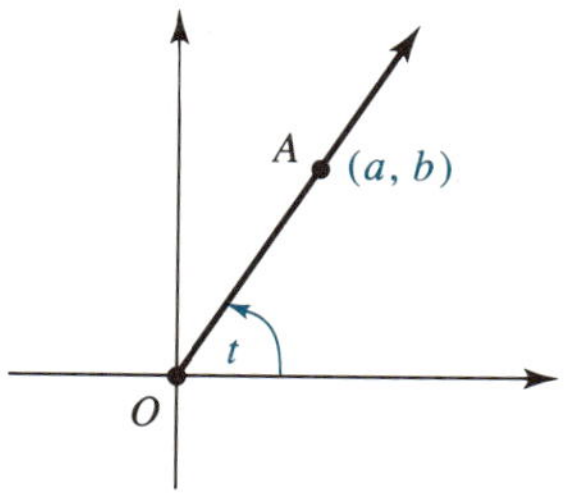

We know the coordinates of A can also be described as $r \cos t$ and $r \sin t$, where r is the length of $\overline{OA}$,

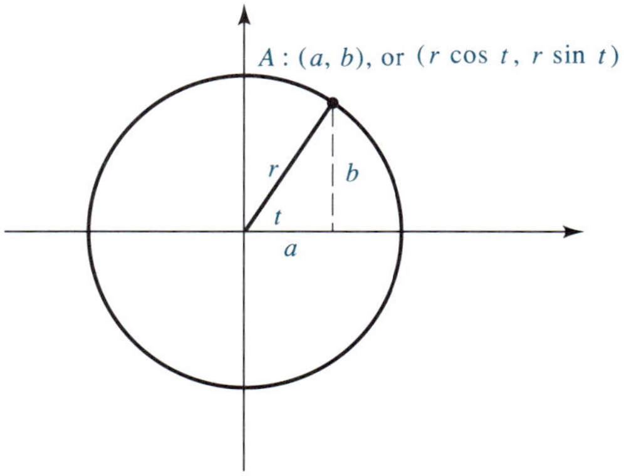

$$r = \sqrt{(a - 0)^2 + (b - 0)^2}, \quad \text{or} \quad \sqrt{a^2 + b^2}$$

and

$$a = r \cos t, \qquad b = r \sin t.$$

Now we can write $f(x) = a \sin x + b \cos x$ as $f(x) = r \cos t \sin x + r \sin t \cos x$, $r = \sqrt{a^2 + b^2}$, or $f(x) = r \sin(x + t)$, $r = \sqrt{a^2 + b^2}$.
Expressions of the form $a \sin x - b \cos x$ or $a \cos x - b \sin x$ can be rewritten in a similar fashion as a multiple of $\sin(x - t)$ or $\cos(x + t)$.

One advantage of this single-term form is that you can determine quickly where maximum or minimum values occur.

EXAMPLE Find the maximum value of $f(x) = 2 \cos x - \sin x$.
We can view $2 \cos x - \sin x$ as resembling $\sin(t - x)$. Let $t = \text{Tan}^{-1}(\frac{2}{1})$, $r = \sqrt{2^2 + 1^2} = \sqrt{5}$, $2 = r \sin t$, and $1 = r \cos t$.

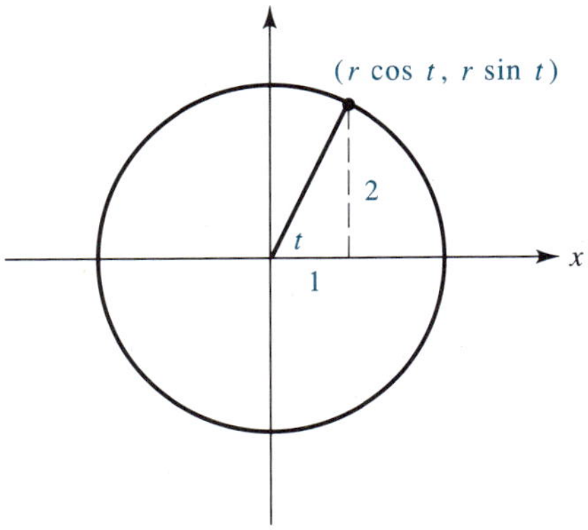

$$f(x) = 2 \cos x - \sin x$$
$$\Leftrightarrow \quad f(x) = \sqrt{5} \sin t \cos x - \sqrt{5} \cos t \sin x$$
$$\Leftrightarrow \quad f(x) = \sqrt{5} \sin(t - x), \qquad \text{where } t = \text{Tan}^{-1}\left(\frac{2}{1}\right).$$

The function f will have a maximum value of $\sqrt{5}$ when $\sin(t - x) = 1$, since this is the largest value sine can have.

Note: It is not necessary to recognize the final form of the single term before calculating r and t. Just be sure to label your sketch carefully. You can identify the single term at the second stage. ■

EXAMPLE For a particular harbor, the height of the tide, h inches, above the average level is given by $h(t) = 20\cos(\pi/6)t + 15\sin(\pi/6)t$, where t is the number of hours after noon. Find the greatest height of the tide and the time at which it occurs in the afternoon, correct to the nearest tenth of an hour.

$$h(t) = 20\cos\frac{\pi}{6}t + 15\sin\frac{\pi}{6}t$$

Let $u = \text{Tan}^{-1}(\frac{20}{15})$ and $r = \sqrt{20^2 + 15^2}$, or 25.

We make a sketch of the data.

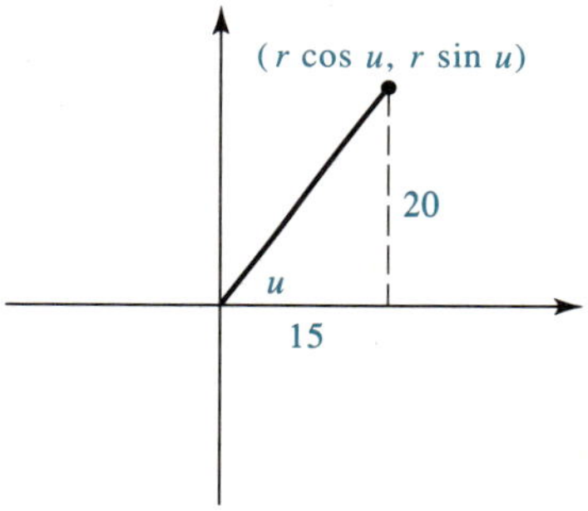

The sketch shows $15 = 25\cos u$ and $20 = 25\sin u$.

$$h(t) = 25\sin u\cos\frac{\pi}{6}t + 25\cos u\sin\frac{\pi}{6}t$$

$$\Leftrightarrow \quad h(t) = 25\left(\sin u\cos\frac{\pi}{6}t + \cos u\sin\frac{\pi}{6}t\right)$$

$$\Leftrightarrow \quad h(t) = 25\sin\left(u + \frac{\pi}{6}t\right)$$

The tide rises 25 inches over the average height because $\sin(u + (\pi/6)t)$ has a maximum value of 1.

$$\sin\left(u + \frac{\pi}{6}t\right) = 1 \Leftrightarrow u + \frac{\pi}{6}t = \frac{\pi}{2} + 2k\pi, \qquad k \in I$$

$$\Leftrightarrow \quad t = \frac{6}{\pi}\left(\frac{\pi}{2} + 2k\pi - \text{Tan}^{-1}\left(\frac{20}{15}\right)\right), \qquad k \in I$$

We also need $0 \leq t \leq 6$ because we want an afternoon time. For $k = 0$,

$$t = \frac{6}{\pi}\left(\frac{\pi}{2} - \text{Tan}^{-1}\frac{20}{15}\right), \quad \text{or} \quad t \approx 1.2.$$

For $k = 1$, $t \approx 13.2$, but this is greater than 6. Only $k = 0$ produces a feasible answer.

The tide reaches its maximum height of 25 inches above average at approximately 1.2 hours after noon. ■

EXERCISES 7.4

In exercises 1 through 6, use trigonometric identities to rewrite the expression without inverse trigonometric functions.

1. $\sin(2\,\text{Sin}^{-1}x)$
2. $\cos(2\,\text{Cos}^{-1}x)$
3. $\sin(\text{Sin}^{-1}\frac{3}{4} + \text{Sin}^{-1}\frac{5}{8})$
4. $\sin(\text{Sin}^{-1}\frac{3}{4} - \text{Sin}^{-1}\frac{5}{8})$
5. $\cos(\text{Cos}^{-1}\frac{12}{13} - \text{Sin}^{-1}\frac{2}{3})$
6. $\cos(\text{Cos}^{-1}\frac{12}{13} + \text{Sin}^{-1}\frac{2}{3})$

In exercises 7 through 12, rewrite each sum or difference as a product.

7. $\cos 6\theta - \cos 2\theta$
8. $\sin 14t + \sin 8t$
9. $\sin 11\alpha - \sin 3\alpha$
10. $\cos 21\beta + \cos 19\beta$
11. $\sin\theta - \sin 5\theta$
12. $\cos 7\theta - \cos 13\theta$

In exercises 13 through 16, rewrite each product as a sum.

13. $(\cos 4\alpha)(\sin\alpha)$
14. $4(\sin 7\theta)(\sin 11\theta)$
15. $3(\cos 2t)(\cos 4t)$
16. $1.9(\sin 5x)(\cos 3x)$
17. Rewrite $(\sin 4x)(\cos 3x)$ as a sum using sum and product formula (c). Also rewrite $(\sin 4x)(\cos 3x)$ as a sum using sum and product formula (d) to demonstrate the results are the same.
18. Develop the formula $\sin(t + u) + \sin(t - u) = 2\sin t\cos u$.
19. Develop the formula $\sin(t + u) - \sin(t - u) = 2\cos t\sin u$.
20. Your friend Eric complains that he cannot memorize formulas easily. You know that he is very good at deriving them, however. Help Eric construct a short list of trigonometric identities that he can memorize and then use to derive quickly the others he needs in this section.

In exercises 21 through 30, solve the equation or inequality.

21. $\sin 2\theta + 2\sin\theta = \cos\theta + 1$
22. $\dfrac{2\tan\theta}{1 + \tan^2\theta} = 1$
23. $(\sin\theta)(\cos\theta) = \dfrac{1}{2}$
24. $(\sin x)(\cos^3 x) - (\sin x)(\cos x) = 0$
25. $\cos 6\theta - \cos 2\theta = 0$
26. $(\sec^2 t)(\csc^2 t) > 0$
27. $\dfrac{2\tan\theta}{1 + \tan^2\theta} \geq 1$
28. $(\sin x)(\cos^3 x) - (\sin x)(\cos x) < 0$
29. $(\sec^2 t)(\csc^2 t) < 0$
30. $(\sin\theta)(\cos\theta) \leq \dfrac{1}{2}$
31. Suppose the end of a spring is oscillating and its position in inches is approximated by $f(t) = \sin t + 2\cos t$, with t measured in seconds. Approximate to two decimal places the maximum distance the end of the spring moves in an oscillation.
32. Suppose the end of a spring is oscillating and its position in inches is approximated by $f(t) = 1.1\sin t + 2.6\cos t$, with t measured in seconds. Approximate to one decimal place the maximum distance the end of the spring moves in an oscillation.
33. (Graphing calculator) Obtain the solution to exercise 31 by graphing f and using the solving routine of your calculator to find the approximate maximum distance.
34. (Graphing calculator) Obtain the solution to exercise 32 by graphing f and using the solving routine of your calculator to find the approximate maximum distance.

7.5 Complex-Number Representation

Toolkit

Several times we've noted that there is no place on the real number line for the complex numbers. We can represent them geometrically in the plane, though, by pairing a complex number in standard form, $a + bi$, with the point (a, b). Geometric definitions for addition and subtraction of complex numbers are discussed in the exercises.

Illustration

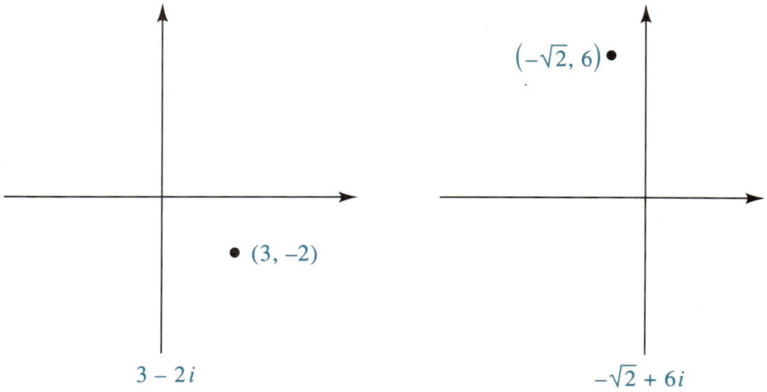

□

A slightly different way of identifying a point in the plane gives us a way to settle a question about complex numbers from Chapter 1: how do you find the standard form for roots of complex numbers?

Instead of locating a point by giving horizontal and vertical coordinates, we can give its location by specifying an angle in standard position and a distance from the origin on the angle's terminal side.

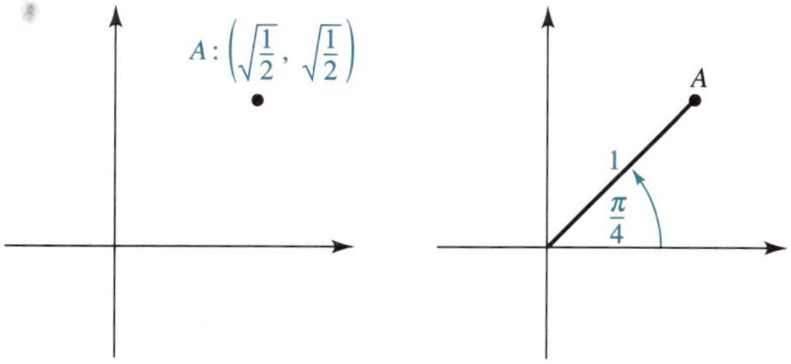

In general, a point (a, b) is $\sqrt{a^2 + b^2}$ units from the origin at an angle θ. The relationships $a = r \cos \theta$ and $b = r \sin \theta$, where $r = \sqrt{a^2 + b^2}$, determine θ. However, θ is not uniquely determined by these statements. The value of θ such that $0 \le \theta < 2\pi$ is called the *principal value*. The ordered pair (r, θ) is called a pair of *polar coordinates* of point A.

If we assign the point (a, b) to the complex number $a + bi$, we can also represent the complex number as $r \cos \theta + r(\sin \theta)i$ or $r(\cos \theta + i \sin \theta)$, where r and θ are as described before. This is called the **trigonometric representation** of $a + bi$.

Definition The real number $\sqrt{a^2 + b^2}$ is called the modulus of $a + bi$. It is the distance from (a, b) to the origin.

It is this style of representation that we need to write the roots of complex numbers in standard form.

EXAMPLE Write $1 + \sqrt{3}i$ in trigonometric form.
Plotting the point $(1, \sqrt{3})$ is a good beginning.

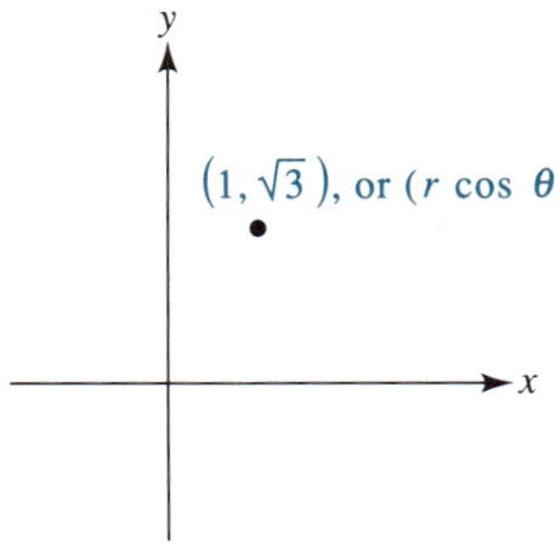

We know that $1^2 + \sqrt{3}^2$, or 4, is r^2, so r is 2. Because $\cos \theta = \frac{1}{2}$ and θ is a first-quadrant number, θ is $\pi/3$.

Therefore, $1 + \sqrt{3}i$ in trigonometric form is $2\left(\cos\frac{\pi}{3} + i \sin\frac{\pi}{3}\right)$. ■

We already have a definition for the product of two complex numbers $a + bi$ and $c + di$, but let's see how the trigonometric representation of the product is related to those of the factors. We can write

$$a + bi = r(\cos \theta + i \sin \theta)$$

and

$$c + di = s(\cos \phi + i \sin \phi).$$

$$(a + bi)(c + di) = [r(\cos \theta + i \sin \theta)][s(\cos \phi + i \sin \phi)]$$

$$\Leftrightarrow \quad (a + bi)(c + di) = rs[(\cos\theta + i \sin \theta)(\cos \phi + i \sin \phi)]$$

$$\Leftrightarrow \quad (a + bi)(c + di) = rs[\cos \theta \cos \phi + i \cos \theta \sin \phi + i \sin \theta \cos \phi + i^2 \sin \theta \sin \phi]$$

$$\Leftrightarrow \quad (a + bi)(c + di) = rs[\cos \theta \cos \phi - \sin \theta \sin \phi + i(\cos \theta \sin \phi + \sin \theta \cos \phi)]$$

$$\Leftrightarrow \quad (a + bi)(c + di) = rs[\cos(\theta + \phi) + i \sin(\theta + \phi)]$$

Identities (12a) and (13)

The product of two complex numbers in trigonometric form is the complex number in trigonometric form whose modulus is the product of the moduli of the factors and whose argument is the sum of the arguments of the factors.

EXAMPLE Write $6\left(\cos\frac{\pi}{3} + i\sin\frac{\pi}{3}\right)\cdot 2\left(\cos\frac{5\pi}{6} + i\sin\frac{5\pi}{6}\right)$ in trigonometric form.

$$6\left(\cos\frac{\pi}{3} + i\sin\frac{\pi}{3}\right)\cdot 2\left(\cos\frac{5\pi}{6} + i\sin\frac{5\pi}{6}\right)$$
$$= 12\left[\cos\left(\frac{\pi}{3} + \frac{5\pi}{6}\right) + i\sin\left(\frac{\pi}{3} + \frac{5\pi}{6}\right)\right], \quad \text{or} \quad 12\left(\cos\frac{7\pi}{6} + i\sin\frac{\pi}{6}\right)$$

In trigonometric form,

$$6\left(\cos\frac{\pi}{3} + i\sin\frac{\pi}{3}\right)\cdot 2\left(\cos\frac{5\pi}{6} + i\sin\frac{5\pi}{6}\right)$$

is

$$\mathbf{12\left(\cos\frac{7\pi}{6} + i\sin\frac{7\pi}{6}\right).}$$

■

You can probably make a very good guess about the quotient of two complex numbers written in trigonometric form. We write the quotient of complex numbers by using the conjugate (Section A.3).

$$\frac{a+bi}{c+di} = \frac{r(\cos\theta + i\sin\theta)}{s(\cos\phi + i\sin\phi)}$$

$$\Rightarrow \quad \frac{a+bi}{c+di} = \frac{r(\cos\theta + i\sin\theta)}{s(\cos\phi + i\sin\phi)}\cdot\frac{(\cos\phi - i\sin\phi)}{(\cos\phi - i\sin\phi)}$$

Note: $\dfrac{a+bi}{c+di}$ is defined only if c and d are not both 0. So $c - di \neq 0$ as well.

$$\Leftrightarrow \quad \frac{a+bi}{c+di} = \frac{r}{s}\cdot\frac{\cos\theta\cos\phi - i\cos\theta\sin\phi + i\sin\theta\cos\phi - i^2\sin\theta\sin\phi}{\cos^2\phi - i^2\sin^2\phi}$$

$$\Leftrightarrow \quad \frac{a+bi}{c+di} = \frac{r}{s}\cdot\frac{(\cos\theta\cos\phi + \sin\theta\sin\phi) + i(\sin\theta\cos\phi - \cos\theta\sin\phi)}{\cos^2\phi + \sin^2\phi}$$

$$\Leftrightarrow \quad \frac{a+bi}{c+di} = \frac{r}{s}\cdot\frac{\cos(\theta - \phi) + i\sin(\theta - \phi)}{1}$$

The quotient of two complex numbers in trigonometric form is the complex number in trigonometric form whose modulus is the quotient of their moduli and whose argument is the difference of the numerator's argument and the denominator's argument.

EXAMPLE Write $\dfrac{3(\cos 50^\circ + i\sin 50^\circ)}{0.1(\cos 10^\circ + i\sin 10^\circ)}$ in trigonometric form.

$$\frac{3(\cos 50° + i \sin 50°)}{0.1(\cos 10° + i \sin 10°)} = 30[\cos(50° - 10°) + i \sin(50° - 10°)]$$
$$= 30(\cos 40° + i \sin 40°)$$

The trigonometric form is 30(cos 40° + *i* sin 40°). ■

The rules in Section A.3 are better for finding the product and quotient of complex numbers in standard form, but trigonometric forms make it easier to develop efficient techniques for finding powers and roots of complex numbers. If you think of raising $a + bi$ to the nth power, n a natural number, as performing repeated multiplications, the following formula seems quite reasonable.

$$(a + bi)^n = r^n(\cos n\theta + i \sin n\theta)$$

if $a + bi = r(\cos \theta + i \sin \theta)$ and $n \in I$. This statement, known as **De Moivre's theorem**, is proved in Chapter 10. It provides a very quick way to find the standard form of integer powers of a complex number, not just natural-number powers.

EXAMPLE Write $\left(\frac{\sqrt{3}}{2} + \frac{1}{2}i\right)^6$ in standard form.

We first write $\frac{\sqrt{3}}{2} + \frac{1}{2}i$ in trigonometric form.

$$r = \sqrt{\left(\frac{\sqrt{3}}{2}\right)^2 + \left(\frac{1}{2}\right)^2}, \quad \text{or} \quad 1$$

$$r \cos \theta = \frac{\sqrt{3}}{2}, \qquad r \sin \theta = \frac{1}{2}, \quad \text{and} \quad r = 1$$

$$\Leftrightarrow \qquad \cos \theta = \frac{\sqrt{3}}{2}, \qquad \sin \theta = \frac{1}{2}$$

$$\Leftrightarrow \qquad \theta = \frac{\pi}{6} + 2k\pi$$

Now we have the trigonometric form we need.

$$\frac{\sqrt{3}}{2} + \frac{1}{2}i = 1\left(\cos \frac{\pi}{6} + i \sin \frac{\pi}{6}\right)$$

$$\left(\frac{\sqrt{3}}{2} + \frac{1}{2}i\right)^6 = 1^6\left(\cos \frac{\pi}{6} \cdot 6 + i \sin \frac{\pi}{6} \cdot 6\right) \qquad \text{De Moivre's theorem}$$

$$\Leftrightarrow \qquad \left(\frac{\sqrt{3}}{2} + \frac{1}{2}i\right)^6 = 1\,(\cos \pi + i \sin \pi)$$

$$\Leftrightarrow \qquad \left(\frac{\sqrt{3}}{2} + \frac{1}{2}i\right)^6 = 1(-1 + i \cdot 0)$$

$$\Leftrightarrow \qquad \left(\frac{\sqrt{3}}{2} + \frac{1}{2}i\right)^6 = -1$$

In standard form, $\left(\frac{\sqrt{3}}{2} + \frac{1}{2}i\right)^6$ is $-1 + 0 \cdot i$. ■

EXAMPLE Write $(-\sqrt{2} - \sqrt{2}i)^{-4}$ in standard form.

The modulus of $-\sqrt{2} - \sqrt{2}i$ is $\sqrt{\sqrt{2}^2 + \sqrt{2}^2}$ or 2. To determine θ, we note $-\sqrt{2} = 2 \cos \theta$ and $-\sqrt{2} = 2 \sin \theta$. We need an angle whose cosine is $-\sqrt{2}/2$ and whose sine is $-\sqrt{2}/2$; $5\pi/4$ is one such angle.

$$(-\sqrt{2} - \sqrt{2}i)^{-4} = 2^{-4}\left\{\cos\left[(-4)\frac{5\pi}{4}\right] + i \sin\left[(-4)\frac{5\pi}{4}\right]\right\}$$

$$\Leftrightarrow \qquad (-\sqrt{2} - \sqrt{2}i)^{-4} = \frac{1}{16}[\cos(-5\pi) + i \sin(-5\pi)]$$

$$\Leftrightarrow \qquad (-\sqrt{2} - \sqrt{2}i)^{-4} = \frac{1}{16}(-1 + i \cdot 0)$$

In standard form, $(-\sqrt{2} - \sqrt{2}i)^{-4}$ is $-\frac{1}{16} + 0 \cdot i$. ■

Is there a square root of $2 + 2i$? Let's explore this question using trigonometric forms. First,

$$2 + 2i = \sqrt{8}\left(\cos \frac{\pi}{4} + i \sin \frac{\pi}{4}\right).$$

We want $r(\cos \theta + i \sin \theta)$ such that

$$[r(\cos \theta + i \sin \theta)]^2 = \sqrt{8}\left(\cos \frac{\pi}{4} + i \sin \frac{\pi}{4}\right)$$

or

$$r^2(\cos 2\theta + i \sin 2\theta) = \sqrt{8}\left(\cos \frac{\pi}{4} + i \sin \frac{\pi}{4}\right).$$

However we write the trigonometric form of this number, its modulus—the distance from the origin—is the same. Hence $r^2 = \sqrt{8}$ and $r = \sqrt{\sqrt{8}}$, or $\sqrt[4]{8}$. The angles must also be equal, but here we need to remember to take into account the fact that $\pi/4$ was not the only choice we could have made for the angle. In fact,

$$2 + 2i = \sqrt{8}\left[\cos\left(\frac{\pi}{4} + 2k\pi\right) + i \sin\left(\frac{\pi}{4} + 2k\pi\right)\right], \qquad k \in I.$$

We actually have $2\theta = \pi/4 + 2k\pi$, $k \in I$. This is equivalent to $\theta = \pi/8 + k\pi$, $k \in I$. Let's look more closely at these values of θ.

k	$\theta = \frac{\pi}{8} + k\pi$	$\cos\theta$
0	$\frac{\pi}{8}$	$\frac{\sqrt{\sqrt{\frac{1}{2}}+1}}{2}$
1	$\frac{9\pi}{8}$	$-\frac{\sqrt{\sqrt{\frac{1}{2}}+1}}{2}$
2	$\frac{\pi}{8} + 2\pi$	$\frac{\sqrt{\sqrt{\frac{1}{2}}+1}}{2}$
3	$\frac{9\pi}{8} + 2\pi$	$-\frac{\sqrt{\sqrt{\frac{1}{2}}+1}}{2}$
-1	$\frac{9\pi}{8} - 2\pi$	$-\frac{\sqrt{\sqrt{\frac{1}{2}}+1}}{2}$
-2	$\frac{\pi}{8} - 2\pi$	$\frac{\sqrt{\sqrt{\frac{1}{2}}+1}}{2}$

Because both sine and cosine have period 2π, only the values $\pi/8$ and $9\pi/8$ produce distinct values for $\cos\theta + i\sin\theta$. There are exactly two square roots of $2 + 2i$,

$$\sqrt[4]{8}\left(\cos\frac{\pi}{8} + i\sin\frac{\pi}{8}\right) \quad\text{and}\quad \sqrt[4]{8}\left(\cos\frac{9\pi}{8} + i\sin\frac{9\pi}{8}\right).$$

Our example should help persuade you of the following:

A complex number $r(\cos\theta + i\sin\theta)$ has exactly ***n n*^th^ roots** ($n \in N$); the roots are

$$\sqrt[n]{r}\left[\cos\left(\frac{\theta + 2k\pi}{n}\right) + i\sin\left(\frac{\theta + 2k\pi}{n}\right)\right]$$

for $k = 0, 1, 2, \ldots, n - 1$.

You may notice the method for finding roots is to use the inverses of the operations used to find a power.

EXAMPLE Find the four fourth roots of -81.

$$\begin{aligned} -81 = -81 + 0i &= \sqrt{(-81)^2 + 0^2}\,(\cos\theta + i\sin\theta) \\ &= 81(\cos\theta + i\sin\theta) \end{aligned}$$

$$\cos\theta = -1 \quad\text{and}\quad \sin\theta = 0 \quad\Leftrightarrow\quad \theta = \pi + 2k\pi$$

The fourth roots are

$$\sqrt[4]{81}\left(\cos\frac{\pi + 2k\pi}{4} + i\sin\frac{\pi + 2k\pi}{4}\right), \qquad k = 0, 1, 2, 3.$$

They may be written as:

$$k = 0: \quad 3\left(\cos\frac{\pi}{4} + i\sin\frac{\pi}{4}\right), \quad \text{or} \quad 3\left(\sqrt{\frac{1}{2}} + \sqrt{\frac{1}{2}}i\right)$$

$$k = 1: \quad 3\left(\cos\frac{3\pi}{4} + i\sin\frac{3\pi}{4}\right), \quad \text{or} \quad 3\left(-\sqrt{\frac{1}{2}} + \sqrt{\frac{1}{2}}i\right)$$

$$k = 2: \quad 3\left(\cos\frac{5\pi}{4} + i\sin\frac{5\pi}{4}\right), \quad \text{or} \quad 3\left(-\sqrt{\frac{1}{2}} - \sqrt{\frac{1}{2}}i\right)$$

$$k = 3: \quad 3\left(\cos\frac{7\pi}{4} + i\sin\frac{7\pi}{4}\right), \quad \text{or} \quad 3\left(\sqrt{\frac{1}{2}} - \sqrt{\frac{1}{2}}i\right)$$

The four fourth roots of -81 are $3(\sqrt{\tfrac{1}{2}} + \sqrt{\tfrac{1}{2}}i)$,

$3(-\sqrt{\tfrac{1}{2}} + \sqrt{\tfrac{1}{2}}i)$, $3(-\sqrt{\tfrac{1}{2}} - \sqrt{\tfrac{1}{2}}i)$, and $3(\sqrt{\tfrac{1}{2}} - \sqrt{\tfrac{1}{2}}i)$. ■

In the special case of the complex number $1 + 0i$, better known as 1, we call its roots the *roots of unity*.

EXAMPLE Find the third roots of unity and graph them.

$1 + 0i = 1(\cos\theta + i\sin\theta)$, with $\cos\theta = 1$ and $\sin\theta = 0$. Hence $1 + 0i = 1(\cos 2k\pi + i\sin 2k\pi)$, and the third roots are

$$\sqrt[3]{1}\left(\cos\frac{0 + 2k\pi}{3} + i\sin\frac{0 + 2k\pi}{3}\right), \qquad k = 0, 1, 2.$$

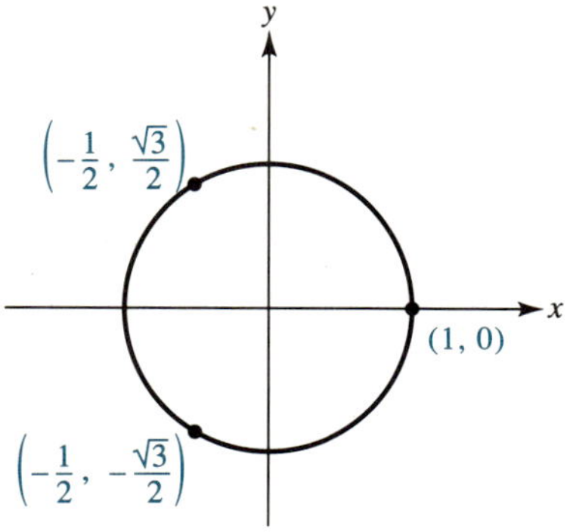

We have for

$$k = 0: \quad 1(\cos 0 + i\sin 0), \quad \text{or} \quad 1 + 0i$$

$$k = 1: \quad 1\left(\cos\frac{2\pi}{3} + i\sin\frac{2\pi}{3}\right), \quad \text{or} \quad -\frac{1}{2} + \frac{\sqrt{3}}{2}i$$

$$k = 2: \quad 1\left(\cos\frac{4\pi}{3} + i\sin\frac{4\pi}{3}\right), \quad \text{or} \quad -\frac{1}{2} - \frac{\sqrt{3}}{2}i$$

The three third roots of unity are $1 + 0i, -\dfrac{1}{2} + (\sqrt{3}/2)i$, and $-\dfrac{1}{2} - (\sqrt{3}/2)i$.

If we graph these roots, each of the third roots of unity lies on a circle of radius 1 with center at the origin. You can also check that the three points are equally spaced around the circle. ■

In Section 1.2 we restricted the coefficients of quadratic equations to the real numbers; now that we have a way to write roots of complex numbers in standard form, this is no longer necessary. First we define $\sqrt{a + bi}$, the principal square root of $a + bi$, $a < 0$ or $b \neq 0$, to be the complex number $c + di$ with positive imaginary part such that $(c + di)^2 = a + bi$.

EXAMPLE Solve $x^2 - 3x + (3 - i) = 0$; write the solutions in standard form.

Unless you quickly see factors for the left-hand side, the quadratic formula should be used. Substituting in the formula, x is either

$$\frac{3 + \sqrt{-3 + 4i}}{2} \quad \text{or} \quad \frac{3 - \sqrt{-3 + 4i}}{2}.$$

To write the solutions in standard form, we need to find the standard form for the principal square root of $-3 + 4i$. In trigonometric form, $-3 + 4i$ is $5(\cos\theta + i\sin\theta)$ with $\theta = \text{Cos}^{-1}(-\frac{3}{5})$.

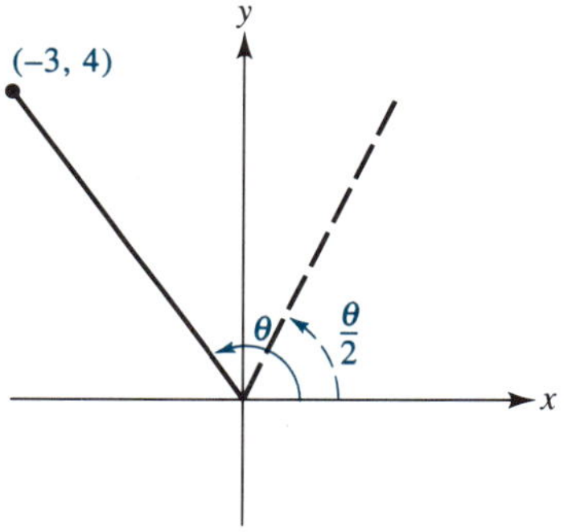

The two square roots are $\sqrt{5}(\cos\theta/2 + i\sin\theta/2)$ and $\sqrt{5}[\cos(\theta/2 + \pi) + i\sin(\theta/2 + \pi)]$. We can use the half-angle formulas to compute $\cos\theta/2$ and $\sin\theta/2$.

$$\cos\frac{\theta}{2} = \sqrt{\frac{1 + \cos\theta}{2}}, \quad \text{or} \quad \sqrt{\frac{1 - \frac{3}{5}}{2}}$$

$$\sin\frac{\theta}{2} = \sqrt{\frac{1 - \cos\theta}{2}}, \quad \text{or} \quad \sqrt{\frac{1 + \frac{3}{5}}{2}}$$

The principal square root of $-3 + 4i$ is $1 + 2i$.

The solution set is $\{2 + i, 1 - i\}$. ■

In Section 1.3 we were not able to check two possible solutions for an equation. Are

$$\left(-\frac{1}{2} + \frac{\sqrt{3}}{2}i\right) \quad \text{and} \quad \left(-\frac{1}{2} - \frac{\sqrt{3}}{2}i\right)$$

solutions for $x + \sqrt{x} = -1$? In trigonometric form, $-\frac{1}{2} + (\sqrt{3}/2)i$ is $\cos(2\pi/3 + 2k\pi) + i\sin(2\pi/3 + 2\pi)$, and $-\frac{1}{2} - (\sqrt{3}/2)i$ is $\cos(4\pi/3 + 2k\pi) + i\sin(4\pi/3 + 2\pi)$. The two square roots of $-\frac{1}{2} + (\sqrt{3}/2)i$ are $\frac{1}{2} + (\sqrt{3}/2)i$ and $-\frac{1}{2} - (\sqrt{3}/2)i$; the two square roots of $-\frac{1}{2} - (\sqrt{3}/2)i$ are $-\frac{1}{2} + (\sqrt{3}/2)i$ and $\frac{1}{2} - (\sqrt{3}/2)i$. The checks are organized in the following table.

x	Principal Square Root $\sqrt{x}$	LHS $x + \sqrt{x}$	RHS -1	Solution
$-\frac{1}{2} + \frac{\sqrt{3}}{2}i$	$\frac{1}{2} + \frac{\sqrt{3}}{2}i$	$\sqrt{3}i$	-1	No
$-\frac{1}{2} - \frac{\sqrt{3}}{2}i$	$-\frac{1}{2} + \frac{\sqrt{3}}{2}i$	-1	-1	Yes

The solution set for $x + \sqrt{x} = -1$ is

$$\left\{-\frac{1}{2} - \frac{\sqrt{3}}{2}i\right\}.$$

EXERCISES 7.5

In exercises 1 through 8, rewrite each ordered pair in polar coordinates.

1. $\left(\sqrt{\frac{5}{2}}, \sqrt{\frac{5}{2}}\right)$
2. $(\sqrt{3}, -1)$
3. $(-5, -5\sqrt{3})$
4. $(-7, 7)$
5. $\left(\frac{-\sqrt{3}}{4}, \frac{1}{4}\right)$
6. $(0, -8)$
7. $(3, 2)$
8. $(-4, -6)$

In exercises 9 through 16, rewrite each complex number in trigonometric form.

9. $0 + 1i$
10. $-1 + 0i$
11. $-1 + 1i$
12. $-15 - 15i$
13. $\sqrt{3} - 1i$
14. $7\sqrt{3} + 7i$
15. $-4 - 4\sqrt{3}i$
16. $-0.5 + 0.5\sqrt{3}i$

In exercises 17 through 20, perform the indicated operation and write the result in trigonometric form.

17. $3\left(\cos\frac{\pi}{3} + i\sin\frac{\pi}{3}\right)\cdot 4\left(\cos\frac{\pi}{2} + i\sin\frac{\pi}{2}\right)$
18. $3(\cos 10° + i\sin 10°)\cdot(-14)(\cos 20° + i\sin 20°)$
19. $\dfrac{3(\cos 10° + i\sin 10°)}{(-14)(\cos 20° + i\sin 20°)}$
20. $\dfrac{4(\cos \pi/2 + i\sin \pi/2)}{3(\cos \pi/3 + i\sin \pi/3)}$

In exercises 21 through 26, rewrite each power of a complex number in trigonometric form and use De Moivre's theorem to convert it to standard complex-number form.

21. $(\sqrt{\tfrac{1}{2}} + \sqrt{\tfrac{1}{2}}i)^5$
22. $(-15 - 15i)^6$
23. $(\sqrt{3} - 1i)^4$
24. $(7\sqrt{3} + 7i)^3$
25. $(-4 - 4\sqrt{3}i)^{-4}$
26. $(-0.5 + 0.5\sqrt{3}i)^{-3}$

In exercises 27 through 32, write the:

27. Four fourth roots of $4 + 4i$
28. Three cube roots of 2
29. Two square roots of $-1 + \sqrt{3}i$
30. Four fourth roots of $\sqrt{3} - 1i$
31. Five fifth roots of -1
32. Five fifth roots of i
33. Find the five fifth roots of unity and plot them on a unit circle. Connect the points to form a star and a regular polygon.
34. **a.** Write -1 in trigonometric form and find its two square roots. Confirm that the definition of principal square root agrees with the usual definition of $\sqrt{-1}$.
    b. Show that if $a + bi$ is a square root of $m + ni$, then $-a - bi$ is also a square root of $m + ni$.

In exercises 35 and 36, solve the equation and write the solutions in standard form.

35. $x^2 - x + (i + 1) = 0$
36. $2ix^2 + (1 + 2i)x + (-1 + i) = 0$

In exercises 37 through 42, plot in the same plane (a) the two complex numbers, (b) the sum of the two complex numbers, and (c) the result of subtracting the second complex number from the first.

37. $2 + i, 3 + 3i$
38. $4 + 3i, 5 + 2i$
39. $4 - i, 5 + 2i$
40. $3 - 2i, 4 + 5i$
41. $-21 - 3i, 6 - 0.2i$
42. $4.3 + 2.1i, -3.4 + 5.7i$

In exercises 43 through 48, use the following definition:

Definition A **vector** in the plane is an ordered pair of real numbers $\langle x, y\rangle$. The sum of two vectors, $\langle a, b\rangle$ and $\langle c, d\rangle$, denoted by $\langle a, b\rangle + \langle c, d\rangle$, is the vector $\langle a + c, b + d\rangle$.

Find the sum of each vector pair. How do these seem to relate to Exercises 37 through 42?

43. $\langle 2, 1\rangle, \langle 3, 3\rangle$ **44.** $\langle 4, 3\rangle, \langle 5, 2\rangle$

45. $\langle 4, -1\rangle, \langle 5, 2\rangle$ **46.** $\langle 3, -2\rangle, \langle 4, 5\rangle$

47. $\langle -21, -3\rangle, \langle 6, -0.2\rangle$ **48.** $\langle 4.3, 2.1\rangle, \langle -3.4, 5.7\rangle$

Another way to define a vector is as a directed line segment.

Definition A vector $\overrightarrow{PQ}$ in the plane is the directed line segment extending from the initial point P to the terminal point Q.

The vector with initial point (0, 0) and terminal point (a, b) is denoted by $\langle a, b\rangle$. The coordinate a is referred to as the x-component, or horizontal component, of the vector $\langle a, b\rangle$, and the coordinate b is referred to as the y-component, or vertical component, of the vector $\langle a, b\rangle$.

In exercises 49 through 54, sketch each of the given vectors as a directed line segment on the same set of axes. Also sketch the vector representing the sum of the given vectors.

49. $\langle 2, 1\rangle, \langle 3, 3\rangle$ **50.** $\langle 4, 3\rangle, \langle 5, 2\rangle$

51. $\langle 4, -1\rangle, \langle 5, 2\rangle$ **52.** $\langle 3, -2\rangle, \langle 4, 5\rangle$

53. $\langle -21, -3\rangle, \langle 6, -0.2\rangle$ **54.** $\langle 4.3, 2.1\rangle, \langle -3.4, 5.7\rangle$

55. Use the results of exercises 49 through 54 to give a geometric method for finding the sum of two vectors.

Chapter 7: Summary

Trigonometric functions have many applications: solving geometric problems involving triangles, modeling real-world situations, and representing complex numbers. Identities can be used to rewrite expressions, and in this chapter they provided an opportunity for more practice with direct proofs.

Key Concepts

radian measure
right triangle definitions of trigonometric functions
laws of cosines and sines
sum and difference formulas
half-angle formulas
identities
trigonometric representation of a complex number *modulus*
De Moivre's theorem $(a + bi)^n = r^n(\cos n\,\theta + i \sin n\,\theta),\ n \in I$
n n^{th} roots of a complex number
vector

Skills

You should be able to:

1. Convert from radian measure to degree measure and vice versa.
2. Compute the sine, cosine, tangent, cotangent, secant, or cosecant of any angle in a right triangle.
3. Solve a right triangle.
4. Use the laws of cosines and sines to solve triangles.

5. Determine whether one triangle, two triangles, or no triangle exists given two sides and an angle not between them.
6. Compute the area of a triangle given two sides and the included angle.
7. Compute the area of a circular sector.
8. Determine whether a statement is an identity.
9. Apply the sum and difference formulas and the half-angle formulas.
10. Graphically represent a complex number, give the modulus of a complex number, and express a complex number in trigonometric form.
11. Use De Moivre's theorem to give an integer power of a complex number and find the n n^{th} roots of a complex number.

SUMMARY EXERCISES

The starred exercises can serve as a practice chapter test. The answers for these are at the end of the answer section at the back of the book.

1. Convert $3\pi/8$ to degree measure.
2. Convert 12° to radian measure.
* 3. Convert 34.7 to degree measure.

In exercises 4 through 6, find the approximate solution to right triangle *ABC* using the given information. Assume standard labeling.

4. $a = 23, b = 42$
5. $\alpha = 55°, c = 13.2$
* 6. $\beta = 13.1°, b = 10{,}004$

In exercises 7 through 14, find the approximate solution(s) for ΔABC using the given information. Assume standard labeling.

7. $a = 12, \alpha = 47°, \gamma = 80°$
8. $a = 53, b = 43, \alpha = 130°$
* 9. $a = 12, b = 30, \gamma = 25°$
10. $b = 4.6, c = 4.1, \gamma = 65°$
11. $a = 13.55, b = 42.84, c = 98.21$
* 12. $b = 12, a = 10, \alpha = 49°$
13. $\beta = 89.9°, a = 13, c = 145$
14. $\alpha = 13°, \beta = 140°, \gamma = 27°$
* 15. Approximate the area of a triangle in which two sides have measured 14 feet and 35 feet and the included angle measures $6\pi/13$.
16. What is the area of a circular sector with radius 11.2 meters and central angle of measure 286°?
17. Compute the exact value of $\sin(\pi/12)$ using $\sin(\pi/6)/2$, $\sin(\pi/3 - \pi/4)$, and $\cos((\pi/6/2) - \pi/2)$.
* 18. Rewrite

$$\frac{\tan(x)\sec(x) - \sin(x)}{\csc^2(x) + 1}$$

in terms of sine and cosine functions only.

In exercises 19 through 25, determine whether the statements are identities. For each identity, state whether the two sides have the same restrictions.

19. $\cos\left(\frac{\pi}{2} + \theta\right) = -\sin(-\theta - \pi)$
20. $\frac{\sec^2\theta - 1}{1/\csc\theta} = (\tan\theta)(\sec\theta)$
* 21. $\sin x + \operatorname{cox} x + 1 = \frac{\csc^2 x + \cot x + (\cot x)(\cos^2 x) + \csc^3 x}{\csc x}$
22. $\frac{\cos\alpha}{\sin\alpha - \cos\alpha} = \frac{(\cos\alpha)(\cos\alpha - \sin\alpha)}{1 - \sin 2\alpha}$
23. $1/\cos(y) = \sec(y)$
* 24. $\sin^6 x + 3(\sin^4 x)(\cos^2 x) + 3(\sin^2 x)(\cos^4 x) + \cos^6 = 1$
25. $\frac{1}{\sec(y)} = \cos(y)$
26. Rewrite $\sin(2\operatorname{Cos}^{-1}\frac{1}{4}3)$ with no inverse trigonometric functions.
* 27. Rewrite $\cos(8x) - \cos(4x)$ as a product.
28. Rewrite $(\sin 2t)(\sin 6t)$ as a sum.

In exercises 29 through 36, solve the equation or inequality.

29. $2(\sin\theta)(\cos\theta) = \frac{\sqrt{3}}{2}$
* 30. $2(\sin\theta)(\cos\theta) = \sqrt{3}$
31. $2(\sin\theta)(\cos\theta) = 0.2$
32. $2(\sin\theta)(\cos\theta) < \sqrt{3}$
* 33. $2(\sin\theta)(\cos\theta) \geq \frac{\sqrt{3}}{2}$
34. $2\sin^2 x - 1 = \sqrt{\frac{1}{2}}$
35. $2\sin^2 x + 5\sin x + 2 = 0$
* 36. $4\sin^2 x - 1 < 0$
37. Rewrite $(3/\sqrt{2}, -3/\sqrt{2})$ in polar coordinates.
38. Write $(\sqrt{3}/2 - \frac{1}{2}i)^5$ in standard form.
* 39. Find the four fourth roots of $\sqrt{3}/2 + \frac{1}{2}i$
40. Write the reciprocal of the sixteenth power of $1 + i$ in standard form.
41. Graph the four fourth roots of unity.

CHAPTER EIGHT

Polynomial Functions

- 8.1 Synthetic Division and Its Uses
- 8.2 Complex and Rational Roots
- 8.3 Graphing Polynomial Functions
- 8.4 Rational and Other Functions Revisited

In this chapter we investigate in more detail the properties of functions defined by one-variable polynomials. A function f whose rule is a polynomial expression is called a *polynomial function*. In Chapters 1 and 2 we saw many important features of first- and second-degree polynomial functions, although they weren't identified as functions yet. In Chapter 4 the simple cubic function, $f(x) = x^3$, was one of the standard functions introduced. Here is a summary of our previous work with polynomial functions.

Function	Degree	Name	Graph	Features
$f(x) = ax + b$	1st	Linear	Line	At most one x-intercept, slope a
$g(x) = ax^2 + bx + c, a \neq 0$	2nd	Quadratic	Parabola	At most two x-intercepts, vertex at $-b/2a$
$h(x) = x^3$	3rd	Cubic	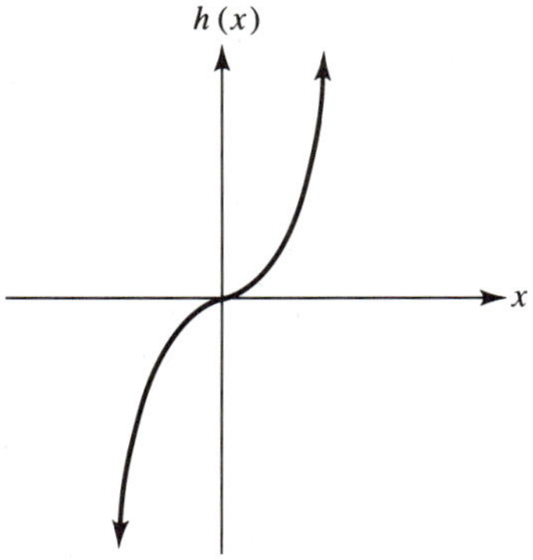	One x-intercept (zero)

This is an appropriate place for a formal definition of a polynomial in one variable.

Definition P is a **polynomial in one variable** if

$$P = a_n x^n + a_{n-1}x^{n-1} + a_{n-2}x^{n-2} + \cdots + a_1 x + a_0,$$

n is a nonnegative integer, and each coefficient a_i is a complex number.

Remember that the real numbers are a subset of the complex numbers. For some of the work in this chapter we'll require that the coefficients be real numbers. Generally we do not write terms with zero coefficients. In particular, if $a_n \neq 0$, we say P has degree n. Also, if $a_n \neq 0$, a_n is called the *leading coefficient* of P.

Illustration The expression $7.8y^5 - 2y^3 + 2.5y^2 + 19y - 3.9$ is a polynomial in one variable. The leading coefficient is 7.8.
The expression $7.8x^5 - 2x^3 + 2.5x^5 + 19x - 3.9 + 4/x$ is not a polynomial in one variable because $1/x$ is x^{-1}, and negative powers of the variable are not allowed.
The expression $7.8x^5y - 2x^3y^3 + 2.5x^2y^4 + 19xy^7$ is not a polynomial in one variable because there are two variables, x and y. □

For the rest of this chapter we use polynomial to mean polynomial in one variable. Two polynomials, P and Q, are equal if and only if

$$P = a_n x^n + a_{n-1}x^{n-1} + a_{n-2}x^{n-2} + \cdots + a_1 x + a_0,$$

$$Q = b_n x^n + b_{n-1}x^{n-1} + b_{n-2}x^{n-2} + \cdots + b_1 x + b_0,$$

and $a_i = b_i$ for each value of $i = 0, 1, \ldots, n$. In other words, $P = Q$ if and only if the corresponding coefficients match each other.

We need a tool, synthetic division, for our investigation of polynomial functions.

8.1 Synthetic Division and Its Uses

Toolkit

You will need division of polynomials from Section A.5.

Pick any two positive integers a and b, so that $a \leq b$. We can always express b as $a \cdot q + r$, with $0 \leq r < a$, by dividing b by a. Here q is the quotient and r the remainder.

Illustration Given 7 and 27, $27 = 7 \cdot \underline{3} + \underline{6}$.
Given 12 and 156, $156 = 12 \cdot \underline{13} + \underline{0}$.
Given 6 and 6, $6 = 6 \cdot \underline{1} + \underline{0}$. □

A similar result holds for polynomials when we measure "size" by the degree of the polynomial. For any two polynomials D and P, D not a constant and degree $D \leq$ degree P, P can be written as $D \cdot Q + R$, where Q is a polynomial of degree equal to degree P minus degree D and $0 \leq$ degree $R <$ degree D.

Illustration Given $x + 3$ and $x^2 - 4x + 7$,

$$x^2 - 4x + 7 = (x + 3) \cdot \underline{(x - 7)} + \underline{28}.$$

Given $x + 3$ and $x^2 - 4x - 21$,

$$x^2 - 4x - 21 = (x + 3) \cdot \underline{(x - 7)} + \underline{0}.$$

Given $x + 3$ and $x - 4$,

$$x - 4 = (x + 3) \cdot \underline{1} + \underline{(-7)}.$$ □

Here we are interested in the special case where D is the polynomial $x - c$. The degree of D is 1, so when a polynomial P is divided by D, the remainder R must have degree 0; that is, R is constant. If R is 0, then D is a factor of P.

Illustration Let P be $x^3 - x^2 - x - 15$ and D be $x - 2$.

The computation

$$\begin{array}{r|l} & x^2 + x + 1 \\ \hline x - 2 & x^3 - x^2 - x - 15 \\ & \underline{x^3 - 2x^2} \\ & \quad x^2 - x \\ & \quad \underline{x^2 - 2x} \\ & \qquad x - 15 \\ & \qquad \underline{x - 2} \\ & \qquad -13 \end{array}$$

shows that $x^3 - x^2 - x - 15 = (x - 2)(x^2 + x + 1) - 13$, where $x^2 + x + 1$ is Q and -13 is R.

Let P be $x^3 - x^2 - x - 15$ and D be $x - 3$. The computation

$$\begin{array}{r|l} & x^2 + 2x + 5 \\ \hline x - 3 & x^3 - x^2 - x - 15 \\ & \underline{x^3 - 3x^2} \\ & \quad 2x^2 - x \\ & \quad \underline{2x^2 - 6x} \\ & \qquad 5x - 15 \\ & \qquad \underline{5x - 15} \end{array}$$

shows that $x^3 - x^2 - x - 15 = (x - 3)(x^2 + 2x + 5) + 0$, where $x^2 + 2x + 5$ is Q and 0 is R. Thus D is a factor of P. □

Division by a polynomial of the form $x - c$ can also tell us about function values for polynomial functions. Suppose

$$P(x) = x^3 - x^2 - x - 15;$$

from our preceding work we can write $P(x) = (x - 2)(x^2 + x + 1) - 13$. In this form we can see very quickly that $P(2) = (2 - 2)(2^2 + 2 + 1) - 13$ is -13, because the first term is 0. This observation won't be much help unless we can find a way to streamline the long-division process. For divisors of the form $x - c$, we can do just that.

In the preceding long-division computations, the coefficients play the most important role. In fact, if you line the coefficients up carefully, the variable factors may be omitted.

$$\begin{array}{r|l} & x^2 + x + 1 \\ \hline x - 2 & x^3 - x^2 - x - 15 \\ & \underline{x^3 - 2x^2} \\ & \quad x^2 - x \\ & \quad \underline{x^2 - 2x} \\ & \qquad x - 15 \\ & \qquad \underline{x - 2} \\ & \qquad -13 \end{array} \qquad \begin{array}{r|rrrr} & 1 & 1 & 1 & \\ \hline x - 2 & \textcircled{1} & -1 & -1 & -15 \\ & \underline{\textcircled{1}} & \underline{-2} & & \\ & & \textcircled{1} & -1 & \\ & & \underline{\textcircled{1}} & \underline{-2} & \\ & & & \textcircled{1} & -15 \\ & & & \underline{\textcircled{1}} & \underline{-2} \\ & & & & -13 \end{array}$$

This isn't much of a savings. Let's analyze our work more closely. We always match the first coefficient exactly at each stage of the division process, so no important information is given by the pairs circled above. We could omit them and write

$$
\begin{array}{r|rrrr}
 & 1 & 1 & 1 & \\ \cline{2-5}
x-2 & 1 & -1 & -1 & -15 \\
 & & \underline{-2} & & \\
 & & 1 & -1 & \\
 & & & \underline{-2} & \\
 & & & 1 & -15 \\
 & & & & \underline{-2} \\
 & & & & -13
\end{array}
$$

Do we need to bring down the coefficients of the second terms? Why not write our work as follows:

$$
\begin{array}{r|rrrr}
 & 1 & \boxed{1} & 1 & \\ \cline{2-5}
x-2 & 1 & -1 & -1 & -15 \\
 & & \underline{-2} & \underline{\bigcirc\!\!\!\!-2} & \underline{-2} \\
 & & 1 & 1 & -13
\end{array}
$$

The circled -2 is the product of the term -2 in the divisor and the boxed 1.

Synthetic division takes advantage of all our attempts to condense the computation and adds one refinement. We have to remember to subtract—that is, to add the opposite of—the second number at each stage. If we began by taking the opposite of the multiplier (-2 in our example), we could just add at each stage. Here is the fully condensed synthetic division version of our computation.

To divide $x^3 - x^2 - x - 15$ by $x - 2$, write the coefficients of the dividend in order of decreasing powers. Set off the computation with a half-box. Write the opposite of the divisor's constant term outside the box.

$$
\begin{array}{r|rrrr}
2 & 1 & -1 & -1 & -15 \\
 & & & & \\ \hline
\end{array}
$$

The leading coefficient of the quotient must be the same as the leading coefficient of the dividend because we are dividing by $x - 2$. Copy the 1 below the line to be the leading coefficient of the quotient. Next repeat the following steps to the end of the row: Multiply the new number below the line by 2, write it under the next coefficient to the right, add these numbers and write the sum below the line.

$$
\begin{array}{r|rrrr}
2 & 1 & -1 & -1 & -15 \\
 & & 2 & 2 & 2 \\ \hline
 & 1 & 1 & 1 & -13
\end{array}
$$

You probably recognize the numbers below the line as the coefficients of the quotient, 1, 1, 1, and the remainder -13. If the numbers in the second row are easily obtained and the additions easily done, you may supress the second line entirely.

$$
\begin{array}{r|rrrr}
2 & 1 & -1 & -1 & -15 \\
 & & 2 & 2 & 2 \\ \hline
 & 1 & 1 & 1 & -13
\end{array}
\qquad
\begin{array}{r|rrrr}
2 & 1 & -1 & -1 & -15 \\ \hline
 & 1 & 1 & 1 & -13
\end{array}
$$

It's useful to mark the remainder clearly; this is commonly done with a vertical bar, as follows:

$$\begin{array}{r|rrrr} 2 & 1 & -1 & -1 & -15 \\ \hline & 1 & 1 & 1 & \vert\, -13 \end{array}$$

EXAMPLE Write $3x^4 - 2x^3 + x^2 - 7x + 4$ as $(x - 4) \cdot Q + R$.
To use synthetic division we write

$$\begin{array}{r|rrrrr} 4 & 3 & -2 & 1 & -7 & 4 \\ \hline \end{array}$$ Coefficients of the dividend

↑ Opposite of the divisor's constant term

$$\begin{array}{r|rrrrr} 4 & 3 & -2 & 1 & -7 & 4 \\ & & 12 & 40 & 164 & 628 \\ \hline & 3 & 10 & 41 & 157 & \vert\, 632 \end{array}$$

We can now read Q as $3x^3 + 10x^2 + 41x + 157$ and R as 632.

$3x^4 - 2x^3 + x^2 - 7x + 4$ can be written as $\mathbf{(x - 4)(3x^3 + 10x^2 + 41x + 157) + 632}$. ■

When using synthetic division, be sure you've ordered the terms of the dividend in decreasing order of powers. Insert a zero in the coefficient list for any missing power.

EXAMPLE Find $f(-3)$ for $f(x) = 2x^4 - 5x^2 + 7$.
We can substitute -3 for x, but synthetic division can also be used to find function values. We saw earlier that the remainder when $2x^4 - 5x^2 + 7$ is divided by $x - (-3)$ is $f(-3)$.

$$\begin{array}{r|rrrrr} -3 & 2 & 0 & -5 & 0 & 7 \\ & & -6 & 18 & -39 & 117 \\ \hline & 2 & -6 & 13 & -39 & \vert\, 124 \end{array}$$

Note the coefficient 0 for the x^3 and x-terms.

The value $f(-3)$ is 124. ■

When the graphs of first-, second-, and third-degree polynomial functions were sketched in Chapters 1, 2, and 4, we used the polynomial principle: there should be no breaks in the graph. We can now state the algebraic version.

> **Intermediate Value Theorem.** If, for a polynomial function f, $f(x_1) = a$ and $f(x_2) = b$, $a \neq b$, then for each number c between a and b, there is a number s between x_1 and x_2 such that $f(s) = c$.

Illustration Let $f(x) = x^3 - x + 2$. Using synthetic division we see that $f(-2) = -4$ and $f(-1)$ is 2.

$$\begin{array}{r|rrrr} -2 & 1 & 0 & -1 & 2 \\ & & -2 & 4 & -6 \\ \hline & 1 & -2 & 3 & \vert\, -4 \end{array} \qquad \begin{array}{r|rrrr} -1 & 1 & 0 & -1 & 2 \\ & & -1 & 1 & 0 \\ \hline & 1 & -1 & 0 & \vert\, 2 \end{array}$$

The function f must have every number between -4 and 2 as a value. In particular, $f(s) = 0$ for some number s. This s must be between -2 and -1.

Note: There may be several numbers such that $f(s) = 0$. The intermediate value theorem says there is at least one. Notice, too, that the theorem tells us nothing about how to find s.

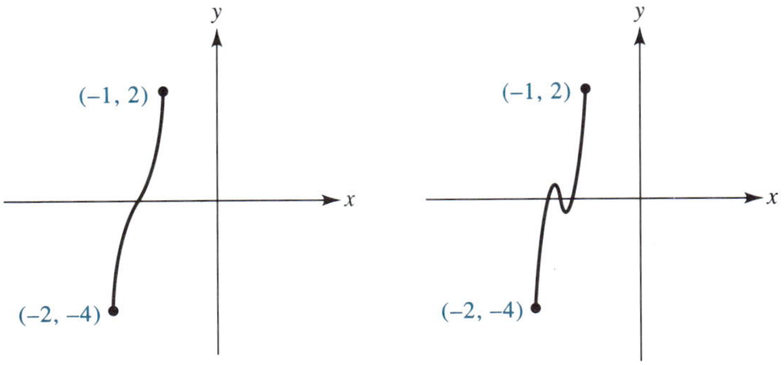

The graph of f between -2 and -1 could be either of these two, based only on the intermediate value theorem. □

EXERCISES 8.1

In exercises 1 through 10, rewrite the polynomial P in the form $D \cdot Q + R$, for each P and D. Indicate when D is a factor of P.

1. $P = x^3 + 3x^2 + 3x + 1, \quad D = x + 1$
2. $P = x^4 - 8x^3 + 24x^2 - 32x + 16, \quad D = x - 2$
3. $P = 2x^5 + 3x^4 - 13x^3 - 70x + 8, \quad D = x + 4$
4. $P = x^4 + 5x^3 - 8x^2 + 5, \quad D = x - 1$
5. $P = x^6 + 40, \quad D = x - 3$
6. $P = 3x^7 - 127, \quad D = x + 2$
7. $P = x^3 + 3x^2 + 3x + 1, \quad D = x + \frac{1}{2}$

   You can choose to use -0.5 as the multiplier in order to use your calculator.
8. $P = 2x^2 + 3x - 13, \quad D = x - \frac{1}{3}$
9. $P = 2ix^2 - (1 - 3i)x + (2 - 5i), \quad D = x - 4i$
10. $P = (5 + i)x^3 - (4 - 3i)x + 2i, \quad D = x + 2i$

In exercises 11 through 20, write P/D as $Q + R/D$.

11. $P = x^3 + 3x^2 + 3x + 1, \quad D = x + 1$
12. $P = x^4 - 8x^3 + 24x^2 - 32x + 16, \quad D = x - 2$
13. $P = 2x^5 + 3x^4 - 13x^3 - 70x + 8, \quad D = x + 4$
14. $P = x^4 + 5x^3 - 8x^2 + 5, \quad D = x - 1$
15. $P = x^6 + 40, \quad D = x - 3$
16. $P = 3x^7 - 127, \quad D = x + 2$
17. $P = x^3 + 3x^2 + 3x + 1, \quad D = x + \frac{1}{2}$
18. $P = 2x^2 + 3x - 13, \quad D = x - \frac{1}{3}$
19. $P = 2ix^2 - (1 - 3i)x + (2 - 5i), \quad D = x - 4i$
20. $P = (5 + i)x^3 - (4 - 3i)x + 2i, \quad D = x + 2i$

In exercises 21 through 30, use synthetic division to find $P(x_0)$ for each x_0.

21. $P(x) = 3x^2 + 4x + 5, \quad x_0 = -1$
22. $P(x) = -3x^3 + 2x^2 - 4x + 8, \quad x_0 = 2$
23. $P(x) = -x^4 - 8x^3 + 9x^2 - 1, \quad x_0 = 3$
24. $P(x) = x^5 - 4x^3 + 2x - 3, \quad x_0 = -2$
25. $P(x) = 2x^3 - 3.1x^2 + 5x - 17.2, \quad x_0 = 4.1$
26. $P(x) = -1.2x^3 + 7.6x^2 - 0.1x + 3.5, \quad x_0 = -2.7$
27. $P(x) = x^2 - 3x + 1, \quad x_0 = i$
28. $P(x) = (2 + 3i)x^2 + (1 - i)x - (5 + 3i), \quad x_0 = -3i$
29. $P(x) = 2x^2 - 4x + 1, \quad x_0 = -\frac{2}{3}$
30. $P(x) = -5x^2 + 3, \quad x_0 = \frac{7}{8}$
31. In which of the exercises 21 through 30 would it have been as easy or easier to evaluate $P(x_0)$ directly as to use synthetic division?
32. In which of the exercises 21 through 30 would it be easier to use a calculator than to do the calculations by hand in the synthetic division process?
33. Describe to a friend how to do synthetic division.
34. If a polynomial expression is divided synthetically using c and a zero remainder is produced, then $x - c$ is ______ and c is ______.

In exercises 35 through 40, find the quotient Q and the remainder R when the polynomial P is divided by D of the form $ax + b$. To do this using synthetic division, first write $ax + b$ as $a(x + b/a)$. You can divide synthetically using $-b/a$ and then divide each term by a because

$$\frac{P}{ax+b} = \frac{P}{a\left(x+\dfrac{b}{a}\right)} = \left[P \div \left(x + \frac{b}{a}\right)\right] \div a.$$

35. $P = 6x^2 - 3x + 20, \quad D = 2x + 1$
36. $P = 3x^2 + 4x + 1, \quad D = 2x - 1$
37. $P = x^3 - 2x^2 - 11x + 18, \quad D = 5x - 2$
38. $P = -2x^3 + x - 5, \quad D = 3x + 4$
39. $P = 8x^3 + 27, \quad D = 2x + 3$
40. $P = 8x^3 + 36x^2 + 54x + 9, \quad D = 2x + 3$

41. You are a teacher constructing a question for a quiz. State a question whose answer could require the use of the following synthetic division.

$$\begin{array}{r|rrrr|r} -2 & 1 & 4 & 3 & -1 & 5 \\ & & -2 & -4 & 2 & -2 \\ \hline & 1 & 2 & -1 & 1 & 3 \end{array}$$

42. State a factoring problem that corresponds to each of the following:

$$\begin{array}{r|rrr|r} a & 1 & 0 & 0 & -a^3 \\ & & a & a^2 & a^3 \\ \hline & 1 & a & a^2 & 0 \end{array}$$

$$\begin{array}{r|rrr|r} -a & 1 & 0 & 0 & a^3 \\ & & -a & a^2 & -a^3 \\ \hline & 1 & -a & a^2 & 0 \end{array}$$

8.2 Complex and Rational Roots

Toolkit

Complex numbers are discussed in Section A.3. If $b = 0$ for a complex number $z = a + bi$, then we say z has no imaginary part, or is a real number.

The values of x for which a polynomial function P has value zero are useful in sketching the graph of P. Such x-values are called **zeros of the function** or **roots of the polynomial** that defines P. In the remainder of this chapter any statement about the zeros of a polynomial function could be made about the roots of the defining polynomial, and vice versa. We will write only one version of such statements but may apply either version in examples.

Synthetic division is a good tool for finding the roots of a polynomial, because we can easily test likely candidates. How do we decide which numbers to test?

Illustration The roots of $6x^3 + 3x^2 - 45x$ can be found by factoring the polynomial.

$$6x^3 + 3x^2 - 45x = 3x(2x^2 + x - 15), \text{ or } 3x(2x - 5)(x + 3)$$

The roots are 0, $\frac{5}{2}$, and -3. We can see that the roots $\frac{5}{2}$ and -3 came from the factors of $2x^2 + x - 15$. If these roots are written in the form a/b, we see that the numerator is a factor of 15 and the denominator is a factor of 2. This suggests where we might look for possible rational roots. Also, the example shows that when we find a root, here 0, we can restrict our search for other roots to a simpler polynomial, here $2x^2 + x - 15$. □

Possible rational roots for the polynomial $a_nx^n + a_{n-1}x^{n-1} + \cdots + a_0$, with a_n, $a_{n-1}, \cdots, a_0$ all integers, are of the form

$$\frac{\text{integer factor of } a_0}{\text{integer factor of } a_n}.$$

EXAMPLE What are the possible rational roots for $6x^4 - 5x^2 + 21$?

Rational roots for this polynomial must be of the form a/b, where a is a factor of 21 and b is a factor of 6. Positive and negative factors should be considered.

factors of 21: $\pm 1, \pm 3, \pm 7, \pm 21$
factors of 6: $\pm 1, \pm 2, \pm 3, \pm 6$

The possible rational roots for $6x^4 - 5x^2 + 21$ are $\pm 1, \pm 3, \pm 7, \pm 21, \pm\frac{1}{2}, \pm\frac{3}{2}, \pm\frac{7}{2}, \pm\frac{21}{2}, \pm\frac{1}{3}, \pm\frac{7}{3}, \pm\frac{1}{6}$, and $\pm\frac{7}{6}$.

Note: Some of the possible pairings produce duplicates of numbers already on the list. ■

We haven't done anything really new here. If you were trying to factor $3x^2 - 5x + 483$ into two binomials, you might try 23 as a constant term in one of them because 23 is a factor of 483. You'd never try 5 because 5 isn't a factor of 483. You know $\frac{23}{7}$ can't be a zero of $f(x) = 3x^2 - 5x + 483$ because if it were, $7x - 23$ would be a factor of $3x^2 - 5x + 483$. On the other hand, $\frac{23}{3}$ is a possible zero because 3 is a factor of 3 and 23 is a factor of 483; thus $3x - 23$ is a possible factor of $3x^2 - 5x + 483$.

The last example produced a lengthy list of possible rational roots. Even with synthetic division, checking these would take a while. It may even be that this polynomial has no rational roots. Clearly we could use some help with this problem. For the rest of this section, we'll restrict the examples to polynomials and functions with short lists of possible rational roots and zeros. In Section 8.3 we'll come back to the question of finding a more efficient method of checking these lists.

EXAMPLE Find all rational zeros of $f(x) = 2x^3 - 7x^2 - 7x + 12$.

The possible rational zeros are $\pm 1, \pm 2, \pm 3, +4, \pm 6, \pm 12, \pm\frac{1}{2}, \pm\frac{3}{2}$.

$$\begin{array}{r|rrrr} 1 & 2 & -7 & -7 & 12 \\ & & 2 & -5 & -12 \\ \hline & 2 & -5 & -12 & 0 \end{array}$$

Note: $0 = f(1)$.

This division shows 1 is a zero for f, and we can restrict our search for other zeros to the polynomial $2x^2 - 5x - 12$. You've had a lot of experience in factoring quadratic polynomials and may choose to finish the problem using earlier factoring methods.

$$2x^2 - 5x - 12 = (2x + 3)(x - 4)$$

The rational zeros of $f(x) = 2x^3 - 7x^2 - 7x + 12$ are 1, $-\frac{3}{2}$, and 4. ■

The rational zeros for $f(x) = 2x^3 - 7x^2 - 7x + 12$ are, in fact, all the zeros of this function. You can see this by remembering that each zero corresponds to a factor.

$$2x^3 - 7x^2 - 7x + 12 = (x - 1)(2x + 3)(x - 4)$$

EXAMPLE Solve $3x^3 - 7x^2 + 14x - 4 = 0$.

Our technique for solving third-degree equations is to make one side zero and factor the other side. The factoring techniques of the appendix fail here, but the solutions we want are the roots of $3x^3 - 7x^2 + 14x - 4$. The possible rational roots are $\pm 1, \pm 2, \pm 4, \pm\frac{1}{3}, \pm\frac{2}{3}, \pm\frac{4}{3}$. Because we haven't discussed how to choose which possibilities to test, we'll just begin at the start of the list.

$$\begin{array}{r|rrrr} 1 & 3 & -7 & 14 & -4 \\ & & 3 & -4 & 10 \\ \hline & 3 & -4 & 10 & 6 \end{array} \qquad \begin{array}{r|rrrr} -1 & 3 & -7 & 14 & -4 \\ & & -3 & 10 & -24 \\ \hline & 3 & -10 & 24 & -28 \end{array}$$

We have similar results until we try $\frac{1}{3}$.

$$\begin{array}{r|rrrr} \frac{1}{3} & 3 & -7 & 14 & -4 \\ & & 1 & -2 & 4 \\ \hline & 3 & -6 & 12 & 0 \end{array}$$

The quotient polynomial $3x^2 - 6x + 12$ can be written as $3(x^2 - 2x + 4)$. Using the quadratic formula to find the factors of $x^2 - 2x + 4$, we have

$$x^2 - 2x + 4 = [x - (1 - \sqrt{3}i)][x - (1 + \sqrt{3}i)].$$

The solution set of $0 = 3x^3 - 7x^2 + 14x - 4$ is $\{\frac{1}{3}, 1 - \sqrt{3}i, 1 + \sqrt{3}i\}$. ■

We are ready to make some general statements about the zeros of a polynomial function.

> **Fundamental Theorem of Algebra:** If $f(x) = a_nx^n + a_{n-1}x^{n-1} + a_{n-2}x^{n-2} + \cdots + a_1x + a_0$, $a_n \neq 0$ and $n \geq 1$, then f has at least one zero.

But look, if f has a zero, say c_1, then $f(x) = (x - c_1)g(x)$, degree $g = n - 1$. If $n - 1 \geq 1$, then g has a zero; call it c_2. Now we have $f(x) = (x - c_1)(x - c_2)(a_nx^{n-2} + \cdots + c)$. If we continue this train of thought, we realize the following must be true.

> **Linear Factor Theorem:** If $f(x) = a_nx^n + a_{n-1}x^{n-1} + a_{n-2}x^{n-2} + \cdots + a_1x + a_0$, $a_n \neq 0$ and $n \geq 1$, then $f(x)$ can be written as
>
> $$a_n(x - c_1)(x - c_2)(x - c_3) \cdots (x - c_n).$$

If the coefficients $a_n, a_{n-1}, \cdots, a_0$ each has no imaginary part, then any zeros with imaginary parts must come in conjugate pairs. You can see this in the earlier example, $g(x) = 3x^3 - 7x^2 + 14x - 4$. The linear factor theorem says that a polynomial of degree n, $n \geq 1$ has exactly n roots. Some of these roots may be repeated; these are called "multiple roots." If $(x - a)^m$ appears in the factored form of the polynomial, then a is a **root of multiplicity m.**

EXAMPLE Confirm the linear factor theorem for $h(x) = x^4 - 4x^3 + 4x^2 + 4x - 5$.
Possible rational zeros of h are $\pm 1, \pm 5$.

1	1	−4	4	4	−5	
		1	−3	1	5	1 is a zero.
	1	−3	1	5	0	

Possible rational roots for $x^3 - 3x^2 + x + 5$ are $\pm 1, \pm 5$.

1	1	−3	1	5	−1	1	−3	1	5	
		1	−2	−1			−1	4	−5	
	1	−2	−1	4		1	−4	5	0	−1 is a root.

We use the quadratic formula to factor $x^2 - 4x + 5$ as $[x - (2 - i)][x - (2 + i)]$.
Putting our information together we have $h(x) = (x - 1)(x + 1)[x - (2 - i)][x - (2 + i)]$, which confirms the linear factor theorem for $h(x)$.

Note: The coefficients of $h(x)$ are real numbers (have no imaginary parts), and there is a pair of conjugate complex zeros, $2 - i$ and $2 + i$. ■

EXAMPLE Find all roots of $x^4 - x^3 - 3x^2 + 5x - 2$.
Possible rational roots are $\pm 1, \pm 2$.

1	1	−1	−3	5	−2	1 is a root of
		1	0	−3	2	$x^4 - x^3 - 3x^2 + 5x - 2$.
1	1	0	−3	2	0	1 is a root of
		1	1	−2		$x^3 - 3x + 2$.
1	1	1	−2	0		1 is a root of
		1	2			$x^2 + x - 2$.
	1	2	0			The root of $x + 2$ is -2.

$$x^4 - x^3 - 3x^2 + 5x - 2 = (x - 1)^3(x + 2).$$

The roots of $x^4 - x^3 - 3x^2 + 5x - 2$ are 1 and −2. The root 1 is a multiple root, with multiplicity 3. ■

EXAMPLE Use the fact that $3 + 2i$ is a root of $x^4 - 2x^3 - 17x^2 + 88x - 78$ to find all its roots.
We know several facts immediately. The polynomial must have 4 roots; if $3 + 2i$ is a root, so is $3 - 2i$ because the coefficients of the polynomial are real numbers; $x - (3 + 2i)$ and $x - (3 - 2i)$ are factors, and the remaining roots must come from the result of dividing $x^4 - 2x^3 - 17x^2 + 88x - 78$ by both $[x - (3 + 2i)]$ and $[x - (3 - 2i)]$. What is this result?

$3 + 2i$	1	−2	−17	88	−78
		$3 + 2i$	$-1 + 8i$	$-70 - 12i$	78
$3 - 2i$	1	$1 + 2i$	$-18 + 8i$	$18 - 12i$	0
		$3 - 2i$	$12 - 8i$	$-18 + 12i$	
	1	4	−6	0	

Using the quadratic formula on $x^2 + 4x - 6 = 0$ is probably the most efficient technique for finding the remaining roots.

$$x = \frac{-4 \pm \sqrt{40}}{2}, \quad \text{or} \quad -2 \pm \sqrt{10}$$

The roots of $x^4 - 2x^3 - 17x^2 + 88x - 78$ are $3 - 2i$, $3 + 2i$, $-2 - \sqrt{10}$, and $-2 + \sqrt{10}$. ■

You know that if you can factor $x^4 - 4x^3 + 4x^2 + 4x - 5$, then you can factor $\sin^4 x - 4\sin^3 x + 4\sin^2 x + 4\sin x - 5$ or $x^{4/3} - 4x^{3/3} + 4x^{2/3} + 4x^{1/3} - 5$ because the coefficient patterns are the same.

EXAMPLE Solve $4\cos^3 x - 4\cos^2 x = 5\cos x - 3$.

Be sure to try old-fashioned factoring first. If we can factor $4\cos^3 x - 4\cos^2 x - 5\cos x + 3 = 0$, then we will almost have the solutions. Can you factor $4z^3 - 4z^2 - 5z + 3$? Possible rational roots are $-3, -\frac{3}{2}, -1, -\frac{3}{4}, -\frac{1}{4}, \frac{1}{4}, \frac{3}{4}, 1, \frac{3}{2}, 3$.

$$\begin{array}{r|rrrr} -1 & 4 & -4 & -5 & 3 \\ & & -4 & 8 & -3 \\ \hline & 4 & -8 & 3 & 0 \end{array}$$

$\cos x + 1$ is a factor.

Since $4\cos^2 x - 8\cos x + 3 = (2\cos x - 1)(2\cos x - 3)$,

$$4\cos^3 x - 4\cos^2 x - 5\cos x + 3 = 0$$

$$\Leftrightarrow \quad (\cos x + 1)(2\cos x - 1)(2\cos x - 3) = 0.$$

Solutions are $\pi + 2k\pi$, $\pi/3 + 2k\pi$, $5\pi/3 + 2k\pi$, k an integer.
The solution set is $\{\pi + 2k\pi, \pi/3 + 2k\pi, 5\pi/3 + 2k\pi, k \in I\}$. ■

EXERCISES 8.2

In exercises 1 through 4, state all possible rational roots for each of the polynomials given.

1. $x^2 - 4x + 5$
2. $4x^2 - 3x + 7$
3. $-3x^3 + 2x^2 - 4x - 12$
4. $-2x^3 - 17x + 16$

In exercises 5 through 10, find all the rational zeros of each polynomial function.

5. $f(x) = x^3 + 4x^2 + x - 6$
6. $h(x) = 2x^3 + x^2 - 18x - 9$
7. $p(x) = x^3 - 2x^2 - 14x - 5$
8. $g(x) = 2x^4 + x^3 + 18x^2 + 9x$
9. $q(x) = 3x^4 + 2x^3 - 37x^2 + 12x$
10. $P(x) = x^4 - 2x^3 - 11x^2 + 12x + 36$

In exercises 11 through 16, find all the roots for each of the following polynomials.

11. $x^2 + 4x + 6$
12. $3x^2 - 2x + 7$
13. $x^3 + 2x - 3$
14. $x^3 - 2x^2 + 16$
15. $x^5 + 3x^3 + 2x$
16. $x^5 + x^3 - 8x^2 - 8$
17. For each of the polynomials in exercises 11 through 16, find the sum of the roots of the polynomial.
18. Can you make any predictions or draw any conclusions from the results you obtained in exercise 17?
19. For each of the polynomials in exercises 11 through 16, find the product of the roots of the polynomial.
20. Can you make any predictions or draw any conclusions from the results you obtained in exercise 19?

In exercises 21 through 30, write an nth-degree polynomial that has the n stated roots. From the sets N, W, I, Q, J, $\mathbb{R}$, and $\mathbb{C}$, choose the first to which the coefficients of the expanded form for each polynomial belong.

21. $P(x)$ has roots $1, -3, 5$.
22. $Q(x)$ has roots $2, 2, 5, -5$.
23. $R(x)$ has roots $0, -\frac{1}{2}, \frac{1}{3}, 2$.
24. $S(x)$ has roots $0, 0, 0.3, 0.1, -0.2$.

25. $T(x)$ has roots $-\sqrt{3}, \sqrt{3}, -1, 1$.
26. $U(x)$ has roots $1 - \sqrt{2}, 1 + \sqrt{2}, 0, 4, 5$.
27. $V(x)$ has roots $i, -i, 3 + i, 3 - i$.
28. $W(x)$ has roots $5 - 3i, 5 + 3i, \sqrt{2}, -1$.
29. $Y(x)$ has roots $0.4, 2 + 3i, 2 - 3i, \sqrt{5}, -2$.
30. $X(y)$ has roots $-i, \sqrt{3}, -1, 0$.
31. Write three third-degree polynomial functions, f_1, f_2, f_3, that have zeros $1, -3, 5$.
32. Write two polynomial functions g_1, g_2, of different degrees that have the set of zeros $\{-i, i, -\sqrt{3}, \sqrt{3}\}$.

In exercises 33 through 36, write the nth-degree polynomial function f with the n stated zeros that contains the ordered pair $(x_1, f(x_1))$.

33. Zeros are $-1, 0, 2$, and $f(1) = 6$.
34. Zeros are $-\sqrt{2}, \sqrt{2}$, and $f(2) = -3$.
35. Zeros are $-1, -1, -1, 6$, and $f(0.5) = -148.5$.
36. Zeros are $3 - i, 3 + i, 4$, and $f(0) = -40$.

In exercises 37 through 42, use the techniques of factoring and factoring by synthetic division to solve the equations or inequalities that are polynomial in form.

37. $2\sin^3 x - 7\sin^2 x + 2\sin x + 3 = 0$
38. $\cos^4 t - 3\cos^3 t + 3\cos^2 t - 3\cos t + 2 = 0$
39. $2x^{4/3} - x^{3/3} + 18x^{2/3} - 9x^{1/3} \le 0$
40. $x^{3/2} > 5x^{2/2} - 2x^{1/2} - 8$
41. $x^3 - (1 + c)x^2 - (6 - c)x + 6c = 0$
42. $e^{2x} - e^x + k = 0$

8.3 Graphing Polynomial Functions

Toolkit

We'll use the polynomial graphing principle from Section 4.5. The graph of a polynomial function is smooth and continuous. It has no vertical asymptotes, no holes, and no cusps. Remember $t \to +\infty$ means t increases without bound, and $t \to -\infty$ means t decreases without bound.

We can now continue the discussion of graphing polynomial functions. Real zeros of a polynomial function give x-intercepts for the function's graph. Remember that our xy-coordinate plane uses real-number lines as axes, so zeros with imaginary parts do not produce x-intercepts.

Illustration The graph of $f(x) = x^2 - 17$ is a parabola with vertex $(0, -17)$ and x-intercepts $(-\sqrt{17}, 0)$ and $(\sqrt{17}, 0)$, corresponding to the zeros $-\sqrt{17}$ and $\sqrt{17}$.

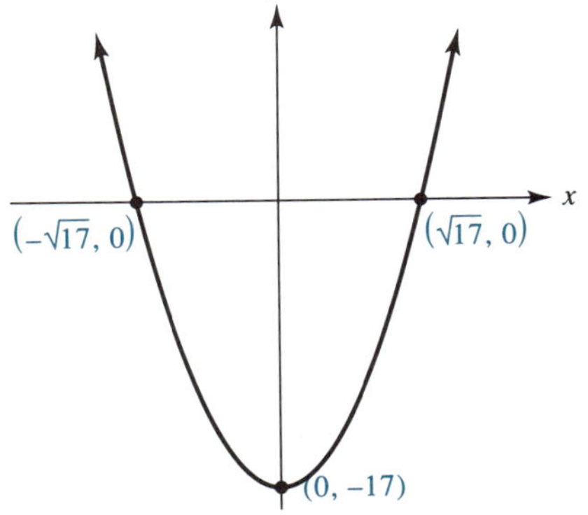

The graph of $g(x) = x^2 + 15$ is a parabola with vertex (0, 15) and no x-intercepts. The zeros of g are $-\sqrt{15}i$ and $\sqrt{15}i$.

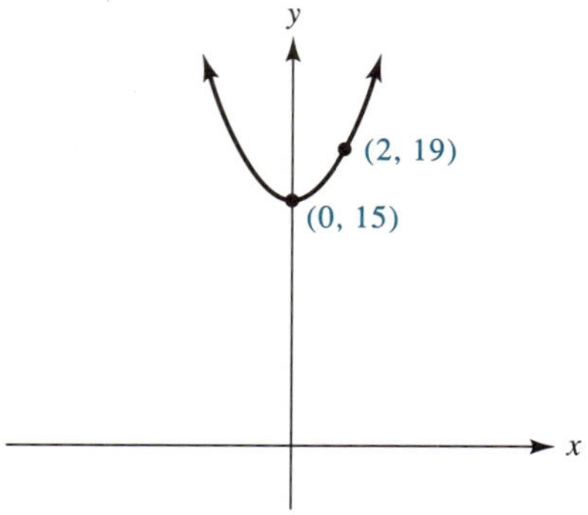

□

At this stage you know that if a function f is defined by a polynomial of degree n, then f has n zeros. If each of these zeros is a real number, you can use these roots to analyze the behavior of the function's graph. Let's look at

$$F(x) = x^3 - 2x^2 + x.$$

We write

$$F(x) = x(x - 1)^2$$

and see that the zeros are 0 and 1 (with multiplicity 2). What does the graph look like between the x-intercepts? A number-line chart similar to those we used in Section 4.5 can aid our analysis.

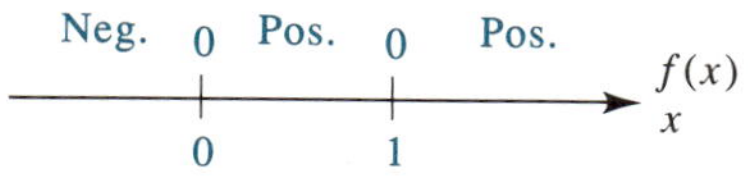

For $x > 1$, $F(x) = x(x - 1)^2 > 0$. For $0 < x < 1$, $F(x)$ is positive; $x < 0$ implies $F(x) < 0$. Using the polynomial graphing principle, we sketch the graph of F.

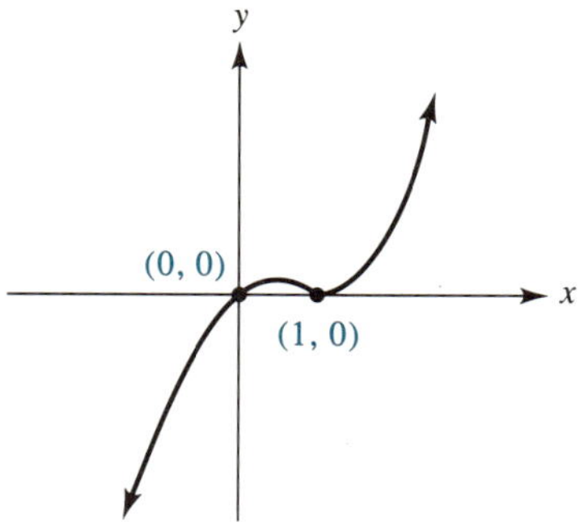

This graph may raise more questions than it answers. Exactly where is the high point between (0, 0) and (1, 0)? How many points should be labeled to identify the graph? How do we know the graph increases without bound as x increases beyond

1? Should the segment in the third quadrant look like Figure 8.3a or Figure 8.3b? How is this graph related to the standard graph of $y = x^3$?

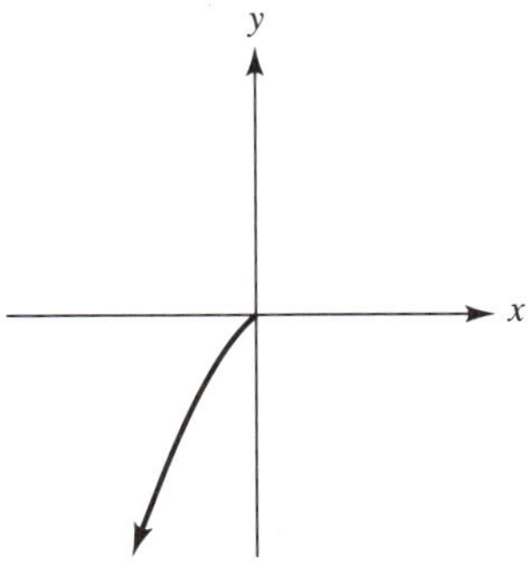

FIGURE 8.3a

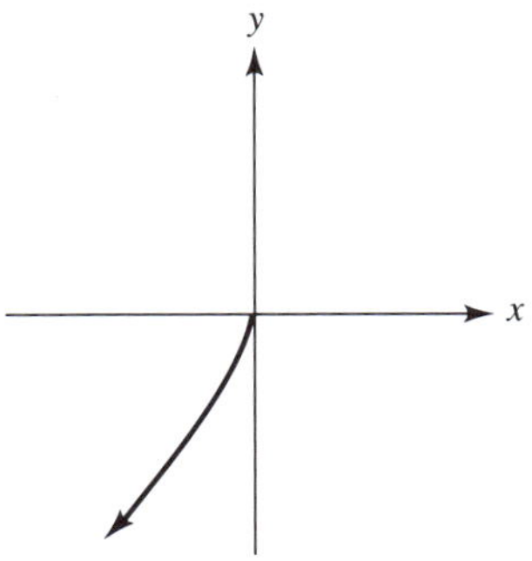

FIGURE 8.3b

You may or may not be relieved to read that the answers to most of these questions lie beyond the scope of this book. The general question of high or low points is one we can't answer here. We can give you an answer about the number of points, but we delay its explanation until Chapter 9. The number of points needed to identify the graph of a polynomial function is one plus the degree of the polynomial. Whether a function has range that is of the form $(-\infty, +\infty)$, $(-\infty, a)$, or $(b, +\infty)$ can be determined by investigating its rule. As x increases to the right of 1, each of the factors of the right-hand side of $F(x) = x(x - 1)^2$ increases without bound. The curving of the graph is not a facet we'll study. As you sketch the graph of a polynomial function, choose curvatures that will not create spikes. If several choices are possible, we have no guidelines now for the selection. The relationship between the graphs of $F(x) = x^3 - 2x^2 + x$ and $y = x^3$ is not one of those we've studied. However, if you can sketch the graph of F, you can sketch the graph of $G(x) = (x - 5)^3 - 2(x - 5)^2 + (x - 5)$ and that of $H(x) = x^3 - 2x^2 + x + 4$ using what you know about translations.

So, what can we do about efficiently sketching the graphs of polynomial functions? We can do a reasonable job if all the zeros are real numbers. The following theorem gives us some help in deciding if the zeros are real even without knowing what they are.

Descartes' Rule of Signs:

If $f(x) = a_n x^n + a_{n-1}x^{n-1} + \cdots + a_0$ with real coefficients and constant term that is not zero, then

1. **The number of positive real zeros of f is the number of times the coefficients change sign or this number minus a positive even integer.**
2. **The number of negative real zeros of f is the number of sign changes in the coefficients of the polynomial $f(-x)$ or this number minus a positive even integer.**

Illustration Since $f(x) = x^4 - x^3 - 3x^2 + 5x - 2$ is a polynomial function with real coefficients and constant term not zero, Descartes' rule of signs applies. There are three sign changes as we read the coefficients from left to right, so f has either 3 or 1 positive real zeros. Because

$$f(-x) = (-x)^4 - (-x)^3 - 3(-x)^2 + 5(-x) - 2,$$
$$\text{or} \quad x^4 + x^3 - 3x^2 - 5x - 2$$

has one sign change in the coefficients, f has one negative real zero. The polynomial defining f was an example in section 8.2; it can be factored as $(x - 1)^3(x + 2)$. The zeros of f are 1, counted three times, and -2. □

Is it possible to decide between the possibilities of 3 or 1 positive real zeros above without knowing the zeros? Sometimes the following analysis will decide this for

$$f(x) = a_nx^n + a_{n-1}x^{n-1} + \cdots + a_1x + a_0, \qquad a_n > 0.$$

Use synthetic division to compute $a_nx^n + a_{n-1}x^{n-1} + \cdots + a_1x + a_0$ divided by $x - c$, $c > 0$.

If the numbers under the line are all nonnegative, c is an **upper bound** on the real **zeros** of f. This means no real zero is larger than c.

If $a_nx^n + a_{n-1}x^{n-1} + \cdots + a_1x + a_0$ divided by $x - c$, $c < 0$, produces numbers under the line that alternate nonnegative and nonpositive values (or vice versa), c is a **lower bound** on the real zeros.

EXAMPLE Sketch the graph of $g(x) = 6x^3 - 27x^2 + 3x + 36$.

The zeros of g are the same as those of $h(x) = 2x^3 - 9x^2 + x + 12$. Descartes' rule of signs tells us h has two or zero positive real zeros and one negative real zero. Possible rational zeros are $-12, -6, -4, -3, -2, -\frac{3}{2}, -1, -\frac{1}{2}, \frac{1}{2}, 1, \frac{3}{2}$, 2, 3, 4, 6, and 12. Let's make choices that divide the list roughly into halves and then quarters if necessary.

$-\frac{1}{2}$	2	-9	1	12	No information except
		-1	5	-3	$h(-\frac{1}{2}) = 9$.
	2	-10	6	9	

-3	2	-9	1	12	There are no real zeros smaller than -3,
		-6	45	-138	but there must be one between $-\frac{1}{2}$ and -3.
	2	-15	46	-126	This is because $h(-\frac{1}{2}) > 0$ and $h(-3) < 0$.

-1	2	-9	1	12	We have the one negative zero, -1.
		-2	11	-12	The other zeros are those of $k(x) =$
	2	-11	12	0	$2x^2 - 11x + 12$.

The remaining zeros are $\frac{3}{2}$ and 4.

$$g(x) = 3(x + 1)(2x - 3)(x - 4)$$

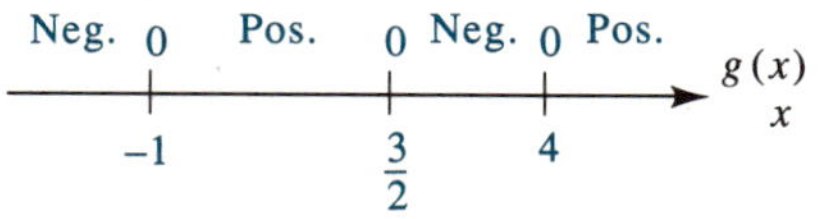

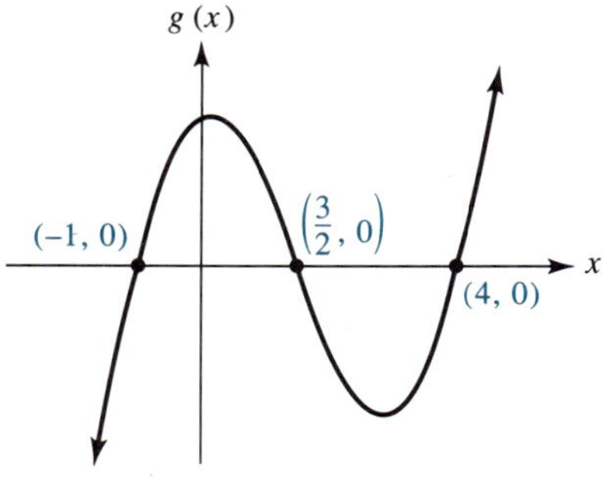

■

EXAMPLE Sketch the graph of $g(x) = 4x^4 + 12x^3 - 7x^2 - 4x + 2$.

There are two or zero positive real zeros and two or zero negative real zeros. Possible rational zeros are $-2, -1, -\frac{1}{2}, -\frac{1}{4}, \frac{1}{4}, \frac{1}{2}, 1, 2$.

$\frac{1}{4}$	4	12	-7	-4	2	
		1	$\frac{13}{4}$	$-\frac{15}{16}$	$-\frac{79}{64}$	
	4	13	$-\frac{15}{4}$	$-\frac{79}{16}$	$\frac{49}{64}$	
1	4	12	-7	-4	2	1 is an upper bound; there is no need to test 2.
		4	16	9	5	
	4	16	9	5	7	
$\frac{1}{2}$	4	12	-7	-4	2	$\frac{1}{2}$ is a zero. There must be another positive real zero. We need look only at $2x^3 + 7x^2 - 4$.
		2	7	0	-2	
	4	14	0	-4	0	
$\frac{1}{2}$	2	7	0	-2		$\frac{1}{2}$ is a double root of g.
		1	4	2		
	2	8	4	0		

Because $2x^2 + 8x + 4 = 2(x^2 + 4x + 2)$ or

$$2(x - (-2 + \sqrt{2}))(x - (-2 - \sqrt{2})),$$

$$g(x) = 8\left(x - \frac{1}{2}\right)^2\left((x - (-2 + \sqrt{2})\right)\left(x - (-2 - \sqrt{2})\right).$$

■

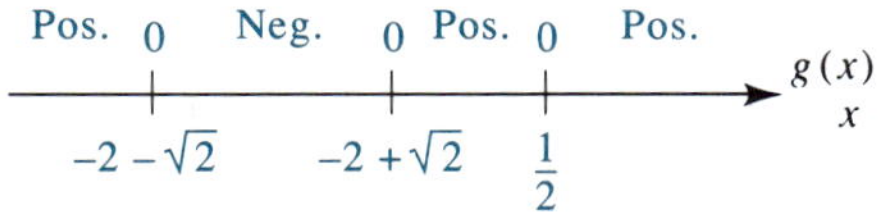

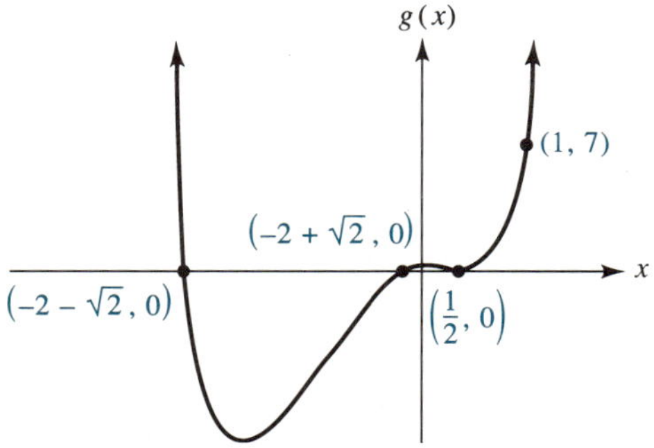

Sometimes it's as important to know what a statement doesn't say as what it does say. Consider what happens if you shift the graph of g in the last example up one unit. Let

$$h(x) = (4x^4 + 12x^3 - 7x^2 - 4x + 2) + 1.$$

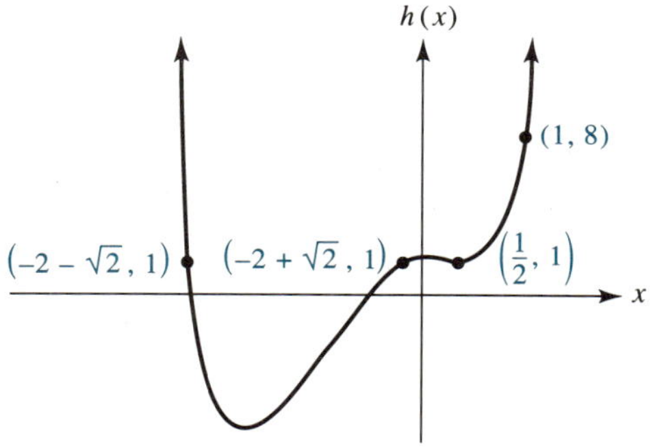

Clearly $\frac{1}{4}$ is an upper bound on the real zeros of h because no zero is larger then $\frac{1}{4}$, but consider the following:

$\frac{1}{4}$	4	12	-7	-4	3
		1	$\frac{13}{4}$	$-\frac{15}{16}$	$-\frac{79}{64}$
	4	13	$-\frac{15}{4}$	$-\frac{79}{16}$	$\frac{113}{64}$

This row is not all nonnegative numbers.

Our statement about upper bounds does not say the third row *must* be nonnegative numbers for c to be an upper bound. A similar comment can be made about the statement concerning lower bounds.

Here's an illustration of how you can adapt your earlier work with translations to graphing a polynomial function.

Illustration For $F(x) = 2x^3 - 3x^2 - 4x + 7$, the possible rational zeros are ± 1, ± 7, $\pm\frac{1}{2}$, $\pm\frac{7}{2}$. The rule of signs tells us that F has two or zero positive zeros and one negative zero. Testing each of the possible rational zeros tells us that none of these is a zero, there is a negative zero between -1 and -7, and 7 is an upper bound on any positive real zeros. This isn't much to go on. Consider the function $G(x) = 2x^3 - 3x^2 - 4x$; the graph of F will be a vertical shift of the graph of G. The zeros for $G(x) = 2x^3 - 3x^2 - 4x$ are

$$0, \quad \frac{3+\sqrt{41}}{4}, \quad \frac{3-\sqrt{41}}{4}.$$

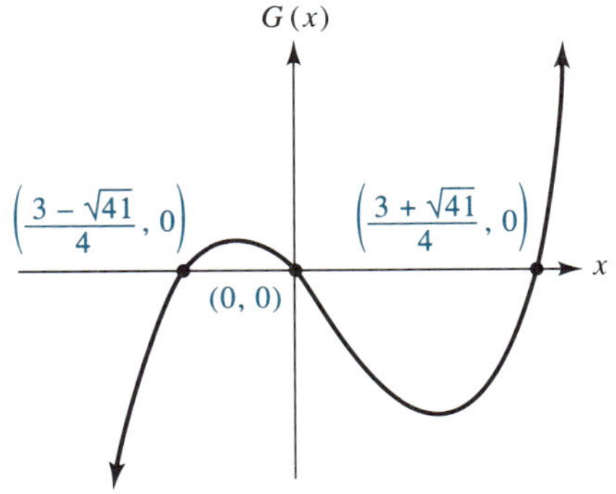

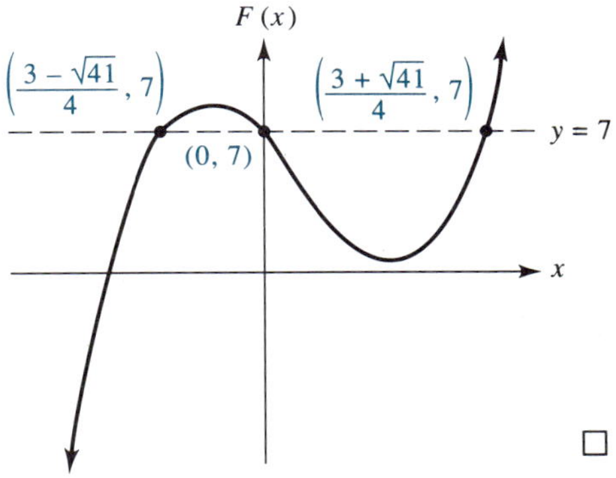

□

Now we can investigate further the location of the zeros of F. Does the graph dip below the x-axis between 1 and $(3 + \sqrt{41})/4$? Because $(3 + \sqrt{41})/4$ is slightly larger than $\frac{9}{4}$, we might test $1, \frac{5}{4}, \frac{6}{4}, \frac{7}{4}, \frac{8}{4}$, and $\frac{9}{4}$ to see if we can detect a sign change in $F(x)$. Because each of these produces a positive value of F, we're inclined to believe that there are no positive zeros. It would take other methods to be certain. If your calculator does graphing, you could also use a geometric (visual location of a test number) approach to estimate zeros.

Returning to the question of the range of a polynomial function $f(x) = a_n x^n + a_{n-1}x^{n-1} + \cdots + a_0$, we state explicitly some facts you've probably noticed already. The behavior of the graph and the range of a polynomial function are determined by the sign of the leading coefficient and the degree of the polynomial. Let's consider first the case $a_n > 0$. If n is even, then as $x \to +\infty$, x^n increases without bound, and when $x \to -\infty$, x^n also increases without bound. There is some lower bound on the values, $f(x)$. The models here are $y = x^2$ and $y = x^4$. The range of f is of the form $(c, +\infty)$.

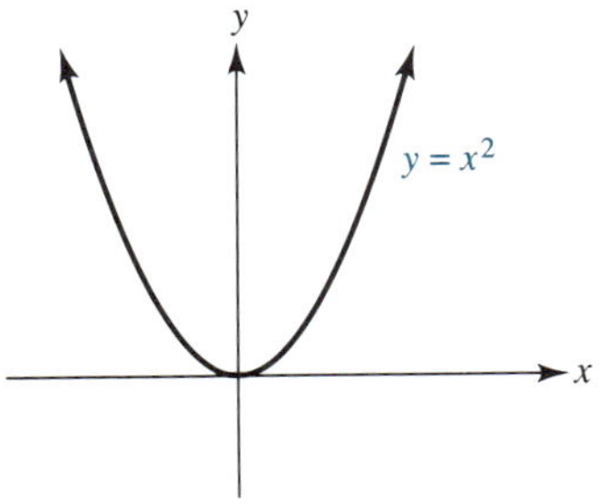

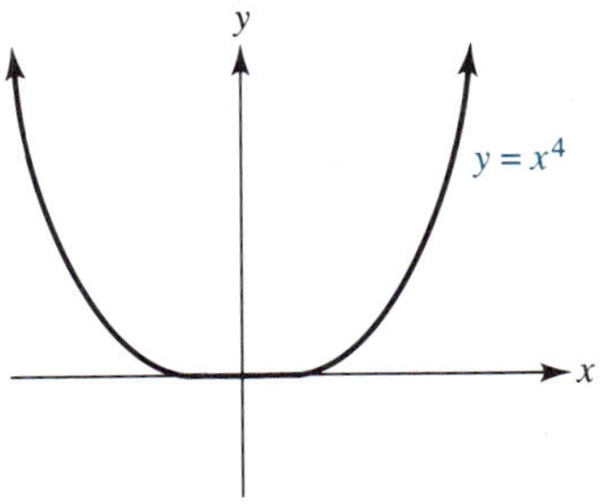

If n is odd, x^n will increase without bound as $x \to +\infty$, and x^n will decrease without bound as $x \to -\infty$. The models here are $y = x^3$ and $y = x^5$. The range of f is of the form $(-\infty, +\infty)$.

If $a^n < 0$ and n is even, the range of f is $(-\infty, c)$. Think of $y = -x^2 + 5$. If $a_n < 0$ and n is odd, the range of f is still $(-\infty, +\infty)$.

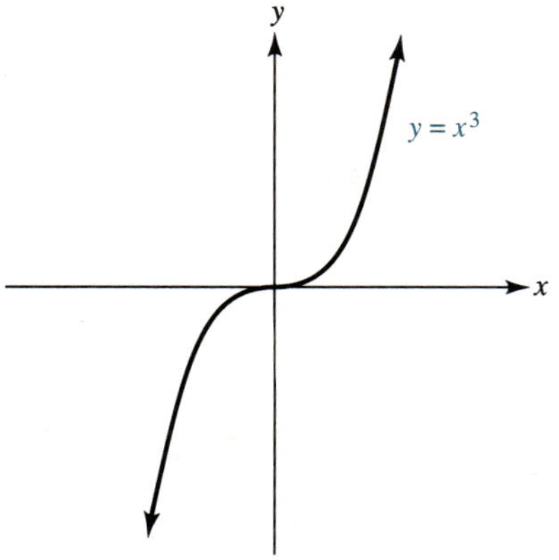

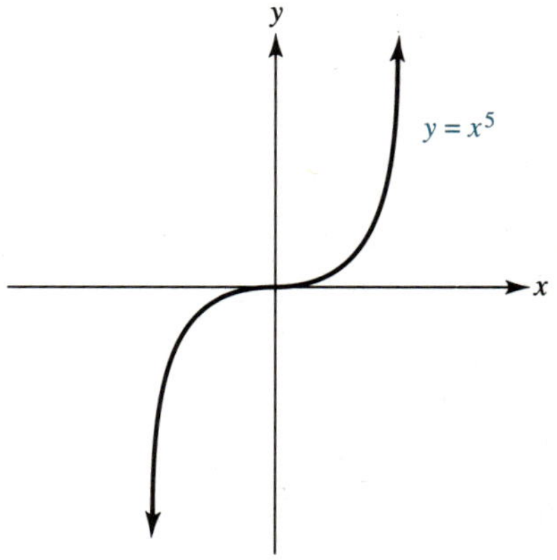

Illustration The function $f(x) = x^3 + 6x^2 - 12x + 13$ has only one real zero. Its approximate value is -7.76.

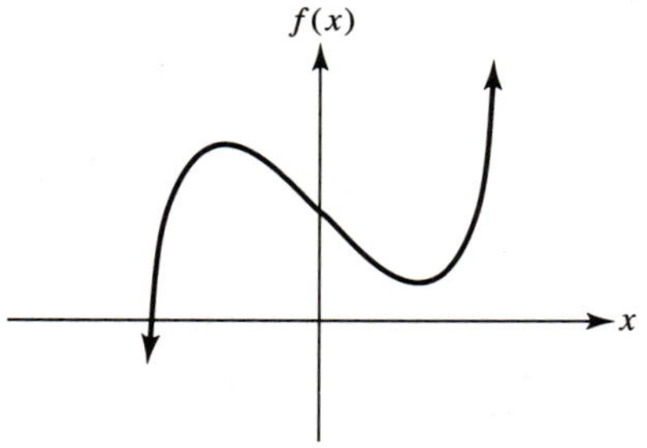

The function $g(x) = x^3 - 5x^2 - 8x + 48$ has two distinct real zeros, -3 and 4.

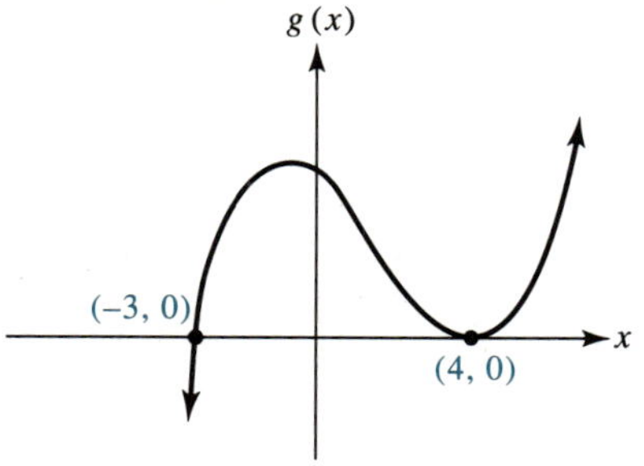

The function $h(x) = -3x^3 - 13x^2 + x + 6$ has three zeros, $\frac{(-5 - \sqrt{13})}{2}$, $\frac{(-5 + \sqrt{13})}{2}$, and $\frac{2}{3}$.

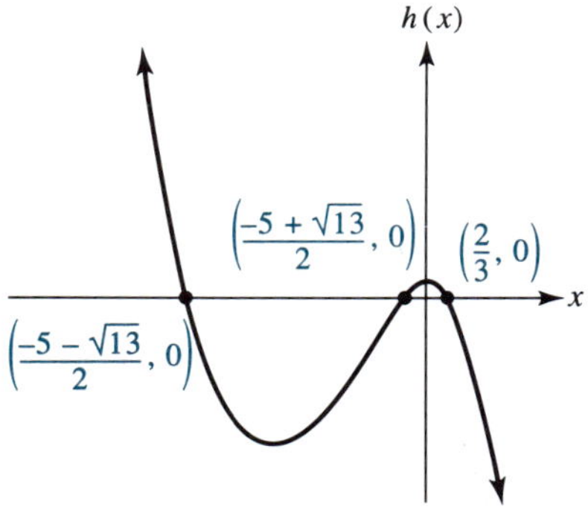

Note: All three functions have range $(-\infty, +\infty)$. □

Illustration The function $p(x) = x^4 - 4x^3 + 12x + 15$ has no real zeros.

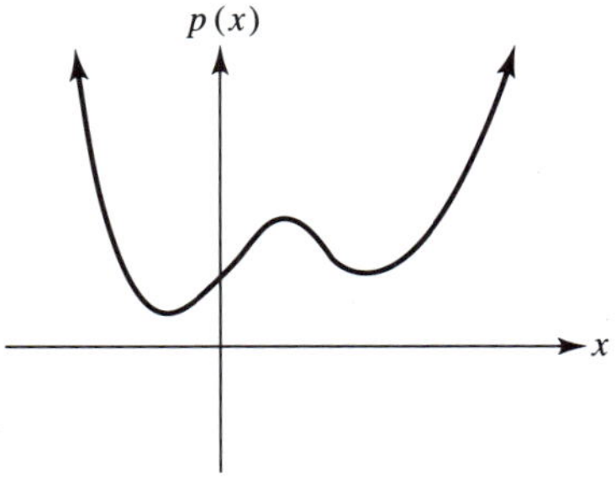

The function $q(x) = -x^4 - 4x^3 - 6x^2 - 4x - 1$ has one real zero, -1.

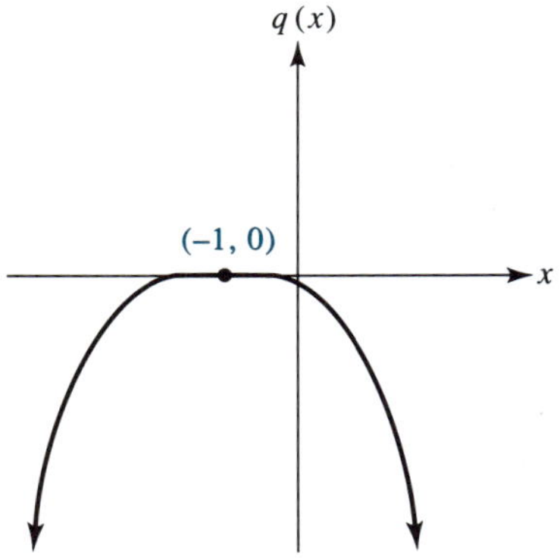

The function $r(x) = -4x^4 + 4x^3 + 59x^2 - 30x - 225$ has two distinct real zeros, $-\frac{5}{2}$ and 3.

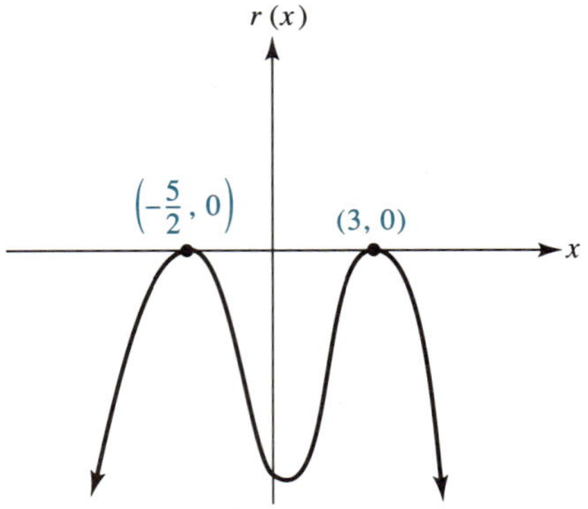

The function $s(x) = -9x^4 + 42x^3 - 34x^2 + 10x - 1$ has three distinct real zeros, $2 - \sqrt{3}$, $\frac{1}{3}$, and $2 + \sqrt{3}$.

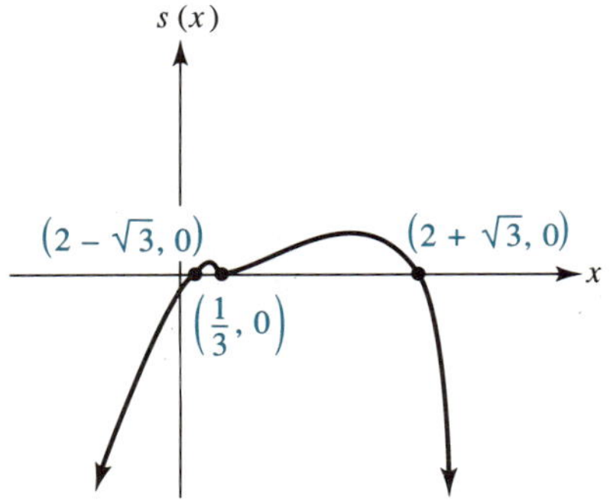

The function $t(x) = x^4 - 4x^3 + 12x - 9$ has four zeros, $-\sqrt{3}$, 1, $\sqrt{3}$, and 3.

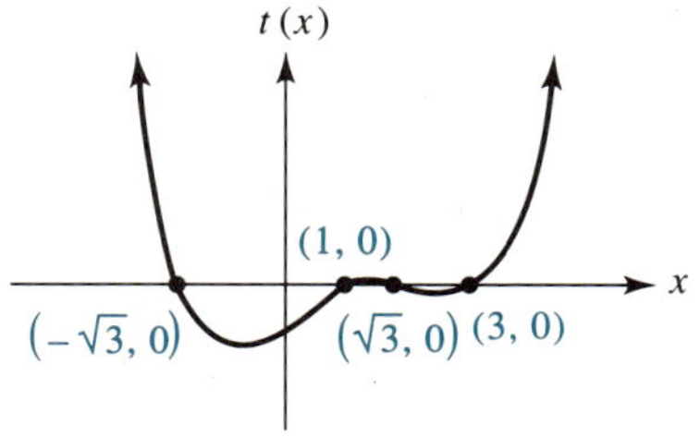

Note: Each of these fourth-degree polynomial functions has a range of the form $(-\infty, c)$ or $(c, +\infty)$. □

EXERCISES 8.3

In exercises 1 through 10, sketch the graph of each function.

1. $f(x) = (1 - x)(x + \sqrt{2})(x - \sqrt{2})$
2. $g(x) = (1 - x)^2(x^2 + 4)$
3. $h(x) = (2x + 3)^2(x - \sqrt{8})^2(x + \sqrt{90})$
4. $m(x) = -(x^2 + 25)(x + \sqrt{3})(x - \sqrt{3})$
5. $n(x) = (5i + x)(5i - x)(x + \sqrt{3})^2(x - \sqrt{3})^3$
6. $p(x) = (5i + x)(5i - x)(x + \sqrt{3})(x - \sqrt{3})^2$
7. $q(x) = (6x - 1)(x - \sqrt{0.03})(x - 0.17)^2$
8. $r(x) = (1 - 6x)(x - \sqrt{0.03})^3(0.17 - x)^3$
9. $s(x) = (0.02x + \pi)(x - 2i)(x + 2i)$
10. $t(x) = (0.02x + \pi)(2i - x)(2i + x)$

If you have a graphing calculator, use your calculator for exercises 11 through 32.

In exercises 11 through 20, sketch the graph of each function. Label the *x*-intercepts. In these exercises an nth-degree polynomial will have *n* rational zeros.

11. $f(x) = x^3 + 4x^2 - 7x - 10$
12. $g(x) = x^3 - 6x^2 - 19x + 24$
13. $h(x) = -x^4 - 4x^3 + 7x^2 + 10x$
14. $p(x) = x^4 - 6x^3 - 19x^2 + 24x$
15. $q(x) = x^4 - 3x^3 - 3x^2 + 11x - 6$
16. $F(x) = x^4 + 10x^3 + 37x^2 + 60x + 36$
17. $G(x) = -6x^3 - 5x^2 + 3x + 2$
18. $H(x) = -9x^4 + 54x^3 - 80x^2 - 6x + 9$
19. $P(x) = -6(2x - 1)^3 - 5(2x - 1)^2 + 3(2x - 1) + 2$
20. $Q(x) = 9(4x + 3)^5 - 54(4x + 3)^4 + 80(4x + 3)^3 + 6(4x + 3)^2 - 9(4x + 3)$

In exercises 21 through 30, sketch the graph of each function. Label the *x*-intercepts.

21. $f(x) = x^3 + 5x^2 - 9x - 10$
22. $g(x) = x^3 + 5x^2 - 19x + 10$
23. $h(x) = -2x^3 + 9x^2 - 7x - 3$
24. $p(x) = -x^4 + 8x^2 - 7$
25. $q(x) = 3x^4 - 2x^3 - 2x^2 + 2x - 1$
26. $F(x) = x^5 - 5x^3 - x^2 + 5$
27. $G(x) = 2x^5 + 5x^4 - 3x^3 - 6x^2 + x + 1$
28. $H(x) = -2x^5 - 5x^4 + 3x^3 + 6x^2 - x - 1$
29. $P(x) = (5x + 1)^5 - 5(5x + 1)^3 - (5x + 1)^2 + 5$
30. $Q(x) = -(x - a)^3 - 5(x - a)^2 + 19(x - a) - 10$
31. Approximate the irrational zero of $f(x) = x^4 + 3x^3 - 9x^2 - 15x + 20$ between 2 and 3 correct to two decimal places.
32. Approximate the irrational zero of $g(x) = x^4 + 10x^3 - 192x^2 - 2130x - 4473$ between 0 and 20 correct to the nearest integer.

In exercises 33 through 36, write a function whose graph could be the one given.

33.

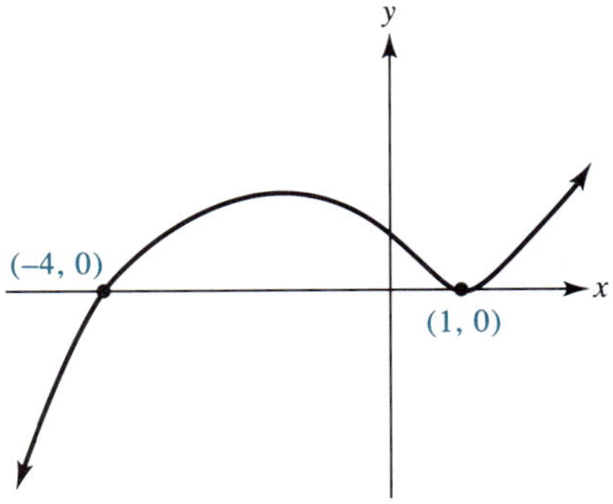

34.

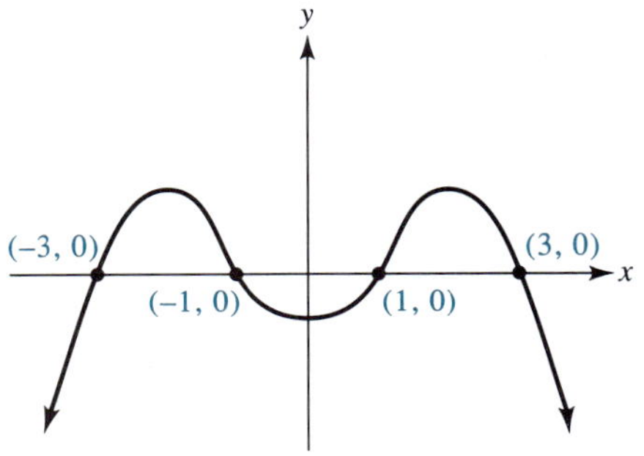

35.

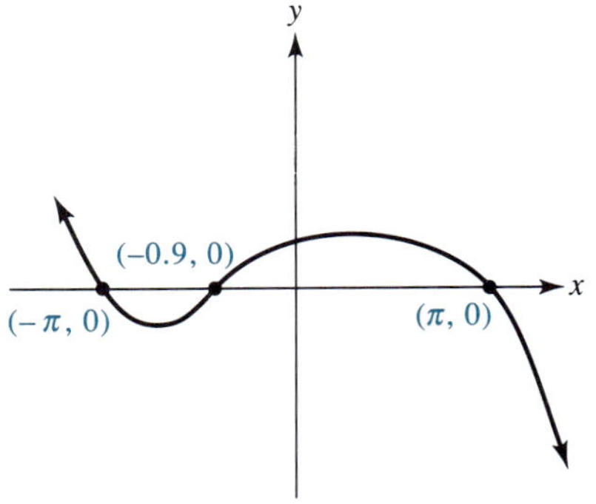

36.

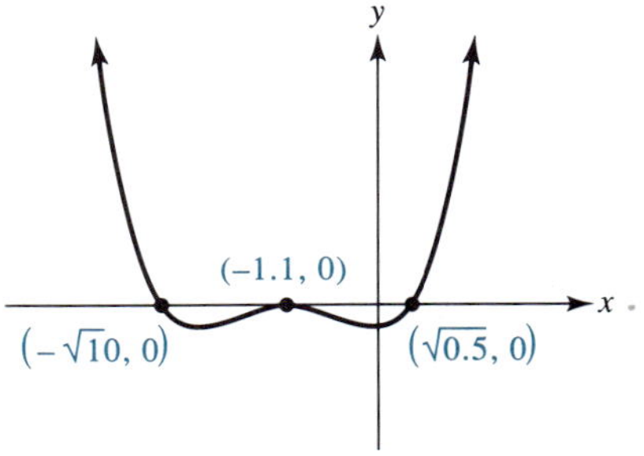

In exercises 37 through 39, sketch the graph of the given third-degree function.

37. $y = x^3 + 2x^2 + 6x + 5$
38. $y = x^3 - 3x^2 - 9x - 5$
39. $y = x^3 - 2x^2 - 13x - 10$

In exercises 40 through 44, sketch the graph of the given fourth-degree function.

40. $y = x^4 + 2x^2 + 1$
41. $y = x^4 + x^2$
42. $y = x^4 + 2x^3 + x^2$
43. $y = x^4 - 3x^2 - 2x$
44. $y = x^4 - 2x^3 - x^2 + 2x$
45. **a.** What can you tell your friend about the graph of $g(x) = -5x^6 + ax^5 + bx^4 + cx^3 + dx^2 + ex + f$?
b. What can you tell your friend about the graph of $h(x) = 2.13x^{17} + a_{16}x^{16} + a_{15}a^{15} + \cdots + a_0$?
46. Explain to your friend how many x-intercepts the graph of an nth-degree polynomial function will have if
a. n is even.
b. n is odd.
47. Summarize what this section and previous ones tell us about the graph of an nth-degree polynomial function. Discuss as many features as you can.
48. For a polynomial function f, what must happen when the rule for f is written in factored form for the graph of f to have this piece?
a.

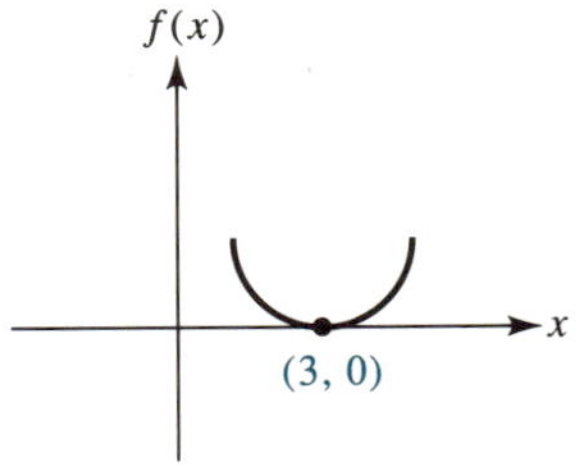

b.

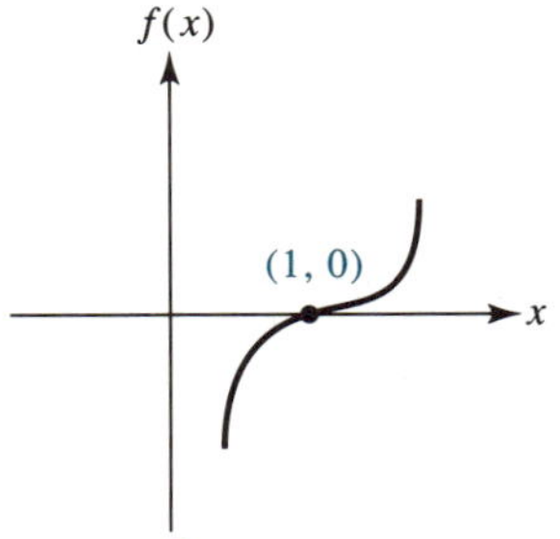

8.4 Rational and Other Functions Revisited

Toolkit

A rational expression is proper if the degree of the numerator is less than the degree of the denominator.

Building on our work in Section 8.3, we can sketch graphs of rational functions that were not manageable before. In this section we will also investigate asymptotes again.

EXAMPLE Sketch the graph of $f(x) = \dfrac{2x^3 - 7x^2 - 7x + 12}{x^4 - x^3 - 3x^2 + 5x - 2}$.

The polynomials that define f were examples in Section 8.2. Retrieving that information we have

$$f(x) = \frac{(x-1)(2x+3)(x-4)}{(x-1)^3(x+2)}, \text{ or } \frac{(2x+3)(x-4)}{(x-1)^2(x+2)}.$$

The graph of f has x-intercepts $(-\frac{3}{2}, 0)$ and $(4, 0)$; it has vertical asymptotes

$x = 1$ and $x = -2$. Using the methods of Section 4.3, you can confirm that the graph has a horizontal asymptote $y = 0$ to the left and to the right.

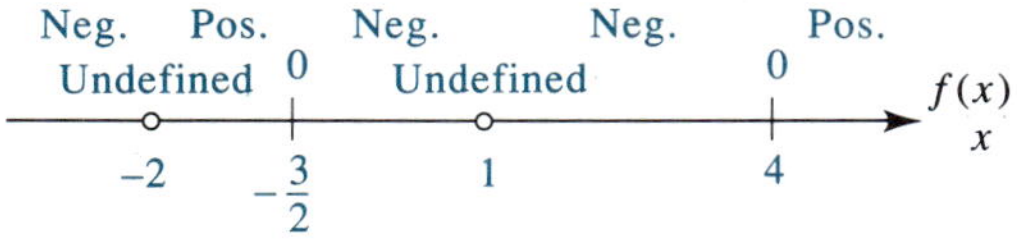

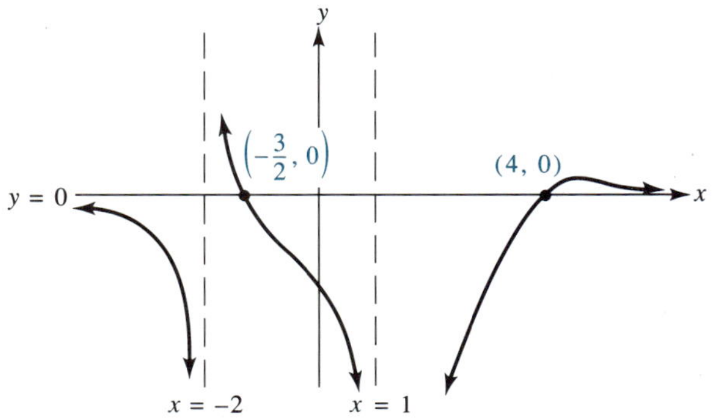

What we've gained in the last two sections are some powerful techniques for factoring polynomials. You can use these skills to find zeros and vertical asymptotes for the graphs of rational functions. Let's reconsider the question of a rational function's behavior for extreme values of the independent variable. Some exercises in Section 4.3 asked you to examine the relation between the degrees of the numerator and denominator and the existence of a horizontal asymptote. If

$$f(x) = \frac{a_m x^m + a_{m-1}x^{m-1} + \cdots + a_0}{b_n x^n + b_{n-1}x^{n-1} + \cdots + b_0} \quad \text{and} \quad m \leq n,$$

when we divide numerator and denominator by x^n we have, if x is not zero,

$$f(x) = \frac{a_m x^{m-n} + a_{m-1}x^{m-n-1} + \cdots + a_0 x^{-n}}{b_n + b_{n-1}x^{-1} + \cdots + b_0 x^{-n}}, \qquad x \neq 0.$$

Each term except b_n in the denominator and possibly $a_n x^{m-n}$ in the numerator has a negative exponent. Every term with a negative exponent will approach 0 as x increases without bound or decreases without bound. For extreme values of x, $f(x) \approx \dfrac{a_m x^{m-n}}{b_n}$. If $m = n$, $f(x) \approx \dfrac{a_m}{b_n}$ and $y = \dfrac{a_m}{b_n}$ is a horizontal asymptote for the graph of f. If $m < n$, $f(x) \approx \dfrac{a_m}{b_n x^{n-m}}$ and $y = 0$ is a horizontal asymptote for the graph of f.

What happens if

$$f(x) = \frac{a_m x^m + a_{m-1}x^{m-1} + \cdots + a_0}{b_n x^n + b_{n-1}x^{n-1} + \cdots + b_0}$$

and $m > n$? Dividing each term by x^n and considering f's behavior for extreme values of x, we have

$$f(x) \approx \frac{a_m x^{m-n}}{b_n}.$$

Terms in the numerator with nonnegative exponents less than $m - n$ will grow more slowly than $a_m x^{m-n}$, and the shape of the graph for extreme values of x is determined by $y = (a_m/b_n)x^{m-n}$. It would be natural to call such a curve an asymptote for the graph of f. If we need a description of the particular **asymptotic curve,** we have to do a bit more work.

Let's examine

$$f(x) = \frac{5x^5 - 3x^3 + 2}{2x^3 + x^2 - 7}.$$

If $x \neq 0$, then

$$f(x) = \frac{5x^2 - 3 + 2x^{-3}}{2 + x^{-1} - 7x^{-3}}.$$

If x is very large or very small, the terms $2x^{-3}$, x^{-1}, and $-7x^{-3}$ are very close to zero and have little effect on the value of $f(x)$. In fact, the -3 in the numerator will also have little effect compared to the term $5x^2$. The shape of f's curve approaches that of a parabola. To find the parabola it approaches, we rewrite $f(x)$ as $Q(x) + R(x)/D(x)$ by carrying out the long division.

$$\begin{array}{r|l}
 & \tfrac{5}{2}x^2 - \tfrac{5}{4}x - \tfrac{7}{8} \\ \hline
2x^3 + x^2 - 7 & 5x^5 \qquad\qquad - 3x^3 \qquad\qquad\qquad + 2 \\
 & 5x^5 + \tfrac{5}{2}x^4 \qquad\qquad - \tfrac{35}{2}x^2 \\ \cline{2-2}
 & \quad -\tfrac{5}{2}x^4 - 3x^3 + \tfrac{35}{2}x^2 \\
 & \quad -\tfrac{5}{2}x^4 - \tfrac{5}{4}x^3 \qquad\qquad + \tfrac{35}{4}x \\ \cline{2-2}
 & \qquad -\tfrac{7}{4}x^3 + \tfrac{35}{2}x^2 - \tfrac{35}{4}x \\
 & \qquad -\tfrac{7}{4}x^3 - \tfrac{7}{8}x^2 \qquad\qquad + \tfrac{49}{8} \\ \cline{2-2}
 & \qquad\qquad \tfrac{147}{8}x^2 - \tfrac{35}{4}x - \tfrac{33}{8}
\end{array}$$

$$f(x) = \frac{5x^5 - 3x^3 + 2}{2x^3 + x^2 - 7}, \quad \text{or} \quad \frac{5}{2}x^2 - \frac{5}{4}x - \frac{7}{8} + \frac{\frac{147}{8}x^2 - \frac{35}{4}x - \frac{33}{8}}{2x^3 + x^2 - 7}$$

The parabola $y = \frac{5}{2}x^2 - \frac{5}{4}x - \frac{7}{8}$ is the curve that the graph of f approaches for extreme values of x.

Definition A rational function f has an asymptotic curve if the degree of the numerator is greater than the degree of the denominator. To find the rule for the asymptotic curve of $f(x) = \dfrac{P(x)}{D(x)}$, write $P(x)$ as $Q(x) + \dfrac{R(x)}{D(x)}$. The graph of $y = Q(x)$ is the asymptotic curve.

EXAMPLE Sketch the graph of $g(x) = \dfrac{x^4 - 4x^2 + 4}{3x^2 - 2x - 1}$.

Factored form would make zeros and vertical asymptotes easy to pick out. We've chosen an example for which that is easy to do in order to concentrate on the asymptotic behavior of the graph, but you might have to use the techniques of this chapter for other functions.

$$g(x) = \frac{(x^2 - 2)^2}{(3x + 1)(x - 1)}$$

Zeros: $-\sqrt{2}, \sqrt{2}$
Vertical asymptotes: $x = -\frac{1}{3}, x = 1$

We use a number line to organize information about the sign of $g(x)$.

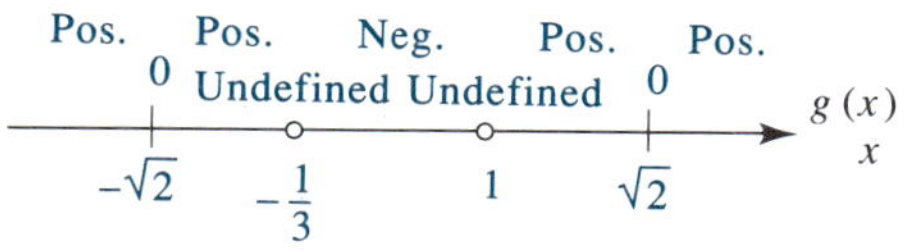

Thus far we have

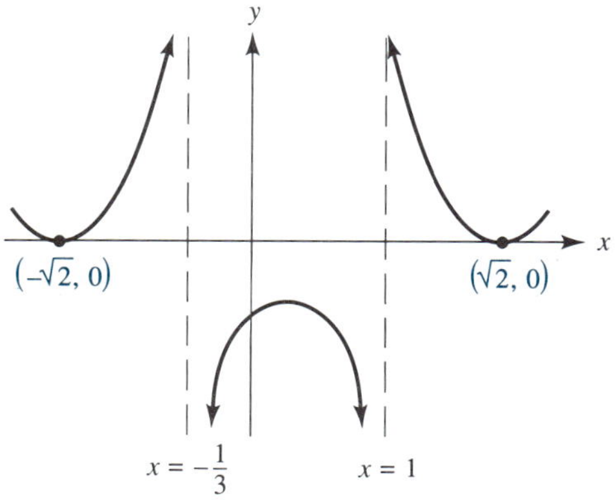

To find the asymptotic curve, we write $g(x)$ in mixed-number form. For extreme values of x, $g(x) \approx \frac{1}{3}x^2 + \frac{2}{9}x - \frac{29}{27}$ and the parabola defined by $y = \frac{1}{3}x^2 + \frac{2}{9}x - \frac{29}{27}$ approximates the graph of g for extreme values of x.

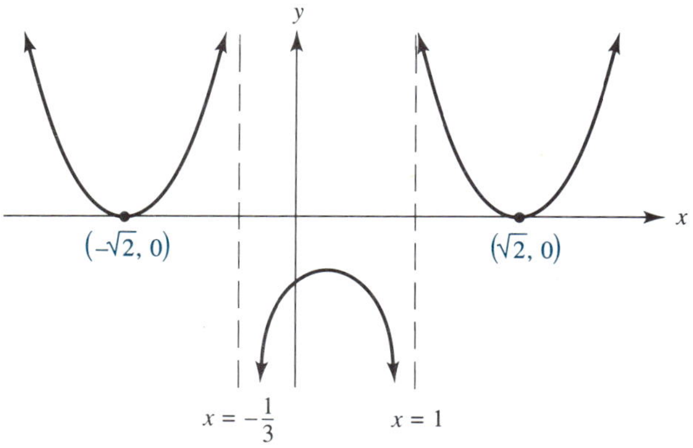

■

We use the equation of the asymptotic curve to shape our sketch on the left- and right-hand ends of our picture. We choose not to show the asymptotic curve itself because it wouldn't contribute much to the geometric information.

In the special case that the asymptotic curve is $y = ax + b$, $a \neq 0$, the line $y = ax + b$ is called an *oblique asymptote* for the graph of f.

Remember that earlier a horizontal asymptote was described as a line that approximated the graph of the function for extreme values of the independent variable. In the same way, if the graph of a rational function approaches a non-horizontal line or curve, this curve may be used to approximate values of the function.

EXAMPLE Approximate $f(257)$ for $f(x) = \dfrac{2x^{15} - 4x^{13} + 5}{10x^{13} - 8x^2 + 2x - 7}$ using the asymptotic curve.

We find the asymptotic curve by dividing.

$$
\begin{array}{r|l}
 & 0.2x^2 - 0.4 \\
\hline
10x^{13} - 8x^2 + 2x - 7 & 2x^{15} - 4x^{13} \qquad\qquad\qquad\qquad + 5 \\
 & 2x^{15} - 1.6x^4 + 0.4x^3 - 1.4 \\
\cline{2-2}
 & -4x^{13} + 1.6x^4 - 0.4x^3 + 6.4 \\
 & -4x^{13} + 3.2x^2 - 0.8x + 2.8 \\
\cline{2-2}
 & 1.6x^4 - 0.4x^3 - 3.2x^2 + 0.8x + 3.6
\end{array}
$$

$$f(x) \approx 0.2x^2 - 0.4.$$

An approximation for $f(257)$ is $0.2(257)^2 - 0.4$, or **13,209.4.**

Note: Direct substitution gives $f(x) \approx 13{,}209.4000001$, so our approximation is very good. ■

In Sections 4.3 and 4.4 we extended the graphing techniques to functions whose rules use fractional exponents. The new factoring methods of this chapter give us tools for making some decisions needed in sketching the graphs of functions that we could not do before.

Illustration In order to sketch the graph of $f(x) = x^{5/3} - x^{4/3} - 6x + 2x^{2/3} - 2x^{1/3} - 12$, we need to know the zeros of the function, where the function values are positive, where the function values are negative, and where the function is not defined. We can rewrite f as $f(x) = (x^{1/3})^5 - (x^{1/3})^4 - 6(x^{1/3})^3 + 2(x^{1/3})^2 - 2x^{1/3} - 12$ and treat its rule as a fifth-degree polynomial in order to find zeros. Possible rational roots of this fifth-degree polynomial are $-12, -6, -4, -3, -2, -1, 1, 2, 3, 4, 6$, and 12.

$$
\begin{array}{r|rrrrr|r}
3 & 1 & -1 & -6 & 2 & -2 & -12 \\
 & & 3 & 6 & 0 & 6 & 12 \\
\hline
-2 & 1 & 2 & 0 & 2 & 4 & 0 \\
 & & -2 & 0 & 0 & -4 & \\
\hline
 & 1 & 0 & 0 & 2 & 0 &
\end{array}
$$

3 is a zero; $x^{1/3} - 3$ is a factor.

-2 is a zero; $x^{1/3} + 2$ is a factor.

The factors of $(x^{1/3})^3 + 2$ are $(x^{1/3} + 2^{1/3})$ and $(x^{2/3} - 2^{1/3}x^{1/3} + 2^{2/3})$. The real zeros of f are 27 from $x^{1/3} - 3$; -8 from $x^{1/3} + 2$; and -2 from $x^{1/3} + 2^{1/3}$. We know that the graph of f has x-intercepts $(-8, 0)$, $(-2, 0)$, and $(27, 0)$. We

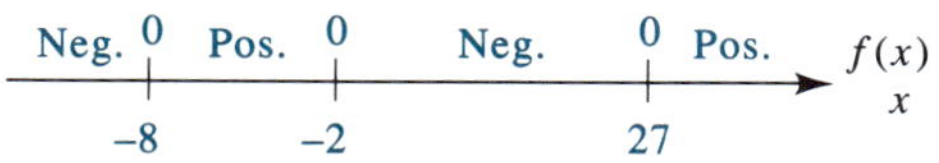

also know that the graph lies above the x-axis over the intervals $(-8, -2)$ and $(27, +\infty)$ and below the x-axis over the intervals $(-\infty, -8)$ and $(-2, 27)$. We cannot sketch the graph because we do not have the properties available for graphs of polynomial functions. We cannot use the polynomial graphing principle; we do not know that the graph will be smooth and continuous. □

EXAMPLE Given $g(t) = \dfrac{6t^{5/2} - 11t^2 + 6t^{3/2} - t}{6t^{3/2} - 15t - 42t^{1/2} + 24}$, where does the graph of g cross the t-axis? Identify the invervals where the graph lies above the t-axis and the intervals where the graph lies below the t-axis.

We need to rewrite the rule for g in factored form; again we treat $6t^{5/2} - 11t^2 + 6t^{3/2} - t$ as we would a fifth-degree polynomial.

$$6t^{5/2} - 11t^2 + 6t^{3/2} - t = t(6t^{3/2} - 11t + 6t^{1/2} - 1),$$
$$\text{or}\quad t[6(t^{1/2})^3 - 11(t^{1/2})^2 + 6(t^{1/2})^1 - 1]$$

You can see that 0 is a zero for g, and possible rational roots for the third-degree polynomial are ± 1, $\pm\frac{1}{2}$, $\pm\frac{1}{3}$, $\pm\frac{1}{6}$.

1	6	−11	6	−1
		6	−5	1
	6	−5	1	0

One factor is $t^{1/2} - 1$.

The other factors are $(3t^{1/2} - 1)(2t^{1/2} - 1)$.

$$6t^{5/2} - 11t^2 + 6t^{3/2} - t = t(t^{1/2} - 1)(3t^{1/2} - 1)(2t^{1/2} - 1)$$

For the denominator we find

$$6t^{3/2} - 15t - 42t^{1/2} + 24 = 3[2(t^{1/2})^3 - 5(t^{1/2})^2 - 14t^{1/2} + 8],$$
$$\text{or}\quad 3(t^{1/2} - 4)(t^{1/2} + 2)(2t^{1/2} - 1)$$

because

4	2	−5	−14	8
		8	12	−8
−2	2	3	−2	0
		−4	2	
	2	−1	0.	

$$g(t) = \frac{t(t^{1/2} - 1)(3t^{1/2} - 1)(2t^{1/2} - 1)}{3(t^{1/2} - 4)(t^{1/2} + 2)(2t^{1/2} - 1)},$$
$$\text{or}\quad \frac{t(t^{1/2} - 1)(3t^{1/2} - 1)}{3(t^{1/2} - 4)(t^{1/2} + 2)}, \qquad t \neq \tfrac{1}{4}$$

Now we can construct a number line and check regions. **The graph of g crosses the t-axis at 0, $\frac{1}{9}$, and 1. It lies above the t-axis on the intervals $(\frac{1}{9}, \frac{1}{4})$, $(\frac{1}{4}, 1)$, and $(16, +\infty)$; it lies below the t-axis on the intervals $(0, \frac{1}{9})$ and $(1, 16)$.** ■

EXERCISES 8.4

In exercises 1 through 4, sketch the graph of each function.

1. $f(x) = \dfrac{(x-1)(2.3-x)}{(x+\sqrt{5})(x-\sqrt{10})}$
2. $g(x) = \dfrac{(3x-1)^2(2.3-x)}{(x+\sqrt{5})(x-\sqrt{16})^2}$
3. $h(x) = \dfrac{x(7-2x)(x+1)(\sqrt{3}-x)}{(x+\pi)^2(x+4)(x^2-2x)}$
4. $j(x) = \dfrac{(x+\sqrt{8})^4}{(3x+7)^2(x-5)^3}$

In exercises 5 through 14, sketch the graph of the rational function. Give the equation for any oblique asymptote or nonlinear asymptotic curve.

5. $k(x) = \dfrac{x^2-3x+1}{x^3-3x^2+2x}$
6. $m(x) = \dfrac{3x^2+5x-3}{-2x^3+x^2+7x}$
7. $n(x) = \dfrac{x^3+3x^2+4x}{-2x^3+x^2+7x}$
8. $p(x) = \dfrac{-x^3+5x}{x^3+4x^2+4x}$
9. $q(x) = \dfrac{x^3}{x^2-5}$
10. $r(x) = \dfrac{2x^3-5x^2-5x+3}{x^2-9}$
11. $s(x) = \dfrac{x^4+1}{x^2-4x-6}$
12. $t(x) = \dfrac{x-x^5}{10x^3+21x^2+5x-6}$
13. $u(x) = \dfrac{x^6-2x^3+1}{5+19x+x^2-x^3}$
14. $v(x) = \dfrac{x^4-x^3-2x^2-6x-3}{10x-2}$

In exercises 15 through 18, use your calculator to approximate $f(473)$ to three decimal places and use the asymptotic curve f_{asym} to approximate $f(473)$.

15. $f(x) = \dfrac{4x^8-7x^3+3x^2-5x+2}{x^7-4x^6+3x^5-2x^4-x^3+x^2-5x-11}$,
$f_{asym}(x) = 4x+16$
16. $f(x) = \dfrac{12x^6+3x^5-x^4+3x^3-2x-45}{3x^5-2x^4-3x^3-8x^2-4x+2}$,
$f_{asym}(x) = 4x+\dfrac{11}{3}$
17. $f(x) = \dfrac{5x^5+2x^4-11x^3-14x+2}{4x^3+7x-2}$,
$f_{asym}(x) = \dfrac{5}{4}x^2+\dfrac{1}{2}x+\dfrac{79}{16}$
18. $f(x) = \dfrac{3x^{10}-x^9-x^8+2x^7+x^6+4x^5+x^4-3x^2}{x^8-x^7+11x^5-2x^4-x^3+10x^2-3}$,
$f_{asym}(x) = 3x^2-2x-3$
19. **a.** Find the oblique asymptote for the graph of $g(x) = \dfrac{2x^{18}-3x^{17}+5x^6-8x}{x^{17}-5x^{16}}$.
b. Use your calculator to approximate $f(1000)$ to four decimal places.
c. Use the oblique asymptote to approximate $f(1000)$.

In exercises 20 through 25, identify the intervals where the graph of each function lies above the x-axis.

20. $h(x) = \dfrac{4x^5-16x^4+9x^3+25x^2-16x-12}{x^2-3x+2}$
21. $j(x) = \dfrac{x^4+6x^3+11x^2-2x-16}{x^4+5x^3-4x^2-44x-48}$
22. $k(x) = \dfrac{(x^{1/2}-3)(x^{1/2}+1)(x^{1/2}-2)}{x^{1/2}(x^{1/2}+2)(x^{1/2}-4)(x^{1/2}-5)}$
23. $m(x) = \dfrac{(x^{1/3}-3)(x^{1/3}+1)(x^{1/3}-2)}{x^{1/3}(x^{1/3}+2)(x^{1/3}-4)(x^{1/3}-5)}$
24. $n(x) = \dfrac{x^{4/3}-4x-6x^{2/3}+4x^{1/3}+5}{x+x^{2/3}-5x^{1/3}+3}$
25. $p(x) = \dfrac{x-4x^{3/4}-6x^{1/2}+4x^{1/4}+5}{x^{3/4}+x^{1/2}-5x^{1/4}+3}$
26. Earlier we found that by dividing the numerator and denominator of the rule for f by x^3 and rewriting $f(x) = \dfrac{5x^5-3x^3+2}{2x^3+x^2-7}$ as $f(x) = \dfrac{5x^2-3+2x^{-3}}{2+x^{-1}-7x^{-3}}$, $x \neq 0$, that $f(x) \approx \dfrac{(5x^2-3)}{2}$ for extreme values of x.
What are some advantages and disadvantages of approximating a function value for f using **(a)** this method, **(b)** the asymptotic curve, and **(c)** the calculator?

Chapter 8: Summary

With the aid of a new tool, synthetic division, we were able to analyze the graphs of polynomial functions in more detail. Descartes' rule of signs and the test for upper and lower bounds help us use synthetic division more efficiently. We applied the new techniques to the graphs of rational and fractional functions also. The idea of asymptotes was extended to that of an asymptotic curve, which can be used to approximate function values.

Key Concepts

polynomial in one variable	*leading coefficient, polynomial function*
synthetic division	*technique, applications*
zero of a function, root of a polynomial	
intermediate value theorem	
fundamental theorem of algebra	
linear factor theorem	
root of multiplicity m	
Descartes' rule of signs	
upper and lower bounds on roots	
asymptotic curve	*oblique asymptote*

Skills

You should be able to:

1. Use synthetic division to find quotients, zeros of a function, roots of a polynomial, factors of a polynomial, or function values.
2. Determine the possible rational roots of a polynomial.
3. Sketch the graph of a polynomial function.
4. Sketch the graph of a rational function.
5. Determine the possible number of positive or negative real roots of a polynomial with real coefficients.
6. Approximate irrational roots of a polynomial with real coefficients.
7. Find an asymptotic curve for the graph of a rational function.
8. Approximate values of rational functions using the asymptotic curve.

SUMMARY EXERCISES

The starred exercises can serve as a practice chapter test. The answers for these are at the end of the answer section at the back of the book.

1. Is $x - 3$ a factor of $2x^5 - 10x^4 + 42x^2 + 15x + 9$? Justify your answer.
2. Find a value of c so that $f(-1)$ will be 3 if $f(x) = -2x^5 + 5x + c$.
* 3. Find a value for k so that -2 will be a zero of $f(x) = 5x^7 - 3x^2 - k$.
4. Write $3x^4 + 10x^3 + 4x^2 + 40x - 32$ as a product of linear factors.
5. Find all the roots of $18x^4 + 9x^3 - 19x^2 - 5x + 5$.
* 6. Write $10x^5 - 91x^4 + 246x^3 - 197x^2 - 52x + 60$ as the product of linear factors.
7. Give the set of all possible rational roots of $6x^4 - 2x^2 + 15x$.
* 8. Use synthetic division to find $f(1 - 3i)$ for $f(x) = x^4 - 2x^2 + ix - 3$.
9. Give the set of all possible rational roots for $4x^3 - 21x - 18$.
* 10. What is the maximum number of negative real roots of $15x^5 - 4x^4 + 31x^3 + 12x^2 - 6x + 14$?
11. What is the maximum number of positive real zeros for $g(x) = -31x^4 - 82x^3 + 75x^2 - 18x + 57$?
* 12. Approximate to the nearest hundredth the one positive real root of $-x^3 + x^2 + 2$.
13. Approximate to the nearest tenth the one negative real zero of $h(x) = x^3 - 2x + 5$.
* 14. Write the polynomial function of smallest degree with real coefficients whose roots include -4, 1, and $2i$ and that has value 12 when the independent variable is 2.
15. Write the polynomial function of smallest possible degree with real coefficients whose zeros include $-\frac{2}{3}$, 1, and $2 - 3i$ and such that $g(-1) = 6$.
16. Sketch the graph of $F(x) = 18x^4 + 9x^3 - 19x^2 - 5x + 5$.
* 17. Write the polynomial function of smallest possible degree with rational coefficients whose zeros include $-\sqrt{3}$, 6.1, and 10 and such that $h(1) = 45.9$.
18. Sketch the graph of $R(x) = \dfrac{x^2 - 1}{x^3 - 1}$.
19. Sketch the graph of $M(x) = \dfrac{x^3 - 1}{x^2 - 1}$.
* 20. Sketch the graph of $F(x) = 10 - 8x + 5x^2 - 4x^3$.
21. On which interval(s) is $t(x) = 5x - 2x^{4/5} - 15x^{3/5} + 6x^{2/5}$ negative?
22. Sketch the graph of $G(x) = x^4 - 2x^3 - 2x^2 + 6x - 3$.
* 23. Find the asymptotic curve for the graph of $S(x) = \dfrac{18x^8 - 3x^6 + 4x^2 - 3}{x^5 - 2x^3}$.
24. Sketch the graph of $H(x) = \dfrac{x^2 + 4x - 3}{(x - 0.5)^2(x + 5)}$.
25. Solve $3\sin^4 x - 8\sin^3 x + 6\sin^2 x = 1$.
* 26. On which interval(s) does $m(x) = 6x^{3/5} - 13x^{2/5} + x^{1/5} + 2$ produce positive values?
27. Solve $4x^{3/2} + 5x - 2x^{1/2} < 3$.
* 28. Approximate $n(2.38)$ for

$$n(x) = \frac{-14x^{12} + 5x^{10} - 17x^5 + 123}{2x^{10} - 3x^5 + 6x - 2}$$

by using the asymptotic curve of its graph.
29. How many x-intercepts could the graph of a fifth-degree polynomial function have? Sketch a graph for each case.
30. Using the solving routine on a graphing calculator, you have obtained the approximate zeros 0.857, 0.577, and -0.577 for $f(x) = 21x^3 - 18x^2 - 7x + 6$. Use this information to guide your choice of possible rational roots in order to find the exact roots quickly.
* 31. Your graphing calculator produces the following graph of $p(x) = 6x^4 - 7x^3 - 21x^2 + 21x + 9$. The distance between marks on the x-axis is 0.5. Use this information to guide your choice of possible rational roots in order to find the exact roots quickly.

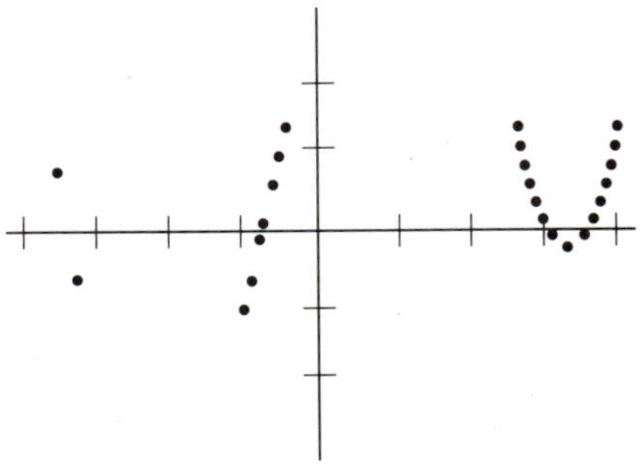

32. Write the equation of the asymptotic curve for

$$r(x) = \frac{4x^8 - 2x^7 - 3x^6 + 2x^4 - x^2 + 4}{x^5 + 3x^3 - x^2 - x}.$$

* 33. Use the asymptotic curve for the function r in exercise 32 to approximate $r(-1021)$.

CHAPTER NINE

Systems of Equations

- 9.1 Linear Systems
- 9.2 Gauss-Jordan Elimination
- 9.3 Matrices
- 9.4 Applications of Matrices
- 9.5 Nonlinear Systems and Inequalities

9.1 Linear Systems

A *system of linear equations* is a set of linear equations for which we want all common solutions.

We refer to a solution that is common to all the equations in the system as a solution of the system. You've found such solutions before in Section 1.1, where you solved pairs of equations involving two variables by using substitution. Also, in Section 5.4 we used a system with three equations and three variables to build a function.

EXAMPLE There are two major ingredients found in the cereal produced by Cereals for All, Inc. These ingredients contribute all the vitamins and minerals to the cereal. Every ounce of ingredient I contains 5 units of vitamin C and 2.5 units of vitamin B; every ounce of ingredient II contains 3 units of vitamin C and 1 unit of vitamin B. If there is a requirement that 15 units of vitamin C and 6 units of vitamin B be present in each box of cereal, how much of each major ingredient must be present?

Let x represent the amount of ingredient I and y represent the amount of ingredient II. The translation produces a system of two linear equations in two variables:

$$\begin{aligned} 5x + 3y &= 15, \\ 2.5x + y &= 6. \end{aligned}$$

You could use substitution to find that the solution is (1.2, 3).
The box of cereal must contain 1.2 ounces of ingredient I and 3 ounces of ingredient II. ■

Now it's time to pull together what you already know about linear systems and find more efficient ways of dealing with them. From a geometric point of view, we see that two lines in a plane may have no points of intersection (Figure 9.1a), one point (Figure 9.1b), or infinitely many points (Figure 9.1c). Thus pairs of linear equations may have 0, 1, or infinitely many common solutions. (What are the possibilities for three lines in a plane?)

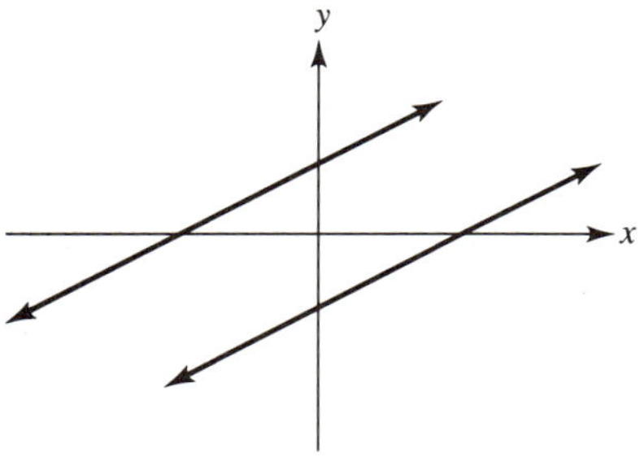

No solution

FIGURE 9.1a

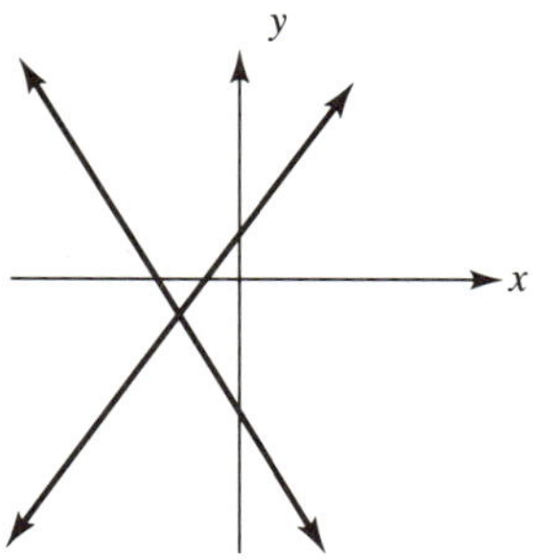

One solution

FIGURE 9.1b

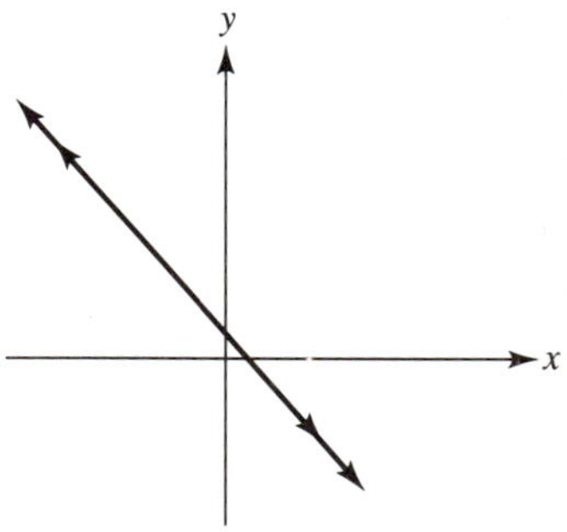

Infinitely many solutions

FIGURE 9.1c

A three-variable linear equation defines a plane in space. How many planes do you need to have a unique solution for a system of three-variable linear equations? Figures 9.1d–9.1i show some possible arrangements for three planes in space.

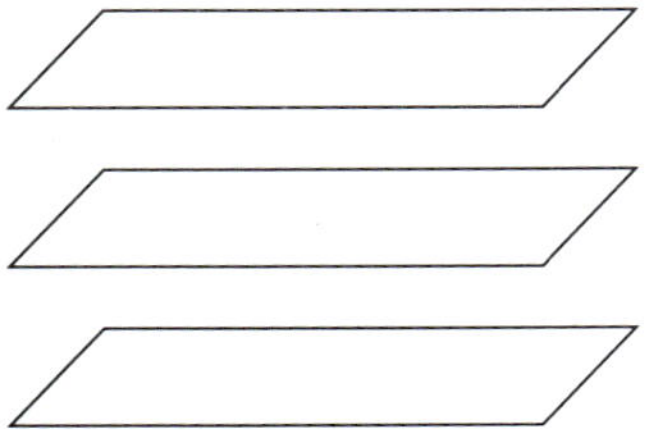

No solution

FIGURE 9.1d

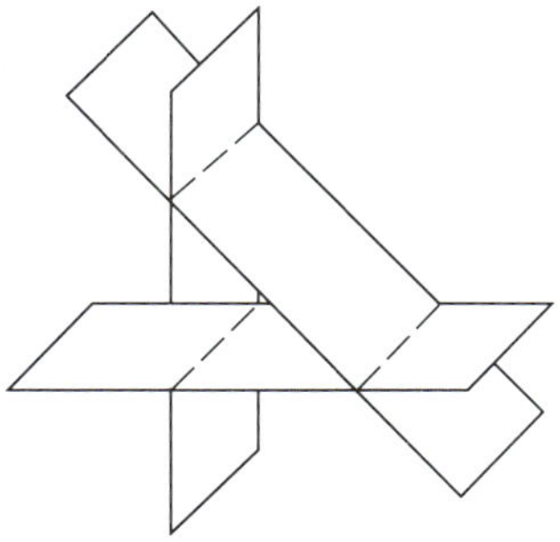

No solution

FIGURE 9.1e

Infintely many solutions

FIGURE 9.1f

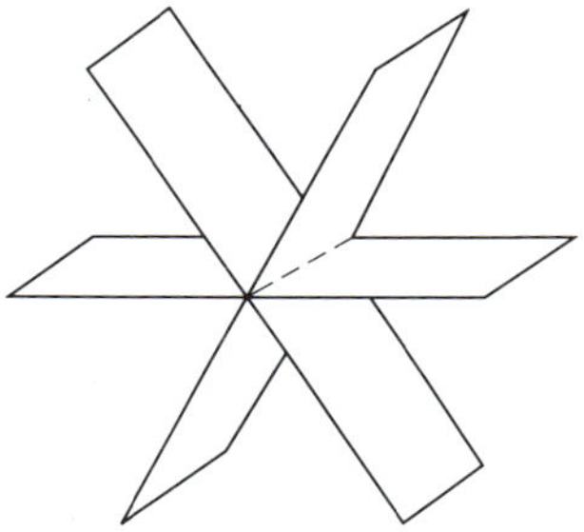

Infinitely many solutions

FIGURE 9.1g

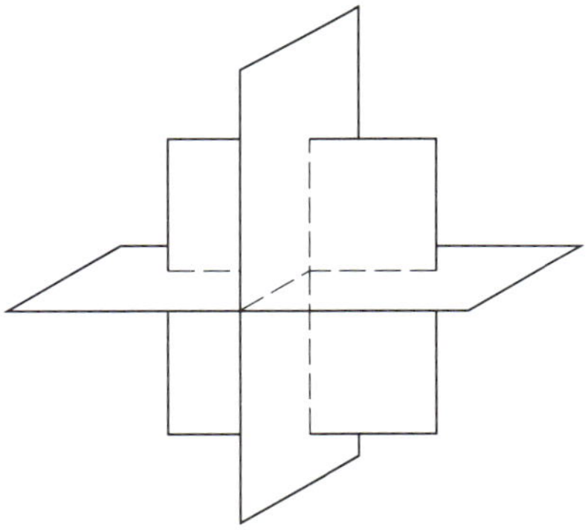

One solution

FIGURE 9.1h

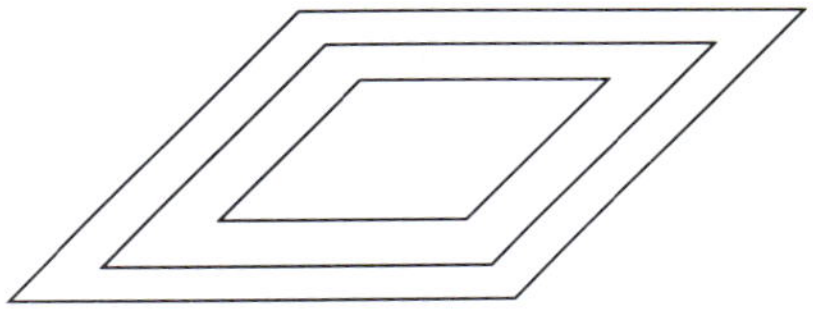

Infinitely many solutions

FIGURE 9.1i

Geometric visualizations fail us when we move to linear systems involving four or more variables, but the algebraic methods and results for two- and three-variable systems extend to larger systems.

1. You must have as many equations as variables if you hope to find a unique solution.
2. A system may have no solution.
3. A system may have an infinite number of solutions. This can happen in various ways, but it always means you can choose a value of at least one of the variables and then find values of the others that will form a solution. This will be easier to see as we look at other methods of solving systems.

Now that we have some idea of the results of solving a system, how do we find solutions? Substitution is not very appealing for large systems. Elimination is a method based on the idea of simplifying the problem. You know that you may multiply both sides of an equation by any nonzero number or add any number to both sides without changing the solution set. Elimination may use both of these rules to eliminate a variable by first performing a multiplication to match coefficients and then adding to remove a variable.

Definition We say two systems are equivalent if they have the same solution set.

We use ⇔ as before to indicate equivalence. Equivalent systems are guaranteed if one can be obtained from the other by replacing one equation with an equivalent equation or by replacing one equation with the result of combining a nonzero multiple of the equation with a multiple of one of the other equations. A single large brace is used to set off equations to be considered as a system.

Illustration
$$\begin{cases} 4x - 3y = 18 \\ 6x + 9y = -24 \end{cases} \Leftrightarrow \begin{cases} 4x - 3y = 18 \\ 2x + 3y = -8 \end{cases}$$

because $6x + 9y = -24 \Leftrightarrow 2x + 3y = -8$. □

Illustration
$$\begin{cases} 8x - 3y = 17 \\ -5x + 2y = -3 \end{cases} \Leftrightarrow \begin{cases} 8x - 3y = 17 \\ y = 61 \end{cases}$$

by replacing $-5x + 2y = -3$ with the result of combining 8 times this equation with 5 times the first equation.

$$-40x + 16y + (\mathbf{40x - 15y}) = -24 + \mathbf{85},$$

or, more simply, $y = 61$.

Note: We could have replaced $-5x + 2y = -3$ with the result of combining this equation with $\frac{5}{8}$ times the first equation.

$$-5x + 2y + \left(5x - \tfrac{15}{8}y\right) = -3 + \tfrac{85}{8}, \quad \text{or} \quad \tfrac{1}{8}y = \tfrac{61}{8}.$$ □

EXAMPLE Solve the system $\begin{cases} 14x - 3y = 19, \\ -5x + 4y = -3. \end{cases}$

It matters not which variable you choose to eliminate. If 5 is used as a multiplier for the first equation and 14 as a multiplier for the second, an equivalent system is produced by replacing the second equation.

$$5(14x - 3y) + \mathbf{14(-5x + 4y)} = 5(19) + \mathbf{14(-3)}$$

$$\Leftrightarrow \qquad 41y = 53$$

$$\begin{cases} 14x - 3y = 19 \\ -5x + 4y = -3 \end{cases} \Leftrightarrow \begin{cases} 14x - 3y = 19 \\ 41y = 53 \end{cases}$$

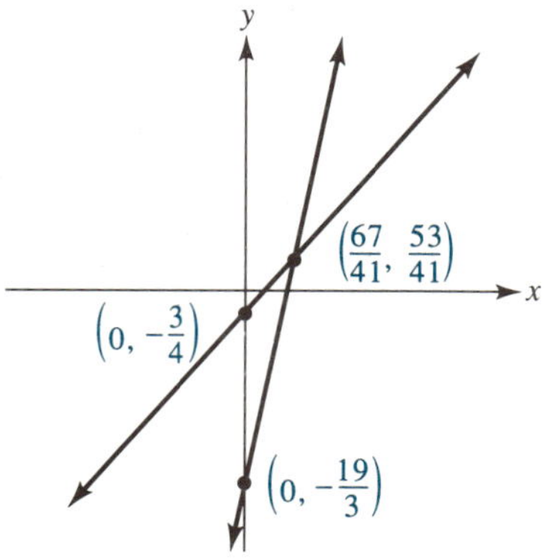

Any common solution must have $\frac{53}{41}$ for y. There are two ways to proceed at this point: Substitute $\frac{53}{41}$ for y in one of the original equations and solve for x or—more appealing for this case—eliminate y from the original system.

$$\begin{cases} 14x - 3y = 19 \\ -5x + 4y = -3 \end{cases} \Leftrightarrow \begin{cases} 14x - 3y = 19 \\ 41x = 67 \end{cases}$$

The solution set is $\left\{\left(\mathbf{\frac{67}{41}}, \mathbf{\frac{53}{41}}\right)\right\}$. ■

Whether to do a double elimination as in the last example or elimination-substitution is a judgment call based on the first coordinate you find. As a first approach, you should examine the original system to see whether substitution or elimination appears to be more efficient.

Illustration

$$\begin{cases} x = 3y + 10 \\ 2x - 18y = 12 \end{cases}$$ Substitution could be more efficient here.

$$\begin{cases} 19x - 15y = 10 \\ 11x + 3y = -2 \end{cases}$$ Elimination could be more efficient here. □

Either substitution or elimination reduces the number of variables. We usually lose one variable with each application of either method. For larger systems, these techniques must be applied repeatedly. In the next section we'll refine elimination, but first we'll use it on some systems with three equations and three variables.

Definition A system with m equations and n variables is said to be m by n, written $m \times n$.

EXAMPLE Find the quadratic function that contains (1, 61), (2, 69), and (5, 45).
Let the function's name be f.

$$f(x) = ax^2 + bx + c \quad \text{and} \quad \begin{aligned} f(1) &= a + b + c = 61 \\ f(2) &= 4a + 2b + c = 69 \\ f(5) &= 25a + 5b + c = 45 \end{aligned}$$

It will be a simple matter to eliminate c.

$$\begin{cases} a + b + c = 61 \\ 4a + 2b + c = 69 \\ 25a + 5b + c = 45 \end{cases}$$

Using the first and second equations, we obtain $3a + b = 8$. Using the first and third equations, we have $24a + 4b = -16 \Leftrightarrow 6a + b = -4$. The system

$$\begin{cases} a + b + c = 61 \\ 3a + b = 8 \\ 6a + b = -4 \end{cases}$$

is equivalent to the original 3×3 system. We have

$$\begin{cases} a + b + c = 61 \\ 3a + b = 8 \\ 6a + b = -4 \end{cases} \Leftrightarrow \begin{cases} a + b + c = 61 \\ 3a + b = 8 \\ 3a = -12 \end{cases}$$

by using the second and third equations to eliminate b. Substitution is easy here. Since a is -4, b is 20, and c is 45.

The function is $f(x) = -4x^2 + 20x + 45$. ■

We need to consider the cases when there may not be a unique solution to a system. For 2×2 and 3×3 systems, we saw the geometric possibilities earlier in this section. How do you recognize these cases in an algebraic setting?

EXAMPLE Solve the system

$$\begin{cases} 6m - 3n + p = 7, \\ 2m + 3n + 5p = -4, \\ -2m + 3n + 2p = 3. \end{cases}$$

It requires no multiplications to eliminate the n-terms.

$$\begin{cases} 6m - 3n + p = 7 \\ 2m + 3n + 5p = -4 \\ -2m + 3n + 2p = 3 \end{cases} \Leftrightarrow \begin{cases} 6m - 3n + p = 7 \\ 8m + 6p = 3 \\ 4m + 3p = 10 \end{cases} \qquad \begin{matrix} Eq_1, Eq_2 \\ Eq_1, Eq_3 \end{matrix}$$

But $8m + 6p = 3 \Leftrightarrow 4m + 3p = \frac{3}{2}$. There can be no solutions for

$$\begin{cases} 6m - 3n + p = 7, \\ 4m + 3p = \frac{3}{2}, \\ 4m + 3p = 10. \end{cases}$$

If you're working on autopilot, you might repeat the elimination process using equation 2 and equation 3 to obtain

$$\begin{cases} 6m - 3n + p = 7 \\ 4m + 3p = \frac{3}{2} \\ 4m + 3p = 10 \end{cases} \Leftrightarrow \begin{cases} 6m - 3n + p = 7 \\ 4m + 3p = \frac{3}{2} \\ 0 = \frac{17}{2} \end{cases}$$

Clearly there can be no solutions for the original system. The graphical representation of this system is Figure 9.1e.
The solution set is { }. ■

EXAMPLE Solve

$$\begin{cases} 2m - 5n + p = -7, \\ 5m + 7n + 4p = -10, \\ -3m + n - 2p = 8. \end{cases}$$

It seems as if it will be about the same work to eliminate any of the variables. Let's begin with m. We note to the side what replacements were made.

$$\begin{cases} 2m - 5n + p = -7 \\ 5m + 7n + 4p = -10 \\ -3m + n - 2p = 8 \end{cases}$$

$$\Leftrightarrow \begin{cases} 2m - 5n + p = -7 \\ -39n + 3p = 15 \\ -3m + n - 2p = 8 \end{cases} \qquad Eq_2 \leftarrow (2Eq_2 - 5Eq_1)$$

$$\Leftrightarrow \begin{cases} 2m - 5n + p = -7 \\ 13n + p = 5 \\ -13n - p = -5 \end{cases} \qquad Eq_3 \leftarrow (2Eq_3 + 3Eq_1)$$

The last pair of equations tells us there are an infinite number of solutions. Choose a value for n or p and solve $13n + p = 5$ to find the other. Notice that if you had continued the elimination process, you would have obtained $0 = 0$. How do we write the solution set in this case? Indicate that one variable, say p, can be replaced by any real number and show how the values for n and m depend on this choice. Let c be any real number; if p is c, then

$$13n + c = 5 \Leftrightarrow n = \frac{5 - c}{13} \qquad \text{Eq}_2$$

$$2m - 5n + c = -7 \Leftrightarrow 2m = -7 - c + 5\left(\frac{5 - c}{13}\right)$$

$$m = \frac{-33 - 9c}{13}. \qquad \text{Eq}_1$$

$$\left\{(a, b, c) \;\middle|\; a = \frac{-33 - 9c}{13}, b = \frac{5 - c}{13}, c \in \mathbb{R}\right\}.$$

Note: The choice to express the solutions in terms of the third coordinate was arbitrary; any of the coordinates could have been used. For example, if we choose n to be any real number b,

$$13b + p = 5 \Leftrightarrow p = 5 - 13b \qquad \text{Eq}_2$$

$$2m + 5b + p = -7 \Leftrightarrow 2m = -7 + 5b - (5 - 13b) \qquad \text{Eq}_1$$

$\{(a, b, c) \mid a = -6 + 9b, b \in \mathcal{R}, c = 5 - 13b\}$ is another representation.

The graphical representation of this system is Figure 9.1g. ■

A system with no solutions is inconsistent. In this case, the equations contradict each other in some way. If there are an infinite number of solutions, we call the system undetermined. This may happen if the system is $n \times n$, as in the previous example, but it *will* happen if there are fewer equations than variables and the system is not inconsistent. A system will have a unique solution only if there are at least as many equations as variables.

Illustration The system $\begin{cases} x + 2z = 3 \\ x - y = -2 \\ x + z - w = -1 \end{cases}$ is 3×4.

We can see that y and z depend only on the value chosen for x. A little thought shows that this is also the case for w. There must be an infinite number of solutions. The solution set is

$$\left\{(a, b, c, d) \;\middle|\; a \in \mathbb{R}, b = a + 2, c = \frac{3 - a}{2}, d = \frac{5 + a}{2}\right\}.$$

As in Section A.2, we sometimes replace this with the less formal

$$\left\{\left(a, a + 2, \frac{3 - a}{2}, \frac{5 + a}{2}\right), a \in \mathbb{R}.\right\}$$ □

EXAMPLE Prove there is no nondegenerate parabola that passes through (1, 4), (1.5, 5), and (2, 6).

Sometimes we can prove a statement by attempting to show its opposite. This type of proof is an example of an *indirect proof*. If we assume there is a nondegenerate parabola through these points, then either $y = ax^2 + bx + c$, $a \neq 0$, or $x = Ay^2 + By + C$ has these points as solutions, with $A \neq 0$. Let's consider the vertical parabola case.

The system

$$\begin{cases} 4 = a(1)^2 + b(1) + c \\ 5 = a(1.5)^2 + b(1.5) + c \\ 6 = a(2)^2 + b(2) + c \end{cases}$$

must be satisfied. But this is equivalent to

$$\begin{cases} 4 = a + b + c \\ 1 = 1.25a + 0.5b & \mathrm{Eq}_2 \leftarrow (\mathrm{Eq}_2 - \mathrm{Eq}_1) \\ 2 = 3a + b & \mathrm{Eq}_3 \leftarrow (\mathrm{Eq}_3 - \mathrm{Eq}_1) \end{cases}$$

or

$$\begin{cases} 4 = a + b + c \\ 1 = 1.25a + 0.5b \\ 0 = 0.5a. & \mathrm{Eq}_3 \leftarrow (\mathrm{Eq}_3 - 2\mathrm{Eq}_2) \end{cases}$$

But this contradicts $a \neq 0$. Hence there is no vertical parabola. Check the case of a horizontal parabola. ■

It may seem awkward to carry along a complete equivalent system at every stage, but there are two reasons we choose to do so. All the information of the system is right there in front of us; we never have to hunt for an original equation to complete a substitution or to start another elimination. The second reason is to make a smooth transition to another technique for solving linear systems.

EXERCISES 9.1

In exercises 1 through 8, solve the system of equations.

1. $\begin{cases} 2x - y = 4 \\ 3x + 5y = -1 \end{cases}$

2. $\begin{cases} x + 4y = 2.1 \\ 4x - 7y = 1.6 \end{cases}$

3. $\begin{cases} 0.3x - 0.4y = 1.8 \\ 0.2x + 0.8y = 3.0 \end{cases}$

4. $\begin{cases} \frac{1}{2}x + \frac{1}{4}y = 2 \\ \frac{1}{2}x - \frac{1}{5}y = 4 \end{cases}$

5. $\begin{cases} 2x - 4y = 8 \\ 4x - 8y = 16 \end{cases}$

6. $\begin{cases} 5x + 6y = 3 \\ 2.5x + 3y = 1.5 \end{cases}$

7. $\begin{cases} 2x - 4y = 8 \\ 4x - 8y = 15 \end{cases}$

8. $\begin{cases} 5x + 6y = 3 \\ 2.5x + 3y = 1 \end{cases}$

In exercises 9 through 16, determine whether there is a unique solution (one ordered pair), no solution, or infinitely many solutions (each an ordered pair).

9. $\begin{cases} 3y - 0.5x = -3 \\ x - 6y = 29 \end{cases}$

10. $\begin{cases} 1.2x + 9y = 15.3 \\ 7x - 4.5y = 3.4 \end{cases}$

11. $\begin{cases} 2x + 4.5y = 12 \\ 4x + 9y = 24 \end{cases}$

12. $\begin{cases} y = -x + 13 \\ x + y = \pi \end{cases}$

13. $\begin{cases} 4.66x - \sqrt{45}y = 290.84 \\ \sqrt{3}x + 2.889y = 897.44 \end{cases}$

14. $\begin{cases} 2x + 3y = 14, \\ 2kx + 3ky = 14k \end{cases}$ k a real constant

15. $\begin{cases} 2jx + 3jy = 14j \\ 2kx + 3ky = 14k \end{cases}$ j and k real constants

16. $\begin{cases} 3jx + 3jy = 14j \\ 2kx + 3ky = 14k \end{cases}$ j and k real constants

Note: **If you had to solve the system to identify the type of solution, look at the systems again after doing problems 23 and 24.**

In exercises 17 through 20, solve the system of equations.

17. $\begin{cases} x + y + z = 3 \\ x - 2y + z = 4 \\ 2x - y + 3z = 8 \end{cases}$

18. $\begin{cases} 5x - y + 2z = 1 \\ 2x + 4y - z = 3 \\ x + y - z = 5 \end{cases}$

19. $\begin{cases} x + y + z = 3 \\ 2x + 2y + 2z = 8 \\ 3x - y + z = 13 \end{cases}$

20. $\begin{cases} x + y + z = 3 \\ 2x + 2y + 2z = 6 \\ 3x - y + z = 13 \end{cases}$

21. Solve $(a + bi)(3 - 5i) = 8 + 7i$, where $i = \sqrt{-1}$. *Hint:* Two complex numbers are equal if their real parts are equal and their imaginary parts are equal.

22. Prove or disprove that there is no nondegenerate horizontal parabola that passes through (1, 4), (1.5, 5), and (2, 6).
(This completes the last example of this section.)

23. Given the system

$$\begin{cases} 2x + 3y = 8 \\ 2x + 3y = k \end{cases}$$

k a constant, what must be true for the solution set to be empty?

24. Given the system

$$\begin{cases} 2x + 3y = 8 \\ 2x + 3y = k \end{cases}$$

k a constant.

a. What must be true for the solution set to be infinite?

b. Give the infinitely many solutions first in terms of x and then in terms of y.

25. Describe the graph of a system of two linear equations in two variables that has infinitely many solutions.

26. Describe the graph of a system of two linear equations in two variables that has no solutions.

In exercises 27 through 30, you are given the last step of a student's solution process for a system of linear equations. Assuming her work is correct, give the solution set.

27. $\begin{cases} 2x + 3y + z = 4 \\ 4x + y = 18 \\ x = 1.3 \end{cases}$

28. $\begin{cases} 2x + 3y + z = 4 \\ 4x + y = 18 \\ 4x + y = 18 \end{cases}$

29. $\begin{cases} 2x + 3y + z = 4 \\ 4x + y = 18 \\ 4x + y = 19 \end{cases}$

30. $\begin{cases} 2x + 3y + z = 4 \\ x = 18 \end{cases}$

(Graphing calculator) Traditionally we've relied on the algebraic methods of solving systems because hand-drawn graphs were not accurate enough for obtaining the intersection points. In exercises 31 through 34, use your graphing calculator to draw the graphs of lines represented by the system and use its solving routine to approximate the coordinates of the point of intersection correct to 2 decimal places.

31. $\begin{cases} 3y - 0.5x = -3 \\ x - 6y = 29 \end{cases}$

32. $\begin{cases} 1.2x + 9y = 15.3 \\ 7x - 4.5y = 3.4 \end{cases}$

33. $\begin{cases} 2x + 4.5y = 12 \\ \sqrt{3}x + 9y = 1.24 \end{cases}$

34. $\begin{cases} y = x + 1 \\ x + y = \pi \end{cases}$

9.2 Gauss-Jordan Elimination

The first observation we make toward refining the techniques for solving systems of linear equations is that the real work is done with the coefficients and constants. As with synthetic division, if we organize our work carefully, we don't need to write the variable factors. We must translate the rules that produce equivalent systems into rules for working with the coefficients and constants that represent the equations.

A method for solving linear systems that incorporates these ideas is **Gauss-Jordan elimination**. To use Gauss-Jordan elimination, write the system in the usual format, variable terms in the same order on the left, constants only on the right side.

Definition The *augmented matrix* associated with the linear system is the block of coefficients, with the constants from the right side as a column adjoined on the right.

Illustration The system

$$\begin{cases} x - 2y + z = -4 \\ -x + 3y - 2z = 5 \\ 2x - y + 3z = -1 \end{cases}$$

is in the required format. The associated augmented matrix is

$$\left[\begin{array}{rrr|r} 1 & -2 & 1 & -4 \\ -1 & 3 & -2 & 5 \\ 2 & -1 & 3 & -1 \end{array}\right].$$

The large brackets set off the augmented matrix. □

How does the method work? We move from the original augmented matrix through a series of augmented matrices that represent equivalent systems until we have

$$\left[\begin{array}{cccccccc|c} 1 & 0 & 0 & 0 & \cdots & 0 & 0 & & r_1 \\ 0 & 1 & 0 & 0 & \cdots & 0 & 0 & & r_2 \\ 0 & 0 & 1 & 0 & \cdots & 0 & 0 & & r_3 \\ \vdots & \vdots & \vdots & \vdots & & \vdots & \vdots & & \vdots \\ 0 & 0 & 0 & 0 & \cdots & 0 & 1 & & r_n \end{array}\right]$$

or we know this form is not posssible to reach. Let us wait a bit to explain how you might know it was not possible. Why stop at

$$\left[\begin{array}{ccc|r} 1 & 0 & 0 & 3 \\ 0 & 1 & 0 & 6 \\ 0 & 0 & 1 & -4 \end{array}\right]?$$

Associated with this augmented matrix is the system of equations

$$\begin{cases} x_1 \qquad = 3, \\ \quad x_2 \quad = 6, \\ \qquad x_3 = -4. \end{cases}$$

Its solution set is obviously $\{(3, 6, -4)\}$. Now how can we guarantee each augmented matrix in the series represents a system equivalent to the others? Each transformation should correspond to a legal move for the system of equations.

Although it's common when solving a system of equations to use multiples of two equations to eliminate a variable, a variable can always be eliminated with at most one multiplication.

Illustration

$$\begin{cases} 7x - 3y = 12 \\ 2x + 5y = -3 \end{cases} \Leftrightarrow \begin{cases} 7x - 3y = 12 \\ \frac{41}{7}y = -\frac{45}{7} \end{cases}$$

by replacing $2x + 5y = -3$ with the result of combining it with $-\frac{2}{7}$ times the first equation.

$$2x + 5y + \left(-2x + \frac{6}{7}y\right) = -3 - \frac{24}{7}, \quad \text{or} \quad \frac{41}{7}y = -\frac{45}{7}$$

□

To solve

$$\begin{cases} x - 2y + z = -4 \\ -x + 3y - 2z = 5 \\ 2x - y + 3z = -1 \end{cases}$$

a natural beginning would be to eliminate the x's by using the first equation and the second and then using -2 times the first equation and the third one. For the augmented matrices, those are operations on rows. *Note:* In the following, R_n represents row n from the previous augmented matrix.

$$\left[\begin{array}{rrr|r} 1 & -2 & 1 & -4 \\ -1 & 3 & -2 & 5 \\ 2 & -1 & 3 & -1 \end{array}\right]$$

$$\rightarrow \left[\begin{array}{rrr|r} 1 & -2 & 1 & -4 \\ 0 & 1 & -1 & 1 \\ 2 & -1 & 3 & -1 \end{array}\right] \qquad R_2 + R_1$$

$$\rightarrow \left[\begin{array}{rrr|r} 1 & -2 & 1 & -4 \\ 0 & 1 & -1 & 1 \\ 0 & 3 & 1 & 7 \end{array}\right] \qquad R_3 - 2R_1$$

We want 0s in the second column above and below the 1. You can create zeros by adding a multiple of one row to another.

$$\left[\begin{array}{rrr|r} 1 & -2 & 1 & -4 \\ 0 & 1 & -1 & 1 \\ 0 & 3 & 1 & 7 \end{array}\right]$$

$$\rightarrow \left[\begin{array}{rrr|r} 1 & 0 & -1 & -2 \\ 0 & 1 & -1 & 1 \\ 0 & 3 & 1 & 7 \end{array}\right] \qquad R_1 + 2R_2$$

$$\rightarrow \left[\begin{array}{rrr|r} 1 & 0 & -1 & -2 \\ 0 & 1 & -1 & 1 \\ 0 & 0 & 4 & 4 \end{array}\right] \qquad R_3 - 3R_2$$

Because multiplying each side of an equation by the same nonzero number produces an equivalent equation, multiplying a row by a nonzero number will not change the solution set.

$$\rightarrow \left[\begin{array}{rrr|r} 1 & 0 & -1 & -2 \\ 0 & 1 & -1 & 1 \\ 0 & 0 & 4 & 4 \end{array}\right]$$

$$\rightarrow \left[\begin{array}{rrr|r} 1 & 0 & -1 & -2 \\ 0 & 1 & -1 & 1 \\ 0 & 0 & 1 & 1 \end{array}\right] \qquad \tfrac{1}{4}R_3$$

$$\left[\begin{array}{rrr|r} 1 & 0 & 0 & -1 \\ 0 & 1 & 0 & 2 \\ 0 & 0 & 1 & 1 \end{array}\right] \qquad \begin{array}{l} R_1 + R_3 \\ R_2 + R_3 \end{array}$$

The solution set is $\{(-1, 2, 1)\}$.

To summarize, the new augmented matrix will represent a system equivalent to the previous system if it is the result of the following row operations.

1. Replacing a row with a nonzero multiple of the row (usually done to create a 1).
2. Replacing a row with the sum of the row and a multiple of another row (usually done to create a 0).
3. Interchanging two rows (sometimes done to save steps).

You probably recognize (1) and (2) as adaptations of our rules for producing equivalent equations. Because a system is a set of equations, their order doesn't matter and (3) won't change the solution set.

Illustration The following pairs represent equivalent systems.

$$\left[\begin{array}{cccc|c} 2 & 0 & 3 & 4 & 6 \\ 5 & 1 & 8 & -3 & 1 \\ -3 & 2 & -5 & 0 & -4 \\ 6 & 10 & -2 & 7 & 13 \end{array}\right] \rightarrow \left[\begin{array}{cccc|c} 1 & 0 & 1.5 & 2 & 3 \\ 5 & 1 & 8 & -3 & 1 \\ -3 & 2 & -5 & 0 & -4 \\ 6 & 10 & -2 & 7 & 13 \end{array}\right] \quad \tfrac{1}{2}R_1$$

Multiply the first row by $\frac{1}{2}$.

$$\left[\begin{array}{ccc|c} 1 & 3 & -4 & 0 \\ 0 & 1 & 2 & 3 \\ 0 & 5 & -1 & -4 \end{array}\right] \rightarrow \left[\begin{array}{ccc|c} 1 & 0 & -10 & -9 \\ 0 & 1 & 2 & 3 \\ 0 & 5 & -1 & -4 \end{array}\right] \quad R_1 - 3R_2$$

Add -3 times row 2 to row 1. Row 2 is unchanged.

$$\left[\begin{array}{ccc|c} 3 & -7 & 4 & 14 \\ -5 & 3 & 1 & 0 \\ 1 & -2 & 6 & 5 \end{array}\right] \rightarrow \left[\begin{array}{ccc|c} 1 & -2 & 6 & 5 \\ -5 & 3 & 1 & 0 \\ 3 & -7 & 4 & 14 \end{array}\right] \quad \begin{array}{c} R_3 \\ \\ R_1 \end{array}$$

Interchange rows one and three. □

It may not be possible to produce

$$\left[\begin{array}{cccccccc|c} 1 & 0 & 0 & 0 & \cdot & \cdot & \cdot & 0 & 0 & r_1 \\ 0 & 1 & 0 & 0 & \cdot & \cdot & \cdot & 0 & 0 & r_2 \\ 0 & 0 & 1 & 0 & \cdot & \cdot & \cdot & 0 & 0 & r_3 \\ \cdot & \cdot & \cdot & \cdot & \cdot & \cdot & \cdot & \cdot & \cdot & \cdot \\ \cdot & \cdot & \cdot & \cdot & \cdot & \cdot & \cdot & \cdot & \cdot & \cdot \\ \cdot & \cdot & \cdot & \cdot & \cdot & \cdot & \cdot & \cdot & \cdot & \cdot \\ 0 & 0 & 0 & 0 & \cdot & \cdot & \cdot & 0 & 1 & r_n \end{array}\right] \qquad (1)$$

using these three operations. Consider what it means if you are applying these operations and produce a row, 0 0 0 0 | 4. The associated equation is $0x_1 + 0x_2 + 0x_3 + 0x_4 = 4$. This is impossible; there can be no solutions to the system, and you stop work on the augmented matrix. Another possibility is that a row entirely of 0s is created: 0 0 0 0 0 | 0. This tells us there may be an infinite number of solutions. If this is the case, we need to work out a convenient way to describe these solutions. An example from Section 9.1 for which you have already seen several ways to describe the infinite solution set may be a good starting place.

Definition The process of applying the three row operations to produce an augmented matrix of form (1)—or as close to it as possible—is called Gauss-Jordan elimination.

EXAMPLE Solve

$$\begin{cases} 2m - 5n + p = -7 \\ 5m + 7n + 4p = -10 \\ -3m + n - 2p = 8 \end{cases}$$

using Gauss-Jordan elimination.

$$\left[\begin{array}{rrr|r} 2 & -5 & 1 & -7 \\ 5 & 7 & 4 & -10 \\ -3 & 1 & -2 & 8 \end{array}\right]$$

$$\to \left[\begin{array}{rrr|r} 1 & -\frac{5}{2} & \frac{1}{2} & -\frac{7}{2} \\ 5 & 7 & 4 & -10 \\ -3 & 1 & -2 & 8 \end{array}\right] \qquad \frac{1}{2}R_1$$

$$\to \left[\begin{array}{rrr|r} 1 & -\frac{5}{2} & \frac{1}{2} & -\frac{7}{2} \\ 0 & \frac{39}{2} & \frac{3}{2} & \frac{15}{2} \\ 0 & -\frac{13}{2} & -\frac{1}{2} & -\frac{5}{2} \end{array}\right] \qquad \begin{array}{l} \\ R_2 - 5R_1 \\ R_3 + 3R_1 \end{array}$$

$$\to \left[\begin{array}{rrr|r} 1 & -\frac{5}{2} & \frac{1}{2} & -\frac{7}{2} \\ 0 & 1 & \frac{1}{13} & \frac{5}{13} \\ 0 & -\frac{13}{2} & -\frac{1}{2} & -\frac{5}{2} \end{array}\right] \qquad \frac{2}{39}R_2$$

$$\to \left[\begin{array}{rrr|r} 1 & -\frac{5}{2} & \frac{1}{2} & -\frac{7}{2} \\ 0 & 1 & \frac{1}{13} & \frac{5}{13} \\ 0 & 0 & 0 & 0 \end{array}\right] \qquad R_3 + \frac{13}{2}R_2$$

We see now there are infinitely many solutions, but we continue the elimination process one more step.

$$\left[\begin{array}{rrr|r} 1 & 0 & \frac{9}{13} & -\frac{33}{13} \\ 0 & 1 & \frac{1}{13} & \frac{5}{13} \\ 0 & 0 & 0 & 0 \end{array}\right] \qquad R_1 + \tfrac{5}{2}R_2$$

The simplest way to write the solutions is to express each variable associated with a 1 that begins a row in terms of variables that are not associated with a leading 1. Here m and n will be expressed in terms of p. The first row tells us $m = -\frac{33}{13} - \frac{9}{13}p$; the second row says $n = \frac{5}{13} - \frac{1}{13}p$.

The solution set is

$$\left\{\left(-\frac{33}{13}-\frac{9}{13}r, \frac{5}{13}-\frac{1}{13}r, r\right), r \in \mathbb{R}\right\}.$$ ■

EXAMPLE Write the solution set for the linear system associated with each of the following.

a. $\left[\begin{array}{cccc|c} 1 & 0 & 0 & 0 & -1.5 \\ 0 & 1 & 0 & 0 & 2.8 \\ 0 & 0 & 1 & 0 & 0.2 \\ 0 & 0 & 0 & 1 & -1.0 \end{array}\right]$ **b.** $\left[\begin{array}{cc|c} 1 & 0 & 6 \\ 0 & 1 & -2 \\ 0 & 0 & 3 \end{array}\right]$

c. $\left[\begin{array}{cccc|c} 1 & 0 & 2 & -1 & \sqrt{3} \\ 0 & 1 & 0 & 3 & -\sqrt{2} \\ 0 & 0 & 0 & 0 & 0 \\ 0 & 0 & 0 & 0 & 0 \end{array}\right]$

For the system associated with **(a)**, the solution set is $\{(-1.5, 2.8, 0.2, -1.0)\}$.
For the system associated with **(b)**, we see there are no solutions, $\{\ \}$.
For the system associated with **(c)**, there are infinitely many solutions, and we have two variables for which we can make arbitrary choices. The first two variables can be expressed in terms of these choices.
The solution set is

$$\{(\sqrt{3}-2r+s, -\sqrt{2}-3s, r, s), r \text{ and } s \text{ real numbers}\}.$$ ■

To apply Gauss-Jordan elimination effectively, work from left to right, creating the 1 in a column first and then the zeros. Swap rows if necessary to place any all-zero rows below rows with leading ones. You may also find in a particular case that interchanging rows may save steps. As always, be alert to possibilities for cutting down the number of steps needed.

Illustration In order to postpone dealing with fractions in the augmented matrix

$$\left[\begin{array}{cccc|c} 1 & 4 & -3 & 2 & -7 \\ 0 & 3 & -1 & 6 & 1 \\ 0 & -11 & 8 & 19 & -4 \\ 0 & 6 & -2 & 3 & 12 \end{array}\right],$$

we replace the last row with the sum of the last row and -2 times the second row before dividing the second row by 3 to create the 1. Be sure that you use one of the three legal operations and don't undo a 0 or 1 you've created to the left. □

You may have noticed that in a sense Gauss-Jordan elimination does more than we need. If we had replaced an original augmented matrix with

$$\left[\begin{array}{ccc|c} 1 & -2 & 3 & 1 \\ 0 & 1 & 4 & 0 \\ 0 & 0 & 1 & -2 \end{array}\right],$$

we could find any solutions at this point. Consider the equations the rows represent and use substitution.

$$1z = -2; \quad 1y + 4z = 0 \Rightarrow y + 4(-2) = 0 \Leftrightarrow y = 8;$$
$$y = 8, \quad z = -2, \quad 1x - 2y + 3z = 1 \Rightarrow x - 2(8) + 3(-2) = 1$$
$$\Leftrightarrow x = 23.$$

The solution set is $\{(23, 8, -2)\}$. Actually the work done for these substitutions and the work to finish the Gauss-Jordan elimination are about the same.

We've seen before that techniques designed for one situation can often be applied in a slightly different case. This is true of Gauss-Jordan elimination as well.

Definition **A linear-in-form system of equations** can be viewed in terms of variable expressions to the first power, even though the expressions themselves may not be of first degree, with the same expressions in each equation.

Some examples will clarify this idea.

Illustration The system

$$\begin{cases} \dfrac{2}{x} - \dfrac{3}{y-1} = 4 \\ \dfrac{1}{x} + \dfrac{5}{y-1} = -2 \end{cases}$$

is linear in form for the expressions $1/x$ and $1/(y - 1)$.
The system

$$\begin{cases} 6(x^2 - 1) + 3\log y - z = 8 \\ -2(x^2 - 1) + \log y + 7z = -13 \\ (x^2 - 1) - 5\log y - 2z = 17 \end{cases}$$

is linear in form for the expressions $x^2 - 1$, $\log y$, and z. □

To apply Gauss-Jordan elimination to a linear-in-form system, we work with the coefficients as before. The final right column will give values for the expressions, not the simple variables.

EXAMPLE Solve

$$\begin{cases} \dfrac{2}{x} - \dfrac{3}{y-1} = 4 \\ \dfrac{1}{x} + \dfrac{5}{y-1} = -2 \end{cases}$$

as a linear-in-form system.
The expressions used are $1/x$ and $1/(y - 1)$.

$$\left[\begin{array}{cc|c} 2 & -3 & 4 \\ 1 & 5 & -2 \end{array}\right]$$

$$\rightarrow \left[\begin{array}{cc|c} 1 & 5 & -2 \\ 2 & -3 & 4 \end{array}\right] \quad \begin{array}{l} R_2 \\ R_1 \end{array}$$

$$\to \left[\begin{array}{cc|c} 1 & 5 & -2 \\ 0 & -13 & 8 \end{array}\right] \quad R_2 - 2R_1$$

$$\to \left[\begin{array}{cc|c} 1 & 5 & -2 \\ 0 & 1 & -\frac{8}{13} \end{array}\right] \quad -\frac{1}{13}R_2$$

$$\to \left[\begin{array}{cc|c} 1 & 0 & \frac{14}{13} \\ 0 & 1 & -\frac{8}{13} \end{array}\right] \quad R_1 - 5R_2$$

The final augmented matrix tells us $1/x = \frac{14}{13}$ and $1/(y - 1) = -\frac{8}{13}$. The solution set is $\{(\frac{13}{14}, -\frac{5}{8})\}$. ■

EXERCISES 9.2

In exercises 1 through 4, write an augmented matrix associated with each system.

1. $\begin{cases} 3x - 4y + z = -8 \\ 7x + 2y - 3z = 13 \\ -x - y = 24 \end{cases}$

2. $\begin{cases} 14x = y - 8.1z + 3 \\ 0.1x - 3y + z = 7.2 \\ 13 - 1.2x = 32z - y \end{cases}$

3. $\begin{cases} 2x - 4y = 1 - z \\ x - 13 = 7z + \pi y \\ (1 - \sqrt{3})z = 0 \end{cases}$

4. $\begin{cases} y - 7.32 = 2.41z \\ -x = 13.8 \\ 28.25z = 8.51x - 2.08 \end{cases}$

In exercises 5 through 8, obtain a 1 in the position marked and record the operation you chose to use.

5. $\left[\begin{array}{ccc|c} 1 & -3 & 8 & 14 \\ 0 & \enclose{circle}{-2} & 5 & 7 \\ 0 & 6 & 4 & -3 \end{array}\right]$

6. $\left[\begin{array}{ccc|c} 1 & -3 & 8 & 14 \\ 0 & -2 & 5 & 7 \\ 0 & 6 & \enclose{circle}{4} & -3 \end{array}\right]$

7. $\left[\begin{array}{cccc|c} 1 & 0 & 4 & 8 & 3 \\ 0 & 1 & -2 & 1 & 1 \\ 0 & 0 & \enclose{circle}{6} & 2 & -3 \\ 0 & 0 & -5 & 5 & 6 \end{array}\right]$

8. $\left[\begin{array}{cccc|c} 1 & 0 & 4 & 8 & 3 \\ 0 & 1 & -2 & 1 & 1 \\ 0 & 0 & 6 & 2 & -3 \\ 0 & 0 & -5 & \enclose{circle}{5} & 6 \end{array}\right]$

In exercises 9 through 12, obtain a 0 in the position marked and record the operation you chose to use.

9. $\left[\begin{array}{ccc|c} 1 & -3 & 8 & 14 \\ 0 & 1 & \enclose{circle}{5} & 7 \\ 0 & 6 & 4 & -3 \end{array}\right]$

10. $\left[\begin{array}{ccc|c} 1 & -3 & \enclose{circle}{8} & 14 \\ 0 & -2 & 5 & 7 \\ 0 & 6 & 1 & -3 \end{array}\right]$

11. $\left[\begin{array}{cccc|c} 1 & 0 & 4 & 8 & 3 \\ 0 & 1 & -2 & 1 & 1 \\ 0 & 0 & 1 & 2 & -3 \\ 0 & 0 & \enclose{circle}{-5} & 5 & 6 \end{array}\right]$

12. $\left[\begin{array}{cccc|c} 1 & 0 & \enclose{circle}{4} & 8 & 3 \\ 0 & 1 & -2 & 1 & 1 \\ 0 & 0 & 1 & 2 & -3 \\ 0 & 0 & -5 & 5 & 6 \end{array}\right]$

In exercises 13 through 22, give the solution set of the linear system associated with the augmented matrix.

13. $\left[\begin{array}{cc|c} 1 & 0 & 4.2 \\ 0 & 1 & -3.1 \end{array}\right]$

14. $\left[\begin{array}{ccc|c} 1 & 0 & 0 & -7.3198 \\ 0 & 1 & 0 & -6.2114 \\ 0 & 0 & 1 & 2.4028 \end{array}\right]$

15. $\left[\begin{array}{ccc|c} 1 & 0 & 0 & \frac{2}{5} \\ 0 & 1 & 0 & \frac{4}{5} \\ 0 & 0 & 1 & 0 \end{array}\right]$

16. $\left[\begin{array}{ccc|c} 1 & 0 & 0 & -\frac{7}{3} \\ 0 & 1 & 0 & \frac{1}{3} \\ 0 & 0 & 1 & -\frac{4}{3} \end{array}\right]$

17. $\left[\begin{array}{ccc|c} 1 & 0 & 0 & 4 \\ 0 & 1 & 3 & 8 \\ 0 & 0 & 0 & 0 \end{array}\right]$

18. $\left[\begin{array}{ccc|c} 1 & 0 & -3 & 3 \\ 0 & 0 & 0 & -1 \\ 0 & 1 & 7 & 0 \end{array}\right]$

19. $\left[\begin{array}{ccc|c} 1 & 0 & 0 & 4 \\ 0 & 1 & 3 & 8 \\ 0 & 0 & 0 & 1 \end{array}\right]$

20. $\left[\begin{array}{ccc|c} 1 & 0 & -3 & 3 \\ 0 & 1 & 7 & -1 \\ 0 & 0 & 0 & 0 \end{array}\right]$

21. $\left[\begin{array}{ccc|c} 0 & 0 & 0 & 0 \\ 0 & 1 & \frac{3}{8} & \frac{7}{8} \\ 1 & 0 & \frac{1}{8} & -\frac{5}{8} \end{array}\right]$

22. $\left[\begin{array}{ccc|c} 1 & 0 & 0 & -4 \\ 0 & 1 & 0 & 3 \\ 0 & 0 & 0 & 1 \end{array}\right]$

In exercises 23 through 30, write an associated augmented matrix for each system, use Gauss-Jordan elimination to solve the system, and state the solution set.

23. $\begin{cases} 2x - 3y = 18 \\ x + 6y = 4 \end{cases}$

24. $\begin{cases} 5x - 3y = -2 \\ 2x - y = 8 \end{cases}$

25. $\begin{cases} x + y - z = 4 \\ 2x - y + z = 8 \\ -x + y + 3z = 5 \end{cases}$

26. $\begin{cases} x + 3z = -1 \\ 2x - y + z = 4 \\ 3y - 6z = 7 \end{cases}$

27. $\begin{cases} y = \frac{2}{3}x + 6 \\ 2x - 3y = 18 \end{cases}$

28. $\begin{cases} y = \frac{2}{3}x - 6 \\ 2x - 3y = 18 \end{cases}$

29. $\begin{cases} 5x - 4y + z = 3 \\ 4x + y - 4z = -2 \\ y - z = 1 \end{cases}$

30. $\begin{cases} 0.5y = 1 + x \\ 3x = z + 2 \\ z = y \end{cases}$

31. State the rule for producing an equivalent system of equations that corresponds to row operation 1.

32. State the rule for producing an equivalent system of equations that corresponds to row operation 2.

33. In solving the system of the first illustration of this section, the following statement appeared: ". . . a natural beginning would be to eliminate the x's . . . ." Since this approach is not an absolute requirement, solve the same system by eliminating the y's first, the z's second, and the x's third. Discuss your results.

34. Lew chooses to multiply the first row of

$$\left[\begin{array}{ccc|c} -2 & 5 & -1 & 7 \\ 5 & 7 & 4 & -10 \\ 3 & 1 & -2 & 8 \end{array}\right]$$

by $-\frac{1}{2}$ to obtain a 1 in the first row, first column. Sue chooses to replace row 1 with the sum of rows 1 and 3. Discuss the advantages and disadvantages of each approach.

In exercises 35 through 38, solve each linear-in-form system using Gauss-Jordan elimination.

35. $\begin{cases} \frac{2}{x} - \frac{3}{y} = 18 \\ \frac{1}{x} + \frac{6}{y} = 4 \end{cases}$

36. $\begin{cases} 5x^2 - 3y^2 = -2 \\ 2x^2 - y^2 = 8 \end{cases}$

37. $\begin{cases} \ln x - 0.5 \ln y = -1 \\ 3 \ln x - \ln z = 2 \\ \ln z = \ln y \end{cases}$

38. $\begin{cases} \frac{2}{x+1} + \frac{5}{y} + \frac{25}{z-2} = 10 \\ \frac{5}{y} + \frac{10}{z-2} = 2 \\ \frac{1}{x+1} + \frac{10}{z-2} = 4 \end{cases}$

39. A method commonly used to solve a linear system of two equations in two unknowns,

$$\begin{cases} ax + by = e, \\ cx + dy = f, \end{cases}$$

is as follows:

1. Compute D, which is $ad - bc$.
2. Compute D_x, which is $de - bf$.
3. Compute D_y, which is $af - ce$.
4. Write the solution set $\{(D_x/D, D_y/D)\}$.

Use Gauss-Jordan elimination on the system and compare your solution set to that obtained in step 4.

In exercises 40 through 43, use Cramer's rule, the method just outlined, to solve the system.

40. $\begin{cases} 3x - 4y = 5.1 \\ x + 7y = -8.3 \end{cases}$

41. $\begin{cases} 12.5x + 6.1y = 104.2 \\ 8.3x + 9.9y = -51.3 \end{cases}$

42. $\begin{cases} 2.1x - 3.4y = 6.2 \\ 9.66x - 15.64y = 28.52 \end{cases}$

43. $\begin{cases} 2.1x - 3.4y = 6.2 \\ 9.66x - 15.64y = 20.52 \end{cases}$

44. Give a verbal description of how to form D, D_x, and D_y for the system

$$\begin{cases} mx + ny = p, \\ tx + sy = q. \end{cases}$$

45. (Graphing calculator) Verify your solutions to exercises 23, 24, 27, and 28 by sketching the system in the same coordinate plane and using a solving routine to find the solution.

9.3 Matrices

Toolkit

The sum of additive inverses a and $-a$ is the additive identity: $a + (-a) = 0$. The product of multiplicative inverses is the multiplicative identity: $a(1/a) = 1, a \neq 0$.

Matrices arise naturally in settings other than as descriptions of linear systems. In this section we'll investigate matrices as mathematical objects, and in the next we'll see some applications of matrices, including yet another way to solve linear systems.

Suppose you run a small regional airline, West-Southeast Air, serving five cities, Acia, Bayville, Calvin, Dryton, and Eden. The company is young and doesn't yet have flights between all cities. The following diagram represents the flights currently available.

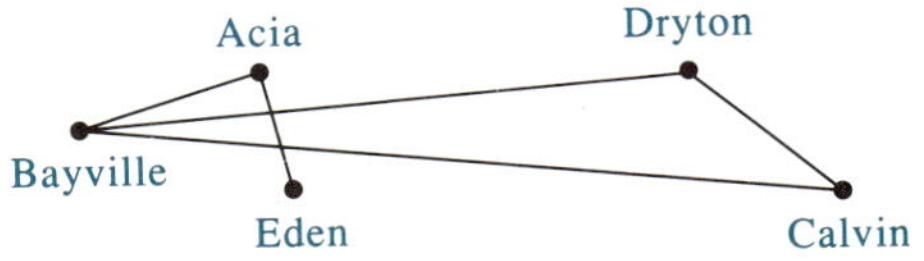

Another way to indicate direct flights between cities is a matrix with rows and columns corresponding to the cities and an entry of 1 if there is a flight between the cities and a 0 if no flight exists.

West-Southeast Flights

	Acia	Bayville	Calvin	Dryton	Eden
Acia	0	1	0	0	1
Bayville	1	0	1	1	0
Calvin	0	1	0	1	0
Dryton	0	1	1	0	0
Eden	1	0	0	0	0

Employee work records could also be kept in matrix form. (Every employee at West-Southeast Air has to be able to handle several jobs.)

Hours Worked During Week of May 17-21

Task:	Sanchez	Chen	Ferguson
Flight Attendant	21	10	11
Baggage Handler	14	7	13
Ticket Agent	5	23	16

The bookkeeper may have a matrix showing hourly wage rates for different jobs. When everything is going well, you may declare a bonus week.

Hourly Wage Rates in Dollars (Regular and Bonus Weeks)

	Flight Attendant	Baggage Handler	Ticket Agent
Regular	10.75	8.37	6.79
Bonus	12.04	9.54	7.81

Some terms and notation are needed.

Definition A **matrix** is a rectangular block of numbers. Its size, or *dimension,* is the number of rows and number of columns.

The flight matrix is 5 by 5, written 5×5. The wage rate matrix is 2×3.

Definition The numbers in a matrix are called *entries,* or *elements.*

An element is often denoted by using double subscripts, the first for the row and the second for the column. In the hourly wage record matrix, the element 7.81 might be named $h_{2,3}$ or h_{23}. Matrices themselves are commonly named by capital letters. For the remainder of this section, we refer to the flight matrix as F, the work matrix as W, and the hourly wage matrix as H.

$$F = \begin{bmatrix} 0 & 1 & 0 & 0 & 1 \\ 1 & 0 & 1 & 1 & 0 \\ 0 & 1 & 0 & 1 & 0 \\ 0 & 1 & 1 & 0 & 0 \\ 1 & 0 & 0 & 0 & 0 \end{bmatrix}$$

$$W = \begin{bmatrix} 21 & 10 & 11 \\ 14 & 7 & 13 \\ 5 & 23 & 16 \end{bmatrix}$$

$$H = \begin{bmatrix} 10.75 & 8.37 & 6.79 \\ 12.04 & 9.54 & 7.81 \end{bmatrix}$$

If we wish to talk about a general matrix, we may use the description A is $[a_{ij}]_{m \times n}$ to indicate A is of size $m \times n$ with typical entry a_{ij} in the ith row and the jth column.

Two matrices $[a_{ij}]_{m \times n}$ and $[b_{ij}]_{m \times n}$ are equal if and only if they are of the same size and corresponding entries are equal; $a_{ij} = b_{ij}$ for each i, j pair.

If you think about how to construct a matrix to represent the hours worked by your airline employees for the month of May, it seems natural to do so by adding the corresponding entries for each of the weekly matrices for May.

Definition The *sum* of two matrices, $[a_{ij}]_{m\times n} + [b_{ij}]_{m\times n}$, is defined to be the matrix $[c_{ij}]_{m\times n}$, where $c_{ij} = a_{ij} + b_{ij}$ for $1 \le i \le m$ and $1 \le j \le n$.

Addition is not defined for matrices of different sizes. Similarly, the *difference* $[a_{ij}]_{m\times n} - [b_{ij}]_{m\times n}$ is the matrix $[c_{ij}]_{m\times n}$, where $c_{ij} = a_{ij} - b_{ij}$ for $1 \le i \le m$ and $1 \le j \le n$. Some of the exercises ask you to explore how these operations behave. If A is of size $m \times n$, then $A - A$ is a matrix of size $m \times n$, all of whose entries are zero. Such a matrix is called a *zero matrix;* we can write it as $[0]_{m\times n}$. A zero matrix behaves as an additive identity for all the matrices of its size; $[a_{ij}]_{m\times n} + [0]_{m\times n} = [a_{ij}]_{m\times n}$ for any $m \times n$ matrix $[a_{ij}]_{m\times n}$.

What about some sort of multiplication for matrices? Actually there are two useful sorts of multiplication. If, as the owner of West-Southeast Air, you decide to increase all the hourly wage rates by 10%, how do you change H? Each entry should be multiplied by 1.10 to give the new rate in each category.

Definition *Scalar multiplication* between a real number, r, and a matrix A, $[a_{ij}]$, is defined by $rA = [ra_{ij}]$; each element of A is multiplied by r.

Illustration If each hourly wage rate at West-Southeast Air is increased by 10%, the new hourly wage matrix, H_{new}, is $1.10H$.

$$H_{\text{new}} = 1.10\begin{bmatrix} 10.75 & 8.37 & 6.79 \\ 12.04 & 9.54 & 7.81 \end{bmatrix}$$

$$H_{\text{new}} = \begin{bmatrix} 11.825 & 9.207 & 7.469 \\ 13.244 & 10.494 & 8.591 \end{bmatrix}$$ □

Is there a useful multiplication for two matrices? Creating a new matrix for the products of corresponding entries is of limited use. By considering our example matrices, H and W, we'll find a way of multiplying matrices that is useful. At the end of each week the bookkeeper must use the information in W and H to compute the gross wages for each worker.

To find Sanchez's wages, the bookkeeper must calculate

$$10.75(21) + 8.37(14) + 6.79(5) \quad \text{and} \quad 12.04(21) + 9.54(14) + 7.81(5).$$

She must multiply each entry in the first row of H by the corresponding entry in the first column of W and add these products, then repeat the process using the second row of H. If she uses the first and second rows of H and the second column of W, she will have the wages for Chen. When finished, there is a new matrix, G of gross wages.

$$H \cdot W = G$$

$$\begin{array}{c} \\ R \\ B \end{array}\begin{array}{c} \begin{array}{ccc} FA & BH & TA \end{array} \\ \begin{bmatrix} 10.75 & 8.37 & 6.79 \\ 12.04 & 9.54 & 7.81 \end{bmatrix} \end{array} \cdot \begin{array}{c} FA \\ BH \\ TA \end{array} \begin{array}{c} \begin{array}{ccc} \text{Sanchez} & \text{Chen} & \text{Ferguson} \end{array} \\ \begin{bmatrix} 21 & 10 & 11 \\ 14 & 7 & 13 \\ 5 & 23 & 16 \end{bmatrix} \end{array}$$

$$= \begin{array}{c} \\ R \\ B \end{array}\begin{array}{c} \begin{array}{ccc} \text{Sanchez} & \text{Chen} & \text{Ferguson} \end{array} \\ \begin{bmatrix} 376.88 & 322.26 & 335.70 \\ 425.45 & 366.81 & 381.42 \end{bmatrix} \end{array}$$

We define H times W, HW, to be G.

Definition If A, $[a_{ij}]_{m \times n}$, and B, $[b_{jk}]_{n \times p}$, are two matrices, *the product, AB*, is $[c_{ik}]_{m \times p}$, where $c_{ik} = a_{i1}b_{1k} + a_{i2}b_{2k} + a_{i3}b_{3k} + \cdots + a_{in}b_{nk}$ for $1 \le i \le m$ and $1 \le k \le p$.

EXAMPLE Calculate

$$\begin{bmatrix} -2 & 3 \\ 0 & 1 \\ 4 & -5 \end{bmatrix} \begin{bmatrix} -1 & 0 & 7 \\ 2 & 1 & 6 \end{bmatrix}.$$

The product is

$$\begin{bmatrix} -2(-1) + 3(2) & -2(0) + 3(1) & -2(7) + 3(6) \\ 0(-1) + 1(2) & 0(0) + 1(1) & 0(7) + 1(6) \\ 4(-1) - 5(2) & 4(0) - 5(1) & 4(7) - 5(6) \end{bmatrix},$$
$$\begin{bmatrix} \mathbf{8} & \mathbf{3} & \mathbf{4} \\ \mathbf{2} & \mathbf{1} & \mathbf{6} \\ \mathbf{-14} & \mathbf{-5} & \mathbf{-2} \end{bmatrix}.$$

■

Using the ith row of A and the jth column of B will produce the i, j^{th} entry of AB. The products of corresponding entries of the ith row of A and j^{th} column of B are added to produce the i, j^{th} entry of AB. In order to carry out matrix multiplication for A and B, the number of columns in A must equal the number of rows of B. In applications, the column names of A should be the same as the row names of B for the multiplication to be meaningful. The product will inherit row names from A and column names from B.

matrix:	A	B	AB
dimension:	$m \times n$	$n \times p$	$m \times p$

EXAMPLE Compare

$$\begin{bmatrix} -1 & 0 & 7 \\ 2 & 1 & 6 \end{bmatrix} \begin{bmatrix} -2 & 3 \\ 0 & 1 \\ 4 & -5 \end{bmatrix}$$

with the result in the previous exercise.

$$\begin{bmatrix} -1 & 0 & 7 \\ 2 & 1 & 6 \end{bmatrix} \begin{bmatrix} -2 & 3 \\ 0 & 1 \\ 4 & -5 \end{bmatrix} = \begin{bmatrix} 30 & -38 \\ 20 & -23 \end{bmatrix}$$

The product in the previous example is a 3 × 3 matrix, whereas the product in this example is a 2 × 2 matrix. Even if it is possible to compute both AB and BA, the products may not be the same. ■

Our examples show that matrix multiplication is not commutative. Some of the exercises ask you to investigate whether this operation is associative. It would be appropriate to ask now about a multiplicative identity and the possibility of multiplicative inverses. For this part of the discussion, we consider only square matrices, those that have the same number of rows as columns. An identity matrix, I, of size $n \times n$ should leave any $n \times n$ matrix A unchanged by multiplication on either side,

$$IA = AI = A.$$

Using the first row of I and the first column of A should yield the entry a_{11} of A. To be sure this happens, the first row of I needs to be 1 0 0 . . . 0.

Definition The $n \times n$ identity matrix , I_n, is

$$\begin{bmatrix} 1 & 0 & 0 & \cdot & \cdot & \cdot & 0 \\ 0 & 1 & 0 & \cdot & \cdot & \cdot & 0 \\ 0 & 0 & 1 & \cdot & \cdot & \cdot & 0 \\ \cdot & \cdot & \cdot & \cdot & \cdot & \cdot & \cdot \\ \cdot & \cdot & \cdot & \cdot & \cdot & \cdot & \cdot \\ \cdot & \cdot & \cdot & \cdot & \cdot & \cdot & \cdot \\ 0 & 0 & 0 & \cdot & \cdot & \cdot & 1 \end{bmatrix},$$

a matrix with 1s on the diagonal and 0s elsewhere.

Are there inverses for matrix multiplication—that is, for A, an $n \times n$ matrix, is there a matrix B, also $n \times n$, such that $AB = BA = I$? If so, B is the *inverse* of A and is denoted by A^{-1}.

Let's see if we can find an inverse for

$$\begin{bmatrix} 2 & 5 \\ 1 & 3 \end{bmatrix}.$$

We want

$$\begin{bmatrix} 2 & 5 \\ 1 & 3 \end{bmatrix} \begin{bmatrix} x & y \\ z & w \end{bmatrix} = \begin{bmatrix} 1 & 0 \\ 0 & 1 \end{bmatrix};$$

this requires

$$\begin{aligned} 2x + 5z &= 1, \qquad & 2y + 5w &= 0, \\ x + 3z &= 0, \quad \text{and} & y + 3w &= 1. \end{aligned}$$

Notice that the augmented matrices needed to solve these two pairs of equations will require us to use the same steps to carry out Gauss-Jordan elimination. Why not handle both at once? Instead of using

$$\left[\begin{array}{cc|c} 2 & 5 & 1 \\ 1 & 3 & 0 \end{array}\right] \text{ and } \left[\begin{array}{cc|c} 2 & 5 & 0 \\ 1 & 3 & 1 \end{array}\right],$$

we could perform Gauss-Jordan elimination on

$$\left[\begin{array}{cc|cc} 2 & 5 & 1 & 0 \\ 1 & 3 & 0 & 1 \end{array}\right].$$

$$\left[\begin{array}{cc|cc} 2 & 5 & 1 & 0 \\ 1 & 3 & 0 & 1 \end{array}\right] \rightarrow \left[\begin{array}{cc|cc} 1 & 3 & 0 & 1 \\ 2 & 5 & 1 & 0 \end{array}\right] \begin{matrix} R_2 \\ R_1 \end{matrix}$$

$$\rightarrow \left[\begin{array}{cc|cc} 1 & 3 & 0 & 1 \\ 0 & -1 & 1 & -2 \end{array}\right] R_2 - 2R_1 \rightarrow \left[\begin{array}{cc|cc} 1 & 3 & 0 & 1 \\ 0 & 1 & -1 & 2 \end{array}\right] -R_2$$

$$\rightarrow \left[\begin{array}{cc|cc} 1 & 0 & 3 & -5 \\ 0 & 1 & -1 & 2 \end{array}\right] R_1 - 3R_2$$

The values 3, -5, -1, 2 for x, y, z, w, respectively, satisfy the equations. More simply,

$$\begin{bmatrix} 3 & -5 \\ -1 & 2 \end{bmatrix} \text{ is the inverse of } \begin{bmatrix} 2 & 5 \\ 1 & 3 \end{bmatrix}.$$

$$\begin{bmatrix} 3 & -5 \\ -1 & 2 \end{bmatrix}\begin{bmatrix} 2 & 5 \\ 1 & 3 \end{bmatrix} = \begin{bmatrix} 6-5 & 15-15 \\ -2+2 & -5+6 \end{bmatrix}, \text{ or } \begin{bmatrix} 1 & 0 \\ 0 & 1 \end{bmatrix}$$

In general, to find the inverse, A^{-1}, of an $n \times n$ square matrix A, adjoin $I_{n\times n}$ and perform Gauss-Jordan elimination. If you can obtain $I_{n\times n}$ on the left of the dividing line, the matrix to the right is A^{-1}.

$$[A|I] \to \cdots \to [I|A^{-1}] \qquad \text{By Gauss-Jordan elimination}$$

You know from Section 9.2 that it may not be possible to create I on the left; if you can't, A does not have an inverse.

There are other methods for calculating A^{-1} if it exists, but this modification for our augmented matrix method will serve our purposes.

EXAMPLE Find A^{-1} if A is

$$\begin{bmatrix} 3 & -1 & 0 \\ 1 & 2 & -4 \\ 3 & 1 & 3 \end{bmatrix}.$$

Augment A with $I_{3\times 3}$ and use Gauss-Jordan elimination.

$$\left[\begin{array}{ccc|ccc} 3 & -1 & 0 & 1 & 0 & 0 \\ 1 & 2 & -4 & 0 & 1 & 0 \\ 3 & 1 & 3 & 0 & 0 & 1 \end{array}\right] \to \left[\begin{array}{ccc|ccc} 1 & 2 & -4 & 0 & 1 & 0 \\ 3 & -1 & 0 & 1 & 0 & 0 \\ 3 & 1 & 3 & 0 & 0 & 1 \end{array}\right]$$

$$\to \left[\begin{array}{ccc|ccc} 1 & 2 & -4 & 0 & 1 & 0 \\ 0 & -7 & 12 & 1 & -3 & 0 \\ 0 & -5 & 15 & 0 & -3 & 1 \end{array}\right] \to \left[\begin{array}{ccc|ccc} 1 & 2 & -4 & 0 & 1 & 0 \\ 0 & -5 & 15 & 0 & -3 & 1 \\ 0 & -7 & 12 & 1 & -3 & 0 \end{array}\right]$$

$$\to \left[\begin{array}{ccc|ccc} 1 & 2 & -4 & 0 & 1 & 0 \\ 0 & 1 & -3 & 0 & \frac{3}{5} & -\frac{1}{5} \\ 0 & -7 & 12 & 1 & -3 & 0 \end{array}\right] \to \left[\begin{array}{ccc|ccc} 1 & 0 & 2 & 0 & -\frac{1}{5} & \frac{2}{5} \\ 0 & 1 & -3 & 0 & \frac{3}{5} & -\frac{1}{5} \\ 0 & 0 & -9 & 1 & \frac{6}{5} & -\frac{7}{5} \end{array}\right]$$

$$\to \left[\begin{array}{ccc|ccc} 1 & 0 & 2 & 0 & -\frac{1}{5} & \frac{2}{5} \\ 0 & 1 & -3 & 0 & \frac{3}{5} & -\frac{1}{5} \\ 0 & 0 & 1 & -\frac{1}{9} & -\frac{2}{15} & \frac{7}{45} \end{array}\right]$$

$$\to \left[\begin{array}{ccc|ccc} 1 & 0 & 0 & \frac{2}{9} & \frac{1}{15} & \frac{4}{45} \\ 0 & 1 & 0 & -\frac{1}{3} & \frac{1}{5} & \frac{12}{45} \\ 0 & 0 & 1 & -\frac{1}{9} & -\frac{2}{15} & \frac{7}{45} \end{array}\right]$$

The inverse is

$$\begin{bmatrix} \frac{2}{9} & \frac{1}{15} & \frac{4}{45} \\ -\frac{1}{3} & \frac{1}{5} & \frac{12}{45} \\ -\frac{1}{9} & -\frac{2}{15} & \frac{7}{45} \end{bmatrix}.$$

■

EXAMPLE Find the inverse of

$$\begin{bmatrix} 2 & -3 & 1 \\ 2 & 2 & 6 \\ -1 & 4 & 2 \end{bmatrix},$$

if it exists.

Adjoin the 3×3 identity matrix and use Gauss-Jordan elimination.

$$\left[\begin{array}{rrr|rrr} 2 & -3 & 1 & 1 & 0 & 0 \\ 2 & 2 & 6 & 0 & 1 & 0 \\ -1 & 4 & 2 & 0 & 0 & 1 \end{array}\right] \rightarrow \left[\begin{array}{rrr|rrr} 1 & -\frac{3}{2} & \frac{1}{2} & \frac{1}{2} & 0 & 0 \\ 2 & 2 & 6 & 0 & 1 & 0 \\ -1 & 4 & 2 & 0 & 0 & 1 \end{array}\right]$$

$$\rightarrow \left[\begin{array}{rrr|rrr} 1 & -\frac{3}{2} & \frac{1}{2} & \frac{1}{2} & 0 & 0 \\ 0 & 5 & 5 & -1 & 1 & 0 \\ 0 & \frac{5}{2} & \frac{5}{2} & \frac{1}{2} & 0 & 1 \end{array}\right] \rightarrow \left[\begin{array}{rrr|rrr} 1 & -\frac{3}{2} & \frac{1}{2} & \frac{1}{2} & 0 & 0 \\ 0 & 1 & 1 & -\frac{1}{5} & \frac{1}{5} & 0 \\ 0 & \frac{5}{2} & \frac{5}{2} & \frac{1}{2} & 0 & 1 \end{array}\right]$$

$$\rightarrow \left[\begin{array}{rrr|rrr} 1 & 0 & 2 & \frac{1}{5} & \frac{3}{10} & 0 \\ 0 & 1 & 1 & -\frac{1}{5} & \frac{1}{5} & 0 \\ 0 & 0 & 0 & 1 & -\frac{1}{2} & 1 \end{array}\right]$$

We can't produce $I_{3\times 3}$ on the left.

The original matrix has no inverse. ■

EXERCISES 9.3

Exercises 1 through 14, refer to the matrices A, B, C. Perform the operations, where possible, or give the indicated value.

$$A = \begin{bmatrix} 2 & -1 & 3 \\ 4 & 8 & 7 \\ 1 & 4 & 0 \end{bmatrix}, \quad B = \begin{bmatrix} 1 & 0 \\ -1 & 4 \\ 2 & -2 \end{bmatrix},$$

$$C = \begin{bmatrix} -0.5 & 6.0 & 0.0 \\ 1.3 & 0.0 & -4.2 \end{bmatrix}$$

1. a_{13}
2. b_{22}
3. c_{23}
4. $\dim_A$
5. $\dim_B$
6. $\dim_C$
7. $A + B$
8. $A \cdot B$
9. $7C$
10. B^{-1}
11. d_{12}, for $D = B \cdot C$
12. e_{22}, for $E = C \cdot B$
13. $B \cdot C$
14. $C \cdot B$

Exercises 15 through 30 refer to matrices A, B, C, D, and E. Perform the indicated operations, where possible.

$$A = \begin{bmatrix} 2 & 0 & 1 \\ 0 & 1 & 4 \\ -1 & 3 & 2 \end{bmatrix}, \quad B = \begin{bmatrix} 0 & 4 & 3 \\ -2 & 4 & 1 \\ 1 & 4 & -1 \end{bmatrix},$$

$$C = \begin{bmatrix} 1 & -4 \\ 3 & 2 \\ 5 & -1 \end{bmatrix}, \quad D = \begin{bmatrix} 0 & 4 \\ 2 & 3 \end{bmatrix}, \quad E = [1 \quad 2 \quad 3]$$

15. $A + B$
16. $B - C$
17. $2B - 3A$
18. EC
19. CE
20. D^{-1}
21. $(A + B)C$
22. $C(A + B)$
23. E^{-1}
24. $CD + C$
25. $C(D + I)$, where I is the 2×2 identity matrix
26. A^{-1}
27. B^{-1}
28. AB^{-1}
29. $(AB)^{-1}$
30. $A^{-1}B$

The transpose of matrix M, denoted by M^t, is formed by writing each row of M as the corresponding column of M^t. For example, for

$$M = \begin{bmatrix} 2 & 3 & 4 \\ 2 & 1 & 7 \end{bmatrix}, \quad M^t = \begin{bmatrix} 2 & 2 \\ 3 & 1 \\ 4 & 7 \end{bmatrix}.$$

Exercises 31 through 34 refer to the definition of transpose, M and M' above, and N, where

$$N = \begin{bmatrix} 2 & 3 \\ -1 & 0 \end{bmatrix}.$$

31. Show $(M')' = M$.
32. Demonstrate that $(NM)' = M'N'$.
33. Compute $(M'N')'$.
34. What is the dimension of A' if $\dim_A$ is $p \times q$?
35. Construct three matrices A, B, and C so that $(AB)C$ exists. Compute $A(BC)$. Do you think matrix multiplication is associative?
36. Construct an example to demonstrate to a friend that matrix multiplication is not commutative.

A building contractor uses a computer program that performs matrix operations to store information and aid in decision-making. Exercises 37 through 43 illustrate the decisions that must be made to store and produce the required information. These problems are obviously contrived. Much more detailed information is necessary for an actual model, but we can get an idea of what is involved in modeling from these exercises.

The contractor has accepted orders for 5 ranch-style houses, 7 Cape Cod houses, and 12 colonial houses. He chooses X to represent the order matrix

	Ranch	Cape Cod	Colonial
Number ordered	[5	7	12].

The houses require steel, wood, glass, paint, and labor. The amounts of each of these needed for a ranch-style house are 5, 20, 16, 7, and 17 units, respectively; for the Cape Cod house, 7, 18, 12, 9, and 21 units, respectively; and for the colonial house 6, 25, 8, 5, and 13 units, respectively.

37. Put this information in a matrix R, where r_{ij} is the amount j needed to build house i. Preserve the order of the materials listed.
38. Compute and interpret XR.

The price for one unit of each of the materials is given by the matrix Y.

	Units of				
	Steel	Wood	Glass	Paint	Labor
Price per unit in dollars	[15	8	5	1	10].

39. Compute and interpret RY'. (Y' is defined just before Exercise 31.)
40. Compute and interpret XRY'.

The contractor decides to take into account the cost of transporting the materials as well as purchasing them. He replaces the matrix Y with a matrix Q, where q_{i1} is the price of material i and q_{i2} is the price of transporting material i. The prices for transporting the materials are 4.5, 2, 3, 0.5, and 0 (dollars), respectively. In exercises 41 through 43, compute and interpret the indicated matrix.

41. RQ
42. XRQ
43. $XRQZ$, where Z is $\begin{bmatrix} 1 \\ 1 \end{bmatrix}$
44. Some calculators compute the inverse of a matrix. If you have access to one of these, use it to redo exercises 20, 21, 27, and the last example of this section. What conclusions do you draw about what your calculator shows?

9.4 Applications of Matrices

Toolkit

The coordinates of A: (x, y) can also be written as $(r \cos\theta, r \sin\theta)$, $r = \sqrt{x^2 + y^2}$, $\theta = \tan^{-1}\left(\frac{y}{x}\right)$.

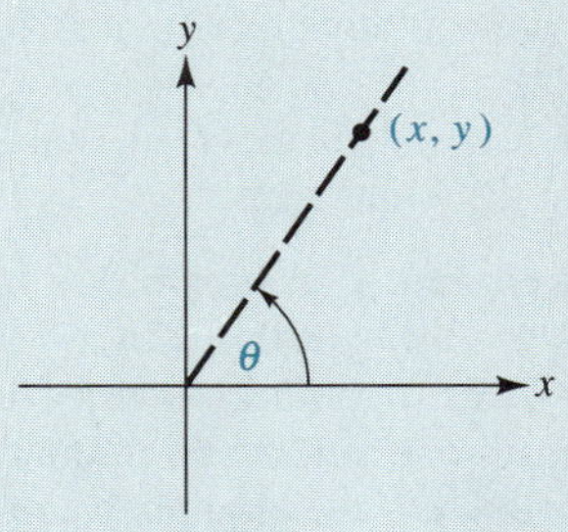

You'll need the double-angle and half-angle identities $\sin 2\theta = 2 \sin\theta \cos\theta$, $\cos 2\theta = \cos^2\theta - \sin^2\theta$,

$$\sin\theta = \pm\sqrt{\frac{1 - \cos 2\theta}{2}}, \text{ and } \cos\theta = \pm\sqrt{\frac{1 + \cos 2\theta}{2}}.$$

By now it seems natural to use inverses of one sort or another to undo operations in solving equations.

Illustration To solve $5x = 7$, we can multiply both sides of the equation by the multiplicative inverse of 5;

$$(\tfrac{1}{5})5x = (\tfrac{1}{5})7 \quad \Leftrightarrow \quad x = \tfrac{7}{5}.$$

To solve $y + 1.8 = 3.2$, we can add the additive inverse of 1.8 to both sides:

$$y + 1.8 + (-1.8) = 3.2 + (-1.8) \quad \Leftrightarrow \quad y = 1.4.$$ □

A system of linear equations of size $n \times n$ can be modeled by a matrix equation, and if a unique solution exists, we can find it by using an inverse matrix. Consider

$$\begin{cases} 3x - y = 6, \\ x + 2y - 4z = -3, \\ 3x + y + 3z = 1. \end{cases}$$

The matrix equation

$$\begin{bmatrix} 3 & -1 & 0 \\ 1 & 2 & -4 \\ 3 & 1 & 3 \end{bmatrix} \begin{bmatrix} x \\ y \\ z \end{bmatrix} = \begin{bmatrix} 6 \\ -3 \\ 1 \end{bmatrix}$$

describes the same relationships between x, y, and z. For example, the entry in the first row, first column of the product on the left side is $3x - y$. This must be the same as the first row, first column entry on the right, 6. In Section 9.3 we found

$$\begin{bmatrix} 3 & -1 & 0 \\ 1 & 2 & -4 \\ 3 & 1 & 3 \end{bmatrix}^{-1} = \begin{bmatrix} \frac{2}{9} & \frac{1}{15} & \frac{4}{45} \\ -\frac{1}{3} & \frac{1}{5} & \frac{12}{45} \\ -\frac{1}{9} & -\frac{2}{15} & \frac{7}{45} \end{bmatrix}.$$

If we multiply both sides of the matrix equation by this inverse, we "undo" the matrix multiplication.

$$\begin{bmatrix} \frac{2}{9} & \frac{1}{15} & \frac{4}{45} \\ -\frac{1}{3} & -\frac{1}{5} & \frac{12}{45} \\ -\frac{1}{9} & -\frac{2}{15} & \frac{7}{45} \end{bmatrix} \begin{bmatrix} 3 & -1 & 0 \\ 1 & 2 & -4 \\ 3 & 1 & 3 \end{bmatrix} \begin{bmatrix} x \\ y \\ z \end{bmatrix} = \begin{bmatrix} \frac{2}{9} & \frac{1}{15} & \frac{4}{45} \\ -\frac{1}{3} & \frac{1}{5} & \frac{12}{45} \\ -\frac{1}{9} & -\frac{2}{15} & \frac{7}{45} \end{bmatrix} \begin{bmatrix} 6 \\ -3 \\ 1 \end{bmatrix}$$

$$\Leftrightarrow \quad \begin{bmatrix} 1 & 0 & 0 \\ 0 & 1 & 0 \\ 0 & 0 & 1 \end{bmatrix} \begin{bmatrix} x \\ y \\ z \end{bmatrix} = \begin{bmatrix} \frac{55}{45} \\ -\frac{105}{45} \\ -\frac{5}{45} \end{bmatrix}$$

$$\Leftrightarrow \quad \begin{bmatrix} x \\ y \\ z \end{bmatrix} = \begin{bmatrix} \frac{11}{9} \\ -\frac{7}{3} \\ -\frac{1}{9} \end{bmatrix}$$

The solution to the original linear system is $(\frac{11}{9}, -\frac{7}{3}, -\frac{1}{9})$.

For linear systems of size $n \times n$ this technique of solving the related matrix equation is less efficient than using Gauss-Jordan elimination. However, if you need to solve several systems with the same coefficients, it is more efficient than the Gauss-Jordan method.

Illustration Each day the head baker at Longbread's Bakery decides how many loaves of bread, dozens of rolls, and dozens of cookies to make based on how much flour, sugar, and yeast she has and the following information. A loaf of bread requires 3 cups of flour, $\frac{1}{2}$ cup of sugar, and 2 yeast cakes; a dozen rolls require 3 cups of flour, $\frac{1}{4}$ cup of sugar, and 2 yeast cakes; a dozen cookies take $2\frac{1}{2}$ cups of flour and 1 cup of sugar. The baker solves

$$\begin{cases} 3b + 3r + 2.5c = f \\ 0.5b + 0.25r + c = s \\ 2b + 2r = y \end{cases}$$

each day; b is the number of loaves of bread, r is the number of dozens of rolls, c is the number of dozens of cookies, f is the number of cups of flour, s is the number of cups of sugar, and y is the number of yeast cakes. She solves the related matrix equation because she needs to compute the inverse of the coefficient matrix only once, and then she multiplies the inverse matrix by the matrix of constants each day.

If B is

$$\begin{bmatrix} 3 & 3 & 2.5 \\ 0.5 & 0.25 & 1 \\ 2 & 2 & 0 \end{bmatrix},$$

then the baker needs to solve

$$B\begin{bmatrix} b \\ r \\ c \end{bmatrix} = \begin{bmatrix} f \\ s \\ y \end{bmatrix}.$$

$$\begin{aligned} B\begin{bmatrix} b \\ r \\ c \end{bmatrix} = \begin{bmatrix} f \\ s \\ y \end{bmatrix} \quad &\Leftrightarrow \quad B^{-1}B\begin{bmatrix} b \\ r \\ c \end{bmatrix} = B^{-1}\begin{bmatrix} f \\ s \\ y \end{bmatrix} \\ &\Leftrightarrow \quad \begin{bmatrix} b \\ r \\ c \end{bmatrix} = B^{-1}\begin{bmatrix} f \\ s \\ y \end{bmatrix} \end{aligned}$$

□

The matrix equation technique has two disadvantages. It can handle only systems that are $n \times n$; if it fails, we don't know if there are no solutions at all or an infinite number of them.

The definition of matrix multiplication was made to allow matrices to be used to model many situations. Here's another example of its use.

EXAMPLE At DeKalb College a course grade of A counts as 4 quality points per quarter credit hour, a B as 3 points, a C as 2 points, a D as 1 point, and an F as 0 points. Takeshi has taken a 5-hour English class, a 3-hour music class, a 5-hour calculus class, and a 2-hour P.E. class this quarter. His grades were B in English, A in music, A in calculus, and C in P.E. Define a quality point matrix Q to represent his grades and a credit hour matrix H so that $\frac{1}{15}QH$ represents Takeshi's grade point average for this quarter. What is his GPA this quarter?

For QH to be defined, Q must have as many columns as H has rows. Let Q be

English	Music	Calculus	P.E.
[3	4	4	2]

The product QH is meaningful only if the row names for H match the column names for Q, so H must be

$$\begin{bmatrix} 5 \\ 3 \\ 5 \\ 2 \end{bmatrix} \begin{matrix} \text{English} \\ \text{Music} \\ \text{Calculus} \\ \text{P.E.} \end{matrix}$$

$$\begin{aligned} \frac{1}{15}QH &= \frac{1}{15}[3 \quad 4 \quad 4 \quad 2]\begin{bmatrix} 5 \\ 3 \\ 5 \\ 2 \end{bmatrix} \\ \Leftrightarrow \quad \frac{1}{15}QH &= \begin{bmatrix} \frac{51}{15} \end{bmatrix} \end{aligned}$$

Takeshi's GPA for this quarter is 3.4. ■

As another type of application of matrices, suppose a customer at West-Southeast Air wants to know if she can fly from Calvin to Acia. Consulting the flight matrix, F, in Section 9.3, you tell her she can fly from Calvin to Bayville to Acia.

$$F: \quad \begin{array}{c} \\ A \\ B \\ C \\ D \\ E \end{array} \begin{array}{c} \begin{array}{ccccc} A & B & C & D & E \end{array} \\ \begin{bmatrix} 0 & 1 & 0 & 0 & 1 \\ 1 & 0 & 1 & 1 & 0 \\ 0 & 1 & 0 & 1 & 0 \\ 0 & 1 & 1 & 0 & 0 \\ 1 & 0 & 0 & 0 & 0 \end{bmatrix} \end{array}$$

This trip is possible because there is a 1 in the second position of the Calvin row and a 1 in the second position of the Acia column. Can she fly from Calvin to Eden?

Definition A collection of points (vertices) and links (edges) like our route map is called a **graph.**

This is a different use of the word graph than the graph of a function. We can use a matrix to represent a graph by naming a row and column for each vertex and putting a 1 in the ij^{th} position if there is an edge between the ith and j^{th} vertices and a zero if there is not. Such a representation is called the *incidence matrix* of the graph.

Definition A matrix whose entries are zeros and ones is called a **Boolean matrix.**

A new operation for Boolean matrices gives us a tool for answering customers' route questions at West-Southeast Air.

Definition The *Boolean product* of two Boolean matrices is defined as follows:

$$[a_{ij}]_{m \times n} \odot [b_{ij}]_{n \times k} = [c_{ij}]_{m \times k}, \text{ where}$$

$$c_{ij} = \begin{cases} 1 & \text{if } a_{i1}b_{1j} + a_{i2}b_{2j} + \cdots + a_{in}b_{nj} \geq 1, \\ 0 & \text{otherwise.} \end{cases}$$

That is, the Boolean product has a 1 at the ij^{th} spot if there was at least one position k where there was a 1 in both the ith row of $[a_{ij}]$ and the jth column of $[b_{ij}]$. This is just what you needed to check to see if the customer could fly from Calvin to Acia. The entries of $F \odot F$ represent one-stop trips from one city to another.

$$F \odot F = \begin{bmatrix} 1 & 0 & 1 & 1 & 0 \\ 0 & 1 & 1 & 1 & 1 \\ 1 & 1 & 1 & 1 & 0 \\ 1 & 1 & 1 & 1 & 0 \\ 0 & 1 & 0 & 0 & 1 \end{bmatrix}$$

You can tell the customer now that there is no one-stop trip possible from Calvin to Eden. Is there a two-stop trip for her?

We can also use matrices to help answer a question from Section 2.5: is the graph of $y = 1/x$ a hyperbola? We'll show it is a variation of the graph of $x^2 - y^2 = 1$.

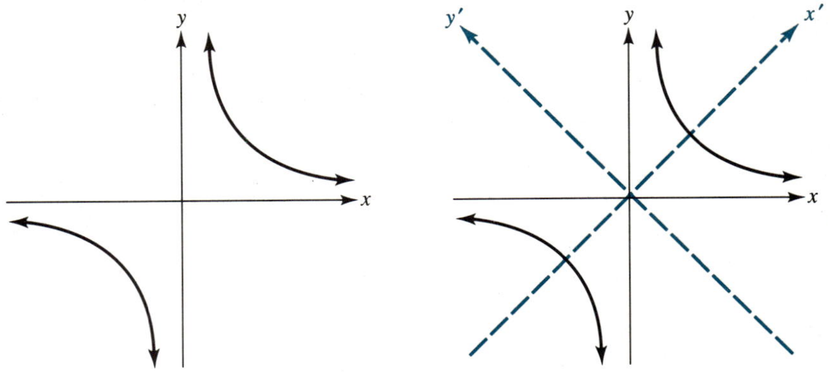

FIGURE 9.4a

In Figure 9.4a it looks as if a new coordinate system with x'- and y'-axes that are the result of rotating the x- and y-axes $\pi/4$ radians counterclockwise would make the graph look like that of $x'^2 - y'^2 = 1$. We need a relationship between the coordinates of a point in the $x'y'$-system and its coordinates in the xy-system. A point on the x'-axis, $(x', 0)$, has coordinates $(x'\cos \pi/4,\ x'\sin \pi/4)$ in the xy-system. A point $(0, y')$ on the y'-axis has coordinates $(y'\cos(\pi/4 + \pi/2),\ y'\sin(\pi/4 + \pi/2))$, or $(-y'\sin \pi/4,\ y'\cos \pi/4)$. Hence a point labeled (x', y') in the $x'y'$-system is

$$\left(x'\cos\frac{\pi}{4} - y'\sin\frac{\pi}{4},\ x'\sin\frac{\pi}{4} + y'\cos\frac{\pi}{4}\right)$$

in the xy-system. We can write this compactly in a matrix equation

$$\begin{bmatrix} x \\ y \end{bmatrix} = \begin{bmatrix} \cos \pi/4 & -\sin \pi/4 \\ \sin \pi/4 & \cos \pi/4 \end{bmatrix} \begin{bmatrix} x' \\ y' \end{bmatrix}.$$

If we translate the relationship $y = \frac{1}{x}$, or $xy = 1$, into $x'y'$-coordinates, we have

$$\begin{aligned}
& xy = \left(x'\cos\frac{\pi}{4} - y'\sin\frac{\pi}{4}\right)\left(x'\sin\frac{\pi}{4} + y'\cos\frac{\pi}{4}\right) \\
\Leftrightarrow\quad & xy = \left(x'\right)^2 \cos\frac{\pi}{4}\sin\frac{\pi}{4} + x'y'\left(\cos\frac{\pi}{4}\right)^2 - x'y'\left(\sin\frac{\pi}{4}\right)^2 \\
& \qquad - (y')^2 \sin\frac{\pi}{4}\cos\frac{\pi}{4} \\
\Leftrightarrow\quad & xy = (x')^2\left(\frac{1}{\sqrt{2}}\right)^2 + x'y'\left(\frac{1}{\sqrt{2}}\right)^2 - x'y'\left(\frac{1}{\sqrt{2}}\right)^2 - (y')^2\left(\frac{1}{\sqrt{2}}\right)^2 \\
\Leftrightarrow\quad & xy = \frac{(x')^2}{2} - \frac{(y')^2}{2} \\
\Leftrightarrow\quad & 1 = \frac{(x')^2}{2} - \frac{(y')^2}{2}.
\end{aligned}$$

This relationship does determine a hyperbola that differs from $x'^2 - y'^2 = 1$ by an expansion factor, so the graph of $y = 1/x$ is a hyperbola.

This last example introduces the idea of another variation of a graph, namely, the rotation by θ radians. The angle θ in the example was $\pi/4$ radians; it was chosen by inspecting the graph and guessing. It's certainly possible to do rotations with other angles, and we'll look at a method to determine exactly the value of the angle needed to describe any parabola, ellipse, or hyperbola as a rotation, reflection, expansion, or translation of a standard graph. Expansions, reflections, and translations can also be described using matrices. For example, a reflection in the x-axis can be seen as the result of creating a new $x'y'$-coordinate system, where

$$\begin{bmatrix} x' \\ y' \end{bmatrix} = \begin{bmatrix} 1 & 0 \\ 0 & -1 \end{bmatrix} \begin{bmatrix} x \\ y \end{bmatrix}.$$

Writing rotations, reflections, or expansions as matrix statements gives an efficient way to generate the algebraic description for the variation of a function or relation. It also provides a convenient method for reversing the variation by using an inverse matrix.

In Section 2.5, we showed that any expansion, reflection, or translation of a basic conic graph can be described by an equation of the form

$$\mathcal{A}x^2 + \mathcal{B}x + \mathcal{C}y^2 + \mathcal{D}y + \mathcal{E} = 0, \tag{1}$$

and vice versa. In this section we saw that a rotation of axes transformed $y = 1/x$ (or $xy = 1$) into an equation of form (1). In general, any equation of the form

$$Ax^2 + Bxy + Cy^2 + Dx + Ey + F = 0 \tag{2}$$

describes a conic section, possibly degenerate, and vice versa. The graph of a conic section is a reflection, rotation, expansion, or translation of the Standard Graphs (1)–(5). We need to show that for an equation of form (2) with $B \neq 0$, it is possible to choose an angle θ, $0 < \theta < \pi/2$, so that rotating the xy-axes through θ gives a new set of axes $x'y'$ and expressing the equation in terms of x' and y' produces an equation of form (1). Roughly speaking, expressing the equation in terms of new coordinates removes the xy-term.

We'll mimic what we did for $y = 1/x$, using θ instead of $\pi/4$. The equation

$$\begin{bmatrix} x \\ y \end{bmatrix} = \begin{bmatrix} \cos\theta & -\sin\theta \\ \sin\theta & \cos\theta \end{bmatrix} \begin{bmatrix} x' \\ y' \end{bmatrix}$$

shows the relationship between old and new coordinates.

$$Ax^2 + Bxy + Cy^2 + Dx + Ey + F = 0, \qquad B \neq 0$$

$$\Leftrightarrow \quad A(x'\cos\theta - y'\sin\theta)^2 + B(x'\cos\theta - y'\sin\theta)(x'\sin\theta + y'\cos\theta) + C(x'\sin\theta + y'\cos\theta)^2 + D(x'\cos\theta - y'\sin\theta) + E(x'\sin\theta + y'\cos\theta) + F = 0, \qquad B \neq 0.$$

Renaming coordinates

We want to choose θ so the $x'y'$-terms vanish. Let's pick out the coefficients of the $x'y'$-terms; they can be produced only by the first three terms. The sum of these coefficients must be 0; we need to solve

$$-2A\cos\theta\sin\theta + B(\cos^2\theta - \sin^2\theta) + 2C\sin\theta\cos\theta = 0, \qquad B \neq 0.$$

Parts of these terms look like parts of the double-angle formula.

$$
\begin{aligned}
& -2A\cos\theta\sin\theta + B(\cos^2\theta - \sin^2\theta) + 2C\sin\theta\cos\theta = 0, && B \neq 0 \\
\Leftrightarrow\quad & -A\sin 2\theta + B\cos 2\theta + C\sin 2\theta = 0, && B \neq 0 \\
\Leftrightarrow\quad & B\cos 2\theta + (C - A)\sin 2\theta = 0, && B \neq 0
\end{aligned}
$$

We wish to choose θ to be a first-quadrant number; this means $\sin 2\theta$ cannot be 0 for our choice of θ.

$$
\begin{aligned}
& B\cos 2\theta + (C - A)\sin 2\theta = 0, \quad B \neq 0 \\
\Leftrightarrow\quad & \frac{\cos 2\theta}{\sin 2\theta} = \frac{A - C}{B} \\
\Leftrightarrow\quad & \cot 2\theta = \frac{A - C}{B} \\
\Leftrightarrow\quad & \theta = \frac{1}{2}\operatorname{Cot}^{-1}\left(\frac{A - C}{B}\right).
\end{aligned}
$$

The last equation gives us a tool for choosing θ.

EXAMPLE Find the angle of rotation needed to remove the xy-term in $2x^2 + \sqrt{3}xy + y^2 + 2x - 4 = 0$.

Using our general solution,

$$
\begin{aligned}
& \theta = \frac{1}{2}\operatorname{Cot}^{-1}\frac{2 - 1}{\sqrt{3}} \\
\Leftrightarrow\quad & \theta = \frac{1}{2}\operatorname{Cot}^{-1}\frac{1}{\sqrt{3}} \\
\Leftrightarrow\quad & \theta = \left(\frac{1}{2}\right)\left(\frac{\pi}{3}\right).
\end{aligned}
$$

The angle of rotation needed is $\pi/6$. ■

If we wish to rewrite $Ax^2 + Bxy + Cy^2 + Dx + Ey + F = 0$ in the new coordinates in order to graph it or determine what kind of conic it is, we need to complete the expansion of $A(x'\cos\theta - y'\sin\theta)^2 + B(x'\cos\theta - y'\sin\theta)(x'\sin\theta + y'\cos\theta) + C(x'\sin\theta + y'\cos\theta)^2 + D(x'\cos\theta - y'\sin\theta) + E(x'\sin\theta + y'\cos\theta) + F = 0$ and to compute values for $\sin\theta$ and $\cos\theta$. We can do both jobs more efficiently by tackling the general case once more. In the same way that we pulled out the coefficients of $x'y'$ to develop the equation for θ, we can identify the coefficients of the remaining terms and use these as conversion formulas.

When $A(x'\cos\theta - y'\sin\theta)^2 + B(x'\cos\theta - y'\sin\theta)(x'\sin\theta + y'\cos\theta) + C(x'\sin\theta + y'\cos\theta)^2 + D(x'\cos\theta - y'\sin\theta) + E(x'\sin\theta + y'\cos\theta) + F = 0$ is written as $\mathcal{A}(x')^2 + \mathcal{B}x' + \mathcal{C}(y')^2 + \mathcal{D}y' + \mathcal{E} = 0$,

$$
\begin{aligned}
\mathcal{A} &= A\cos^2\theta + B\sin\theta\cos\theta + C\sin^2\theta \\
\mathcal{B} &= D\cos\theta + E\sin\theta \\
\mathcal{C} &= A\sin^2\theta - B\sin\theta\cos\theta + C\cos^2\theta
\end{aligned}
$$

$$\mathcal{D} = -D\sin\theta + E\cos\theta$$
$$\mathcal{E} = F$$

To compute $\sin\theta$ and $\cos\theta$, we use the half-angle formulas

$$\sin\theta = \pm\sqrt{\frac{1-\cos 2\theta}{2}} \quad \text{and} \quad \cos\theta = \pm\sqrt{\frac{1+\cos 2\theta}{2}}.$$

Remember $\cot 2\theta = \dfrac{(A-C)}{B}$; sketch 2θ in standard position to determine $\cos 2\theta$.

EXAMPLE Determine the angle of rotation needed to remove the xy-term in $19x^2 + 16xy + \frac{43}{3}y^2 = 25$ and rewrite the equation in terms of the new coordinates. Graph this conic section.

First we see that $\cot 2\theta = \dfrac{(19 - \frac{43}{3})}{16}$, or $\frac{7}{24}$, and $\theta = \frac{1}{2}\,\mathbf{Cot^{-1}(\tfrac{7}{24})}$.
To find $\sin\theta$ and $\cos\theta$, draw 2θ in standard position.

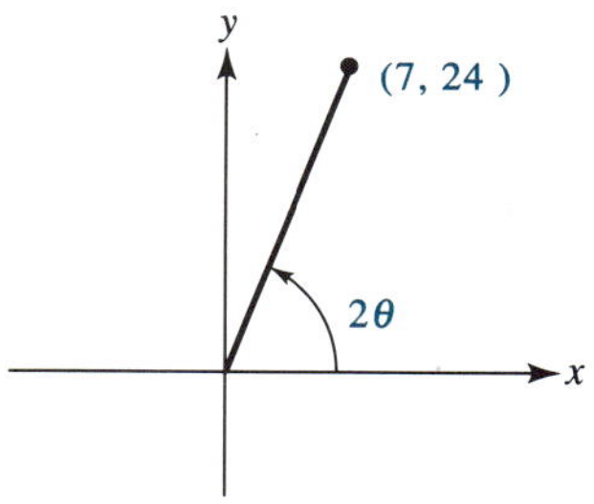

Since $x = r\cos 2\theta$, $7 = 25\cos 2\theta$, $\cos 2\theta = \frac{7}{25}$,

$$\sin\theta = \sqrt{\frac{1-\frac{7}{25}}{2}} \quad \text{and} \quad \cos\theta = \sqrt{\frac{1+\frac{7}{25}}{2}}.$$

Using $A = 19$, $B = 16$, $C = \frac{43}{3}$, $D = 0$, $E = 0$, $F = -25$, and the conversions, we have:

$$\mathcal{A} = 19\left(\frac{16}{25}\right) + 16\left(\frac{12}{25}\right) + \left(\frac{43}{3}\right)\left(\frac{9}{25}\right) = \frac{625}{25}, \text{ or } 25,$$
$$\mathcal{B} = 0,$$
$$\mathcal{C} = 19\left(\frac{9}{25}\right) - 16\left(\frac{12}{25}\right) + \left(\frac{43}{3}\right)\left(\frac{16}{25}\right) = \frac{25}{3},$$
$$\mathcal{D} = 0,$$
$$\mathcal{E} = -25,$$

$$19x^2 + 16xy + 43y^2 - 25 = 0$$
$$\Leftrightarrow \quad 25(x')^2 + \frac{25}{3}(y')^2 - 25 = 0$$

$$\Leftrightarrow \qquad (x')^2 + \frac{(y')^2}{3} = 1.$$

We see the equation defines an ellipse.

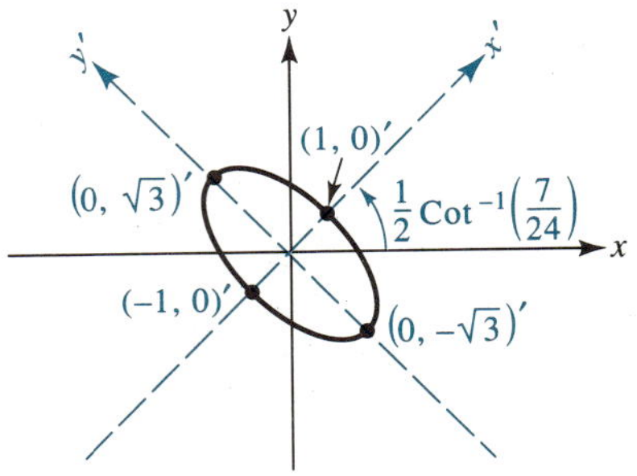

■

Certain features of the graphs of nondegenerate conic sections remain unchanged if the graphs are translated, reflected, or rotated. For example, these actions will not change the radius of the circle defined by $x^2 + y^2 = 16$. We say the radius is *invariant* under rotation, reflection, and translation. What features are invariant for the other conic sections? Roughly speaking, the shape remains unchanged, so any measurement related to the shape is invariant. The case of a parabola requires a new definition. Remember the geometric definition of a parabola as the set of points equidistant from a given point and a given line. If we use the point $(0, p)$ and the line $y = -p$, then the graph has equation $y = x^2/4p$ (see section 2.4).

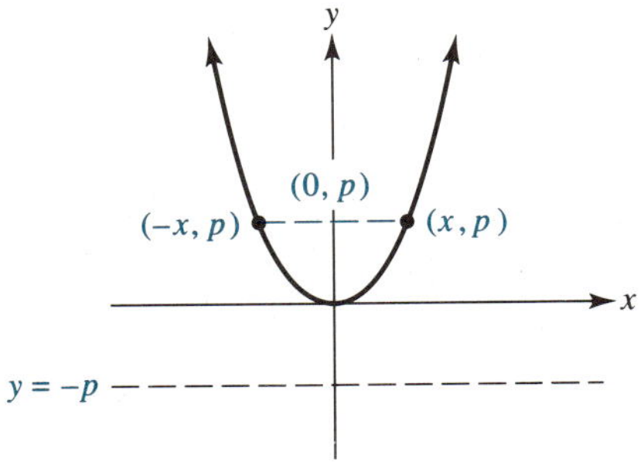

We can find the length of the horizontal line segment through $(0, p)$ with endpoints on the parabola to be $4p$ by solving $p = x^2/4p$ for x. This length is an invariant of the graph under translation, rotation, and reflection. For the conics, we have the following.

Conic Section		*An Invariant Measurement*
Circle		Radius

Conic Section		*An Invariant Measurement*
Ellipse		Maximum chord length
Hyperbola		Minimum distance between arms
Parabola	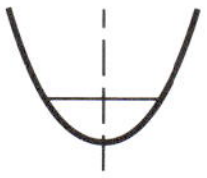	"Width" at focus

This is not a complete catalog of invariant measurements for these objects, but it gives a measurement for each nondegenerate conic that you can obtain by writing the equation as a translation, rotation, or reflection of a basic conic.

Illustration After rewriting $19x^2 + 16xy + \frac{43}{3}y^2 - 25 = 0$ by rotating the axes, we obtained a new equation $(x')^2 + (y')^2/3 = 1$, which defines an ellipse whose maximum length is measured on the y'-axis. The maximum length is $2\sqrt{3}$ units. □

EXAMPLE Rewrite $3x^2 - 2\sqrt{3}xy + y^2 - 12x - 12\sqrt{3}y = 0$ to remove the xy-term and find the width at the focus of the parabola.

$$\cot 2\theta = \frac{3-1}{-2\sqrt{3}} = -\frac{1}{\sqrt{3}} \quad \Rightarrow \quad 2\theta = \frac{2\pi}{3} \quad \Leftrightarrow \quad \theta = \frac{\pi}{3}$$

Recall that $\cos(\pi/3) = \frac{1}{2}$, and $\sin(\pi/3) = \sqrt{3}/2$. Using the conversions

$$\mathcal{A} = 3\left(\frac{1}{4}\right) - 2\sqrt{3}\left(\frac{\sqrt{3}}{4}\right) + 1\left(\frac{3}{4}\right),$$

$$\mathcal{B} = -12\left(\frac{1}{2}\right) - 12\sqrt{3}\left(\frac{\sqrt{3}}{2}\right),$$

$$\mathcal{C} = 3\left(\frac{3}{4}\right) + 2\sqrt{3}\left(\frac{\sqrt{3}}{4}\right) + 1\left(\frac{1}{4}\right),$$

$$\mathcal{D} = 12\left(\frac{\sqrt{3}}{2}\right) - 12\sqrt{3}\left(\frac{1}{2}\right),$$

the $x'y'$-equation is $-24x' + 4(y')^2 = 0$.

$$-24x' + 4(y')^2 = 0 \quad \Leftrightarrow \quad x' = \frac{(y')^2}{6}.$$

The width at the focus is 6 units. ■

EXERCISES 9.4

In exercises 1 through 6, write the matrix equation that corresponds to the given system of equations.

1. $\begin{cases} 5x - 6y = 18 \\ 3x + 2y = 9 \end{cases}$

2. $\begin{cases} 4.3x + 2.1y = 13.0 \\ -7.4y = -12.8 \end{cases}$

3. $\begin{cases} 0.4x - 0.3y + 0.2z = 1.5 \\ x - z = -3 \quad = -3 \\ 0.7y - z = 10 \end{cases}$

4. $\begin{cases} 5x + 13y = 104 \\ x + 20y + 30z = 217 \\ 17x + 2y + 15z = 310 \end{cases}$

5. $\begin{cases} 7x - 3y = 11 \\ 14z - 3x = 11 + y \\ 13y = 4z - 8 \end{cases}$

6. $\begin{cases} 7x - 3y = 11 \\ 3y - z = x + 4 \\ 12 - x = 2y + 5z \end{cases}$

In exercises 7 through 12, give the linear system corresponding to the matrix equation and also give its solution set.

7. $\begin{bmatrix} 4 & 3 \\ 2 & 1 \end{bmatrix} \begin{bmatrix} x \\ y \end{bmatrix} = \begin{bmatrix} -1 \\ 3 \end{bmatrix}$

8. $\begin{bmatrix} 4 & 3 \\ 2 & 1 \end{bmatrix} \begin{bmatrix} x \\ y \end{bmatrix} = \begin{bmatrix} 7 \\ 4 \end{bmatrix}$

9. $\begin{bmatrix} 1 & 0 & -1 \\ 2 & 3 & 5 \\ -1 & -1 & 0 \end{bmatrix} \begin{bmatrix} x \\ y \\ z \end{bmatrix} = \begin{bmatrix} -1 \\ 1 \\ -2 \end{bmatrix}$

10. $\begin{bmatrix} 2 & -1 & 0 \\ 0 & 1 & 6 \\ -1 & -1 & 2 \end{bmatrix} \begin{bmatrix} x \\ y \\ z \end{bmatrix} = \begin{bmatrix} 12 \\ -3 \\ 4 \end{bmatrix}$

11. $\begin{bmatrix} 1 & 1 & 1 \\ 2 & 3 & 0.5 \\ 1 & -1 & -1 \end{bmatrix} \begin{bmatrix} x \\ y \\ z \end{bmatrix} = \begin{bmatrix} 11 \\ -7 \\ 10 \end{bmatrix}$

12. $\begin{bmatrix} 2 & 1 & -5 \\ -3 & 3 & 10 \\ 1 & 0 & 1 \end{bmatrix} \begin{bmatrix} x \\ y \\ z \end{bmatrix} = \begin{bmatrix} 3 \\ -6 \\ 6.2 \end{bmatrix}$

In exercises 13 through 16, write the solution to each matrix equation $CX = K$ given:

13. $C^{-1} = \begin{bmatrix} 2 & 4 \\ 3 & 7 \end{bmatrix}$ and

 a. $K = \begin{bmatrix} -1 \\ 2 \end{bmatrix}$ **b.** $K = \begin{bmatrix} 3.4 \\ 7.1 \end{bmatrix}$ **c.** $K = \begin{bmatrix} 4 \\ 0 \end{bmatrix}$

14. $C^{-1} = \begin{bmatrix} -1 & 3 \\ 2.5 & 1.5 \end{bmatrix}$ and

 a. $K = \begin{bmatrix} 1.2 \\ -3.2 \end{bmatrix}$ **b.** $K = \begin{bmatrix} \frac{1}{4} \\ \frac{3}{4} \end{bmatrix}$ **c.** $K = \begin{bmatrix} 0.315 \\ 2.435 \end{bmatrix}$

15. $C^{-1} = \begin{bmatrix} 1 & -1 \\ 0 & 1 \end{bmatrix}$ and

 a. $K = \begin{bmatrix} 1.2 \\ -3.2 \end{bmatrix}$ **b.** $K = \begin{bmatrix} \frac{1}{4} \\ \frac{3}{4} \end{bmatrix}$ **c.** $K = \begin{bmatrix} 0.315 \\ 2.435 \end{bmatrix}$

16. $C^{-1} = \begin{bmatrix} \frac{3}{8} & -\frac{1}{8} \\ -\frac{2}{8} & \frac{5}{8} \end{bmatrix}$ and

 a. $K = \begin{bmatrix} \frac{1}{3} \\ \frac{2}{3} \end{bmatrix}$ **b.** $K = \begin{bmatrix} 0.02 \\ -0.01 \end{bmatrix}$ **c.** $K = \begin{bmatrix} 178 \\ 243 \end{bmatrix}$

17. **a.** Write the Boolean matrix that represents this graph.
 b. Use Boolean multiplication to find all possible one-stop trips on this graph. List the trips.

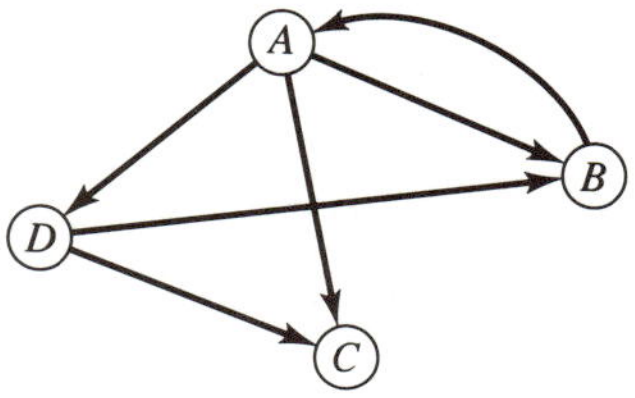

18. Donna is a first-year high school mathematics teacher at Donaldson High School. She is having difficulty making up systems of linear equations that are not too difficult for her students to solve. A veteran teacher gives her the following advice: Make up a system with an obvious solution, such as

$$\begin{cases} x = 2 \\ y = 3 \\ z = 4 \end{cases}$$

Then use the rules for producing an equivalent system to transform this system to one where the solution is not obvious.

$$\begin{cases} x = 2 \\ y = 3 \\ z = 4 \end{cases}$$

$$\to \begin{cases} 2x + y + z = 11 \\ x - y - z = -5 \\ x + z = 6 \end{cases} \qquad \begin{aligned} Eq_1 &\leftarrow 2Eq_1 + Eq_2 + Eq_3 \\ Eq_2 &\leftarrow Eq_1 - Eq_2 - Eq_3 \\ Eq_3 &\leftarrow Eq_1 + Eq_3 \end{aligned}$$

Determine whether this process works by using a corresponding matrix equation to solve the system produced.

19. Use the procedure described in exercise 18 to produce an equivalent system for

$$\begin{cases} x = 1.2 \\ y = -2.3 \\ z = 3.1 \end{cases}$$

20. Many calculators are capable of performing matrix operations but use decimals exclusively. Explain a problem this might create.

21. When $\begin{bmatrix} 2 & 4 \\ -1 & -2 \end{bmatrix}$ is entered in a calculator and the inverse key is pressed, the message "Infinite Result" is obtained. Explain what has happened.

In exercises 22 through 25, determine C and K for the matrix equation $CX = K$ when

22. $C^{-1} = \begin{bmatrix} 1 & 5 \\ -3 & 9 \end{bmatrix}; \quad X = \begin{bmatrix} 4 \\ 2 \end{bmatrix}$

23. $C^{-1} = \begin{bmatrix} 3 & -6 \\ 1 & 9 \end{bmatrix}; \quad X = \begin{bmatrix} 4 \\ 2 \end{bmatrix}$

24. $C^{-1} = \begin{bmatrix} 2 & -1 & 0 \\ 0 & 1 & 6 \\ -1 & -1 & 2 \end{bmatrix}; \quad X = \begin{bmatrix} 1 \\ 4 \\ 8 \end{bmatrix}$

25. $C^{-1} = \begin{bmatrix} 2 & 1 & -5 \\ -3 & 3 & 10 \\ 1 & 0 & 1 \end{bmatrix}; \quad X = \begin{bmatrix} 2.3 \\ -4.8 \\ 1.1 \end{bmatrix}$

26. You are using your calculator to compute a matrix inverse in order to solve a system of linear equations. What do you do if the calculator will not produce an inverse?

In exercises 27 through 30, determine the angle of rotation needed to remove the xy-term in each equation. Rewrite each equation in terms of the new coordinates. Graph the conic section.

27. $2x^2 + \sqrt{3}xy + y^2 = 1$

28. $x^2 + 5xy + y^2 = 1$

29. $5x^2 + 10xy + 5y^2 + x - y = 0$

30. $73x^2 - 72xy + 52y^2 = 100$

In exercises 31 through 34, determine the length of the appropriate invariant measurement (radius, maximum length, minimum distance between arms, or width at focus) for each conic.

31. $2x^2 + \sqrt{3}xy + y^2 = 1$

32. $x^2 + 5xy + y^2 = 1$

33. $5x^2 + 10xy + 5y^2 + x - y = 0$

34. $73x^2 - 72xy + 52y^2 = 100$

9.5 Nonlinear Systems and Inequalities

Toolkit

To produce an equivalent nonlinear system, we apply the same rules as for a linear system.

Substitution is a common method for solving nonlinear systems; not all our other techniques for solving linear systems are adaptable to the nonlinear case.

EXAMPLE Solve

$$\begin{cases} x^2 + y^2 = 4, \\ xy = 2. \end{cases}$$

Since $xy = 2 \Leftrightarrow x = 2/y$, we can substitute $2/y$ for x in the first equation.

$$\begin{cases} x^2 + y^2 = 4 \\ x = \dfrac{2}{y} \end{cases} \Leftrightarrow \begin{cases} \left(\dfrac{2}{y}\right)^2 + y^2 = 4 \\ x = \dfrac{2}{y} \end{cases}$$

$$\Leftrightarrow \begin{cases} \dfrac{4}{y^2} + y^2 = 4 \\ x = \dfrac{2}{y} \end{cases} \Leftrightarrow \begin{cases} 4 + y^4 = 4y^2, \quad y \neq 0 \\ x = \dfrac{2}{y} \end{cases}$$

$$\Leftrightarrow \begin{cases} y^4 - 4y^2 + 4 = 0, \quad y \neq 0 \\ x = \dfrac{2}{y} \end{cases}$$

$$\Leftrightarrow \begin{cases} (y^2 - 2)^2 = 0, \quad y \neq 0 \\ x = \dfrac{2}{y} \end{cases}$$

Solutions for the first equation are $-\sqrt{2}$ and $\sqrt{2}$.

$$\{(\sqrt{2}, \sqrt{2}), (-\sqrt{2}, -\sqrt{2})\}.$$

■

If the equations in the system have familiar graphs, a sketch of these graphs can be useful. We can obtain an idea of the number of solutions and their locations from the graphs.

Illustration

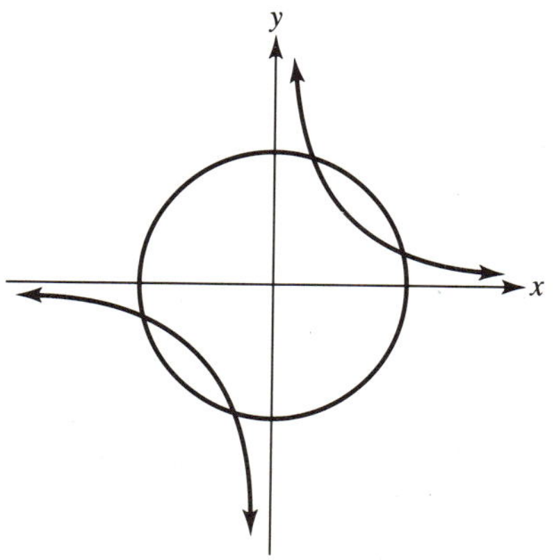

The system

$$\begin{cases} x^2 + y^2 = 16 \\ y = 6/x \end{cases}$$

has four solutions.

□

Occasionally the graphs may yield the exact solutions. In some cases graphical approximation may be the best we can do. Consider

$$\begin{cases} y = 2\cos x, \\ y = e^x. \end{cases}$$

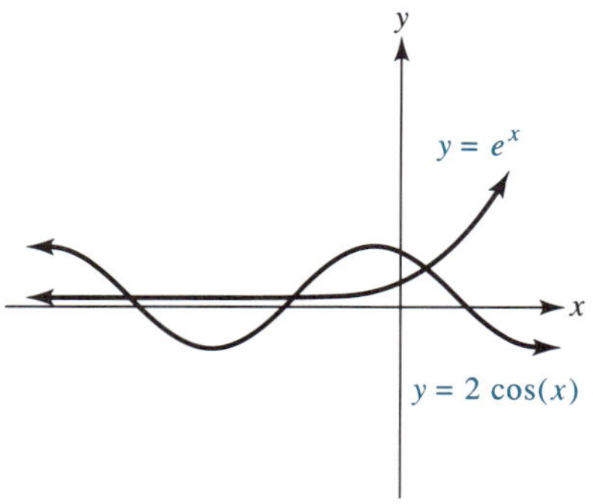

A sketch of the graphs suggests that there are infinitely many distinct solutions with exactly one in the first quadrant, but we lack methods for finding exact solutions. Some graphing calculators will give excellent approximations.

We saw in Section 9.2 that nonlinear systems are sometimes linear in form and can be solved by elimination or one of the other techniques for solving linear systems. It's also possible to use elimination effectively in some cases where the system is not quite linear in form.

EXAMPLE Solve

$$\begin{cases} x^2 + 2y^2 - z^2 = 8, \\ x^2 - y^2 + 3z = 6, \\ y^2 - 5z = -6. \end{cases}$$

This system doesn't appear to be linear in form because each equation doesn't contain the same type of terms. We could solve the third equation for z and apply substitution. Let's try an elimination method instead. First, we pad the equations so the system is linear in form.

$$\begin{cases} x^2 + 2y^2 - z^2 = 8 \\ x^2 - y^2 + 3z = 6 \\ y^2 - 5z = -6 \end{cases}$$

$$\Leftrightarrow \begin{cases} x^2 + 2y^2 - z^2 + 0z = 8 \\ x^2 - y^2 + 0z^2 + 3z = 6 \\ 0x^2 + y^2 + 0z^2 - 5z = -6 \end{cases}$$

We can now use Gauss-Jordan elimination.

$$\left[\begin{array}{cccc|c} 1 & 2 & -1 & 0 & 8 \\ 1 & -1 & 0 & 3 & 6 \\ 0 & 1 & 0 & -5 & -6 \end{array}\right] \to \left[\begin{array}{cccc|c} 1 & 2 & -1 & 0 & 8 \\ 0 & -3 & 1 & 3 & -2 \\ 0 & 1 & 0 & -5 & -6 \end{array}\right]$$

$$\to \left[\begin{array}{cccc|c} 1 & 0 & -1 & 10 & 20 \\ 0 & 1 & 0 & -5 & -6 \\ 0 & 0 & 1 & -12 & -20 \end{array}\right] \to \left[\begin{array}{cccc|c} 1 & 0 & 0 & -2 & 0 \\ 0 & 1 & 0 & -5 & -6 \\ 0 & 0 & 1 & -12 & -20 \end{array}\right]$$

This matrix tells us $z^2 = -20 + 12z$, $y^2 = -6 + 5z$, and $x^2 = 2z$. But $z^2 = -20 + 12z \Leftrightarrow z^2 - 12z + 20 = 0 \Leftrightarrow (z - 2)(z - 10) = 0$. Possible values for z are 2 and 10. Substituting 2 for z in $y^2 = -6 + 5z$ and $x^2 = 2z$, we obtain four solutions: (2, 2, 2), (−2, 2, 2), (2, −2, 2), and (−2, −2, 2). We still have to investigate the situation when $z = 10$. There are four more solutions in this case. The solution set is

$\{(2, 2, 2), (-2, 2, 2), (2, -2, 2), (-2, -2, 2), (2\sqrt{5}, 2\sqrt{11}, 10), (-2\sqrt{5}, 2\sqrt{11}, 10), (2\sqrt{5}, -2\sqrt{11}, 10)\ (-2\sqrt{5}, -2\sqrt{11}, 10)\}$. ■

Techniques for solving nonlinear systems are usually approximation methods. In this section the exercises with nonlinear systems serve mainly to give a review of all our equation-solving methods.

EXAMPLE Approximate, to the nearest thousandth, any solutions of

$$\begin{cases} \ln xy = 6.2, \\ \ln x^2y = 4.1. \end{cases}$$

Using properties of the natural logarithmic function, we rewrite the system to obtain a system that is linear in form.

$$\begin{cases} \ln x + \ln y = 6.2 \\ 2\ln x + \ln y = 4.1 \end{cases}$$

Elimination yields

$$\begin{cases} 2\ln x + \ln y = 4.1, \\ \ln x = -2.1. \end{cases}$$

There is one solution $(e^{-2.1}, e^{8.3})$; approximated to the nearest thousandth, it is (0.122, 4023.872). ■

EXAMPLE Approximate to one decimal place any solutions of $\frac{1}{2}x = \sin(x)$.
We can model this equation with a system

$$\begin{cases} y = \frac{1}{2}x \\ y = \sin(x) \end{cases}$$

and consider the graphs of these functions. It looks as if there are three solutions.

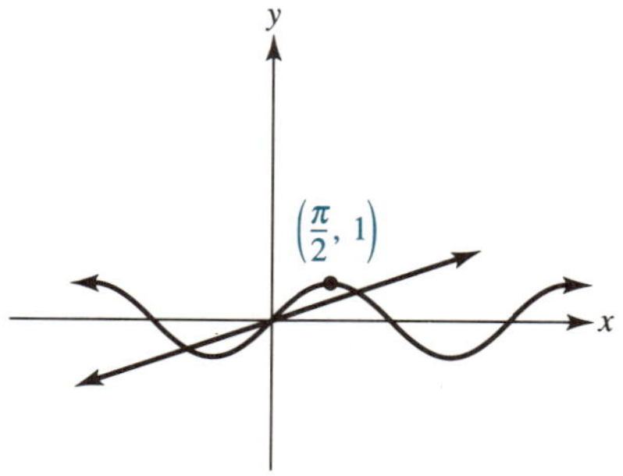

One is (0, 0); $\sin(0) = 0 = \frac{1}{2}(0)$. Another one is near (2, 1). Using a calculator, we approximate $\sin(2)$ and compare it with $\frac{1}{2}(2)$; $\sin(2) \approx 0.909$. This is not quite close enough, so we try again; $\sin(1.9) \approx 0.946$, and $\frac{1}{2}(1.9) = 0.95$. We chose a smaller value of x because this will give a larger value of $\sin(x)$ in this neighborhood. Because the sine function and $y = \frac{1}{2}x$ are both symmetric with respect to the origin, the other solution is approximately -1.9.

The solutions are 0, 1.9 (approx.), and −1.9 (approx.). ■

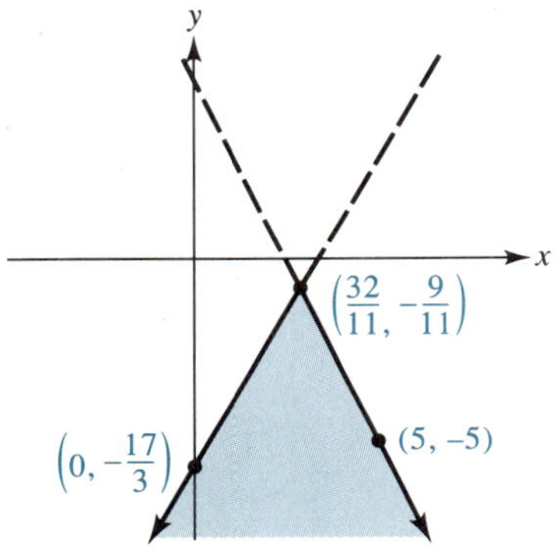

Another variation on linear systems of equations is linear systems of inequalities. For two-variable systems, our general technique of finding boundaries and testing regions adapts easily to these situations. Quite often it's the graph of the solution set that is of interest because the system itself may be the best algebraic description of the solution set. The set solution of

$$\begin{cases} 5x - 3y \geq 17 \\ 2x + y \leq 5 \end{cases}$$

is $\{(x, y) | 5x - 3y \geq 17 \text{ and } 2x + y \leq 5\}$. Remember that a solution for a system is a solution common to all statements in the system; a system solution belongs to the intersection of the solution sets of the individual statements.

The graphs of $5x - 3y = 17$ and $2x + y = 5$ form the boundaries of the solution set. The boundaries are lines. Taking each in turn and testing, we find that points below the line $5x - 3y = 17$ satisfy the first inequality and that points below the line $2x + y = 5$ satisfy the second. The graph of the solution set is the intersection of the graphs of the individual solution sets.

Systems of linear inequalities appear in a very nice application of mathematics called **linear programming.** In actual linear programming problems, hundreds of variables and inequalities may be involved, but we'll use a slightly smaller example.

At Longbread's Bakery the pastry section specializes in large and small sticky buns. The pastry cook can make at most 36 ounces of sweet dough at a time because of the size of the mixer. For each batch of sweet dough she makes, the baker uses a 30-ounce packet of special filling. A large bun uses 4.5 ounces of dough and 3 ounces of filling; a small one uses 2 ounces of dough and 2 ounces of filling. The pastry cook's problem is to decide how many large and small sticky buns to make from each batch of dough. She wants to choose these numbers so that the profit made is the largest possible. At today's prices, the bakery makes 25 cents on each large sticky bun and 15 cents on each small one.

We can model the constraints with which she has to work with inequalities. Let x be the number of large buns and y be the number of small ones per batch.

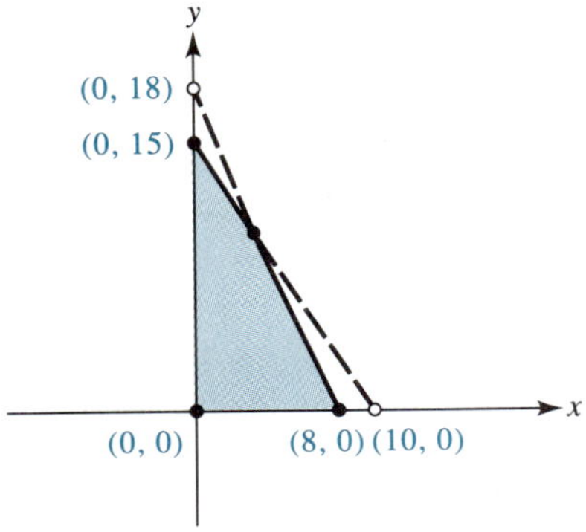

$4.5x + 2y \leq 36$	Dough constraint
$3x + 2y \leq 30$	Filling constraint
$x \geq 0$	
$y \geq 0$	

The profit is $25x + 15y$, measured in cents. A sketch of the solution set for the system of inequalities shows there are infinitely many solutions. Even if the cook must choose whole numbers, there are many solutions. How can she decide if there is a best choice among all of these to maximize profit?

Here's the nice piece of mathematics: The maximum for $25x + 15y$ must occur at one of the corners of the shaded region. We have one corner to identify; it is the intersection of the lines defined by $4.5x + 2y = 36$ and $3x + 2y = 30$.

Corner	(0, 0)	(8, 0)	(4, 9)	(0, 15)
$25x + 15y$	0	200	235	225

The cook should make 4 large sticky buns and 9 small buns from each 36-ounce batch of dough in order to maximize the profit.

This technique can also be used to make a choice to minimize an expression whose variables are subject to linear constraints. The minimum value must also occur at a corner. Consider the following picture. The shaded area is the set of solutions for a system of constraint inequalities.

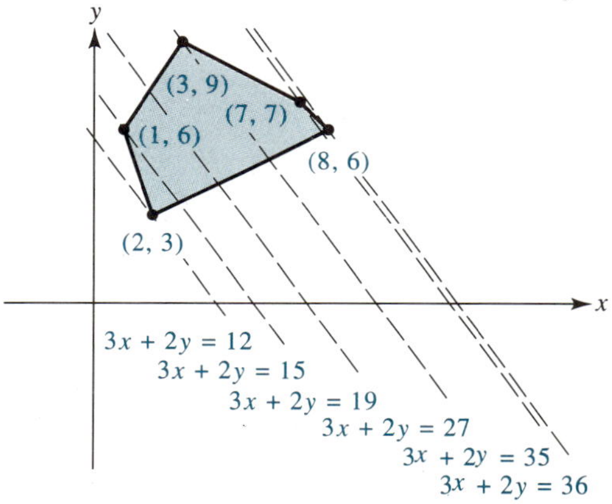

Whether we want a minimum or a maximum value for $4x + 3y$, we see that this happens at a corner.

Our example just gives the flavor of this type of problem; we haven't spelled out all the details necessary for the technique to work. For large-scale problems, there are several very efficient methods of finding and testing corner points that can be carried out on a computer.

EXAMPLE Graph the solution set of

$$\begin{cases} x^2 + y^2 < 4, \\ y < \sqrt{3}x, \\ 3x + y \geq 2. \end{cases}$$

We sketch all the boundaries first and then test regions to find the solution set for each inequality. The boundaries for the first two inequalities should be dashed and that for the third solid, showing whether the boundaries are part of the solution set or not.

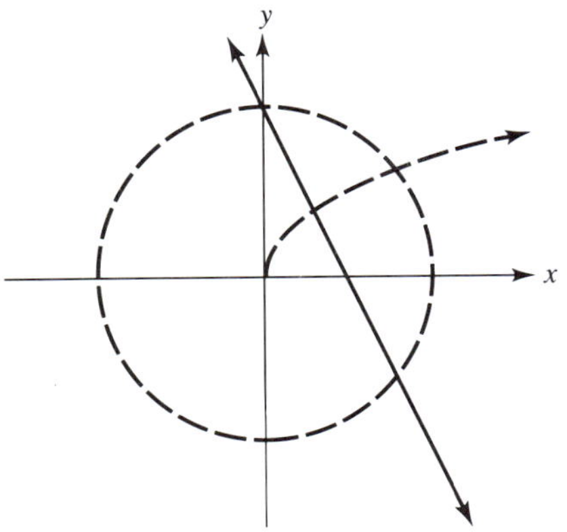

For the first inequality, solutions occur inside the dashed circle. The second inequality restricts solutions to the first and fourth quadrants and below the dashed half-parabola. To satisfy the linear inequality, solutions must lie above (to the right of) the line. The solution set for the system is the intersection of these solution sets. The graph of the solution set is

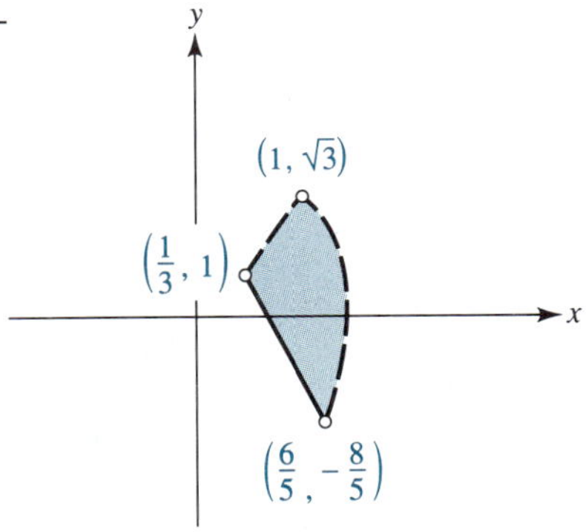

EXAMPLE (Graphing calculator): Graph the solution set of

$$\begin{cases} \sec x \leq y + 1, \\ -\pi/2 < x < \pi/2, \\ y < 3^x. \end{cases}$$

The boundary defined by $\sec x = y + 1$ is the graph of the secant function shifted down one unit.

$$\sec(x) = y + 1 \quad \Leftrightarrow \quad y = \sec x - 1$$

The third inequality has the graph of $y = 3^x$ as its boundary.

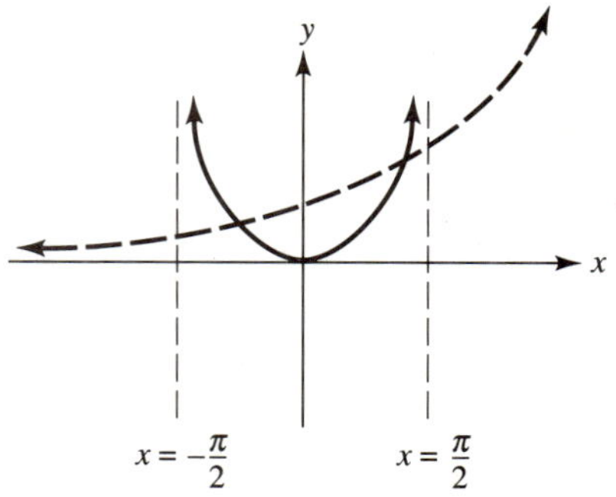

Using the solving routine of the graphing calculator, we approximate the points of intersection. Testing regions to find the solutions for the individual inequalities and forming their intersection, we have the following graph of the solution set.

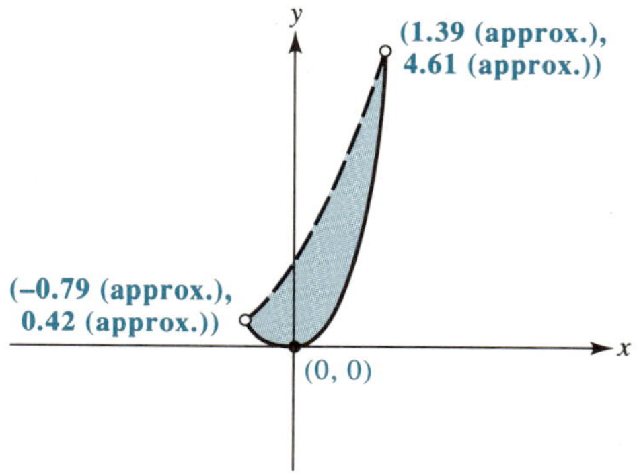

EXERCISES 9.5

In exercises 1 through 12, solve each nonlinear system.

1. $\begin{cases} x^2 + y^2 = 6 \\ y = x^2 \end{cases}$

2. $\begin{cases} x^2 + y^2 = 6 \\ y^2 = x \end{cases}$

3. $\begin{cases} x^2 + y^2 = 3 \\ x + y = 0 \end{cases}$

4. $\begin{cases} x^2 + y^2 = 6 \\ x - y = 0 \end{cases}$

5. $\begin{cases} 2x^3 - 50x = y \\ y = 0 \end{cases}$

6. $\begin{cases} x^3 - 4x + x^2 = y \\ y = 4 \end{cases}$

7. $\begin{cases} y = x^3 \\ y = x(x - 3)(x + 1) \end{cases}$

8. $\begin{cases} y = -x^3 \\ y = x(x - 3)(x + 1) \end{cases}$

9. $\begin{cases} y = 2x^2 - 9 \\ y = \dfrac{2}{x} \end{cases}$

10. $\begin{cases} y = \sqrt{x} \\ y = x - 4 \end{cases}$

11. $\begin{cases} y = \tan x \\ y = 5 \end{cases}$

12. $\begin{cases} y = \ln(x + 2) \\ y = -1 \end{cases}$

In exercises 13 through 24, sketch the graph of the solution set of each system.

13. $\begin{cases} x^2 + y^2 \le 6 \\ y \ge x^2 \end{cases}$

14. $\begin{cases} x^2 + y^2 < 6 \\ y^2 > x \end{cases}$

15. $\begin{cases} x^2 + y^2 \ge 3 \\ x + y > 0 \end{cases}$

16. $\begin{cases} x^2 + y^2 > 6 \\ x - y \le 0 \end{cases}$

17. $\begin{cases} 2x^3 - 50x \ge y \\ y \le 0 \end{cases}$

18. $\begin{cases} x^3 - 4x + x^2 < y \\ y > 4 \end{cases}$

19. $\begin{cases} y > x^3 \\ y \le x(x - 3)(x + 1) \end{cases}$

20. $\begin{cases} y > -x^3 \\ y \le x(x - 3)(x + 1) \end{cases}$

21. $\begin{cases} y < 2^x \\ y > 5 \end{cases}$

22. $\begin{cases} y < \sqrt{x} \\ y \ge x - 4 \end{cases}$

23. $\begin{cases} y \ge \tan x \\ y < 5 \end{cases}$

24. $\begin{cases} y \le \ln(x + 2) \\ y \ge -1 \end{cases}$

In exercises 25 through 30, treat the system as linear in form and use Gauss-Jordan elimination to find the solution set.

25. $\begin{cases} \ln(xy) = 5 \\ \ln(xy^2) = 4 \end{cases}$

26. $\begin{cases} \ln(xy) = 5 \\ \ln(x^2y^2) = 4 \end{cases}$

27. $\begin{cases} x^2 + 2y^2 - 2z^2 - 10x = -8 \\ x^2 + y^2 - 7x = -3 \\ x^2 + z^2 - 4x = -1 \end{cases}$

28. $\begin{cases} x^2 + y^2 + 2x = -3 \\ y^2 - 3x = 1 \\ x^2 + z^2 + 5x = 3 \end{cases}$

29. $\begin{cases} \frac{1}{x} + \frac{1}{y} + \frac{1}{z} = 4 \\ \frac{3}{x} - \frac{1}{z} = 2 \\ \frac{2}{y} + \frac{3}{z} = 8 \end{cases}$

30. $\begin{cases} \frac{1}{x} + \frac{1}{y} + \frac{1}{z} = 3 \\ \frac{3}{x} - \frac{1}{z} = -2 \\ \frac{2}{y} + \frac{3}{z} = 1 \end{cases}$

31. Jane is the manager of a company that assembles and packages two types of computers. Model 279P requires 4 hours of assembly time and $\frac{1}{2}$ hour of packaging time per computer; model 345Q requires 2 hours of assembly time and 1 hour of packaging time per computer. The company charges \$35 for each Model 279P computer and \$25 for each Model 345Q computer it assembles and packages. If at most 800 worker-hours are available per day for assembly and at most 250 worker-hours are available per day for packaging, how should Jane set up the work schedule so that maximum revenue is produced?

32. A metal company produces two types of alloy. One order of the first requires 4 tons of iron, 8 tons of zinc, and 5 tons of copper; one order of the second requires 3 tons of iron, 6 tons of zinc, and 6 tons of copper. The maximum numbers of tons of iron, zinc, and copper available in a month are 480, 540, and 600, respectively. If the company makes a profit of \$248.34 on one order of the first alloy and \$320.98 on one order of the second alloy, how many orders of each should be filled in order to maximize profit?

33. Doris and Tom have set up a linear programming model for production of brooms, dusters, and mops for their company. The cost function is $C(b, d, m) = 2.51b + 1.23d + 2.29m$, where $C(b, d, m)$ is the cost in dollars of producing b brooms, d dusters, and m mops. From the constraint set they have determined the complete set of corner points to be $\{(2, 3, 5), (4, 1, 8), (10, 0, 0), (0, 7, 0)\}$. What decision should they make about production in order to minimize cost?

34. Doris decides to investigate the revenues generated by producing the brooms, dusters, and mops from exercise 33 d. The revenue function is $R(b, d, m) = 7.34b + 4.99d + 8.49m$, where $R(b, d, m)$ is the revenue generated in dollars by producing b brooms, d dusters, and m mops. What decision should Doris make about production in order to maximize revenue? What would be a better procedure than looking at cost only or revenue only?

(Graphing calculator) In exercises 35 through 42, approximate to three decimal places the solution(s) of each equation. (Use the graphs of y = left-hand side and y = right-hand side to estimate the solutions first.)

35. $x^2 = \cos x$

36. $x^2 = \sin x$

37. $(x - 2)^2 = \ln x$

38. $(x + 2)^2 = \ln x$

39. $0.6x = \sin x$

40. $6x = \sin x$

41. $0.6x = e^x$

42. $6x = e^x$

Chapter 9: Summary

A system of linear equations can be solved in a variety of ways: substitution, elimination, Gauss-Jordan elimination, and inverse matrix. Some of these methods can be adapted to solving systems of linear-in-form or nonlinear equations or inequalities.

Key Concepts

systems of linear equations — *equivalent systems, $m \times n$ system*

linear-in-form systems of equations

Gauss-Jordan elimination — *augmented matrix*

matrix	*dimension, sum, difference, scalar multiplication, multiplication, inverse*
graph	*incidence matrix*
Boolean matrix	*Boolean product*
nonlinear systems of equations	
linear programming	

Skills

You should be able to:

1. Solve a linear or linear-in-form system by elimination-elimination, substitution, or elimination-substitution, as appropriate.
2. Solve linear or linear-in-form systems by Gauss-Jordan elimination.
3. Identify the dimension of a matrix and the position of an element of a matrix.
4. Perform matrix addition, subtraction, multiplication, and scalar multiplication.
5. Perform Boolean multiplication.
6. Find the inverse of square matrix or determine none exists.
7. Solve a linear matrix equation.
8. Determine the angle of rotation and sketch the graph of a rotated conic.
9. Solve a nonlinear system by substitution or by padding to use Gauss-Jordan elimination.
10. Graph the solution set of a system of nonlinear inequalities.

SUMMARY EXERCISES

The starred exercises can serve as a practice chapter test. The answers for these are at the end of the answer section at the back of the book.

In exercises 1 through 8, solve the system.

1. $\begin{cases} 5x + 12y = 7 \\ 2x - 4y = 18 \end{cases}$

2. $\begin{cases} 5.0x - 1.2y = 7.3 \\ 0.3x - 1.4y = 1.8 \end{cases}$

* **3.** $\begin{cases} x - 2y + 7z = 12 \\ 3x + y = -4 \\ 3y - 4z = 6 \end{cases}$

4. $\begin{cases} x - 2y + 7z = 12 \\ 3x + y + z = -4 \\ 2x + 3y - 4z = 6 \end{cases}$

5. $\begin{cases} \ln(x^2y^3) = -1 \\ \ln(xy) = 4 \end{cases}$

* **6.** $\begin{cases} \ln(x) + e^y = 1 \\ 3\ln(x) - 2e^y = -1 \end{cases}$

7. $\begin{cases} \dfrac{5}{x} + \dfrac{12}{y} = 7 \\ \dfrac{2}{x} - \dfrac{4}{y} = 18 \end{cases}$

8. $\begin{cases} \dfrac{2}{a} + \dfrac{3}{b} - \dfrac{1}{c} = 8 \\ \dfrac{1}{a} - \dfrac{2}{b} + \dfrac{1}{c} = -4 \\ \dfrac{5}{a} + \dfrac{1}{b} - \dfrac{3}{c} = -2 \end{cases}$

In exercises 9 through 12, determine whether there is a unique solution, no solution, or infinitely many solutions.

* **9.** $\begin{cases} 4x - 2y = 16 \\ 2x - y = 9 \end{cases}$

10. $\begin{cases} 4x - 2y = 16 \\ 2x - y = 8 \end{cases}$

11. $\begin{cases} 4x - 2y = 16 \\ x - y = 8 \end{cases}$

*** 12.** $\begin{cases} x + y + z = 16 \\ 2x - y + z = 8 \end{cases}$

In exercises 13 through 18, give the solution set of the linear system associated with the augmented matrix.

13. $\left[\begin{array}{cc|c} 1 & 0 & -\frac{1}{2} \\ 0 & 1 & \frac{3}{4} \end{array}\right]$

14. $\left[\begin{array}{ccc|c} 1 & 0 & 0 & 304 \\ 0 & 1 & 0 & -56 \\ 0 & 0 & 1 & 123 \end{array}\right]$

15. $\left[\begin{array}{ccc|c} 1 & 0 & 0 & 304 \\ 0 & 0 & 0 & -56 \\ 0 & 0 & 1 & 123 \end{array}\right]$

16. $\left[\begin{array}{ccc|c} 1 & 0 & 0 & 304 \\ 0 & 0 & 0 & 0 \\ 0 & 0 & 1 & 123 \end{array}\right]$

17. $\left[\begin{array}{ccc|c} 1 & 0 & 3 & 5 \\ 0 & 1 & -2 & 0 \end{array}\right]$

*** 18.** $\left[\begin{array}{ccc|c} 1 & 0 & 2 & 304 \\ 0 & 1 & 3 & -56 \\ 0 & 0 & 0 & 0 \end{array}\right]$

In exercises 19 through 28, use matrices *A*, *B*, and *C* to find

$$A = \begin{bmatrix} 2 & 3 \\ -5 & 1 \\ 0 & 6 \end{bmatrix}, \quad B = \begin{bmatrix} -1 & 0 & 3 \\ 2 & -4 & 1 \\ 0 & 5 & -3 \end{bmatrix}, \quad C = \begin{bmatrix} -8 & 10 \\ 3 & 1 \end{bmatrix}$$

19. $\dim_A$

20. $\dim_B$

21. AB

22. BA

23. AC

*** 24.** $AC + BA$

25. $0.5C$

26. C^{-1}

27. B^{-1}

28. CA^t

29. Solve $\begin{bmatrix} -8 & 10 \\ 3 & 1 \end{bmatrix}\begin{bmatrix} x \\ y \end{bmatrix} = \begin{bmatrix} 9.2 \\ -1.1 \end{bmatrix}$.

30. Solve $\begin{bmatrix} -1 & 0 & 3 \\ 2 & -4 & 1 \\ 0 & 5 & -3 \end{bmatrix}\begin{bmatrix} x \\ y \\ z \end{bmatrix} = \begin{bmatrix} -23 \\ 46 \\ 69 \end{bmatrix}$.

31. Use Boolean multiplication to find all the one-step trips for the airline whose flight matrix is

	A	*B*	*C*	*D*	*E*	*F*
A	0	1	1	0	0	1
B	1	0	1	0	0	0
C	1	1	0	0	1	0
D	0	0	0	0	1	1
E	0	0	1	1	0	1
F	1	0	0	1	1	0

In exercises 32 and 33, solve the system.

32. $\begin{cases} x^2 + y^2 = 1 \\ xy = 1 \end{cases}$

*** 33.** $\begin{cases} x^2 + y^2 = 9 \\ xy = 1 \end{cases}$

In exercises 34 and 35, approximate the solutions to two decimal places.

34. $\begin{cases} x^2 + y^2 = 1 \\ y = 0.31 \end{cases}$

*** 35.** $\begin{cases} y = -x^2 + 5 \\ y = 1.5x \end{cases}$

In exercises 36 through 39, sketch the graph of each system.

*** 36.** $\begin{cases} x^2 + y^2 \le 1 \\ xy < 1 \end{cases}$

37. $\begin{cases} x^2 + y^2 \le 9 \\ xy > 1 \end{cases}$

38. $\begin{cases} x^2 + y^2 > 1 \\ y \le 0.31 \end{cases}$

*** 39.** $\begin{cases} y \le -x^2 + 5 \\ y \le 1.5x \end{cases}$

In exercises 40 and 41, determine the angle of rotation and sketch the graph of the rotated conic.

40. $5x^2 - 16\sqrt{3}xy - 11y^2 = 4$

*** 41.** $13x^2 - 6\sqrt{3}xy + 7y^2 = 4$

CHAPTER 10

Headstarts

- 10.1 Counting Techniques
- 10.2 Mathematical Induction
- 10.3 The Binomial Theorem
- 10.4 Sequences
- 10.5 Sigma Notation and Series

In this chapter you'll find some topics and methods that will be of use in your next math course. Even if this is your last collegiate math course, you may find them interesting and applicable elsewhere. Although these ideas are side trails from our themes of graphing and solving equations and inequalities, we'll show connections with many earlier topics.

10.1 Counting Techniques

Suppose there are six 9 o'clock classes for which you and your twenty closest friends are qualified and four 10 o'clock classes you all can take. Can each of you have a different schedule of morning classes? Since any one of the 9:00 classes can be paired with any of the 10:00 classes, there are $6 \cdot 4$, or 24, different schedules. Yes, you and your 20 friends can each have different schedules if you like. This is an illustration of the **multiplication principle** of counting: If there are m ways to perform the first task and n ways to perform the second, there are $m \cdot n$ ways to do the pair.

An extended version of the multiplication principle says if there are k tasks and the ith task can be done n_i ways, then the k tasks may be performed in $n_1\, n_2 \cdots n_k$ ways.

EXAMPLE Lisa's favorite lunch spot features a soup-and-sandwich lunch special. If there are 6 kinds of soup and 14 different sandwiches today, how many different lunch specials can they advertise?

Each of the 6 soups can be paired with any of the 14 sandwiches.
There are $6 \cdot 14$, or 84, different lunch specials. ■

EXAMPLE How many ways can Sam arrange the photos of his five girlfriends on his mantel? There are 5 ways he can choose the first picture, 4 ways to choose the second one, and so on.

There are $5 \cdot 4 \cdot 3 \cdot 2 \cdot 1$, or 120, ways to arrange the 5 photos. ■

This last example is a bit special in that we keep making the choices from the same set of photos and do not have any repeats.

Definition An ordered arrangement of r items chosen from a set of n items is called a **permutation** of n items taken r at a time.

In the example, n and r were both 5.

Illustration Sam actually has only one girlfriend, but he has photos of eight girls in all and space for 5 of them on the mantel. He can make $8 \cdot 7 \cdot 6 \cdot 5 \cdot 4$, or 6720, permutations of 5 photos chosen from his collection of 8.

In computing the number of permutations of n items taken r at a time, $P(n, r)$, products of the form $n(n-1)(n-2) \cdots (n-k)$ appear. Mathematicians have a compact way to represent these expressions.

Definition The product $n(n-1)(n-2) \cdots 3 \cdot 2 \cdot 1$ is called n **factorial** and is written $n!$.

(Your calculator may have an $x!$ key; perhaps it also has a $P(n, r)$ key, often labeled ${}_nP_r$.) $P(n, r) = n!$ if $r = n$; if $r < n$, then $P(n, r)$ should have r factors, beginning with n: $n(n-1)(n-2)\cdots(n-(r-1))$. Another way to express this is

$$P(n, r) = \frac{n!}{(n-r)!}$$

$$\frac{n!}{(n-r)!} = \frac{n(n-1)(n-2)\cdots[n-(r-1)](n-r)\cdots 3\cdot 2\cdot 1}{(n-r)(n-r-1)(n-r-2)\cdots 3\cdot 2\cdot 1},$$

$$\text{or}\quad n(n-1)(n-2)\cdots(n-r+1)$$

We can also use this formula when $r = n$ if we define 0! to be 1.

Sometimes the word permutation just means an ordered arrangement. If there are n items from which to choose to fill r positions and unlimited repeats *are* allowed, there are

$$\underbrace{n\cdot n\cdot n\cdots n}_{r\text{ factors}} = n^r$$

possible permutations, by the multiplication principle. In applications of this principle it will be clear whether repeats are allowed or not.

EXAMPLE The digits 0 through 9 can be used to form part numbers for the Widget line of tools. If 7-digit part numbers are used, how many different part numbers are available?

There are 10 digits available, and each can be used in any position of the 7-digit part number.

There are 10 · 10 · 10 · 10 · 10 · 10 · 10, or 10,000,000, part numbers available, ranging from 0000000 to 9999999. ■

EXAMPLE The digits 0 through 9 can be used to form part numbers for the Gidget line of tools. If 7-digit part numbers are used with no digit repeated, how many different part numbers are available?

There are 10 digits available, and each can be used in only one position of the 7-digit part number.

There are 10 · 9 · 8 · 7 · 6 · 5 · 4, or 604,800, part numbers available. ■

A related problem is one where order matters and limited repeats are allowed. As an illustration we'll use a popular one. How many different "words" (ordered arrangements) can be made using the letters in MISSISSIPPI? Begin by imagining tags on each of the letters so that the first I is distinguishable from the third I, and so on. We know there are 11! permutations in this case. These permutations can be grouped into batches, where all the permutations in a particular batch are exactly the same except for the arrangement of S's in them. For example,

$$S_1S_2I_1I_2I_3I_4MP_1P_2S_3S_4 \quad\text{and}\quad S_3S_1I_1I_2I_3I_4MP_1P_2S_2S_4$$

are grouped together. How many permutations belong to each batch? There are as many as there are ways to arrange four S's, 4!. Without the subscripts, these arrangements are the same. If there are no tags on the S's, there ae 11!/4! distinguishable arrangements. Treating the I's and P's in the same way, we see there are 11!/(4!4!2!), or 34,650, permutations of the letters in Mississippi.

If we have k different elements with the jth element occuring i_j times for a total of

$$i_1 + i_2 + i_3 + \cdots + i_k, \text{ or } n, \text{ items,}$$

there are

$$\frac{n!}{i_1!i_2! \cdots i_k!}$$

permutations of the n items.

EXAMPLE How many words can you make using the letters in WALLABY?

Of the 7 letters available, there are 2 L's and 2 A's. **We can make $\frac{7!}{2!2!1!1!1!}$, or 1260, words.** ■

We need different counting rules if the ordering of the items is irrelevant; for example, in how many ways can you choose 3 members of a 22-person club to form a committee?

A **combination** of r items chosen from n items, $C(n, r)$, is the formal name for such a choice without regard to order.

(Your calculator may have a key labeled ${}_nC_r$ for computing $C(n, r)$.)

EXAMPLE How many two-element subsets can be made for the set $\{a, b, c, d, e\}$?
This is a question about combinations; the elements of a subset are not ordered. Let's list the subsets of size 2.

$$\{a, b\}, \{a, c\}, \{a, d\}, \{a, e\}, \{b, c\}, \{b, d\},$$
$$\{b, e\}, \{c, d\}, \{c, e\}, \{d, e\}$$

There are 10 two-element subsets of $\{a, b, c, d, e\}$. ■

Clearly, direct counting—as in the last example—is not always feasible. A formula for $C(n, r)$ would be useful; we base it on the formula for $P(n, r)$. The problem of counting the number of ordered arrangements of r items chosen from n items can be viewed as consisting of two tasks: select r items and then arrange them all possible ways. That is, $P(n, r)$ is $C(n, r)$ times the number of arrangements we can make of r items using all of them.

$$P(n, r) = C(n, r) \cdot r!, \quad \text{or} \quad C(n, r) = \frac{P(n, r)}{r!} = \frac{n!}{(n - r)!r!}$$

As a check we can see that the number of two-element subsets of $\{a, b, c, d, e\}$ is $C(5, 2)$ using the formula.

$$\frac{5!}{(5 - 2)!2!} = \frac{5!}{3!2!}, \text{ or } 10$$

This is exactly the number we found by direct counting. Notice that $C(n, r)$ counts the number of ways to choose r items from a set of n items *without* repeats. If unlimited repeats are allowed and the order of the elements chosen is not relevant, in how many ways can we choose r items from n items? To answer this question we use a common problem-solving technique: make a model of the problem that is related to an already-solved problem.

Little Frankie, 5 years old, is flying today from Eden to Dryton on West-Southeast Air. To keep him occupied, Chris (the flight attendant) offers him a choice of 3 small toys from a collection of 7 different types of toys. The order in which Frankie chooses doesn't matter, and *he can have two or three of the same toy if he likes*. Frankie is cranky and refuses to say which he wants; he'll only nod yes or no. Chris offers the toys one at a time, handing one to Frankie if he nods yes and offering a different toy only after Frankie has nodded no. How can this "conversation" go? Frankie will nod yes three times. To make our job easier, Chris will record an N every time Frankie nods no and a Y for yes. Here are some possible outcomes.

Outcome	*Toy 1*	*Toy 2*	*Toy 3*	*Toy 4*	*Toy 5*	*Toy 6*	*Toy 7*
1	N	YN	N	YN	N	Y	
2	N	N	N	YYY			
3	N	N	N	N	YYN	N	Y

We need 3 Y's and 6 N's to represent all possible conversations. Outcomes 1 and 2 can be padded with N's to fit the pattern. Try out a few more outcomes yourself to see that all the possibilities are covered. You already know how to count the number of strings of 9 letters made from 3 Y's and 6 N's. There are $9!/(3!6!)$ such strings, so there are 84 choices Frankie can make. Chris hopes Frankie likes the one he makes.

Let's analyze where the numbers in our result $9!/(3!6!)$ came from in order to build a general formula. The number of Y's is easy; Frankie had 3 choices. Outcome 3 illustrates the type of situation that will require the most N's; the final Y is at the end of the string. If there are n items, we'll need $n - 1$ N's in this case, one for each item but the last. Now we know that we need strings $n - 1 + r$ long, where r is the number of choices to be made.

The general formula for choosing r items from n items without regard to order and with unlimited repeats is

$$\frac{(n + r - 1)!}{(n - 1)!r!}.$$

EXAMPLE You've won 4 albums from the Top Ten Album list. How many different ways could you claim your prize?

Order doesn't matter, and repeats are all right.

There are $\frac{(10 + 4 - 1)!}{9!4!}$, or 715, ways. ■

Another useful counting principle is the *addition principle*. If one task can be done in n ways and another task in m ways, there are $n + m$ ways to do one task or the other but not both.

EXAMPLE How many 6-person subcommittees can be chosen from a committee of 8 engineers and 6 scientists if there are to be exactly 2 engineers or exactly 3 scientists on the subcommittee?

To use the addition principle we first count the subcommittees with exactly 2 engineers and then those with exactly 3 scientists. The multiplication principle implies we may choose 2 engineers and then 4 scientists for the first case. The number of subcommittees with exactly 2 engineers is

$$\frac{8!}{2!6!} \cdot \frac{6!}{4!2!}, \quad \text{or} \quad 420.$$

Similarly, the number of subcommittees with exactly 3 scientists is

$$\frac{6!}{3!3!} \cdot \frac{8!}{3!5!} \quad \text{or} \quad 1120.$$

The total number of subcommittees as described is 1540. ■

Sometimes it's more convenient to count what you don't want than what you do want.

EXAMPLE As president of the local chapter of Preserve the Whales, you must appoint a program committee of 3 people from your 57 members. You're required to pull names out of a hat to do the selecting. How many different 3-person committees can be chosen?

Dr. Zuk, a prominent member, is difficult to work with, and you hope he's not selected. How many of the possible three-person committees are Zuk-free?

We've seen the problem of choosing 3 people from 57 without regard to order. **There are $C(57, 3)$ different 3-person committees;** $C(57, 3)$ is $57!/(54!3!)$, or 29,260.

To find the number of committees without Dr. Zuk, it may be easier first to count the committees of which he is a member; there are only $C(56, 2)$ ways to fill the other two slots and $C(56, 2)$ is 1540. **If 1540 committees have Dr. Zuk on them, then 29,260 − 1540, or 27,720, committees don't have him.** Your chances look pretty good. ■

EXERCISES 10.1

In exercises 1 through 16, evaluate each of the given expressions.

1. $7!$
2. $9!$
3. $P(4, 2)$
4. $P(6, 4)$
5. $P(6, 6)$
6. $P(6, 0)$
7. $\dfrac{14!}{11!}$
8. $\dfrac{7!}{4!}$
9. $\dfrac{P(6, 4)}{2!}$
10. $\dfrac{P(8, 5)}{5!}$
11. $\dfrac{P(6, 0)}{0!}$
12. $\dfrac{P(6, 6)}{6!}$
13. $C(10, 3)$
14. $C(10, 7)$
15. $C(15, 15)$
16. $C(15, 0)$
17. Arnold frequently dresses for school while not fully awake. On Monday morning Arnold's closet contains 9 shirts and 10 pairs of slacks. If he is particularly sleepy and pays no attention to color or style, in how many ways can he select a shirt and pair of slacks?
18. Arnold has a choice of 8 pairs of socks and 4 pairs of shoes. In how many ways can he select socks and shoes?

19. The extension numbers at the new office complex consist of four digits. How many different numbers are there if repeats are allowed and
  a. any digit can be used?
  b. the first digit cannot be zero?
  c. no digit can be zero?

20. The extension numbers at the state department office bulding consist of five digits. How many different numbers are there if repeats are not allowed and
  a. any digit can be used?
  b. the first digit cannot be zero?
  c. no digit can be zero?

21. Kris and Tom are arranging books on a new book shelf. They decide to reserve the third shelf for novels. If there are 25 novels in their collection, in how many ways can they arrange
  a. all 25 on the shelf?
  b. 10 of the 25 books?
  c. 7 of the 25 books if the tallest book must be first?

22. After the novels on the third shelf are arranged, Tom decides to arrange his sporting books on the top shelf. He has 20 books, but notices there are 2 copies of one and 4 copies of another. How many distinguishable arrangements of these 20 books can he make?

23. Kris and Tom separate, and Kris wins the toss for who gets to select 13 of the 25 novels.
  a. In how many ways can Kris make her selections?
  b. In how many ways can Kris select 3 of her 13 books to take for her summer vacation reading?
  c. In how many ways can Tom select 2 of his 12 books to give to his new girlfriend?
  d. Kris decides to borrow two books from Tom to take on her vacation, along with the three she has selected from her collection. How many different choices does she have? Remember, Tom has only 10 books now.

24. In a psychology experiment, four-year olds are allowed to select 3 toys from a box of 7 toys.
  a. How many ways can a child make a selection of 3 toys without regard to order?
  b. The experiment is changed slightly. The child is allowed to select 1 toy and play with it for 5 minutes. That toy is replaced and the child chooses again. After another 5 minutes the toy is replaced and the final selection is made. In how many ways can the choices be made?

25. A set consists of the first four letters of the Greek alphabet, $\{\alpha, \beta, \gamma, \delta\}$. How many subsets can be formed if the subset contains
  a. 4 letters?
  b. 3 letters?
  c. 2 letters?
  d. 1 letter?
  e. 0 letters?

26. How many subsets in all can be formed for the set
  a. in exercise 25?
  b. of 7 toys in exercise 24?
  c. of 9 shirts in Arnold's closet?

27. Prove $C(n, n) = 1$ for all $n \in W$.

28. Prove $C(n, r) = C(n, n - r)$ for all $n, r \in W$ with $r \leq n$.

29. Prove $C(n, r) + C(n, r - 1) = C(n + 1, r)$ for all n, $r \in W$ with $r \leq n$.

30. Hoke has won a shopping spree at the local discount store. He may choose 10 items from a section of the store containing 20 different products. There are well over 10 items of each of the 20 different products. In how many ways can he make his selections?

31. The president of a college decides to form a 4-member committee of the faculty to review the budget. How many committees can be formed if the president chooses from
  a. the entire faculty of 40?
  b. the tenured faculty of 26?
  c. the business faculty of 5?

32. **a.** How many of the faculty committees of exercise 31b have exactly one business faculty member if 3 of the business faculty are tenured?
  b. How many have exactly 2 members from the business faculty?
  c. How many have exactly 3 members from the business faculty?
  d. How many have at least one member from the business faculty?
  e. How many have no members from the business faculty?

10.2 Mathematical Induction

Toolkit

You'll need these identities from Chapter 7: $\cos(\alpha + \beta) = \cos\alpha\cos\beta - \sin\alpha\sin\beta$, and $\sin(\alpha + \beta) = \sin\alpha\cos\beta + \cos\alpha\sin\beta$. You'll also use these sets from the appendix: $N = \{1, 2, 3, \ldots\}$ and $W = \{0, 1, 2, 3, \ldots\}$. In Section 5.1, we first discussed direct proofs.

Sometimes it's difficult to fit a counting problem into one of the models in Section 10.1, but we can make a guess based on experiments. These guesses need to be verified in some way before we can use them in other situations. Here's a puzzle problem to consider. Mark n points on a circle and connect these points with line segments; count the maximum number of nonoverlapping regions produced inside the circle. If there are n points, how many regions are created? Let's experiment.

Number of points	Illustration	Maximum number of regions
1		1
2		2
3		4
4		8

Are you ready to make a guess about how the maximum number of regions is related to the number of points? Draw a picture for 6 points and check your guess. Was your guess confirmed? This problem illustrates nicely why mathematicians look for proof to verify patterns instead of relying on wishful thinking.

In this section we explain a common method of proof for statements that involve the whole numbers. The technique is called **mathematical induction**; each proof by induction consists of two parts, a basis step and an induction step. The purpose of the basis step is to confirm that the statement is true for the smallest whole number that makes sense in the statement. The induction step shows that if the statement is true for some whole number, then it's true for the next one. You can think of an induction proof this way: The statement you want to prove is a shorthand way of making an infinite number of statements, one for each whole number. For example, the statement

A set with n elements has exactly 2^n subsets.

represents

A set with 0 elements has exactly 2^0 subsets.
A set with 1 element has exactly 2^1 subsets.
A set with 2 elements has exactly 2^2 subsets.
⋮

Picture these statements as dominos standing on their short ends. The basis step in an induction proof is done to check if there is a beginning domino that can be pushed over. The induction step confirms that the dominos are close enough together so that when the first one is pushed over, all the dominos, an infinite number, will fall—that is, that all the statements are true.

Let's use mathematical induction to prove that a set with n elements has exactly 2^n subsets, where n represents a whole number.

Basis step Zero is the smallest whole number for which the statement makes sense. If it has 0 elements, the set is the empty set and its only subset is itself. Thus the statement is confirmed for 0.

Induction step The key to this step is to use what the statement says for a whole number k to show that the statement is true for the next whole number, $k + 1$. The statement associated with the whole number k is called $P(k)$, and the statement associated with the whole number $k + 1$ is called $P(k + 1)$. (The letters P and Q are frequently used to name statements.) We wish to show that if $P(k)$ is true, then $P(k + 1)$ is true. It helps to write out each version.

$P(k)$: A set with k elements has exactly 2^k subsets.
$P(k + 1)$: A set with $k + 1$ elements has exactly 2^{k+1} subsets.

Because $P(k)$ tells us about a set with k elements, let's create such a set from the set with $k + 1$ elements by choosing one element, call it z, at random and setting it aside. The new set with k elements has exactly 2^k subsets. Think about a list of these subsets, $S_1, S_2, \ldots, S_{2^k}$. Each of these is a subset of the original set with $k + 1$ elements. We can produce another batch of subsets by including z in each: $S_1 \cup \{z\}, S_2 \cup \{z\}, \ldots, S_{2^k} \cup \{z\}$. Altogether we have $2(2^k)$ subsets of the set with $k + 1$ elements.

You might be worried about whether these are all the subsets. Yes, because if there is another subset, say S, either it contains z or it doesn't. If S doesn't contain z, it is one of the $S_1, S_2, \ldots, S_{2^k}$. If it does contain z, removing z must give one of the S_i, and S is $S_i \cup \{z\}$. We have shown a set with $k + 1$ elements has exactly $2(2^k)$, or 2^{k+1}, subsets.

Mathematical induction is a subtle idea and takes some getting used to. In the rest of the section we give examples of its use in different settings. There's always an element of creativity in a proof, so there aren't any hard-and-fast rules on how to construct one, but this doesn't mean you can't learn to write one. Working through examples helps you find ways to build proofs of your own, just as reading lots of essays helps you learn to write essays of your own.

Where do statements to prove come from? Often conjectures come from the kind of experimenting you did earlier connecting points on a circle. Consider the following sums of positive even numbers.

Number of terms	1	2	3	4	5
Expanded sum	2	2 + 4	2 + 4 + 6	2 + 4 + 6 + 8	2 + 4 + 6 + 8 + 10
Sum	2	6	12	20	30

Make a guess about the relationship between the number of terms and the sum before reading further.

EXAMPLE Prove $2 + 4 + 6 + \cdots + 2n = n(n + 1)$ for $n \in N$.

For practice, we'll use induction, but Exercise 33 outlines a direct proof. Because this statement is made in terms of the natural numbers, a subset of the whole numbers, it's reasonable to try an induction proof.

Basis step: The smallest value of n for which the statement is defined is 1. In this case, the left-hand side is 2; the right-hand side is $1(1 + 1)$, or 2, so the first domino can be pushed over.

Induction step: We write the statement for two consecutive values of n.

$$P(k): \quad 2 + 4 + 6 + \cdots + 2k = k(k + 1)$$
$$P(k + 1): \quad 2 + 4 + 6 + \cdots + 2(k + 1) = (k + 1)[(k + 1) + 1]$$

Remember we want to use $P(k)$ to show $P(k + 1)$ is true. To do this we try to find either the left-hand side or the right-hand side of $P(k)$ in $P(k + 1)$. The left-hand side of $P(k + 1)$ is the expanded sum of the first $k + 1$ positive even numbers; it is the sum of the first k positive even numbers plus $2(k + 1)$. Replacing the left-hand side of $P(k)$ as it appears in the left-hand side of $P(k + 1)$,

$$\mathbf{2 + 4 + \cdots + 2k} + 2(k + 1) = \mathbf{k(k + 1)} + 2(k + 1)$$
$$\Leftrightarrow \quad 2 + 4 + \cdots + 2(k + 1) = (k + 1)(k + 2)$$

Factoring the right-hand side

But this last expression is the right-hand side of $P(k + 1)$. The induction step is finished. Each falling domino pushes over the next one. ■

In some earlier sections you've used statements that can now be proved by mathematical induction. One of these is De Moivre's theorem from Section 7.5:

$$[r(\cos\theta + i\sin\theta)]^n = r^n[\cos n\theta + i\sin n\theta], \quad n \in I.$$

EXAMPLE Give a proof of De Moivre's theorem,
$[r(\cos\theta + i\sin\theta)]^n = r^n[\cos n\theta + i\sin n\theta], \quad n \in I.$

The proof has two parts, an induction proof for $n \geq 0$ and a direct proof for $n < 0$. The induction proof begins with the basis step. If n is 0, the left and right sides are the same: $[r(\cos 0 + i\sin 0)]^0$ is 1 and $r^0(\cos 0 + i\sin 0)$ is $1(1 + 0)$, or 1.

For the induction step, consider

$$P(k): \quad [r(\cos\theta + i\sin\theta)]^k = r^k(\cos k\theta + i\sin k\theta)$$

and

$$P(k+1): \quad [r(\cos\theta + i\sin\theta)]^{k+1} = r^{k+1}[\cos(k+1)\theta + i\sin(k+1)\theta].$$

It's not too hard to see the relation between the left-hand sides of $P(k)$ and $P(k + 1)$.

$$[r(\cos\theta + i\sin\theta)]^{k+1} = [r(\cos\theta + i\sin\theta)]^k \cdot r(\cos\theta + i\sin\theta)$$

$$\Leftrightarrow \quad [r(\cos\theta + i\sin\theta)]^{k+1} = r^k(\cos k\theta + i\sin k\theta) \cdot r(\cos\theta + i\sin\theta)$$

Replacing the left-hand side of $P(k)$ with the right-hand side of $P(k)$

$$\Leftrightarrow \quad [r(\cos\theta + i\sin\theta)]^{k+1} = r^{k+1}[(\cos k\theta + i\sin k\theta)(\cos\theta + i\sin\theta)]$$

$$\Leftrightarrow \quad [r(\cos\theta + i\sin\theta)]^{k+1} = r^{k+1}[\cos k\theta\cos\theta + i\cos k\theta\sin\theta + i\sin k\theta\cos\theta - \sin k\theta\sin\theta]$$

$$\Leftrightarrow \quad [r(\cos\theta + i\sin\theta)]^{k+1} = r^{k+1}[\cos k\theta\cos\theta - \sin k\theta\sin\theta + i\cos k\theta\sin\theta + i\sin k\theta\cos\theta]$$

$$\Leftrightarrow \quad [r(\cos\theta + i\sin\theta)]^{k+1} = r^{k+1}[\cos(k\theta + \theta) + i\sin(k\theta + \theta)]$$

$$\Leftrightarrow \quad [r(\cos\theta + i\sin\theta)]^{k+1} = r^{k+1}[\cos(k+1)\theta + i\sin(k+1)\theta]$$

This establishes $P(k + 1)$ and completes the induction part of the proof.
If $n < 0$, let $m = -n$. We have

$$[r(\cos\theta + i\sin\theta)]^n = \frac{1}{[r(\cos\theta + i\sin\theta)]^m}$$

$$\Leftrightarrow \quad [r(\cos\theta + i\sin\theta)]^n = \frac{1}{[r^m(\cos m\theta + i\sin m\theta)]}$$

Since $m > 0$

$$\Leftrightarrow \quad [r(\cos\theta + i\sin\theta)]^n = \frac{1}{[r^m(\cos m\theta + i\sin m\theta)]} \cdot \frac{\cos m\theta - i\sin m\theta}{\cos m\theta - i\sin m\theta}$$

$$\Leftrightarrow \quad [r(\cos\theta + i\sin\theta)]^n = \frac{r^{-m}(\cos m\theta - i\sin m\theta)}{\cos^2 m\theta + \sin^2 m\theta}$$

$$\Leftrightarrow \quad [r(\cos\theta + i\sin\theta)]^n = r^n[\cos(-n\theta) - i\sin(-n\theta)]$$

$$\Leftrightarrow \quad [r(\cos\theta + i\sin\theta)]^n = r^n[\cos(n\theta) + i\sin(n\theta)]$$

Since cosine is even and sine is odd ■

In Section 4.4 a function is defined as periodic with period p if p is the smallest positive number for which $f(x) = f(x + p)$. In that discussion it was claimed that if f has period p, $f(x) = f(x + np)$ for every integer n that gives $x + np$ a value in the domain of f. Mathematical induction can be used to prove the following:

If f has period p, then for every whole number n such that $x + np$ belongs to the domain of f, $f(x + np) = f(x)$.

Basis Step: If n is 0, certainly $f(x + 0 \cdot p) = f(x)$.

Induction Step: We will use $P(k)$, which is $f(x + kp) = f(x)$, to show $P(k + 1)$: $f(x + (k + 1)p) = f(x)$. Because f has period p, $f(x + (k + 1)p) = f((x + kp) + p)$, or $f(x + kp)$. But $P(k)$ says $f(x + kp) = f(x)$, so $f(x + (k + 1)p) = f(x)$.

Here's one more induction proof to study before you try the exercises. This last example is presented in a more formal format to give you some experience with this style. Since $7 - 1$, $7^2 - 1$, and $7^3 - 1$ are multiples of 6, we might guess that $7^n - 1$, $n \in N$, is always divisible by 6.

EXAMPLE Use an induction argument to prove or disprove $7^n - 1$ is divisible by 6 for all natural numbers n.

Basis step: For $n = 1$, $7^n - 1$ is $7 - 1$, or 6. Certainly 6 divides itself evenly.

Induction step: For reference we write $P(k)$: $7^k - 1$ is divisible by 6 and $P(k + 1)$: $7^{k+1} - 1$ is divisible by 6. We need to express $7^{k+1} - 1$ in a way that allows us to make use of $P(k)$. Rewriting $7^{k+1} - 1$ as $7 \cdot 7^k - 1$ doesn't quite give what we want, but it is fixable.

$$7^{k+1} - 1 = 7 \cdot 7^k - 1 = 7 \cdot 7^k - 7 + 6 = 7(7^k - 1) + 6$$

The number $7(7^k - 1) + 6$ is divisible by 6 because each term is divisible by 6. Hence $7^{k+1} - 1$ (same number expressed differently) is divisible by 6.

Note: There is a direct proof of this statement; however, the preceding serves as another example of mathematical induction. ■

EXERCISES 10.2

In exercises 1 through 12, first rewrite each statement in the form $P(n)$ and then write the basis statement. Is the basis statement true or false?

1. The sum of the first n whole numbers is half the product of n and $(n + 1)$.
2. The sum of the first n whole numbers is half the product of n and $(n - 1)$.
3. The term containing the factor x^{n-1} in the expansion of $(x + y)^n$, n a whole number, has coefficient n.
4. The product of a number c and the sum of n numbers is the sum of the products of c and each of the n numbers.
5. The sum of the cubes of the first i counting numbers is the square of half the product of i and $(i + 1)$.
6. The sum of the cubes of the first j whole numbers is the square of half the product of j and $(j + 1)$.
7. There are at most 2^n regions created when n points are marked on an ellipse and each pair is connected with a line segment.
8. There are at most 2^{n-1} regions created when n points are marked on an ellipse and each pair is connected with a line segment.
9. There are at most 2^n regions created when n points are marked on a square and each pair is connected with a line segment.
10. There are at most 2^{n-1} regions created when n points are marked on a square and each pair is connected with a line segment.
11. The sum of the angles of a closed figure with n sides of equal length is $(180n - 360)°$.
12. The sum of the angles of a closed figure with $(n + 1)$ sides of equal length is $(180n - 360)°$.

13.–24. Write at least three statements $P(k)$ for each statement in exercises 1 through 12 using specific values of k. Do you think, after examining each of these, that $P(k)$ might be true for all k?.

In exercises 25 through 32, use mathematical induction to prove each statement.

25. The sum of the first n counting numbers is half the product of n and $(n + 1)$.

26. $5 + 6 + 7 + \cdots + n + (n + 1) = \dfrac{(n + 6)(n - 3)}{2}$.

27. The expression $4^n - 1$ is divisible by 3.

28. The expression $a^n - 1$ is divisible by $a - 1$, $a \neq 1$.

29. The sum of the squares of the first n counting numbers is one-sixth the product of n, $n + 1$, and $2n + 1$.

30. The sum of the squares of the first n whole numbers is one-sixth the product of $n - 1$, n, and $2n - 1$.

31. $(x - 1)(x^n + x^{n-1} + x^{n-2} + \cdots + x + 1) = x^{n+1} - 1$.

32. $(x - a)(x^{n-1} + x^{n-2}a + x^{n-3}a^2 + \cdots + xa^{n-2} + a^{n-1}) = x^n - a^n$.

33. Prove $2 + 4 + \cdots + 2n = n(n + 1)$ by considering $S = 2 + 4 + 6 + \cdots + 2n$, $S = 2n + 2(n - 1) + 2(n - 2) + \cdots + 4 + 2$, and $S + S$.

34. Can you determine a pattern for the sum of the first n counting numbers that are multiples of 3? If so, give the statement $P(n)$ and prove it using induction.

35. Can you determine a pattern for the sum of the first n whole numbers that are multiples of 7? If so, give the statement $P(n)$ and prove it using induction.

36. Let's return to the problem that opens this chapter. The method of finite differences (Section 5.4) can be used to predict the maximum number of regions determined by n points.

Number of points	*Number of regions*	*Difference*		
1	1			
		> 1		
2	2		> 1	
		> 2		> 1
3	4		> 2	
		> 4		> 2
4	8		> 4	
		> 8		> 3
5	16		> 7	
		> 15		
6	31			

Use the pattern established to make a guess for the maximum number of reigons there are for 7 points and for 8 points. (*Hint:* Backtrack.)

37. a. Use the table in exercise 36 to determine the degree of the polynomial function that gives the maximum number of regions.

b. Find the polynomial function.

10.3 The Binomial Theorem

Toolkit

A palindrome is a number, word, or list that reads the same backward or forward.

Sherlock Holmes' nemesis Moriarity was said to be an expert on the binomial theorem. In this section you'll meet Moriarity's specialty, use combinations, and apply mathematical induction.

The numbers $C(n, r)$ are sometimes called binomial coefficients because they count the number of $x^r y^{n-r}$-terms in the expanded version of $(x + y)^n$, $n \in W$. Why is this so? There are n factors in $(x + y)^n$, each $x + y$; an $x^r\,y^{n-r}$-term is created by r of the factors contributing an x and $n - r$ of them contributing a y. The order of these contributions is irrelevant because of the commutative property of multiplication, so the number of ways this can happen is a combinations question.

Illustration $(x + y)^3 = (x + y)(x + y)(x + y)$. How are the x^2y-terms produced? Two factors contribute an x, and one contributes a y. This happens three ways:

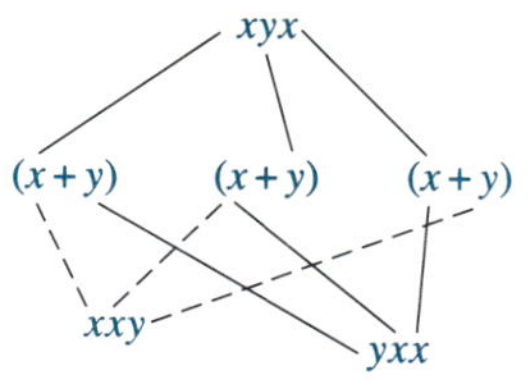

$$C(3, 2) = \frac{3!}{2!1!}, \text{ or } 3.$$

□

EXAMPLE What is the coefficient of x^7 in the expansion of $(x + 2)^{10}$?
An x^7-term is created as the product of 7 factors of x and 3 factors of 2. The number of ways this can happen is $C(10, 7)$, or 120. Hence there are 120 $x^7(2)^3$-terms.
The coefficient of x^7 is 960. ■

You're already familiar with some special cases of the binomial theorem. If $a + b \neq 0$,

$$(a + b)^0 = 1$$
$$(a + b)^1 = a + b$$
$$(a + b)^2 = a^2 + 2ab + b^2$$
$$(a + b)^3 = a^3 + 3a^2b + 3ab^2 + b^3.$$

A good guess for the types of terms in the expanded version of $(a + b)^n$, $n \in W$, is a^n, $a^{n-1}b$, $a^{n-2}b^2$, . . . , a^2b^{n-2}, ab^{n-1}, b^n. The difficulty before now was in describing the coefficients. The binomial theorem puts the pieces together.

The Binomial Theorem: For $n \in W$,

$$(a + b)^n = C(n, n)a^n + C(n, n - 1)a^{n-1}b + C(n, n - 2)a^{n-2}b^2 + \cdots + C(n, 2)a^2b^{n-2} + C(n, 1)ab^{n-1} + C(n, 0)b^n.$$

A few comments are in order before we look at a proof for the binomial theorem. The value of $C(n, n)$ and the value of $C(n, 0)$ are each 1; the list of coefficients forms a palindrome because $C(n, k) = C(n, n - k)$ for each $k \leq n$, k and n whole numbers. You may have used Pascal's triangle before to find these coefficients. We'll discuss Pascal's triangle later in the section. (If you haven't seen it before, don't worry; we'll begin with its definition.)

We need a property of binomial coefficients in the proof: $C(n, r) + C(n, r - 1) \equiv C(n + 1, r)$. Although this statement is stated in terms of whole numbers, a direct proof rather than an induction proof is used. Remember, we try direct proofs first. An induction proof here would have to be a double one because the statement depends on both n and r. The direct proof begins with the definitions of $C(n, r)$ and $C(n, r - 1)$. (You may have done this in exercise 29 of Section 10.1.)

$$C(n, r) + C(n, r - 1) = \frac{n!}{(n - r)!r!} + \frac{n!}{[n - (r - 1)]!(r - 1)!}$$

$$\Leftrightarrow \quad C(n, r) + C(n, r - 1) = \frac{n!(n - r + 1)}{(n - r)!r!(n - r + 1)} + \frac{n!r}{(n - r + 1)!(r - 1)!r}$$

$$\Leftrightarrow \quad C(n, r) + C(n, r-1) = \frac{n!(n-r+1)}{(n-r+1)!r!} + \frac{n!r}{(n-r+1)!r!}$$

$$\Leftrightarrow \quad C(n, r) + C(n, r-1) = \frac{n!(n-r+1+r)}{(n-r+1)!r!}$$

$$\Leftrightarrow \quad C(n, r) + C(n, r-1) = \frac{n!(n+1)}{(n+1-r)!r!}$$

$$\Leftrightarrow \quad C(n, r) + C(n, r-1) = \frac{(n+1)!}{[(n+1)-r]!r!}$$

$$\Leftrightarrow \quad C(n, r) + C(n, r-1) = C(n+1, r)$$

The statement of the binomial theorem suggests an induction proof. A direct proof based on the comments in the second paragraph of this section is a bit wordier. It is sometimes a matter of taste whether one uses a direct, indirect, or induction proof.

The Binomial Theorem: For $n \in W$, $(a + b) \neq 0$,

$$(a+b)^n = C(n, n)a^n + C(n, n-1)a^{n-1}b + C(n, n-2)a^{n-2}b^2 + \cdots + C(n, 2)a^2b^{n-2} + C(n, 1)ab^{n-1} + C(n, 0)b^n.$$

PROOF OF THE BINOMIAL THEOREM

(proof by induction)

Basis Step: The smallest n is 0. The left-hand side then is 1. The right-hand side is $C(0, 0)a^0$, or 1.

Induction Step: We need to use $P(k)$ to show $P(k + 1)$.

$$P(k): \quad (a+b)^k = C(k, k)a^k + C(k, k-1)a^{k-1}b + C(k, k-2)a^{k-2}b^2 + \cdots + C(k, 2)a^2b^{k-2} + C(k, 1)ab^{k-1} + (k, 0)b^k.$$

$$P(k+1): \quad (a+b)^{k+1} = C(k+1, k+1)a^{k+1} + C(k+1, k)a^kb + C(k+1, k-1)a^{k-1}b^2 + \cdots + C(k+1, 2)a^2b^{k-1} + C(k+1, 1)ab^k + C(k+1, 0)b^{k+1}.$$

The left-hand side of $P(k + 1)$, $(a + b)^{k+1}$, factors as $(a + b)(a + b)^k$, and we can replace $(a + b)^k$ with the right-hand side of $P(k)$.

$$(a+b)(\boldsymbol{a} + \boldsymbol{b})^k = (a+b)\mathbf{[C(k, k)}\boldsymbol{a}^{\boldsymbol{k}} + \mathbf{C(k, k-1)}\boldsymbol{a}^{\boldsymbol{k-1}}\boldsymbol{b} + \cdots + \mathbf{C(k, 1)}\boldsymbol{ab}^{\boldsymbol{k-1}} + \mathbf{C(k, 0)}\boldsymbol{b}^{\boldsymbol{k}}\mathbf{]}.$$

Expanding the right-hand side, we have

$$C(k, k)a^{k+1} + C(k, k-1)a^kb + \cdots + C(k, 1)a^2b^{k-1} + C(k, 0)ab^k + C(k, k)a^kb + C(k, k-1)a^{k-1}b^2 + \cdots + C(k, 1)ab^k + C(k, 0)b^{k+1}.$$

We know $C(k, k)$ and $C(k, 0)$ are each 1. What about the other terms? Combining like terms, we have $a^{k+1} + [C(k, k) + C(k, k - 1)]a^kb + [C(k, k - 1) +$

$C(k, k-2)]a^{k-1}b^2 + \cdots + [C(k, 1) + C(k, 0)]ab^k + b^{k+1}$. But we know how to combine these binomial coefficients: $C(n, r) + C(n, r-1) = C(n+1, r)$, for $n \geq 0$, $r \leq n$. Finally, replacing 1 with $C(k+1, k+1)$ in the first term and $C(k+1, 0)$ in the last, we have

$$(a+b)^{k+1} = C(k+1, k+1)a^{k+1} + C(k+1, k)a^k b + \cdots + C(k+1, 1)ab^k + C(k+1, 0)b^{k+1},$$

and the induction step is complete. □

EXAMPLE Use the binomial theorem to expand $(2x-3)^6$.

Here $2x$ is playing the role of a and -3 plays the role of b. By the theorem, we have
$C(6, 6)(2x)^6 + C(6, 5)(2x)^5(-3) + C(6, 4)(2x)^4(-3)^2 + C(6, 3)(2x)^3(-3)^3 + C(6, 2)(2x)^2(-3)^4 + C(6, 1)(2x)(-3)^5 + C(6, 0)(-3)^6$.
Carrying out the arithmetic operations, $(2x-3)^6$ is $64x^6 - 576x^5 + 2160x^4 - 4320x^3 + 4860x^2 - 2916x + 729$. ■

Pascal's triangle is a triangular array of binomial coefficients that has many interesting properties. Although named for Blaise Pascal (1623–62), Pascal's triangle was known as early as 1303 to the Chinese. Let's begin by writing the coefficients for $(a+b)^n$, $n \in \{0, 1, 2, \ldots\}$, with those for each n on a separate row.

$(a+b)^0 = 1$	1	row 0
$(a+b)^1 = a + b$	1 1	row 1
$(a+b)^2 = a^2 + 2ab + b^2$	1 2 1	row 2
$(a+b)^3 = a^3 + 3a^2b + 3ab^2 + b^3$	1 3 3 1	row 3

The triangle can be extended to find the coefficients of $(a+b)^n$ for values of n greater than 3 by adding other rows to the triangle as follows. A row always begins and ends with a 1. Other entries are found by adding the two entries above and to the right and left of the position.

```
        1
      1   1
    1   2   1
  1   3   3   1
    \ /  \ /  \ /
1   4   6   4   1
```

Although this method is easy to carry out, it would be very tedious if you needed the coefficients for the twelfth power of some binomial.

These sums produce the next row because adjacent entries are $C(n, r)$ and $C(n, r-1)$, and we know $C(n, r) + C(n, r-1) \equiv C(n+1, r)$. Some of the other properties of the triangle are also easily explained when we view the entries as values of $C(n, r)$. Here are some examples.

1. Each row of the triangle begins and ends with a 1, because $C(n, n)$ and $C(n, 0)$ are 1 for every $n \in W$.
2. Each row is a palindrome. This happens because $C(n\ r) = C(n, n-r)$ for each $0 \leq r \leq n$.

3. The sum of entries in the nth row is 2^n. In Section 10.1 we saw that $C(n, r)$ counts the number of r-element subsets with n elements. But a set with n elements has a total of 2^n subsets.

Let's investigate two extensions of the binomial theorem. First, suppose you want to expand a power of a trinomial, instead of a binomial. We'll experiment on $(x + y + z)^3$. Each term in the expansion will have degree 3; in fact, all possible terms of degree 3 using x, y, and z will be produced. So there will be x^3, x^2y, x^2z, xyz, y^3, xy^2, y^2z, z^3, xz^2, and yz^2-terms. We now need their coefficients. Certainly x^3, y^3, and z^3 each occurs only once—namely, when each of the three factors $(x + y + z)$ contributes the same variable. How many times can x^2y, x^2z, xy^2, y^2z, xz^2, and yz^2 each appear? To produce x^2y, we need two factors to contribute x, one to contribute y, and none to contribute z. This is like the MISSISSIPPI word problem; x^2y is created $3!/(2!1!0!)$ times. The same argument applies to the other terms with one variable squared. This leaves only the xyz-terms; each of the three factors must contribute a different variable. This can be done $3!/(1!1!1!)$ or $3!$ ways.

$$
\begin{aligned}
(x + y + z)^3 = {} & \frac{3!}{3!0!0!}x^3 + \frac{3!}{2!1!0!}x^2y + \frac{3!}{2!1!0!}x^2z + \frac{3!}{1!1!1!}xyz \\
& + \frac{3!}{3!0!0!}y^3 + \frac{3!}{2!1!0!}xy^2 + \frac{3!}{2!1!0!}y^2z + \frac{3!}{3!0!0!}z^3 \\
& + \frac{3!}{2!1!0!}xz^2 + \frac{3!}{2!1!0!}yz^2
\end{aligned}
$$

The exercises provide more practice with this extension of the binomial theorem.

As another variation, what happens if n is a negative integer in $(a + b)^n$? Here is one special case that has connections with the next sections. We can write an expanded version of $(1 - x)^{-1}$ by beginning the long-division process for $1 \div (1 - x)$.

$$
\begin{array}{r|l}
 & 1 + x + x^2 + x^3 + x^4 + x^5 + \cdots \\
\hline
1 - x & 1 \\
 & \underline{1 - x} \\
 & \quad x \\
 & \quad \underline{x - x^2} \\
 & \qquad x^2 \\
 & \qquad \underline{x^2 - x^3} \\
 & \qquad\quad x^3 \\
 & \qquad\quad \underline{x^3 - x^4} \\
 & \qquad\qquad x^4 \\
 & \qquad\qquad \underline{x^4 - x^5} \\
 & \qquad\qquad\quad \vdots
\end{array}
$$

Let's check. If we accept an infinite version of the distributive property,

$$(1 - x)(1 + x + x^2 + x^3 + x^4 + x^5 + \cdots)$$
$$= (1 + x + x^2 + x^3 + \cdots) + (-x - x^2 - x^3 - \cdots), \text{ or } 1.$$

EXERCISES 10.3

In exercises 1 through 10, write each expression in an expanded form.

1. $(x - 1)^4$
2. $(x + 1)^6$
3. $(x + 3)^5$
4. $(y - 1.1)^3$
5. $(2a + b)^7$
6. $(3x - y)^6$
7. $(-x - 4y)^4$
8. $(-4x - y)^5$
9. $(3x - 2y)^7$
10. $(1 - x)^{14}$

The expression

$$\frac{f(x + h) - f(x)}{(x + h) - x}$$

can frequently be rewritten. In exercises 11 through 16, expand $f(x + h)$ and then rewrite the expression in lowest terms for the given function.

11. $f(x) = x^2$
12. $f(x) = x^5$
13. $f(x) = -x^4$
14. $f(x) = 3x^6$
15. $f(x) = 2x^3 - 4x^2 + x - 3$
16. $f(x) = -2.66x^4 + 5.93x + 3.40$

In exercises 17 through 22, use the functions of exercises 11 through 16 to write the expression

$$\frac{f(x + h) - f(x)}{(x + h) - x}$$

in lowest terms by first factoring the numerator, $f(x + h) - f(x)$. Do not expand $f(x + h)$.

In exercises 23 through 36, give the nth term in the expansion of each expression.

23. $(x - 2y)^{16}$, $n = 4$
24. $(x - 2y)^{16}$, $n = 13$
25. $(2x + 5y)^{41}$, $n = 22$
26. $(2x + 5y)^{41}$, $n = 21$
27. $(2.4x - 1.1)^{103}$, $n = 99$
28. $(1.1x - 2.4)^{103}$, $n = 99$
29. $(\sqrt{3x} - 1)^{33}$, $n = 8$
30. $(\sqrt{3}x - 1)^{33}$, $n = 8$
31. $(5a^2 + 4b^3)^{63}$, $n = 50$
32. $[(5a)^2 + (4b)^3]^{63}$, $n = 50$
33. $(x^{-4} + y^{-1})^{10}$, $n = 4$
34. $(x^{-4} + x^{-1})^{11}$, $n = 6$
35. $(x^5 + x^{-5})^{11}$, $n = 6$
36. $(x^{-1/3} - x^{1/3})^{10}$, $n = 5$
37. What is the constant term in the expansion of $(x^{-6} - 2x^6)^{20}$?
38. Explain to a friend how to determine $C(n, r)$ in the ith term in the expansion of $(x + y)^n$.
39. Think of $C(n + 1, r)$ as the number of ways to choose r-element subsets from a set of $n + 1$ elements. Give an explanation of why $C(n + 1, r) = C(n, r) + C(n, r - 1)$.

In exercises 40 and 41, rewrite the expression in expanded form.

40. $(x - 2y + z)^3$
41. $(3x - 4y + 2z)^2$

In exercises 42 through 46, determine the coefficient of the given term in the expansion of $(x - y + 2z)^6$.

42. x^5y-term
43. xy^5-term
44. x^3y^3-term
45. $x^2y^2z^2$-term
46. xy^2z^3-term

10.4 Sequences

Toolkit

In Section 4.3 the notation $y \to b$ as $x \to +\infty$ was used to indicate that as x grows larger without bound, y approaches b.

The word *term* is used differently in this section than previously.

Definition A **sequence** is a function whose domain is the natural numbers.

For sequences, it's common to write f_n for $f(n)$. A sequence can be described by listing enough values of the dependent variable to establish a pattern for the function values.

Illustration The sequence 1, 2, 4, 8, 16, 32, . . . may be the function $t(n) = 2^{n-1}$, $n \in N$.
The sequence -2, 1, 4, 7, 10, . . . may be the function $m(n) = 3n - 5$, $n \in N$.

Note: We've said these sequences *may* be described by the rules given because there may be other rules that also produce these lists. The technique of successive differences in Section 5.4 can help find a rule that works. □

If you want to be sure there is no misunderstanding about the function, write a functional description: $f(n)$ = expression. Common practice is to combine both descriptive methods.

Illustration The sequence 1, 2, 4, 8, . . . is described with no ambiguity by $a_n = 2^{n-1}$, $n \in N$. Actually, since a sequence has domain N, we needn't write $n \in N$.
The sequence -2, 1, 4, 7, . . . is clearly identified by $a_n = 3n - 5$. □

This descriptive style is explicit; it tells us exactly how to compute the nth term or entry of the list. Another way to give a sequence is recursively: give a starting value and define each value in terms of previous values on the list.

EXAMPLE List the first five terms of the sequence defined by $b_1 = 1$, $b_n = \frac{1}{3}b_{n-1}$.

Each term is one-third times the previous term.
The first five terms are 1, $\frac{1}{3}$, $\frac{1}{9}$, $\frac{1}{27}$, $\frac{1}{81}$. ■

Certain types of sequences are very common, and their properties have been studied. Sequences such as -2, 1, 4, . . . , $3n - 5$, . . . and $b_1 = 4$, $b_n = b_{n-1} - 1.5$ are examples of arithmetic sequences.

Definition In an *arithmetic* sequence there is a fixed difference between successive terms. This common difference is usually denoted by d.

Illustration An example of an arithmetic sequence with common difference 3 is -2, 1, 4, . . ., $-5 + 3n$, ...
An arithmetic sequence with $d = -1.5$ is $b_1 = 4$, $b_n = b_{n-1} - 1.5$. □

Each arithmetic sequence can be defined explicitly by a first-degree expression, $a_n = a_1 + d(n - 1)$, where d is the common difference and a_1 is the first term.

Clearly an arithmetic sequence can be defined recursively using a fixed value for a_1 and $a_n = a_{n-1} + d$.

EXAMPLE Bill has just been hired for his first after-school job. The manager told Bill he'll begin at \$3.35 an hour, and if his work is good, he'll receive an increase of \$0.05 an hour every 2 weeks. If all goes well for him, what will Bill's hourly rate be in 24 weeks?

This is essentially an arithmetic problem, but it is an example of an arithmetic sequence. The first term is \$3.35, the beginning rate; the common difference is \$0.05. In 24 weeks, 12 two-week periods will have passed, **so Bill will be making \$3.35 + (12 − 1)(\$0.05), or \$3.90 per hour.** ■

The next common sequence type we look at is the geometric.

Definition A *geometric* sequence is a sequence with each term except the first obtained by multiplying the previous term by a fixed constant.

In the geometric sequence $1, 2, 4, \ldots, 2^{n-1}, \ldots$, the fixed constant is 2.

Definition For a geometric sequence, the fixed constant, r, is called the *common ratio* for the sequence.

Again, it's easy to see that any geometric sequence can be described either explicitly or recursively.

Illustration We can write $1, 2, 4, \ldots$, where a_n is 2^{n-1}, as $a_1 = 1$, $a_n = 2a_{n-1}$. The common ratio r is 2. We can express the sequence $b_1 = 1$, $b_n = \frac{1}{3}b_{n-1}$ as $1, \frac{1}{3}, \frac{1}{9}, \ldots$, where b_n is $\left(\frac{1}{3}\right)^{n-1}$; r is $\frac{1}{3}$. □

Each geometric sequence can be defined explicitly by $gn = g, (r)^{n-1}$.

EXAMPLE On Saturday afternoon Bill came in to watch the ball game after cutting grass. He decided to get something to drink, but noticed there was only one can of cola. He developed the following strategy: at any break in the game he will pour half of the contents of the can into his glass and consume only that much until the next break. If Bill continued to watch the game forever and there are breaks throughout, give the sequence that serves as a model for the amounts consumed at the breaks.

The mathematical model for this problem is an infinite geometric sequence

$$\mathbf{\frac{1}{2}, \frac{1}{4}, \frac{1}{8}, \frac{1}{16}, \ldots, \frac{1}{2^n}, \ldots}$$

with $a_1 = \frac{1}{2}$, $a_n = \frac{1}{2}a_{n-1}$, and $r = \frac{1}{2}$. ■

Arithmetic and geometric sequences are simple structures. If we know any 2 terms of such a sequence and the type, we can obtain complete information about the sequence.

EXAMPLE If $a_1 = 14$ and $a_{30} = 217$ are terms of an arithmetic sequence, what is a_{20}?

Since the sequence is arithmetic, a_{30} must be $a_1 + (30 - 1)d$; $14 + 29d = 217 \Leftrightarrow d = 7$.

$\mathbf{a_{20} = 14 + (20 - 1)7}$**, or 147.** ■

EXAMPLE If the geometric sequence b has terms $b_2 = 333$ and $b_5 = \frac{37}{3}$, what is b_8?
Since the sequence is geometric, $b_5 = r^3 \cdot b_2$.
$\frac{37}{3} = r^3 \cdot 333 \Leftrightarrow \frac{37}{999} = r^3 \Leftrightarrow \sqrt[3]{\frac{37}{999}} = r.$

Therefore, $b_8 = (\frac{37}{3})(\sqrt[3]{\frac{37}{999}})^3$, or $37^2/2997$. ■

Another type of sequence is defined as follows.

Definition If the reciprocals of the terms of a sequence form an arithmetic sequence, the original sequence is a **harmonic sequence.**

Illustration The sequence $1, \frac{1}{2}, \frac{1}{3}, \frac{1}{4}, \ldots$, where $a_n = 1/n$, is harmonic because the reciprocals of the terms are $1, 2, 3, \ldots, n, \ldots$, an arithmetic sequence.
The sequence $\frac{2}{3}, \frac{2}{6}, \frac{2}{9}, \ldots$, with $a_n = 2/(3n)$, is also harmonic. □

Because arithmetic sequences can be written either explicitly or recursively, harmonic sequences can also be written either way.

Illustration The sequence $\frac{2}{3}, \frac{2}{6}, \frac{2}{9}, \ldots$ can be described by

$$g_1 = \frac{2}{3}, \qquad g_n = \frac{2}{\dfrac{2}{g_{n-1}} + 3}.$$ □

EXAMPLE Alice is idly swinging in a hammock. If she pushes off strongly, the ends of the supports rise 8 inches from the horizontal on the first rise and less on each successive swing. Alice wonders just how these distances are related. She measures and finds this sequence 8 inches, 2 inches, $\frac{8}{7}$ inches, $\frac{4}{5}$ inch, $\frac{8}{13}$ inch, $\frac{1}{2}$ inch, $\frac{8}{19}$ inch. Is this the beginning of a harmonic sequence? If so, what is its tenth term?

A sequence is harmonic if the reciprocals of terms form an arithmetic sequence, so let's look at the reciprocals of Alice's measurements, $\frac{1}{8}, \frac{1}{2}, \frac{7}{8}, \frac{5}{4}, \frac{13}{8}, 2, \frac{19}{8}$. We could probably compare terms more easily if they all had the same denominator.

$$\frac{1}{8}, \frac{4}{8}, \frac{7}{8}, \frac{10}{8}, \frac{13}{8}, \frac{16}{8}, \frac{19}{8}$$

Yes, this is an arithmetic sequence with common difference $\frac{3}{8}$; its tenth term is $\frac{1}{8} + (10 - 1)\frac{3}{8}$, or $\frac{28}{8}$. **The tenth term of Alice's sequence will be $\frac{8}{28}$ inch, or $\frac{2}{7}$ inch.** Will she be aware of the lift in the tenth swing? ■

It isn't always easy to write a sequence in both explicit and recursive formats. As an example, consider Fibonacci-style sequences, where each term after the first two is the sum of the previous two. Those sequences are named for the original Fibonacci sequence: $1, 1, 2, 3, 5, \ldots$, where $f_n = f_{n-2} + f_{n-1}$. Leonardo Fibonacci, a thirteenth-century Italian mathematician, used this sequence to model the growth of a rabbit population under certain ideal conditions.

For Fibonacci's rabbits, assume that you begin with one pair of baby rabbits, a female and a male. When the pair is a month old, they can produce another male-female pair, and they do this on the first day of every ensuing month. Assuming none of the rabbits dies, count the pairs of rabbits every month for 6 months.

At the end of month	0	1	2	3	4	5	6
Number of rabbit pairs	1	1	2	3	5	8	13

An explicit form for Fibonacci's sequence was not developed for 500 years (see exercise 39).

A question of interest regarding any type of sequence is what happens to the terms of the sequence as n becomes larger and larger. For an arithmetic sequence, terms either increase or decrease without bound, depending on whether d is positive or negative. There are more possibilities for the terms of a geometric sequence. Consider the following illustrations.

Illustration $0.1, 0.01, 0.001, \ldots, (0.1)^n, \ldots$ with $n \in N$ is a geometric sequence with common ratio 0.1. As n becomes larger, a_n approaches 0 but is never 0.
$f(n) = 2.3^n$, $n \in N$ is a geometric sequence with common ratio 2.3. For larger values of n, f produces larger numbers. The terms of the sequence grow without bound.
$c_1 = 3$, $c_n = (-0.2)c_{n-1}$ has common ratio -0.2. The terms will be alternately positive and negative as n progresses through the natural numbers. The terms will approach 0 as n gets larger.
$-\pi, \pi^2, -\pi^3, \pi^4, \ldots, (-\pi)^n, \ldots$ with n a natural number, has alternate positive and negative terms. The positive terms increase without bound, and the negative terms decrease without bound as n gets larger. The common ratio is $-\pi$. □

These examples show that the behavior of the nth term as n gets larger for a geometric sequence depends on the value of the common ratio, r. If $-1 < r < 1$, or $|r| < 1$, the terms approach 0; if $|r| > 1$, the terms are unbounded. What happens if $|r| = 1$?

EXAMPLE For the sequence $1, x, x^2, x^3, \ldots, x^n, \ldots$, describe the behavior of the nth term as n grows without bound.
This is a geometric sequence with common ratio x. **If $|x| < 1$, $x^n \to 0$ as $n \to +\infty$. If $|x| > 1$, the terms are unbounded in size. For $|x| = 1$, $|x^n| = 1$ for all n.** ■

EXERCISES 10.4

In exercises 1 through 6, write the first six terms of each sequence with $n \in N$.

1. $t(n) = (-2)^n$
2. $s(n) = (-2)^{n-1}$
3. $m(n) = 8 - 3n$
4. $p_n = 3(n + 1) - 8$
5. $q_n = -(2 + n)^n$
6. $r_n = (2 + n)^{n+2}$

In exercises 7 through 12, write a function t with domain the natural numbers that describes the arithmetic sequence.

7. $1, 6, 11, 16, 21, \ldots$
8. $7, 10, 13, 16, 19, \ldots$
9. $-8, -10, -12, -14, \ldots$
10. $-8, -6, -4, -2, \ldots$
11. $10{,}098, 10{,}129, 10{,}160, 10{,}191, \ldots$
12. $-734, -626, -518, -410, \ldots$

In exercises 13 through 20 write g_n for each geometric sequence.

13. $\frac{1}{4}, \frac{1}{8}, \frac{1}{16}, \frac{1}{32}, \ldots$
14. $-\frac{1}{4}, -\frac{1}{8}, -\frac{1}{16}, -\frac{1}{32}, \ldots$
15. $\frac{1}{4}, -\frac{1}{8}, \frac{1}{16}, -\frac{1}{32}, \ldots$

16. $-\frac{1}{4}, \frac{1}{8}, -\frac{1}{16}, \frac{1}{32}, \ldots$

17. 11, 55, 275, 1375, 6875, . . .

18. $11, \frac{11}{5}, \frac{11}{25}, \frac{11}{125}, \ldots$

19. $-\frac{2}{3}, -\frac{2}{7}, -\frac{6}{49}, -\frac{18}{343}, \ldots$

20. $-\frac{2}{3}, \frac{2}{7}, -\frac{6}{49}, \frac{18}{343}, \ldots$

In exercises 21 through 30, identify each of the following as an arithmetic sequence, a geometric sequence, a harmonic sequence, a Fibonacci-type sequence, or none of these. If arithmetic, state the common difference *d*; if geometric, state the common ratio, *r*.

21. 3, 3.3, 3.63, 3.993, 4.3923, . . .

22. 3, 3.3, 3.6, 3.9, 4.2, . . .

23. $\frac{1}{3}, \frac{1}{5}, \frac{1}{7}, \frac{1}{9}, \frac{1}{11}, \ldots$

24. $3, -\frac{3}{10}, \frac{3}{100}, -\frac{3}{1000}, \ldots$

25. −4.33, −6.32, −8.31, −10.3, . . .

26. 1, 5, 6, 11, 17, 28, . . .

27. 25, 36, 49, 64, . . .

28. $-\frac{1}{4}, -\frac{1}{12}, -\frac{1}{36}, -\frac{1}{108}, \ldots$

29. −6, 4, −2, 2, 0, 2, 2, 4, 6, . . .

30. −6, 4, 14, 24, . . .

In exercises 31 through 38, write the given term for each sequence.

31. a_{10} of $\frac{1}{4}, -\frac{1}{8}, \frac{1}{16}, -\frac{1}{32}, \ldots$

32. a_{109} of $\frac{1}{4}, -\frac{1}{8}, \frac{1}{16}, -\frac{1}{32}, \ldots$

33. a_{24} of 7, −1, −9, −17, . . .

34. a_{17} of $\frac{1}{7}, -1, -\frac{1}{9}, -\frac{1}{17}, \ldots$

35. a_{10} of the arithmetic sequence with $a_1 = 3$, $a_{34} = 85.5$

36. g_{10} of the geometric sequence with $g_1 = 1$, $g_{23} =$ 4,194,304

37. a_{42} of the harmonic sequence associated with the arithmetic sequence of exercise 35

38. a_3 of the arithmetic sequence with $a_{47} =$ 10,325 and $d = -2$

39. An explicit formula for the nth Fibonacci number is

$$f_n = \frac{1}{\sqrt{5}}\left[\frac{1+\sqrt{5}}{2}\right]^n - \frac{1}{\sqrt{5}}\left[\frac{1-\sqrt{5}}{2}\right]^n.$$

Use this to calculate f_3 and compare with f_3 obtained recursively.

40. List the first 20 terms of the Fibonacci sequence; below each term, beginning with the second term, write f_n/f_{n-1}. Using a calculator to investigate these ratios, what do you conclude?

10.5 Sigma Notation and Series

Definition Any sequence can be used to generate a new sequence whose n^{th} term is the sum of the first n terms of the original sequence. The new sequence is called a **series.**

The original sequence is referred to as the *base sequence* for that series. To emphasize that we are viewing a sequence as a series, we call the terms *partial sums*.

Base sequence	*Series*	*Third partial sum*
$1, 2, 4, 8, \ldots, 2^{n-1}, \ldots$	1, 3, 7, 15, . . .	1 + 2 + 4, or 7
$-2, 1, 4, 7, \ldots, -5 + 3n, \ldots$	−2, −1, 3, 10, . . .	−2 + 1 + 4, or 3
$f_1 = 1, f_2 = 1,$ $f_n = f_{n-2} + f_{n-1}$	1, 2, 4, 7, 12, . . .	1 + 1 + 2, or 4

A recursive definition of a series $A_1, A_2, A_3, \ldots$ with base sequence $a_1, a_2, a_3, \ldots$ is straightforward: $A_1 = a_1$, $A_n = A_{n-1} + a_n$.

Illustration The sequence $\frac{1}{2}, \frac{1}{4}, \frac{1}{8}, \frac{1}{16}, \ldots$ appeared in the last section to describe the amounts of cola Bill drank at each break in the ball game. Each term of the associated series,

$\frac{1}{2}, \frac{1}{2} + \frac{1}{4}, \frac{1}{2} + \frac{1}{4} + \frac{1}{8}, \frac{1}{2} + \frac{1}{4} + \frac{1}{8} + \frac{1}{16}, \ldots$ gives the total amount of cola consumed by the end of the nth break. □

To describe explicitly the nth term in a series, it's useful to employ sigma notation. This is just a shorthand for writing sums. We write a capital sigma, Σ, as the signal to add, provide a generic description of the summands, and give a counter that tells where to begin and end the summing. The counter is called the *index of summation*. With sigma notation we can write $1 + 3 + 5 + 7 + 9 + \cdots + 107$ as

$$\sum_{i=1}^{54} (2i - 1).$$

The generic description of each summand is $2i - 1$; i, the index, counts from 1 to 54 by 1s. Notice that when i is 54, $2i - 1$ is 107, the last term to be added.

EXAMPLE Write the expanded version of

$$\sum_{j=2}^{17} \frac{1}{j - 1}.$$

The index j takes on integer values from 2 through 17, producing

$$\frac{1}{2-1} + \frac{1}{3-1} + \frac{1}{4-1} + \cdots + \frac{1}{17-1}; \quad \text{or} \quad \mathbf{1 + \frac{1}{2} + \frac{1}{3} + \cdots + \frac{1}{16}}. \blacksquare$$

Sigma notation is very flexible. Even though the index always counts by 1s, adjusting the description of the summands can produce the effect of counting by other size steps. The index may be used merely as a subscript if there is not a convenient way to express the summands as a function of the index. For example, the terms of the series generated by the Fibonacci sequence could be given by

$$F_n = \sum_{k=1}^{n} f_k, \qquad \text{where } f_1 = f_2 = 1, \quad f_k = f_{k-2} + f_{k-1}.$$

Here are some useful properties for dealing with sigma notation.

$$\sum_{i=1}^{k} a_i + \sum_{i=k+1}^{n} a_i = \sum_{i=1}^{n} a_i \qquad \text{for all } k \text{ such that } 1 \le k \le n$$

$$\sum_{i=1}^{k} (a_i + b_i) = \sum_{i=1}^{k} a_i + \sum_{i=1}^{k} b_i$$

$$\sum_{i=k}^{k} a_i = a_k \qquad \sum_{i=1}^{n} c = nc \qquad \sum_{i=1}^{n} ca_i = c\sum_{i=1}^{n} a_i$$

In Section 10.4 you saw four different types of sequences. Let's investigate the types of series these sequences generate and look for explicit forms. We will name the series according to the base sequence type.

The sequence $3, 10, 17, 24, \ldots, 3 + 7(n - 1), \ldots$, is arithmetic with common difference 7. The associated arithmetic series is $3, 13, 30, 54, \ldots$. The nth partial sum A_n should be $3n$ plus some number of sevens; how many sevens?

$$A_1 = 3, \qquad A_2 = 3 + 10, \qquad A_3 = 3 + 10 + 17,$$

$$A_n = \sum_{j=1}^{n} [3 + 7(j - 1)]$$

Using the properties of sigma notation, we have

$$A_n = \sum_{j=1}^{n} (3 + 7(j - 1)) \Leftrightarrow A_n = \sum_{j=1}^{n} 3 + \sum_{j=1}^{n} 7(j - 1)$$

$$\Leftrightarrow A_n = 3n + 7\sum_{j=1}^{n} (j - 1).$$

There are several ways to get a short form for

$$\sum_{j=1}^{n} (j - 1).$$

Here's one; another is described just before exercise 35. In fact, we first saw this problem in exercise 46 of Section 4.1 and then again in exercise 25 of Section 10.2. Certainly

$$\sum_{j=1}^{n} (j - 1) = 0 + 1 + 2 + 3 + \cdots + (n - 2) + (n - 1),$$

and

$$\sum_{j=1}^{n} (j - 1) = (n - 1) + (n - 2) + \cdots + 3 + 2 + 1 + 0,$$

so

$$2\sum_{j=1}^{n} (j - 1) = \underbrace{(n - 1) + (n - 1) + (n - 1) + \cdots + (n - 1)}, \text{ } n \text{ terms}$$

or

$$(n - 1)n \quad \text{and} \quad \sum_{j=1}^{n} (j - 1) = \frac{(n - 1)n}{2}.$$

Our conclusion is that the series associated with

$$3, 10, 17, 24, \ldots, 3 + 7(n - 1), \ldots$$

is

$$3, 13, 30, \ldots, 3n + 7\frac{(n - 1)n}{2}, \ldots.$$

We never used any special facts about 3 or 7, so we can generalize the pattern.

If $A_1, A_2, A_3, \ldots$ is the arithmetic series based on a_1, $a_n = a_{n-1} + d$, then

$$A_n = a_1 n + d\frac{(n - 1)n}{2}.$$

Another form for A_n may be useful, too.

$$a_1 n + d\frac{(n-1)n}{2} = \frac{2a_1 n + d(n-1)n}{2}$$
$$= \frac{n[a_1 + a_1 + d(n-1)]}{2}, \quad \text{or} \quad \frac{n(a_1 + a_n)}{2}.$$

The nth partial sum, A_n, is half the product of n and the sum of the first and n^{th} terms of the base sequence.

EXAMPLE This week's grocery special is canned tomato soup. Karen has made a triangular display of soup cans by putting one can less in each successive row. If the bottom row has 25 cans, how many cans are in the display?
The numbers of cans in each row form part of an arithmetic sequence: 25, 24, . . . , 1. The total number of cans is the 25th partial sum or $25(25 + 1)/2$.
There are 325 cans in all. ■

A geometric series $G_1, G_2, G_3, \ldots$ is based on a geometric sequence g_1, $g_n = rg_{n-1}$. We have defined G_n to be $g_1 + g_2 + g_3 + \cdots + g_n$ or $g_1 + rg_1 + r^2g_1 + \cdots + r^{n-1}g_1$. To find a short form for G_n, we do some fancy arithmetic somewhat in the style of what we did earlier. We look at

$$G_n = g_1 + rg_1 + r^2g_1 + \cdots + r^{n-1}g_1$$

and

$$rG_n = rg_1 + r^2g_1 + r^3g_1 + \cdots + r^ng_1,$$

so

$$G_n - rG_n = g_1 - r^ng_1 \quad \text{and} \quad G_n(1-r) = g_1(1-r^n).$$

The nth partial sum in a geometric series, G_n, is

$$g_1\frac{1-r^n}{1-r}$$

if the base sequence is g_1, $g_n = rg_{n-1}$.

EXAMPLE A ball bounces to a height that is $\frac{2}{3}$ the height from which it fell. If the ball falls from a height of 5 feet, how far will it have traveled at the time it bounces the sixth time?
At the first bounce the ball has traveled 5 feet, at the second bounce, $2 \cdot (\frac{2}{3}) \cdot 5$ feet (because it travels up and back down), at the third bounce, $2 \cdot (\frac{2}{3})^2 \cdot 5$ feet and so on. The sum is

$$5 + 2\cdot\left(\frac{2}{3}\right)\cdot 5 + 2\cdot\left(\frac{2}{3}\right)^2\cdot 5 + 2\cdot\left(\frac{2}{3}\right)^3\cdot 5 + 2\cdot\left(\frac{2}{3}\right)^4\cdot 5 + 2\cdot\left(\frac{2}{3}\right)^5\cdot 5.$$

This is 5 plus the fifth partial sum of the base sequence $g_1 = 2(\frac{2}{3})5$, $g_n = \frac{2}{3}g_{n-1}$.

Using the preceding formula, the fifth partial sum is

$$2\left(\tfrac{2}{3}\right)(5)\frac{1-\left(\frac{2}{3}\right)^5}{1-\frac{2}{3}}.$$

The ball has traveled $22\frac{89}{243}$ feet at the time it bounces the sixth time. ■

Earlier we investigated what happened to terms of a sequence as $n \to +\infty$. What happens to G_n as $n \to +\infty$? Since

$$G_n = g_1\frac{1-r^n}{1-r},$$

the answer must depend on the behavior of r^n. We know that for $|r| < 1$, $r^n \to 0$ as $n \to +\infty$, so G_n approaches

$$g_1\frac{1}{1-r}, \quad \text{or} \quad \frac{g_1}{1-r}.$$

We could also express this as

$$\sum_{i=1}^{+\infty} g_i = \frac{g_1}{1-r}, \qquad |r| < 1.$$

EXAMPLE As Bill's game goes on and on, what happens to the total amount of cola he drinks?

The base geometric sequence has a common ratio of $\frac{1}{2}$. **The amount of cola he drinks will approach**

$$\frac{\frac{1}{2}}{1-\frac{1}{2}},$$

or **1 can of cola.** ■

If $r > 1$, G_n will grow without bound. This tells us something about an example from Section 10.3, $(1-x)^{-1} = 1 + x + x^2 + x^3 + \cdots$. The right-hand side will represent a number if and only if $|x| < 1$, because its value is the number the nth partial sum approaches as $n \to +\infty$.

The next type of sequence discussed was harmonic. In general there is not a short form for the nth partial sum, H_n, of a harmonic series. Nor can we easily describe the behavior of H_n as $n \to +\infty$. Even if $h_n \to 0$ as $n \to +\infty$, H_n may not have a bound.

The Fibonacci-type sequences generate series with some elegant patterns. We examine the Fibonacci series based on the original, 1, 1, 2, 3, 5, 8, 13, ...

n	1	2	3	4	5	6	7
F_n	1	2	4	7	12	20	33

Compare these partial sums with the terms of the base sequence. It seems that F_n is $f_{n+2} - 1$. Let's see if we can prove this. The conjecture uses the natural numbers as subscripts, so an induction proof is appropriate.

> **Prove:** The nth partial sum, F_n, in the Fibonacci series is $f_{n+2} - 1$.

Proof: ***Basis step:*** The smallest value of n we can use is 1. F_1 is f_1; the right-hand side is $f_{n+2} - 1$, or $f_3 - 1$. But f_1 is 1 and f_3 is 2, so the basis step is confirmed.

Induction step: $P(k)$: $F_k = f_{k+2} - 1$; $P(k+1)$: $F_{k+1} = f_{k+3} - 1$. Because $F_{k+1} = F_k + f_{k+1}$, we have $F_{k+1} = (f_{k+2} - 1) + f_{k+1}$, or $F_{k+1} = f_{k+3} - 1$, by the definition of f_{k+3}. The induction step is complete. □

Is there a similar pattern for any Fibonacci-type series? The induction step of the proof would work with any constant in place of the 1. In the basis step, the 1 is special. It's there because $f_1 = f_3 - f_2$ and f_2 is 1. This suggests that for a general Fibonacci-type sequence, $\mathscr{f}_1, \mathscr{f}_2, \mathscr{f}_n = \mathscr{f}_{n-1} + \mathscr{f}_{n-2}$, its nth partial sum, $\mathscr{F}_n$ is $\mathscr{f}_{n+2} - \mathscr{f}_2$. Providing the proof of this is one of the exercises. This short form for $\mathscr{F}_n$ tells us that as $n \to +\infty$, $\mathscr{F}_n$ increases or decreases without bound as $\mathscr{f}_{n+2}$ does.

EXERCISES 10.5

In exercises 1 through 8, write the first five terms S_n (the first five partial sums) of the series based on the sequence s_n.

1. $s_n = 2 + 4n$
2. $s_n = 2s_{n-1}$, $s_1 = 5$
3. $s_n = \dfrac{1}{n}$
4. $s_n = s_{n-2} + s_{n-1}$, $s_1 = s_2 = 2$
5. $s_1 = 3.1$, $r = 0.1$
6. $s_n = \frac{1}{3}s_{n-1}$, $s_1 = \frac{2}{3}$
7. 7, 10, 13, 16, . . .
8. $-\frac{1}{4}, \frac{1}{8}, -\frac{1}{16}, \frac{1}{32}, \ldots$

In exercises 9 through 14, first expand the expression to find the sum and then use the properties of sigma notation to evaluate the expression.

9. $\sum_{i=1}^{10} (2i)$
10. $\sum_{i=2}^{10} (-0.02i)$
11. $\sum_{i=1}^{10} (1 + 2i)$
12. $\sum_{i=3}^{10} (1 - 2i)$
13. $\sum_{i=1}^{5} (1 + 3i) + \sum_{i=6}^{10} (1 + 3i)$
14. $\sum_{i=1}^{5} (2 + 3i) + \sum_{i=6}^{10} (1 + 3i)$

In exercises 15 through 20, use the formula developed in this section to compute the nth term of the arithmetic series A_n associated with the arithmetic sequence a_n.

15. $a_n = 3 + 2n$
16. $a_n = 7n$
17. $-8, -10, -12, -14, \ldots$
18. $-8, -6, -4, -2, \ldots$
19. $a_n = 13.5 - 2n$
20. $a_n = -13.5 + 2n$

In exercises 21 through 26, use the formula developed in this section to compute the nth term of the geometric series G_n associated with the geometric sequence g_n.

21. $g_n = 4 \cdot \left(\frac{1}{2}\right)^{n-1}$
22. $g_n = \left(\frac{2}{3}\right)^n$
23. 11, 55, 275, 1375, . . .
24. $-\frac{2}{3}, -\frac{2}{7}, -\frac{6}{49}, -\frac{18}{343}, \ldots$
25. $g_n = 3\left(-\frac{1}{10}\right)^n$
26. $g_n = 3\left(-\frac{1}{10}\right)^{n-1}$

Exercises 27 and 28 deal with series formed from sequences of the form $p_n = n^k$, where k is a constant. The series formed is called a power series.

27. In Section 10.2, exercise 29, you were asked to prove that $1^2 + 2^2 + 3^2 + \cdots + n^2 = \left(\frac{1}{6}\right)(n)(n+1)(2n+1)$. Use this to evaluate the eighteenth partial sum of the power series P_n based on the sequence $p_n = n^2$.
28. In Section 10.2, exercise 30, you were asked to prove that $0^2 + 1^2 + 2^2 + \cdots + (n-1)^2 = \left(\frac{1}{6}\right)(n-1)(n)(2n-1)$. What do you think the formula is for $8^2 + 9^2 + 10^2 + \cdots + (n+7)^2$? Try out your formula. Does it seem to work?

In exercises 29 through 34, identify each as the nth term of an arithmetic, geometric, harmonic, or power sequence. Where possible, give a formula for the nth partial sum of each series based on the sequence.

29. $p_n = 4(-\frac{1}{3})^n$

30. $q_n = \dfrac{1}{n-2}$

31. $s_n = -7 + \frac{1}{2}n$

32. $r_n = [n(n-1)]^2$

33. $t_n = 4 + 3n$

34. $u_n = n^3$

Another method for arriving at the formula for the nth partial sum of an arithmetic series is attributed to Gauss. Supposedly, as a child he noticed that the sum of the first and last terms of the partial sum were the same as the sum of the second and next-to-last terms, and so on. He concluded that there were half as many of these sums as there were terms in the partial sum. Use this method to find each sum in exercises 35 through 38.

35. $1 + 2 + 3 + 4 + \cdots + 100$

36. $1 + 2 + 3 + 4 + \cdots + 101$

(Confirm that this method works with an odd number of summands.)

37. $3.01 + 3.03 + 3.05 + \cdots + 9.13$

38. $2908 + 2911 + 2914 + \cdots + 3070$

39. How does the verbal description used in exercises 35 through 38 compare to the formula developed in this section?

In exercises 40 through 43, evaluate the expression.

40. $8 + 10 + 12 + \cdots + 464$

41. $\dfrac{5}{6} + \dfrac{5}{12} + \dfrac{5}{24} + \cdots + \dfrac{5}{12{,}228}$

42. $-\dfrac{1}{3} + \dfrac{1}{9} - \dfrac{1}{27} + \cdots + \dfrac{1}{4{,}782{,}969}$

43. $-7.2 - 6.9 - 6.6 + \cdots + 18.0$

44. In Section A.5, exercise 63, you are asked to develop a pattern for $(x^n - y^n)/(x - y)$. Replacing x with 1 and y with r, compare your pattern with G_n for a geometric series whose base sequence is g_1, $g_n = r^{n-1}g_1$.

Chapter 10: Summary

This chapter begins with some counting techniques, most importantly those for permutations and combinations. Mathematical induction, a powerful proof technique, was used to prove some of our counting facts and other statements from earlier chapters. For sequences and series, we attempted to find both explicit and recursive definitions of terms and investigated the behavior of terms for large values of n.

Key Concepts

multiplication principle	
addition principle	
permutation	*with or without repeats*
factorial	
combination	*with or without repeats*
mathematical induction	*basis step, induction step*
binomial theorem	
sequences	*arithmetic, geometric, harmonic, Fibonacci-type*
series	*term, partial sum*

Skills

You should be able to:

1. Apply the multiplication principle or the addition principle of counting.
2. Count permutations with or without repeats.
3. Count combinations with or without repeats.
4. Give a proof using mathematical induction.
5. Use the binomial theorem.
6. Identify arithmetic, geometric, harmonic, or Fibonacci-type sequences.
7. Determine the nth term of a sequence.
8. Determine the nth partial sum of a series.
9. Use sigma notation.

SUMMARY EXERCISES

The starred exercises can serve as a practice chapter test. The answers to these are at the end of the answer section at the back of the book.

In exercises 1 through 9, evaluate each expression.

1. $6!$

2. $\dfrac{9!}{2!3!4!}$

* **3.** $P(10, 5)$

4. $P(10, 1)$

5. $P(10, 10)$

* **6.** $C(10, 5)$

7. $C(10, 1)$

8. $C(10, 10)$

* **9.** $\dfrac{(n + r - 1)!}{(n - 1)!r!}$, $n = 10$, $r = 4$

10. A breakfast menu contains 3 choices of fruit, 4 main choices (eggs, pancakes, French toast, croissants), and 6 beverage choices. How many breakfasts are there if one choice must be made from each of the three categories?

11. If you are not required to make a choice from each of the three categories, how many breakfast orders can be made?

* **12.** Monique has purchased 15 different bedding plants and decides to use 7 of them in her window box. How many arrangements can she make?

13. James also purchases 15 bedding plants, but 3 are geraniums and 5 are marigolds. How many different arrangements can he make of the 15 plants if he can't tell the geraniums apart and he can't tell the marigolds apart?

14. David and Elaine must choose 3 of their 7 dogs to enter in the local dog show. In how many ways can they make the choice?

* **15.** In the student council elections, 4 of the candidates are freshmen, 6 are sophomores, 5 are juniors, and 4 are seniors. If 6 representatives are to be elected, in how many ways could 1 freshman, 1 sophomore, 2 juniors, and 2 seniors be elected?

16. Write the statement $P(5)$ if $P(n)$ states that the sum of the cubes of the first n counting numbers is one-fourth the product of the square of n and the square of one more than n. Is $P(5)$ true?

17. Write the basis step and the induction step for a proof by mathematical induction of $P(n)$ from exercise 16.

* **18.** Prove or disprove $P(n)$ from exercise 16.

19. Expand $(2a - 3b)^5$.

20. What is the coefficient of the x^6y^{31}-term in the expansion of $(x + y)^{37}$?

* **21.** What is the constant term in the expansion of $(2x - x^{-1})^{14}$?

22. What is the 32nd term in the expansion of $(0.33c^{1/3} - 1.29d^{1/2})^{55}$?

In exercises 23 through 29, identify each sequence as arithmetic, geometric, harmonic, Fibonacci-type, or none of these.

23. $s_n = \frac{3}{2}s_{n-1}$, $s_1 = 6$

* **24.** $s_n = 14 - 3(n - 1)$

25. $s_n = \dfrac{6}{2n - 1}$

26. 3, 4, 7, 11, 18, 29, . . .

* **27.** 1, 8, 27, 64, 125, . . .

28. −0.2, 0.08, −0.032, 0.0128, . . .

29. −0.2, −0.08, 0.04, 0.16, . . .

In exercises 30 through 32, write the first five terms of each sequence.

* **30.** $s_n = \frac{3}{2}s_{n-1}, s_1 = 6$
31. $s_n = 14 - 3(n - 1)$
32. $s_n = \dfrac{6}{2n - 1}$

In exercises 33 through 36, write the nth term of each sequence.

* **33.** 3, 4, 7, 11, 18, 29, . . .
34. 1, 8, 27, 64, 125, . . .
35. −0.2, 0.08, −0.032, 0.0128, . . .
* **36.** −0.2, −0.08, 0.04, 0.16, . . .
37. What is the 30th term of an arithmetic series with $a_1 = 34$ and $a_{54} = 35.06$?
38. What is the behavior of the nth term of $1, \frac{1}{3}, \frac{1}{9}, \frac{1}{27}, \ldots$ as n grows without bound?

In exercises 39 through 43, write the nth partial sum of the series S_n based on the given sequence.

* **39.** 1, 2, 4, 8, . . .
40. −2, 1, 4, 7, . . .
41. 1, −2, 4, −8, . . .
* **42.** 2, −1, −4, −7, . . .
43. 1, 1, 2, 3, 5, 8, . . .

In exercises 44 through 49, use the properties of sigma notation and any formulas that have been developed in the chapter to evaluate the expressions.

44. $\sum_{i=1}^{5} 4 + \sum_{i=6}^{9} 4$

* **45.** $\sum_{i=1}^{18} i$

46. $\sum_{i=1}^{18} ai$

47. $\sum_{i=1}^{15} i^2$

* **48.** $\sum_{i=1}^{11} (3i - 5)$

49. $\sum_{i=1}^{11} (2i^2 + 3i - 5)$

APPENDIX

Preview and Review

- A.1 Introduction
- A.2 Sets
- A.3 Numbers
- A.4 Exponents and Radicals
- A.5 Expressions and Operations

Toolkit

For this appendix you will need to know about arithmetic with real numbers—addition, subtraction, multiplication, division, and powers for signed numbers, fractions, and decimals.

A.1 Introduction

In this appendix we will review some of the basic ideas needed throughout the book and consider some of the extensions of these ideas.

Definition A number whose value is not specified may be represented by a letter. This letter is called a **variable.** We refer to a number whose value is specified as a **constant.**

Occasionally a constant is denoted by a letter for convenience—for example, π, pi. In writing expressions involving variables, we generally denote multiplication with juxtaposition (writing factors next to each other) and division with a vinculum (the long bar in a fraction).

Illustration

$3x$ means three times x.

kxy means k times x times y.

$\dfrac{14}{m_2}$ means fourteen divided by m_2 (read "m sub 2"). □

Expressions that use a $-$ sign should be read carefully. If it is between two numbers, whatever their form, the $-$ indicates subtraction and is read "minus." If it appears in front of a simple constant, the $-$ is read "negative." In any other case, it should be read "the opposite of" to avoid confusion. Is $-x$ a positive or a negative number?

Illustration

$2x - y$	2 times x minus y
-147	Negative 147
-3^2	The opposite of the square of 3
$-a$	The opposite of a
$-(3m + 2)$	The opposite of the sum of 3 times m and 2 □

This seems a good place for a reminder about the order in which we perform basic operations in evaluating an expression.

1. Symbols of grouping are a signal to carry out the operations within them first. Symbols of grouping include parentheses, brackets, braces and the vinculum, which groups both the numerator and denominator. These symbols may be doing double duty by also indicating an operation to be carried out next, for example, the vinculum.
2. Exponents are used next.

3. Multiplications and divisions follow. These are done in "reading order," from left to right.
4. The operation of taking the opposite of a number is done next.
5. Additions and subtractions are now done in reading order, left to right.

You should test your calculator to see if it follows the same rules.

A more complete description of the order of operations appears in Section 4.1.

EXERCISES A.1

These exercises are designed to help you to become familiar with your calculator and how it handles various operations and the order of operations.

In exercises 1 through 14, approximate the value of the expression correct to three decimal places. To represent the approximation for $\sqrt{2}$ correct to three decimal places, we will write 1.414 (approx.).

1. $-2.456\left[4.119 - \left(314.005 + \frac{4.999}{7.375}\right)\right]$
2. $\frac{6.4887 - 98.3245}{32.9811} - \left(-56.0998\right)\left(0.0034\right)$
3. 8.567^2
4. -8.567^2
5. $(-8.567)^2$
6. $(-8.567)^3$
7. $\sqrt[3]{8.1}$
8. $\sqrt[4]{0.283 + 7.2^2}$
9. $-34.879\sqrt{34.997 - 2\sqrt{0.493}}$
10. $14^{2/5} + \frac{1}{14^2 - 3}$
11. $\frac{58.2985}{[3.8566 - (2.4498)(-34.8821)]^{3.1}}$
(Try evaluating the denominator and then using the $1/x$ key.)
12. $\frac{7 + 2\sqrt{5}}{4 - 3\sqrt{2}}$
13. $13.333\pi + 4^{3.1}$
14. $\frac{7.32^{5/2}}{2\pi - 6.24} + 7.003$
15. Arrange the following numbers in order from smallest to largest.

$$\sqrt[3]{-43.894},\ 472.39,\ 2\pi - 11,\ -3.52751,\ 14^{7/3}$$

16. Use your calculator to approximate the value of each expression correct to 6 decimal places.

$$\frac{1}{-0.9}, \frac{1}{-0.99}, \frac{1}{-0.999}, \frac{1}{-0.9999}, \frac{1}{-0.99999}, \frac{1}{-0.999999}$$

Is there a number to which your answers are coming closer and closer?

17. Use your calculator to approximate the value of each expression to five decimal places.

$$1.7^2,\ 1.73^2,\ 1.732^2,\ 1.7321^2,\ 1.73205^2,\ 1.732051^2$$

Is there a number to which your answers are coming closer and closer? Based on this, for what is 1.732051 a good approximation?

18. The formula for computing the volume of a can is $V = \pi r^2 h$, where

V represents the volume,
r represents the radius,
h represents the height.

What is the three-decimal-place approximation of the volume of a can with radius 2.843 inches and height 4.113 inches?

19. The formula for computing the surface area of a can is $SA = 2\pi rh + 2\pi r^2$, where

SA represents the surface area,
r represents the radius,
h represents the height.

What is the three-decimal-place approximation of the surface area of a can with radius 2.843 and height 4.113?

20. The formula for computing the resistance of a circuit wired in parallel is

$$\frac{1}{R} = \frac{1}{R_1} + \frac{1}{R_2}$$

where

R is the resistance of the circuit,
R_1 is the resistance of one wire,
R_2 is the resistance of the other wire.

What is the approximate resistance, correct to three decimal places, of a circuit wired in parallel if the resistance of one wire is 3.875 ohms and the resistance of the other wire is 5.399 ohms?

21. Jane and Jim decide to deposit \$3,500 in a money-market account that earns 8.02% interest compounded monthly. The formula for computing the amount of an investment is

$$A = P\left[1 + \frac{i}{m}\right]^{mn}$$

where

P represents the principal or initial investment,

i represents the quoted yearly interest rate as a decimal,

m represents the number of compounding periods per year,

n represents the number of years,

A represents the amount at the end of mn periods.

If they make no other deposits or withdrawals for 3 years, how much will be in their account?

22. Before Jane and Jim make the deposit, they find that another savings institution offers 8.08% interest compounded quarterly. Would they earn more money investing with this institution or the one in exercise 21?

23. In statistics z-scores are used to compare means (averages) on a standard scale. To compute a z-score for two populations, the following formula is used:

$$z = \frac{\overline{x_1} - \overline{x_2}}{\sqrt{(s_1^2/n_1) + (s_2^2/n_2)}},$$

where

z represents the z-score,

n_1 represents the size of the first sample,

$\overline{x_1}$ represents the sample mean (average) of the first sample,

s_1 represents the standard deviation of the first sample,

and n_2, $\overline{x_2}$, s_2 represent the corresponding values of the second sample.

What is the approximate z-score, correct to two decimal places, for a study comparing the following two groups of students? There are 42 students in the first group with a mean grade-point average of 2.56 and standard deviation 1.13. There are 37 students in the second group with a mean grade point average of 2.77 and standard deviation 1.02.

24. In statistics if the sample sizes are less than 30, a t-score is computed instead of a z-score. The formula is

$$t = \frac{\overline{x_1} - \overline{x_2}}{s_p\sqrt{(1/n_1) + (1/n_2)}},$$

where

$$s_p = \sqrt{\frac{(n_1 - 1)s_1^2 + (n_2 - 1)s_2^2}{n_1 + n_2 - 2}}$$

and the variables represent the same values as those in exercise 23. What is the approximate t-score, correct to two decimal places, for the following study, which differs from that of exercise 23 only in the sample sizes of the two groups? There are 24 students in the first sample and 15 students in the second sample.

A.2 Sets

To solve an equation or inequality means to find the set of numbers that make the statement true. Such numbers are called *the solutions of the equation or inequality*. In order to have a convenient way of describing solutions for equations and inequalities, you need to be familiar with some simple ideas about sets.

Definition A **set** is a well-defined collection of objects.

"Well-defined" means we can always decide if an object belongs to the set or not. For example, the collection of tall people in our math class is not well defined. The collection of people in our math class that are 6 feet or taller is well defined.

Sets may be represented in a variety of ways. If the set is small or has a simple pattern, the objects or elements may be listed within braces. A set may be given a name, often a single letter, for easy reference.

Illustration $A = \{4, 7, a, \text{Tom}\}$
$B = \{0, 3, 6, 9, \ldots, 180\}$
$C = \{-5, 0, 5, 10, \ldots\}$ □

This style of describing a set is called enumeration. The three dots (an ellipsis) in sets B and C represent unwritten elements that follow the pattern established. Never use an ellipsis unless the pattern is clear.

A set whose elements are numbers that conform to a pattern may be described with *set-builder notation*.

Illustration $D = \{x \mid x = 3n - 2, n \text{ is a counting number}\}$
This description of D is read "the set of all x such that $x = 3n - 2$, where n is a counting number." Here we could also write $D = \{1, 4, 7, 10, \ldots\}$, but if we hadn't restricted n to be a counting number, it might not be possible to use enumeration. □

The set E below has no simple enumeration.

Illustration $E = \{m \mid m = 1.5r + 7, r \text{ is a real number}\}$ □

Although set-builder notation should include a description of what can be used to build the elements as well as the pattern to follow, we'll often omit the description. In such cases it's assumed that any number that makes sense in the pattern is allowable. We discuss this again when we talk about functions in Chapter 4.

Another common way to represent a set is with a picture, geometrically. We use this most often when the set elements form a piece of the real number line. Consider $F = \{q \mid q < 7\}$ and $G = \{t \mid t \geq 12\}$. The boundary points, 7 and 12, must be distinguished in some way to show that one belongs to the set described and the other does not. In this text we use a square bracket,], to mark a boundary point that is an element and a parenthesis,), to mark one that is not an element. (You may have used hollow dots, ○, and solid dots, ●, before.) Either symbol should open towards the rest of the set. The rest of the set is shaded on the number line.

Illustration

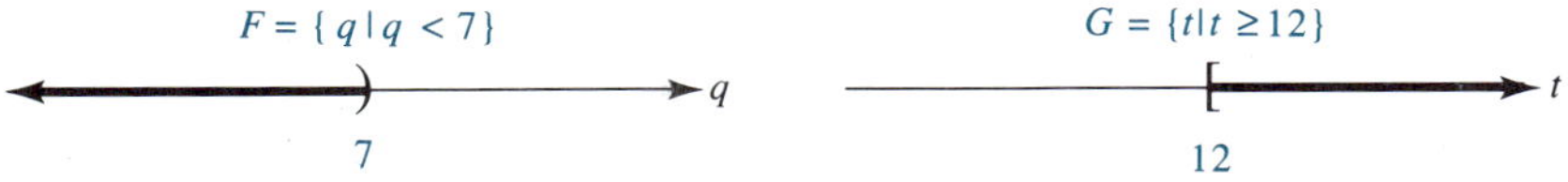

□

Just as we sometimes abbreviate set-builder notation, we can sometimes condense the geometric description of a set into interval notation. An **interval** is a connected piece of the real number line. An interval is determined by its boundary points. If the interval extends without bound to the right, we use the symbol $+\infty$ in place of the right boundary. If the interval has no left boundary, we use $-\infty$. These symbols $+\infty$ and $-\infty$ are read "positive infinity," and "negative infinity." Since neither $+\infty$ nor $-\infty$ can belong to the set, they will be enclosed with a parenthesis. Boundary points are enclosed with a square bracket or a parenthesis as they would be in a number line picture.

Illustration $F = \{q \mid q < 7\}$ can be represented as $(-\infty, 7)$.
$G = \{t \mid t \geq 12\}$ can be represented as $[12, +\infty)$. □

If we want to discuss a particular element of a set, the notation $a \in A$, read "a is an element of A," is useful. If b is not an element of A, the notation can be modified to $b \notin A$.
We'll find that each time we discuss a new mathematical idea, we'll generally develop notation to go with it.

Occasionally our technique for solving a problem will require that we solve parts of it and then combine the solution sets. The two most common ways of combining sets to produce new sets are union and intersection.

Definition The *union* of sets A and B, written $A \cup B$, is the set of elements that belong to either A or B. Symbolically, $A \cup B = \{x \mid x \in A \text{ or } x \in B\}$.

EXAMPLE Find the union of each pair of sets:

$M = \{2, 4, 9\}$ and $N = \{3, 7\}$
$S = \{1, 3, 5, \ldots, 29\}$ and $T = \{17, 19, 21\}$
$V = (-1, 7]$ and $W = [-9, 0)$.

$M \cup N = \mathbf{\{2, 3, 4, 7, 9\}}$

$S \cup T = \mathbf{\{1, 3, 5, \ldots, 29\}}$, or S

$V \cup W = \mathbf{[-9, 7]}$

Note: V and W are given in interval notation. ■

Illustration $H = \{x \mid x \leq 5 \text{ or } x > 10.3\}$ can be represented by $(-\infty, 5] \cup (10.3, +\infty)$. □

Definition The *intersection* of sets A and B, written $A \cap B$, is the set of elements that belong to both A and B. Symbolically, $A \cap B = \{x \mid x \in A \text{ and } x \in B\}$.

It may happen that no elements belong to both sets; then $A \cap B = \{\ \}$, the empty set.

Illustration If $M = \{2, 4, 9\}$ and $N = \{3, 7\}$, $M \cap N = \{\ \}$.
If $S = \{1, 3, 5, \ldots, 29\}$ and $T = \{17, 19, 21\}$,
$S \cap T = \{17, 19, 21\}$, or T.
If $V = (-1, 7]$ and $W = [-9, 0)$, $V \cap W = (-1, 0)$.
Note: V and W are given in interval notation. □

EXAMPLE Write $\{y \mid y \leq -3 \text{ and } y < -17\}$ using interval notation.

In interval notation $\{y \mid y \leq -3$ and $y < -17\}$ is $(-\infty, -17)$. ■

Another method for producing new sets from an existing set is to build subsets of it.

Definition M is a *subset* of N, written $M \subseteq N$, if each element of M belongs to N.

Illustration If $A = \{4, 7, a, \text{Tom}\}$ and $B = \{4, \text{Tom}\}$, then $B \subseteq A$.
A is not a subset of B. This can be written $A \nsubseteq B$.
Note: $\{\ \} \subseteq A$ and $\{\ \} \subseteq B$. In fact, $\{\ \}$ is a subset of any set. □

EXERCISES A.2

In exercises 1 through 5, use enumeration to give the set described in set-builder notation.

1. $A = \{a \mid a = 5n - 2, n \text{ is a counting number}\}$
2. $B = \{b \mid b = -n^2, n \text{ is a positive even number}\}$
3. $C = \{c \mid c = k^3 - k, k \text{ is a positive odd number}\}$
4. $D = \left\{d \mid d = \dfrac{2}{m} + 3, m \text{ is a counting number}\right\}$
5. $E = \{e \mid e = 4k^2 \div 7, k \text{ is a positive even number}\}$

In exercises 6 through 10, use set-builder notation to describe the set.

6. $J = \{8, 11, 14, 17, \ldots\}$
7. $K = \{1, 3, 5, 7, \ldots\}$
8. $L = \{1, 8, 27, 64, \ldots\}$
9. $M = \{2, 8, 18, 32, 50, \ldots\}$
10. $N = \{-3, 9, -27, 81, \ldots\}$

In exercises 11 through 20, write interval notation for the set pictured.

11. 

12. 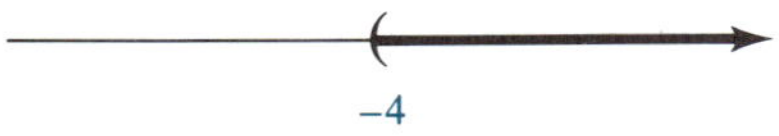

13. 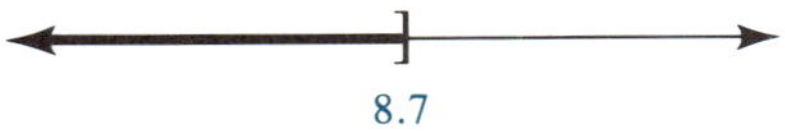

14. 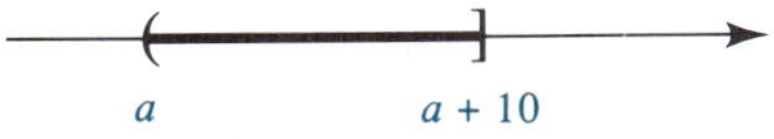

15. 

16. 

17. 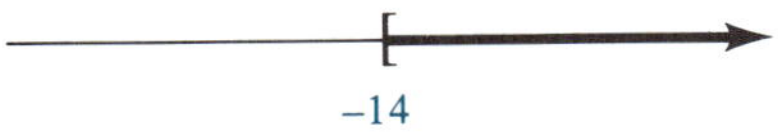

18. 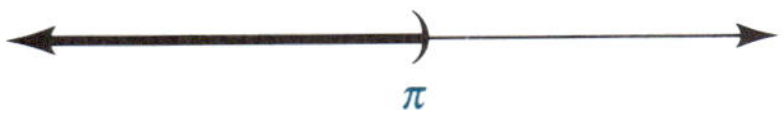

19. 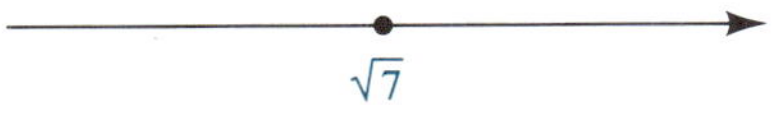

20. 

In exercises 21 through 30, draw the set represented by the inverval given.

21. $(-3, 5]$
22. $[7, 21]$
23. $(-\infty, 2.3)$
24. $[6.2, 9) \cup (10, +\infty)$
25. $[4, 7) \cap (5.1, 10]$
26. $[3\frac{2}{3}, +\infty)$
27. $(-\infty, a - 10] \cup [a - 3, a)$
28. $[-12.3, -8] \cap (-10, +\infty)$
29. $(-\infty, -2\frac{1}{8}) \cup [0, +\infty)$
30. $(-\infty, b] \cup (b + 5, +\infty)$

In exercises 31 through 38, $A = \{2, 4, 6, 8, \ldots\}$, $B = \{3, 6, 9, 12, \ldots\}$, $C = \{4, 8, 12\}$, $D = \{3, 9, 27, \ldots\}$. Perform the indicated set operations.

31. $A \cup C$
32. $A \cap B$
33. $B \cap D$
34. $B \cup D$
35. $B \cap C$
36. $A \cap D$
37. $(A \cup C) \cap D$
38. $A \cup (C \cap D)$
39. Complete the table. Use the results of this experiment to find a formula for the number of subsets a set with k elements has.

Number of elements	0	1	2	3	4	. . .	k
Number of subsets	1	2				. . .	

40. Are there cases when the order in which we do $\cap$ and $\cup$ give different results? Let A, B, C represent any sets.
    a. Is $A \cup B = B \cup A$ always true?
    b. Does $A \cap B = B \cap A$?
    c. Does $(A \cup B) \cup C = A \cup (B \cup C)$?
    d. Does $A \cap (B \cap C) = (A \cap B) \cap C$?
    e. Does $(A \cup B) \cap C = A \cup (B \cap C)$?
41. Is there a distributive property for $\cap$ and $\cup$? Experiment with $A \cap (B \cup C)$, $(A \cap B) \cup (A \cap C)$, $A \cup (B \cap C)$, $(A \cup B) \cap (A \cup C)$. Try writing conclusions for your experiments.
42. A survey of 100 customers at the Eatemups Drive-In has the following results: Susan reported that 47 customers liked catsup on their hamburgers, and 19 liked catsup and mustard. Dan reported that 36 liked mustard on their hamburgers, and 12 liked mustard and relish. Jackie found 29 customers who liked relish on their hamburgers, 15 who liked relish and catsup, and 6 who liked all three. Surely there's been some double counting. If each customer is reported in all the categories to which his or her preferences correspond, how many customers like only relish on their hamburgers?

A.3 Numbers

When talking about set-builder notation we used the descriptions counting numbers and real numbers for two common sets of numbers. If you were a bit fuzzy about their definitions, it wouldn't have seriously affected your understanding of those examples. Now it's time to be sure we're speaking exactly the same language about numbers.

Definition The *natural numbers* are 1, 2, 3, ... They are sometimes called the counting numbers. The usual name for the set of natural numbers is N.

Definition The *whole numbers* are 0, 1, 2, 3, ... A common name for the set of whole numbers is W.

Definition The set of *integers*, I, is $\{\ldots, -2, -1, 0, 1, 2, \ldots\}$. This is the set of whole numbers and their opposites.

Definition The set of *rational numbers* can be described with set-builder notation as $Q = \{q \mid q = a/b, a \in I, b \in I, b \neq 0\}$.

It would be misleading to call rational numbers fractions because the definition includes the integers and mixed numbers also. For example, -17 and $3\frac{2}{3}$ satisfy the definition because they can be written as $-\frac{17}{1}$ and $\frac{11}{3}$, respectively. Sometimes an alternate definition is helpful: $Q = \{q \mid q$ has a finite or repeating decimal representation$\}$. We explore the equivalence of these definitions in the exercises.

Definition The *irrational numbers* are those that have infinite nonrepeating decimal representations. A common name for this set is J.

Some common irrational numbers are π and the square roots of natural numbers that aren't perfect squares. For the time being we'll assume that these are irrational without proof. Here are some other numbers that clearly satisfy the definition.

Illustration These infinite nonrepeating decimal numbers are irrational numbers.

$$-12.012345678910111213\ldots \qquad 8.323323332\ldots$$

Are the patterns clear? □

Most of our work will be done with the real number system.

Definition The set of *real numbers*, $\mathbb{R}$, is the union of Q, the set of rational numbers, and J, the set of irrational numbers.

On a number line each point has a real number assigned to it, and vice versa. Before discussing one last set of numbers, there are some properties of the real numbers for which we need names. The associative, commutative, and distributive properties describe how the usual order of operations may be changed.

Associative Properties: If two additions are to be done, the order of doing them does not change the final result. If two multiplications are to be done, the order in which they are done does not change the final result.

Commutative Properties: Changing the order of two real numbers to be added doesn't change the result. Changing the order of two real numbers to be multiplied doesn't change the result.

Identity Properties: Zero added to any real number leaves the number unchanged. Zero is called the *additive identity*. The product of 1 and any real number is the real number. One is called the *multiplicative identity*.

Inverse Properties: Each real number has an additive inverse, which is another real number such that their sum is 0. An additive inverse is often called an *opposite*. Each nonzero real number has a multiplicative inverse, which is another real number such that their product is 1. A multiplicative inverse is often called a *reciprocal*.

Distributive Property: The product of a real number and the sum of two real numbers may be calculated either by adding first and then multiplying or by multiplying each summand by the number and then adding the products.

These properties are often described symbolically as follows:

Let a, b, and c represent any real numbers.

Associative Properties: $(a + b) + c = a + (b + c)$

$(ab)c = a(bc)$

Commutative Properties: $a + b = b + a$

$ab = ba$

Identity Properties: $0 + a = a$

$1a = a$

Inverse Properties: For each real number a, there is an opposite, written $-a$, such that $a + (-a) = 0$.
For each nonzero real number b, there is a reciprocal, written $1/b$, such that $b(1/b) = 1$.

Distributive Property: $a(b + c) = ab + ac$

There are some rules for working with equations and inequalities that should be included in our review, since they are basic tools for dealing with equations and inequalities.

Rules for Equations: Let a, b, and c be any real numbers.

1. If $a = b$, then $a + c = b + c$.
2. If $a = b$, then $ac = bc$.
3. $ab = 0$ if and only if either $a = 0$ or $b = 0$.

Rules for Inequalities: Let a, b, and c be any real numbers.

1. If $a < b$, then $a + c < b + c$.
2. If $a < b$ and $c > 0$, then $ac < bc$.
3. If $a < b$ and $c < 0$, then $ac > bc$.
4. If $a < b$ and $ab > 0$, then $1/a > 1/b$.

Since any subtraction can be replaced by adding the opposite of the second number, the two rules involving addition also cover the case of subtraction. Similarly, any division can be done by multiplying by the reciprocal of the second number. Also, because $a < b$ is equivalent to $b > a$, the inequality rules could be written in several different ways.

One last set of properties for the real numbers is included here for later reference.

Definition The **absolute value** of a number is its distance from zero on the number line. The absolute value of a is written $|a|$. An algebraic definition of $|a|$ is given by

$$|a| = \begin{cases} a & \text{if } a \geq 0, \\ -a & \text{if } a < 0. \end{cases}$$

Be careful of the notation here. The opposite of a is $-a$, and without knowledge about a, we can't say whether $-a$ is positive, negative, or zero. Reading $-a$ as "negative a" seems to cause a great deal of confusion. Remember the suggestion made in Section A.1 about reading $-$ as "the opposite of." It's helpful to have the geometric definition of $|a|$ also.

How does the operation of producing the absolute value behave when other operations are involved? In addition to indicating an operation, the absolute value bars are symbols of grouping, like parentheses.

Properties of the absolute value operation

Addition: The absolute value of a sum is less than or equal to the sum of the absolute values of the summands.

Multiplication: The absolute value of a product is equal to the product of the absolute values of the two factors.

Illustration Here is an illustration of each property.

$$|6 + (-2)| \leq |6| + |-2|$$

$$\left|\frac{-6}{3}\right| = \frac{|-6|}{|\ 3|}$$ □

To finish our catalog of numbers, we define one last set, the complex numbers. For every real number r, $r^2 \geq 0$. Sometimes we need numbers whose squares are negative.

Definition We define i to be a number such that $i^2 = -1$.

This is not a real number; it's something different. However, it does obey certain familiar rules:

1. $\underbrace{i + i + i + \cdots + i}_{(n \text{ terms})} = n \cdot i$
2. $0 \cdot i = 0$
3. $\underbrace{i \cdot i \cdot i \cdots i}_{(n \text{ factors})} = i^n$
4. $-n \cdot i = n(-i)$

On the other hand, i is not paired with any point on the number line, so it is neither positive nor negative. This also means i does not have an absolute value (distance from 0), and we can't say i is larger or smaller than a real number.

Definition The set of *complex numbers*, $\mathbb{C}$, is $\{z | z = r + si;\ r, s \text{ in } \mathbb{R},\ i^2 = -1\}$.

This is a definition in the style of the first definition of Q; it tells how the numbers should look. Because $0 \cdot i = 0$, each real number r is also a complex number; $r =$

$r + 0i$ fits the pattern. In fact, now is a good time to point out that the sets of numbers we've defined have the following set relationships.

$$N \subseteq W \subseteq I \subseteq Q \subseteq \mathbb{R} \subseteq \mathbb{C}; \qquad J \subseteq \mathbb{R} \subseteq \mathbb{C};$$
$$Q \cap J = \{\ \}, \qquad Q \cup J = \mathbb{R}$$

Pictorially we have Figure A.3a.

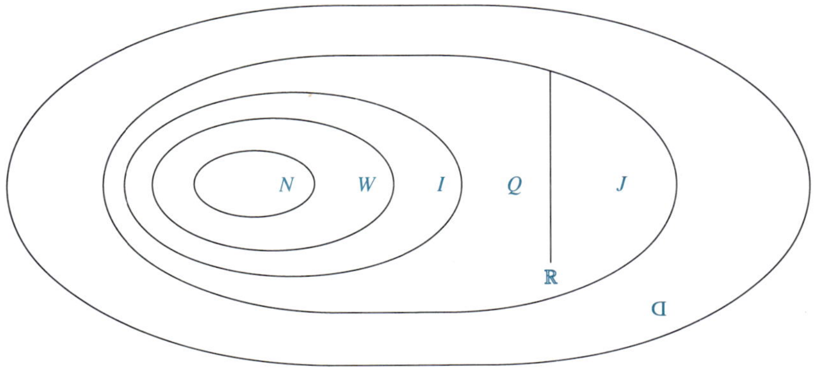

FIGURE A.3a

Using only the sets in Figure A.3a, we can speak of the smallest subset of $\mathbb{R}$ or $\mathbb{C}$ to which a number belongs. For example, the smallest set to which -2 belongs is I; the smallest set to which $-\frac{1}{2}$ belongs is Q.

You probably expect that we should be able to do arithmetic with the complex numbers and that for those complex numbers whose smallest set is $\mathbb{R}$, the results should agree with known facts. We want $(5 + 0i) + (3 + 0i) = 8 + 0i$ because $5 + 3 = 8$. It's easy to see how to define addition and subtraction so that this is true.

$$(a + bi) + (c + di) \text{ is } (a + c) + (b + d)i.$$
$$(a + bi) - (c + di) \text{ is } (a - c) + (b - d)i.$$

These definitions are easy to remember; we've just done what comes naturally using the associative and distributive properties, even though we have stated these only for operations on real numbers. In fact, when dealing with complex numbers, you may use all the properties of operations on real numbers we have discussed *except* those that deal with absolute value, size, or positive or negative values. Multiplication is defined by using the distributive and associative properties.

$$(a + bi)(c + di) = ac + adi + bci + bdi^2$$

Is this a complex number? We rewrite it to see if it matches the pattern $r + si$.

$$\begin{aligned} &ac + adi + bci + bdi^2 \\ &= ac + (ad + bc)i + bd(-1) \\ &= (ac - bd) + (ad + bc)i \end{aligned}$$

Multiplying two complex numbers produces a complex number.

Definition We call the $(r + si)$ form the *standard form of a complex number*.

The number r is the *real part* of $r + si$; the number s is called the *imaginary part*.

EXAMPLE Perform the indicated operations.

a. $(3 - \sqrt{5}i) + (-2 - 3\sqrt{5}i)$
b. $(8 - 5i)(8 + 5i)$

a. $(3 - \sqrt{5}i) + (-2 - 3\sqrt{5}i) = [3 + (-2)] + [-\sqrt{5} + (-3\sqrt{5})]i$, or **$1 - 4\sqrt{5}i$**

b. $(8 - 5i)(8 + 5i) = 64 - 40i + 40i - 25i^2$, or **89** ■

The powers of i form an interesting pattern that can be used as a shortcut in some problems. Consider

$$i; \quad i^2 = -1; \quad i^3 = i^2 \cdot i = -1 \cdot i = -i;$$

$$i^4 = i^2 \cdot i^2 = -1 \cdot -1 = 1; \quad i^5 = i^4 \cdot i = i$$

The list is beginning to repeat itself. Think about how this observation could give a way to put i^{17} in standard form.

What about the quotient of two complex numbers? Is it a complex number? Let's look at an example,

$$\frac{3 + i}{8 + 5i}.$$

If we try to mimic division for the real numbers,

$$8 + 5i\overline{)3 + i},$$

we find it seems to have no end. Another approach is needed. What is it we want? We'd like a complex number $a + bi$ such that

$$a + bi = \frac{3 + i}{8 + 5i}.$$

This means $(a + bi)(8 + 5i) = 3 + i$; if you know how to solve a system of linear equations, you can find a and b (see Section 9.1). There is another method for finding $a + bi$. In a preceding example, we saw $(8 + 5i)(8 - 5i)$ is 89, a real number. Consider the following:

$$\frac{3 + i}{8 + 5i} = \frac{3 + i}{8 + 5i} \cdot \frac{8 - 5i}{8 - 5i} = \frac{24 - 7i - 5i^2}{89} = \frac{29}{89} - \frac{7}{89}i.$$

We have found the standard form of

$$\frac{3 + i}{8 + 5i}$$

by multiplying numerator and denominator by $8 - 5i$.

Definition A pair of complex numbers $r + si$ and $r - si$ are called **conjugates.**

The product of a pair of conjugates is always a real number. Division of complex numbers is carried out by multiplying numerator and denominator by the conjugate of the denominator.

$$\frac{a + bi}{c + di} = \frac{a + bi}{c + di} \cdot \frac{c - di}{c - di} = \frac{ac + bd}{c^2 + d^2} + \frac{bc - ad}{c^2 + d^2} \cdot i$$

This is the formal definition of complex-number division. Don't memorize it; just carry out the multiplications each time.

EXERCISES A.3

In exercises 1 through 12, list all of the sets N, W, I, Q, J, $\mathbb{R}$, $\mathbb{C}$ to which each number belongs.

1. $\pi + 0i$
2. 0
3. $-1.333\ldots$
4. $\frac{24}{9}i$
5. $2 - \sqrt{3}$
6. 2.94
7. $1 - \pi$
8. $\sqrt{17} + 2i$
9. $-4.686686668\ldots$
10. $\sqrt{0.25}$
11. $-\sqrt[3]{27}$
12. $-3\frac{1}{8}$
13. Locate the real numbers in exercises 1 through 6 on a real-number line.
14. Locate the real numbers in exercises 7 through 12 on a real-number line.

Exercises 15 through 21 explore the two ways of defining the set of rational numbers.

15. Write each of the following as a/b, $\quad a, b, \in I, b \neq 0$:

    $0.125, -3.01, 17.005, 0.c_1c_2c_3, 0.d_1d_2d_3 \ldots d_n$.
16. If we let n represent $0.2828\ldots$, then $100n - n$, or $99n$, is $28.2828\ldots - 0.2828\ldots$, or 28, and n is $\frac{28}{99}$. Use the method of this example to write each of the following as a/b, $\quad a, b \in I, b \neq 0$:

    $-0.31331331331\ldots, -1.7272\ldots, 1.999\ldots,$ $2.1474747\ldots, 0.cdecdecde\ldots$.
17. Write instructions that tell a friend how to write any repeating decimal in the form a/b, $\quad a, b \in I, b \neq 0$.
18. Write as a decimal: $4\frac{2}{5}, -\frac{3}{2}, \frac{17}{20}, -2\frac{13}{40}$.
19. Write as a decimal: $\frac{2}{7}, -\frac{1}{3}, \frac{5}{27}, 1\frac{3}{13}, \frac{43}{6}$.
20. **a.** Calculate $1/n$ for $n = 2,3,4,5,\ldots, 20$. Examine your results as well as those for exercises 18 and 19 and tell when a/b will produce a finite decimal.

    b. If a/b produces a repeating decimal, what is the maximum size of the block that repeats?
21. What is it about the grade-school algorithm (method) for long division that guarantees a/b will be a finite or repeating decimal?

In exercises 22 through 29, determine a correct order symbol so that applying the properties of equations and inequalitites to the first statement produces the second.

22. $\frac{3}{5} - 7y \geq 2y + \frac{1}{2}; \quad 9y \ ? \ \frac{1}{10}$
23. $-3(1.4 + 9z) \leq 14z + 5.2; \quad -18.8 \ ? \ 82z$
24. $\frac{2x - 3}{-4} > \frac{3x + 12}{2}; \quad \frac{-21}{8} \ ? \ x$
25. $\frac{8}{y^4} - 11 \geq \frac{2}{y^4} - 3; \quad y^4 \ ? \ \frac{3}{4}$
26. $y = -3; \quad y \ ? \ -5$
27. $a > \frac{2}{3}; \quad a^2 \ ? \ \frac{4}{9}$
28. $a < \frac{2}{3}; \quad a^2 \ ? \ 0$
29. $a < -\frac{2}{3}; \quad a^2 \ ? \ 0$
30. Write the verbal versions of rule 1 for equations and rule 2 for inequalities given symbolically in this section.
31. Write the symbolic versions of the properties of the absolute value operation given in this section.
32. Compare $|a - b|$ and $|a| - |b|$ for various choices of a and b. What conclusion can you draw?
33. Compare

    $$\left|\frac{a}{b}\right| \quad \text{and} \quad \frac{|a|}{|b|}$$

    for various choices of a and b. What conclusion can you draw?

In exercises 34 through 39, replace the ? with = if the two quantities are identical; otherwise replace the ? with $\leq$ or $\geq$ to make a true statement.

34. $|x - 2| \ ? \ |2 - x|$
35. $-17|x| \ ? \ |-17x|$

36. $|-135x| \ ? \ 135x$

37. $|x^2 + 2| \ ? \ x^2 + 2$

38. $|4x - 5| \ ? \ |4x| + 5$

39. $|a| \ ? \ a$

40. Write directions for a friend for a shortcut to write i^n, $n \in N$, in standard complex number form.

In exercises 41 through 52, write the expression in standard complex-number form.

41. $(10 - 8i)(0.5 - 6i)$

42. $(18 + 2i) + 3(7 - 5i)$

43. i^{149}

44. $(13 + 21i) - (10 - 12i)$

45. $(6 + \frac{1}{2}i)^2$

46. $(2 - 5i)(2 + 5i)$

47. $(2.7 + 3i) + (-5 + 7i)(11 - 6i)$

48. $(3i)^5$

49. $\dfrac{7}{9i}$

50. $\dfrac{3.1 - 9i}{1 - 0.2i}$

51. $\dfrac{14}{3 + 4i}$

52. $\dfrac{1}{i^7}$

A.4 Exponents and Radicals

What can we do with these numbers that were defined in the last section? An important operation on a single number is raising it to a power. The first numbers used as powers, or exponents, were the natural numbers.

Definition a^n is the product of n factors of a: $\underbrace{(a)(a)(a) \cdots (a)}_{n \text{ factors}}$, $n \in N$. The number n is the **exponent**; a is the *base*.

Any number may be used as the base. Remember exponentiation, raising to a power, has precedence over the arithmetic operations and the operation of taking the opposite of a number. *Note:* This means $4(3)^2$ is 36 and -5^2 is -25. Since this first definition is in terms of multiplication, it is not surprising that there are special rules for exponentiation when combined with multiplication and division. What is surprising is that so many people try to invent rules for addition or subtraction combined with exponentiation. The rules that do connect powers and addition are quite nice—but not as simple as some students would like them to be.

We have five rules for exponents that hold for our first definition of exponents and all the extensions we make later of this definition. We list them now.

Rules for Exponents:

1. $a^m a^n = a^{m+n}$
2. $a^n b^n = (ab)^n$
3. $\dfrac{a^m}{a^n} = a^{m-n}, \ a \neq 0$
4. $\dfrac{a^n}{b^n} = \left(\dfrac{a}{b}\right)^n$
5. $(a^m)^n = a^{mn}$

The third rule prompts our first extension of what kind of number can be an exponent. What happens if $m \leq n$? Let's look at a few examples. They all fit the pattern of rule 3, but they can also be viewed as fractional expressions, and we can rewrite them in lowest terms.

Illustration

Original	By rule 3	Lowest terms
$\frac{12^{14}}{12^{14}}$	12^0	1
$\frac{x^5}{x^7}$	x^{-2}	$\frac{1}{x^2}$
$\frac{(3x)^4}{(3x)^5}$	$(3x)^{-1}$	$\frac{1}{3x}$
$\frac{(7-5z)^4}{(7-5z)^4}$	$(7-5z)^0$	1, if $7-5z \neq 0$

□

These examples suggest natural definitions for a^{-n} and for a^0.

Definition When n is a positive integer and $a \neq 0$; $\boldsymbol{a^{-n}} = 1/a^n$; if $a \neq 0$, $\boldsymbol{a^0} = 1$.

Notice that 0^0 is undefined; a^0 is the result of a division, and division by 0 is undefined. Again be careful of notation. The definition $a^{-n} = 1/a^n$ means $3^{-2} = 1/3^2$, or $1/9$ and $-7^{-2} = -1/7^2$, or $-1/49$. Remember, -7^{-2} is the opposite of 7^{-2}. We can also interpret a^{-n} as meaning take the reciprocal of the base and then compute the nth power: $a^{-n} = (1/a)^n$.

The fifth rule for exponents suggests a way to extend the definition of exponent to rational numbers. According to rule 5, what exponent would we need to make this statement true: $(16^t)^2 = 16$? One-half is a solution of this equation, but notice that there are already two numbers, 4 and -4, whose square is 16. For a definition of a rational-number exponent to be consistent with the existing facts of arithmetic, we must choose $16^{1/2}$ to be either 4 or -4. There are also questions to be settled about negative bases. For example, how should we choose t so that $[(-4)^t]^4 = -4$? For the moment we'll restrict our definition to real-number bases and distinguish two cases, depending on whether the base is negative or nonnegative.

Definition Let n be a positive integer and a be a real number.
If n is odd, $\boldsymbol{a^{1/n}}$ is the real number whose n^{th} power is a.
If n is even and $a \geq 0$, then $\boldsymbol{a^{1/n}}$ is the nonnegative real number whose n^{th} power is a.

So $16^{1/2}$ should be 4, since $4^2 = 16$ and 4 is positive.

If n is even and $a < 0$, $a^{1/n}$ is not a real number. We present a full discussion of this case in Section 7.5. For now we can handle the special case of $n = 2$ and $a < 0$ by using the fact that $i^2 = -1$. We define $(-1)^{1/2}$ to be i, since $i^2 = -1$. We can rewrite any $a^{1/2}$, $a < 0$, as $a^{1/2} = (-a)^{1/2}\, i$. So $(-49)^{1/2}$ is $49^{1/2} \cdot i$, or $7i$.

EXAMPLE Evaluate $8^{1/3}$, $(-27)^{1/3}$, $25^{1/2}$, and $(-64)^{1/2}$.

The value of $8^{1/3}$ is **2** because 2^3 is 8, the value of $(-27)^{1/3}$ is $\mathbf{-3}$ because $(-3)^3$ is -27, the value of $25^{1/2}$ is **5** because 5^2 is 25, and the value of $(-64)^{1/2}$ is $\mathbf{8i}$ because $(8i)^2$ is -64. ■

Another notation has also been developed for $a^{1/n}$. The number $a^{1/n}$ can be written $\sqrt[n]{a}$, read as "the nth root of a."

Definition In $\sqrt[n]{a}$, n is the *index*, a is the *radicand*, and the whole expression is the *radical*.

The radical symbol $\sqrt[n]{}$ is a symbol of grouping, so operations in the radicand are carried out first. The index 2 is usually omitted.

Not every rational number is the reciprocal of a positive integer, so we are not finished with this extension of the exponent definition.

Definition For any positive rational number m/n, if $a^{1/n}$ is a real number, $\boldsymbol{a^{m/n}} = (a^{1/n})^m$, or $(\sqrt[n]{a})^m$

Generally it's more convenient to evaluate roots before powers, but knowing $(\sqrt[n]{a})^m = \sqrt[n]{a^m}$ may be useful. *Note:* This equation is true only if on the left-hand side, $\sqrt[n]{a}$ is a real number. For example, $(\sqrt{-4})^2 \neq \sqrt{(-4)^2} = \sqrt{16} = 4$. One consequence of this fact is that the operations of squaring a real number and taking the square root of it are not precise opposites of each other in the way that doubling and halving a number are. Actually,

$$\sqrt{x^2} = |x|,$$

and $|x|$ may not be x. More generally, if n is even, raising a number to the nth power and taking the nth root are not operations that will exactly undo each other. We are interested in operations that undo each other for solving equations and inequalities.

For negative rational numbers, we combine our last two definitions. If p/q is a negative rational number, it can always be expressed as $-m/n$, where m and n are positive. We can define $a^{p/q}$ as $a^{-m/n}$ or $1/(a^{1/n})^m$ if $a^{1/n}$ is a real number.

It's also possible to extend the definition of exponents to cover irrational numbers. We discuss this in Section 4.3.

EXAMPLE Evaluate $32^{2/5}$, $\sqrt[4]{81^2}$, and $(\sqrt[4]{-16})^3$.

$32^{2/5}$ is 4 because $(\sqrt[5]{32})^2$ is 2^2, or 4.
$\sqrt[4]{81^2}$ is 9 because $81^{2/4}$ is $81^{1/2}$, or 9.
$(\sqrt[4]{-16})^3$ is not a real number because $\sqrt[4]{-16}$ is $(-16)^{1/4}$, which is not real. ■

We need to be able to write numbers using exponential or radical forms.

EXAMPLE Rewrite $(-4)^{5/2}$, $2a^{1/3}$, and $8^{1/2}$ in radical form.

$(-4)^{5/2}$ is $\sqrt{(-4)^5}$, which is not a real number,
$2a^{1/3}$ is $2\sqrt[3]{a}$, and $8^{1/2}$ is $\sqrt{8}$. ■

EXAMPLE Rewrite $\sqrt{a^3}$, $\sqrt[3]{7^{12}}$, and $\sqrt[6]{(3-a)^5}$ in exponential form.

$\sqrt{a^3}$ is $a^{3/2}$, $\sqrt[3]{7^{12}}$ is $7^{12/3}$, or 7^4, and $\sqrt[6]{(3-a)^5}$ is $(3-a)^{5/6}$. ■

EXAMPLE Express $(27/4)^{-2/3}$ with positive exponents using the definition and rules of exponents.

$(27/4)^{-2/3}$ is $(4/27)^{2/3}$ because a negative exponent means to take the reciprocal of the base with the opposite exponent. We can write $(4/27)^{2/3}$, using the definition and rules of exponents.

$$\frac{4^{2/3}}{27^{2/3}} = \frac{4^{2/3}}{(27^{1/3})^2} = \frac{4^{2/3}}{3^2} = \frac{4^{2/3}}{9}$$

We express $(27/4)^{-2/3}$ as $\dfrac{4^{2/3}}{9}$. ■

EXAMPLE Find the approximate value of $\left(\dfrac{256}{125}\right)^{-3/5}$ correct to four decimal places. If your calculator cannot handle negative exponents, rewrite the expression with a positive exponent before doing the computation.

$$\left(\frac{256}{125}\right)^{-3/5} = \left(\frac{125}{256}\right)^{3.5} \approx 0.6504$$

The approximate value of $\left(\dfrac{256}{125}\right)^{-3/5}$ is 0.6504. ■

We prefer to work with the exact form of radical expressions. Do not give an approximation for a radical unless specifically asked to do so.

Some uses of radicals require a form where the numerator or denominator contains no radicals. The method used to rationalize numerators or denominators of the form $a + b\sqrt{c}$ is very similar to that used in producing a real denominator for a quotient of complex numbers. Numbers such as $a + b\sqrt{c}$ and $a - b\sqrt{c}$ are called *conjugates* just as the complex number pairs $a + bi$, $a - bi$ are.

EXAMPLE Rewrite $\dfrac{2\sqrt{x} - 3}{2\sqrt{x} + 3}$ with no radicals in the numerator.

$$\frac{2\sqrt{x} - 3}{2\sqrt{x} + 3} = \frac{2\sqrt{x} - 3}{2\sqrt{x} + 3} \cdot \frac{2\sqrt{x} + 3}{2\sqrt{x} + 3} = \frac{4x - 9}{4x + 12\sqrt{x} + 9}$$

Rationalizing the numerator of $\dfrac{2\sqrt{x} - 3}{2\sqrt{x} + 3}$ produces $\dfrac{4x - 9}{4x + 12\sqrt{x} + 9}$. ■

EXERCISES A.4

In exercises 1 through 26, use the rules of exponents and carry out the indicated operations. Assume each variable represents a real number and specify any other restrictions needed to produce real number results.

1. $x^{1/2} \cdot x^{1/3}$
2. $y^{2/5} \cdot y$
3. $(a + 3)^2(a + 3)^{2/3}$
4. $(6b)^{4/7}(6b)^{-1/5}$
5. $(x - 2y)^{-1/3}(x - 2y)$
6. $\dfrac{x^{-2}}{x^5}$
7. $\dfrac{y^7}{y^{-4}}$
8. $\dfrac{a^{3/8}}{a^{-1/8}}$
9. $\dfrac{b^{4/5}}{b^{1/2}}$
10. $\dfrac{(r + 3s)^{-4}}{(r + 3s)^{-12}}$
11. $(x^{2/3})^4$
12. $(y^{12})^{5/6}$
13. $(a^{-2/5})^{15}$
14. $[(2a + 7b)^{3/7}]^{14/27}$
15. $(y^{4n})^{3/n}$
16. $(-125x^6y^5)^{1/3}$
17. $\dfrac{x^{4/5}y^{-1}}{x^{-1/5}y^2}$
18. $\dfrac{a^{-1/n}b^{5/n}}{a^{-1}b^{-7/n}}$
19. $(3a^{-1/2} - b^{-1/2})^2$
20. $(5a^{-1} - a)^2$
21. $(x^{-1} - y^{-1})^{-2}$
22. $(9a^4b^{-1}c^{-2})^{3/2}(36a^{-2}bc^{-4})^{-5/2}$
23. $(27a^3b^{-2}c)^{2/3}(-8a^{-6}b^2c^{-3})^{-4/3}$
24. $(5^{-1}x - y^{-1})(10x + 3y^{-1})$
25. $(2^{-1}x + 3y^{-1})(5x - 3^{-1}y^{-1})$
26. $[(5x - y)^4(y - 3x)^8]^{1/4}$

In exercises 27 through 32, perform the indicated operations and write the answer in $a + bi$ form.

27. $(3i)^5$

28. $(1 + i)^{10}$

29. $\left(\sqrt{\frac{3}{2}} + \sqrt{\frac{1}{2}}i\right)^2$

30. $\left(-\frac{1}{2} + \frac{\sqrt{3}}{2}i\right)^3$

31. $\left(-\frac{1}{2} - \frac{\sqrt{3}}{2}i\right)^3$

32. $\left(-1 + \sqrt{3}i\right)^3$

33. Try to extend the pattern for i^n to negative integral exponents. (See exercise 40 in Section A.3.)

34. JoAnne and Janet are comparing their answers to homework problems to those in the back of the book. They have $(\sqrt{5} - 8)/(\sqrt{3} + 5)$, but the answer in the book is $(8\sqrt{3} + 5\sqrt{5} - \sqrt{15} - 40)/22$. Are the answers the same? If so, which form is better? Why?

35. Write the verbal versions of exponent rules 1 and 3.

In problems 36 through 41, approximate each of the following to four decimal places.

36. $\dfrac{2}{1 + \sqrt{3}}$

37. $\dfrac{5 + \sqrt{2}}{3 - \sqrt[3]{7}}$

38. $\dfrac{6}{7 - \sqrt{5}}$

39. $\dfrac{12 - \sqrt[3]{6}}{\sqrt[3]{7}}$

40. $10 - \dfrac{4}{2 - 3\sqrt{8}}$

41. $\sqrt{\dfrac{1}{4} + 32 - \dfrac{14.8}{\sqrt[3]{9}}}$

42. Use your calculator to estimate

$$\sqrt{12 + \sqrt{12 + \sqrt{12 + \cdots}}}.$$

43. In evaluating x^{m^n} does the order in which the powers are done matter?

44. We know $(-8)^{2/3}$ is 4, because

$$(-8)^{2/3} = \left[(-8)^{1/3}\right]^2 = (-2)^2 = 4.$$

Try raising -8 to the $\frac{2}{3}$ power using your calculator. What happens? How can you work around this problem?

45. We've defined rational exponents. Will your calculator accept an irrational number as an exponent? For example, can you approximate 5^π and $3^{\sqrt{2}}$? If your machine gives answers to these questions, what do you think it uses as a definition of irrational exponent?

A.5 Expressions and Operations

We've used the word expression informally in the previous sections. Here is a formal definition for use from now on.

Definition An algebraic **expression** is a number written using variables, constants, exponents, and any of the four arithmetic operations.

We generally assume that any variables represent numbers that would give a meaningful expression. Radicals may be used in place of rational exponents. We often shorten the phrase algebraic expression to expression.

Definition If it involves only the operations of exponentiation, multiplication or division, the expression is a *term*. The constant factor in a term is its *coefficient*.

If in an expression the variable bases have only positive integral exponents and no denominator contains a variable, the expression is a **polynomial.** The *degree* of a polynomial is calculated by finding the sum of the exponents on variable bases in each term and taking the largest of these sums.

Illustration A polynomial with degree 3 is $1.2x^2y$. It contains one term. Its coefficient is 1.2.
The polynomial 24 has degree 0.
The polynomial $1.5xy^2z - \sqrt{3}x^2y + 12z^2$ has degree 4 and contains three terms. □

Polynomials may be also classified by the number of terms they contain.

Definition A one-term polynomial is called a *monomial*; a two-term polynomial is a *binomial*; a three-term polynomial is a *trinomial*.

Polynomials are discussed in more detail in Chapter 8, but this description will do for now.

Definition If an expression is the quotient of two polynomials, it is a **rational expression**. We'll call other expressions involving division by a variable **fractional expressions**.

Remember there may be values for which a denominator is zero that are not explicitly noted.

Since expressions are numbers, the four arithmetic operations can be carried out with them. We'll postpone the discussion of these operations with rational expressions and fractional expressions until after we talk about factoring.

For addition and subtraction we need the idea of like terms.

Definition Terms are *like terms* if they have the same variables with exactly the same exponents.

Illustration Like terms are $-15x^3yz$ and $3x^3yz$.
$4\sqrt[4]{xy^2}$ and $-3\sqrt[4]{xy^2}$ are like terms.
The terms $-15x^3yz^2$ and $3x^3yz$ are not like terms. □

The distributive property can be used to combine like terms that are added or subtracted.

Illustration

$$(3.8x^2y - 4y) - (2.5y^2 - 5.1x^2y) = [3.8 - (-5.1)]x^2y - 4y - 2.5y^2, \text{ or } 8.9x^2y - 4y - 2.5y^2$$

$$\sqrt{12x} + 6\sqrt{3x} = 2\sqrt{3x} + 6\sqrt{3x} = (2 + 6)\sqrt{3x}, \text{ or } 8\sqrt{3x}$$ □

Multiplication of expressions involves the commutative, associative, and distributive properties as well as the rules of exponents.

Illustration Using the distributive property and rules of exponents, $x^2(x^{3/2} - 4x^{1/2} - 2)$ can be written as $x^{7/2} - 4x^{5/2} - 2x^2$. Similarly,

$$(m^2 - n)(7m^2 + 8n) = 7m^4 + 8m^2n - 7m^2n - 8n^2, \text{ or } 7m^4 + m^2n - 8n^2.$$ □

For division we'll discuss only the most common cases. Division by a single term is done by using the distributive property and the rules of exponents.

Illustration

$$\frac{15z - 3z^{1/3}}{3z^{1/3}} = \frac{15z}{3z^{1/3}} - \frac{3z^{1/3}}{3z^{1/3}}, \text{ or } 5z^{2/3} - 1, \qquad z \neq 0$$

$$\frac{21ab^2 - 10ab}{-6a^2b} = \frac{21ab^2}{-6a^2b} - \frac{10ab}{-6a^2b}, \text{ or } \frac{-7b}{2a} + \frac{5}{3a}, \qquad b \neq 0$$

Note: We must explicitly state the restrictions that were implicit in the factors removed. □

Division of a polynomial by a multiterm polynomial can be done using the long-division algorithm. We can do this more easily if we arrange the terms of the divisor and those of the dividend both in either ascending or descending powers of the same variable. The division terminates when the degree of the variable chosen for the ordering is smaller in the remainder than it is in the divisor.

EXAMPLE Compute $(x^3 + 8y^3 - 3xy^2 + 5x^2y) \div (x - 2y)$.

If $x \neq 2y$, this can be computed using

$$\begin{array}{r|l} & x^2 + 7xy \quad + 11y^2 \\ \hline x - 2y & x^3 + 5x^2y - 3xy^2 \quad + 8y^3 \\ & \underline{x^3 - 2x^2y} \\ & \qquad 7x^2y - \;\; 3xy^2 \\ & \qquad \underline{7x^2y - 14xy^2} \\ & \qquad\qquad 11xy^2 + \;\; 8y^3 \\ & \qquad\qquad \underline{11xy^2 - 22y^3} \\ & \qquad\qquad\qquad 30y^3 \end{array}$$

The division terminates, since the degree of x in the divisor is 1 and no x's appear in the last difference, $30y^3$.

So $(x^3 + 8y^3 - 3xy^2 + 5x^2y) \div (x - 2y) = x^2 + 7xy + 11y^2 + \dfrac{30y^3}{x - 2y}$. ■

Illustration If $a \neq 4$, $(a^3 - 64) \div (a - 4)$ can be computed using

$$\begin{array}{r|l} & a^2 + 4a \quad + 16 \\ \hline a - 4 & a^3 \qquad\qquad\quad - 64 \\ & \underline{a^3 - 4a^2} \\ & \qquad 4a^2 \\ & \qquad \underline{4a^2 - 16a} \\ & \qquad\qquad 16a - 64 \\ & \qquad\qquad \underline{16a - 64} \end{array}$$

So $(a^3 - 64) \div (a - 4) = a^2 + 4a + 16$, $a \neq 4$. □

This method of long division can sometimes be used for expressions that are not polynomials. Practicing on polynomials will give you the skills needed to use the technique in other settings.

Solving equations and inequalities and performing some operations may require that we rewrite an algebraic expression as a product of factors. The problem will dictate how much factoring needs to be done. For practice, we want to do this factoring "completely." In this chapter we will restrict the coefficients of the factors to the smallest subset of $\mathbb{C}$ that contains the coefficients of the expression.

To factor an expression completely, we first determine if there are any factors common to all terms. If there are, we use the distributive property to rewrite the expression. When one variable appears in all terms, we factor it out as a common factor using the smallest of the exponents with which it appears.

Illustration With all the common factors factored out,

$$\begin{aligned} 8m^2n - 32m^2n^3 &= 8m^2n(1 - 4n^2), \\ x^{-1/2} + 3x^{-1} - 2x^{-3/2} &= x^{-3/2}(x + 3x^{1/2} - 2), \\ 8x^2 - 6x^{-3} + 2x^{-1} &= 2x^{-3}(4x^5 - 3 + x^2), \\ 5x\sqrt[3]{x^2y} - \sqrt[3]{x^2yz} &= \sqrt[3]{x^2y}(5x - \sqrt[3]{z}). \end{aligned}$$

□

We next examine the other factor created by factoring out the common factors. (This would be the original expression if 1 were the only common factor.) Count the terms in this factor. We consider the possible cases separately.

If there are two terms, we check to see if they form one of the special patterns: the difference of squares, the sum of cubes, or the difference of cubes. There are other special patterns for two terms, but these will suffice for now.

If the two terms form the difference of squares, the expression can be factored as the product of the sum and difference of the square roots of the terms.

EXAMPLE Factor $y^2 - 4$.

Writing $y^2 - 4$ as the product of the sum and difference of the square roots of the terms gives $y^2 - 4 = (|y| + 2)(|y| - 2)$ because $\sqrt{y^2} = |y|$. If $y \geq 0$, then $|y|$ is y and $y^2 - 4 = (y + 2)(y - 2)$. On the other hand if $y < 0$, then $|y|$ is $-y$ (the opposite of y) and

$$\begin{aligned} y^2 - 4 &= (-y + 2)(-y - 2) = [-(y - 2)][-(y + 2)] \\ &= (y + 2)(y - 2). \end{aligned}$$

In either case, $y^2 - 4 =$ **$(y + 2)(y - 2)$**, so, yes, you can still factor the difference of squares as you've probably done before. Symbolically,

$$A^2 - B^2 = (A + B)(A - B). \qquad \blacksquare$$

Illustration Factored completely,

$$\begin{aligned} 4m^2 - 1/9 &= (2m + 1/3)(2m - 1/3), \\ (a + b)^2 - 100 &= (a + b + 10)(a + b - 10), \\ z^{-2} - 36 &= (z^{-1} - 6)(z^{-1} + 6). \end{aligned} \qquad \square$$

EXAMPLE Factor completely $x^4 - 25$.

The complete factorization is **$(x^2 - 5)(x^2 + 5)$** because the smallest set to which the coefficients 1 and -25 belong is the integers.

There are times when we write $x^4 - 25$ as $(x + \sqrt{5})(x - \sqrt{5})(x^2 + 5)$. There are also times we need to rewrite $x^4 - 25$ as $(x + \sqrt{5})(x - \sqrt{5})(x + \sqrt{5}i)$ $(x - \sqrt{5}i)$. We need the restrictions for this appendix only to provide guidelines for our practice. In other situations, the problem for which we need to factor dictates how we choose the factors, and the definition of complete factoring in this appendix does not apply. $\blacksquare$

The factoring pattern for the sum of cubes and for the difference of cubes is the same when described verbally. The factors are a two-term expression formed from the cube root of the first term and the cube root of its second term combined with the same operation as that of the original expression and a three-term expression built from the two terms of the first factor as follows: Square the first term, combine it with the product of the terms using the inverse of the operation in the original, and add the square of the second term.

Illustration Factored completely,

$$x^3 - 8 \text{ is } (\sqrt[3]{x^3} - \sqrt[3]{8})[(\sqrt[3]{x^3})^2 + \sqrt[3]{x^3}\sqrt[3]{8} + (\sqrt[3]{8})^2],$$

$$\text{or} \quad (x - 2)(x^2 + 2x + 4),$$
$$n^6 + 27 = (n^2 + 3)(n^4 - 3n^2 + 9),$$
$$a^{3/2} + \frac{1}{64} = \left(a^{1/2} + \frac{1}{4}\right)\left(a^{2/2} - \frac{1}{4}a^{1/2} + \frac{1}{16}\right),$$
$$(m + 2n)^3 - 1 = [(m + 2n) - 1][(m + 2n)^2 + (m + 2n) + 1],$$
$$125 - (2b + c)^3 = [5 - (2b + c)][25 + 5(2b + c) + (2b + c)^2]. \quad \square$$

As in Section A.3, we've described a pattern verbally to provide practice in reading mathematical terminology and to avoid notation that might seem to restrict the pattern. The verbal form gives us a set of directions to follow. Try writing the pattern for factoring the sum or difference of cubes symbolically.

For the case of three terms, we present one possible factoring scheme, one that produces two two-term factors. This is often described as a trial-and-error method. Candidates for this type of factoring must be *quadratic in form*: the exponent on each variable in the middle term must be half the exponent on the same variable in the first or last term. Terms may need to be rearranged to match this pattern.

Illustration The following are quadratic in form.

$x^2 - 3x + 7$

$y^{-4} + 10y^{-2} + 8$

$a^2c^2 + 12 - ac$

$6m + m^{1/2}n^2 + 5n^4$

$(t + 3)^2 - 5(t + 3) + 7$

The following are not quadratic in form.

$x^3 - 4x + 3$

$y^{-6}x^2 + 10y^{-2}x + 8$ $\square$

Illustration Here are some quadratic-in-form expressions written as the product of two two-term expressions.

$$x^2 - 7x + 12 = (x - 3)(x - 4)$$
$$m^{-4} + 11m^{-2} + 18 = (m^{-2} + 9)(m^{-2} + 2)$$
$$2(a - b)^2 + (a - b) - 15 = [2(a - b) - 5][(a - b) + 3] \quad \square$$

For four or more terms, the basic factoring technique is to group the terms into blocks that fit one of the previous factoring methods and then to carry out a second-level factoring.

Illustration The following are factored using grouping.

$$\underbrace{2tm + 10tn} - \underbrace{3ms - 15ns} = 2t(m + 5n) - 3s(m + 5n), \text{ or } (m + 5n)(2t - 3s)$$
$$\underbrace{25y^2 - 10y + 1} - x^2 = (5y - 1)^2 - x^2, \text{ or } [(5y - 1) - x][(5y - 1) + x]$$

$$\underbrace{m^2 + 4mn + 4n^2}\; - \;\underbrace{a^2 - 6a - 9} = (m + 2n)^2 - (a + 3)^2, \text{ or}$$
$$(m + 2n + a + 3)(m + 2n - a - 3)$$

□

One use of factoring is in rewriting rational expressions and other quotients in lowest terms. In order to remove all common factors from the numerator and denominator of an expression, we need to know what the factors are.

EXAMPLE Write $\dfrac{x^2 - y^2}{2x - 2y}$ in lowest terms.

$$\frac{x^2 - y^2}{2x - 2y} = \frac{(x + y)(x - y)}{2(x - y)}, \text{ or } \frac{x + y}{2}, \qquad \text{if } x \neq y.$$

In lowest terms, $\dfrac{x^2 - y^2}{2x - 2y}$ is $\dfrac{x + y}{2}$, $x \neq y$. ■

EXAMPLE Write $\dfrac{x^{8/3} - 9x^{5/3} + 20x^{2/3}}{x^2 + 3x - 28}$ in lowest terms.

$$\frac{x^{8/3} - 9x^{5/3} + 20x^{2/3}}{x^2 + 3x - 28} = \frac{x^{2/3}(x^2 - 9x + 20)}{(x + 7)(x - 4)}$$
$$= \frac{x^{2/3}(x - 4)(x - 5)}{(x + 7)(x - 4)} = \frac{x^{2/3}(x - 5)}{x + 7}, \qquad \text{if } x \neq 4.$$

In lowest terms, $\dfrac{x^{8/3} - 9x^{5/3} + 20x^{2/3}}{x^2 + 3x - 28}$ is $\dfrac{x^{2/3}(x - 5)}{x + 7}$, $x \neq 4$.

Note: We must explicitly state the restrictions that were implicit in the factors removed. ■

Remember that we're now dealing with signed numbers, and we may be able to find a pair of opposites to remove.

EXAMPLE Write $\dfrac{27y^3 - 8}{2 - 3y}$ in lowest terms.

$$\frac{27y^3 - 8}{2 - 3y} = \frac{(3y - 2)(9y^2 + 6y + 4)}{2 - 3y}, \text{ or}$$
$$-(9y^2 + 6y + 4), \qquad y \neq \frac{2}{3}$$

The expressions $3y - 2$ and $2 - 3y$ are a pair of opposites; their quotient is -1 if y is not $\frac{2}{3}$.

In lowest terms, $\dfrac{27y^3 - 8}{2 - 3y}$ is $-(9y^2 + 6y + 4)$, $y \neq \dfrac{2}{3}$. ■

Illustration In lowest terms,

$$\frac{4 - 25x^2}{5x^{7/3} + 13x^{4/3} - 6x^{1/3}} \text{ is } -\frac{2 + 5x}{x^{1/3}(x + 3)}, \qquad x \neq \frac{2}{5}.$$

□

When we multiply rational or fractional expressions, it's generally more efficient to begin by removing any factors common to the numerators and denominators.

EXAMPLE Perform the indicated operation and write the expression in lowest terms:

$$\frac{x^2 - 2x - 15}{2x^2 - 24x + 70} \cdot \frac{x^2 - 3x - 28}{x^2 + 9x + 18}.$$

$$\frac{x^2 - 2x - 15}{2x^2 - 24x + 70} \cdot \frac{x^2 - 3x - 28}{x^2 + 9x + 18} = \frac{(x + 3)(x - 5)}{2(x - 7)(x - 5)} \cdot \frac{(x - 7)(x + 4)}{(x + 3)(x + 6)}, \text{ or}$$

$$\frac{x + 4}{2x + 12}, \qquad x \neq -3, 5, 7$$

$\frac{x^2 - 2x - 15}{2x^2 - 24x + 70} \cdot \frac{x^2 - 3x - 28}{x^2 + 9x + 18}$ is $\frac{x + 4}{2x + 12}$, $x \neq -3, 5, 7$. ■

EXAMPLE Perform the indicated operation and write the expression in lowest terms:

$$\frac{a^2 - 4}{a + 3} \div \frac{2 - a}{3 + a}.$$

$$\frac{a^2 - 4}{a + 3} \div \frac{2 - a}{3 + a} = \frac{a^2 - 4}{a + 3} \cdot \frac{3 + a}{2 - a}$$

$$= \frac{(a + 2)(a - 2)}{a + 3} \cdot \frac{3 + a}{2 - a} = -(a + 2),\ a \neq -3, 2$$

$\frac{a^2 - 4}{a + 3} \div \frac{2 - a}{3 + a}$ is $-(a + 2)$, $a \neq -3, 2$. ■

Factoring denominators can be useful in determining the least common denominator needed to add or subtract rational or fractional expressions.

EXAMPLE Perform the indicated operation and write the expression in lowest terms:

$$\frac{4x}{x^2 - 9} - \frac{3x - 11}{x^2 - 7x + 12}.$$

$$\frac{4x}{x^2 - 9} - \frac{3x - 11}{x^2 - 7x + 12} = \frac{4x}{(x - 3)(x + 3)} - \frac{3x - 11}{(x - 4)(x - 3)}$$

$$= \frac{4x(x - 4)}{(x - 3)(x + 3)(x - 4)} - \frac{(3x - 11)(x + 3)}{(x - 3)(x + 3)(x - 4)}$$

Remember to multiply the numerator by the same nonzero number as the denominator in order to have an expression equivalent to the original.

$$= \frac{x^2 - 14x + 33}{(x-3)(x+3)(x-4)} = \frac{(x-11)(x-3)}{(x-3)(x+3)(x-4)}$$

$$= \frac{x-11}{x^2 - x - 12}, \qquad x \neq 3$$

$$\frac{4x}{x^2-9} - \frac{3x-11}{x^2-7x+12} \text{ is } \frac{x-11}{x^2-x-12}, x \neq 3.$$ ■

EXAMPLE Perform the indicated operations and write the expression in lowest terms:

$$\frac{-4}{x^{12/5} - 16x^{2/5}} - \frac{4}{3x^{9/5} - 12x^{4/5}}.$$

$$\frac{-4}{x^{2/5}(x+4)(x-4)} - \frac{4}{3x^{4/5}(x-4)}$$

$$= \frac{-4(3)x^{2/5}}{3x^{4/5}(x+4)(x-4)} - \frac{4(x+4)}{3x^{4/5}(x-4)(x+4)}$$

$$= \frac{-12x^{2/5} - 4x - 16}{3x^{4/5}(x+4)(x-4)} = \frac{-12x^{2/5} - 4x - 16}{3x^{14/5} - 48x^{4/5}}$$

$$\frac{-4}{x^{12/5} - 16x^{2/5}} - \frac{4}{3x^{9/5} - 12x^{4/5}} \text{ is } \frac{-12x^{2/5} - 4x - 16}{3x^{14/15} - 48x^{4/5}}.$$ ■

EXERCISES A.5

In exercises 1 and 2, state the degree and the number of terms, and list the coefficients for each polynomial.

1. $-12x^5 + 4x^6 - 5$
2. x^2y^3z
3. Write a four-term polynomial of degree 7 with coefficients -3, 4.22, π, $\sqrt{2}$.
4. Write a monomial of degree 6 with variables w and z.

In exercises 5 through 15, carry out the indicated operations.

5. $(x^{1/2} + 3x - 5) - (4x - 2\sqrt{x} - 8)$
6. $16 - 4(3xy - 7x^2y + xy^2) + (9x^2y + 6xy - 14xy^2)$
7. $(2 \cdot 3^n + 8x) + (-3x + 3^n)$
8. $(6x^{-2} - 4x^{-1} + 22) - (12x^{-1} + 18 - 10x^{-2})$
9. $(4.6a^{3/2} + 5.1a - 2a^{1/2}) - (6a - 3.1a^{3/2} + 1.7a^{1/2})$
10. $\left(\frac{1}{8}xy^{1/2} + \frac{2}{3}x^{1/2} - \frac{2}{3}xy\right) + \left(\frac{1}{6}x^{1/2} - \frac{5}{12}xy^{1/2} - \frac{2}{15}xy\right)$
11. $(4x - 3)(5x + 2)(4x + 3)$
12. $x^{5/3}(x^2 - 3x - 1)$
13. $y^{-4/5}(y^2 + 7y + 9)$
14. $(3x - 2)^3(3x + 2)^3$
15. $\sqrt{2x+1}\,[7\sqrt{2x+1} - (2x + 1) + 3]$

In exercises 16 through 20, assume each variable represents a real number and describe any restrictions needed on the variables in each expression so that it represents a real number.

16. $\frac{a}{b}$
17. $\sqrt{5 - x}$
18. $\sqrt{9 - x^2}$
19. $\sqrt[3]{x + 8}$
20. $\frac{4}{x - y}$

In exercises 21 through 24, carry out the indicated division. State all restrictions on the variables in your quotient.

21. $\frac{x^2 - 3x + 4}{x^{2/3}}$
22. $\frac{5a^2 + 1.2a}{2\sqrt{a}}$
23. $\frac{9ab^2 - 3a^2b + 15a^2b^2}{-3a^2b}$
24. $\frac{30xyz - 5x^2y^{1/2} + 40x^{3/2}z}{15x^{1/2}yz^2}$

In exercises 25 through 28, write the expression in lowest terms; record any restrictions that must be written explicitly as a result of the rewriting.

25. $\dfrac{6x^2 + 5x - 4}{6 - 13x + 2x^2}$

26. $\dfrac{5x^{14/5} - 23x^{9/5} + 12x^{4/5}}{2x^{12/5} - x^{7/5} - 28x^{2/5}}$

27. $\dfrac{x^2 - xy - 6y^2}{x^2 - 4xy + 3y^2}$

28. $\dfrac{x^5 + x^4 - 156x^3}{144x^2 - x^4}$

In exercises 29 through 46, factor the expression completely under the restrictions of this section.

29. $3x + 10x^{1/2} + 3$

30. $(6y - 7)^3 + 64x^3$

31. $4 - x^{2/3}$

32. $2x^{-6} + x^{-3} - 15$

33. $5am - 3bn + 15an - bm$

34. $x^3 + 3x^2 - x - 3$

35. $x^{-1} - 4x^{-2} - 21x^{-3}$

36. $x^{-4} - 9$

37. $3.2x^2 + 9.6x + 6.4$

38. $x^2 + 6x - 4y^2 + 9$

39. $(a - 2b)^2 + 5(a - 2b) + 6$

40. $5x^2 - 35xy + 60y^2$

41. $\frac{7}{5}x^{2/5} - \frac{2}{5}x^{-3/5}$

42. $\frac{1}{36} - 144x^2$

43. $x^3 - 27y^3$

44. $250x^3 + 16y^3$

45. $16a^4 - 81b^4$

46. $(x - 4)^2 - y^2$

In exercises 47 through 54, perform the indicated operations and write the result in lowest terms.

47. $\dfrac{10 - 5x}{15x} \cdot \dfrac{x^2 + 5x + 6}{x^2 - 4}$

48. $\dfrac{2y^2 - 11y + 5}{4y^2 - 9} \cdot \dfrac{3 + 2y}{1 - 2y}$

49. $\dfrac{m^{3/2}n^{1/2} + m^{1/2}n^{3/2}}{m^2n^{1/2} - n^{5/2}} \div \dfrac{m^{1/2} + n^{1/2}}{m^2 - 2mn + n^2}$

50. $\dfrac{a^{2/3} - 4}{a^{2/3} - 4a^{1/3} + 3} \div \dfrac{3a^{2/3} + 6a^{1/3}}{a^{2/3} - 9}$

51. $\dfrac{x - y}{(x - y)^2} \div \left(\dfrac{x^3 + y^3}{x - y} \cdot \dfrac{y^2 - x^2}{x^2 + 2xy + y^2}\right)$

52. $\dfrac{\sqrt{x}}{5 - 3\sqrt{x}} + \dfrac{x}{25 - 9x}$

53. $\dfrac{3.2}{x^{2/3}y^{2/3}} + \dfrac{6.9}{xy} - \dfrac{7.5}{x^{1/3}y^{2/3}}$

54. $\dfrac{x^2 + 2}{2x^4 - x^3 - 15x^2} - \dfrac{x - 3}{20 + 3x - 2x^2}$

In exercises 55 and 56, write the expression as a simple fraction.

55. $1 + \dfrac{x}{1 + \dfrac{x}{1 + x}}$

56. $x - \dfrac{2}{x - \dfrac{2}{x - 2}}$

In exercises 57 through 62, carry out the indicated division.

57. $(x^3 + 5x^2 - 7x + 2) \div (x - 3)$

58. $(y^4 - 3y^3 + 8y^2 - 7y + 12) \div (y^2 + y - 2)$

59. $(a^4 + a^2b^2 - 5ab^3) \div (a + 2b)$

60. $(n^4 - 625) \div (n + 5)$

61. $(x^{3/2} - 4x + 5x^{1/2} - 34) \div (x^{1/2} + 2)$

62. $(y^{4/5} + 7y^{2/5} - 6) \div (y^{1/5} - 1)$

63. Try to find a pattern for $(x^n - y^n) \div (x - y)$, $n \in N$, by finding the result in some cases—for example, $(x^2 - y^2) \div (x - y)$ and $(x^3 - y^3) \div (x - y)$.

64. Try to find a pattern for $(x - y) \div (x^{1/n} - y^{1/n})$, $n \in N$, by finding the result in some cases, such as $(x - y) \div (x^{1/2} - y^{1/2})$, $(x - y) \div (x^{1/3} - y^{1/3})$, and $(x - y) \div (x^{1/4} - y^{1/4})$.

65. In arithmetic an improper fraction is defined as a fraction whose numerator is greater than or equal to its denominator. An improper fraction can be rewritten as a mixed number by carrying out the division. Define an improper rational expression and an improper fractional expression; for instance, how will you describe "greater" in this case?

Appendix: Summary

This appendix contains many topics used throughout the text such as sets, interval notation, and basic operations with polynomials and other expressions. You may have seen most or all of the topics before, but in each section, we have provided some extensions or previews of ideas discussed elsewhere in the book.

Key Concepts

variable	
constant	
set	*ways to describe, union, intersection, subset*
interval	
numbers	*natural, whole, integer, rational, irrational, real, complex*
properties of operations on the real numbers	*associative, commutative, identity, inverse, distributive*
rules for equations and inequalities	
absolute value	*properties*
exponents	*properties, definition of rational exponents*
expression	*term, like terms*
polynomial	*degree, coefficients, arithmetic*
rational or fractional expressions	*lowest terms, arithmetic*
factoring	*common factors, difference of squares, sum or difference of cubes, quadratic in form, grouping*

Skills

You should be able to:

1. Form the union or intersection of sets.
2. Use interval notation and number line graphs.
3. Classify numbers.
4. Perform complex-number arithmetic.
5. Recognize real-number properties.
6. Apply rules for equations, inequalities, and absolute value.
7. Apply rules of exponents.
8. Perform arithmetic with polynomials, rational expressions, and fractional expressions.
9. Factor completely with the coefficient restrictions of this chapter.
10. Identify quadratic-in-form expressions.

SUMMARY EXERCISES

The starred exercises can serve as a practice chapter test. The answers are in the answer section at the back of the book.

In exercises 1 through 3, enumerate the set.

1. $\{q \mid q = n/3, n \in N\}$
2. $\{m \mid m = i + i^2, i \in I\}$

* 3. $\{t \mid t = |2 - 5k|, k \in W\}$

In exercises 4 through 6, use $A = \{1, 3, 5, 7, \ldots\}$, $B = \{3, 6, 9, \ldots\}$, and $C = \{1, 4, 7, 10, \ldots\}$ to find the set described.

4. $B \cap (A \cup C)$
5. $(C \cap A) \cup (C \cap B)$

* 6. $A \cup (B \cap C)$

In exercises 7 through 9, graph and give a simplified interval to describe the result.

7. $[4.2, 7] \cup (4.2, 7)$
8. $(-\infty, -2.8) \cup (\sqrt[3]{-27}, +\infty)$

* 9. $[5, +\infty) \cap (-7, 8)$

10. List the elements of E in order from smallest to largest, where $E = \left\{-1.57557\ldots, \sqrt[3]{7}, -\frac{52}{33}, (1 + \sqrt{2})^2, -\frac{12}{16}, 2.05, -1.575, \frac{45}{8}, 3 + \sqrt{6}, \left(\frac{1}{8}\right)^{-1/3}\right\}$. Give the smallest interval that will contain all the integers in E.
11. If $F = \left\{\sqrt{5}, -3.1, \sqrt[3]{8}, 2.14114\ldots, \sqrt{3} + 1, -2 - \sqrt{2}, 2.14114, 15/8\right\}$, give the smallest interval that will contain all the rationals in F.

* 12. Graph $G = \left\{\sqrt[4]{80}, \sqrt[3]{-27}, -6 + \sqrt{3}, 3.23223\ldots, 3.23, 0.29, \sqrt{1.44}, -32^{2/5}\right\}$. Give the smallest interval that will contain all the irrationals in G.

In exercises 13 through 15, give the property of the real numbers that guarantees the right-hand side is equivalent to the left-hand side.

13. $8x(3x - 2)(3x - 2) = 8x(3x - 2)^2$
14. $14.5(4.1 + 6.8x) = (4.1 + 6.8x)(14.5)$

* 15. $\sqrt{2}(6 + \sqrt{3}) + 12(6 + \sqrt{3}) = (\sqrt{2} + 12)(6 + \sqrt{3})$

In exercises 16 through 21, fill in the blank with $<$ or $>$ to make the statement true.

16. If $4(3 - x) > 12 + 6(x - 3)$, then $x - 3$ ______ $-\frac{12}{10}$.
17. If $-3(2x + 5) < 15$, then $3x$ ______ -15.

* 18. If $6/(y - 4) > 2$, then $y - 2$ ______ 5.

19. If $\sqrt{x} + 5 = 7\sqrt{x} - 2$, then $\sqrt{x} + 8$ ______ $7\sqrt{x}$.
20. $|m - 7|$ ______ $2|m| + |7|$.

* 21. If $6|8x - 4| < 96x$, then $|2 - x|$ ______ $4x$.

In exercises 22 through 27, perform the indicated operations and write the result in standard form.

22. $(6 - 3i)^2 - 2(1 - \frac{1}{2}i)$
23. $\dfrac{11 + 10i}{3 - 2i} + 7i(4 + i)$

* 24. $27i^{15}(8 - 5i) + 2 - 3i(4 + 9i)$

25. $\dfrac{\left(\frac{1}{2} - \frac{3}{2}i\right)^2}{i^{51}} \div \dfrac{(1 - 3i)^5}{-(1 - i)^4}$
26. $\dfrac{7 - 4i \div 2 \cdot 9i}{1 + 2i}$

* 27. $\dfrac{(1 - i)^6}{3i} \cdot (5 + 2i)$

In exercises 28 through 33, factor completely subject to the restrictions of this chapter.

28. $(6m + 7)^3 - 125$
29. $4y^{-4/5} - 16y^{-2/5} + 15$

* 30. $6(2a - 1)^{1/2}(a + 4)^{1/2} + (2a - 1)^{3/2}(a + 4)^{-1/2}$

31. $x^{-2} + 11x^{-3} + 30x^{-4}$
32. $a^3 - 5ac + 12cb - 4ab - 3a^2c + 15c^2$

* 33. $y^{11/2} - 64y^{-1/2}$

In exercises 34 through 36, carry out the division indicated.

34. $(-30xy^2z + 25xz^4 - 15x^2yz^3) \div (15x^2yz^3)$
35. $(y - 8) \div (y^{1/3} - 2)$

* 36. $(x^5 + 3x^3 + 4x + 1) \div (x^2 + x + 1)$

In exercises 37 through 39, rewrite each expression in lowest terms; state *all* restrictions on the variables necessary for the result to represent a real number equal to the original.

37. $\dfrac{3x^2 + 14x - 24}{12x^2 + 5x - 28}$
38. $\dfrac{x^3 + 8}{6\sqrt{x - 2} + 3x\sqrt{x - 2}}$

* 39. $\dfrac{x^{-4/5} - 6x^{-2/5} + 5}{x^{-4/5} - 25}$

In exercises 40 through 42, perform the operations; write the result in lowest terms with only positive exponents.

40. $\dfrac{(x - y)^2}{4x^{-2} - 9y^{-2}} \cdot \dfrac{8x^{-3} - 27y^{-3}}{x^2 - y^2}$
41. $\dfrac{x^4 - 2x^2 + 1}{x^3 + 1} \div \dfrac{x^2 + xy - x - y}{2x^2 - 2x + 2}$

* 42. $\dfrac{2\sqrt[3]{x^2}}{x + 27} - \dfrac{2}{\sqrt[3]{x} + 3}$

Solutions to Odd-Numbered Exercises and Chapter Tests

Chapter One

1.1

1. $\left\{\frac{1}{4}\right\}$
3. $\left\{\frac{22.18}{0.27}\right\}$, or $\left\{\frac{2218}{27}\right\}$; the solution is approximately 82.148.
5. {all reals except -2}
7. $\left\{\frac{21}{13}\right\}$; the solution is approximately 1.615.
9. $\left\{\frac{1.3 + 2\sqrt{7}}{2 + \sqrt{2}}\right\}$; the solution is approximately 1.931.
11. $\left\{\frac{7}{4}\right\}$
13. $\left\{-\frac{9}{4}\right\}$
15. $\{-12, 22\}$
17. $\{0\}$
19. $\{\ \}$
21. $\{3\}$
23. $\{2, 6\}$
25. $\{1 + i\}$
27. $\left\{-3 + \frac{3}{2}i\right\}$
29. yes
31. no
33. The least interest will be paid for 10% compounded semiannually for 4 years.
35. Luis will be 30 years old.
37. The width is approximately 4.5 feet and the length is approximately 7.5 feet.
39. She earned 25 A's.
41. The relationship must be $h = \frac{2r}{r - 2}$.
43. a. She needs $\frac{22.4}{3}$ ounces of 40% solution and $\frac{19.6}{3}$ ounces of 10% solution.
    b. She needs 9.1 ounces of 40% solution and 4.9 ounces of water.
45. She ships 7 flats every three days.

1.2

1. $\{2, 4\}$
3. $\{2, 4\}$
5. The solutions are $\frac{-35 - \sqrt{1233}}{2}$ and $\frac{-35 + \sqrt{1233}}{2}$; the approximate solutions are -35.057 and 0.057.
7. $\left\{\frac{8}{3}, 5\right\}$
9. $\left\{\frac{4}{11}\right\}$
11. $\{-\sqrt{65}i, \sqrt{65}i$
13. $\left\{\frac{0.01 - \sqrt{0.0399}i}{0.0002}, \frac{0.01 + \sqrt{0.0399}i}{0.0002}\right\}$,

or $\{500 - 50\sqrt{3.99}i, 500 + 50\sqrt{3.99}i\}$

15. $\{1 - 2i, 1 + 2i\}$

17. $\{-\sqrt{2}i, \sqrt{2}i, -2\sqrt{2}, 2\sqrt{2}\}$ or $\{-\sqrt{2}i, \sqrt{2}i, -\sqrt{8}, \sqrt{8}\}$

19. $\{\ \}$

21. One possible answer is $\left(x - \frac{7}{3}\right)\left(x + \frac{1}{4}\right) = 0$.

23. One possible answer is $16x^2 - 4x - 159 = 0$.

25. $\left(x - \frac{5 - \sqrt{21}}{2}\right)\left(x - \frac{5 + \sqrt{21}}{2}\right)$

27. $3\left(x - \frac{15 - \sqrt{201}}{6}\right)\left(x - \frac{15 + \sqrt{201}}{6}\right)$

29. $(x - 1)^2 + 3$

31. $\frac{7}{10}\left(y - \frac{11}{14}\right)^2 - \frac{121}{280}$

33. rational

35. complex and not real

37. irrational

39. complex and not real

41. The formula gives $-\frac{1}{3}$ and 2. $-3(2)^2 + 5(2) + 2$ is 0 and $-3\left(-\frac{1}{3}\right)^2 + 5\left(-\frac{1}{3}\right) + 2$ is 0.

43. $\left\{\frac{a - \sqrt{a^2 + 24}}{6}, \frac{a + \sqrt{a^2 + 24}}{6}\right\}$

45. The trapezoid's upper base is $4\sqrt{21}$ meters, its lower base is $2\sqrt{21}$ meters, and its height is $\frac{2}{3}\sqrt{21}$ meters.

47. Tracy runs exactly 15 miles per hour, and she should definitely consider turning professional.

49. Tim must drive approximately 7.9 miles to the lighthouse.

1.3

1. $\{28\}$
3. $\{\ \}$
5. $\{343\}$
7. $\{-4\}$
9. $\left\{\frac{4094}{3}\right\}$
11. $\{\ \}$
13. $\{1\}$
15. $\{0, 5\}$
17. $\{0, 5\}$
19. $\{3 - \sqrt{10}, 3 + \sqrt{10}\}$
21. $\{4, i, -i\}$
23. $\{4\}$
25. $\{\ \}$
27. $\{1 - \sqrt{3}, 1 + \sqrt{3}\}$
29. $\{0\}$
31. $\{-1, 0\}$
33. $\left\{\frac{7}{2}\right\}$
35. $\{33 + 8\sqrt{19}\}$ or $\left\{\frac{66 + \sqrt{4864}}{2}\right\}$
37. $\{2\}$
39. $\{4\}$

1.4

1. $\{-2, 0, 5\}$
3. $\{-\sqrt{2}, \sqrt{2}\}$
5. $\left\{1, \frac{-1 + \sqrt{3}i}{2}, \frac{-1 - \sqrt{3}i}{2}, -2, 1 + \sqrt{3}i, 1 - \sqrt{3}i\right\}$
7. $\left\{\frac{1}{\sqrt[3]{5}}, 1, \frac{-1 - \sqrt{3}i}{2}, \frac{-1 + \sqrt{3}i}{2}, \frac{-\sqrt[3]{\frac{1}{5}} - \sqrt{3\sqrt[3]{\frac{1}{25}}}}{2}, \frac{-\sqrt[3]{\frac{1}{5}} + \sqrt{3\sqrt[3]{\frac{1}{25}}}}{2}\right\}$
9. $\{-3\}$
11. $\{-7, 2\}$
13. $\left\{\frac{5 - \sqrt{37}}{6}, \frac{5 + \sqrt{37}}{6}\right\}$
15. $\left\{\frac{-1}{8}, 1\right\}$
17. $\left\{\frac{31}{32}, 2\right\}$
19. $\{1\}$
21. $\{0, 8\}$
23. $\left\{\frac{8}{27}\right\}$
25. $\left\{\frac{5}{3}\right\}$
27. $\left\{\frac{-1}{3}\right\}$
29. $\left\{\frac{-1}{\sqrt[5]{27}}\right\}$
31. $\{0\}$

33. $\left\{\frac{5}{4-\sqrt{44}i}+\frac{3}{2}, \frac{5}{4+\sqrt{44}i}+\frac{3}{2}\right\}$

35. $\left\{\frac{-1-\sqrt{3}i}{2}, \frac{-1+\sqrt{3}i}{2}, -1, 6\right\}$

37. An nth-degree polynomial equation will have at most n solutions.

39. $\{-2, 1+\sqrt{3}i, 1-\sqrt{3}i\}$

Summary Exercises

1. $\left\{\frac{8}{5}\right\}$

3. $\left\{\frac{17}{4}\right\}$

5. $\left\{\frac{-4-8i}{17+i}\right\}$, or $\left\{\frac{-38}{145}-\frac{66}{145}i\right\}$

7. $\left\{\frac{128}{35}\right\}$

9. $\{1\}$

11. $\{2-\sqrt{8}i, 0, 4, 2, +\sqrt{8}i\}$

13. $\left\{\frac{7-\sqrt{31}i}{10}, \frac{7+\sqrt{31}i}{10}\right\}$

15. $\{-12, 6\}$

17. $\left\{\frac{9-\sqrt{3}i}{2}, \frac{9+\sqrt{3}i}{2}\right\}$

19. K could be $-\sqrt{456}$ or $\sqrt{456}$.

21. Possible answers are $x^2 - 18x + 81 = 0$ and $x - 9 = 0$.

23. a is 7, h is 2, k is -33.

25. $\{2, 10\}$

27. $\{1\}$

29. $\{2, 6\}$

31. There are 53 sixth-graders and 62 fourth-graders.

33. $\{-5, -2\}$

35. $\{2-\sqrt{5}, 0, 2+\sqrt{5}\}$

37. $\{-i, i, -1, 1\}$

39. $\left\{\frac{1-\sqrt{33}}{2}, -1, 0, 2, \frac{1+\sqrt{33}}{2}\right\}$

41. It took approximately 20.1 minutes.

Chapter Two

2.1

1. and 3.

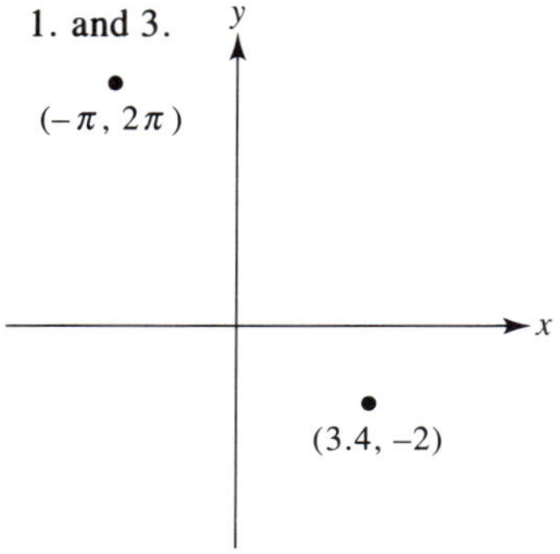

5. 20.4 units

7. 14 units

9. 5 units

11. $\sqrt{129.61}$ units, or approximately 11.3846 units

13. $\sqrt{100+(2+\pi)^2}$, or approximately 11.2444 units

15. $\frac{MD}{AC} = \frac{DB}{CB}$ by similar triangles, so $MD = \dfrac{(y_1 - y_2)\frac{x_2 - x_1}{2}}{(x_2 - x_1)}$ and $MD = \frac{y_2 - y_1}{2}$ because $x_2 \neq x_1$. The x-coordinate of M must be $\frac{x_1 + x_2}{2}$ and $MD = y_m - y_2 = \frac{y_1 - y_2}{2}$, so $y_m = \frac{y_1 + y_2}{2}$. The midpoint of AB has coordinates $\left(\frac{x_1 + x_2}{2}, \frac{y_1 + y_2}{2}\right)$.

17. $(5.5, -6)$

19. $\left(\frac{5}{3}, 2\right)$

21. $\left(-\frac{5}{3}, -\frac{2}{3}\right)$

23. $(0, 2-\sqrt{12}), (0, 2+\sqrt{12})$

25. $(e+6.2, 0), (e-6.2, 0)$

27.

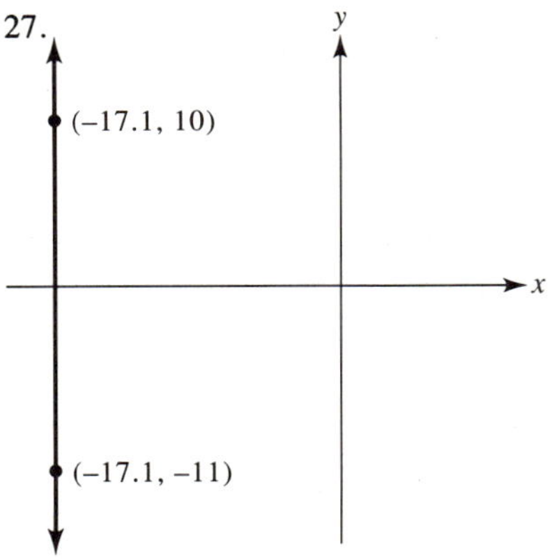

29.

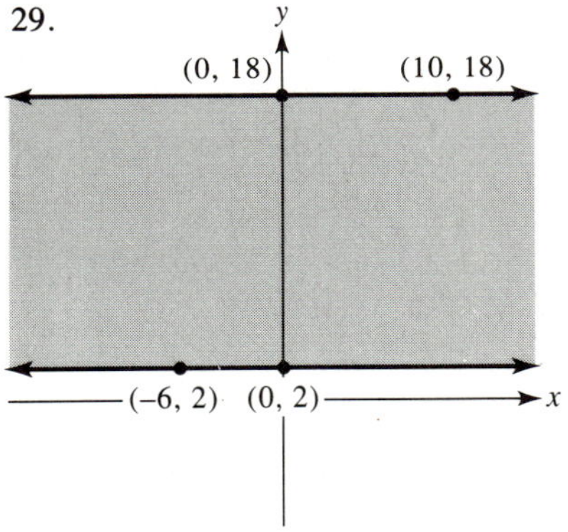

31.

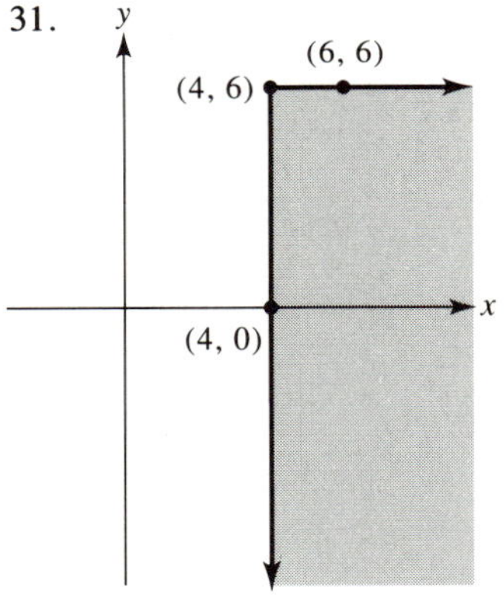

33.

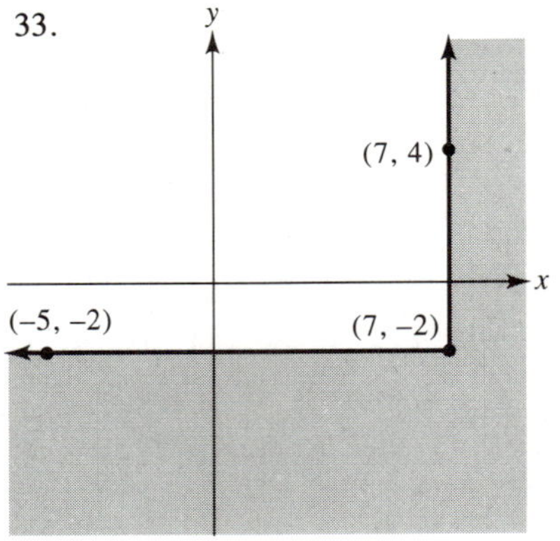

35. It is neither because the diagonal lengths are $\sqrt{26}$ and $\sqrt{50}$ units.

37. a. The shortest is 10 blocks.
    b. The shortest is 7 blocks.

39. There are two ways to connect the stations using $(\sqrt{29} + \sqrt{18} + \sqrt{5})$ blocks.

2.2

1. Linear
3. Nonlinear
5. Linear
7.

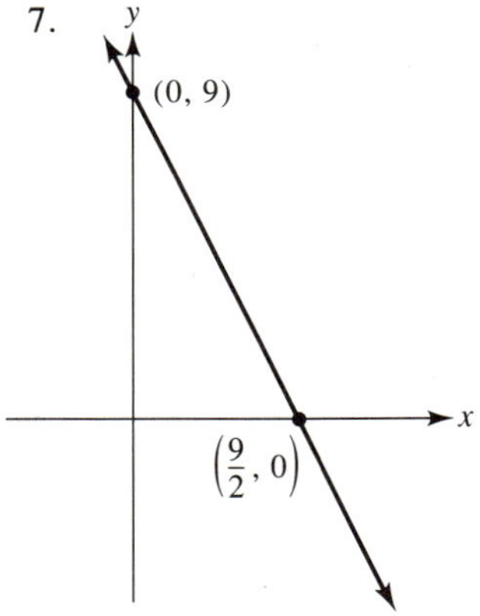

x-intercept: (4.5, 0); y-intercept: (0, 9); slope: -2

9.

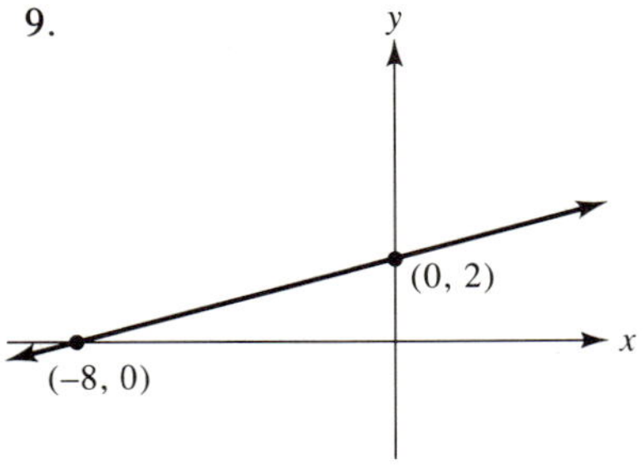

x-intercept: $(-8, 0)$; y-intercept: (0, 2); slope: $\frac{1}{4}$

11.
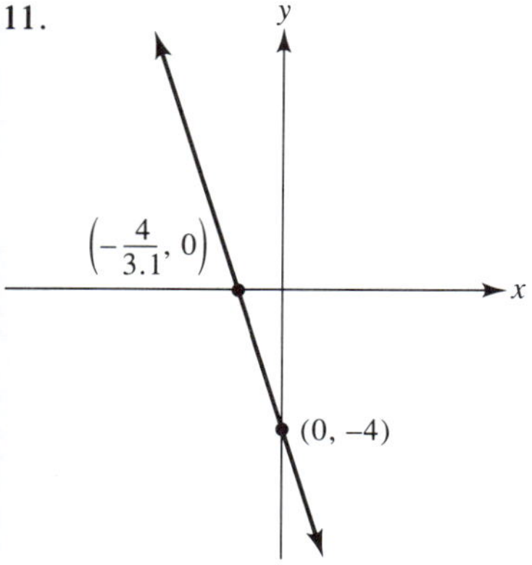

x-intercept: $\left(-\dfrac{4}{3.1}, 0\right)$; y-intercept: $(0, -4)$; slope: -3.1

13.
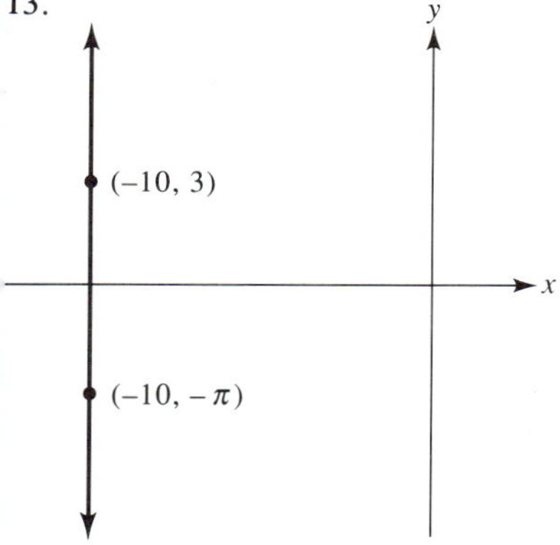

x-intercept: $(-10, 0)$; slope: not defined

15.
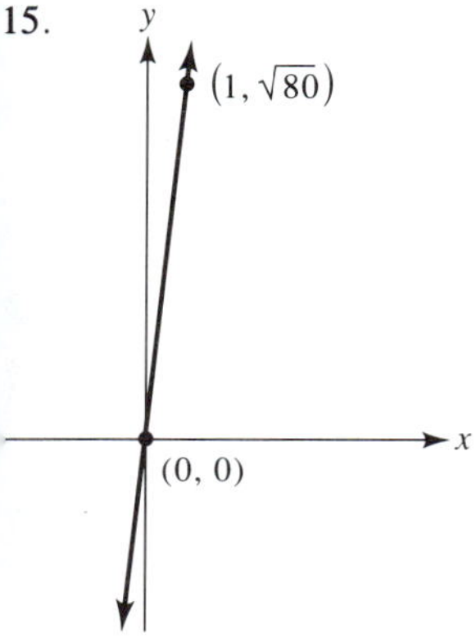

x-intercept: $(0, 0)$; y-intercept: $(0, 0)$; slope: $\sqrt{80}$

17. $-\frac{1}{16}$

19. $(6b)/(5a)$, if $a \neq 0$; slope not defined if $a = 0$

21. $-\frac{3}{11}$

23. 0

25. $-1/45.3$

27. $y = \frac{8}{7}(x + 3)$

29. $x = \sqrt{7}$

31. $y + 1 = -\frac{3}{5}x$

33. $x = -1$

35. $y = -\frac{1}{5}(x - \pi)$

37. $EF = \frac{2}{15}\sqrt{34}$; $FG = \frac{1}{5}\sqrt{34}$; $EG = \frac{1}{3}\sqrt{34}$; $EF + FG = EG$.

39. The slope of l_1 is b/a; the slope of l_2 is d/c. By the Pythagorean theorem, $(\sqrt{a^2 + b^2})^2 + (\sqrt{c^2 + d^2})^2 = (\sqrt{(a - c)^2 + (b - d)^2})^2$. This is equivalent to $a^2 + b^2 + c^2 + d^2 = a^2 - 2ac + c^2 + b^2 - 2bd + d^2 \Leftrightarrow 0 = -2ac - 2bd \Leftrightarrow ac = -bd \Leftrightarrow b/a = -c/d$.

41.
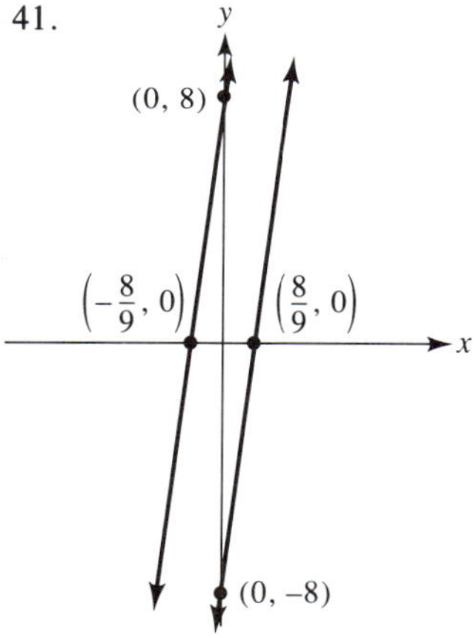

43.
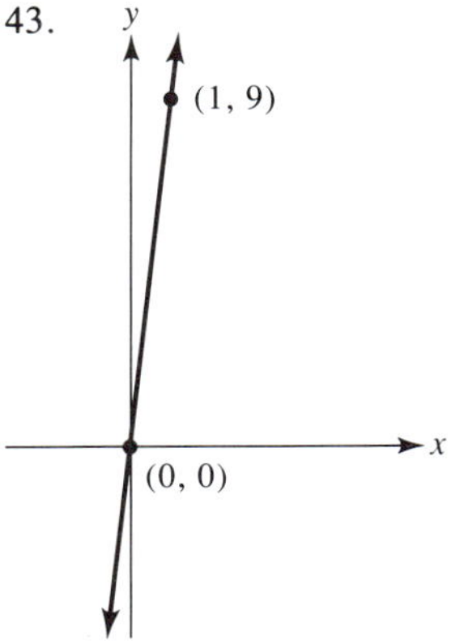

45.

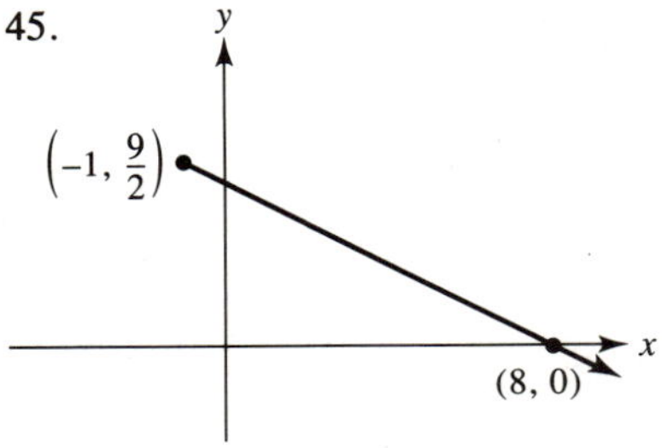

47.

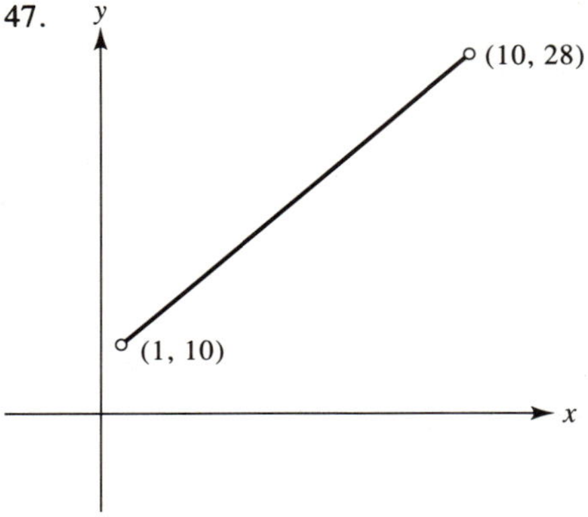

49. No solutions.

51. The exact x-intercept is $(2.3/\sqrt{3}, 0)$.

53. The exact x-intercept is $(\sqrt[3]{7.43/3}, 0)$.

2.3

1.

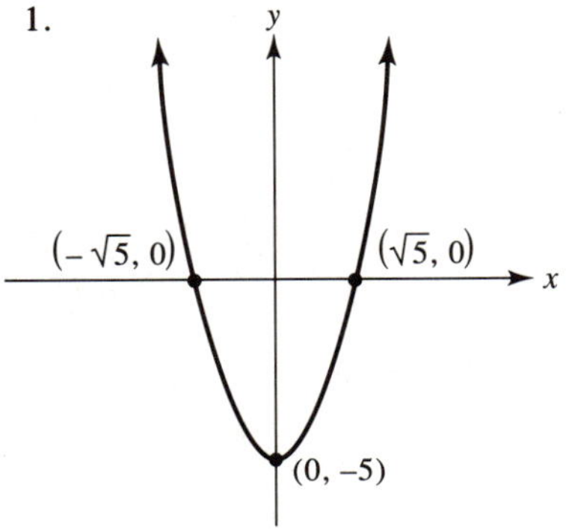

3.

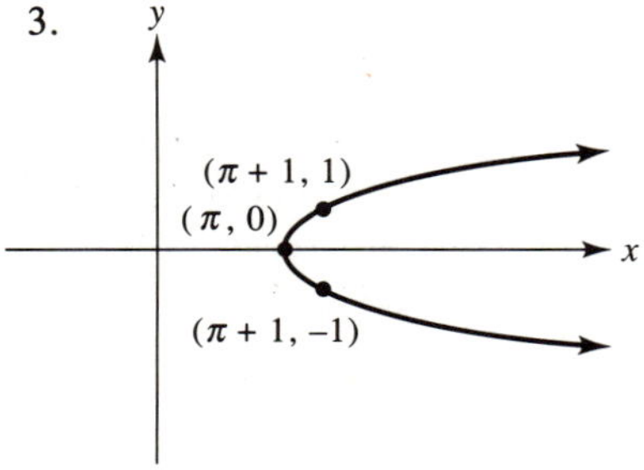

5.

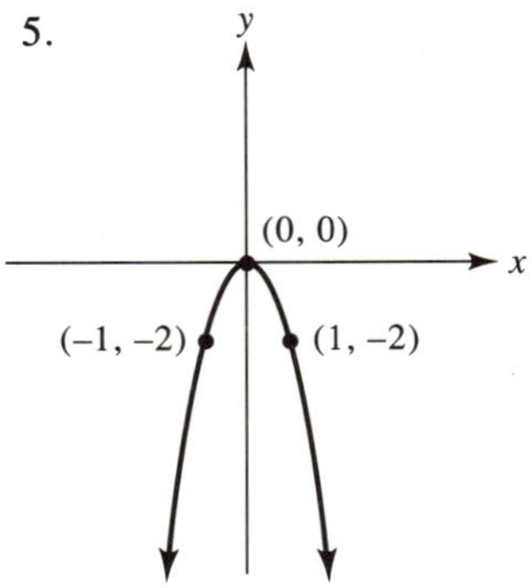

7.

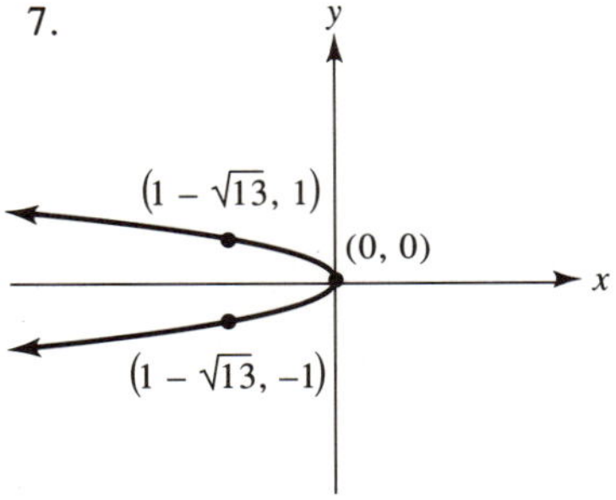

9.

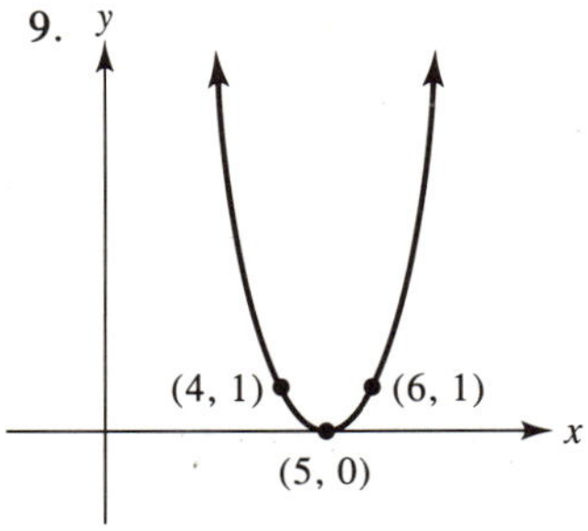

11.

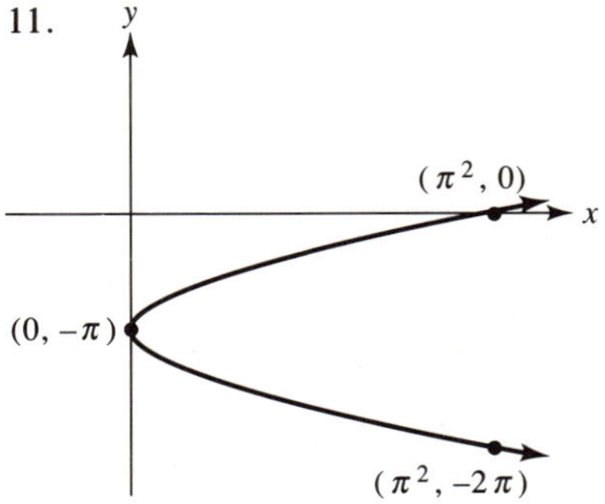

13.

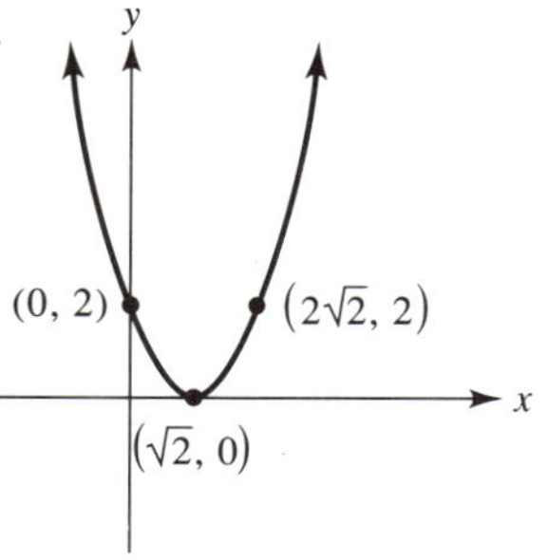

15.

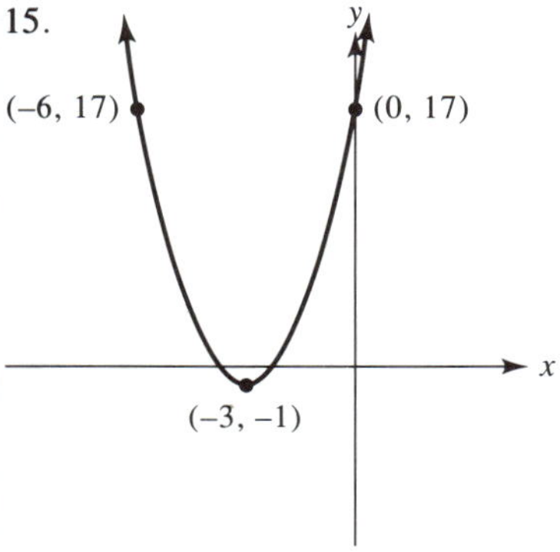

17.

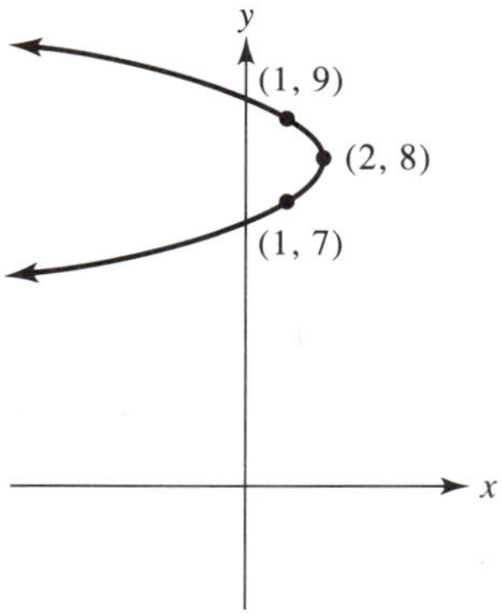

19.

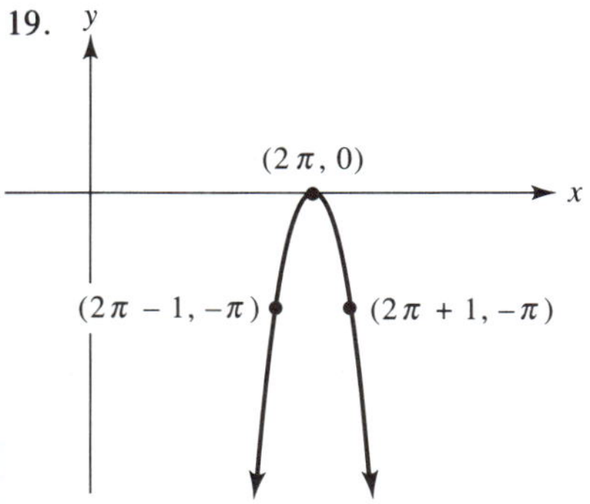

21. Vertex: (4, 1); opens up

23. Vertex: $\left(-\dfrac{13}{4}, -\dfrac{5}{2}\right)$; opens right

25. Vertex: (2, 5); opens up

27. Vertex: $\left(-\frac{1}{6}, \frac{25}{12}\right)$; opens down

29. $(a/4, -3 - a^2/8)$

31. Minimum 5

33. Maximum -5

35. $\frac{25}{4}$

37. Vertex: $\left(\dfrac{-b}{2a}, \dfrac{4ac - b^2}{4a}\right)$;

x-intercepts: $\left(\dfrac{-b - \sqrt{b^2 - 4ac}}{2a}, 0\right)$, $\left(\dfrac{-b + \sqrt{b^2 - 4ac}}{2a}, 0\right)$.

The x-coordinate of the vertex is the x-coordinate of the midpoint of the line segment connecting the x-intercepts.

39.

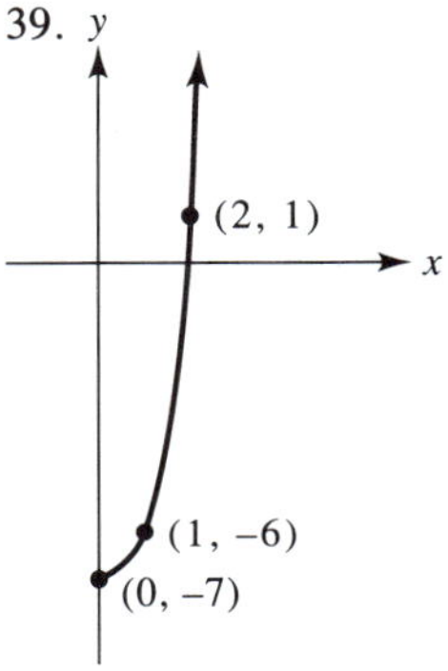

41.

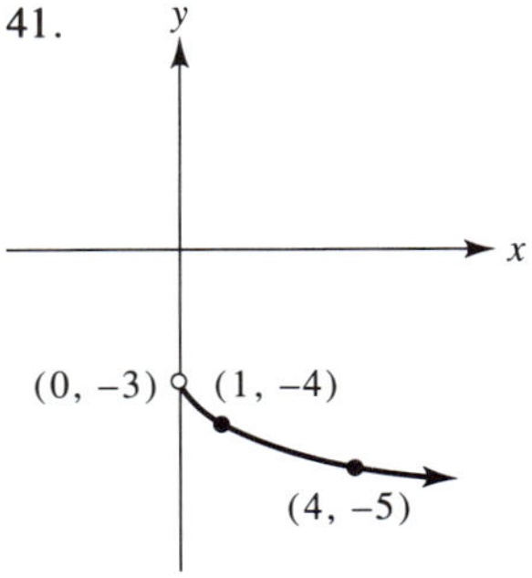

43.

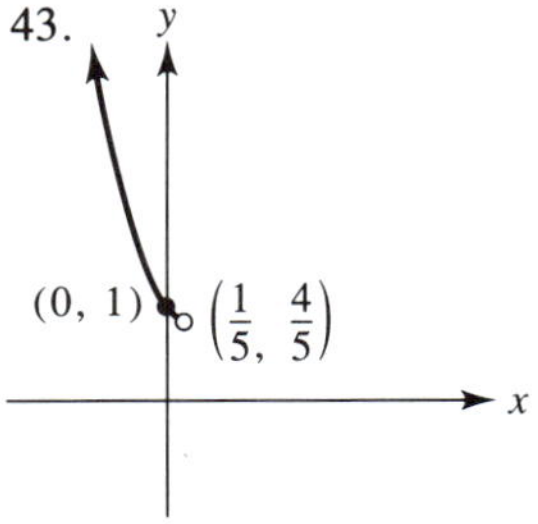

45. $x \geq \frac{7}{2}$

47. The maximum height is 144 feet after 2 seconds.

49. He should plant 39 trees per acre; this would yield 4524 pounds of fruit.

Answers to exercises 51 through 54 may vary, but they should be close to the values given.

51. b. None c. (3.5, 3.14)

53. b. -0.372 and 0.991 c. (0.313, -3.905)

2.4

1.

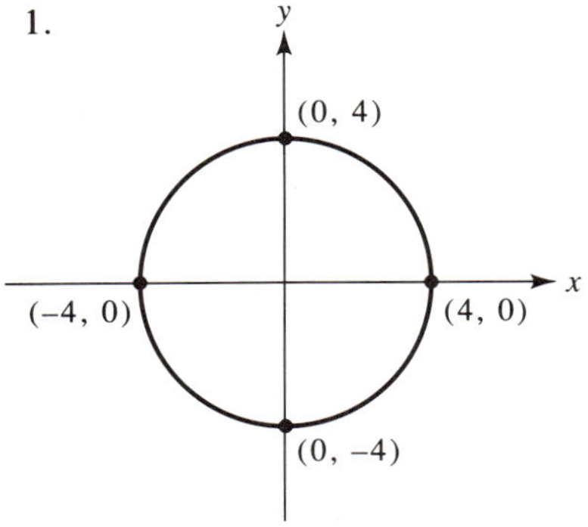

3.

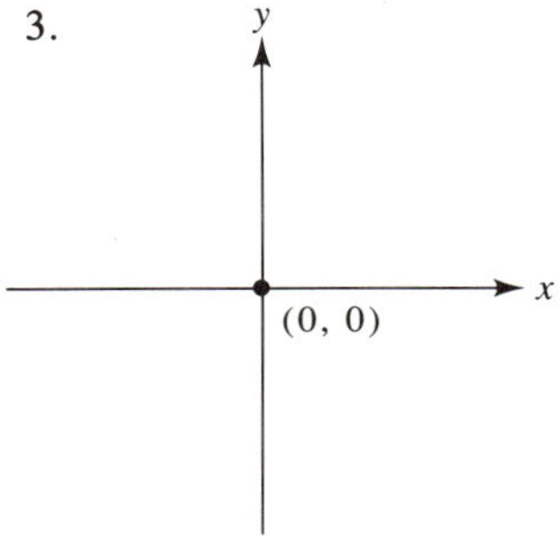

5.

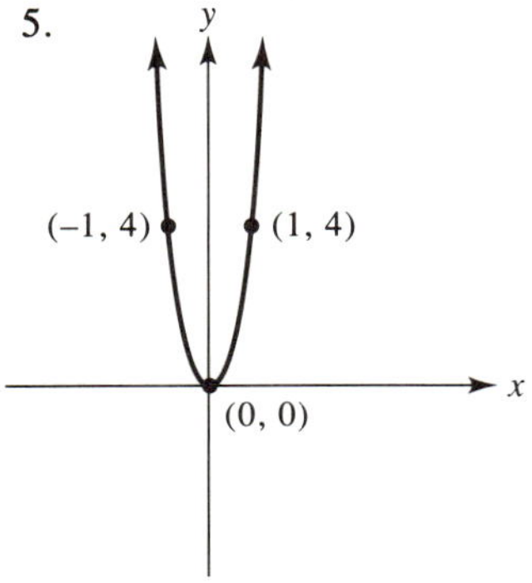

7.

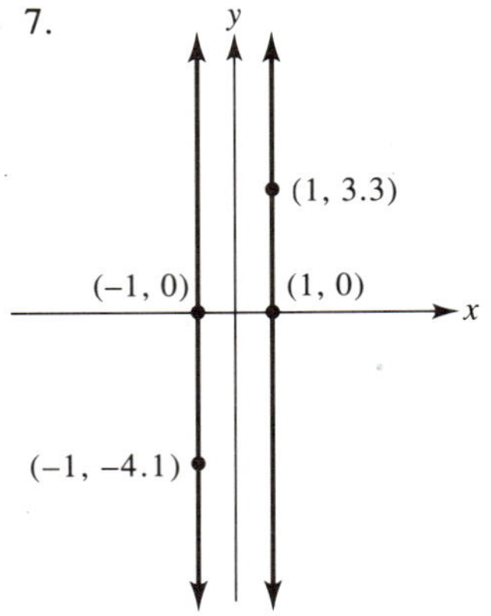

9. There are no real-number solutions.

11.

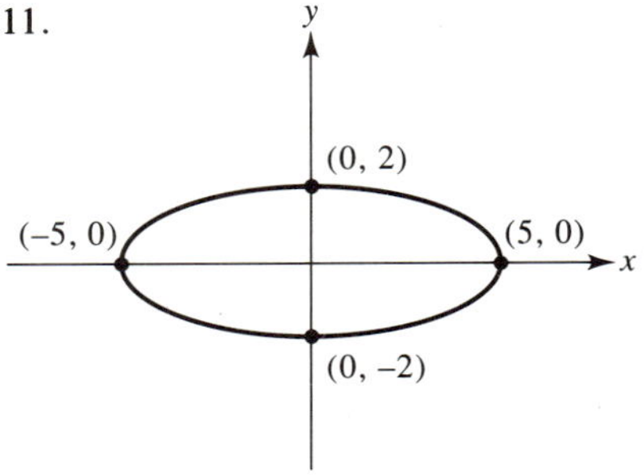

13.

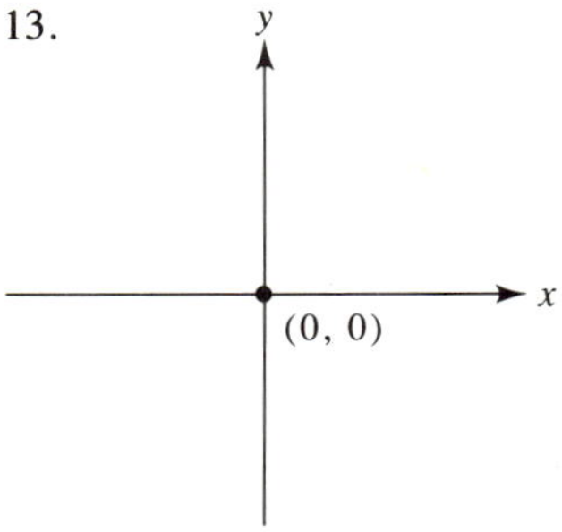

15. There are no real-number solutions.

17.

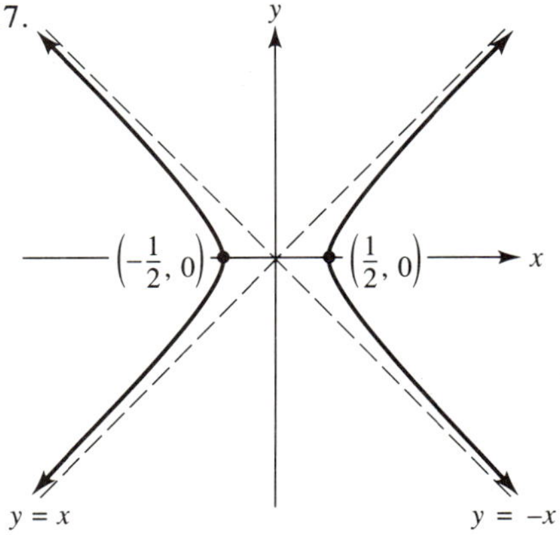

19.

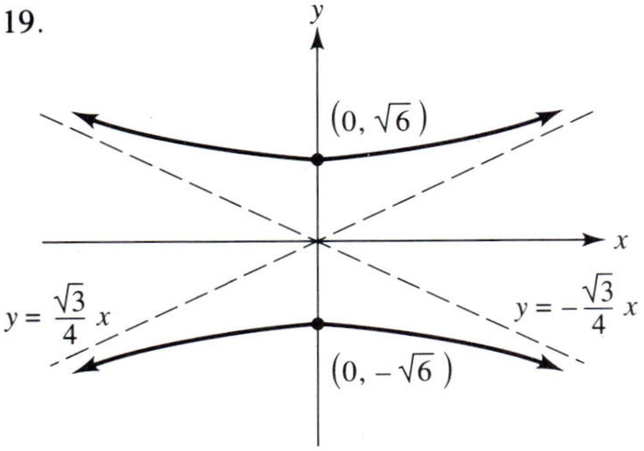

21.

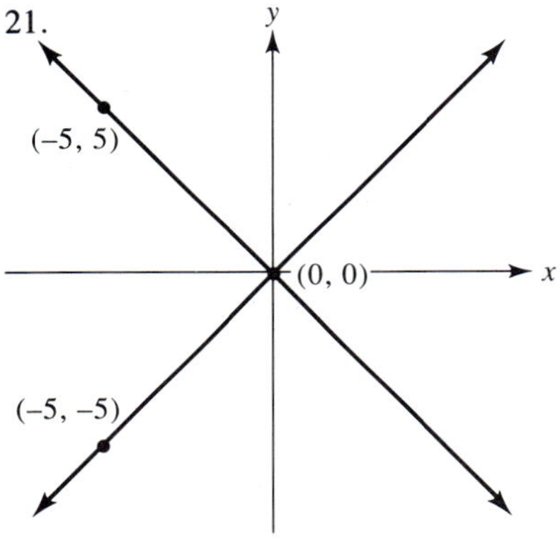

23. 3, 4, 6, 7, 9, 10, 13, 15, 21

25. For any point (x, y) on the parabola, $\sqrt{x^2 + (y - r)^2} = \sqrt{y^2}$. So $x^2 + (y - r)^2 = y^2 \Leftrightarrow x^2 - 2ry + r^2 = 0 \Leftrightarrow y = (x^2 + r^2)/(2)r$.

27. $\dfrac{x^2}{4} + \dfrac{y^2}{16.81} = 1$

29. $\dfrac{y^2}{9} - \dfrac{x^2}{4} = 1$

31. The two foci must coincide at the center.

33. A long flat ellipse beginning to resemble a horizontal line (the x-axis)

35. The branches of the hyperbola are pushed close to the x-axis.

37. $y = \pm\dfrac{b}{a}\sqrt{a^2 + x^2}$. When $|x|$ is very large, $y \approx \pm\dfrac{b}{a}|x|$ or $y \approx \pm\dfrac{b}{a}x$.

39. An ellipsoid with center (0, 0, 0), longest in y-dimension (a football)

41.

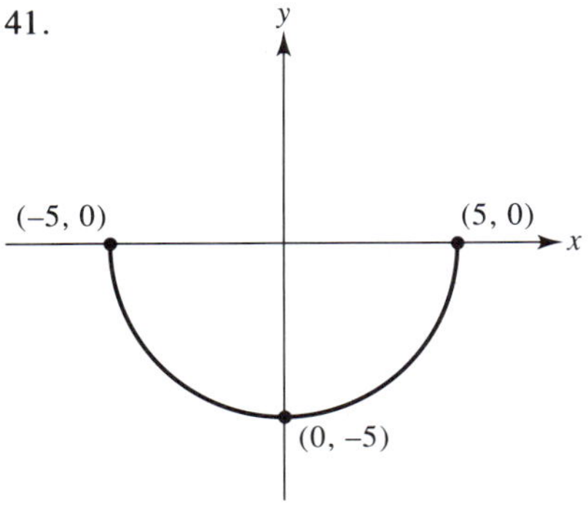

43.

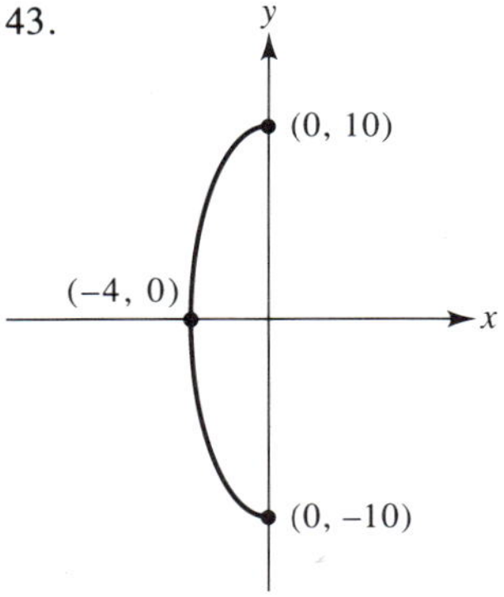

45.

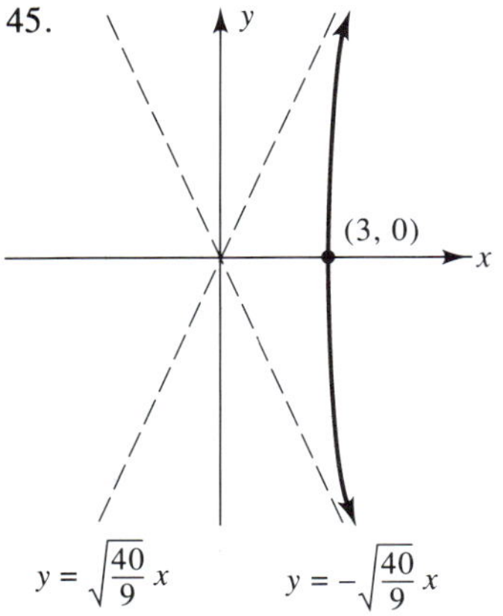

47.

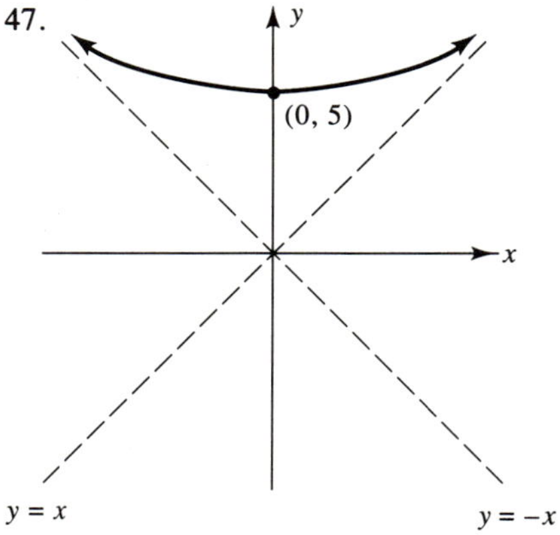

49.

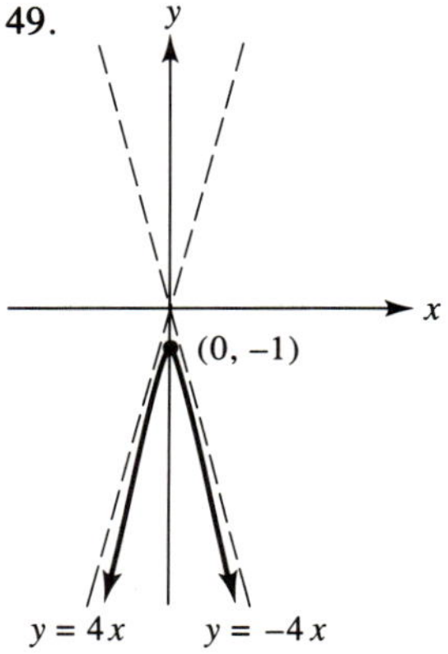

51. If a value for x produces exactly one y-value, the equation can be graphed directly. Others need to be rewritten as $y = \cdots$, perhaps in two pieces.

2.5

1.

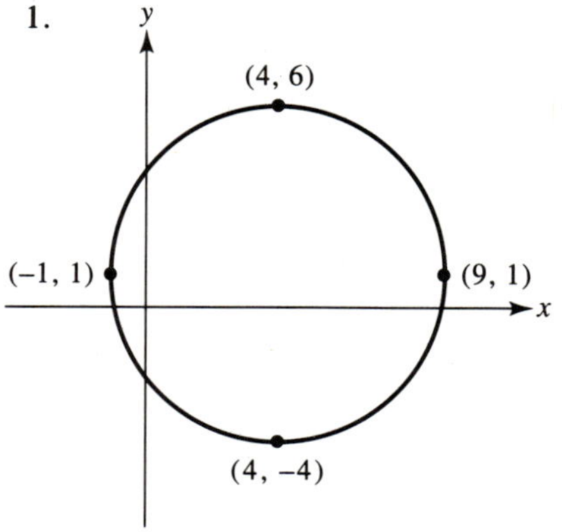

3.

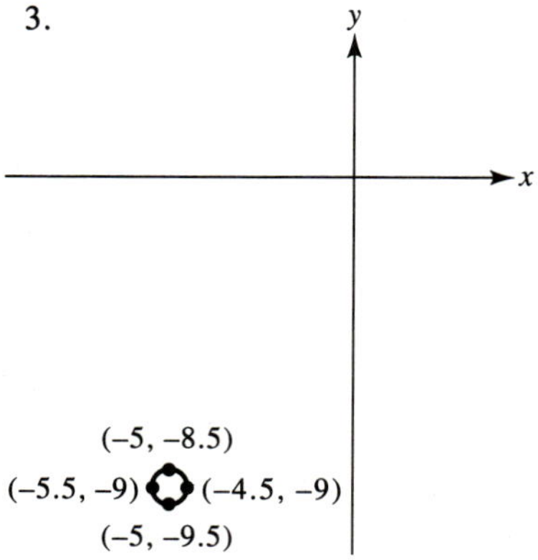

5.

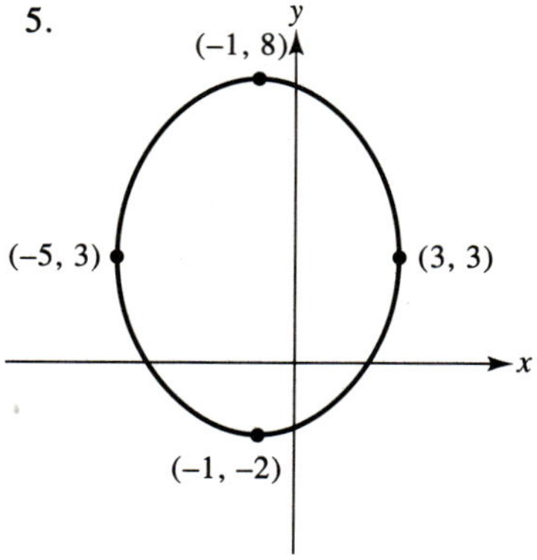

7.

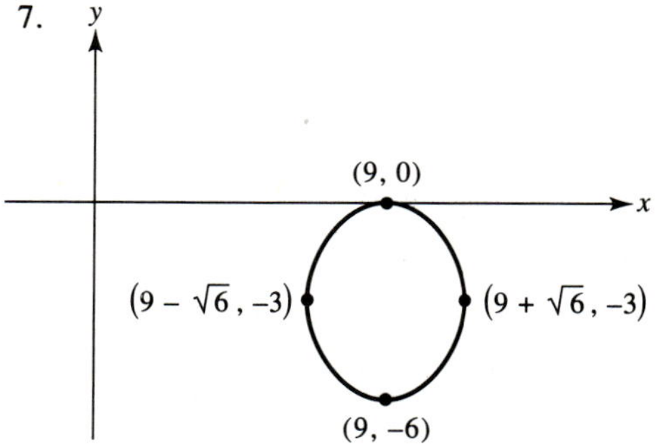

9.

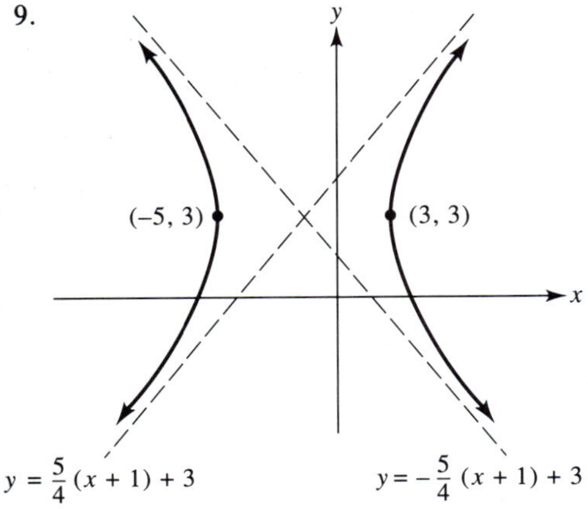

11.

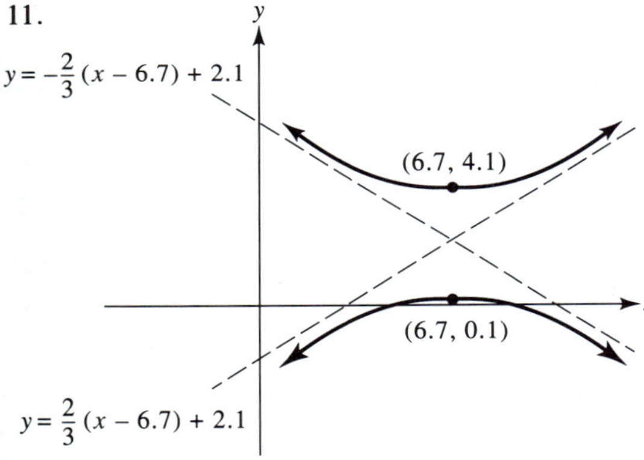

13.

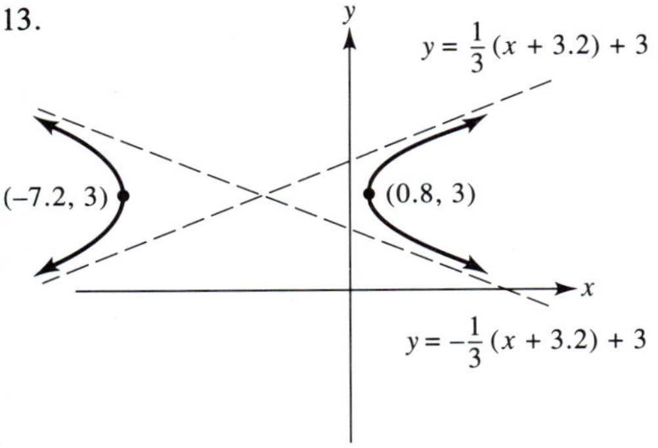

15.

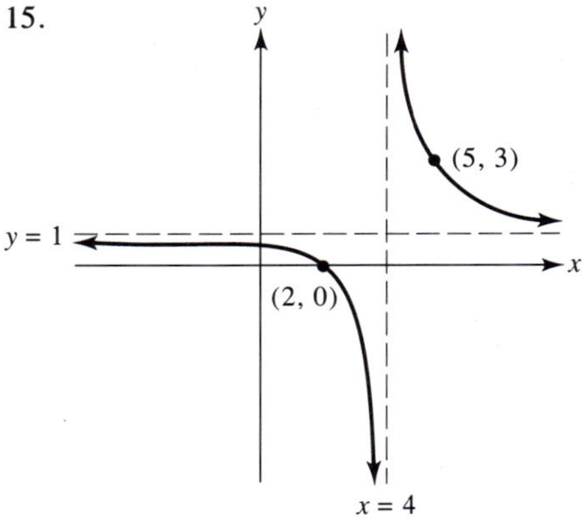

17.

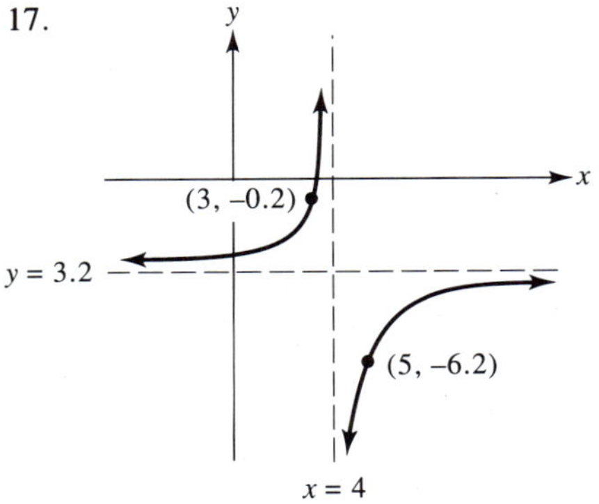

19.

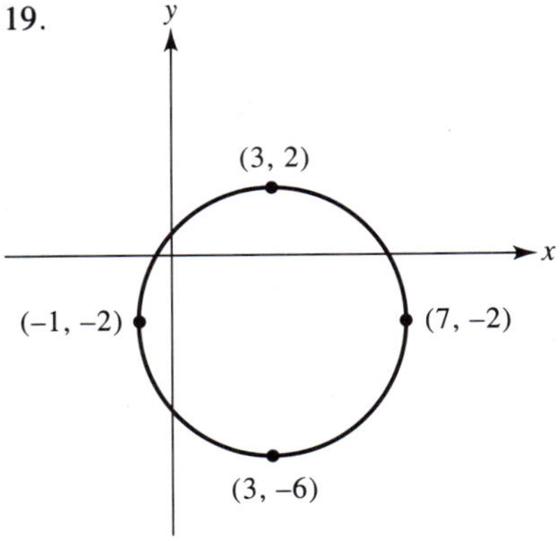

21.

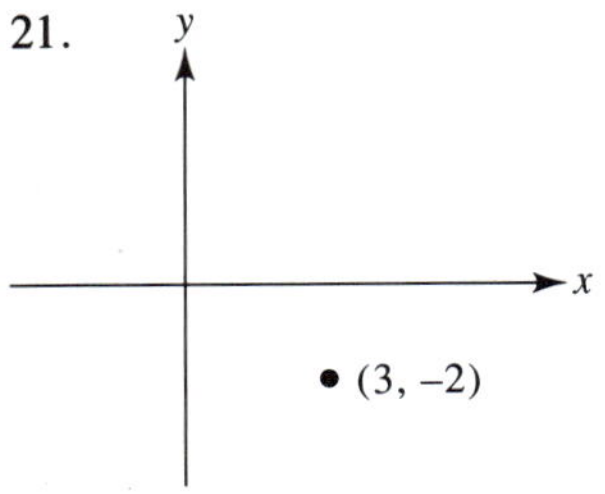

23.

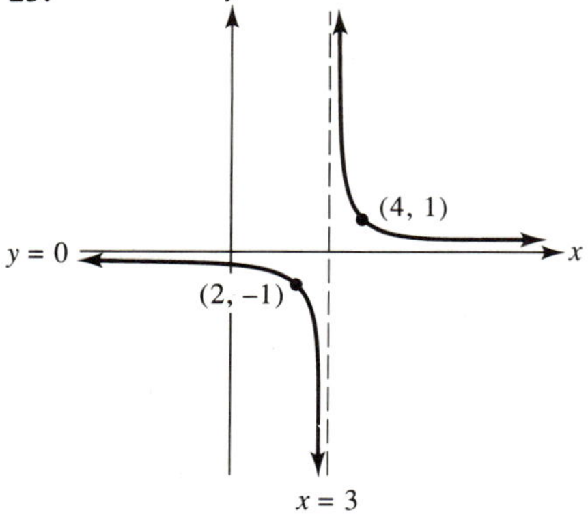

25. There are no real-number solutions.

27. $y = \dfrac{1}{x + 7.3} + 150$

29. $y + \sqrt{2} = -0.3(x + 6)$

31. $(x + \sqrt[3]{7})^2 + (y + \frac{3}{4})^2 = 1$

33. $\dfrac{x^2}{9} + \dfrac{y^2}{4} = 1, y \geq 0$

35. $y = 3(x - 1)^2 + 2$

37. An ellipse taller than it is wide

39. A hyperbola that opens up and down

41. A sphere with center $(2, 0, -5)$ and radius 2

43. A paraboloid with vertex $(0, -\pi, 0)$

45. The answer depends on the individual.

Summary Exercises

1.

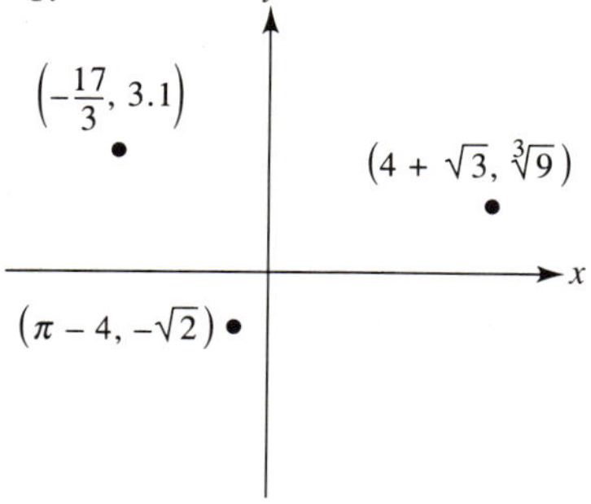

3. $(1.7, -4.595)$

5. 22.69

7.

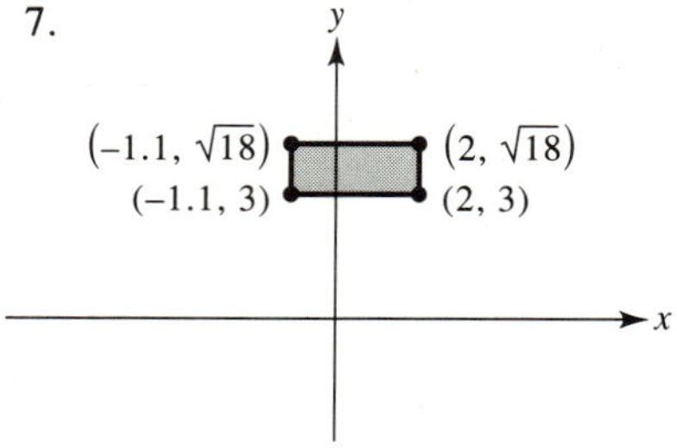

9.

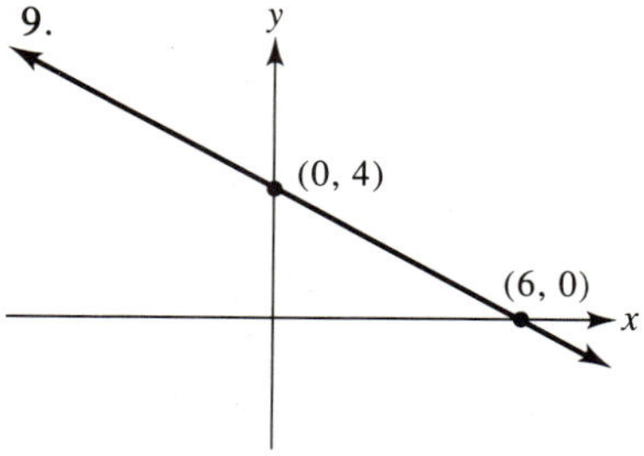

11.

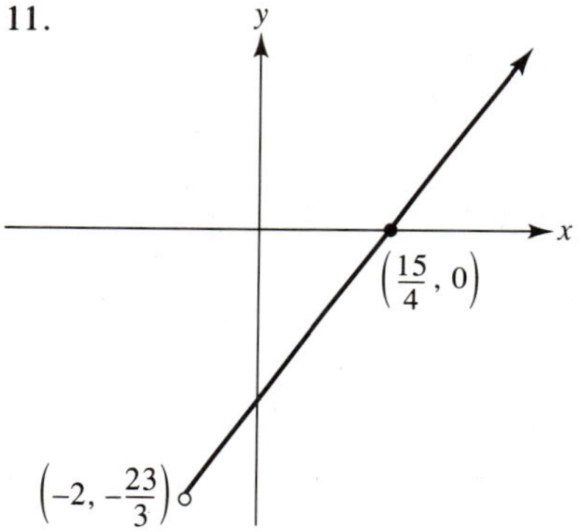

13. $y + 3 = \dfrac{7}{3}(x - 4)$

15. $y = -\dfrac{7}{3}(x - 4)$

17. $x = -3.45$

19. $(-1 + \sqrt{2}, 4 + \sqrt{2})$ and $(-1 - \sqrt{2}, 4 - \sqrt{2})$

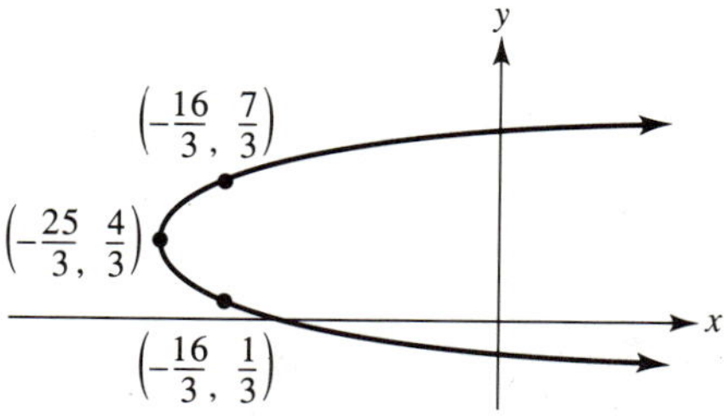

23.

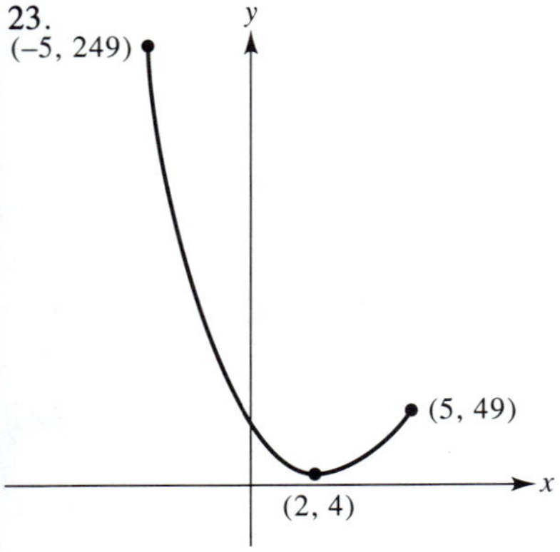

25. $\frac{15}{2}$

27.

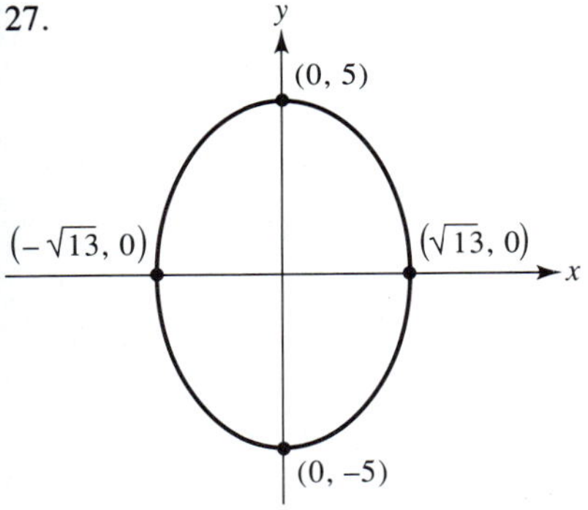

29.

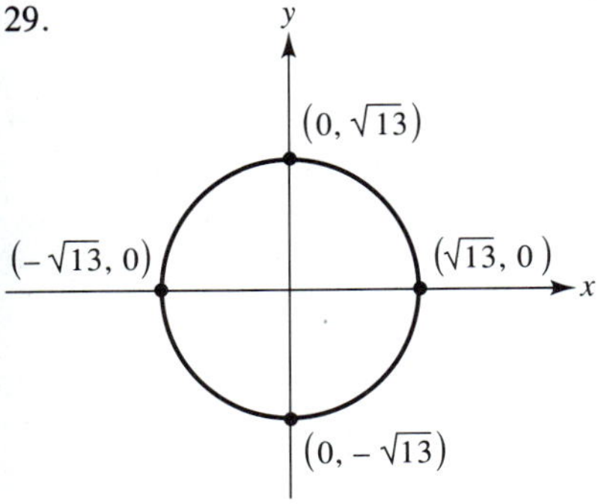

31.

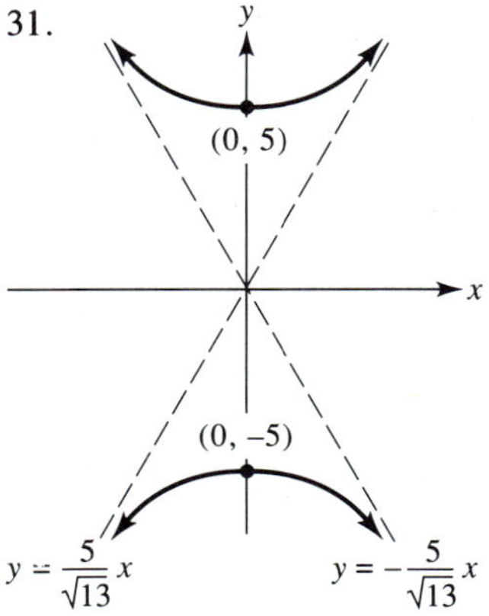

33. If $a = f$ and $a \neq 0$, then the graph is the line segment with endpoints $(-f, 0)$ and $(f, 0)$. If $a = 0$ and $a \neq f$, then the graph is empty. If $a = f = 0$, then the graph is the point $(0, 0)$.

35.

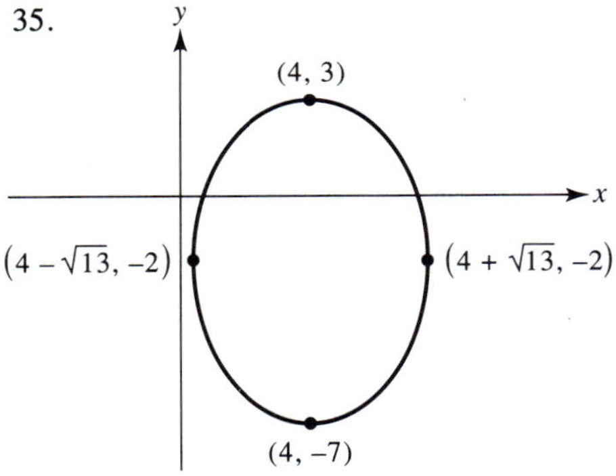

37.

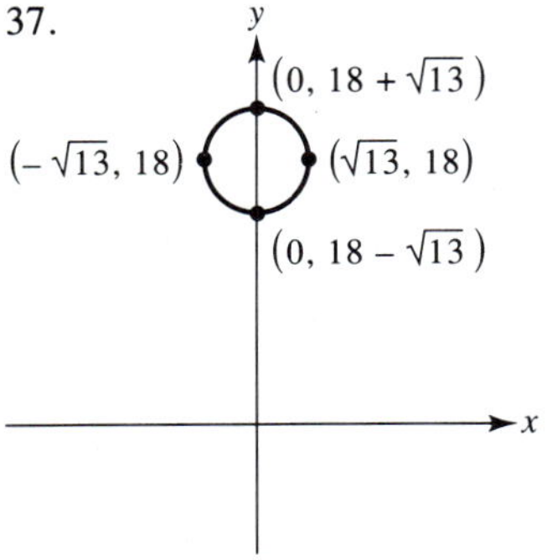

39.

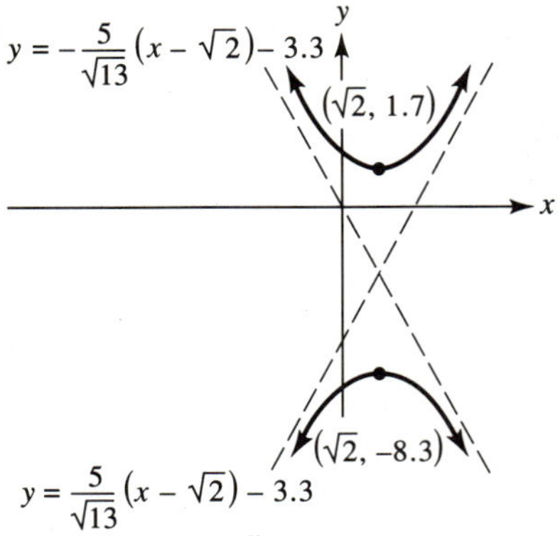

41.

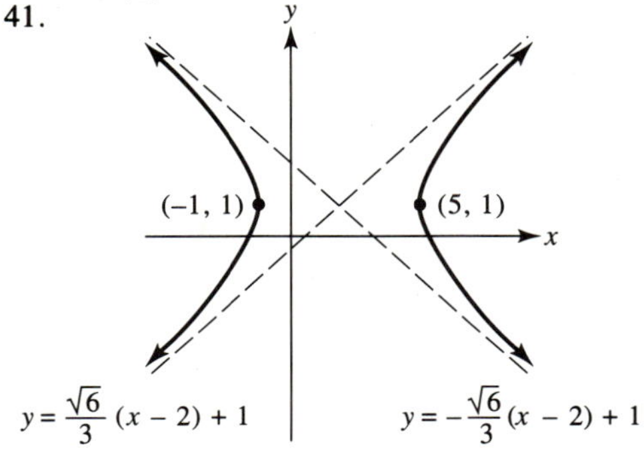

43. a.

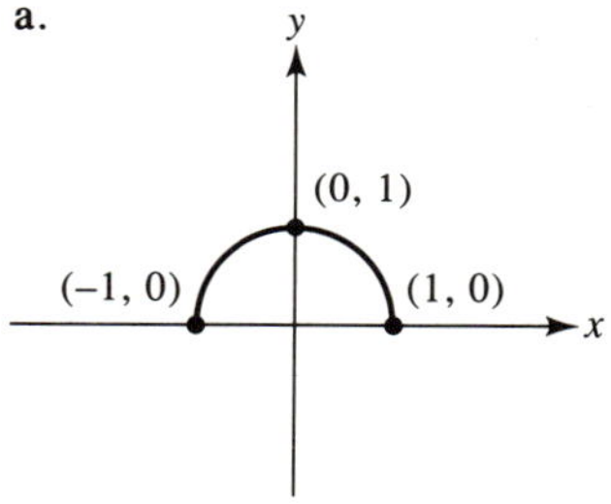

b.

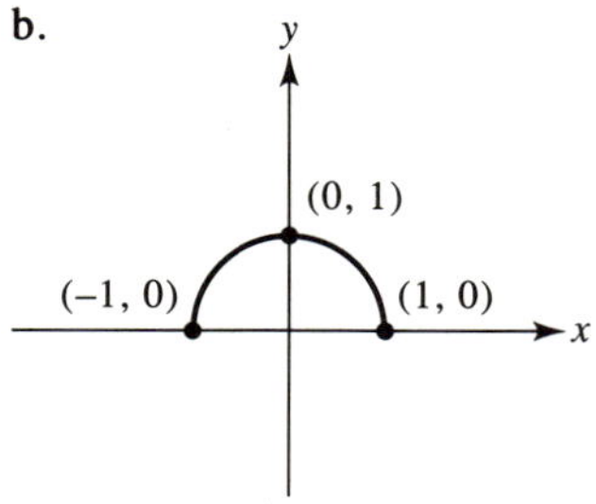

45. $(x + 3)^2 + (y - 7)^2 = 1$ and $(x + 3)^2 + (y - 7)^2 = 4$ are possible answers.

Chapter Three

3.1

1. $\{x \mid x \leq \frac{8}{5}\}$ or $(-\infty, \frac{8}{5}]$
3. $\{y \mid y > 10.91\}$
5. $\left\{w \mid w \geq \dfrac{3 - \sqrt{7}}{2}\right\}$
7. $\{0\}$
9. $\mathbb{R}$
11. $\{x \mid x < \frac{2}{3} \text{ or } x > 2\}$
13. $\left\{x \mid x < \dfrac{20 - \sqrt{13}}{15}, \text{ or } x > \dfrac{20 + \sqrt{13}}{15}\right\}$
15. $\mathbb{R}$
17. $\{\ldots, -6, -5, -4\} \cup \{7, 8, 9, \ldots\}$
19. $\{\ \}$
21. $\{c \mid c \leq -27\}$
23. $\left\{x \mid x < \dfrac{7}{5}\right\}$
25. $|x - 7| \leq 100$
27. $|x + 7| < \pi$
29. $|(a - 3) - 5| \leq \frac{132}{5}$
31. a. $<$ b. $>$ c. $\geq$
33. He must make a 77 or higher. It looks chancey.
35. Todd needs to sell at least $4560 worth of merchandise.
37. He can consume 1850 calories per day.
39.

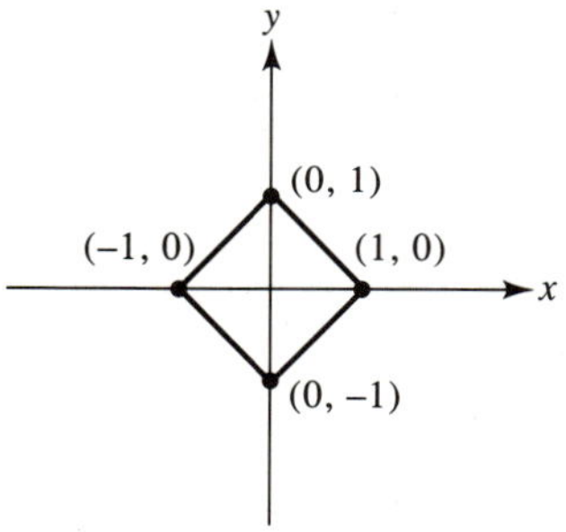

41. a. $\sqrt{2} + 4k\sqrt{2}, k \in W$
    b. $2\sqrt{2} + 4k\sqrt{2}, k \in W$
    c. $3\sqrt{2} + 4k\sqrt{2}, k \in W$

3.2

1. $\{x \mid x < -5 \text{ or } x > 5\}$
3. $\mathbb{R}$
5. $\{x \mid x \leq -\sqrt{17} \text{ or } x \geq \sqrt{17}\}$
7. $\left\{x \mid -\sqrt{\dfrac{39}{12}} \leq x \leq \sqrt{\dfrac{39}{12}}\right\}$
9. $\left\{x \mid x \neq \dfrac{4}{11}\right\}$

11. $\mathbb{R}$
13. $\mathbb{R}$
15. $\{x \mid x < -\sqrt{8} \text{ or } x > \sqrt{8}\}$
17. $\{\ \}$
19. $\mathbb{R}$
21. $\left\{x \mid x \le \dfrac{26 - \sqrt{736}}{10} \text{ or } 5 < x \le \dfrac{26 + \sqrt{736}}{10}\right\}$
23. $\left\{x \mid x < 0 \text{ or } \dfrac{8 - \sqrt{54}}{2} \le x \le \dfrac{8 + \sqrt{54}}{2} \text{ or } 8 < x\right\}$
25. $\{x \mid -2 < x < 2 \text{ or } 3 < x\}$
27. $\{x \mid -2 \le x \le 2\}$
29. $\{x \mid x \le -1 \text{ or } \pi < x\}$
31. $\{w \mid w \le 2\}$
33. $\{x \mid -3 \le x < -2\}$
35. a. $(2, +\infty)$
    b. $(-\infty, 2)$
    c. $[2, +\infty)$
    d. $(-\infty, 2]$
37. a. $(-\infty, 2)$
    b. $(2, +\infty)$
    c. $(-\infty, 2]$
    d. $[2, +\infty)$
39. a. $\left(-\infty, -\frac{1}{3}\right) \cup (5, +\infty)$
    b. $\left(-\frac{1}{3}, 5\right)$
    c. $\left(-\infty, -\frac{1}{3}\right] \cup [5, +\infty)$
    d. $\left[-\frac{1}{3}, 5\right]$
41. a. $\left(-\frac{1}{3}, 5\right) \cup (5, +\infty)$
    b. $\left(-\infty, -\frac{1}{3}\right)$
    c. $\left[-\frac{1}{3}, +\infty\right)$
    d. $\left(-\infty, -\frac{1}{3}\right] \cup [5, 5]$
43. a. $\left(-\infty, -\frac{1}{3}\right) \cup (5, +\infty)$
    b. $\left(-\frac{1}{3}, 5\right)$
    c. $\left(-\infty, -\frac{1}{3}\right] \cup [5, +\infty)$
    d. $\left[-\frac{1}{3}, 5\right]$
45. a. $\left(-\frac{1}{3}, 5\right)$
    b. $\left(-\infty, -\frac{1}{3}\right)$
    c. $\left[-\frac{1}{3}, +\infty\right)$
    d. $\left(-\infty, -\frac{1}{3}\right) \cup [5, 5]$
47. The set is all numbers between $-\frac{1}{2}$ and $\frac{1}{2}$ except 0.
49. Any triple of three consecutive integers from 1 down has a product that is not positive.
51. Possible answers are:
    a. $2\sqrt{2}, -2\sqrt{2}, 6\sqrt{2}, -6\sqrt{2}$
    b. $3\sqrt{2}, -\sqrt{2}, 7\sqrt{2}, -5\sqrt{2}$
    c. $\sqrt{2}, -3\sqrt{2}, 5\sqrt{2}, -7\sqrt{2}$
    d. $4\sqrt{2}, 8\sqrt{2}, -12\sqrt{2}, -4\sqrt{2}$

3.3

1.
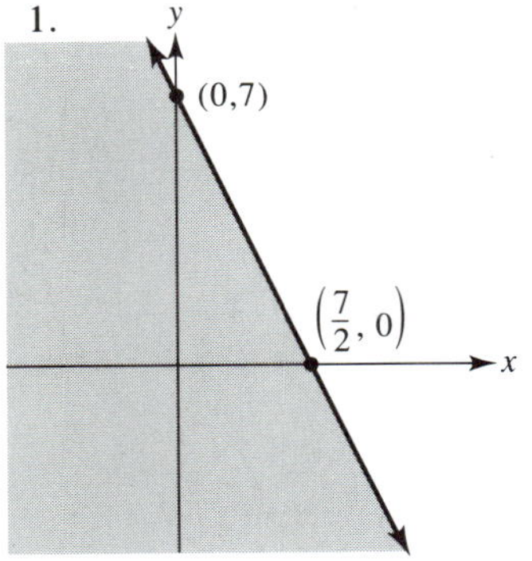

3.
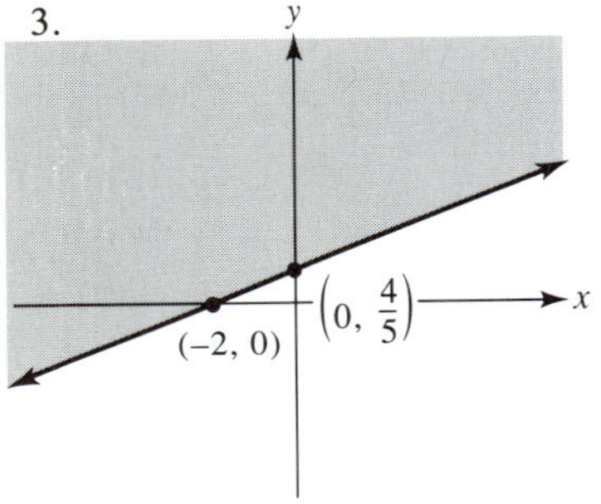

5.
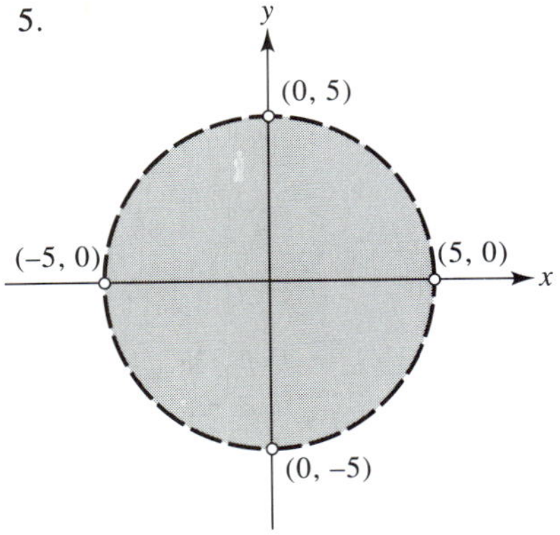

7.
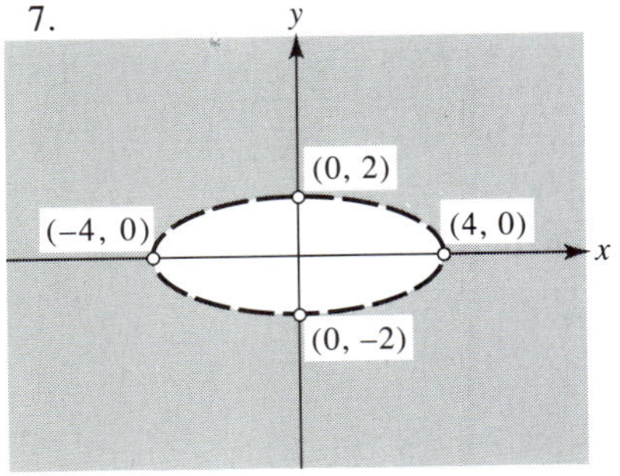

9.
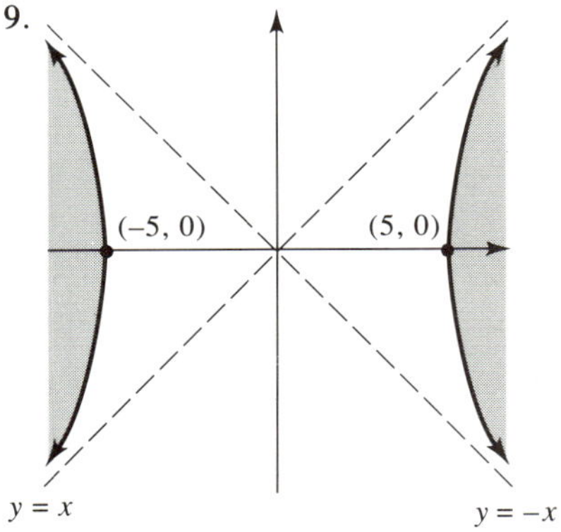

11.
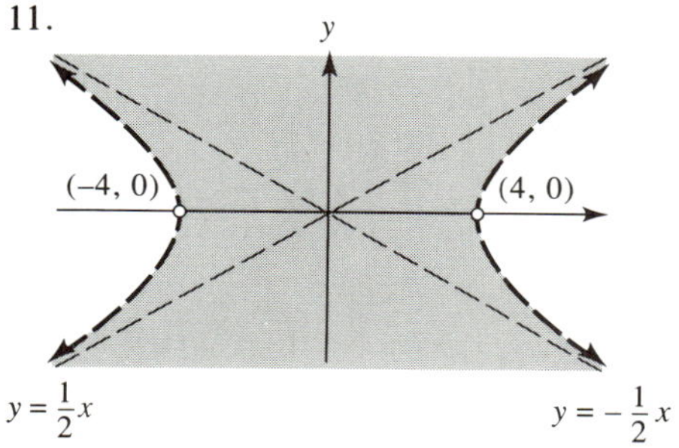

13.
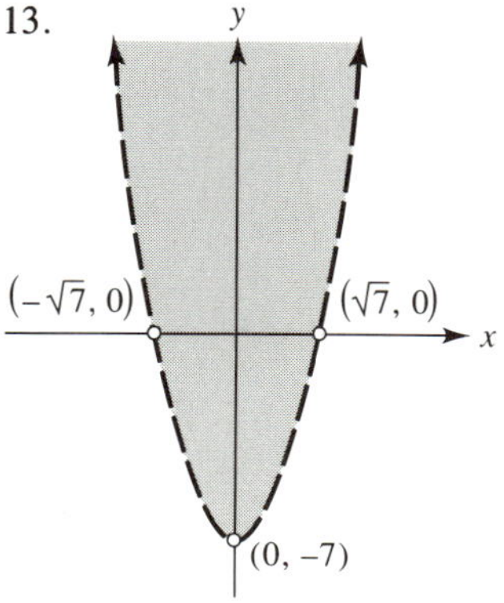

15.
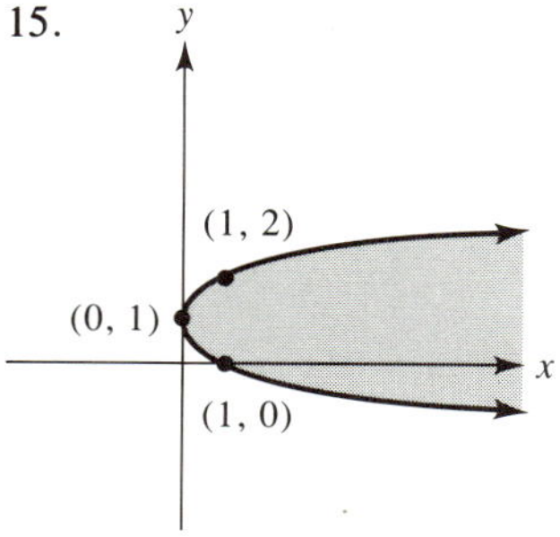

17.
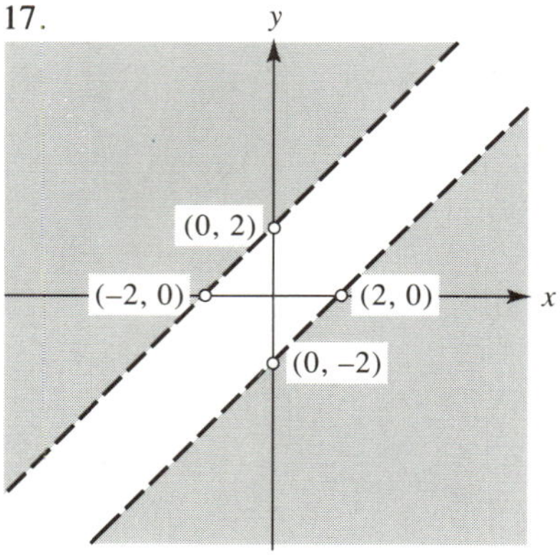

19. There are no real solutions.

21.
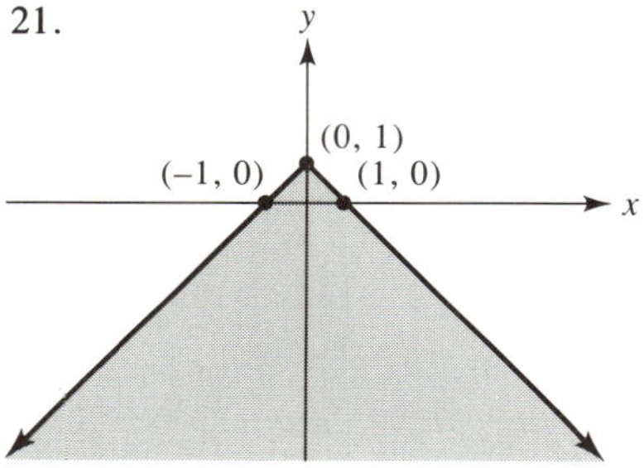

23.
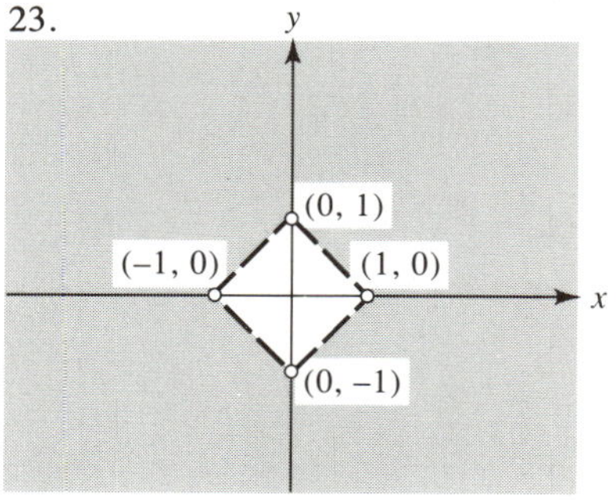

25. $x^2 + y^2 \leq 16$

27. $\dfrac{(x+5)^2}{\left(\frac{3}{2}\right)^2} + \dfrac{\left(y - \frac{1}{2}\right)^2}{4} > 1$

29. $y > -x + 3$

31. $y \geq x^2$ and $x \geq 0$

33. $\dfrac{(x-3)^2}{2^2} + \dfrac{(y-3)^2}{(1.5)^2} \leq 1$

35. The midpoints are $\left(\frac{1}{2}, \frac{1}{2}\right)$, $\left(-\frac{1}{2}, \frac{1}{2}\right)$, $\left(-\frac{1}{2}, -\frac{1}{2}\right)$, $\left(\frac{1}{2}, -\frac{1}{2}\right)$.

37. a. $\left(\frac{1}{2}, -\frac{1}{2}\right)$ b. $\left(-\frac{1}{2}, \frac{1}{2}\right)$ c. $\left(\frac{1}{2}, \frac{1}{2}\right)$ d. $\left(\frac{1}{2}, -\frac{1}{2}\right)$

3.4

1.

3. 

5. 

7. No real solutions

9. 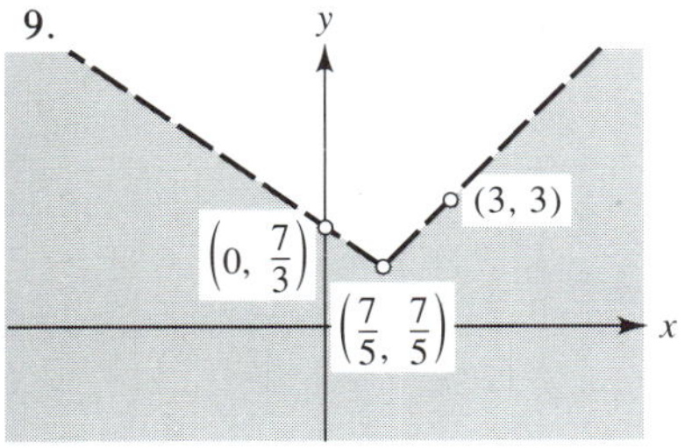

11. 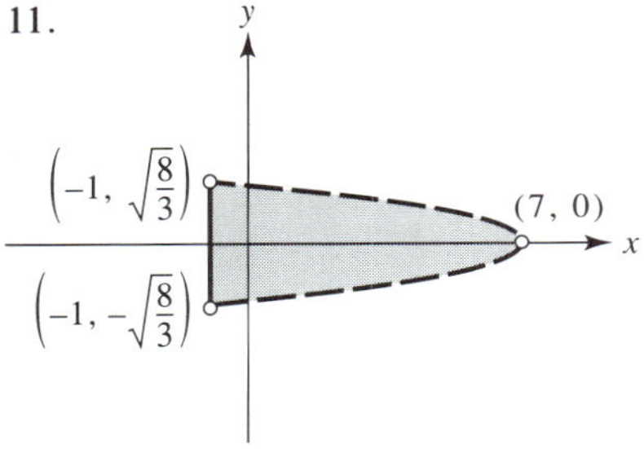

13. 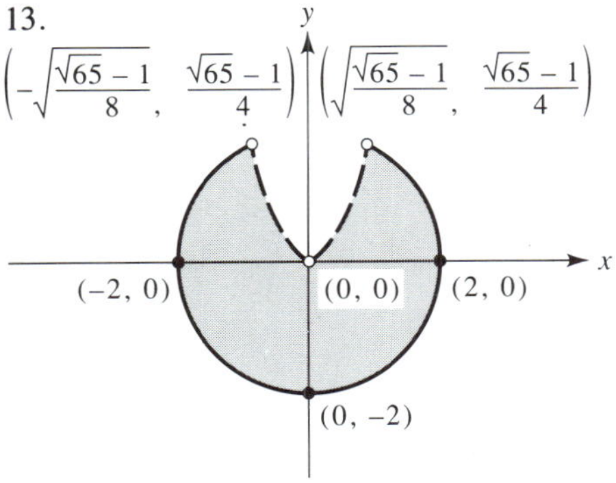

15. 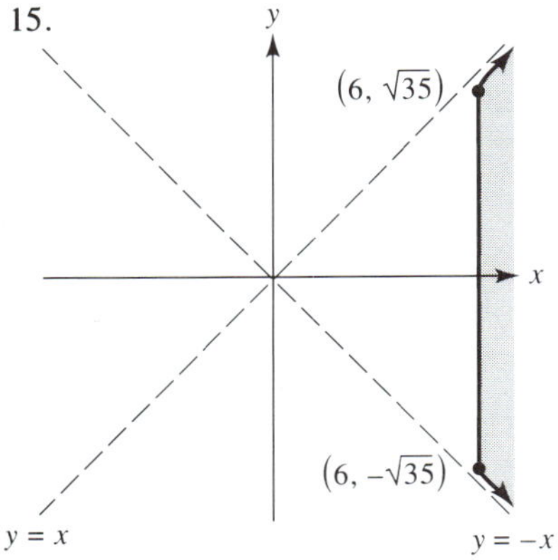

17. 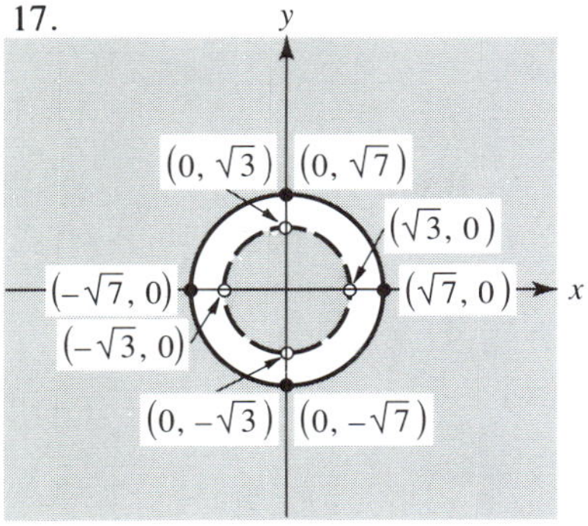

19. 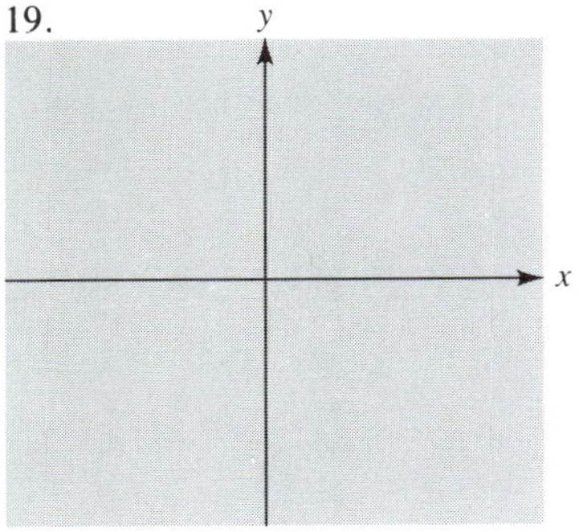

21. 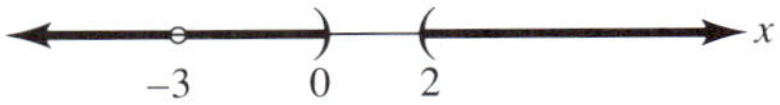

23. 

25. 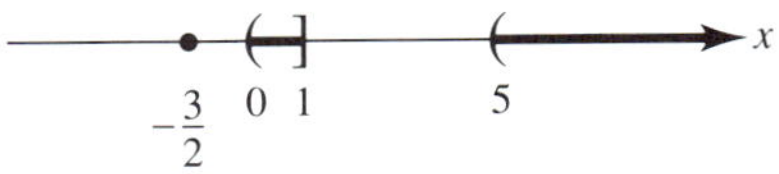

27. 

29. 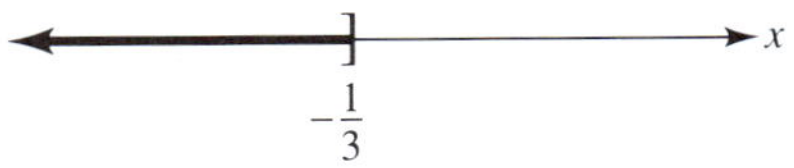

31. 

33. 

35. 

37. 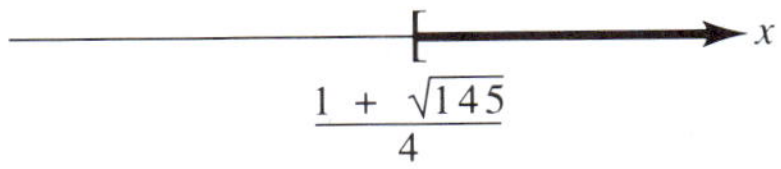

39. There are no real solutions.

41. a. $(2, +\infty)$
 b. None
 c. $[2, +\infty)$
 d. $[2, 2]$

43. a. $(2, +\infty)$
 b. $(-\infty, 2)$
 c. $[2, +\infty)$
 d. $(-\infty, 2]$

45. a. $(-\infty, 2) \cup (2, +\infty)$
 b. None
 c. $(-\infty, +\infty)$
 d. $[2, 2]$

47. One answer is $1 \le x \le 2$ and $2 \le y \le 3$.

49. One answer is $y \ge 0$ and $x^2 + y^2 \ge 7$; another is $y \ge 0$ and $y \ge \sqrt{7 - x^2}$.

51. One answer is

$\frac{3}{4}\lvert x\rvert = y$ and $\lvert x\rvert \le 4$	*chin*
or $\lvert x\rvert = 4$ and $3 \le y \le 10$	*sides*
or $y = 10$ and $-4 \le x \le 4$	*top*
or $(x - 1)^2 + (y - 7.5)^2 \le \frac{1}{16}$	*eye*
or $(x + 1)^2 + (y - 7.5)^2 \le \frac{1}{16}$	*eye*
or $x^2 + (y - 6)^2 \le \frac{1}{64}$	*nose*
or $\dfrac{x^2}{\frac{1}{4}} + \dfrac{(y - 5)^2}{1} \le 1$	*mouth*
or $(x - 4)^2 + (y - 10)^2 \le 1$	*ear*
or $(x + 4)^2 + (y - 10)^2 \le 1$	*ear*

53. a. $\frac{3}{2}\sqrt{2} + 4k\sqrt{2}, k \in I$
 b. $\frac{5}{2}\sqrt{2} + 4k\sqrt{2}, k \in I$

Summary Exercises

1. $\{x \mid x \le \frac{22}{39}\}$
3. $\left\{w \mid w \ge \dfrac{4}{\sqrt{8} - 3}\right\}$
5. $\{u \mid u < 0 \text{ or } u > 1\}$
7. $\{\ \}$
9. $\{x \mid -\frac{7}{3} \le x \le \frac{19}{3}\}$
11. $\{\ \}$
13. $\{w \mid w \le -5 \text{ or } 1 \le w\}$
15. $\{x \mid x \le \frac{49}{11} \text{ or } 5 < x\}$
17. $\{x \mid -7 < x < 0 \text{ or } \frac{9}{4} < x\}$
19. $\{x \mid -8 \le x < 1756\}$
21. $\{\ \}$
23. The numbers are between -7 and 0.
25. The quotient is nonnegative for any two consecutive odd integers from 1, 3, 5, ...
27. $\{x \mid x \le -4 \text{ or } 0 < x \le 3\}$
29. $\{x \mid -\frac{1}{4} \le x < 0 \text{ or } \frac{1}{3} \le x\}$
31. $\{x \mid -4 \le x < 0 \text{ or } 0 < x \le 3\}$

33.

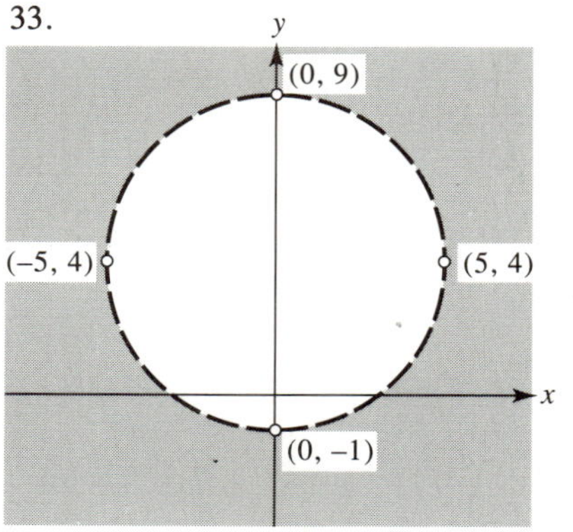

35.

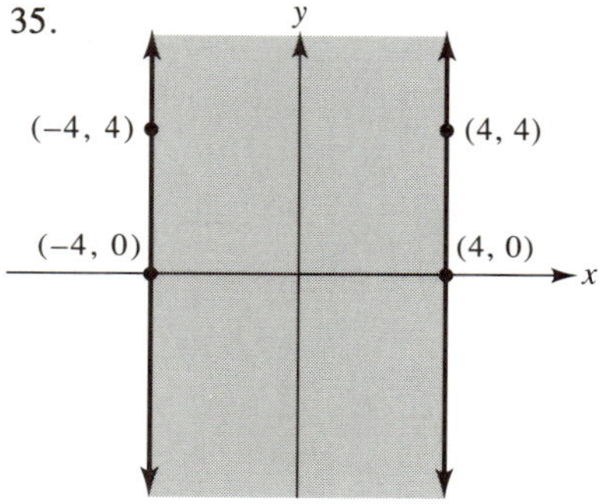

37.

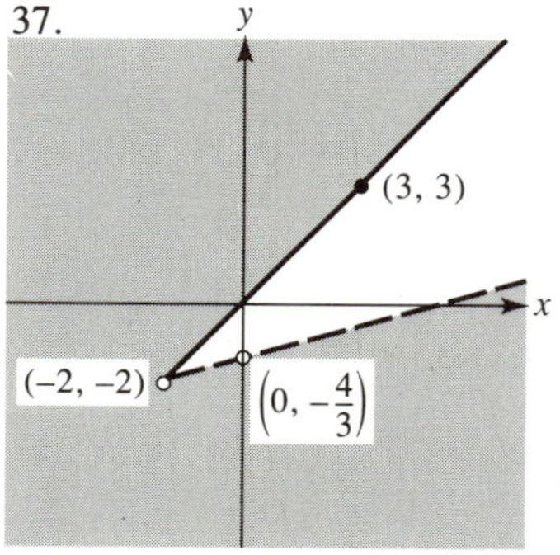

39.

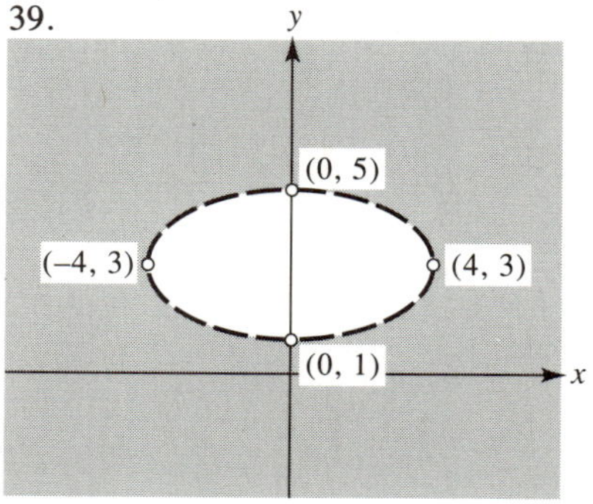

41.

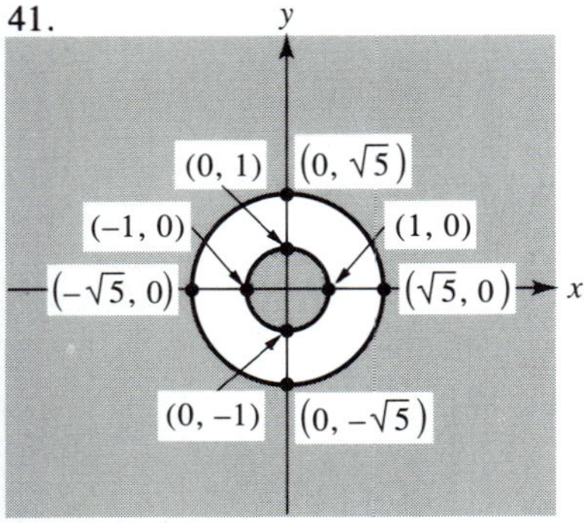

43. $x < 0$ and $y < 0$

45. $y \geq x^2$

47. $(y - \sqrt{2}) < (\sqrt{2} - \sqrt{3})(x - 9)$

49. $x > -\pi$ or $x < -490$

Chapter Four

4.1

1. Function
3. Not a function
5. Function
7. Not a function
9. Not a function
11. Function
13. Function
15. Domain: $\mathbb{R}$; range: $\mathbb{R}$
17. Domain: N; range: $\{0, 1, 2, 3, 4, 5, 6\}$
19. Domain: $(-\infty, -\sqrt{8}] \cup [\sqrt{8}, +\infty)$; range: $[0, +\infty)$
21. a. -3 b. 3 c. $2r^2 - 5$ d. $8s^2 - 5$
    e. $2(r + t)^2 - 5$
23. a. -2
    b. Undefined
    c. $\dfrac{6}{r^2 - 3}, r \neq -\sqrt{3}, \sqrt{3}$
    d. $\dfrac{6}{5s - 3}, s \neq \dfrac{3}{5}$
    e. $\dfrac{6}{2r + t - 3}, 2r + t \neq 3$

25.

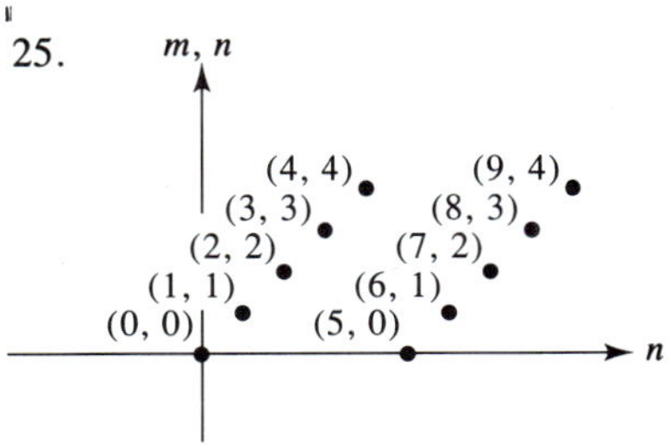

The graph continues to the right.

27.

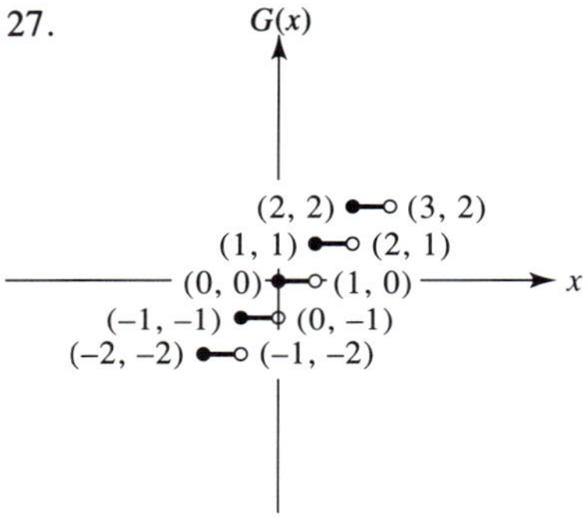

The graph continues to the left and the right.

29.

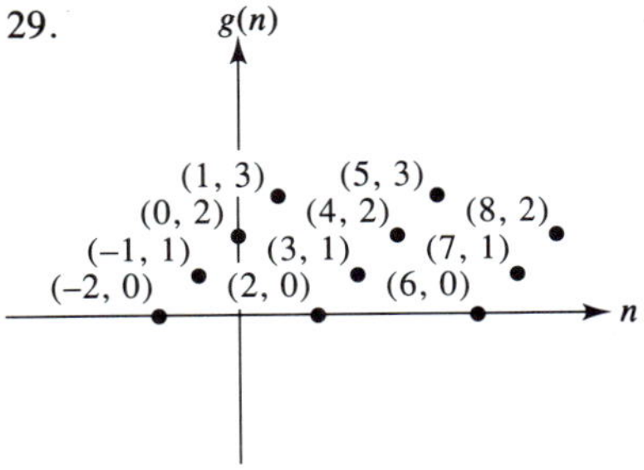

The graph continues to the right.

31. a. 8.57 (approx.)
    b. 5.37 (approx.)
    c. 11.36 (approx.)
    d. 9.37 (approx.)
33. $[-\frac{2}{3}, 2)$
35. $(\sqrt{5}, \sqrt{8}) \cup (-\sqrt{8}, -\sqrt{5})$
37. a. $\mathbb{R}$
    b. $\left(\dfrac{4}{21}, \dfrac{4\pi}{3}\right)$
    c. $\left(-\dfrac{6}{7}, -6\pi\right)$
    d. $\left(\dfrac{\sqrt{6}}{7}, \sqrt{6}\pi\right)$

39. a. $\{(x, y) \mid y \in \mathbb{R}\}$
    b. 2
    c. 0
    d. 4

41. $f(n) = 3n + 5,\ n \in N$ is one possible answer.

43. $h(n) = n^3,\ n \in N$ is one possible answer.

45. a. 0.771
    b. $m = \dfrac{y - 2.457}{0.3}$
    c. The value of y is 2.6883.

4.2

1.

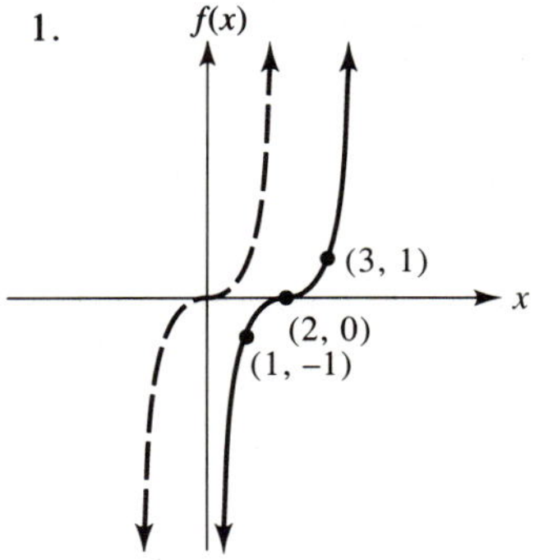

3.

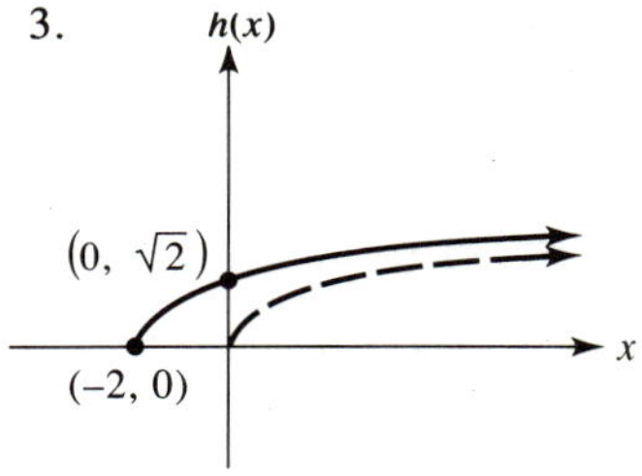

5.

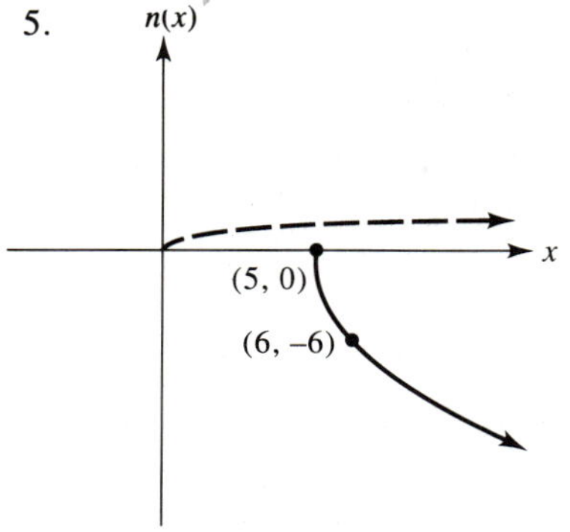

7.

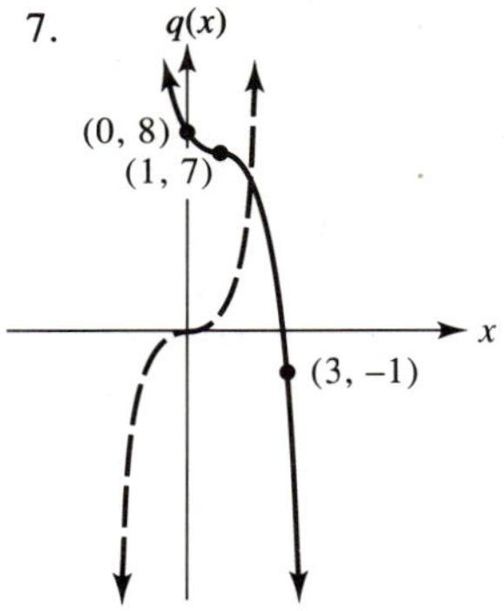

9.

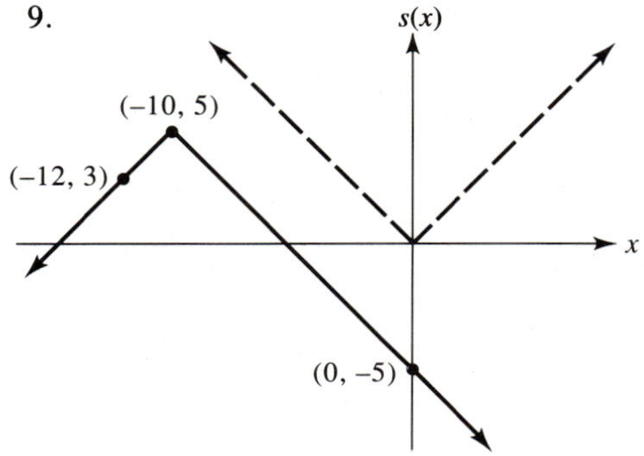

11.

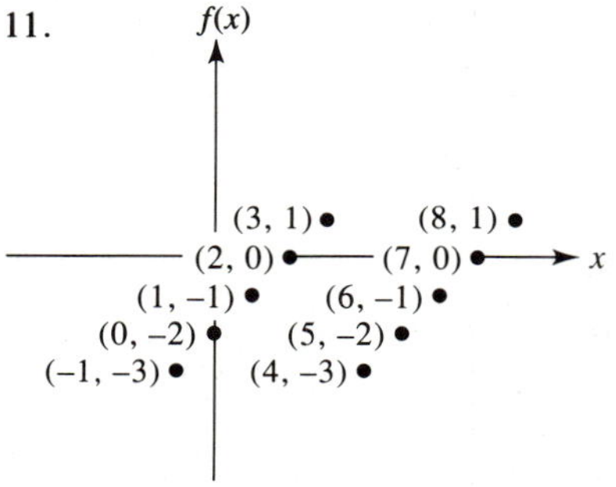

The graph continues to the right.

13.

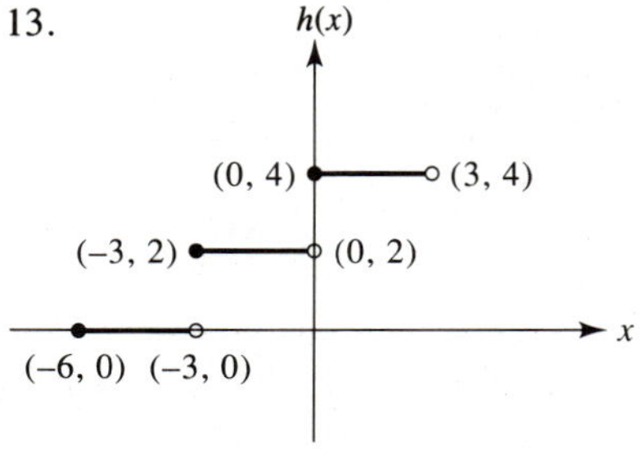

The graph continues to the right.

15.

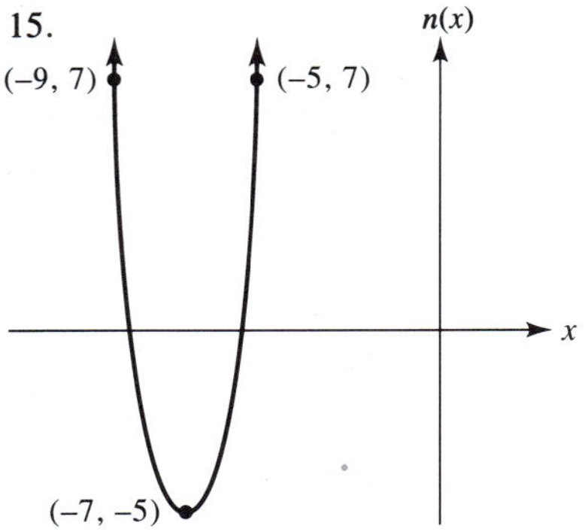

17.

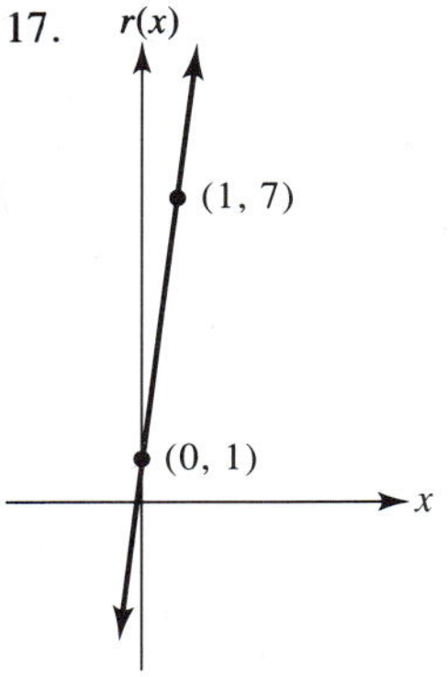

19.

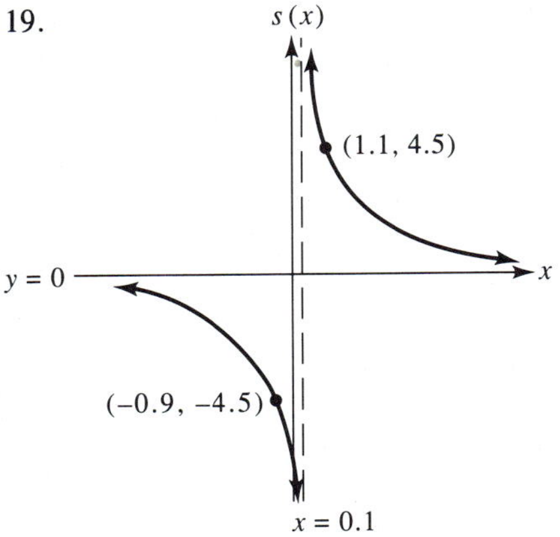

21. $f(x) = (x + 2)^3 + 4$

23. $h(x) = -\text{mod}_4(x - 3)$

25. The graph is reflected in the x-axis, has expansion factor of 5, and is shifted 8 units left and 7 units up.

27. The graph is reflected in the x-axis and is shifted 12 units right and 23 units up.

29. $F(x) = 4\left|x + \frac{7}{4}\right|$: expansion factor 4; translation $\frac{7}{4}$ units left

31. $r(x) = -2^3\left(x - \frac{3}{2}\right)^3$: expansion factor 8; translation $\frac{3}{2}$ units right; reflection in the x-axis

33. $t(x) = (0.7)^2(x + 3)^2$: expansion factor $(0.7)^2$; translation 3 units left

35. $S(x) = \sqrt{5}\sqrt{-(x - 2)}$: expansion factor $\sqrt{5}$; translation 2 units right; reflected in y-axis

37.

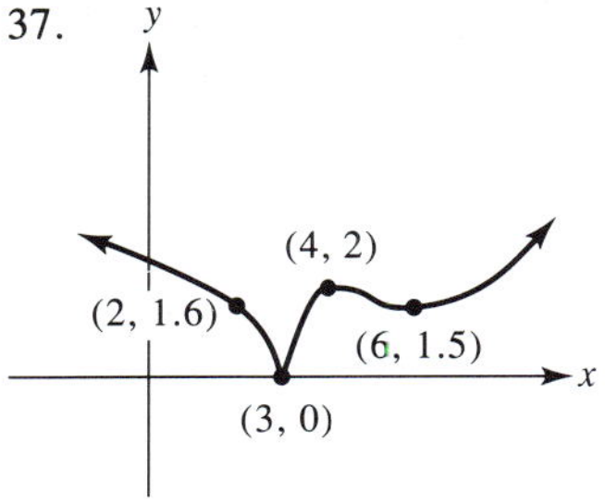

39.

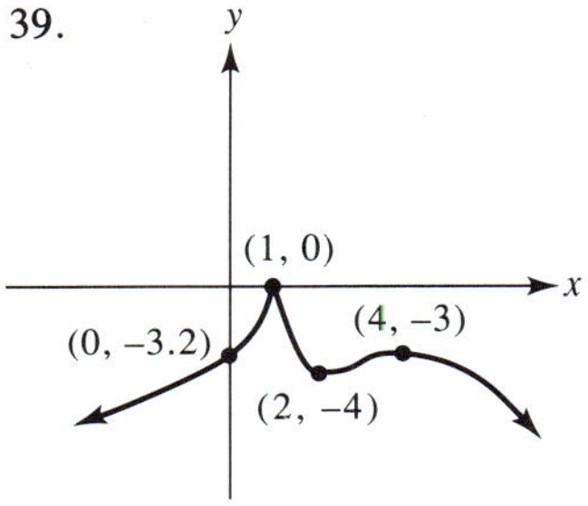

41.

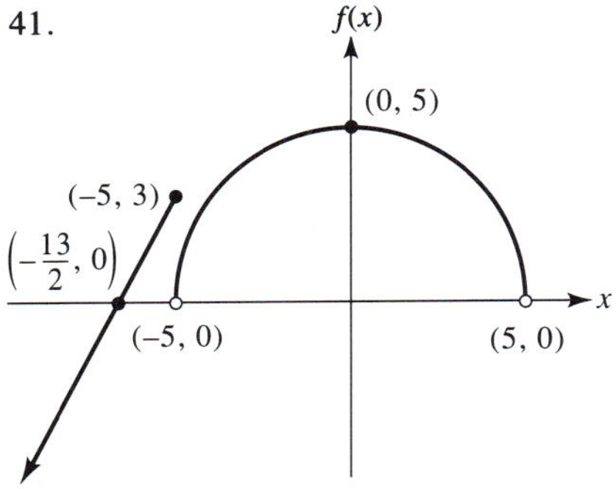

43.

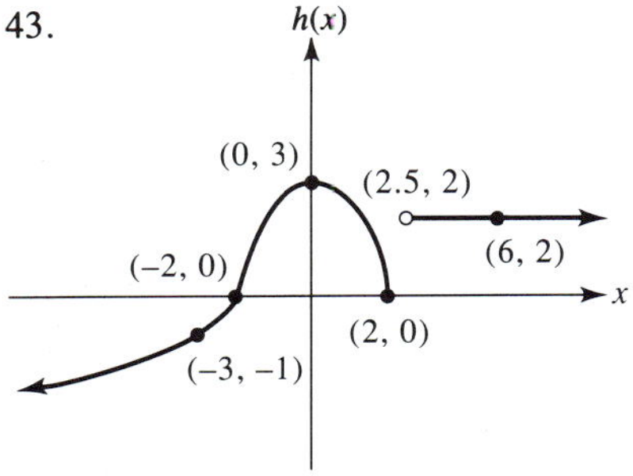

45. Most of the features of the standard graphs show up well, but domain and range cannot be clearly identified. (Check the graph of $y = x^{2/3}$.) It can be difficult to determine what's happening near the origin for $y = |x|$ and $y = \sqrt{x}$. The graphs for the unit circle and an el-

lipse must be produced in two pieces, the top half and the bottom half.

47. a. $\mathbb{R}$ b. $[-1, 1]$

4.3

1.

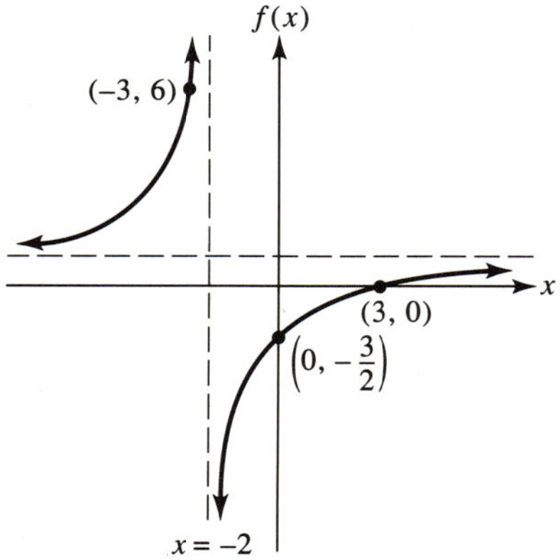

3.

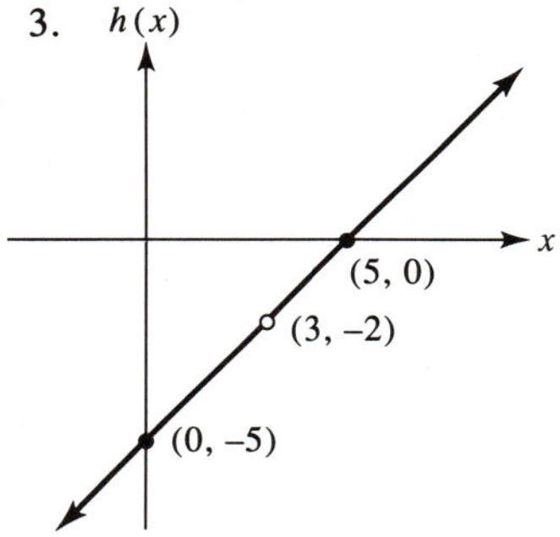

5.

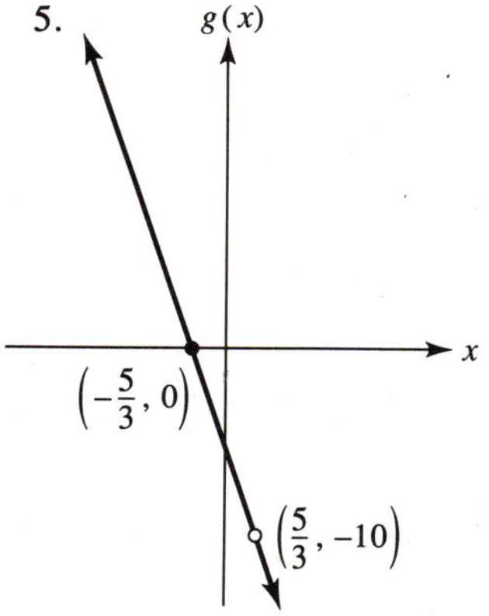

7.

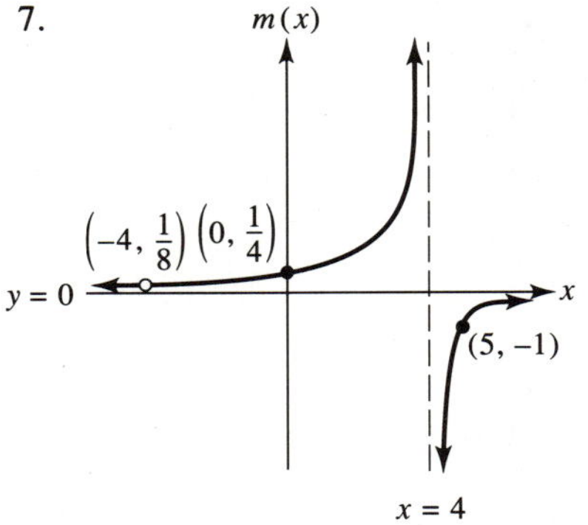

9.

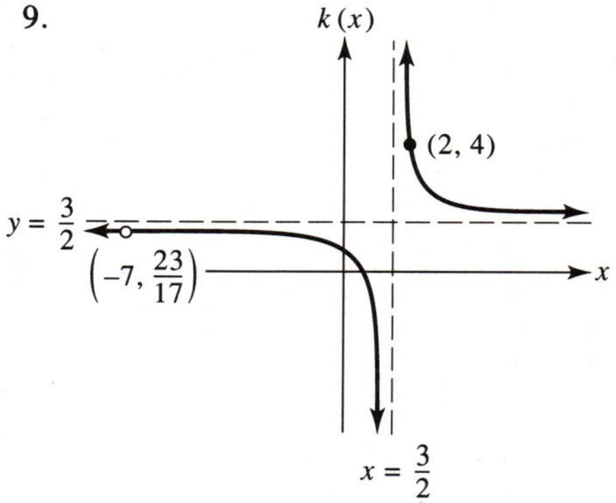

11.

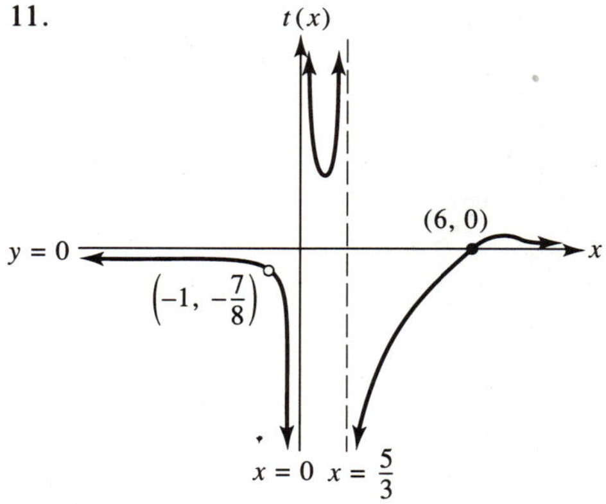

13.

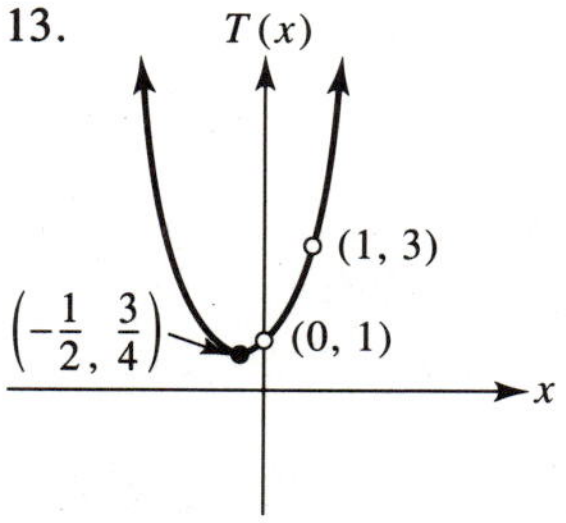

15.

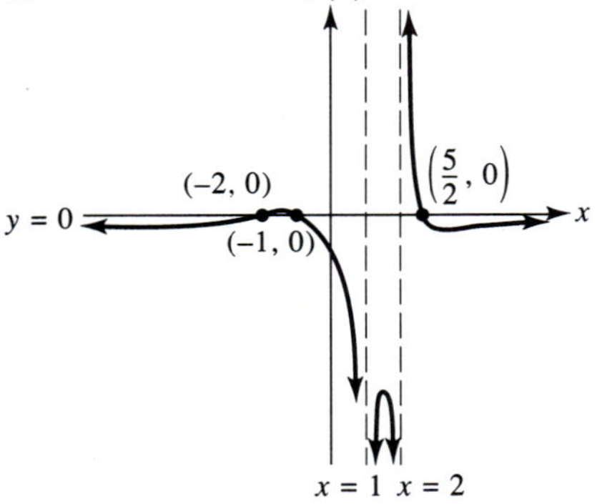

17.

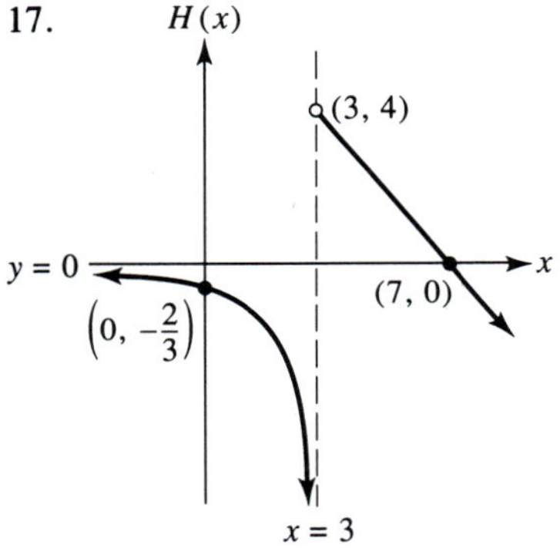

19.

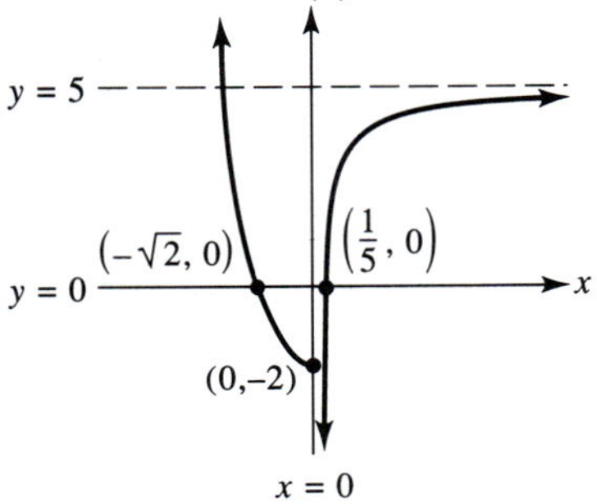

21. $\frac{2}{5}$

23. $\frac{13}{4}$

25. $2^{\sqrt{8}}$

27. $e^{1.8t}$

29. $a^{\sqrt{3}-3}$

31.

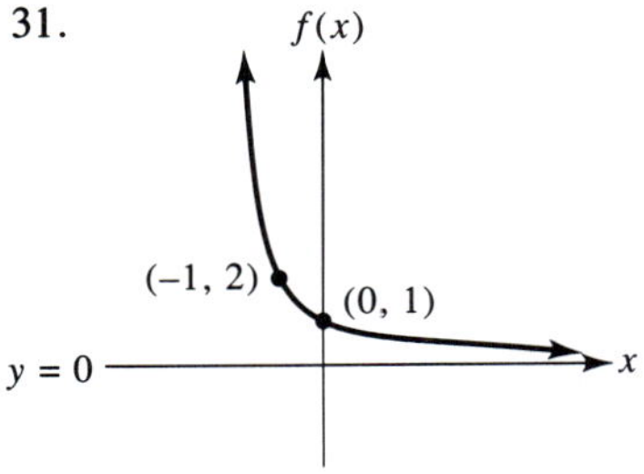

33.

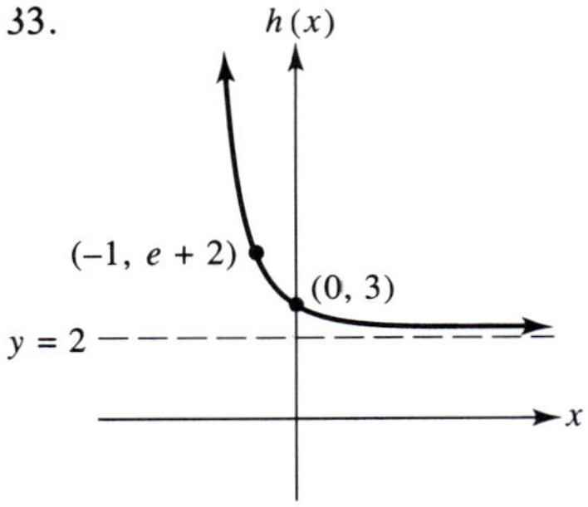

35.

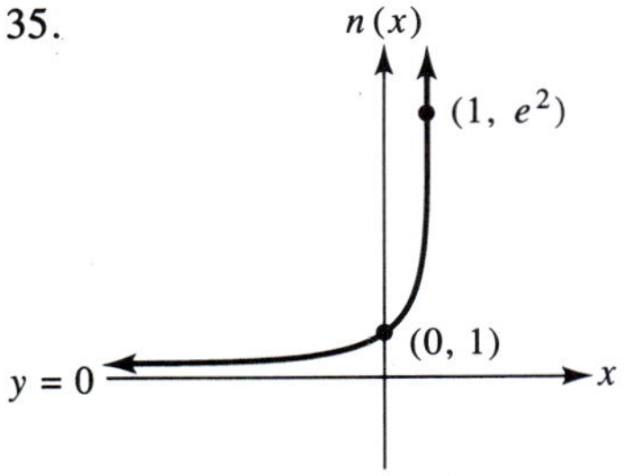

37.

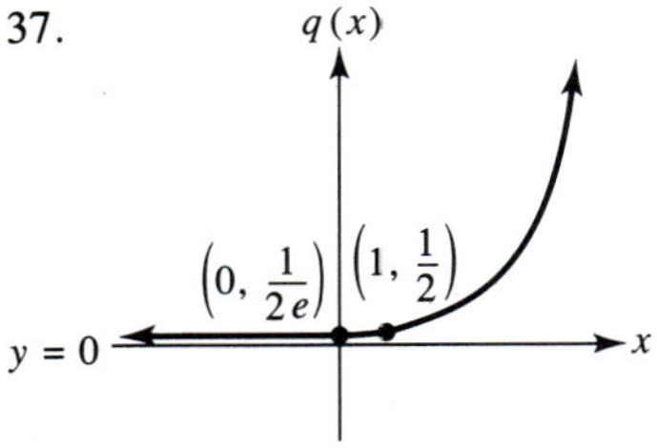

39.

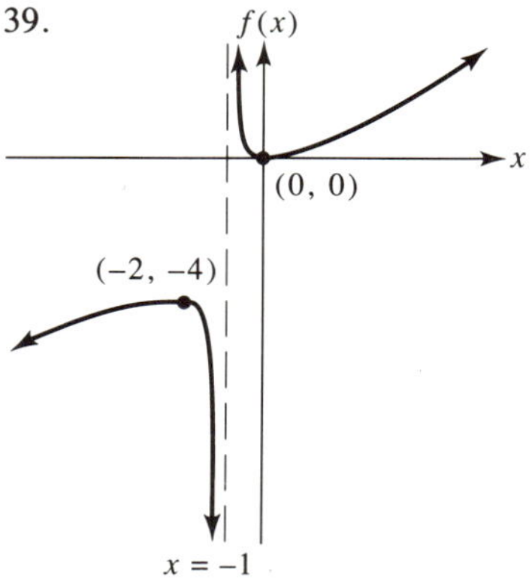

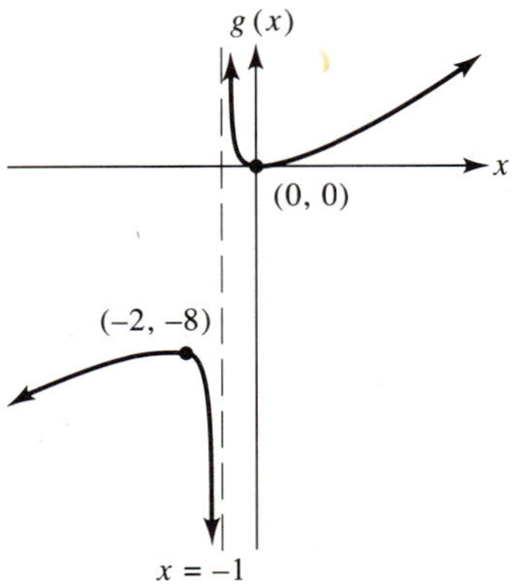

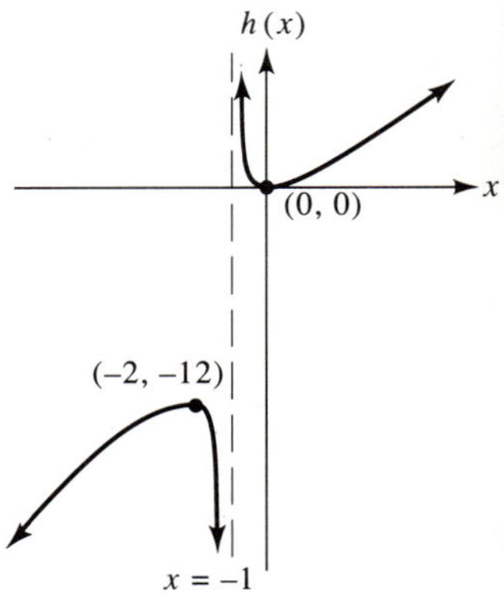

The graphs appear to approach slant lines rather than horizontal lines.

41.

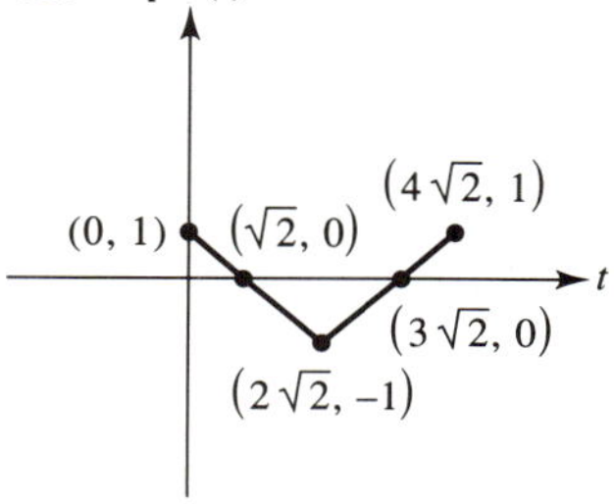

$0 \leq t \leq 4\sqrt{2}$

4.4

1.

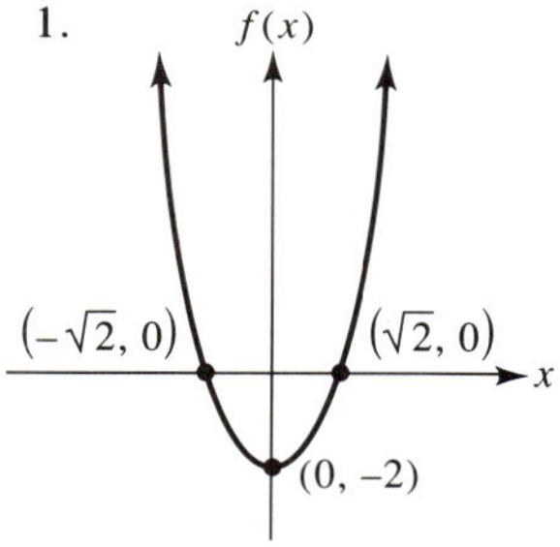

The graph is symmetric with respect to the y-axis.

3.

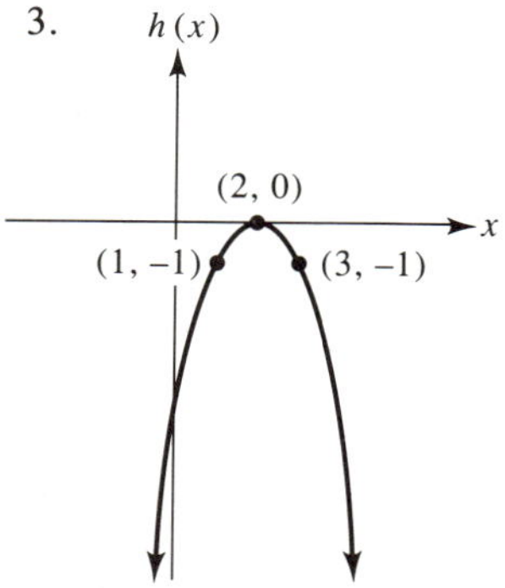

The graph is symmetric with respect to $x = 2$.

5.

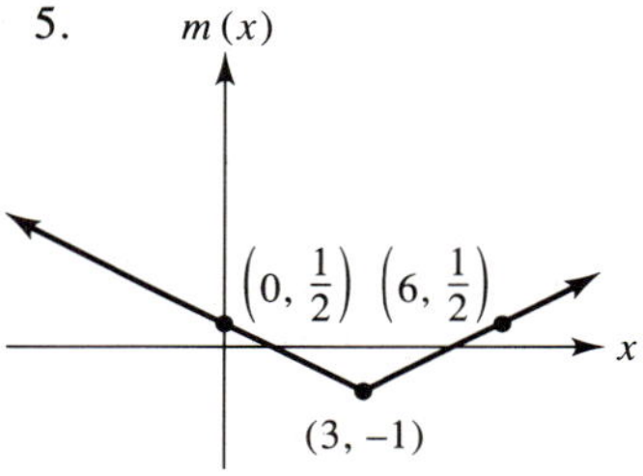

The graph is symmetric with respect to $x = 3$.

7.

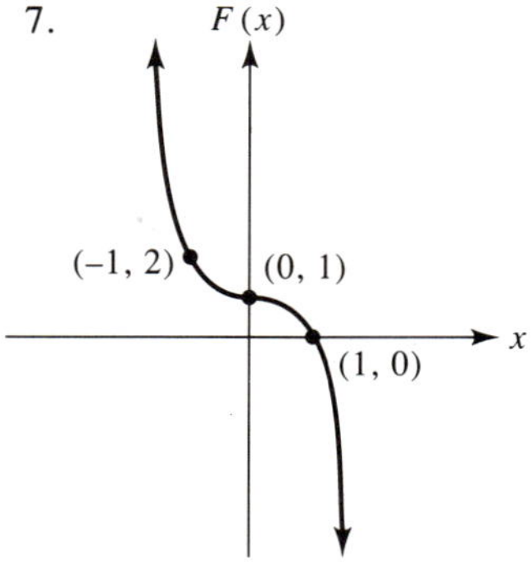

The graph is symmetric with respect to (0, 1).

9.

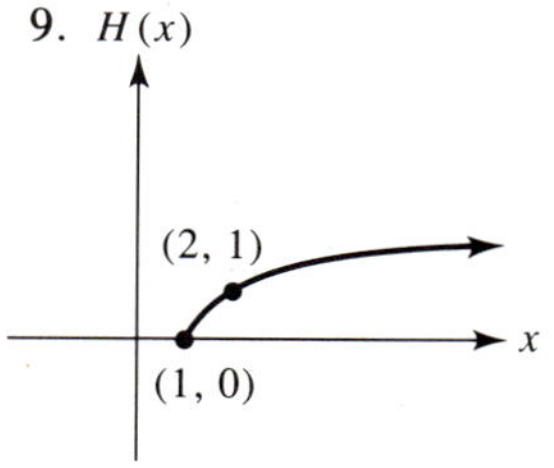

There are no symmetries.

11. 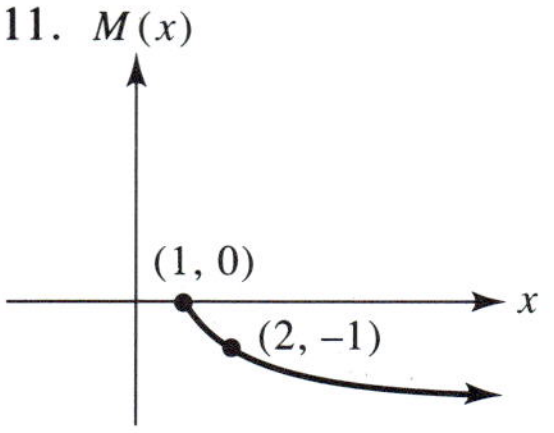

There are no symmetries.

13. $[0, +\infty)$
15. $(-\infty, 2]$
17. $[3, +\infty)$
19. The function is increasing nowhere.
21. $[1, +\infty)$
23. The function is increasing nowhere.
25. $f_1(x) = x^2 - 2, x \geq 0$ or $f_2(x) = x^2 - 2, x \leq 0$
27. $h_1(x) = -(2 - x)^2, x \leq 2$ or $h_2(x) = -(2 - x)^2, x \geq 2$
29. $m_1(x) = \frac{1}{2}|x - 3| - 1, x \leq 3$ or $m_2(x) = \frac{1}{2}|x - 3| - 1,$ $x \geq 3$
31. $f(-x) = -12x^3 + 3x + 4$; neither
33. $h(-x) = -x^{1/3} - 8x^{5/3} = -h(x)$; odd
35. $t(-x) = -38|x| - 7 = t(x)$; even
37. a, c, d
39. $f(n) = 2n, n \in N$ is a possible answer.
41. $h(n) = 5n, n \in N$ is a possible answer.
43. 5
45. 6
47. 4
49. A graph is symmetric with respect to the x-axis if the portion above the x-axis is the mirror image of the portion below the x-axis. A graph is symmetric with respect to the x-axis if whenever (a, b) belongs to the graph, so does $(a, -b)$.
51. The graph of f is symmetric with respect to $x = r$ if the portion to the right of $x = r$ is a mirror image of the portion to the left of $x = r$. The graph of f is symmetric with respect to $x = r$ if whenever (a, b) belongs to the graph, so does $(2r - a, b)$.
53. $f(p) \cdot f(q) = \frac{1}{p} \cdot \frac{1}{q} = \frac{1}{pq} = f(pq)$. In general, no.
55. Neither
57. Increasing
59. Decreasing
61. a. 1.7 b. −3.1 c. 2.6
63. $-2 + 5n, n \in I$
65. $4\sqrt{2}$

4.5

1. 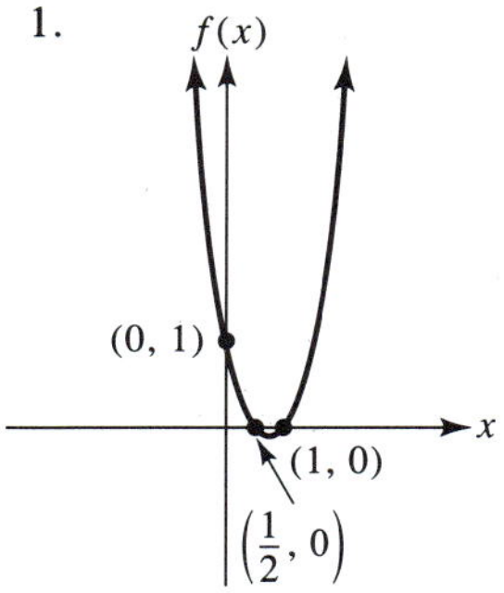

3. 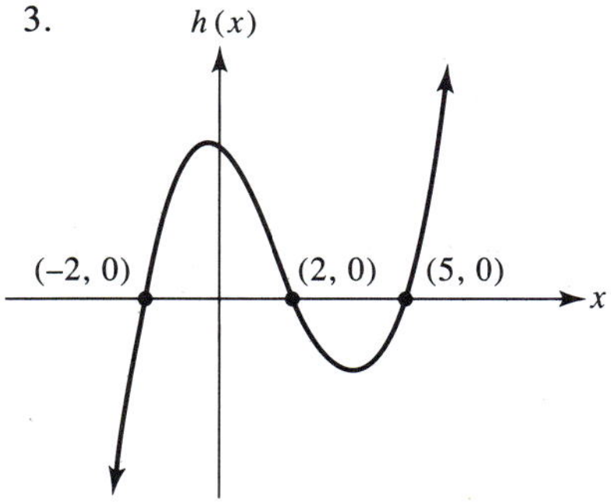

5. 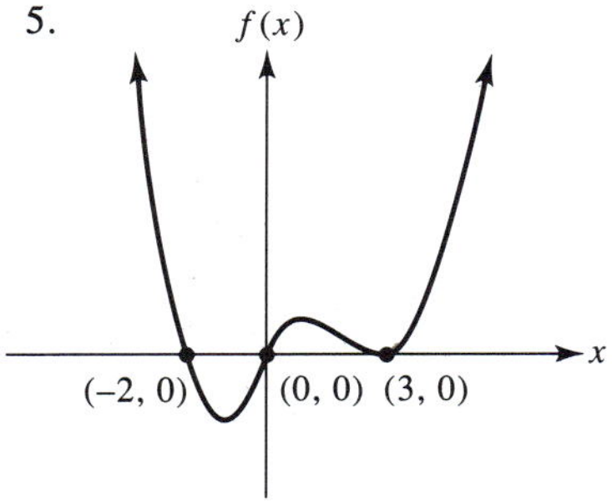

7. 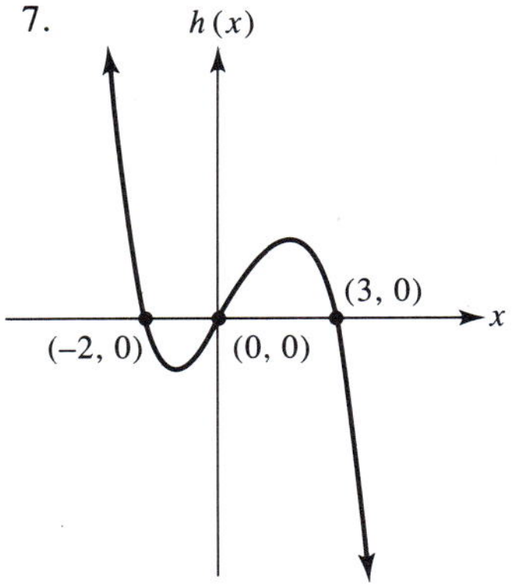

9.

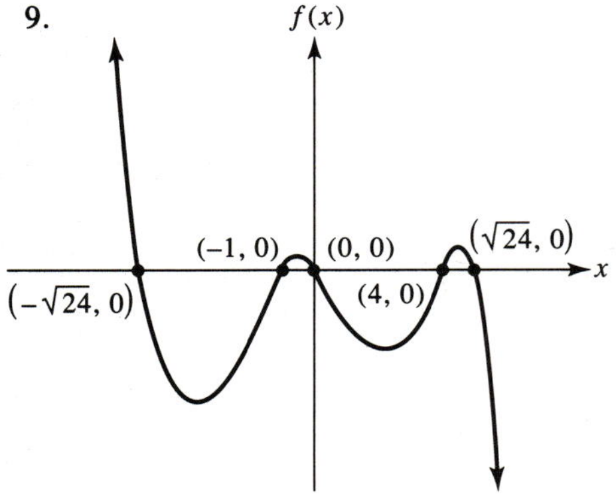

11.

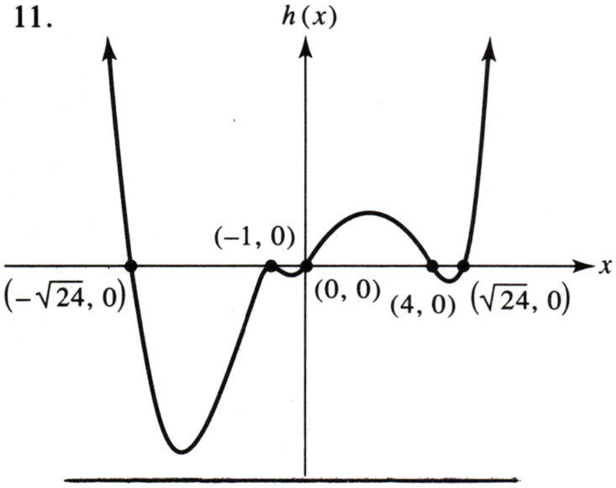

13.

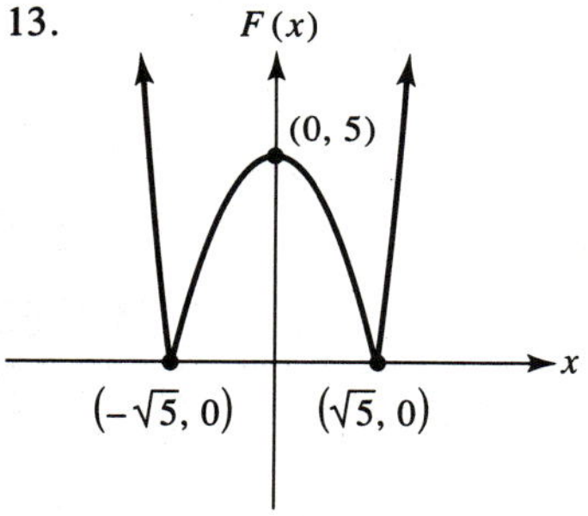

15.

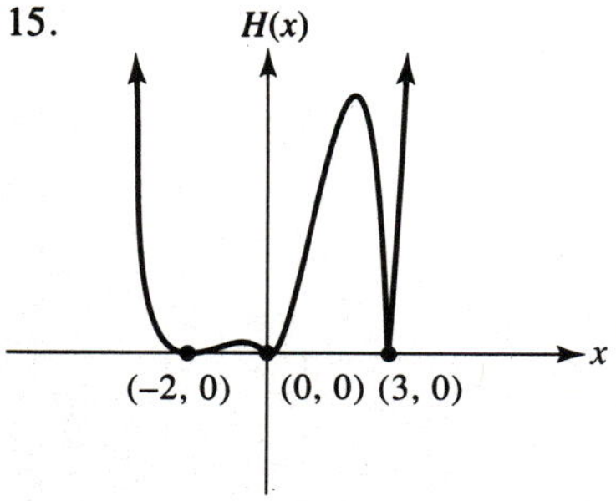

17.

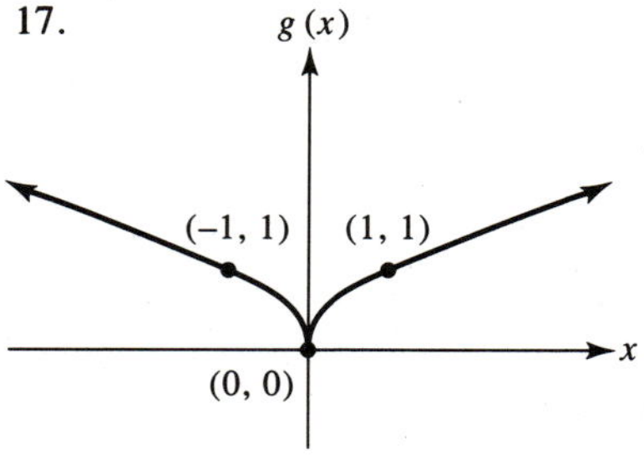

19.

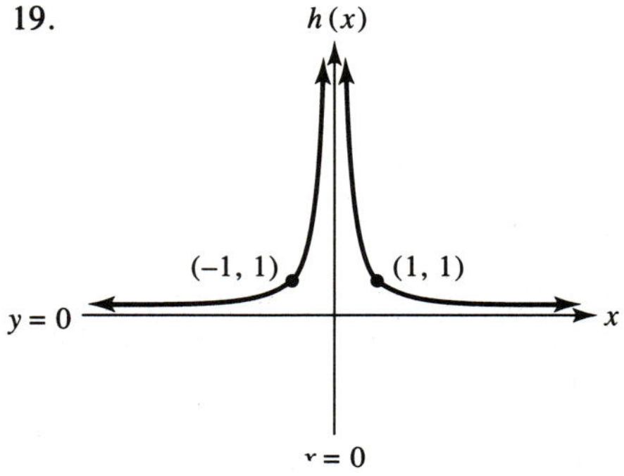

21.

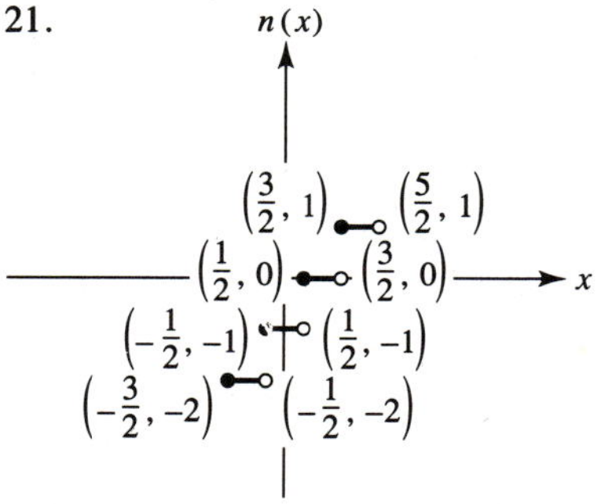

The graph continues to the left and the right.

23.

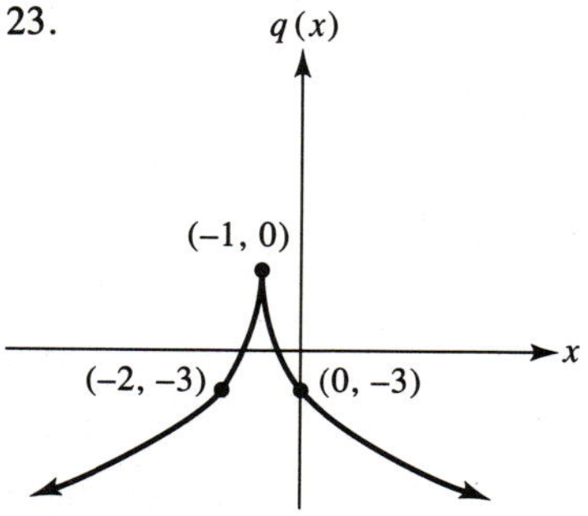

25.

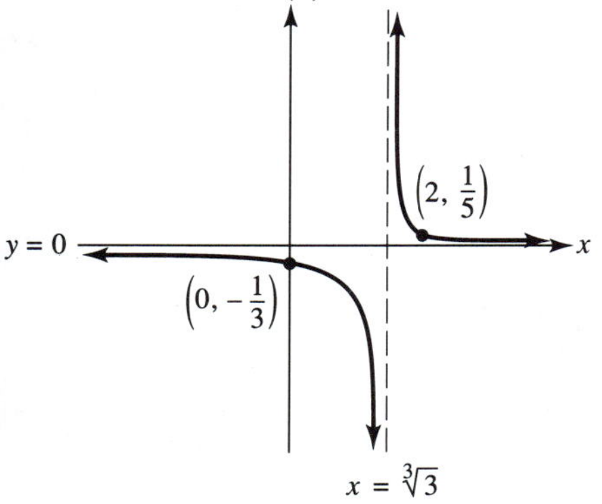

27.

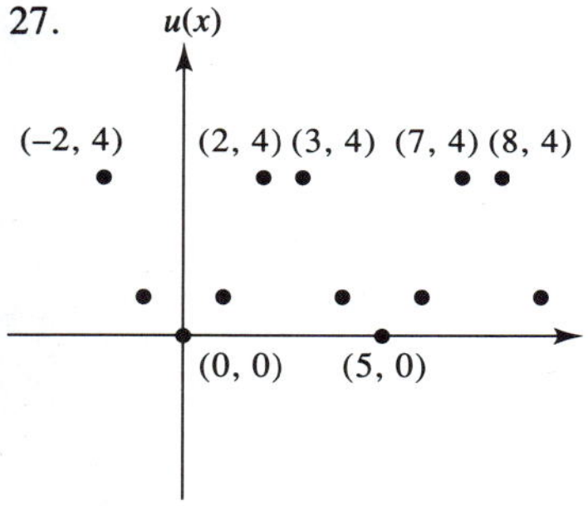

The graph continues to the left and the right.

29. a.

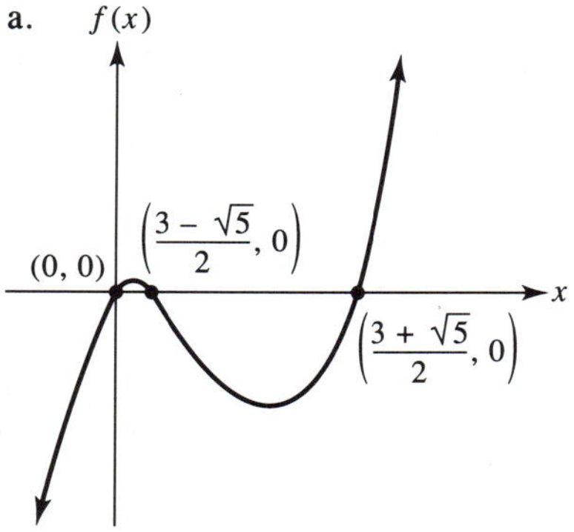

b. The graph of g is the same as the graph of f, shifted 1 unit down.

31.

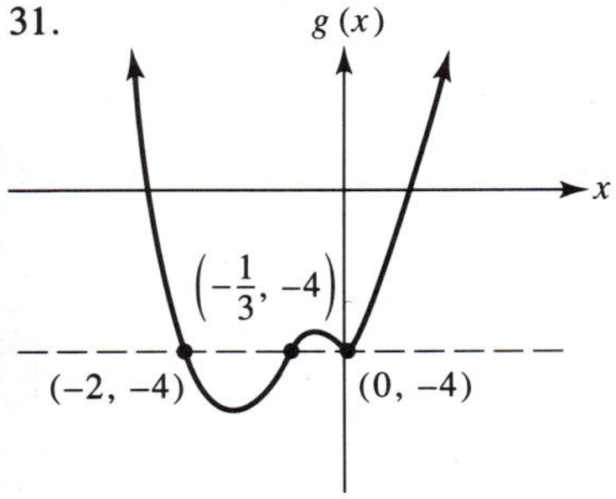

33.

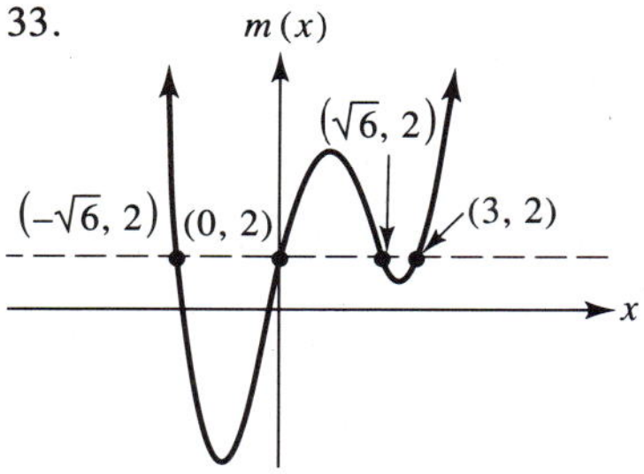

35.

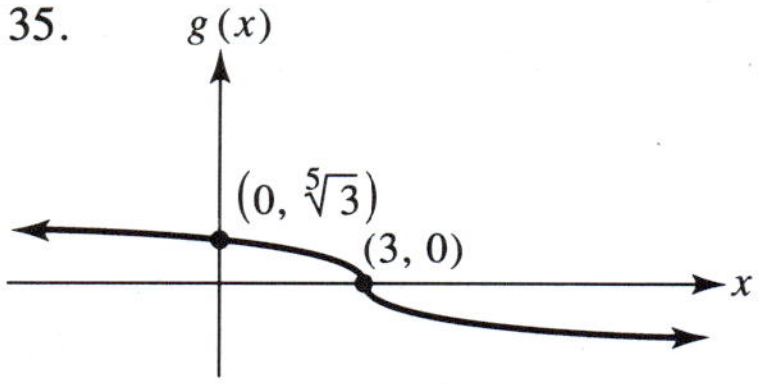

37.

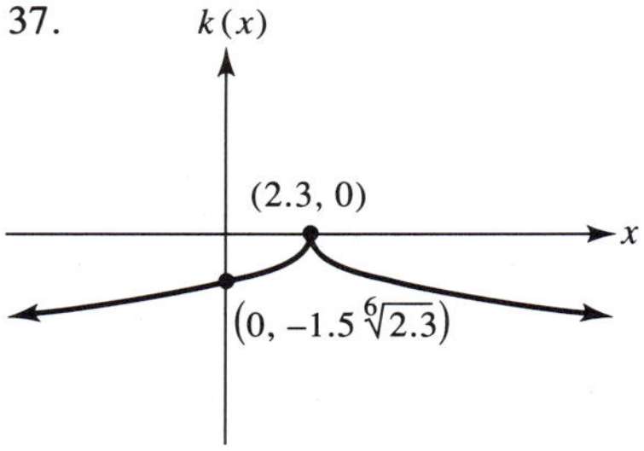

39. $g\left(-\frac{5}{12}\right) \approx -0.069$; $g\left(-\frac{1}{6}\right) \approx 0.025$;
$g\left(-\frac{5}{12} - 0.01\right) \approx -0.080$; $g\left(-\frac{1}{6} - 0.01\right) \approx 0.027$;
$g\left(-\frac{5}{12} + 0.01\right) \approx -0.058$; $g\left(-\frac{1}{6} + 0.01\right) \approx 0.024$
no no

41. a. domain $\mathbb{R}$, range $[-1, 1]$

b. $0, \frac{1}{2}, \frac{1}{2}$

Summary Exercises

1. Not a function
3. Not a function
5. $\dfrac{2}{h+2} + 5,\ h \neq -2$
7.

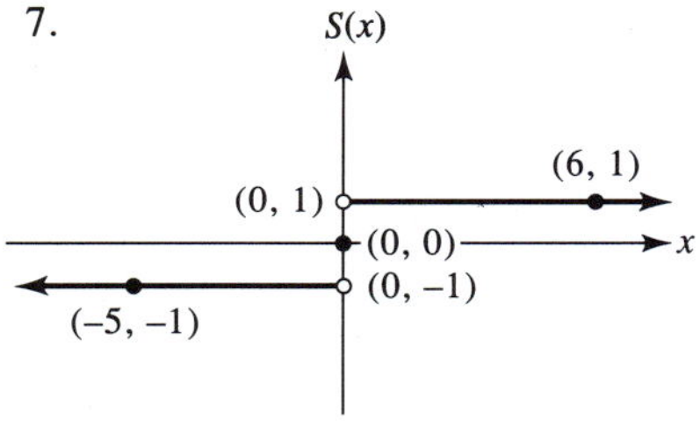

S is odd.

9. The graph of g is that of S shifted one unit right.

11. a. 1.0016 b. 1.2559 c. 1.0941

13.

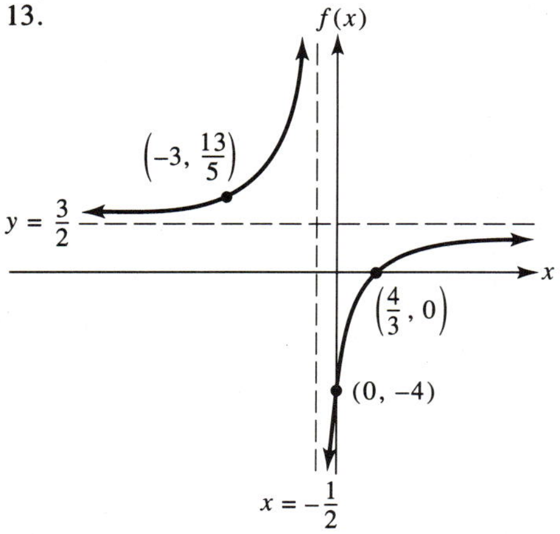

15.

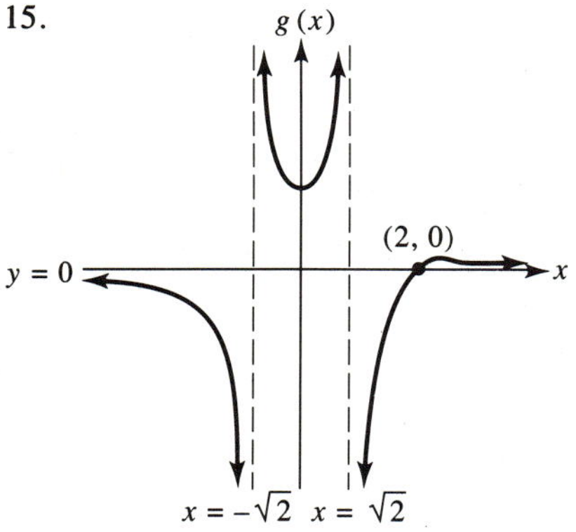

17.

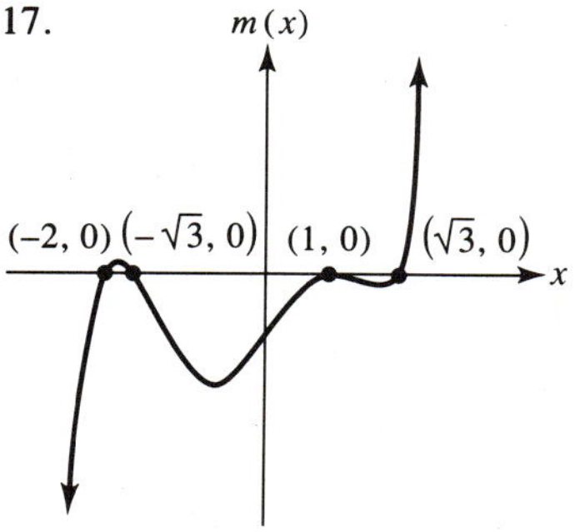

19.

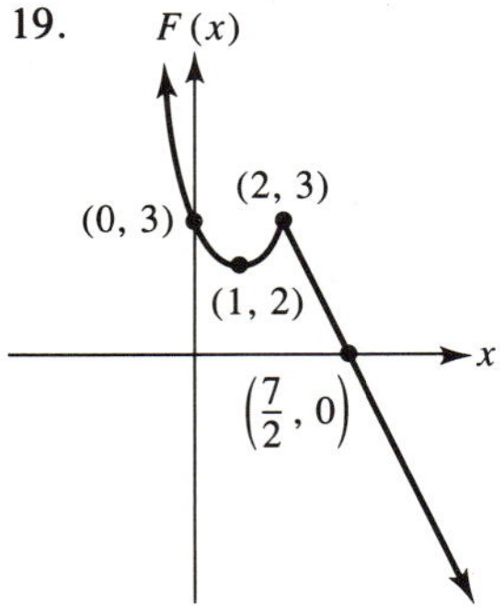

21.

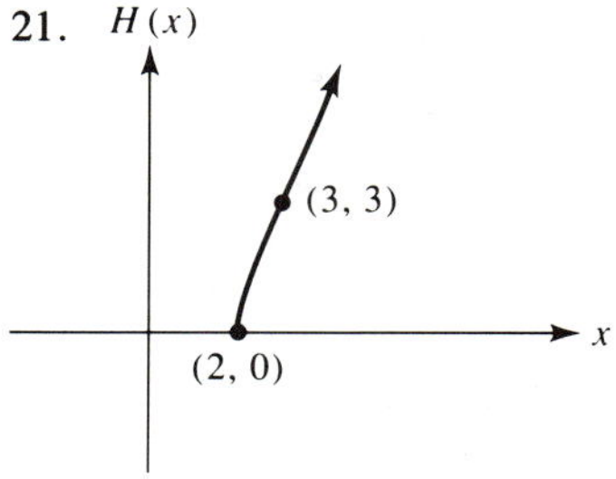

23.

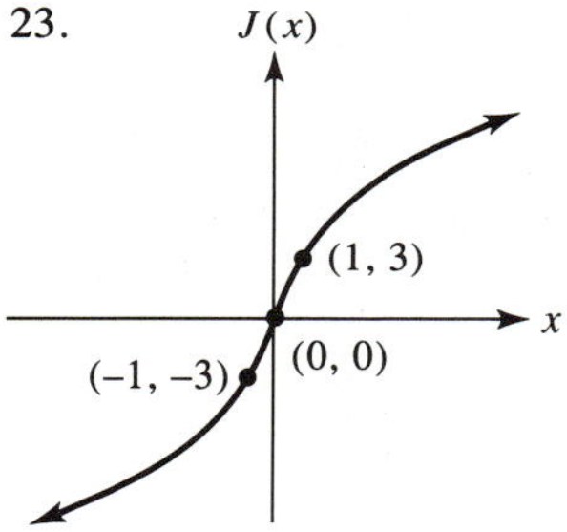

25.

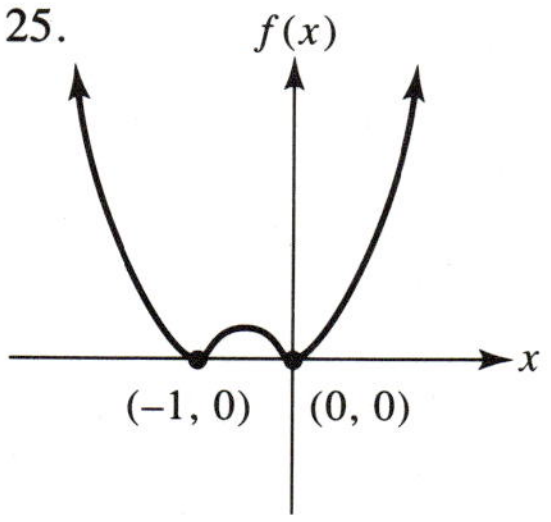

27.

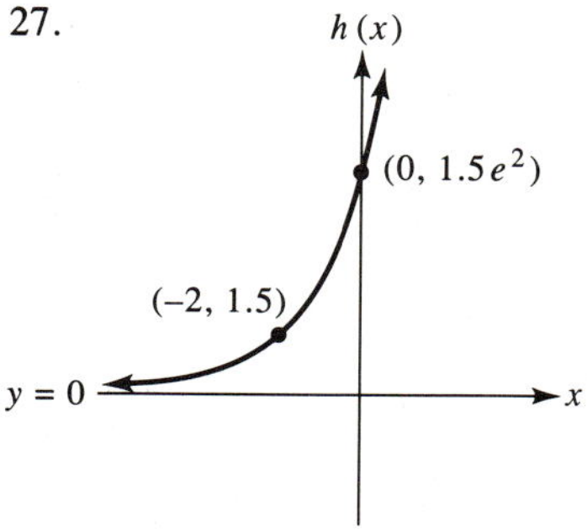

29.

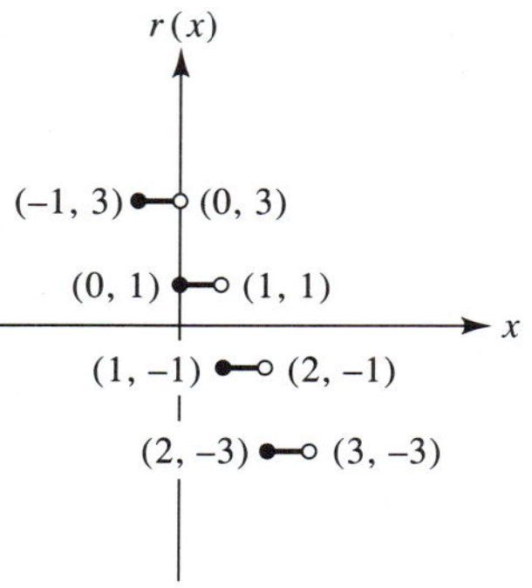

The graph continues to the left and the right.

31.

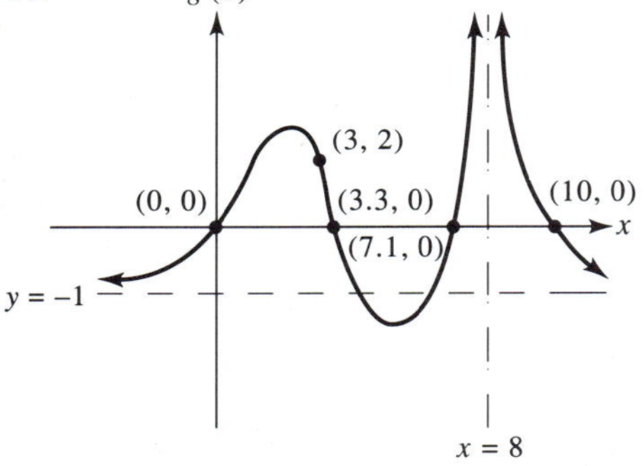

33.

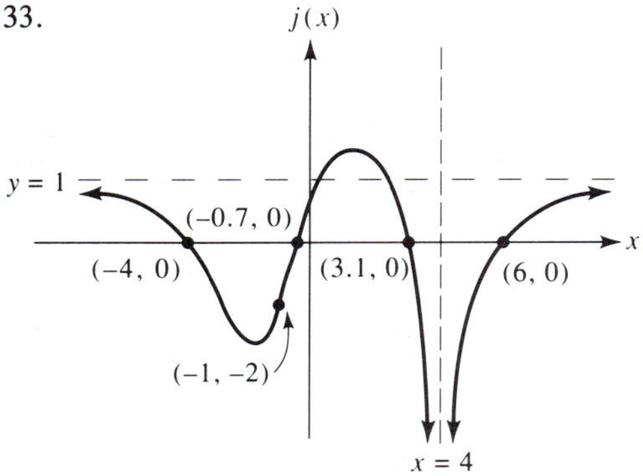

35. $G(x) = -g(x + 5) + 4$

37. 16

39. $f(n) = \text{mod}_7(n)$ is a possible answer.

41. b can be either $-\sqrt{8.9}$ or $\sqrt{8.9}$.

43. $h(x) = 3x^2 - 5, x \in \left(-\infty, \sqrt{\tfrac{5}{3}}\right)$

45. $F_1(x) = -5x^2 + 6, x \in (-\infty, 0]$ or $F_2(x) = -5x^2 + 6,$ $x \in [0, +\infty)$ is an answer.

47. $H(x) = -|x + 1| - 4, x \in [-7, -2]$ is an answer. Yes, it is 1-1.

Chapter Five

5.1

1. $(f + g)(x) = \dfrac{x + 3}{x - 2} + x - 2, (-\infty, 2) \cup (2, +\infty)$
   $(f \cdot g)(x) = x + 3, x \neq 2, (-\infty, 2) \cup (2, +\infty)$

3. $f + g = \left\{(1, -3), \left(2, 9\tfrac{1}{3}\right), (3, 16), (5, 16)\right\}, \{1, 2, 3, 5\}$
   $f \cdot g = \{(1, -28), (2, 3), (3, 63), (5, 64)\}, \{1, 2, 3, 5\}$

5. $(f + g)(x) = x + 4 + \sqrt{\dfrac{x - 2}{x + 2}}, [2, +\infty)$
   $(f \cdot g)(x) = (x + 4)\sqrt{\dfrac{x - 2}{x + 2}}, [2, +\infty)$

7. $(f + g)(n) = \text{mod}_5(n) + n/5, n \in W$
   $(f \cdot g)(n) = \dfrac{n\,\text{mod}_5(n)}{5}, n \in W$

9. $(f + g)\,(x) = \sqrt[3]{x + 3} + \sqrt[3]{4 - 2x^2}, (-\infty, +\infty)$
   $(f \cdot g)(x) = \sqrt[3]{(x + 3)(4 - 2x^2)}, (-\infty, +\infty)$

11. $(f - g)(x) = \dfrac{x^2 + 7x + 12}{x^2 + 8x + 15} - \dfrac{x^2 + 7x + 10}{x^2 + 6x + 8},$
   $(-\infty, -5) \cup (-5, -4) \cup (-4, -3) \cup (-3, -2) \cup (-2, +\infty);$
   $(f/g)(x) = \dfrac{(x + 4)^2}{(x + 5)^2}, (-\infty, -5) \cup (-5, -4) \cup$
   $(-4, -3) \cup (-3, -2) \cup (-2, +\infty)$

13. $(f - g)(x) = (1.01)^{0.5x} - (1.01)^x, (-\infty, +\infty)^{0.5x}$
   $(f/g)(x) = (1.01)^{-0.5x}, (-\infty, 0) \cup (0, +\infty)$

15. $(f - g)(x) = \text{div}_3(n) - 6n, n \in W$
   $(f/g)(x) = \dfrac{\text{div}_3 n}{6n}, n \in N$

17. $(f - g)(x) = \sqrt{2x} - \sqrt[3]{16x}, [0, +\infty)$
   $(f/g)(x) = \dfrac{\sqrt{2x}}{\sqrt[3]{16x}}, (0, +\infty)$

19. $(f - g)(x) = 0, [4, +\infty)$
   $(f/g)(x) = 1, (4, +\infty)$

21. Yes, the reciprocal is $g(x) = 1/e^x$, $\text{dom}_g = \mathbb{R}$.

23. No

25. ct(7) = 1, ct(81) = 4, ct(210) = 4, ct(512) = 9

27. If m is divisible by n, then we can write m as $p_1 p_2 \cdots p_s$ and n as $p_1 p_2 \cdots p_t$, $t \leq s$, each p_i a prime. Then $m/n = p_1 p_2 \cdots (p_t p_{t+1} \cdots p_s)/(p_1 p_2 \cdots p_t)$, or $p_{t+1} p_{t+2} \cdots p_s$. There are $s - t$ primes in $p_{t+1} p_{t+2} \cdots p_s$. ct($m/n$) is ct($m$) − ct($n$).

29. Function multiplication is associative for f, g, and h if $(f \cdot (g \cdot h))(x) = ((f \cdot g) \cdot h)(x)$ for all $x \in \mathbb{R}$. By definition the left-hand side $(f \cdot (g \cdot h))(x) = f(x)(g(x)\,h(x)) = f(x)g(x)h(x)$ because multiplication is associative for the real numbers.

$f(x)\,g(x)\,h(x) = (f(x)\,g(x))\,h(x) =$
$(f \cdot g)(x) \cdot h(x) = ((f \cdot g) \cdot h)(x)$

using the definition of function multiplication, but this is the right-hand side.

31. Function multiplication is commutative for these functions if $(f \cdot g)(x) = (g \cdot f)(x)$ for all $x \in \mathbb{R}$. By the definition of function multiplication, $(f \cdot g)(x) = f(x)g(x)$. Since multiplication is commutative for the reals, $f(x)\,g(x) = g(x)\,f(x)$. But this is $(g \cdot f)(x)$.
33. For $f(x) = x$, $g(x) = 2x$, $(f - g)(x) \neq (g - f)(x)$. This is one counterexample.
35. Because $(b + c)/2 = (c + b)/2$ for all integers b, c, the operation @ is commutative.
37. There is not an integer i for which $(b + i)/2 = b$ for all b.
39. $a * (b * c) = |a - b * c| = |a - |b - c||$. On the other hand, $(a * b) * c = |a * b - c| = ||a - b| - c|$. The operation $*$ is not associative.
41. Without an identity we can't define an inverse.
43. If f and g are odd, $f + g$ and $f - g$ may be even or odd. Given $f(x) = x$, $g(x) = x^3$, $h(x) = -x$, and $k(x) = x^3$; $f + g$ is odd, $f + h$ is even, $g - f$ is odd, and $g - k$ is even.

5.2

1. $(f \circ g)(x) = x, (-\infty, +\infty)$;
$(g \circ f)(x) = x, (-\infty, +\infty)$
3. $(f \circ g)(x) = 5.3(3 - x)^2 - 7.1(3 - x) + 2$, $(-\infty, +\infty)$;
$(g \circ f)(x) = 1 + 7.1x - 5.3x^2, (-\infty, +\infty)$
5. $(f \circ g)(x) = \dfrac{5x + 15}{5x + 4}, \left(-\infty, -\dfrac{4}{5}\right) \cup \left(-\dfrac{4}{5}, +\infty\right)$;
$(g \circ f)(x) = 5\left(\dfrac{x + 7}{x - 4}\right) + 8, (-\infty, 4) \cup (4, +\infty)$
7. $(f \circ g)(x) = x, (-\infty, -1) \cup (-1, +\infty)$;
$(g \circ f)(x) = x, (-\infty, 3) \cup (3, +\infty)$
9. $(f \circ g)(x) = x, (-\infty, 0) \cup (0, +\infty)$;
$(g \circ f)(x) = x, (-\infty, 0) \cup (0, +\infty)$
11. $f(x) = \sqrt[3]{x}$, $g(x) = (x + 1)^2$
13. $f(x) = |x|$, $g(x) = x^2 - 7x + 12$
15. $f(x) = \sqrt{x}$, $g(x) = 4 - x^2$
17. $f(x) = x + 1$, $g(x) = \text{mod}_5(x)$
19. $f(x) = e^x$, $g(x) = 4x^2$
21. $(f \circ (g \circ h))(x) = \dfrac{5}{\sqrt{5 - (x + 1)^2}}$, $((f \circ g) \circ h)(x) = \dfrac{5}{\sqrt{5 - (x + 1)^2}}$; yes
23. $f^{-1} = \{(5, 3), (2, 9), (-9, 19)\}$
25. 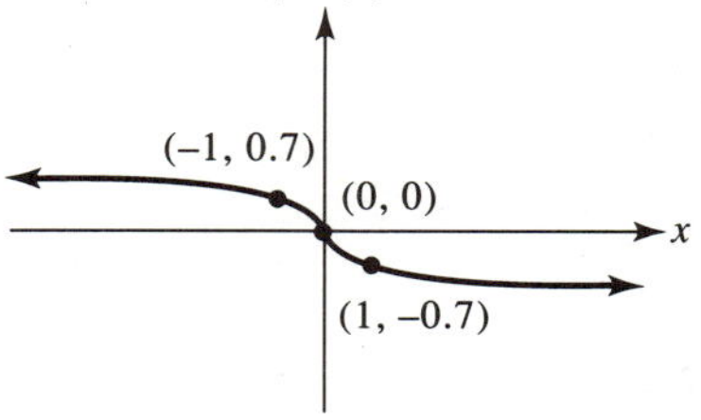

27. 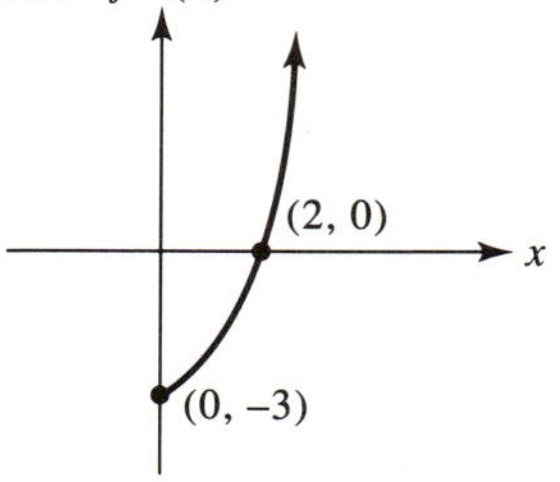

29. $f^{-1}(x) = \dfrac{x - 5}{3}$
31. $f^{-1}(x) = \sqrt[5]{x}$
33. $f^{-1}(x) = \dfrac{1}{x - 2}$
35. $f^{-1}(x) = x^2 + 3, x \geq 0$
37. $f^{-1}(x) = \sqrt{x} - 5$
39. We must show $(f \circ (g \circ h))(x) = ((f \circ g) \circ h)(x)$ and $\text{dom}_{f\circ(g\circ h)} = \text{dom}_{(f\circ g)\circ h}$. $(f \circ (g \circ h))(x) = f(g \circ h(x)) = f(g(h(x))) = (f \circ g)(h(x)) = ((f \circ g) \circ h)(x)$ because of the definition of composition. We have $\text{dom}_{f\circ(g\circ h)} = \{x \mid x \in \text{dom}_{g\circ h} \text{ and } (g \circ h)(x) \in \text{dom}_f\} = \{x \mid x \in \text{dom}_h \text{ and } h(x) \in \text{dom}_g \text{ and } g(h(x)) \in \text{dom}_f\}$.
But this is also a description of $\text{dom}_{(f\circ g)\circ h}$.
41. a, d could not be inverses.
43. 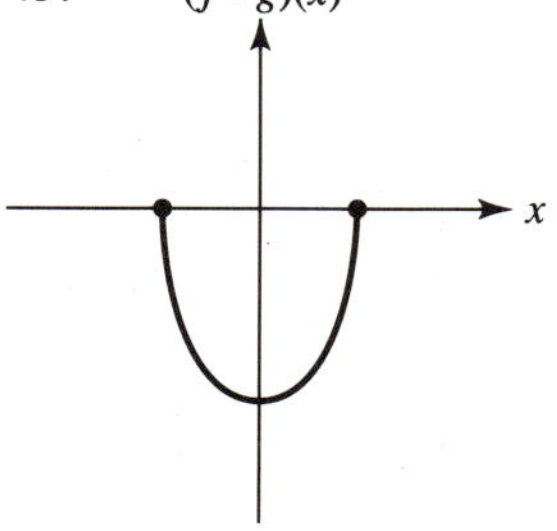

5.3

1.

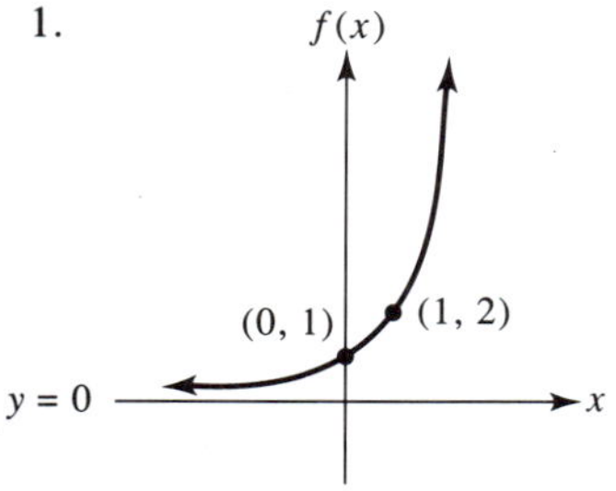

3.

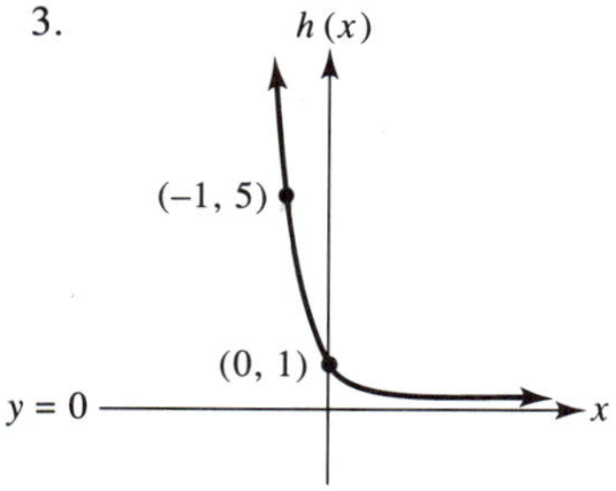

5.

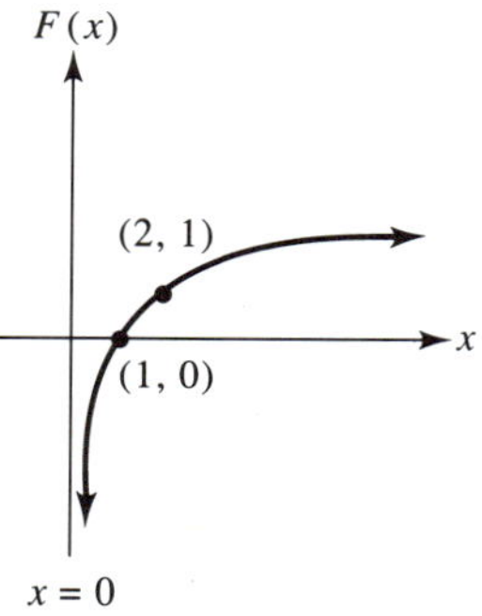

7.

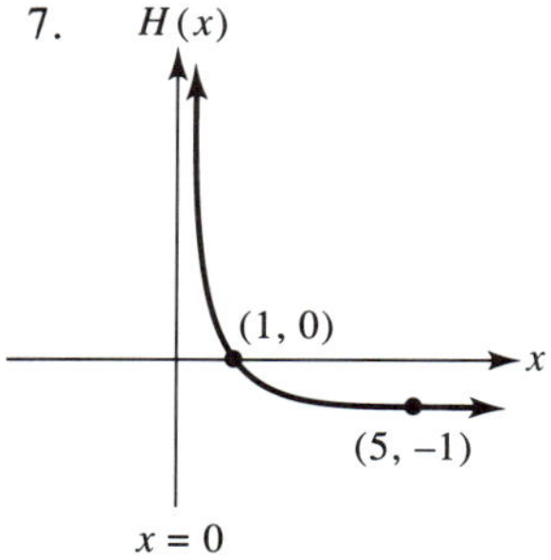

9.

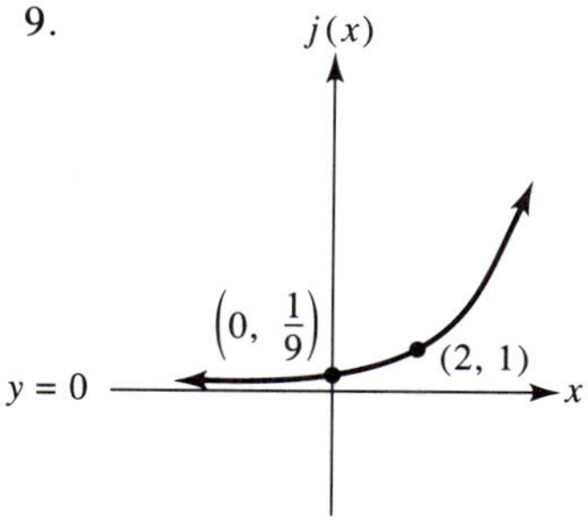

11.

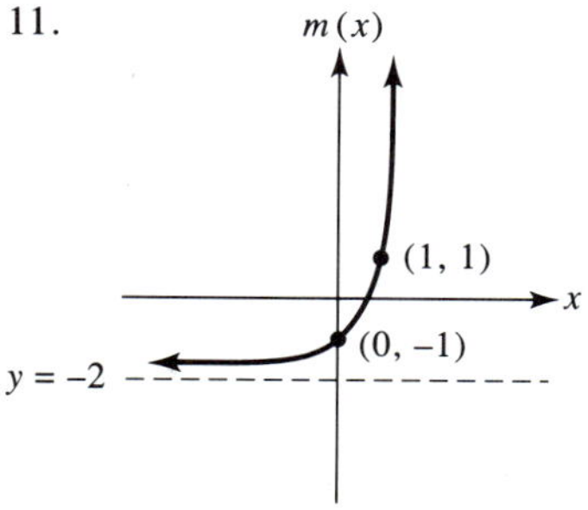

13.

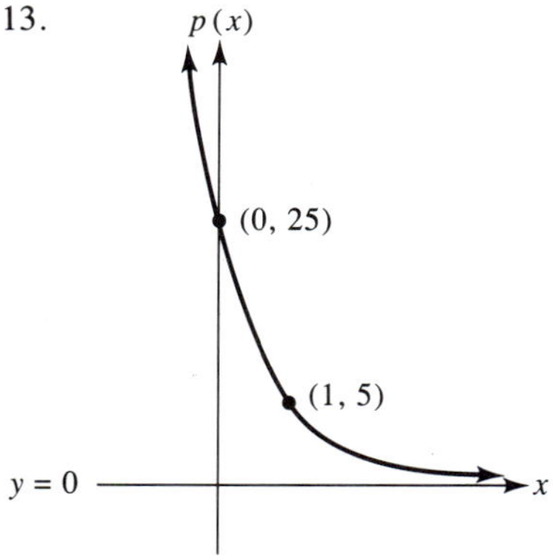

15.

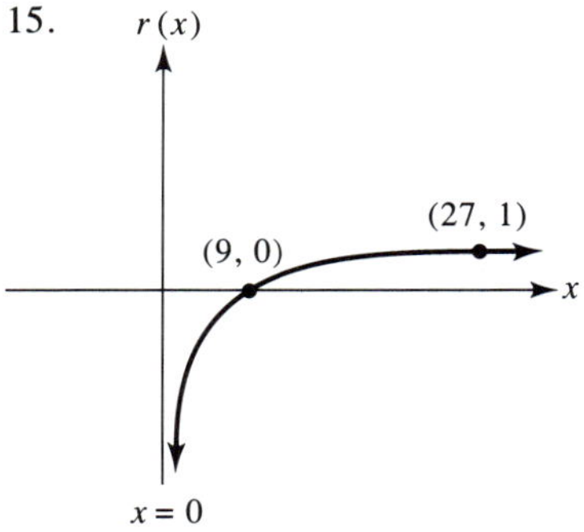

17.

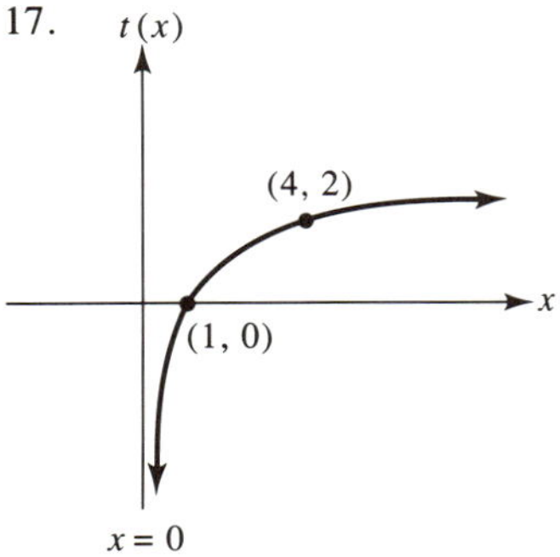

19.

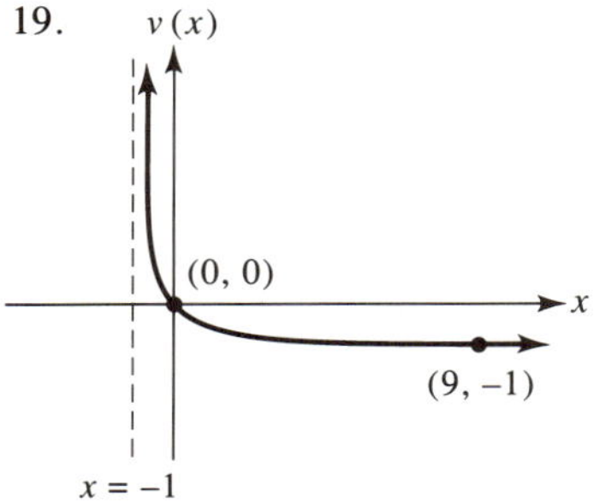

21. They are not the same; $\text{dom}_f = (0, +\infty)$, $\text{dom}_F = (-\infty, 0) \cup (0, +\infty)$.
23. $f^{-1}(x) = \log_7 x$
25. $h^{-1}(x) = (0.5)^x$
27. 0.7738 (approx.)
29. 0.3280 (approx.)
31. 0.4308 (approx.)
33. 1.2851 (approx.)
35. 0.5000 (approx.)
37. $\ln 6 + 2 \ln x + \frac{1}{3}\ln y + \frac{2}{3}\ln z$
39. $\frac{1}{3}\ln 7 + \frac{2}{3}\ln x + \frac{5}{3}\ln z - \ln 2 - \frac{1}{2}\ln y$
41. $\ln xz/y$
43. $\ln \dfrac{\sqrt[3]{7x^2z^5}}{2\sqrt{y}}$
45. She chose to invest at 8%.
47. You'd better move to the neighbors; it will take over 37 hours.
49. I'd report 2.4 as the pH; it is an acidic solution.
51. a. It should have 112 parking places.
    b. $p \approx f(x) = \dfrac{x^{1.261}}{1.441}$
53. It has approximately 213,000 square feet.
55. { }
57. $\left\{\dfrac{\ln 5 + \sqrt{(\ln 5)^2 + 12(\ln 10)^2}}{2 \ln 10}\right\}$
59. $\{\frac{5}{3}\}$
61. $\{e^7 - 1\}$
63. { }
65. $\{-\sqrt{8}, \sqrt{8}\}$
67. ln 900
69. $\dfrac{1}{3}\ln 2$
71. $\ln(15)^{-8}$
73. $\sqrt[9]{2^{-13}}$

5.4

1. $y = -5x + 3$
3. $y = \dfrac{1}{100{,}000}x + 16$
5. Either form is fine; slope-intercept is easier for exercise 1 and exercise 4.
7. $y = -(x - 4)^2$
9. $y = x^2 + 2x + 2$
11. $y = -3.2^{0.1x}$
13. $y = \dfrac{-1}{\ln(e/2)}\ln(2x)$ or $y = \dfrac{1}{\ln 2 - 1}\ln(2x)$
15. $y = -6x$
17. $z = 816\dfrac{x}{y}$
19. $z = \frac{7}{9}x^2\sqrt[3]{y}$
21. y varies inversely as x.
23. $f(x, y)$ varies directly as the square root of x and inversely as y.
25. w varies jointly as the square of x and the fourth power of y and inversely as the cube of z.
27. The pressure will be 7680 lb/ft^2.
29. k is $\dfrac{-\ln 0.5}{120}$; $k \approx 0.00578$
31. $f(x) = 2x^2 + 4x - 1$
33. $h(x) = 0.2 \ln 2x$
35. $F(x) = \begin{cases} 0.5x^2 + 1, & 0 \le x \le 4 \\ 0.25x + 8, & 4 \le x \end{cases}$
37. You'd need 6 data points, each one an ordered triple.
39. It is third degree.
41. a. $p(x) = 0.5x - 322$
    b. $644 \le x \le 2000$; the upper bound is a judgment call.

Summary Exercises

1. $(f + g)(x) = \sqrt{2 - 3x} + x^2 + 2;\ \left(-\infty, \frac{2}{3}\right]$
   $(f/g)(x) = \dfrac{\sqrt{2 - 3x}}{x^2 + 2};\ \left(-\infty, \frac{2}{3}\right]$
   $(f \circ g)(x) = \sqrt{-4 - 3x^2};\ \{\ \}$
   $(g \circ f)(x) = 4 - 3x;\ \left(-\infty, \frac{2}{3}\right]$
   $f^{-1}(x) = \dfrac{2 - x^2}{3};\ [0, +\infty)$
   $g^{-1}(x)$ is not defined.
3. $(f + g)(x) = \ln x + 2x;\ (0, +\infty)$
   $(f/g)(x) = \dfrac{\ln x}{2x};\ (0, +\infty)$
   $(f \circ g)(x) = \ln(2x);\ (0, +\infty)$
   $(g \circ f)(x) = 2 \ln x;\ (0, +\infty)$
   $f^{-1}(x) = e^x;\ \mathbb{R}$
   $g^{-1}(x) = \dfrac{x}{2};\ \mathbb{R}$
5. If f is odd and g is even, $f + g$ and $f - g$ are neither, except if $g(x) = 0$.
   $(f + g)(-x) = f(-x) + g(-x) = -f(x) + g(x)$.
7.

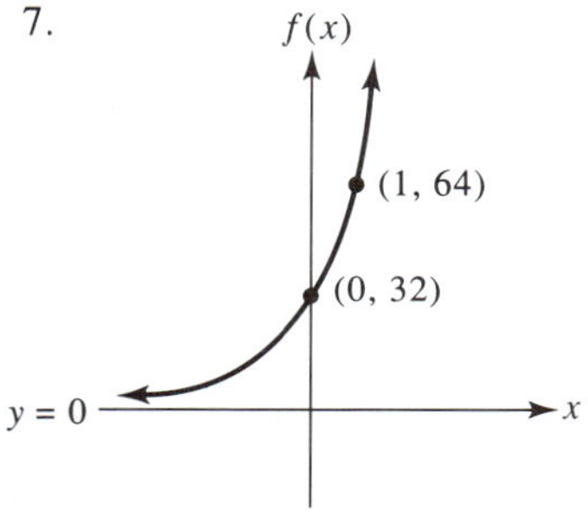

9.

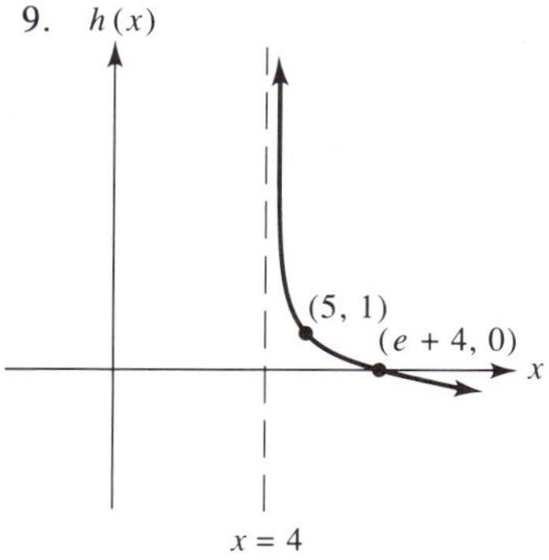

11. $(f \circ g)(x) = x$; dom $= [0, +\infty)$ if n is even.
    $(f \circ g)(x) = x$; dom $= \mathbb{R}$ if n is odd.
    $(g \circ f)(x) = |x|$; dom $= \mathbb{R}$ if n is even.
    $(g \circ f)(x) = x$; dom $= \mathbb{R}$ if n is odd.
    If n is odd, f and g are inverses.
13. $\frac{1}{3}[\ln 4 + 2 \ln c + \ln(a + 3b)]$
15. $\ln\sqrt{\dfrac{x^3}{y}}$
17. $\{\ \}$
19. $\left\{\dfrac{\ln(\pi \cdot 3^2)}{\ln 3}\right\}$, or $\{(\log_3 \pi) + 2\}$
21. $\{-2 + \ln 3.2\}$
23. $\log_6 1.549$
25. $(0.000008)^{1{,}000{,}000}$
27. y varies directly as the fourth power of x and inversely as the cube root of z.
29. $S(t) = -16t^2 + 64t$

Chapter Six

6.1

1. II $\left(-\dfrac{\sqrt{3}}{2}, \dfrac{1}{2}\right)$
3. IV $\left(\dfrac{1}{2}, -\dfrac{\sqrt{3}}{2}\right)$
5. III $\left(-\dfrac{\sqrt{3}}{2}, \dfrac{1}{2}\right)$
7. II $\left(-\sqrt{\dfrac{1}{2}}, \sqrt{\dfrac{1}{2}}\right)$
9. I $\left(\dfrac{\sqrt{3}}{2}, \dfrac{1}{2}\right)$
11. IV $\left(\dfrac{1}{2}, -\dfrac{\sqrt{3}}{2}\right)$
13. No quadrant; $(-1, 0)$
15. I
17. III
19. I
21. I
23. III
25. Some possible answers are $\pi, -\pi, 3\pi, -3\pi$.
27. Some possible answers are $11\pi/6, -\pi/6, 23\pi/6, -13\pi/6$.
29. Answers are of the form $\pi/2 + 2k\pi, k \in I$.
31. Some possible answers include $\dfrac{2\pi}{3}, -\dfrac{4\pi}{3}, \dfrac{8\pi}{3}, -\dfrac{10\pi}{3}$.
33. Answers are of the form $\pi/4 + 2k\pi, k \in I$.
35. $\pi + 2k\pi, k \in I$
37. $11\pi/6 + 2k\pi, k \in I$
39. $[4\pi/3, 3\pi/2]$
41. $[\pi/3, \pi/2]$

43. $[5\pi/6, \pi]$
45. $[-2\pi/3, -5\pi/6]$

6.2

1. $\frac{1}{2}$
3. $-\sqrt{3}/2$
5. $-\frac{1}{2}$
7. $-\sqrt{\frac{1}{2}}$
9. $\frac{1}{2}$
11. $\frac{4}{5}$
13. $-\sqrt{\frac{1}{3}}$
15. Positive
17. Negative
19. Positive
21. 0.43388 (approx.)
23. 0.50596 (approx.)
25. 11.99821 (approx.)
27. 0.75680 (approx.)
29. −8.18724 (approx.)
31. 1.5574 (approx.)
33. $3\pi/2$ is a possible answer.
35. Possible answers include $\pi/6$, $-\pi/6$, $11\pi/6$, $-11\pi/6$.
37. Answers are of the form $\pi/2 + 2k\pi$, $k \in I$
39. $-4 - \sqrt{3}$
41. The term $\sec(-\pi/2)$ is undefined.
43. The value of t is $4\pi/3 + 2k\pi$, $k \in I$ or $5\pi/3 + 2k\pi$, $k \in I$.
45. We must show $\sec(-x) = \sec(x)$, $x \in \text{dom}_{\sec}$. $\sec(-x) = \dfrac{1}{\cos(-x)} = \dfrac{1}{\cos(x)} = \sec(x)$ because cosine is even. The function cosecant is an odd function, because $\csc(-x) = \dfrac{1}{\sin(-x)} = \dfrac{1}{-\sin x} = -\csc x$. The function tangent is an odd function; $\tan(-x) = \dfrac{\sin(-x)}{\cos(-x)} = -\dfrac{\sin x}{\cos x} = -\tan x$. Because cotangent is the reciprocal of tangent, it is also odd.
47. One of the times is $\text{Cos}^{-1}(\frac{1}{6})$ seconds.
49. The sine function and ps are periodic, odd, have domain $\mathbb{R}$, and range $[-1, 1]$.

6.3

1.

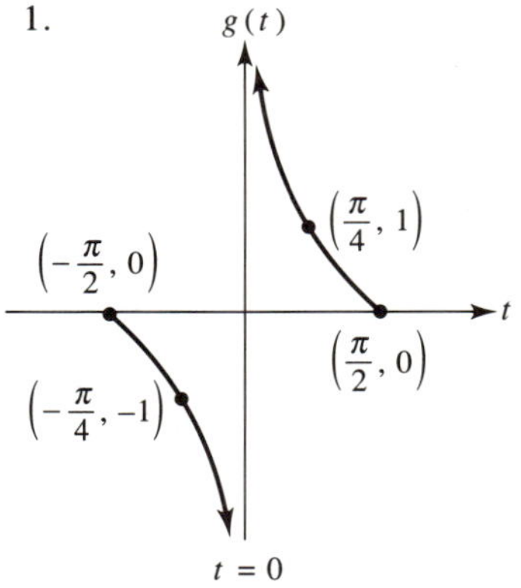

3.

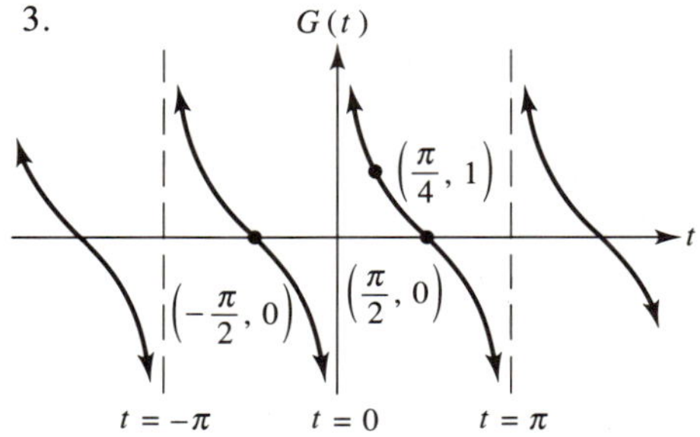

The graph continues to the left and the right.

5. The function f is increasing on every interval $[-\pi/2 + 2k\pi, \pi/2 + 2k\pi]$, $k \in I$; f is decreasing on $[\pi/2 + 2k\pi, 3\pi/2 + 2k\pi]$, $k \in I$.
7. The function h is increasing on every interval $(-\pi/2 + 2k\pi, \pi/2 + 2k\pi)$, $k \in I$; h is never decreasing.
9. The function G is increasing on each of the intervals $[0 + 2k\pi, \pi/2 + 2k\pi]$, $(\pi/2 + 2k\pi, \pi + 2k\pi]$, $k \in I$; G is decreasing on each of the intervals $(-\pi/2 + 2k\pi, 0 + 2k\pi]$, $[\pi + 2k\pi, 3\pi/2 + 2k\pi)$, $k \in I$.
11. a. No b. No c. Yes d. No e. Yes f. No
13. a. No b. No c. Yes d. No e. Yes f. No
15. a. Yes b. No c. No d. Yes e. No f. No
17. $\sec(\pi + 2\pi) = \dfrac{1}{\cos(x + 2\pi)} = \dfrac{1}{\cos x} = \sec x$. No smaller number can be the period because cosine has period 2π.
19. Since the period of tangent is π, $\cot(x + \pi) = \dfrac{1}{\tan(x + \pi)} = \dfrac{1}{\tan x}$. No smaller number will work.
21. They do not have this property.
23. It is the sine graph shifted left $\pi/6$ units, up 2 units, and inverted. The graph continues to the left and the right.

6.4

1.

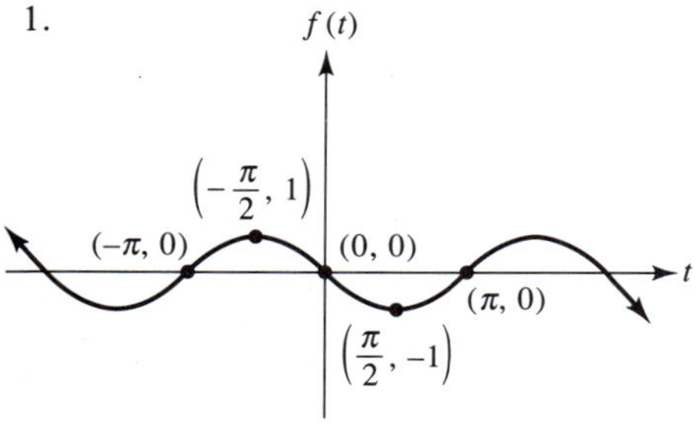

Amplitude: 1; period: 2π; phase shift: 0

3.

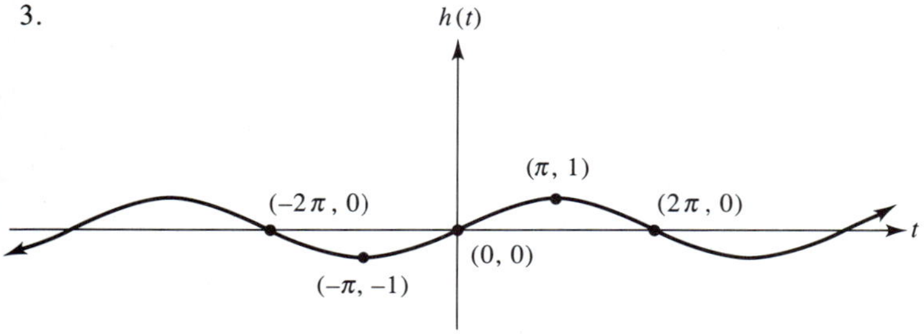

Amplitude: 1; period: 4π; phase shift: 0

5.

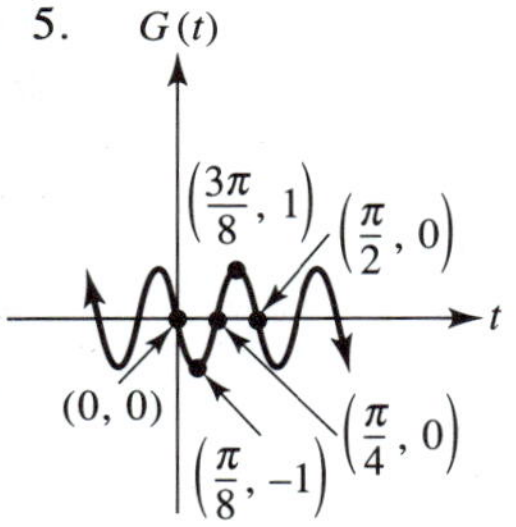

Amplitude: 1; period: $\pi/2$; phase shift: 0

7.

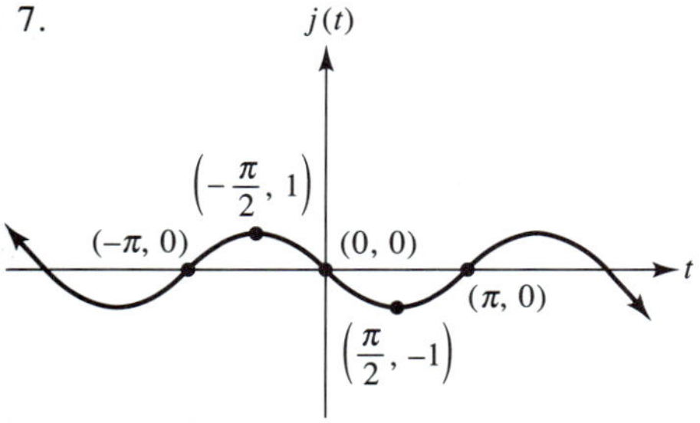

Amplitude: 1; period: 2π; phase shift: 0

9.

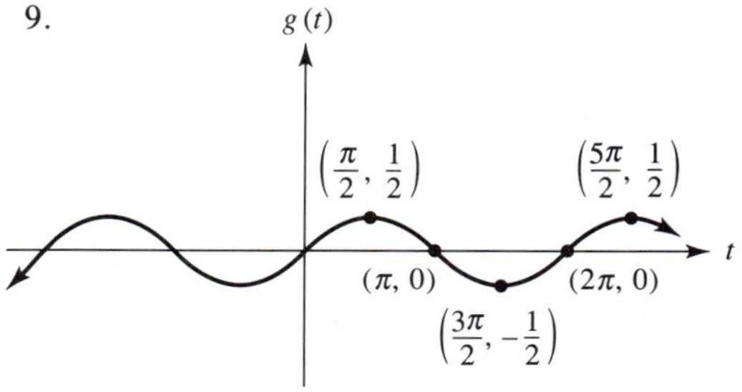

Amplitude: $\frac{1}{2}$; period: 2π; phase shift: π

11.

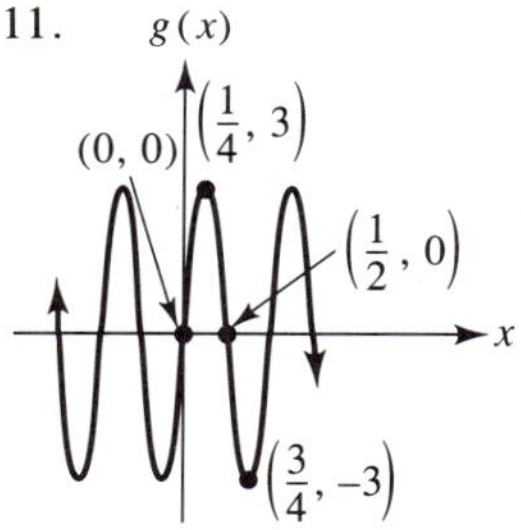

Amplitude: 3; period: 1; phase shift: 0

13.

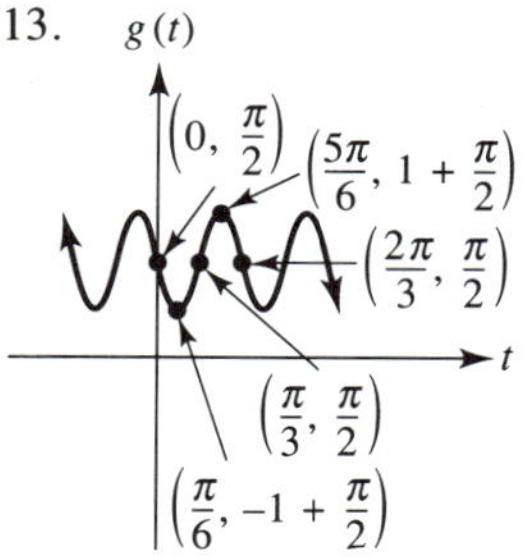

Amplitude: 1; period: $2\pi/3$; phase shift: 0

15.

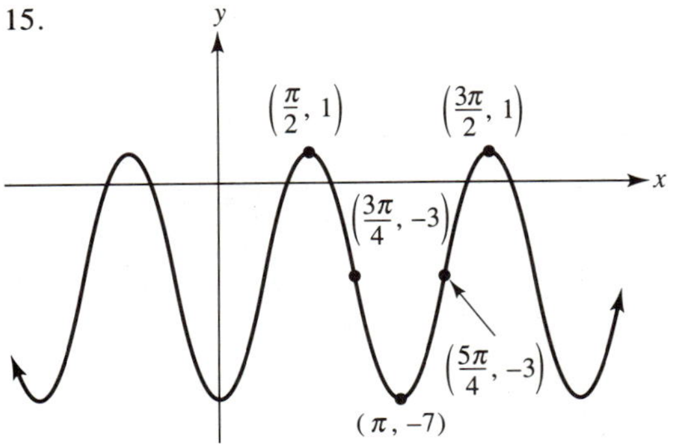

Amplitude: 4; period: π; phase shift: $\pi/2$

17.

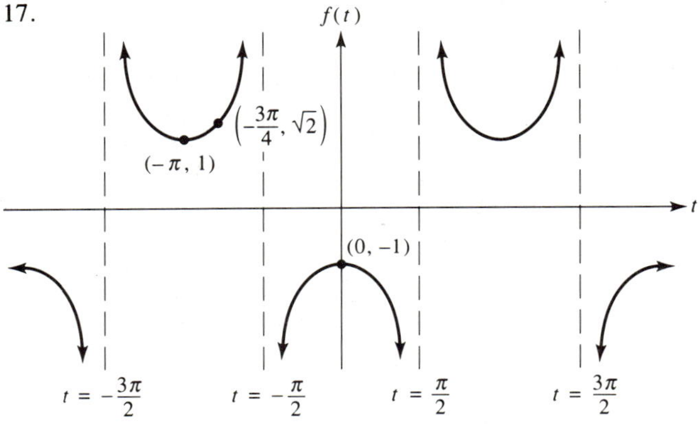

The graph continues to the left and right; period: 2π; phase shift: $-\pi$.

19.

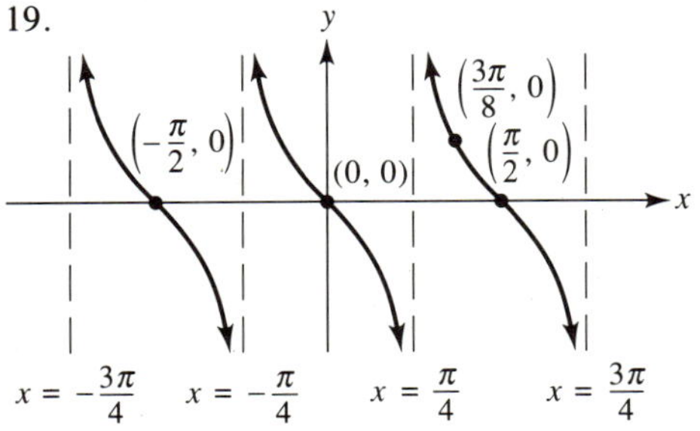

The graph continues to the left and right; period: $\pi/2$; phase shift: $\pi/2$.

21.

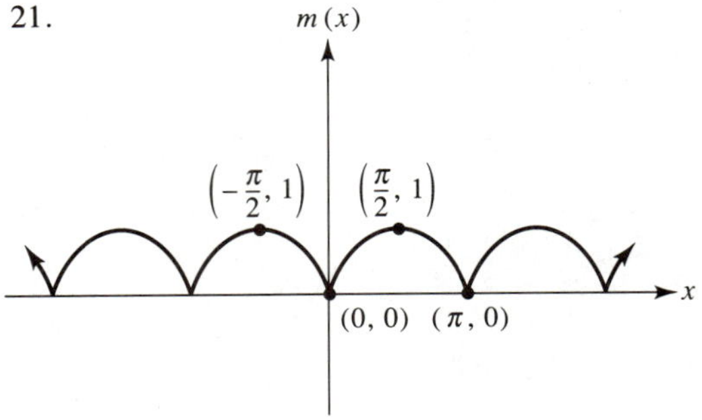

23.

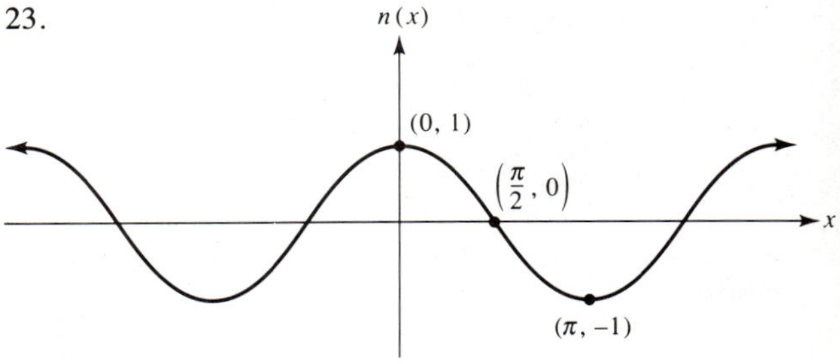

This is the same as the graph of the cosine function.

25.

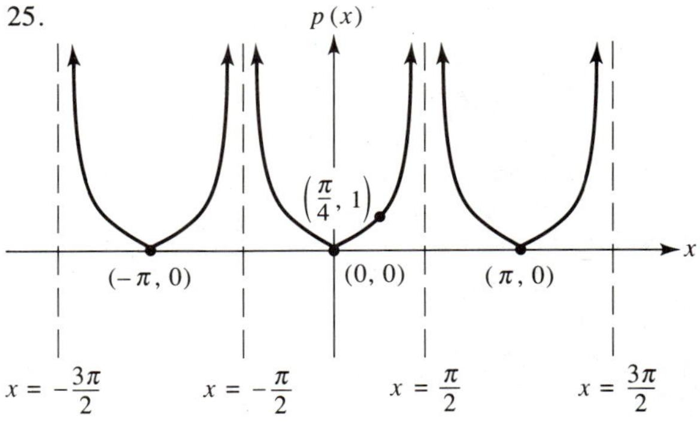

The graph continues to the left and right.

27. $f(x) = 3 \sin \frac{1}{2}(x - \pi)$

29. $h(x) = e \sin 4(x - 5\pi)$

31. $m(x) = a \sin \dfrac{2\pi}{3}(x + e)$

6.5

1. $-\pi/4$
3. $\pi/3$
5. $-\pi/4$
7. $-\pi/2$
9. $\pi/6$
11. 0.33800 (approx.)
13. 1.15928 (approx.)
15. -1.21828 (approx.)
17. Not defined
19. Not defined
21. 0.55107 (approx.)
23. Undefined
25. 0
27. $(1 + \pi)/2$
29. Not defined
31. $\frac{5}{2}$
33. $3\pi/4$
35. Not defined
37. 0.534 (approx.)
39. -0.642 (approx.)
41. $\text{Sec}^{-1}x = y$ if and only if $\text{Sec } y = x$ with $\text{Sec } x = \sec x$, $x \in [0, \pi/2) \cup (\pi/2, \pi]$ is one possible answer.

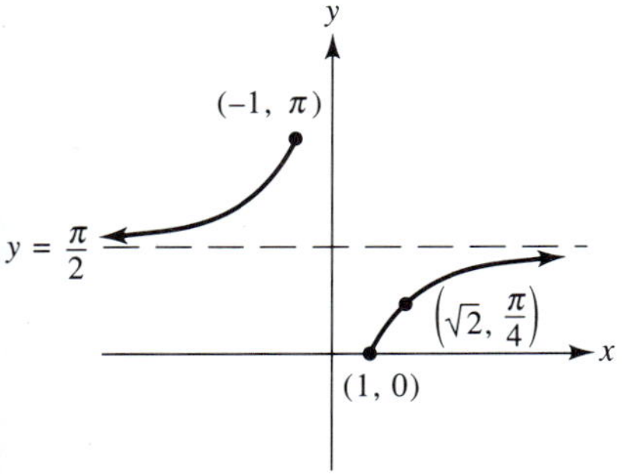

43. $\text{Cot}^{-1}x = y$ if and only if $\text{Cot } y = x$ with $\text{Cot } x = \cot x$, $x \in (0, \pi)$ is one possible answer.

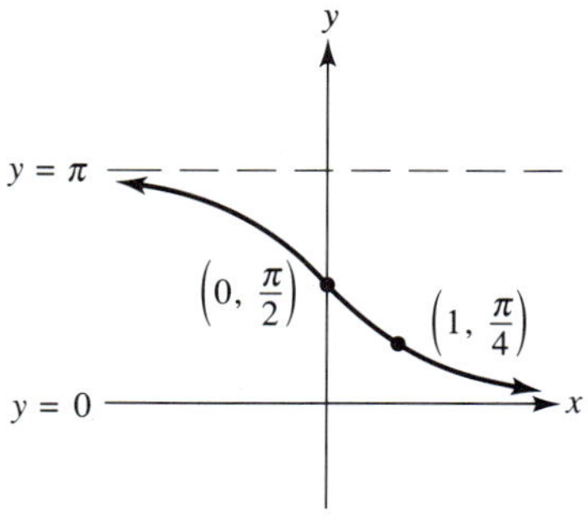

45.

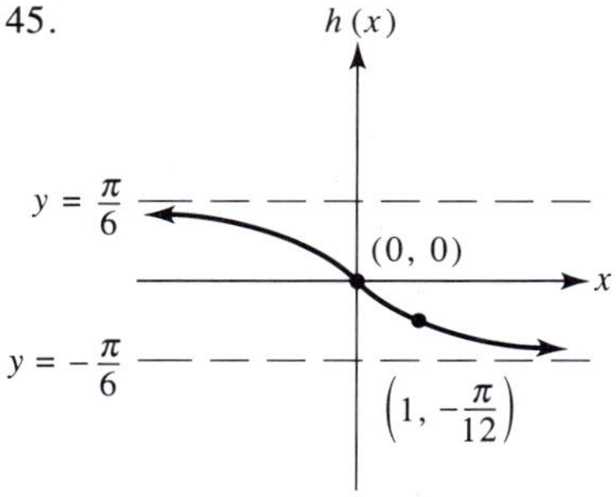

47.

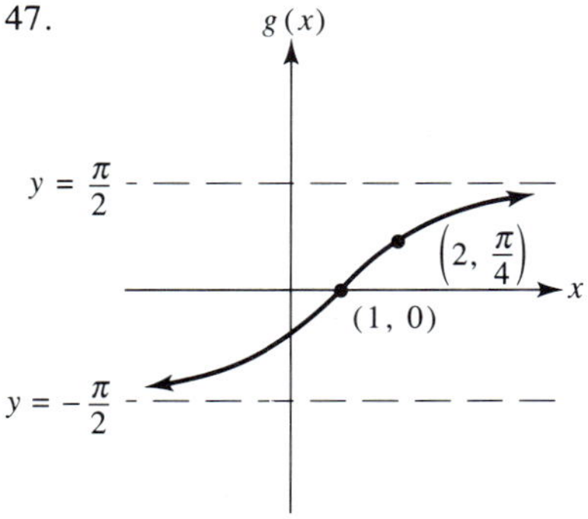

49. The graph is that of Cos^{-1} shifted 3 units left with a vertical expansion factor of 2.

51. The graph is the graph of Cos^{-1} reflected in the y-axis, shifted right 3 units with a vertical expansion of 2.

6.6

1. $\{\sqrt{3}/2\}$
3. $\{\ \}$
5. $\{\pi/3 + k\pi\},\ k \in I$
7. $\{\mathrm{Cos}^{-1}(0.3) + 2k\pi\} \cup \{-\mathrm{Cos}^{-1}(0.3) + 2k\pi\},\ k \in I$
9. $\{\ \}$
11. $\{\sqrt{3}/2\}$
13. $\{\ \}$
15. $\{\pi/15 + k\pi/5\},\ k \in I$
17. $\{\pi/3 - 1 + 2k\pi\} \cup \{-\pi/3 - 1 + 2k\pi\},\ k \in I$
19. $\{\ \}$
21. $\{7\pi/6 + 2k\pi\} \cup \{11\pi/6 + 2k\pi\} \cup \{3\pi/2 + 2k\pi\},\ k \in I$
23. $\{7\pi/6 + 2k\pi\} \cup \{11\pi/6 + 2k\pi\},\ k \in I$
25. $\{k\pi\},\ k \in I$
27. $\left\{\mathrm{Tan}^{-1}\left(\dfrac{-5 - \sqrt{13}}{6}\right) + k\pi\right\} \cup \left\{\mathrm{Tan}^{-1}\left(\dfrac{-5 + \sqrt{13}}{6}\right) + k\pi\right\},\ k \in I$
29. $\{\pi/2 + k\pi\},\ k \in I$
31. $\{\mathrm{Cos}^{-1}(\frac{1}{4}) + 2k\pi\} \cup \{-\mathrm{Cos}^{-1}(\frac{1}{4}) + 2k\pi\},\ k \in I$
33. $[\sqrt{3}/2, 1]$
35. The union of all intervals $(0 + 2k\pi, \pi + 2k\pi),\ k \in I$
37. $\{0.7227 \text{ (approx.)} + 2k\pi, -0.7227 \text{ (approx.)} + 2k\pi\},\ k \in I$
39. One solution is approximately -3.183.

Summary Exercises

1. $\left(-\dfrac{\sqrt{3}}{2}, -\dfrac{1}{2}\right)$
3. $\left(\dfrac{1}{2}, \dfrac{\sqrt{3}}{2}\right)$
5. 1
7. -2
9. $-\sqrt{3}$
11. 0.2679 (approx.)
13. 0.6625 (approx.)
15. 1.0630 (approx.)
17.

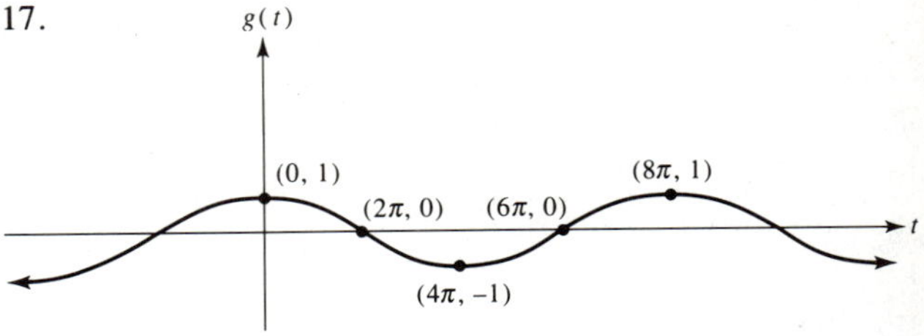

19.

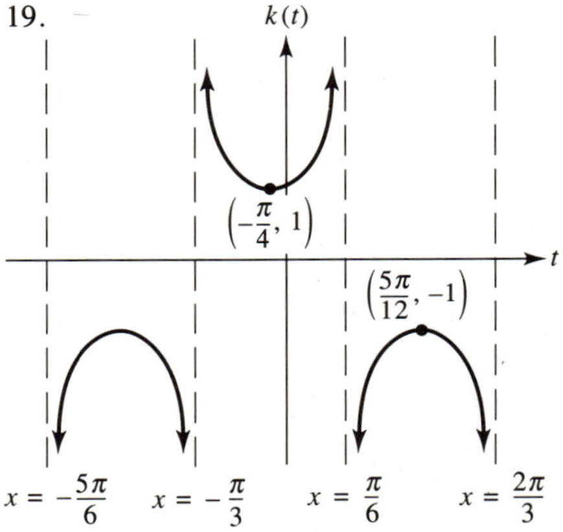

The graph continues to the left and right.

21.

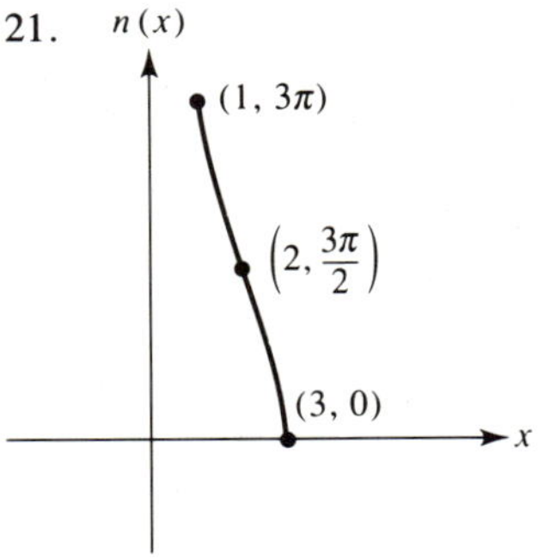

23. 2π; no; yes; no
25. π; no; no; yes
27. 2π; no; no; yes
29. None; yes; no; no
31. None; yes; no; no
33. None; yes; no; yes
35. Amplitude: 3.5; period: $\dfrac{2\pi}{2.3}$; phase shift: $\dfrac{\pi}{(8)(2.3)}$

37. Expansion factor: $\frac{5}{6}$; period: $\frac{8\pi}{3}$; phase shift: $\frac{8\pi}{15}$

39. Expansion factor: $\frac{2}{3}$; period: 10π; phase shift: $\frac{5\pi}{7}$

41. $\pi/7$

43. $\sqrt{3}$

45. $\{\mathrm{Sin}^{-1}(\ln 2) + 1 + 2k\pi\} \cup \{\pi - \mathrm{Sin}^{-1}(\ln 2) + 1 + 2k\pi\}, k \in I$

47. $\{\pi/6 + 2k\pi\} \cup \{5\pi/6 + 2k\pi\} \cup \{3\pi/2 + 2k\pi\}, k \in I$

Chapter Seven

7.1

1. $7\pi/4, -\pi/4$
3. $\pi/6, -11\pi/6$
5. $2\pi/3, -4\pi/3$
7. 0 or 2π
9. $(\sqrt{3}, 1)$
11. $(6\sqrt{2}, -6\sqrt{2})$
13. $-300°$
15. $612°/\pi$
17. 570°
19. $4\pi/3$
21. $13\pi/180$
23. $-5\pi/6$

In answers 25 through 30, triangles are not necessarily drawn to scale.

25.

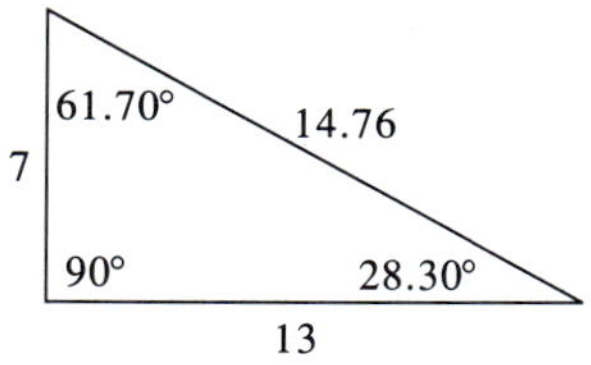

27.

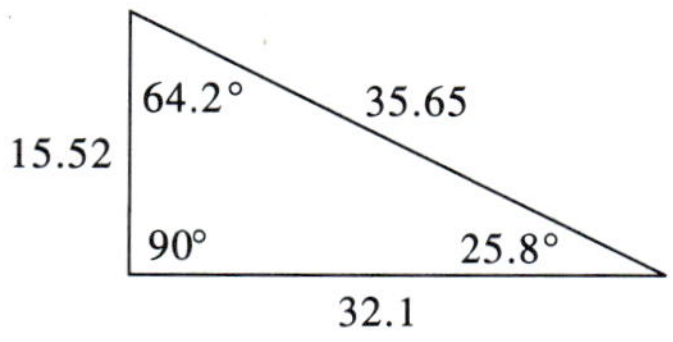

29.

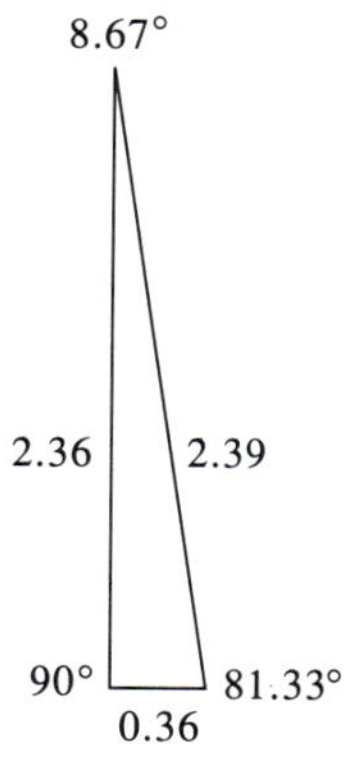

31. 25.20°
33. 488.77 feet
35. 725.13 centimeters
37. 88.02°
39. No, the tree is approximately 45.5 feet tall.
41. The angle is 1084.1 mils, to the nearest hundredth of a mil.
43. It will take 742 square feet of roofing tiles.
45. It's much shorter to travel the outside perimeter (approximately 85 miles).

7.2

In answers 1 through 13, triangles are not necessarily drawn to scale.

1. No such triangle can exist. ($\cos a = 1.15$ has no solution.)

3.

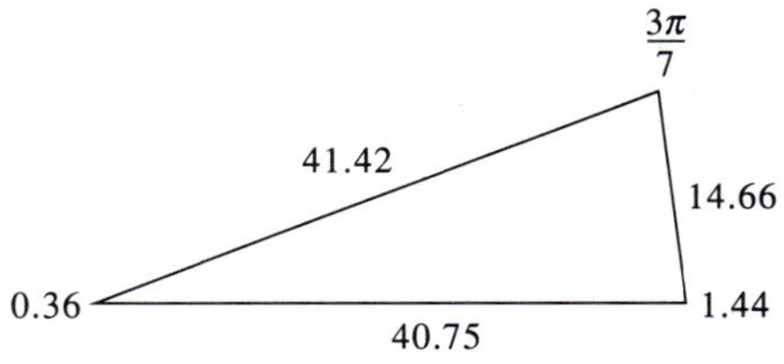

5.

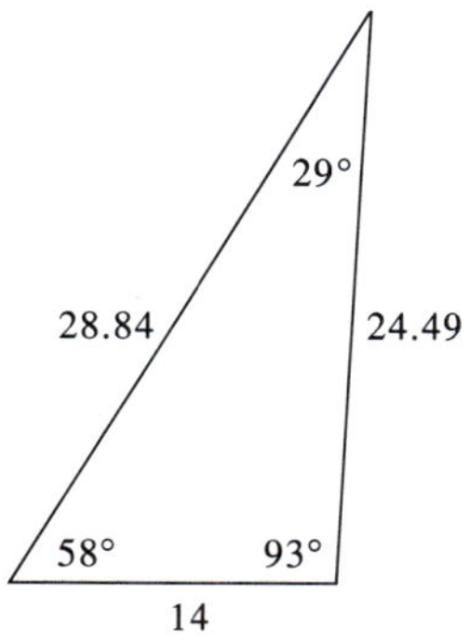

7.

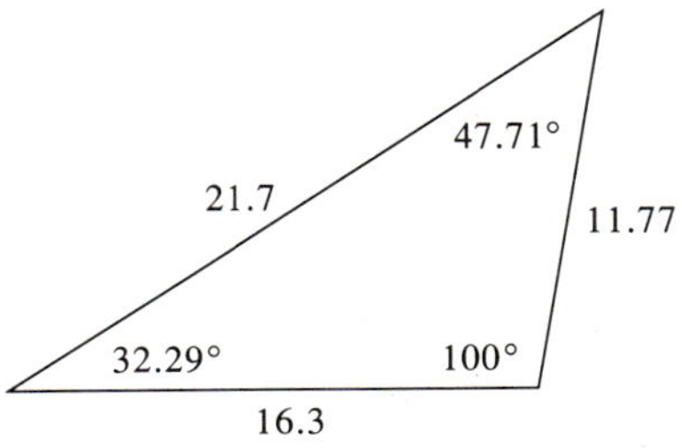

9.

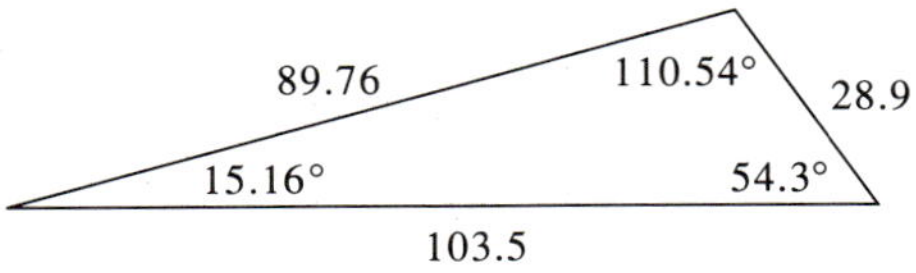

11. No such triangle exists.

13.

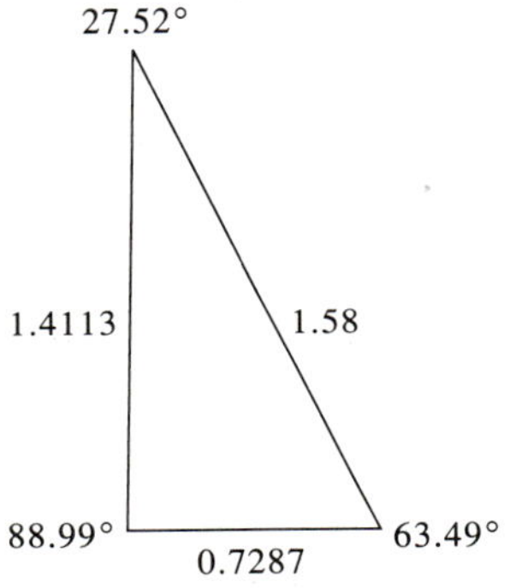

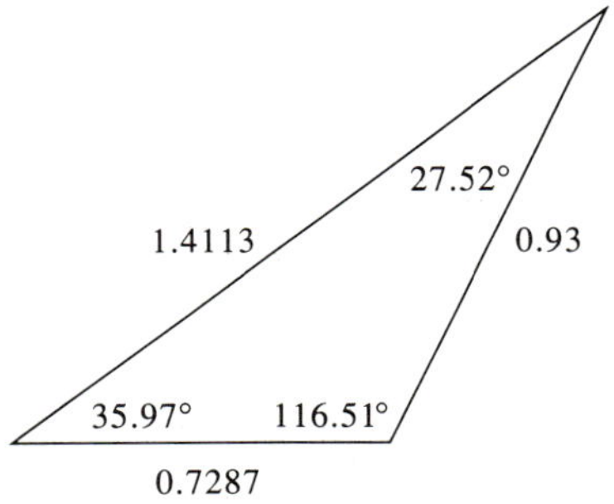

15. Two triangles
17. No such triangle exists.
19. One triangle
21. a. $38.4 \sin 13.5°$ square units
    b. approximately 9.0 square units
23. a. $1495.575 \sin 54.3°$ square units
    b. approximately 1214.5 square units
25. a. $\frac{8\pi}{3}$ square inches
    b. approximately 8.38 square inches
27. a. $\frac{18633.24\pi}{180}$ square meters
    b. approximately 325.21 square meters
29. a. $\frac{86(311.4)\pi}{180}$ square centimeters
    b. approximately 467.41 square centimeters
31. Use the law of cosines if you have three sides or if you have two sides and the angle between them. Use the law of sines if you have two angles and a side or if you have two sides and the angle opposite one of them.
33. With standard labeling, $AB = c$. Using the Pythagorean theorem, $y^2 + (AB - b\cos\alpha)^2 = a^2$.

$$y^2 + (AB - b\cos\alpha)^2 = a^2$$
$$\Leftrightarrow \quad (b\sin\alpha)^2 + (c - b\cos\alpha)^2 = a^2$$
$$\Leftrightarrow \quad b^2\sin^2\alpha + c^2 - 2bc\cos\alpha + b^2\cos^2\alpha = a^2$$
$$\Leftrightarrow \quad b^2 + c^2 - 2bc\cos\alpha = a^2$$

Law of cosines

35. They should buy 39 feet of pipe and trim off approximately $\frac{1}{3}$ foot.
37. The angles are approximately 19°, 145°, and 16°.
39. It will extend just over 25 feet. (25.0904).
41. He sailed approximately 6.566 miles.
43. The area of ΔOAD is $\frac{1}{2}\sin\theta\cos\theta$ square units.

    The area of ΔOBC is $\frac{1}{2}\cdot 1\cdot\tan\theta$, or $\frac{1}{2}\frac{\sin\theta}{\cos\theta}$, square units.

    The area of ΔOBD is $\frac{1}{2}\cdot\theta\cdot 1^2$, or $\theta/2$, square units.

    $\frac{1}{2}\sin\theta\cos\theta < \frac{1}{2}\theta < \frac{1}{2}\frac{\sin\theta}{\cos\theta}$.

7.3

1. $\frac{\cos\theta}{\sin^2\theta}$
3. $\cos(t) + 1$, $\sin(t) \neq 0$
5. $\sin^2\theta + \frac{\sin^4\theta}{\cos^2\theta}$
7. $\frac{\sqrt{6} + \sqrt{2}}{4}$
9. $\frac{\sqrt{2 - \sqrt{2}}}{2}$
11. $\frac{\sqrt{2} - \sqrt{6}}{2}$
13. $\frac{\sqrt{6} - \sqrt{2}}{4}$

15. $\dfrac{\sqrt{2+\sqrt{2}}}{2}$

17. $\dfrac{\sqrt{2-\sqrt{2}}}{2}$; $\dfrac{\sqrt{2+\sqrt{2}}\sqrt{2}-\sqrt{2-\sqrt{2}}\sqrt{2}}{4}$; $1\left(\dfrac{\sqrt{2-\sqrt{2}}}{2}\right)$

19. $\tan(\alpha+\beta) = \dfrac{\tan\alpha + \tan\beta}{1 - \tan\alpha\tan\beta}$

21. $\tan 2\theta = \dfrac{2\tan\theta}{1-\tan^2\theta}$

23. The cosine of the sum of two numbers is the product of the cosines of each number minus the product of the sines of each number.

25. The cosine of half a number is either the square root of one-half the sum of the number's cosine and one, or the opposite of this square root. The sign depends on the quadrant of half the number.

27. a. $\dfrac{\sqrt{6}-\sqrt{2}}{4}$ b. $\dfrac{\sqrt{6}-\sqrt{2}}{4} \approx 0.2588$
c. $\dfrac{\sqrt{2}-\sqrt{3}}{2} \approx 0.2588$

The approximations are the same to 12 decimal places. Since both numbers are positive, $u^2 = v^2 \Leftrightarrow u = v$ and the squares are equal.

29. They are equivalent.

31. They are not equivalent.

33. Identity

35. Identity; the left-hand side is undefined for $t = k\pi/2$, $k \in I$.

37. Identity

39. Identity

41. Identity

43. Identity

45. Neither

47. For the expressions in exercise 45 there are two distinct graphs. For exercises 33 and 40 they appear the same. Holes might not show up.

7.4

1. $2\sqrt{1-x^2}\,x,\ 0 \le x \le 1;\ -2\sqrt{1-x^2}\,x,\ -1 \le x < 0$

3. $\dfrac{3\sqrt{39}}{32} + \dfrac{5\sqrt{7}}{32}$

5. $\dfrac{12\sqrt{5}}{39} + \dfrac{10}{39}$

7. $-2\sin(4\theta)\sin(2\theta)$

9. $2\cos(7\alpha)\sin(4\alpha)$

11. $2\cos(3\theta)\sin(-2\theta)$

13. $\frac{1}{2}(\sin 5\alpha - \sin 3\alpha)$

15. $\frac{3}{2}[\cos 6t + \cos(-2t)]$

17. $\frac{1}{2}(\sin 7x + \sin x)$ (c) Since $\sin(-x) = -\sin(x)$,
$\frac{1}{2}[\sin 7x - \sin(-x)]$ (d) these are the same.

19. $\sin(t+u) - \sin(t-u) = \sin t\cos u + \sin u\cos t - \sin t\cos u + \sin u\cos t$, or $2\sin u\cos t$

21. $\left\{\dfrac{\pi}{6} + 2j\pi\right\} \cup \left\{-\dfrac{\pi}{6} + 2k\pi\right\} \cup \{(2m+1)\pi\}$, $j, k, m \in I$

23. $\left\{\dfrac{\pi}{4} + k\pi\right\}$, $k \in I$

25. $\left\{(2k+1)\dfrac{\pi}{2}, (2k+1)\dfrac{\pi}{4}, k\pi\right\}$, $k \in I$

27. $\left\{\dfrac{\pi}{4} + k\pi\right\}$, $k \in I$

29. $\{\ \}$

31. The maximum distance moved is approximately 4.47 inches.

33. The approximate maximum distance is 2(2.236) inches.

7.5

1. $(\sqrt{5}, \pi/4)$

3. $(10, 7\pi/6)$

5. $\left(\dfrac{1}{2}, \dfrac{2\pi}{3}\right)$

7. $(\sqrt{13}, \mathrm{Tan}^{-1}(\frac{2}{3}))$

9. $\cos(\pi/2) + i\sin(\pi/2)$

11. $\sqrt{2}[\cos(3\pi/4) + i\sin(3\pi/4)]$

13. $2[\cos(5\pi/3) + i\sin(5\pi/3)]$

15. $4[\cos(7\pi/6) + i\sin(7\pi/6)]$

17. $12(\cos(5\pi/6) + i\sin(5\pi/6)]$

19. $\frac{3}{4}[\cos(-10°) + i\sin(-10°)]$

21. $-\sqrt{\frac{1}{2}} - \sqrt{\frac{1}{2}}i$

23. $-8\sqrt{3} + 8i$

25. $\dfrac{1}{4^6} - \dfrac{\sqrt{3}}{4^6}i$

27. $\sqrt[8]{32}[\cos(\pi/16) + i\sin(\pi/16)]$, $\sqrt[8]{32}[\cos(9\pi/16) + i\sin(9\pi/16)]$, $\sqrt[8]{32}[\cos(17\pi/16) + i\sin(17\pi/16)]$, $\sqrt[8]{32}[\cos(25\pi/16) + i\sin(25\pi/16)]$

29. $\dfrac{\sqrt{2}}{2} + i\dfrac{\sqrt{6}}{2}, -\dfrac{\sqrt{2}}{2} - i\dfrac{\sqrt{6}}{2}$

31. $\cos\dfrac{\pi}{5} + i\sin\dfrac{\pi}{5}, \cos\dfrac{3\pi}{5} + i\sin\dfrac{3\pi}{5}, -1, \cos\dfrac{7\pi}{5} + i\sin\dfrac{7\pi}{5}, \cos\dfrac{9\pi}{5} + i\sin\dfrac{9\pi}{5}$

33.

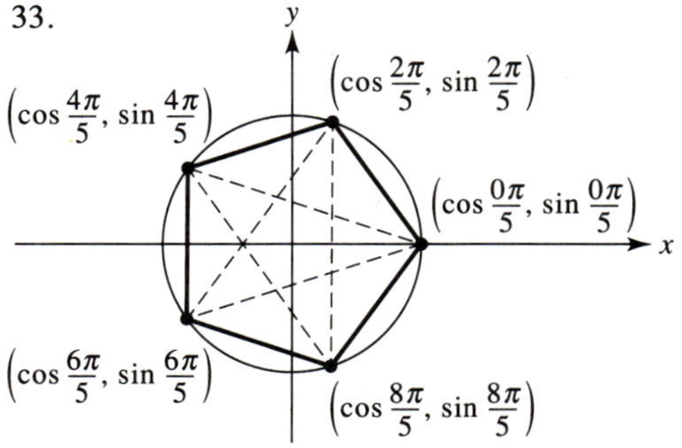

35. $\{i, 1 - i\}$

37.

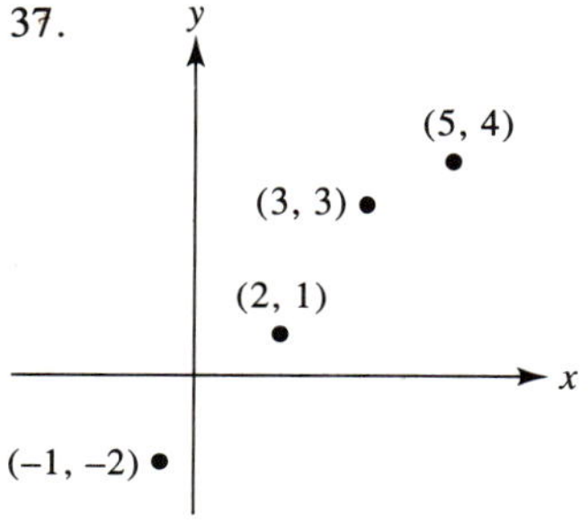

39.

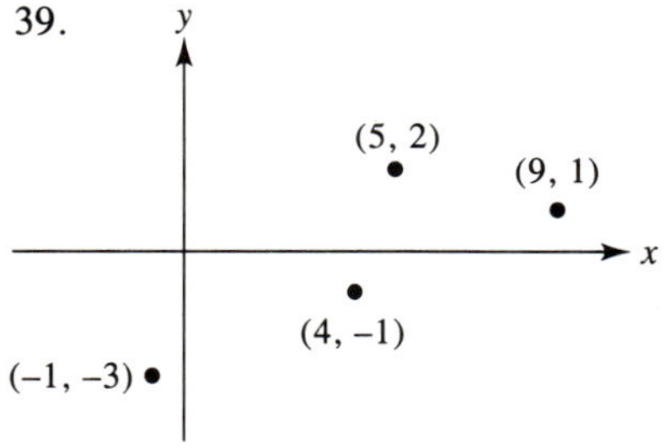

41.

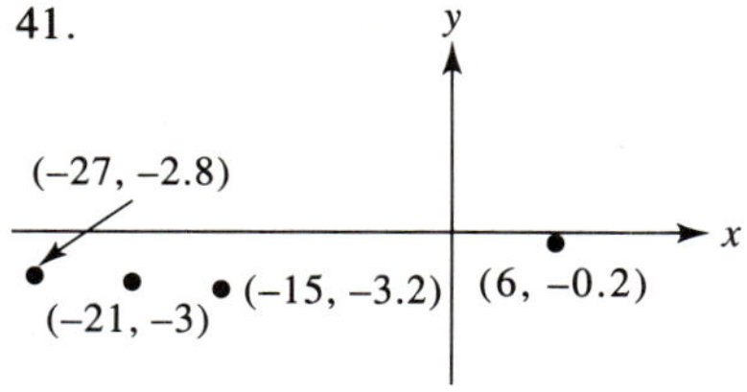

43. <5, 6>

45. <9, 1>

47. <−15, −3.2>

49.

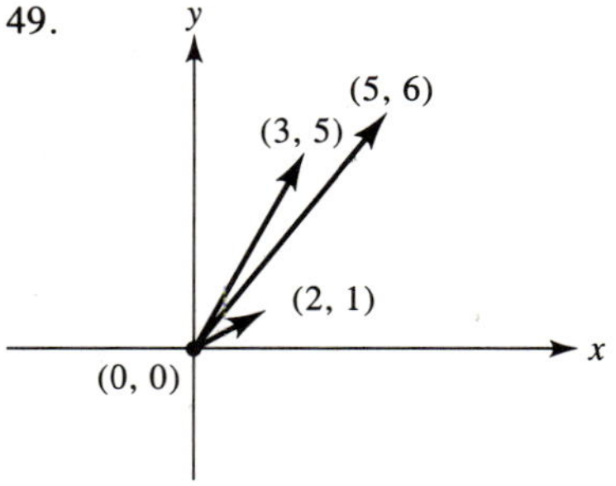

51.

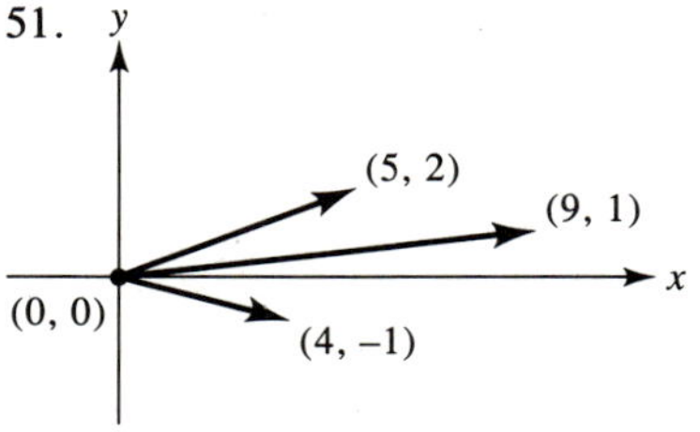

53.

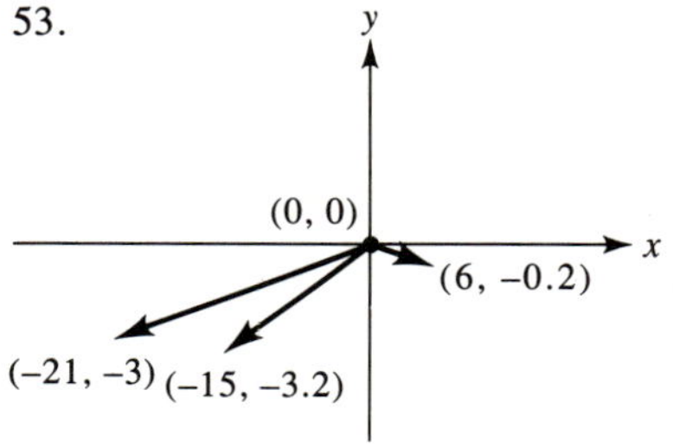

55. If we use the original two vectors as two sides of a parallelogram, the sum vector corresponds to the diagonal of this parallelogram.

Summary Exercises

1. 67.5°

3. $(6246/\pi)^\circ$

5.

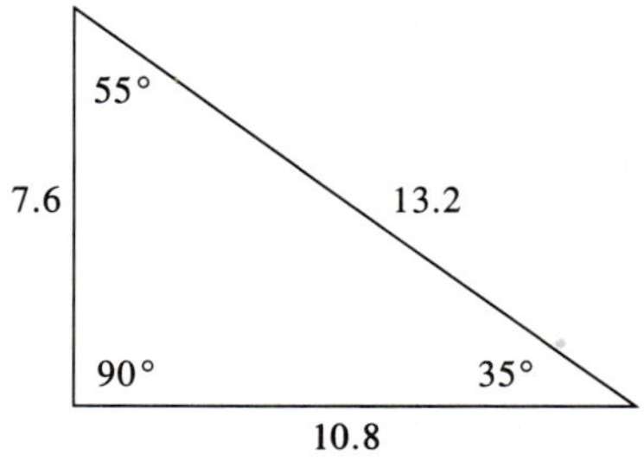

7. 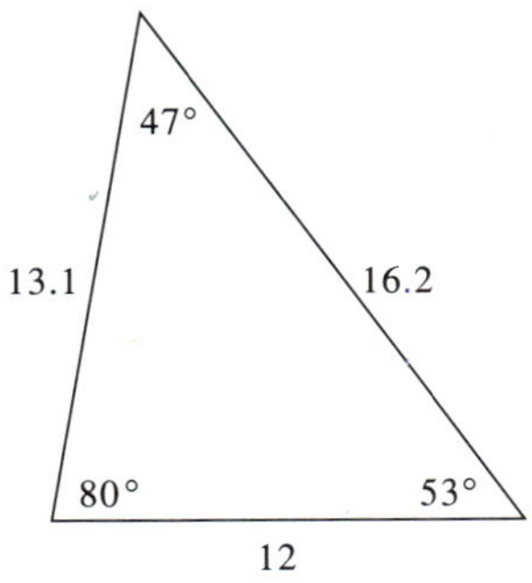

9. 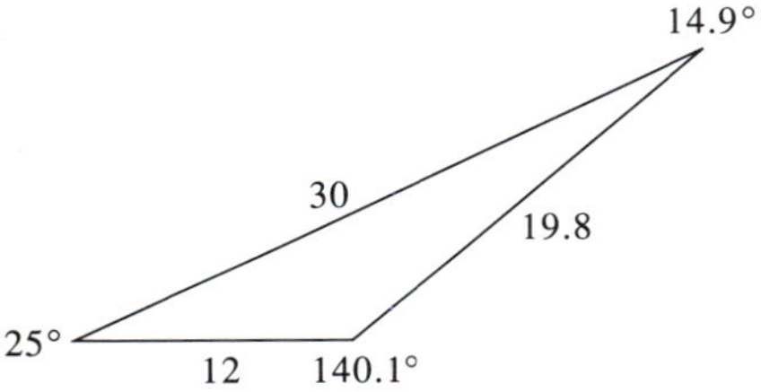

11. No such triangle exists.

13. 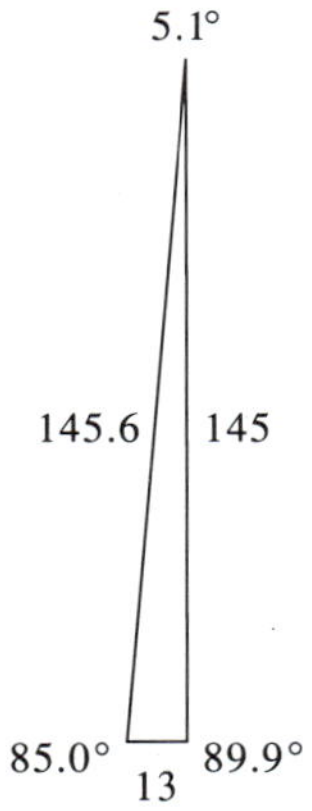

15. 355.2 square feet

17. a. $(\sqrt{6} - \sqrt{2})/4$ b. $\sqrt{2 - \sqrt{3}}/2$ c. $\sin \pi/12$

19. Identity

21. Neither

23. Identity

25. Identity; no

27. $-2(\sin 6x)(\sin 2x)$

29. $\{\pi/6 + k\pi, \pi/3 + k\pi\}, k \in I$

31. $\left\{\dfrac{\text{Sin}^{-1}0.2}{2} + k\pi, \dfrac{\pi}{4} - \dfrac{\text{Sin}^{-1}0.2}{2} + k\pi\right\}, k \in I$

33. The union of the intervals $[\pi/6 + k\pi, 2\pi/6 + k\pi]$, $k \in I$

35. $\{7\pi/6 + 2k\pi, 11\pi/6 + 2k\pi\}, k \in I$

37. $3(\cos 7\pi/4 + i \sin 7\pi/4)$

39. $\cos \pi/24 + i \sin \pi/24$, $\cos 13\pi/24 + i \sin 13\pi/24$, $\cos 25\pi/24 + i \sin 25\pi/24$, $\cos 37\pi/24 + i \sin 37\pi/24$

41. 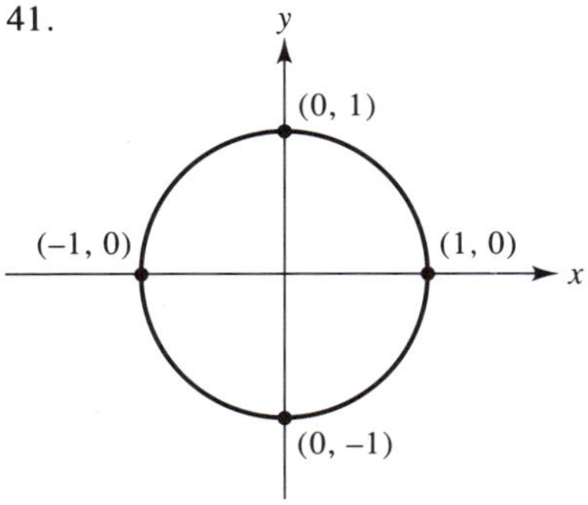

Chapter Eight

8.1

1. $(x + 1)(x^2 + 2x + 1) + 0$; D is a factor of P.
3. $(x + 4)(2x^4 - 5x^3 + 7x^2 - 28x + 42) - 160$
5. $(x - 3)(x^5 + 3x^4 + 9x^3 + 27x^2 + 81x + 243) + 769$
7. $\left(x + \frac{1}{2}\right)\left(x^2 + \frac{5}{2}x + \frac{7}{4}\right) + \frac{1}{8}$
9. $(x - 4i)(2ix + (-9 + 3i)) - 10 - 41i$
11. $x^2 + 2x + 1$
13. $2x^4 - 5x^3 + 7x^2 - 28x + 42 - \dfrac{160}{x + 4}$
15. $x^5 + 3x^4 + 9x^3 + 27x^2 + 81x + 243 + \dfrac{769}{x - 3}$
17. $x^2 + \dfrac{5}{2}x + \dfrac{7}{4} + \dfrac{\frac{1}{8}}{x + \frac{1}{2}}$
19. $2ix + (-9 + 3i) - \dfrac{10 + 41i}{x - 4i}$
21. 4
23. -217
25. 89.031
27. $-3i$
29. $\dfrac{41}{9}$
31. Perhaps exercise 21 and exercise 27
33. Have someone try out your instructions.
35. $Q = 3x - 3$, $R = 23$
37. $Q = \frac{1}{5}x^2 - \frac{8}{25}x - \frac{291}{125}$, $R = \frac{1668}{125}$
39. $Q = 4x^2 - 6x + 9$, $R = 0$
41. Perform $(x^4 + 4x^3 + 3x^2 - x + 5) \div (x + 2)$ or find $f(-2)$ for $f(x) = x^4 + 4x^3 + 3x^2 - x + 5$.

8.2

1. $\pm 1, \pm 5$
3. $\pm 1, \pm 2, \pm 3, \pm 4, \pm 6, \pm 12, \pm \frac{1}{3}, \pm \frac{2}{3}, \pm \frac{4}{3}$
5. $-2, -3, 1$
7. 5
9. $-4, 0, \frac{1}{3}, 3$
11. $-2 - \sqrt{2}i, -2 + \sqrt{2}i$
13. $1, \dfrac{-1 - \sqrt{11}i}{2}, \dfrac{-1 + \sqrt{11}i}{2}$
15. $0, -i, i, -\sqrt{2}i, \sqrt{2}i$
17. $-4; \frac{2}{3}; 0; 2; 0; 0$
19. $6; \frac{7}{3}; 3; -16; 0; 8$
21. $P(x) = (x - 1)(x + 3)(x + 5); I$
23. $R(x) = x(x + \frac{1}{2})(x - \frac{1}{3})(x - 2); Q$
25. $T(x) = (x^2 - 3)(x^2 - 1); I$
27. $V(x) = (x^2 - 10x + 34)(x - \sqrt{2})(x + 1); R$
29. $X(y) = y(y + i)(y - \sqrt{3})(y + 1); C$
31. $f_1(x) = (x - 1)(x + 3)(x - 5)$, $f_2(x) = 5(x - 1)(x + 3)(x - 5)$, and $f_3(x) = -2(x - 1)(x + 3)(x - 5)$ is one possible collection of functions.
33. $f(x) = -3(x + 1)(x - 2)x$
35. $f(x) = -8(x + 1)^3(x - 6)$
37. $\left\{\frac{\pi}{2} + 2k\pi\right\} \cup \left\{\text{Sin}^{-1}\left(\frac{5 - \sqrt{37}}{2}\right) + 2k\pi\right\} \cup \left\{\pi - \text{Sin}^{-1}\left(\frac{5 - \sqrt{37}}{2}\right) + 2k\pi\right\}, k \in I$
39. $(-\infty, 0] \cup [\frac{1}{8}, +\infty)$
41. $\{c, 3, -2\}$

8.3

Your graphs in this chapter may differ slightly from the answers given. However, yours should have the same x-intercepts, asymptotes, and general shape.

1.

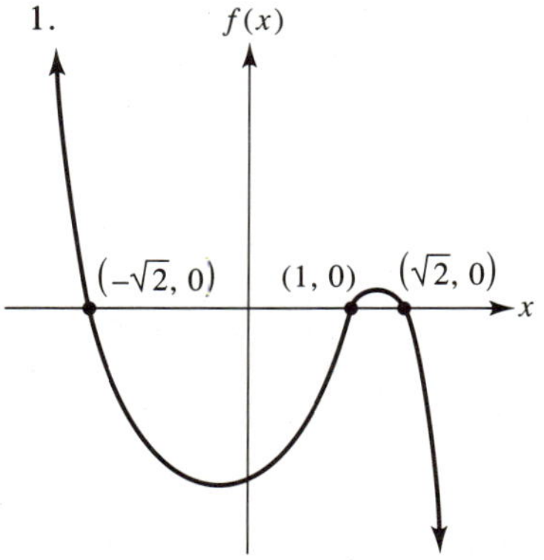

3.

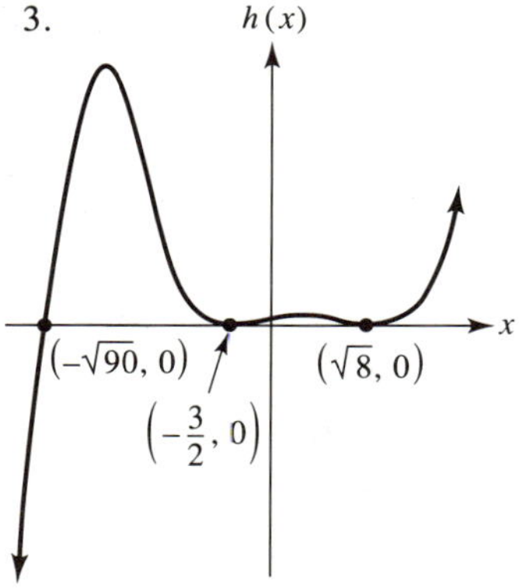

5.

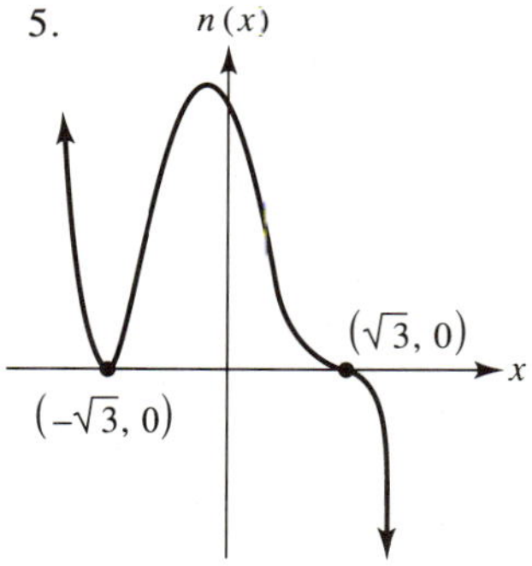

7.

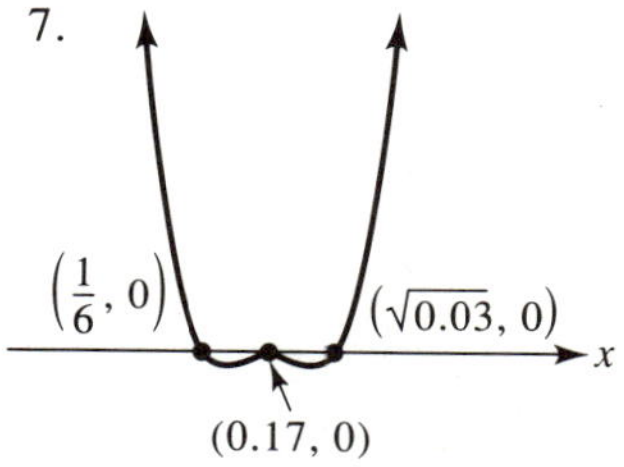

9.

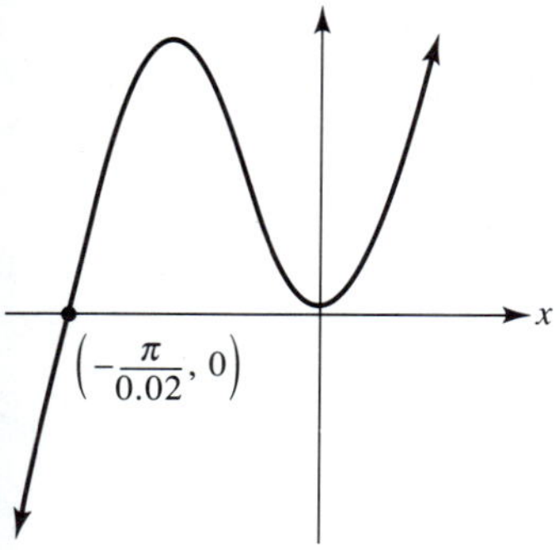

11.

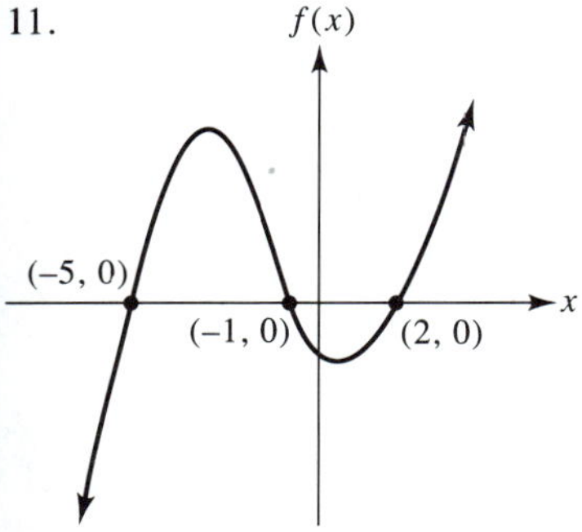

13.

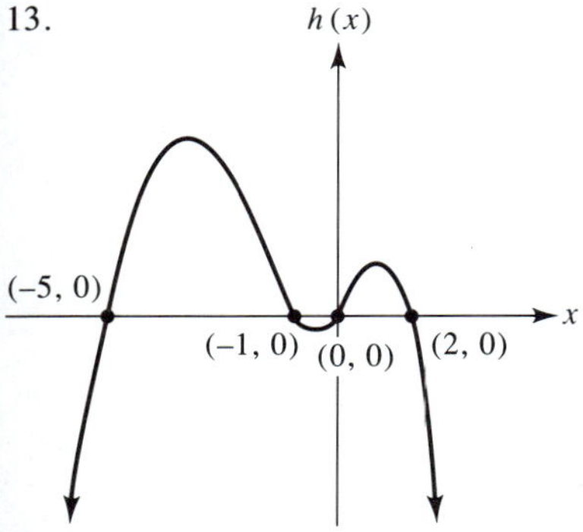

15.

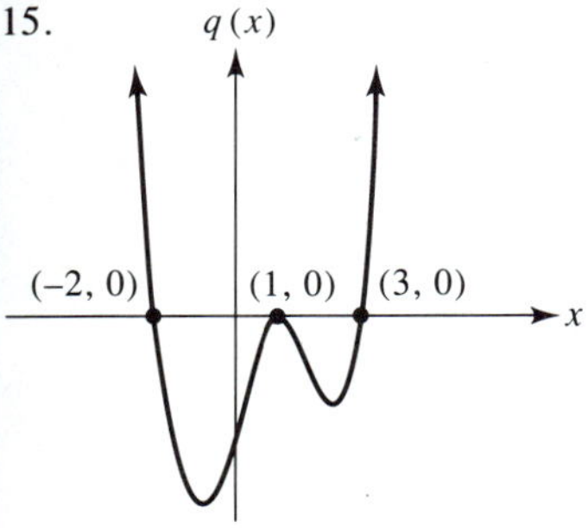

17.

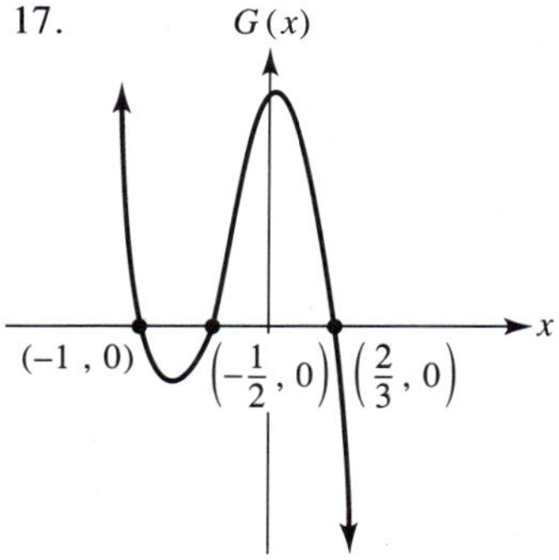

19.

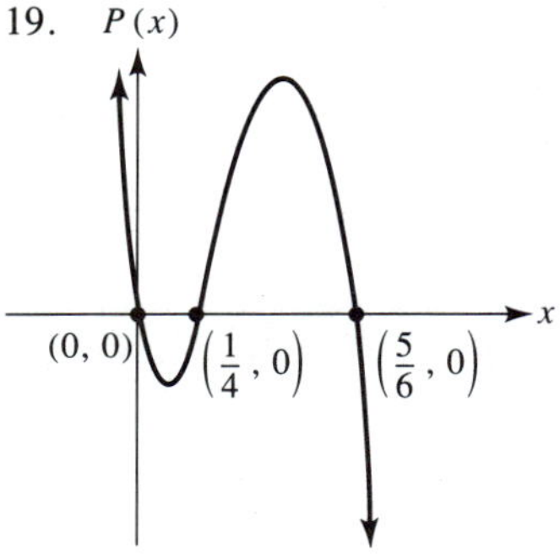

21.

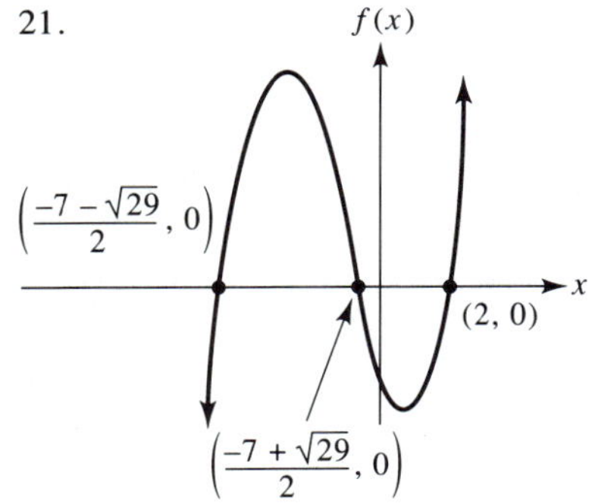

23.

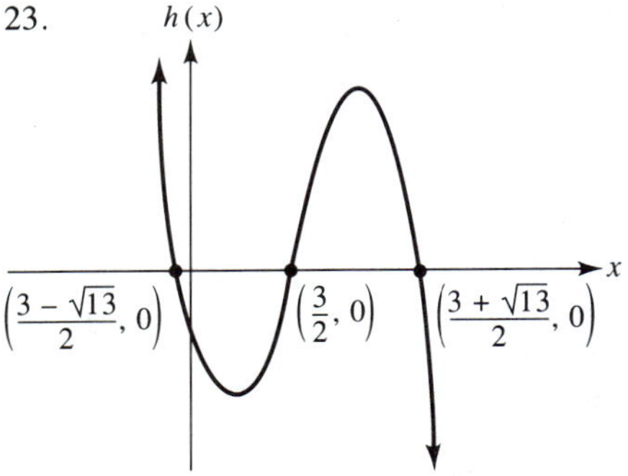

25.

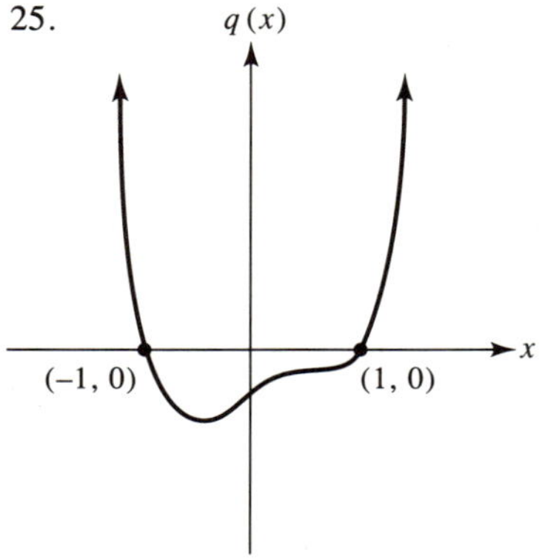

27.

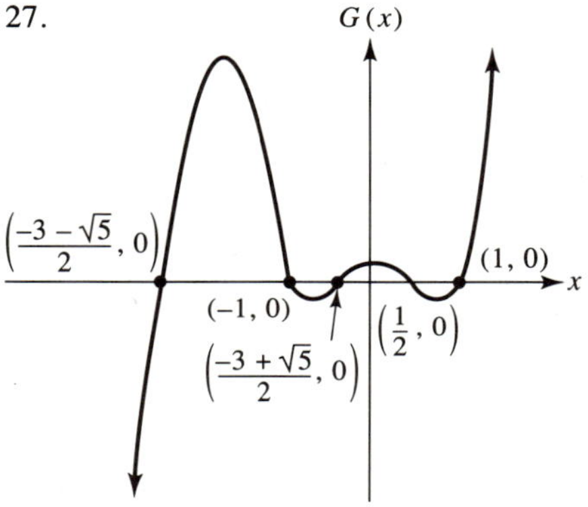

29.

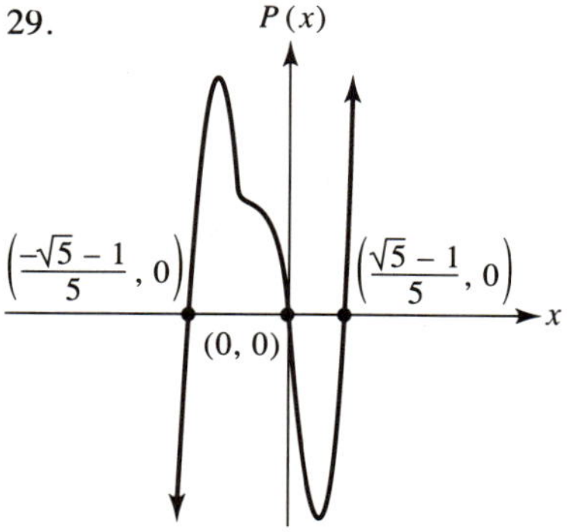

31. 2.24

33. $y = (x + 4)(x - 1)^2$

35. $y = -(x + \pi)(x + 0.9)(x - \pi)$

37.

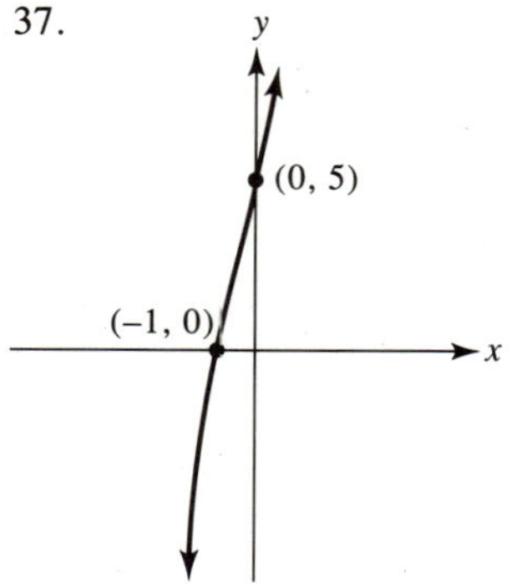

39.

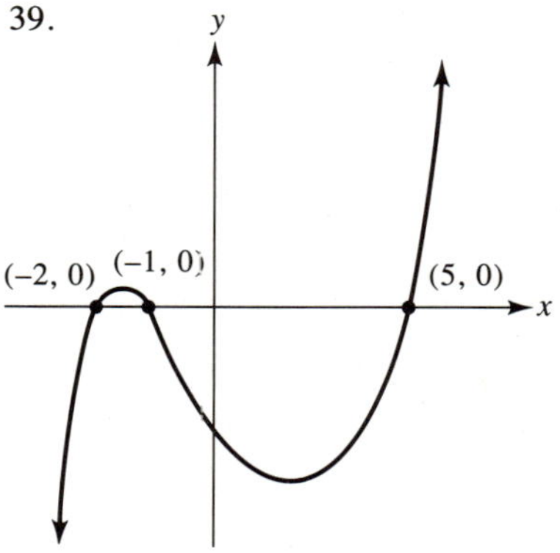

41.

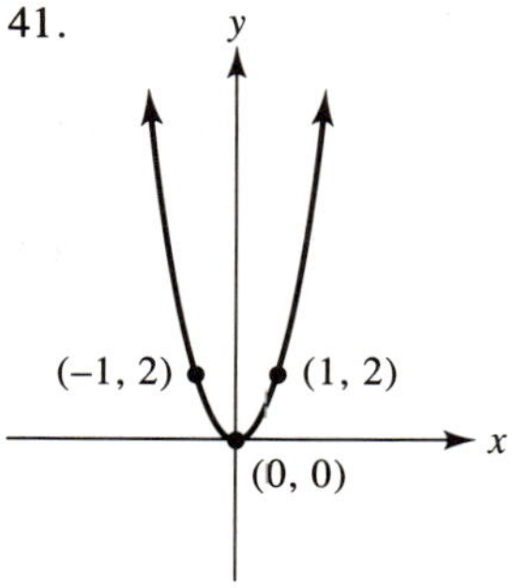

43.

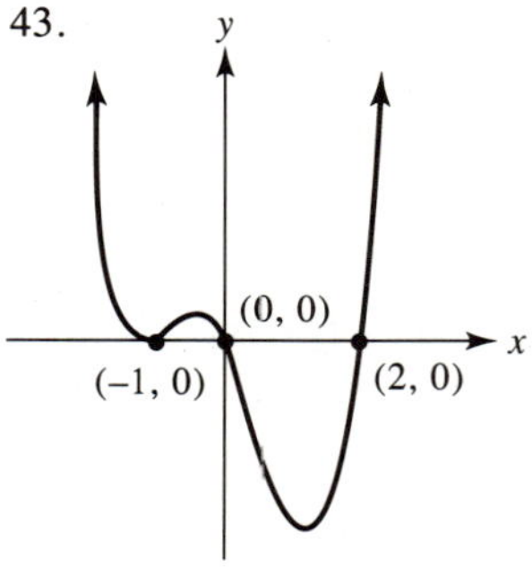

45. a. The graph of g has at most 6 x-intercepts, has range $(-\infty, c)$, changes directions at most 5 times, is smooth and continuous, and opens downward.

b. The graph of h has at most 17 x-intercepts, has range $(-\infty, +\infty)$, changes direction at most 16 times, is smooth and continuous, and goes uphill to the right and downhill to the left.

47. Did you describe the number of x-intercepts, the range, the number of changes in direction, the fact that the graph is smooth and continuous, and when it opens up or down?

8.4

1.

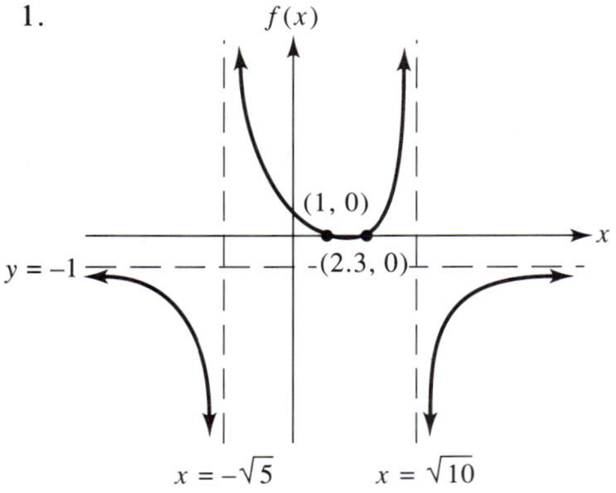

3.

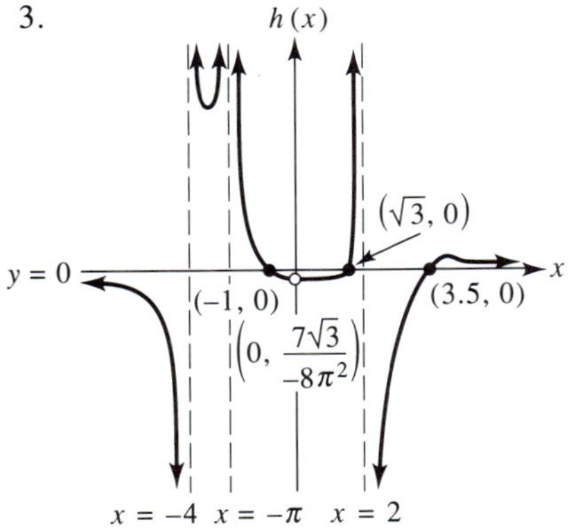

5.

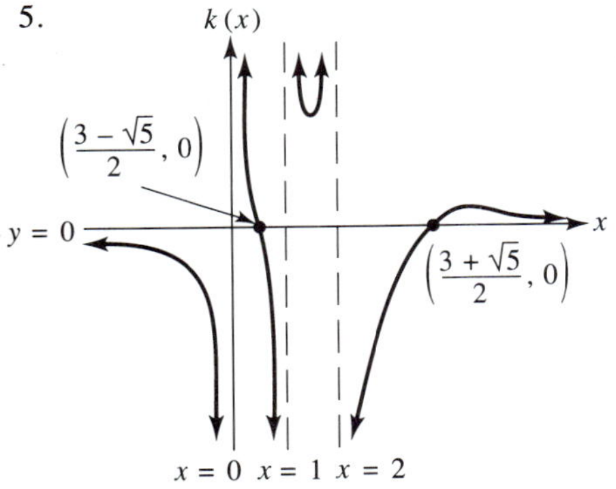

7.

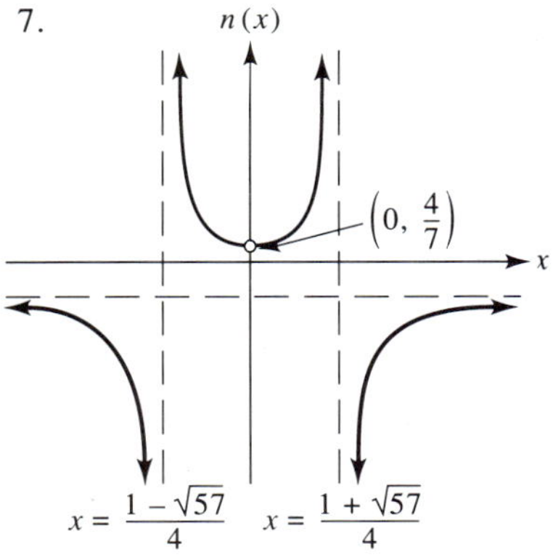

9.

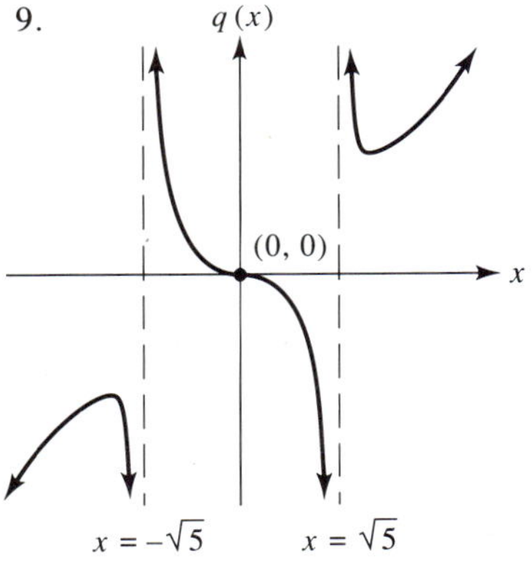

$y = x$

11.

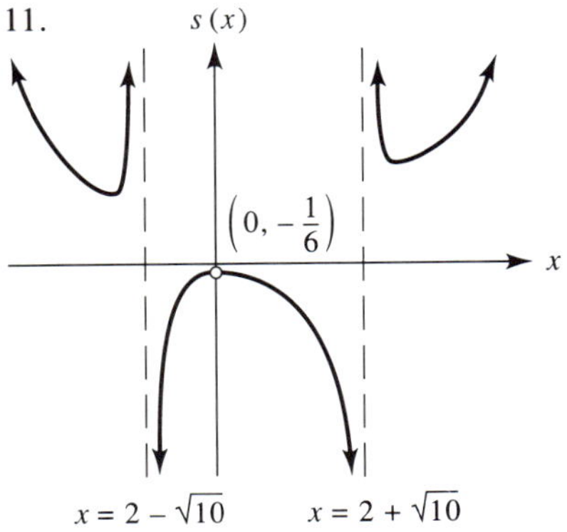

$y = x^2 + 4x + 22$

13.

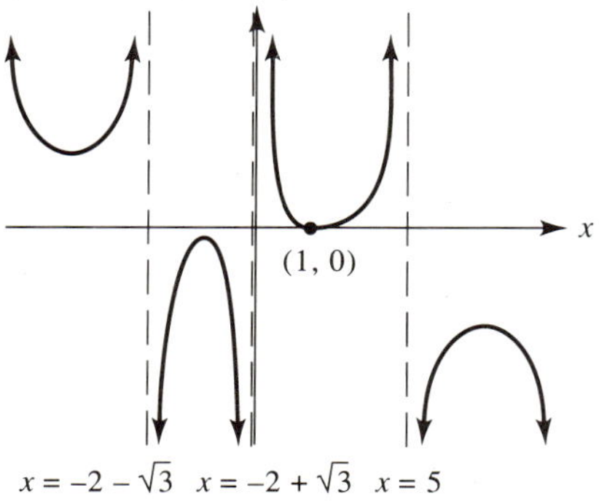

$y = -x^3 - x^2 - 20x - 42$

15. $f(473) \approx 1908.111$

$f_{\text{asym}}(473) = 1908.0$

17. $f(473) \approx 279{,}892.812$

$f_{\text{asym}}(473) = 279{,}902.668$

19. a. $y = 2x + 2$ b. $f(1000) \approx 2007.0352$

c. $y = 2002$

21. $(-\infty, -4), (-2, 1), (3, +\infty)$

23. $(-8, -1), (0, 8), (27, +\infty), (125, +\infty)$

25. $(0, 1), (625, +\infty)$

Summary Exercises

1. No, it does not divide the polynomial evenly.

3. -648

5. $-1, \frac{1}{2}, \sqrt{5}/3, -\sqrt{5}/3$

7. $0, \pm 1, \pm 3, \pm 5, \pm 15, \pm\frac{1}{2}, \pm\frac{3}{2}, \pm\frac{5}{2}, \pm\frac{15}{2}, \pm\frac{1}{3}, \pm\frac{5}{3}, \pm\frac{1}{6}, \pm\frac{5}{6}$

9. $\pm 1, \pm 2, \pm 3, \pm 6, \pm 9, \pm 18, \pm\frac{1}{2}, \pm\frac{3}{2}, \pm\frac{9}{2}, \pm\frac{1}{4}, \pm\frac{3}{4}, \pm\frac{9}{4}$

11. 3

13. -2.1

15. $g(x) = \frac{37}{6}(3x^4 - 13x^3 + 41x^2 - 5x - 26)$

17. $h(x) = -0.5(x^4 - 16.1x^3 + 58x^2 + 48.3x - 183)$

19.

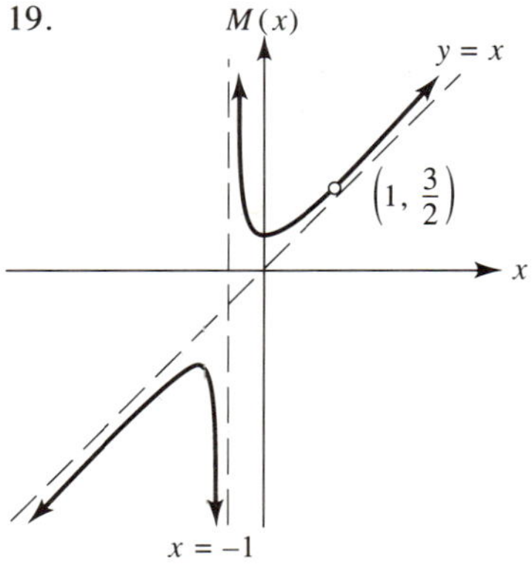

21. $(-\infty, -(\sqrt{3})^5), ((\frac{2}{5})^5, (\sqrt{3})^5)$

23. $y = 18x^3 + 33x$

25. $\{\pi/2 + 2k\pi\} \cup \{\text{Sin}^{-1}(-\frac{1}{3}) + 2k\pi\} \cup \{\pi - \text{Sin}^{-1}(-\frac{1}{3}) + 2k\pi\}, k \in I$

27. $[0, \sqrt{3}/2)$

29. A fifth-degree polynomial function could have 1, 2, 3, 4, or 5 x-intercepts.

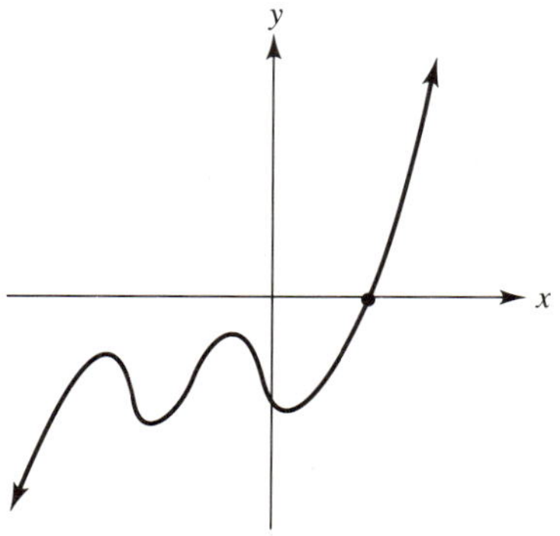

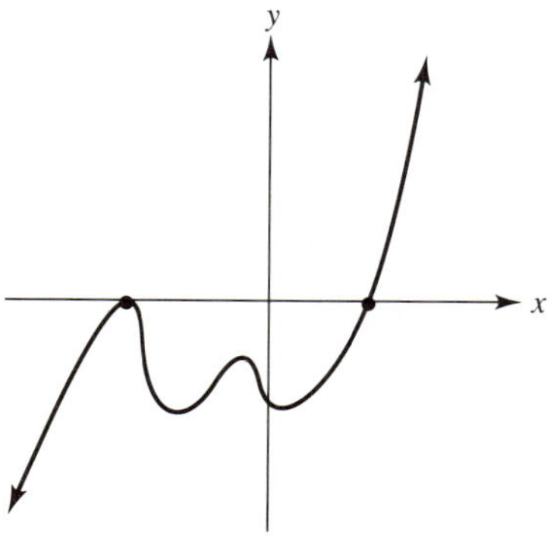

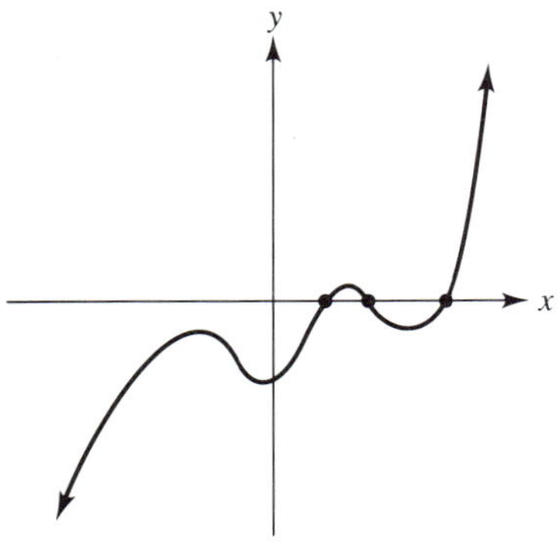

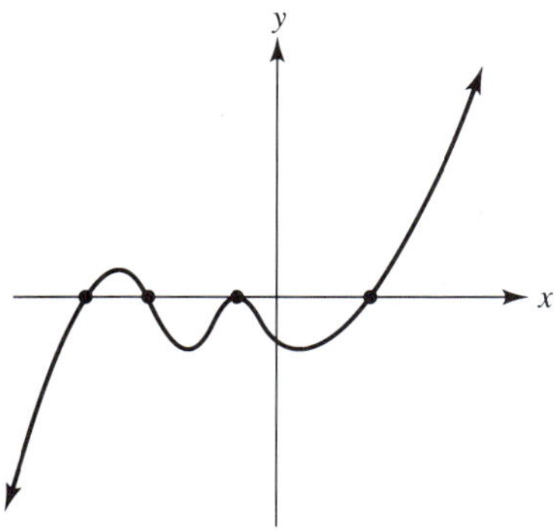

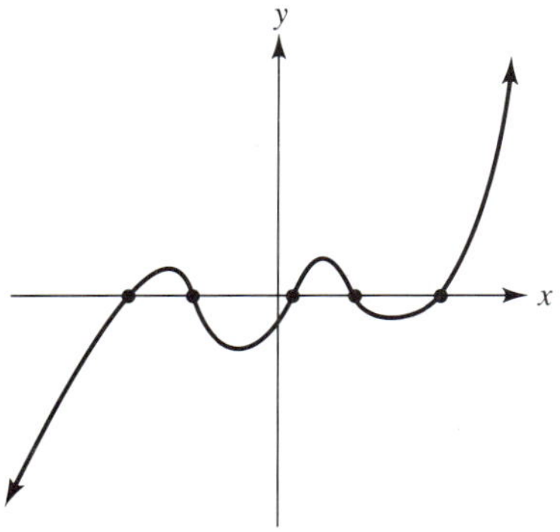

31. $\frac{3}{2}, -\frac{1}{3}, \sqrt{3}, -\sqrt{3}$
33. $-4{,}259{,}398{,}621$

Chapter Nine

9.1

1. $\left\{\left(\frac{19}{13}, -\frac{14}{13}\right)\right\}$
3. $\left\{\left(\frac{33}{4}, \frac{27}{16}\right)\right\}$
5. $\{(a, b) \mid a = 4 + 2b, b \in \mathbb{R}\}$
7. $\{\ \}$
9. No solutions
11. Infinitely many solutions
13. One solution
15. Infinitely many solutions
17. $\left\{\left(\frac{7}{3}, -\frac{1}{3}, \frac{3}{3}\right)\right\}$
19. $\{\ \}$
21. If a is $-\frac{11}{34}$, then b is $\frac{61}{34}$.
23. $k \neq 8$
25. It is a line.
27. $\{(1.3, 12.8, -37.0)\}$
29. $\{\ \}$
31. $\{\ \}$
33. $\{(10.04, -1.79)\}$

9.2

1. $$\left[\begin{array}{rrr|r} 3 & -4 & 1 & -8 \\ 7 & 2 & -3 & 13 \\ -1 & -1 & 0 & 24 \end{array}\right]$$
3. $$\left[\begin{array}{lrr|r} 2 & -4 & 1 & 1 \\ 1 & -\pi & -7 & 13 \\ 0 & 0 & 1-\sqrt{3} & 0 \end{array}\right]$$
5. $$\left[\begin{array}{lrr|r} 1 & -3 & 8 & 14 \\ 0 & 1 & -\frac{5}{2} & \frac{7}{2} \\ 0 & 6 & 4 & -3 \end{array}\right] \quad -\tfrac{1}{2}R_2$$
7. $$\left[\begin{array}{llrr|r} 1 & 0 & 4 & 8 & 3 \\ 0 & 1 & -2 & 1 & 1 \\ 0 & 0 & 1 & 7 & 3 \\ 0 & 0 & -5 & 5 & 6 \end{array}\right] \quad R_3 + R_4$$
9. $$\left[\begin{array}{lrr|r} 1 & -3 & 8 & 14 \\ 0 & -26 & 0 & 43 \\ 0 & 6 & 4 & -3 \end{array}\right] \quad 4R_2 - 5R_3$$
11. $$\left[\begin{array}{llrr|r} 1 & 0 & 4 & 8 & 3 \\ 0 & 1 & -2 & 1 & 1 \\ 0 & 0 & 1 & 2 & -3 \\ 0 & 0 & 0 & 15 & -9 \end{array}\right] \quad R_4 + 5R_3$$
13. $\{(4.2, -3.1)\}$
15. $\left\{\left(\frac{2}{5}, \frac{4}{5}, 0\right)\right\}$
17. $\{(4, 8 - 3r, r), r \in \mathbb{R}\}$
19. $\{\ \}$
21. $\left\{\left(-\frac{5}{8} - \frac{1}{8}r, \frac{7}{8} - \frac{3}{8}r, r\right) r \in \mathbb{R}\right\}$

23. $\left[\begin{array}{rr|r} 2 & -3 & 18 \\ 1 & 6 & 4 \end{array}\right]$; $\{(8, -\frac{2}{3})\}$

25. $\left[\begin{array}{rrr|r} 1 & 1 & -1 & 4 \\ 2 & -1 & 1 & 8 \\ -1 & 1 & 3 & 5 \end{array}\right]$; $\{(4, \frac{9}{4}, \frac{9}{4})\}$

27. $\left[\begin{array}{rr|r} -\frac{2}{3} & 1 & 6 \\ 2 & -3 & 18 \end{array}\right]$; $\{\ \}$

29. $\left[\begin{array}{rrr|r} 5 & -4 & 1 & 3 \\ 4 & 1 & -4 & -2 \\ 0 & 1 & -1 & 1 \end{array}\right]$; $\{(10, \frac{46}{3}, \frac{43}{3})\}$

31. An equivalent equation is produced if each side of an equation is multiplied by a nonzero number.

33. It took longer because fractions were involved.

35. $\{(\frac{1}{8}, -\frac{3}{2})\}$

37. $\{e^4, e^{10}, e^{10}\}$

39. $$\left[\begin{array}{cc|c} a & b & e \\ c & d & f \end{array}\right] \to \left[\begin{array}{cc|c} 1 & \dfrac{b}{a} & \dfrac{e}{a} \\ 0 & d - \dfrac{cb}{a} & f - \dfrac{ce}{a} \end{array}\right] \to$$

$$\left[\begin{array}{cc|c} 1 & 0 & \dfrac{de - bf}{ad - bc} \\ 0 & 1 & \dfrac{af - ce}{ad - bc} \end{array}\right]$$; the result is the same.

41. $\left\{\left(\dfrac{1344.51}{73.12}, -\dfrac{1506.11}{73.12}\right)\right\}$

43. The method fails; $D = 0$. In fact, there are no solutions.

9.3

1. 3

3. -4.2

5. 3×2

7. Not defined

9. $\begin{bmatrix} -3.5 & 42.0 & 0.0 \\ 9.1 & 0.0 & -29.4 \end{bmatrix}$

11. 6

13. $\begin{bmatrix} -0.5 & 6.0 & 0.0 \\ 5.7 & -6.0 & -16.8 \\ -3.6 & 12.0 & 8.4 \end{bmatrix}$

15. $\begin{bmatrix} 2 & 4 & 4 \\ -2 & 5 & 5 \\ 0 & 7 & 1 \end{bmatrix}$

17. $\begin{bmatrix} -6 & 8 & 3 \\ -4 & 5 & -10 \\ 5 & -1 & -8 \end{bmatrix}$

19. Not defined

21. $\begin{bmatrix} 34 & -4 \\ 38 & 13 \\ 26 & 13 \end{bmatrix}$

23. Not defined

25. $\begin{bmatrix} -7 & -12 \\ 7 & 20 \\ 3 & 16 \end{bmatrix}$

27. $\begin{bmatrix} \frac{2}{10} & -\frac{4}{10} & \frac{2}{10} \\ \frac{1}{40} & \frac{3}{40} & \frac{6}{40} \\ \frac{3}{10} & -\frac{1}{10} & -\frac{2}{10} \end{bmatrix}$

29. $\begin{bmatrix} \frac{1}{95} & \frac{13}{95} & -\frac{17}{95} \\ \frac{2}{95} & \frac{9}{380} & \frac{13}{760} \\ \frac{14}{95} & -\frac{8}{95} & -\frac{1}{190} \end{bmatrix}$

31. $(M^t)^t = \begin{bmatrix} 2 & 2 \\ 3 & 1 \\ 4 & 7 \end{bmatrix}^t = \begin{bmatrix} 2 & 3 & 4 \\ 2 & 1 & 7 \end{bmatrix} = M$

33. $\begin{bmatrix} 10 & 9 & 29 \\ -2 & -3 & -4 \end{bmatrix}$

35. If both $(AB)C$ and $A(BC)$ exist, then $(AB)C = A(BC)$.

37.

	S	W	G	P	L
Ranch	5	20	16	7	17
Cape Cod	7	18	12	9	21
Colonial	6	25	8	5	13

39.

	Cost
Ranch	492
Cape Cod	528
Colonial	465

represents the cost of materials for each type of house.

41.

	Material	Transportation
Ranch	492	114
Cape Cod	528	108
Colonial	465	103.5

represents the material and transportation costs for each type of house.

43. [14,304] represents the combined cost of materials and transportation for the orders.

9.4

1. $\begin{bmatrix} 5 & -6 \\ 3 & 2 \end{bmatrix}\begin{bmatrix} x \\ y \end{bmatrix} = \begin{bmatrix} 18 \\ 9 \end{bmatrix}$

3. $\begin{bmatrix} 0.4 & -0.3 & 0.2 \\ 1 & 0 & -1 \\ 0 & 0.7 & -1 \end{bmatrix}\begin{bmatrix} x \\ y \\ z \end{bmatrix} = \begin{bmatrix} 1.5 \\ -3 \\ 1.0 \end{bmatrix}$

5. $\begin{bmatrix} 7 & -3 & 0 \\ -3 & -1 & 14 \\ 0 & 13 & -4 \end{bmatrix}\begin{bmatrix} x \\ y \\ z \end{bmatrix} = \begin{bmatrix} 11 \\ 11 \\ -8 \end{bmatrix}$

7. $\begin{cases} 4x + 3y = -1 \\ 2x + y = 3 \end{cases}$ $\{(5, -7)\}$

9. $\begin{cases} x - z = 1 \\ 2x + 3y + 5z = 1 \\ -x - y = -2 \end{cases}$ $\{(-2.5, 4.5, -1.5)\}$

11. $\begin{cases} x + y + z = 11 \\ 2x + 3y + 0.5z = -7 \\ x - y - z = 10 \end{cases}$ $\{(10.5, -11.3, 11.8)\}$

13. a. $\begin{bmatrix} 6 \\ 11 \end{bmatrix}$ b. $\begin{bmatrix} 35.2 \\ 59.9 \end{bmatrix}$ c. $\begin{bmatrix} 8 \\ 12 \end{bmatrix}$

15. a. $\begin{bmatrix} 4.4 \\ -3.2 \end{bmatrix}$ b. $\begin{bmatrix} -\frac{1}{2} \\ \frac{3}{4} \end{bmatrix}$ c. $\begin{bmatrix} -2.12 \\ 2.435 \end{bmatrix}$

17. a. $\begin{bmatrix} 0 & 1 & 1 & 1 \\ 1 & 0 & 0 & 0 \\ 0 & 0 & 0 & 0 \\ 0 & 1 & 1 & 0 \end{bmatrix}$ b. A to B, B to C, A to C, B to D, A to A, B to B, D to A

19. One possible answer is $\begin{cases} x + y + z = 2 \\ x + 2y - z = -6.5 \\ 3x - y + 2z = 12.1 \end{cases}$

21. The matrix has no inverse.

23. $C = \begin{bmatrix} \frac{9}{33} & \frac{6}{33} \\ -\frac{1}{33} & \frac{3}{33} \end{bmatrix}; K = \begin{bmatrix} \frac{16}{11} \\ \frac{2}{33} \end{bmatrix}$

25. $C = \dfrac{1}{44}\begin{bmatrix} 3 & -1 & 35 \\ 23 & 7 & -25 \\ -3 & 1 & 9 \end{bmatrix}; K = \begin{bmatrix} 52.2 \\ -8.2 \\ -1.8 \end{bmatrix}$

27. $\theta = \dfrac{\pi}{6}$; $2(x')^2 + (y')^2 = 1$

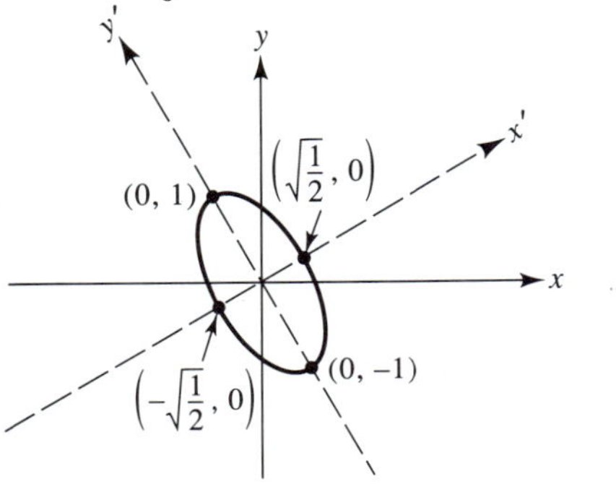

29. $\theta = \pi/4$; $(x')^2 - (y')^2 = 9$

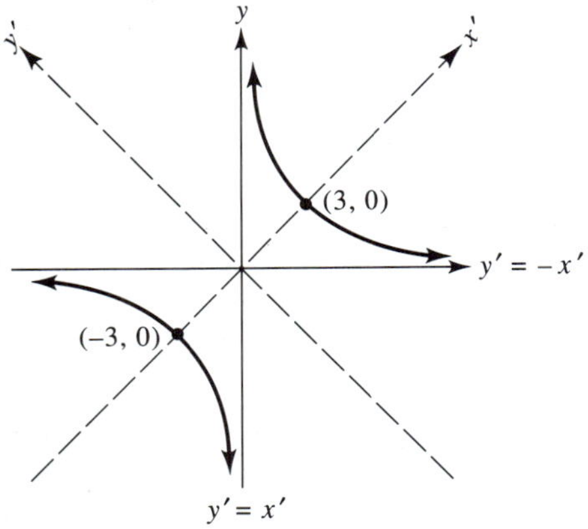

31. The maximum length is 2 units.

33. The minimum distance between arms is 6 units.

9.5

1. $\{(\sqrt{3}i, -3), (-\sqrt{3}i, -3), (\sqrt{2}, 2), (-\sqrt{2}, 2)\}$

3. $\{(\sqrt{\frac{3}{2}}, -\sqrt{\frac{3}{2}}), (-\sqrt{\frac{3}{2}}, \sqrt{\frac{3}{2}})\}$

5. $\{(0, 0), (5, 0), (-5, 0)\}$

7. $\{(0, 0), (-\frac{3}{2}, -\frac{27}{8})\}$

9. $\left\{(-2, -1), \left(\dfrac{2 - \sqrt{6}}{2}, \dfrac{4}{2 - \sqrt{6}}\right), \left(\dfrac{2 + \sqrt{6}}{2}, \dfrac{4}{2 + \sqrt{6}}\right)\right\}$

11. $\{(\text{Tan}^{-1}5 + k\pi, 5)\}, k \in I$

13.

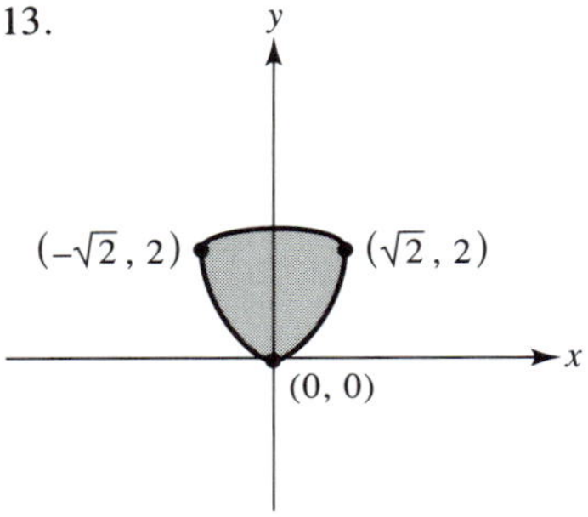

15.

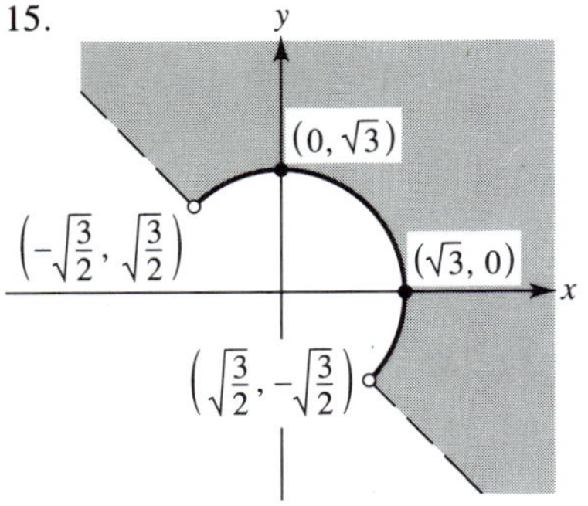

17.

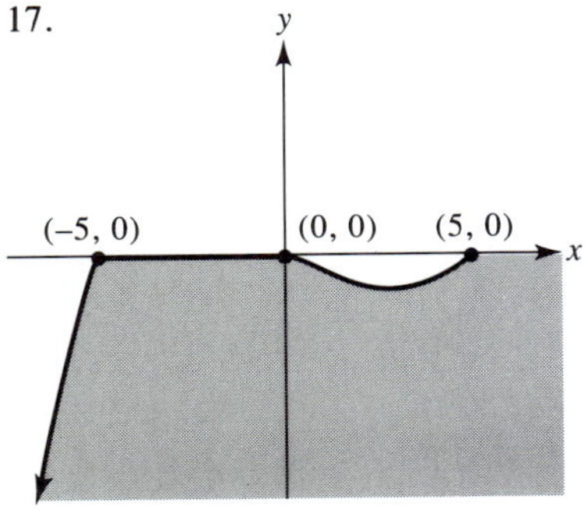

19.

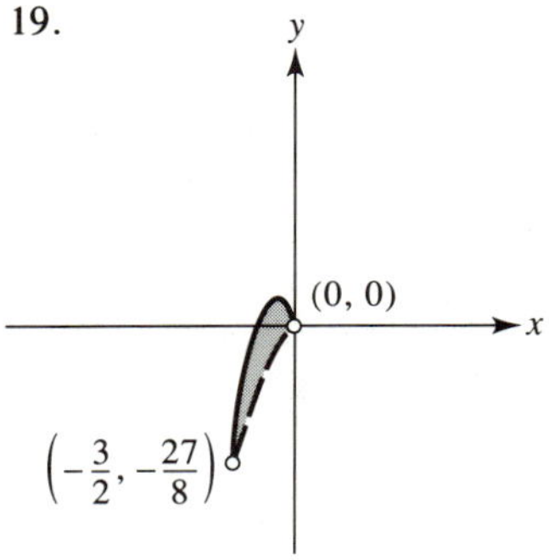

21.

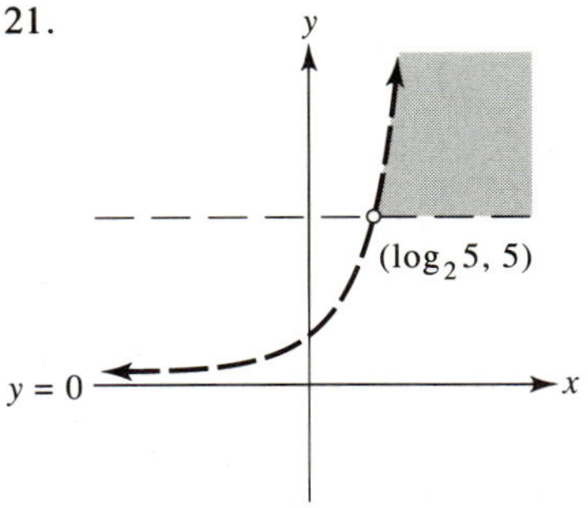

23.

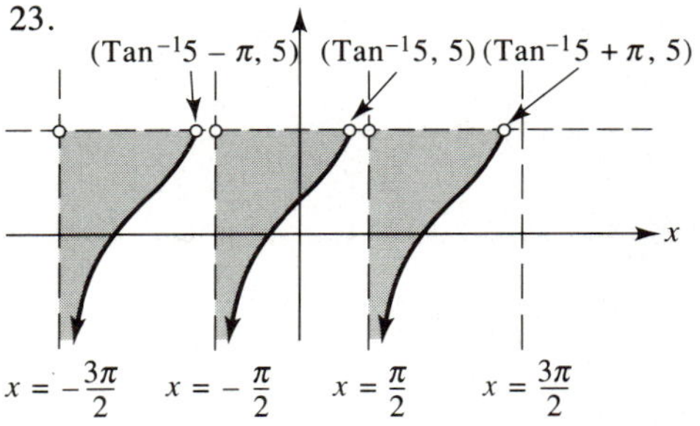

The graph continues to the left and right.

25. $\{(e^6, e^{-1})\}$
27. $\{(2, \sqrt{7}, \sqrt{3}), (2, -\sqrt{7}, \sqrt{3}), (2, -\sqrt{7}, -\sqrt{3}), (2, \sqrt{7}, -\sqrt{3})\}$
29. $\{(\frac{1}{2}, -\frac{1}{2}, \frac{1}{4})\}$
31. Jane should set the schedule to produce $\frac{1100}{3}$ Model 279P computers and $\frac{200}{3}$ Model 345Q computers each day.
33. To minimize cost, Doris and Tom should make 7 dusters per day.
35. {(0.824 (approx.), 0.679 (approx.)), (−0.824 (approx.), 0.679 (approx.))}
37. {(1.412 (approx.), 0.345 (approx.)), (3.057 (approx.), 1.117 (approx.))}
39. {(0, 0), (−1.660 (approx.), −0.996 (approx.)), (1.660 (approx.), 0.996 (approx.))}
41. None

Summary Exercises

1. $\{(\frac{61}{11}, -\frac{19}{11})\}$
3. $\{(-\frac{142}{35}, \frac{286}{35}, \frac{162}{35})\}$
5. $\{(e^{13}, e^{-9})\}$
7. $\{(\frac{11}{61}, -\frac{11}{19})\}$
9. None
11. One
13. $\{(-\frac{1}{2}, \frac{3}{4})\}$
15. $\{\ \}$
17. $\{(5 - 3c, 2c, c), c \in \mathbb{R}\}$
19. 3×2
21. Undefined
23. $\begin{bmatrix} -7 & 23 \\ 43 & -49 \\ 18 & 6 \end{bmatrix}$
25. $\begin{bmatrix} -4 & 5 \\ 1.5 & 0.5 \end{bmatrix}$
27. $\begin{bmatrix} \frac{7}{23} & \frac{15}{23} & \frac{12}{23} \\ \frac{6}{23} & \frac{3}{23} & \frac{7}{23} \\ \frac{10}{23} & \frac{5}{23} & \frac{4}{23} \end{bmatrix}$
29. $\begin{bmatrix} -\frac{10.1}{11} \\ \frac{9.4}{11} \end{bmatrix}$
31. $\begin{bmatrix} 1 & 1 & 1 & 1 & 1 & 0 \\ 1 & 1 & 1 & 0 & 1 & 1 \\ 1 & 1 & 1 & 1 & 0 & 1 \\ 1 & 0 & 1 & 1 & 1 & 1 \\ 1 & 1 & 0 & 1 & 1 & 1 \\ 0 & 1 & 1 & 1 & 1 & 1 \end{bmatrix}$ represents all one-stop trips.
33. $\left\{\left(\sqrt{\frac{9 - \sqrt{77}}{2}}, \sqrt{\frac{9 + \sqrt{77}}{2}}\right), \left(\sqrt{\frac{9 + \sqrt{77}}{2}}, \sqrt{\frac{9 - \sqrt{77}}{2}}\right), \left(-\sqrt{\frac{9 - \sqrt{77}}{2}}, -\sqrt{\frac{9 + \sqrt{77}}{2}}\right), \left(-\sqrt{\frac{9 + \sqrt{77}}{2}}, -\sqrt{\frac{9 - \sqrt{77}}{2}}\right)\right\}$
35. {(−1.61 (approx.), −2.41 (approx.)), (3.11 (approx.), 4.66 (approx.))}

37.

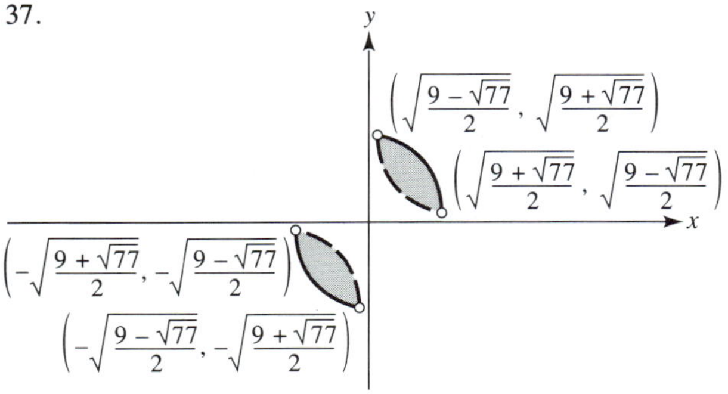

39.

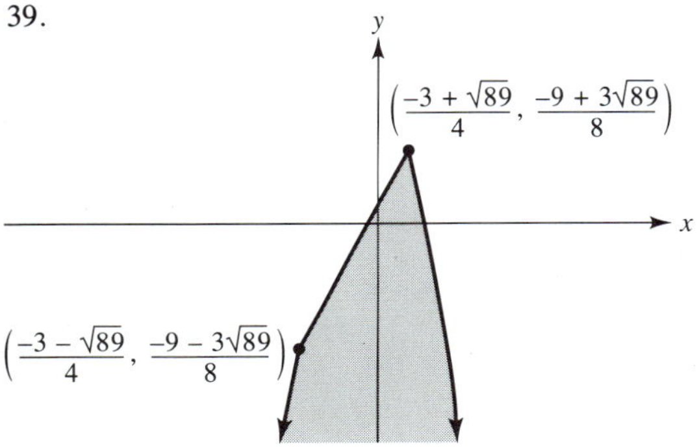

41.

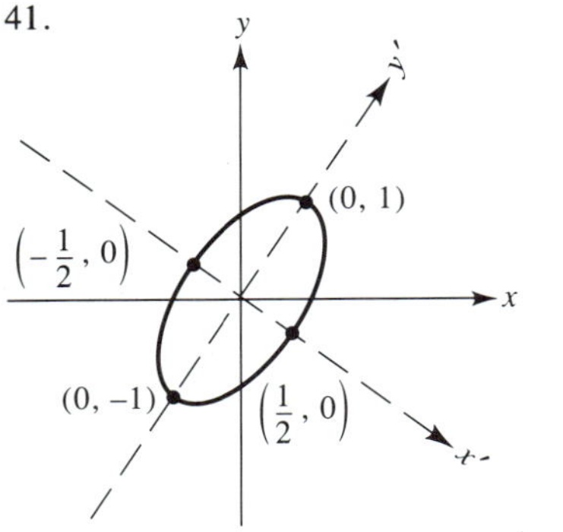

θ is $-\frac{\pi}{6}$; $16(x')^2 + 4(y')^2 = 4$

Chapter Ten

10.1

1. 5040
3. 12
5. 1
7. 2184
9. 180
11. 1
13. 120
15. 1
17. Arnold can select 90 shirt and slacks pairs.
19. a. There are 10,000 numbers.
    b. There are 9000 numbers.
    c. There are 6561 numbers.
21. a. Kris and Tom can arrange the books in 25! ways.
    b. Kris and Tom can arrange the books in approximately 1.3076743×10^{12} ways ($P(25,10)$ ways).
    c. Kris and Tom can arrange the books in 96,909,120 ways.
23. The selections can be made in the following number of ways:
    a. 5,200,300 b. 286 c. 66 d. 12,870
25. There are the following number of subsets:
    a. 1 b. 4 c. 6 d. 4 e. 1
27. $C(n, n) = \frac{n!}{n!(n-n)!} \Leftrightarrow C(n, n) = \frac{1}{0!}$, or 1
29. See Section 10.3 for a proof.
31. a. The number of committees is 91,390.
    b. The number of committees is 14,950.
    c. The number of committees is 5.

10.2

1. $P(k)$: $1 + 2 + 3 + \cdots + k = \frac{k(k+1)}{2}$

   $P(1)$: $1 = \frac{1(1+1)}{2}$ is true.
3. $P(k)$: $(x+y)^k = x^k + kx^{k-1}y + \cdots + y^k$

   $P(1)$: $(x+y)^1 = x^1 + 1x^0y$, or $x + y$, is true.
5. $P(k)$: $1^3 + 2^3 + 3^3 + \cdots + k^3 = \left[\frac{k(k+1)}{2}\right]^2$

   $P(1)$: $1^3 = \left(\frac{1(1+1)}{2}\right)^2$ is true.
7. $P(k)$: There are 2^k regions created when k points are marked on an ellipse, and each pair is connected with a line segment.

   $P(0)$: There is $1(2^0)$ region created when 0 points are marked on an ellipse, and each pair is connected with a line segment. This statement is true.
9. $P(k)$: There are 2^k regions created when k points are marked on a square, and each pair is connected with a line segment.

   $P(0)$: There is $1(2^0)$ region created when 0 points are marked on a square. This statement is true.
11. $P(k)$: The sum of the angles of a closed figure with k sides of equal length is $(180k - 360)°$.

    $P(3)$: The sum of the angles of a closed figure with 3 sides of equal length is $[180(3) - 360]°$. This statement is true.

In the answers for 13 through 24, these are possible choices for k.

13. $P(1)$ is true in exercise 1.

$P(2)$: $1 + 2 = \frac{2(2+1)}{2}$ is true.

$P(5)$: $1 + 2 + 3 + 4 + 5 = \frac{5(6)}{2}$ is true.

Yes

15. $P(1)$ is true in exercise 3.

$P(2)$: $(x + y)^2 = x^2 + 2xy + y^2$ is true.

$P(4)$: $(x + y)^4 = x^4 + 4x^3y + 6x^2y^2 + 4xy^3 + y^4$ is true.

Yes

17. $P(1)$ is true in exercise 5.

$P(2)$: $1^3 + 2^3 = \left(\frac{2(2+1)}{2}\right)^2$ is true.

$P(4)$: $1^3 + 2^3 + 3^3 + 4^3 = \left(\frac{4(4+1)}{2}\right)^2$ is true.

Yes

19. $P(0)$ is true in exercise 7.

$P(1)$: There are 2 regions created when 1 point is marked on an ellipse, and each pair . . . segment. (false)

$P(2)$: There are 2^2 regions created when 2 points are marked on an ellipse, and each pair . . . segment. (false)

No

21. $P(0)$ is true in exercise 9.

$P(1)$: There are 2 regions created when 1 point is marked on a square, and each pair . . . segment. (false)

$P(3)$: There are 2^3 regions created when 3 points are marked on a square, and . . . segment. (false)

No

23. $P(3)$ is true in exercise 11.

$P(4)$: The sum of the angles of a closed figure with 4 sides of equal length is $(180(4) - 360)°$. (true)

$P(6)$: The sum of the angles of a closed figure with 6 sides of equal length is $(180(6) - 360)°$. (true)

Yes

For exercises 25 through 34, we give the basis step, and if the basis step is true, one way to use $P(k)$ to demonstrate $P(k + 1)$. Your proof should contain a full explanation in the style of the examples.

25. The basis step was done in exercise 1.

$P(k + 1)$:

$$\underbrace{1 + 2 + 3 + \cdots + k}_{\text{LHS of } P(k)} + (k + 1) = \frac{(k+1)(k+2)}{2}$$

$\Leftrightarrow$ $P(k + 1)$:

$$\underbrace{\frac{k(k+1)}{2}}_{\text{RHS of } P(k)} + (k + 1) = \frac{(k+1)(k+2)}{2}$$

$\Leftrightarrow$ $P(k + 1)$:

$$\frac{k(k+1)}{2} + \frac{2(k+1)}{2} = \frac{(k+1)(k+2)}{2}$$

$\Leftrightarrow$ $P(k + 1)$: $\frac{(k+1)(k+2)}{2} = \frac{(k+1)(k+2)}{2}$

27. Basis step: $P(0)$: $4^0 - 1$ is divisible by 3 is true.

$P(k + 1)$: $4^{k+1} - 1$ is divisible by 3.

$P(k + 1)$: $4^{k+1} - 1 \equiv 4^{k+1} - 4 + 4 - 1 \equiv 4(4^k - 1) + (4 - 1) \equiv 4(4^k - 1) + 3$

$P(k)$ says the first term is divisible by 3; the sum is divisible by 3.

29. Basis step: $P(1)$: $1^2 = \frac{1}{6}(1)(2)(3)$ is true.

$P(k + 1)$:

$$\underbrace{1^2 + 2^2 + \cdots + k^2}_{\text{LHS of } P(k)} + (k + 1)^2 = \frac{1}{6}(k + 1)(k + 2)[2(k + 1) + 1]$$

31. Basis step: $P(1)$: $(x - 1)(x + 1) = x^{1+1} - 1$ is true.

$P(k + 1)$: $(x - 1)(x^{k+1} + x^k + x^{k-1} + \cdots + x + 1) = x^{k+2} - 1$

$\Leftrightarrow$ $P(k + 1)$: $(x - 1)x^{k+1} + \mathbf{(x - 1)(x^k + x^{k-1} + x^{k-2} + \cdots + x + 1)} = x^{k+2} - 1$

$\Leftrightarrow$ $P(k + 1)$: $x^{k+2} - x^{k+1} + x^{k+1} - 1 = x^{k+2} - 1$

33. $S + S = (2 + 2n) + [4 + 2(n - 1)] + [6 + 2(n - 2)] + \cdots + (2n + 2)$ or $2(n + 1) + 2(n + 1) + 2(n + 1) + \cdots + 2(n + 1) = n[2(n + 1)]$

$\Leftrightarrow$ $2S = n[2(n + 1)]$ $\Leftrightarrow$ $S = n(n + 1)$

35. $P(n)$: $0 + 7 + 14 + \cdots + 7(n - 1) = 7\frac{n(n+1)}{2}$

Basis step: $P(0)$: $0 = 7\frac{0(0+1)}{2}$ is true.

Induction step (sketch): $P(k + 1)$: $0 + 7 + 14 + \cdots + 7(k - 1) + 7k = \frac{7}{2}(k + 1)(k + 2)$

10.3

1. $x^4 - 4x^3 + 6x^2 - 4x + 1$
3. $x^5 + 15x^4 + 90x^3 + 270x^2 + 450x + 243$
5. $128a^7 + 448a^6b + 672a^5b^2 + 560a^4b^3 + 280a^3b^4 + 84a^2b^5 + 14ab^6 + b^7$
7. $x^4 + 16x^3y + 96x^2y^2 + 256xy^3 + 256y^4$
9. $2187x^7 - 10{,}206x^6y + 20{,}412x^5y^2 - 22{,}680x^4y^3 + 15{,}120x^3y^4 - 6048x^2y^5 + 1344xy^6 - 128y^7$
11. $f(x + h) = x^2 + 2xh + h^2$; $x + h$

13. $f(x + h) = -x^4 - 4x^3h - 6x^2h^2 - 4xh^3 - h^4$; $-4x^3 - 6x^2h - 4xh^2 - h^3$
15. $f(x + h) = 2x^3 + 6x^2h + 6xh^2 + 2h^3 - 4x^2 - 8xh - 4h^2 + x + h - 3$; $6x^2 + 6xh + 2h^2 - 8x - 4h + 1$
17. $[(x + h) - x][(x + h) + x]$; $x + h$
19. $[x - (x + h)][x^3 + x^2(x + h) + x(x + h)^2 + (x + h)^3]$; $-4x^3 - 6x^2h - 4xh^2 - h^3$
21. $2[(x + h) - x][(x + h)^2 + (x + h)x + x^2] - 4[(x + h) - x][(x + h) + x)] + [(x + h) - x]$; $6x^2 + 6xh + 2h^2 - 8x - 4h + 1$
23. $-4480x^{13}y^3$
25. $C(41, 21)(2x)^{20}(5y)^{21}$
27. $C(103, 98)(2.4x)^5(-1.1)^{98}$
29. $-4{,}272{,}048(\sqrt{3}x)^{26}$
31. $C(63, 49)(5a^2)^{14}(4b^3)^{49}$
33. $120x^{-28}y^{-3}$
35. $462x^5$
37. 189,190,144
39. Set aside one of the $n + 1$ elements. Each r-element subset of the original set is either an r-element subset of the set of the n remaining elements or can be made from an $(r - 1)$-element subset of the n elements with the element you set aside joined to it.
41. $9x^2 - 24xy + 12xz - 16yz + 16y^2 + 4z^2$
43. -6
45. 360

10.4

1. $-2, 4, -8, 16, -32, 64$
3. $5, 2, -1, -4, -7, -10$
5. $-3, -16, -125, -1296, -16807, -262144$
7. $t(n) = 5n - 4, n \in N$
9. $t(n) = -2n - 6, n \in N$
11. $t(n) = 31n + 10{,}067, n \in N$
13. $g_n = \dfrac{1}{2^{n+1}}$
15. $g_n = (-\frac{1}{2})^{n+1}$
17. $g_n = 11(5)^{n-1}$
19. $g_n = -\frac{2}{3}(\frac{3}{7})^{n-1}$
21. Geometric; r is 1.1
23. Harmonic
25. Arithmetic; d is -1.99.
27. None of these
29. Fibonacci-type
31. $a_{10} = -1/2048$
33. $a_{24} = -177$
35. $a_{10} = 25.5$
37. $a_{42} = 1/105.5$
39. f_3 is 2.

10.5

1. 6, 16, 30, 48, 70
3. $1, \frac{3}{2}, \frac{11}{6}, \frac{25}{12}, \frac{137}{60}$
5. 3.1, 3.41, 3.441, 3.4441, 3.44441
7. 7, 17, 30, 46, 65
9. 110
11. 120
13. 175
15. A_n is $n(4 + n)$.
17. A_n is $-n^2 - 7n$.
19. $\dfrac{n}{2}(25 - 2n)$
21. G_n is $8\left(1 - \dfrac{1}{2^n}\right)$.
23. G_n is $\dfrac{11}{4}(5^n - 1)$.
25. $\frac{3}{11}((-\frac{1}{10})^n - 1)$
27. 2109
29. Geometric; $(-\frac{1}{3})^n - 1$
31. Arithmetic; $\frac{n}{4}(n - 27)$
33. Arithmetic; $\frac{n}{2}(11 + 3n)$
35. 5050
37. 1863.49
39. They are the same.
41. $\dfrac{5}{3}\left(1 - \dfrac{1}{2^{12}}\right)$
43. 459

Summary Exercises

1. 720
3. 30,240
5. 3,628,800
7. 10
9. 715
11. 140
13. 2,335,132,800
15. 1440

17. Basis step: $P(0)$: $0^3 = \frac{1}{4} \cdot 0^2 \cdot (0 + 1)^2$

Induction step: We use $P(k)$: $\sum_{i=1}^{k} i^3 = \frac{1}{4} \cdot k^2 \cdot (k + 1)^2$

to show $P(k + 1)$: $\sum_{i=1}^{k+1} i^3 = \frac{1}{4} \cdot (k + 1)^2 \cdot (k + 2)^2$.

19. $32a^5 - 240a^4b + 720a^3b^2 - 1080a^2b^3 + 810ab^4 - 243b^5$
21. $-56{,}229{,}888$
23. Geometric
25. Harmonic
27. None of these (power)
29. Arithmetic
31. 14, 11, 8, 5, 2
33. a_n is $a_{n-1} + a_{n-2}$ and $a_1 = 3$, $a_2 = 4$.
35. a_n is $-0.2(-0.4)^{n-1}$.
37. 34.58
39. $2^n - 1$
41. $\dfrac{(-2)^n - 1}{-3}$
43. $a_{n+2} - 1$
45. 171
47. 1240
49. 1155

Appendix

A.1

1. 762.745 (approx.)
3. 73.393 (approx.)
5. 73.393 (approx.)
7. 2.008 (approx.)
9. -202.156 (approx.)
11. 0.000 (approx.)
13. 115.404 (approx.)
15. $2\pi - 11$, $\sqrt[3]{-43.894}$, -3.52751, $14^{7/3}$, 472.39
17. 2.89000, 2.99290, 2.999824, 3.00017, 3.00000, 3.00000; yes, 3; the square root of 3
19. The surface area is approximately 124.256 cubic centimeters.
21. They will have $4448.48.
23. The z-score is approximately -0.87.

A.2

1. {3, 8, 13, 18, . . .}
3. {0, 24, 120, 336, . . .}
5. $\{\frac{16}{7}, \frac{64}{7}, \frac{144}{7}, \frac{256}{7}, \ldots\}$
7. One description is $\{k \mid k = 2n - 1, n \text{ is a counting number}\}$.
9. One description is $\{m \mid m = 2t^2, t \text{ is a counting number}\}$.
11. [5, 13)
13. $(-\infty, 8.7]$
15. $(-\infty, 7] \cup [12, +\infty)$
17. $[-14, +\infty)$
19. $[\sqrt{7}, \sqrt{7}]$
21. 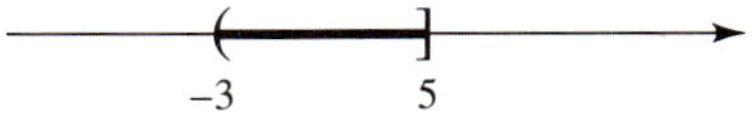

23. 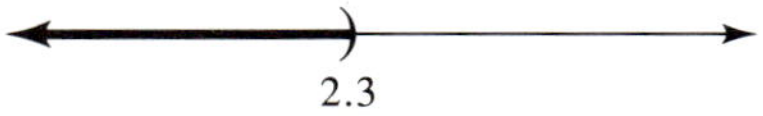

25. 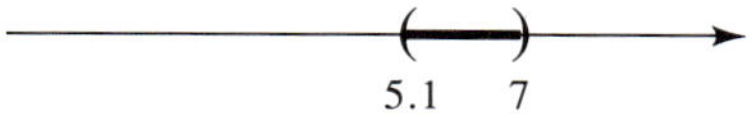

27. 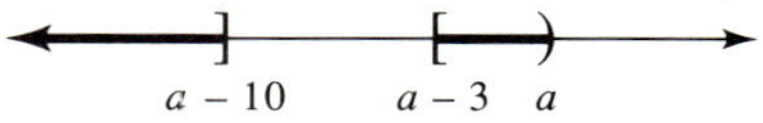

29. 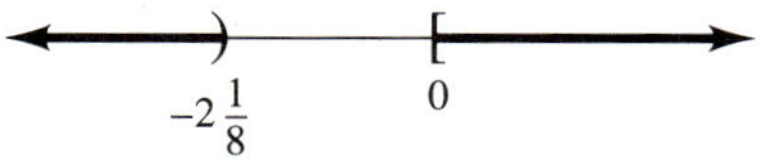

31. {2, 4, 6, 8, . . .}
33. {3, 9, 27, . . .}
35. {12}
37. { }
39.

Number of elements	0	1	2	3	4	k
Number of subsets	1	2	4	8	16	2^k

41. There are two distributive properties:
$A \cap (B \cup C) = (A \cap B) \cup (A \cap C)$
$A \cup (B \cap C) = (A \cup B) \cap (A \cup C)$

A.3

1. J, $\mathbb{R}$, $\mathbb{C}$
3. Q, $\mathbb{R}$, $\mathbb{C}$
5. J, $\mathbb{R}$, $\mathbb{C}$
9. J, $\mathbb{R}$, $\mathbb{C}$
11. I, Q, $\mathbb{R}$, $\mathbb{C}$
13. 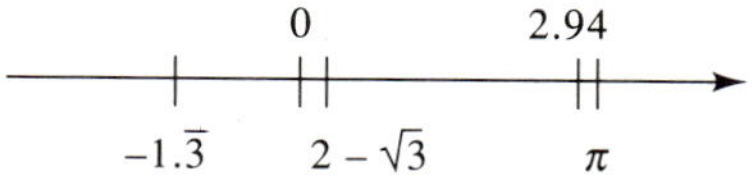

15. $\frac{1}{8}$, $-\frac{301}{100}$, $\frac{3401}{200}$, $\frac{c_1c_2c_3}{1000}$, $\frac{d_1d_2d_3 \ldots d_n}{10^n}$
17. One possible explanation follows: Let n represent the repeating decimal. Form $1n$, $10n$, $100n$, $1000n$, . . . until

you find two that have the same decimal part. Subtract to find a multiple of n that is an integer. Divide this integer by the multiplier of n.

19. $0.\overline{285714}$, $-0.3333\ldots$, $0.\overline{185}$, $1.\overline{230769}$, $7.16666\ldots$

21. Each result of the subtraction step must be less than the divisor. There are only as many such numbers as the size of the divisor. After some number of steps we must get 0 (the division ends) or repeat a number (the digits in the quotient begin to repeat).

23. $\leq$

25. $\leq$

27. $>$

29. $>$

31. $|a + b| \leq |a + b|$; $|ab| = |a| \cdot |b|$

33. $\left|\dfrac{a}{b}\right| = \dfrac{|a|}{|b|}$

35. $\leq$

37. $=$

39. $\geq$

41. $-43 - 64i$

43. $0 + 1i$

45. $\dfrac{143}{4} + 6i$

47. $-10.3 + 110i$

49. $0 - \dfrac{7}{9}i$

51. $\dfrac{42}{25} - \dfrac{56}{25}i$

A.4

1. $x^{5/6}$, $x \geq 0$

3. $(a + 3)^{8/3}$

5. $(x - 2y)^{2/3}$

7. y^{11}, $y \neq 0$

9. $b^{3/10}$, $b > 0$

11. $x^{8/3}$

13. a^{-6} or $1/a^6$, $a \neq 0$

15. y^{12}, $y \neq 0$ if $n < 0$

17. xy^{-3}; $x, y \neq 0$

19. $3a^{-1} - 6a^{-1/2}b^{-1/2} + b^{-1}$; $a, b > 0$

21. $\dfrac{x^2y^2}{y^2 - 2xy + x^2}$; $y \neq x$, $x, y \neq 0$

23. $\dfrac{9a^{10}c^{14/3}}{16b^4}$; $a, b, c \neq 0$

25. $\dfrac{15x^2y^2 + 89xy - 6}{6y^2}$, $y \neq 0$

27. $24\sqrt{3}i$

29. $1 + \sqrt{3}i$

31. 1

33. The pattern does extend.

35. 1. To obtain the product of two numbers with the same base, retain the base and add the exponents.
    2. To obtain the quotient of two numbers with the same base, retain the base and subtract the exponent in the denominator from the exponent in the numerator.

37. 5.9005 (approx.)

39. 5.3232 (approx.)

41. -1.4362 (approx.)

43. Yes; $(3^2)^4 = 3^8$ but $3^{(2^4)} = 3^{16}$.

45. Yes; yes; it may use the decimal approximation of the irrational number.

A.5

1. Degree: 6; number of terms: 3; coefficients: $-12, 4, -5$

3. One possible answer is $-3x^7 + 4.22x^5 + \pi x^3 + \sqrt{2}x$.

5. $3\sqrt{x} - x + 3$

7. $3^{n+1} + 5x$

9. $7.7a^{3/2} - 0.9a - 3.7a^{1/2}$

11. $80x^3 + 32x^2 - 45x - 18$

13. $y^{6/5} + 7y^{1/5} + 9y^{-4/5}$

15. $7(2x + 1) - (2x + 1)^{3/2} + 3(2x + 1)^{1/2}$

17. $x \leq 5$

19. No restrictions are needed.

21. $x^{4/3} - 3x^{1/3} + 4x^{-2/3}$, $x \neq 0$

23. $-3a^{-1}b + 1 - 5b$; $a, b \neq 0$

25. $\dfrac{3x + 4}{x - 6}$, $x \neq \dfrac{1}{2}$

27. $\dfrac{x + 2y}{x - y}$, $x \neq 3y$

29. $(3x^{1/2} + 1)(x^{1/2} + 3)$

31. $(2 + x^{1/3})(2 - x^{1/3})$

33. $(m + 3n)(5a - b)$

35. $x^{-3}(x - 7)(x + 3)$

37. $3.2(x + 2)(x + 1)$

39. $[(a - 2b) + 3][(a - 2b) + 2]$

41. $\frac{1}{5}x^{-3/5}(7x - 2)$

43. $(x - 3y)(x^2 + 3xy + 9y^2)$

45. $(2a + 3b)(2a - 3b)(4a^2 + 9b^2)$

47. $\dfrac{-(x + 3)}{3x}; x \neq -2, 2$

49. $m - m^{1/2}n^{1/2}; n \neq m, -m, 0$

51. $\dfrac{-1}{x^3 - 2x^2y + 2xy^2 - y^2}; x \neq -y$

53. $\dfrac{3.2x^{1/3}y^{1/3} + 6.9 - 7.5x^{2/3}y^{1/3}}{xy}$

55. $\dfrac{x^2 + 3x + 1}{1 + 2x}$

57. $x^2 + 8x + 17 + \dfrac{53}{x - 3}$

59. $a^3 - 2a^2b + 5ab^2 - 15b^3 + \dfrac{30b^4}{a + 2b}$

61. $x - 6x^{1/2} + 17$

63. $(x^n - y^n) \div (x - y) = x^{n-1} + x^{n-2}y + x^{n-3}y^2 + \cdots + y^{n-1}$

65. An improper rational or fractional expression should be defined so that it is possible to carry out the indicated division. We can do this if the degree of the numerator is greater than or equal to the degree of the denominator. An improper rational (or fractional) expression is a rational (or fractional) expression whose numerator has degree greater than or equal to the degree of the denominator.

Summary Exercises

1. $\{\frac{1}{3}, \frac{2}{3}, \frac{3}{3}, \frac{4}{3}, \ldots\}$
3. $\{2, 3, 8, 13, 18, \ldots\}$
5. $\{1, 7, 13, 19, \ldots\}$
7. 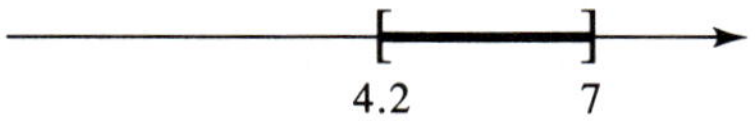

$[4.2, 7]$
9. 

$[5, 8)$
11. $[-3.1, 2.14114]$
13. Associative property for multiplication
15. Distributive property
17. $>$
19. $>$
21. $<$
23. $-6 + 32i$
25. $-\frac{13}{500} + \frac{9}{500}i$
27. $\frac{40}{3} + \frac{16}{3}i$
29. $(2y^{-2/5} - 5)(2y^{-2/5} - 3)$
31. $x^{-4}(x + 6)(x + 5)$
33. $y^{-1/2}(y - 2)(y^2 + 2y + 4)(y + 2)(y^2 - 2y + 4)$
35. $y^{2/3} + 2y^{1/3} + 4$
37. $\dfrac{x + 6}{4x + 7}; x \neq -\dfrac{7}{4}, \dfrac{4}{3}$
39. $\dfrac{x^{-2/5} - 1}{x^{-2/5} + 5}, x \neq 0, 5^{-2/5}$
41. $\dfrac{2x^2 - 2}{x + y}, x \neq -1, 1, -y$

Chapter One Test

3. $\left\{\dfrac{17}{4}\right\}$
6. $\left\{\dfrac{1.6 + 2.7i}{10.4 - 0.5i}\right\}$
9. $\{1\}$
12. $\left\{\dfrac{-1}{2\pi}, \dfrac{9}{4\pi}\right\}$
15. $\{-12, 6\}$
18. $\{3 - \sqrt{2}, 3 + \sqrt{2}\}$
21. Possible answers are $x^2 - 18x + 81 = 0$ and $x - 9 = 0$.
24. The width is $\dfrac{-l + \sqrt{5}l}{2}$ units and the golden ratio is $\dfrac{-1 + \sqrt{5}}{2}$.
27. $\{1\}$
30. $\{6\}$
31. There are 53 sixth-graders and 62 fourth-graders.
33. $\{-5, -2\}$
36. $\left\{-\dfrac{1}{2}, 3\right\}$
39. $\left\{\dfrac{1 - \sqrt{33}}{2}, -1, 0, 2, \dfrac{1 + \sqrt{33}}{2}\right\}$
42. The red part is 22 square kweebs in area.

Chapter Two Test

3. $(1.7, -4.595)$

6.

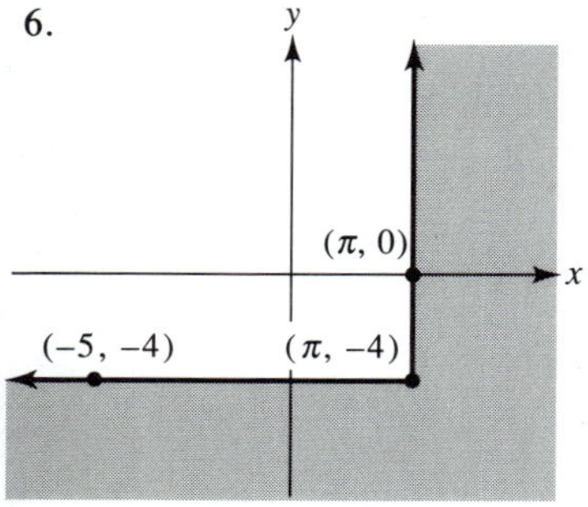

9.

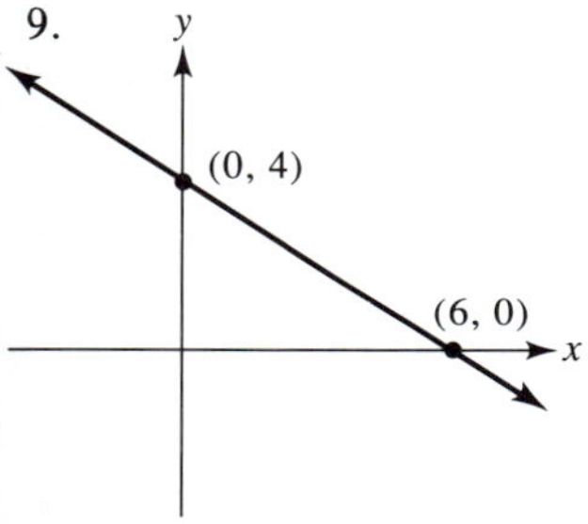

12. Slope: $-a/b$, $b \neq 0$; x-intercept: $(c/a, 0)$, $a \neq 0$; y-intercept: $(0, c/b)$, $b \neq 0$

15. $y = -\frac{7}{3}(x - 4)$

18. They must have the same x-coordinate.

21.

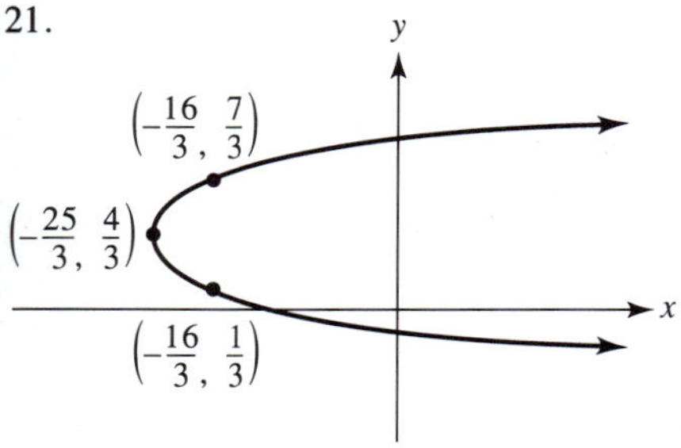

24. They could intersect in 0, 1, 2, 3, or 4 points.

27.

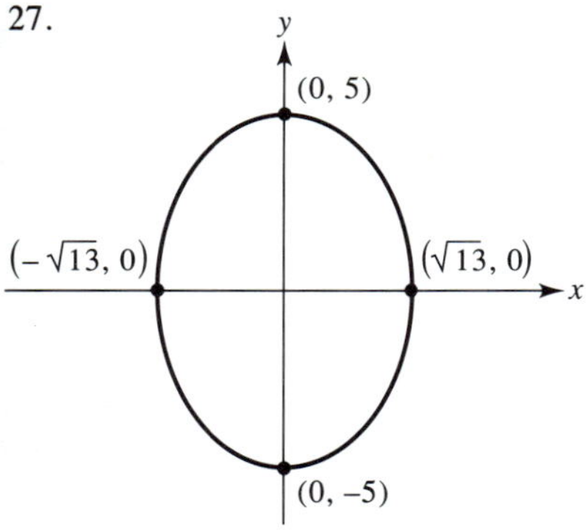

30. There are no real number solutions.

33. If $a = f$ and $a \neq 0$, then the graph is the line segment with endpoints $(-f, 0)$ and $(f, 0)$. If $a = 0$ and $a \neq f$, then the graph is empty. If $a = f = 0$, then the graph is the point $(0, 0)$.

36.

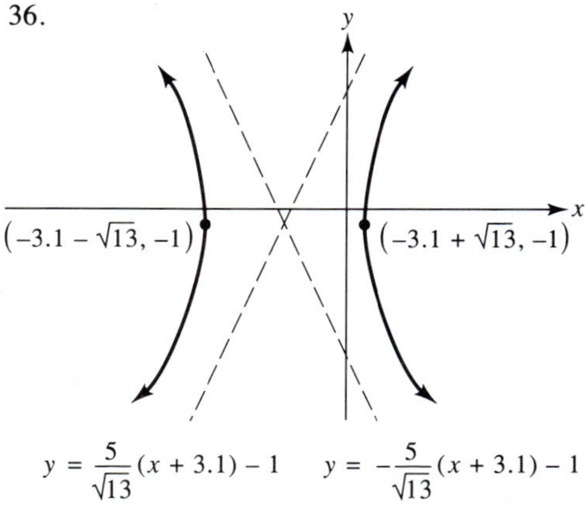

39.

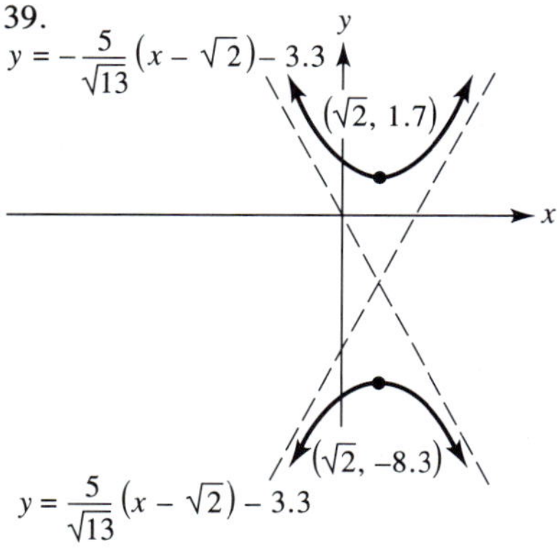

42.

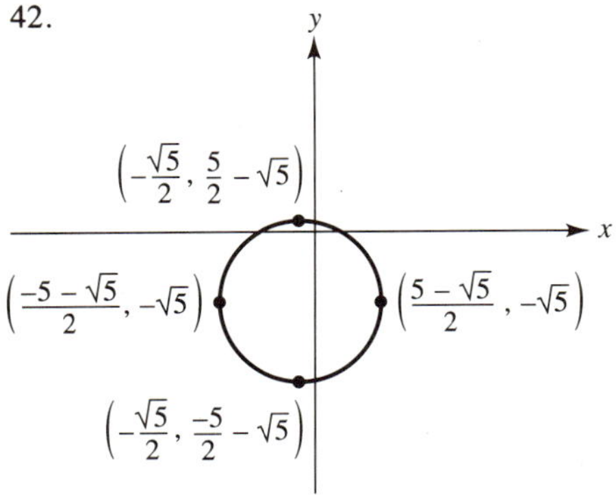

45. $(x + 3)^2 + (y - 7)^2 = 1$ and $(x + 3)^2 + (y - 7)^2 = 4$ are possible answers.

Chapter Three Test

3. $\left\{w \mid w \geq \dfrac{4}{\sqrt{8} - 3}\right\}$

6. $\mathbb{R}$

9. $\{x \mid -\frac{7}{3} \leq x \leq \frac{19}{3}\}$

12. $\{x \mid -3 < x < -1 \text{ or } 1 < x < 3\}$

15. $\{x \mid x \leq \frac{49}{11} \text{ or } 5 < x\}$

18. $\{x \mid -5 \leq x < -\sqrt{5} \text{ or } \sqrt{5} < x < 3 \text{ or } 6 \leq x \text{ or } x = 0\}$

21. $\{\ \}$

24. The border must be at least $12 - \sqrt{76}$ feet wide.

27. $\{x \mid x \leq -4 \text{ or } 0 < x \leq 3\}$

30. $\{x \mid 0 < x \leq 3\}$

33.

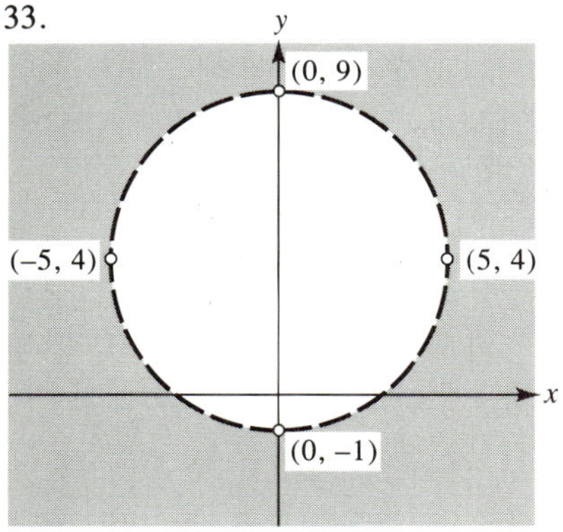

36.

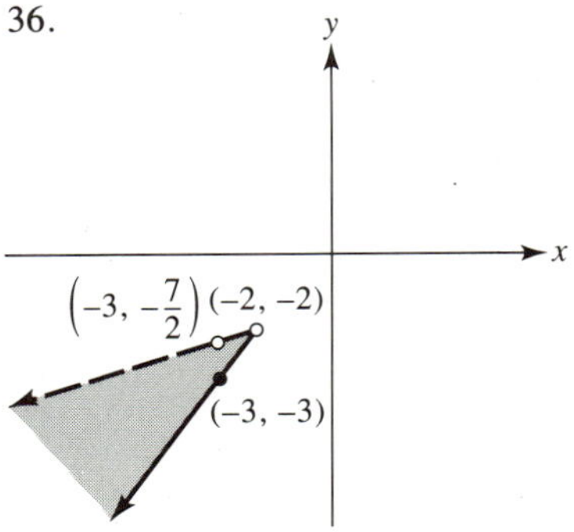

39.

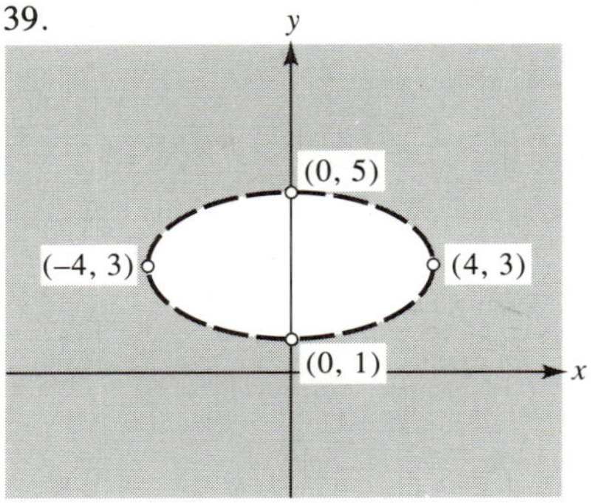

42.

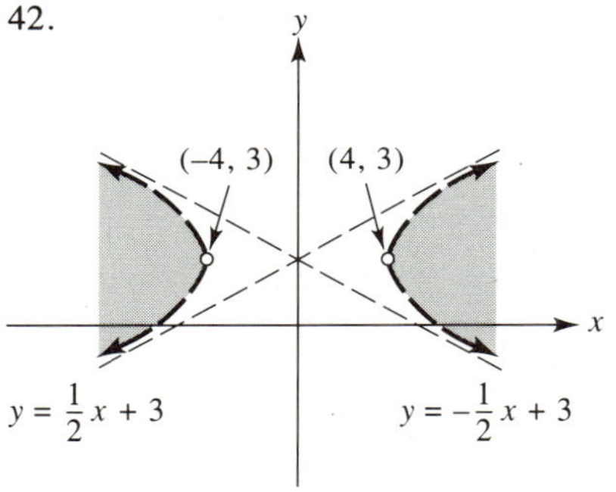

45. $y \geq x^2$

48. $y > 0$ and $x > y^2$; another solution is $0 < y < \sqrt{x}$

Chapter Four Test

3. Non-function

6. $\sqrt{2a + 2},\ a \in [-1, +\infty)$

7. S is odd.

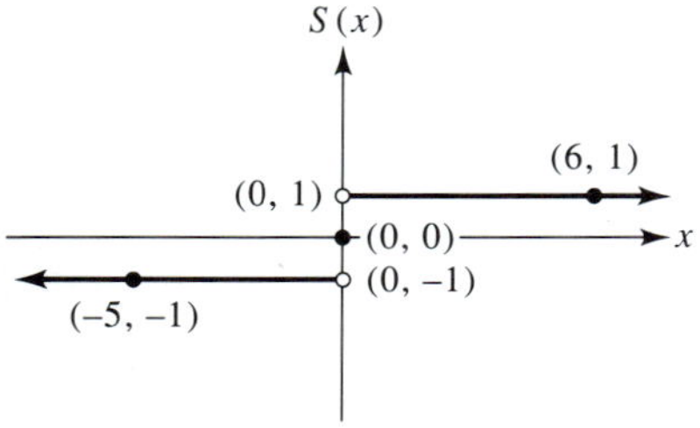

12. a. 6.672 (approx.) b. 2.674 (approx.)
c. 4.233 (approx.)

15.

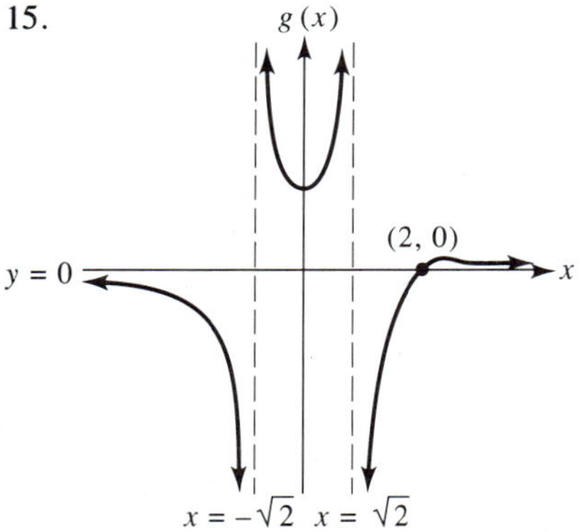

18.

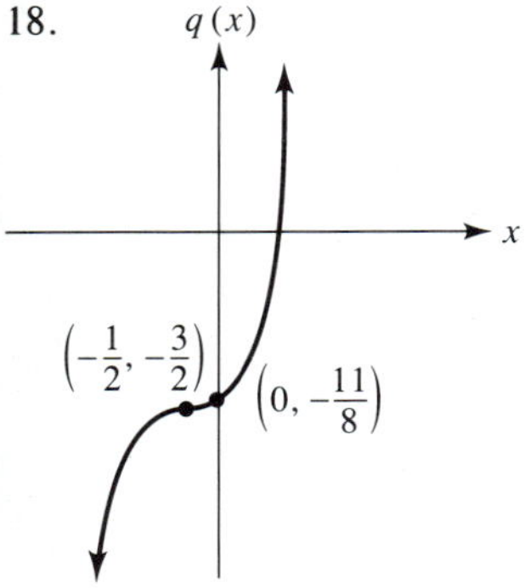

21.

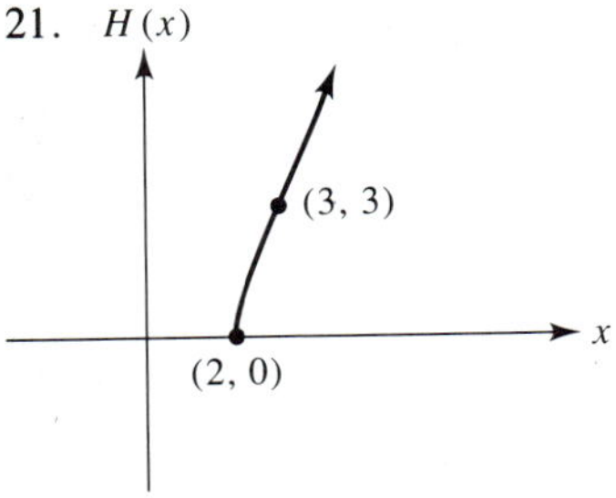

24.

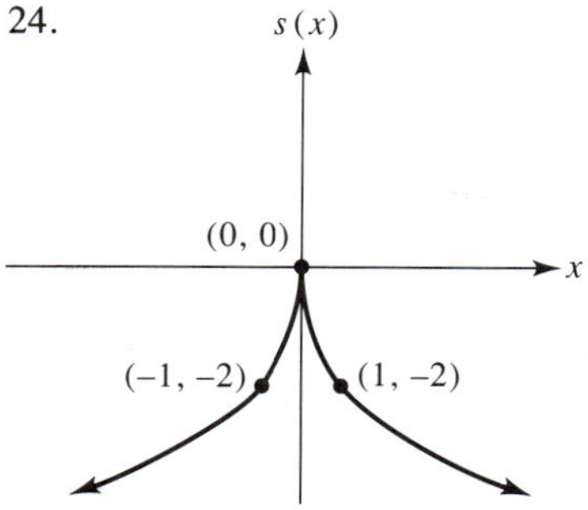

27.

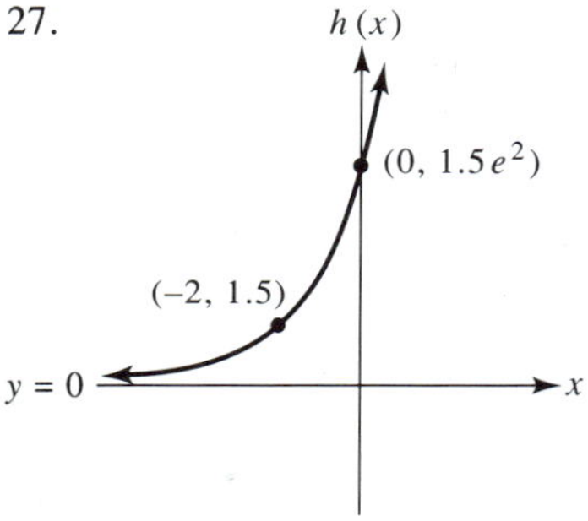

30.

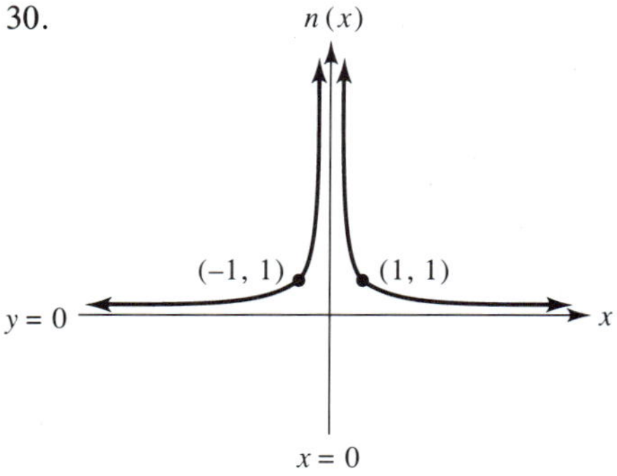

33.

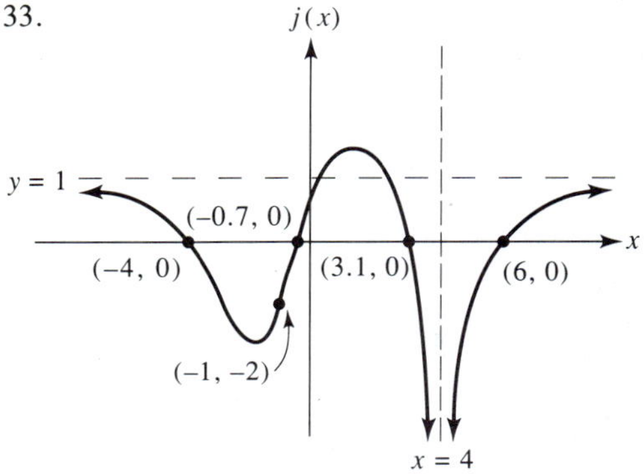

36. $H(x) = (x - 3)^{2/3} - 2$

39. $f(n) = \text{mod}_7(n)$ is a possible answer.

42. $c^3 + 4$; there is only one b.

47. $H(x) = -|x + 1| - 4, x \in [-7, -2]$ is an answer. Yes, it is 1-1.

Chapter Five Test

3. $(f + g)(x) = \ln x + 2x$; $(0, +\infty)$
$\left(\dfrac{f}{g}\right)(x) = \dfrac{\ln x}{2x}$; $(0, +\infty)$
$(f \circ g)(x) = \ln(2x)$; $(0, +\infty)$
$(g \circ f)(x) = 2 \ln x$; $(0, +\infty)$
$f^{-1}(x) = e^x$; $\mathbb{R}$
$g^{-1}(x) = x/2$; $\mathbb{R}$

6. One pair is $f(x) = \sqrt[4]{x}$ and $g(x) = (2x - 1)^3$. Another pair is $f(x) = x^{3/4}$ and $g(x) = 2x - 1$.

9. 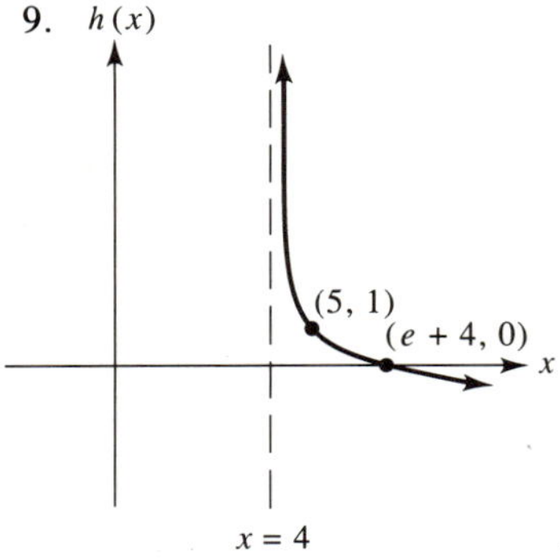

12. $\ln 7 + 6 \ln a - \frac{1}{2}(\ln 2 + \ln b + \ln c)$

15. $\ln\sqrt{x^3/y}$.

18. $\{\frac{25}{4}\}$

21. $\{-2 + \ln 3.2\}$

24. 1000^{204}

27. y varies directly as the fourth power of x and inversely as the cube root of z.

30. It will take approximately $11\frac{1}{4}$ years. The same amount of time is required. $3P = P(1.025)^{4t} \Leftrightarrow 3 = (1.025)^{4t}$.

Chapter Six Test

3. $(1/2, \sqrt{3}/2)$

6. -1

9. $-\sqrt{3}$

12. 1.0613 (approx.)

15. 1.0630 (approx.)

18. 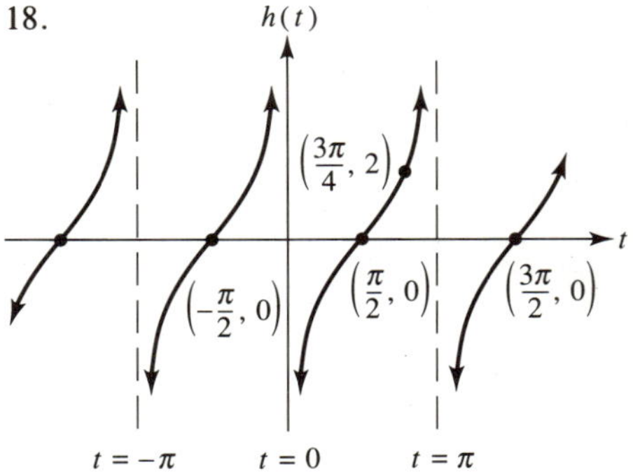

The graph continues to the left and right.

21. 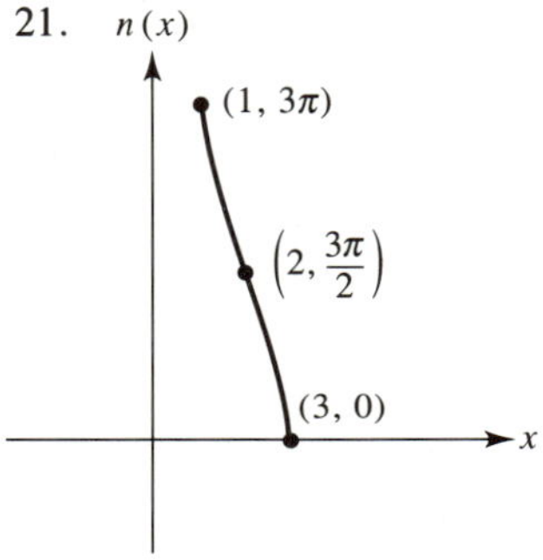

24. π; no; no; yes

27. 2π; no; no; yes

30. None; yes; no; yes

33. None; yes; no; yes

36. Amplitude: 1; period: $2\pi/6.8$; phase shift: $-\frac{29}{68}$

39. Expansion factor: $\frac{2}{3}$; period: 10π; phase shift: $5\pi/7$

42. 0

45. $\{\text{Sin}^{-1}(\ln 2) + 1 + 2k\pi\} \cup$
$\{\pi - \text{Sin}^{-1}(\ln 2) + 1 + 2k\pi\}, k \in I$

48. $\{\pi/3 + k\pi\} \cup \{2\pi/3 + k\pi\}, k \in I$

Chapter Seven Test

3. $(6246/\pi)°$

6. 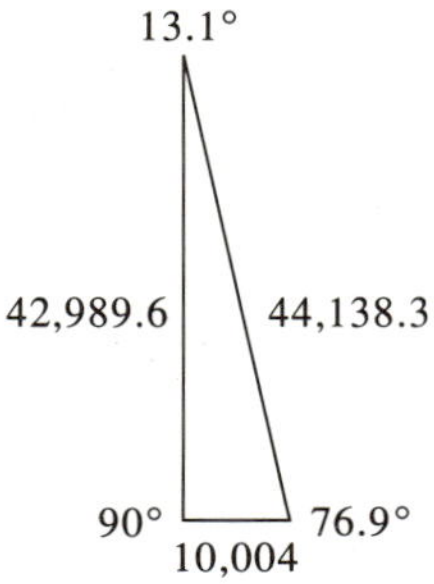

9.

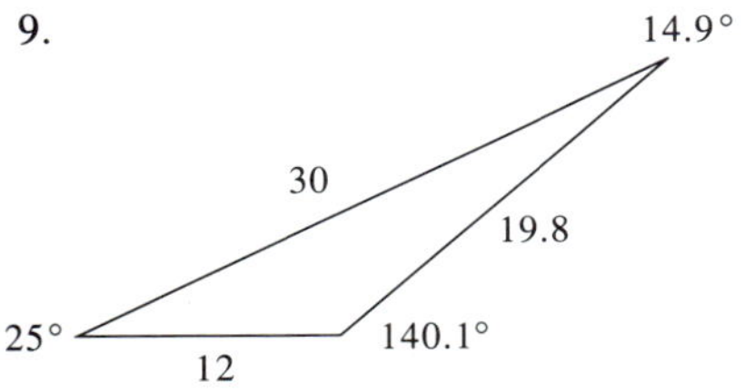

12.

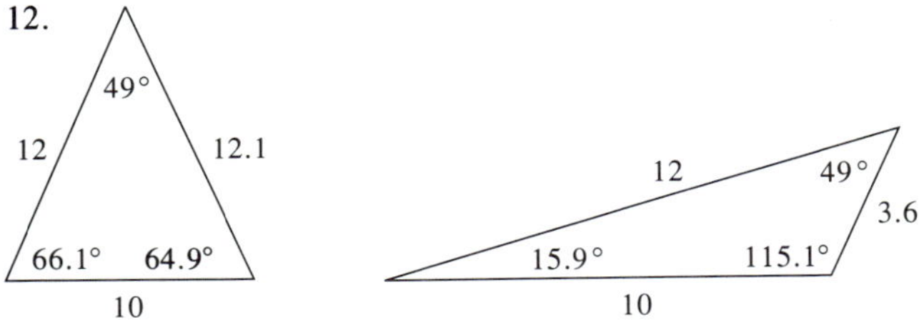

15. 355.2 square feet
18. $\sin^5 x/\cos^4 x$, $\sin x \neq 0$
21. Neither
24. Identity
27. $-2(\sin 6x)(\sin 2x)$
30. $\{\ \}$
33. The union of the intervals $[\pi/6 + k\pi, 2\pi/6 + k\pi]$, $k \in I$
36. The union of the intervals $(5\pi/6 + k\pi, 7\pi/6 + k\pi)$, $k \in I$
39. $\cos \pi/24 + i \sin \pi/24$, $\cos 13\pi/24 + i \sin 13\pi/24$, $\cos 25\pi/24 + i \sin 25\pi/24$, $\cos 37\pi/24 + i \sin 37\pi/24$

Chapter Eight Test

3. -648
6. $-0.5, 0.6, 2, 5$
8. $47 + 109i$
10. 1
12. 1.70
14. $p(x) = \frac{1}{4}(x^4 + 3x^3 + 12x - 16)$
17. $h(x) = -0.5(x^4 - 16.1x^3 + 58x^2 + 48.3x - 183)$
20.

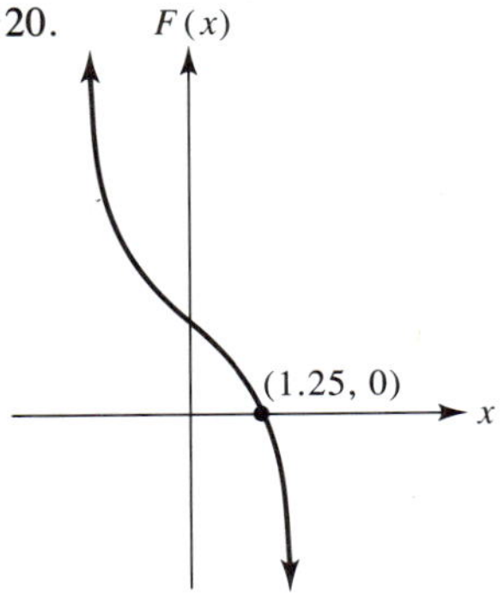

23. $y = 18x^3 + 33x$
26. $\left(-\frac{1}{243}, \frac{1}{32}\right)$, $(32, +\infty)$
28. -37.1508
31. $\frac{3}{2}, -\frac{1}{3}, \sqrt{3}, -\sqrt{3}$
33. $-4{,}259{,}398{,}621$

Chapter Nine Test

3. $\{(-\frac{142}{35}, \frac{286}{35}, \frac{162}{35})\}$
6. $\{(e^{1/5}, \ln \frac{4}{5})\}$
9. None
12. Infinitely many
15. $\{\ \}$
18. $\{(304 - 2c, -56 - 3c, c), c \in \mathbb{R}\}$
21. Undefined
24. $\begin{bmatrix} -9 & 38 \\ 67 & -41 \\ -7 & -7 \end{bmatrix}$
27. $\begin{bmatrix} \frac{7}{23} & \frac{15}{23} & \frac{12}{23} \\ \frac{6}{23} & \frac{3}{23} & \frac{7}{23} \\ \frac{10}{23} & \frac{5}{23} & \frac{4}{23} \end{bmatrix}$
30. $\begin{bmatrix} 59 \\ 21 \\ 12 \end{bmatrix}$
33. $\left\{\left(\sqrt{\frac{9-\sqrt{77}}{2}}, \sqrt{\frac{9+\sqrt{77}}{2}}\right), \left(\sqrt{\frac{9+\sqrt{77}}{2}}, \sqrt{\frac{9-\sqrt{77}}{2}}\right), \left(-\sqrt{\frac{9-\sqrt{77}}{2}}, -\sqrt{\frac{9+\sqrt{77}}{2}}\right), \left(-\sqrt{\frac{9+\sqrt{77}}{2}}, -\sqrt{\frac{9-\sqrt{77}}{2}}\right)\right\}$
35. $\{(-1.61$ (approx.), -2.41 (approx.)), (3.11 (approx.), 4.66 (approx.))}
36.

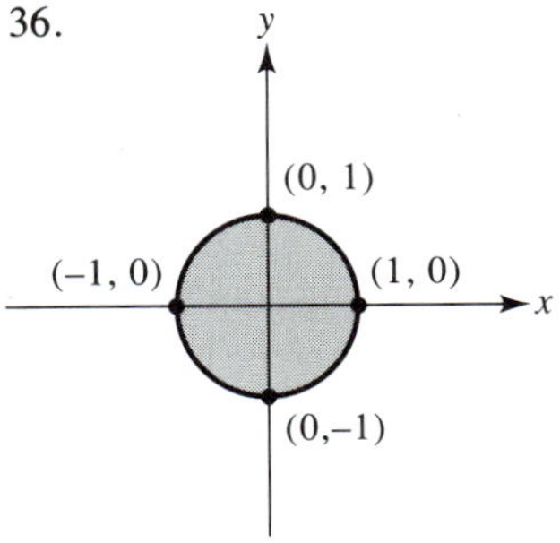

39.

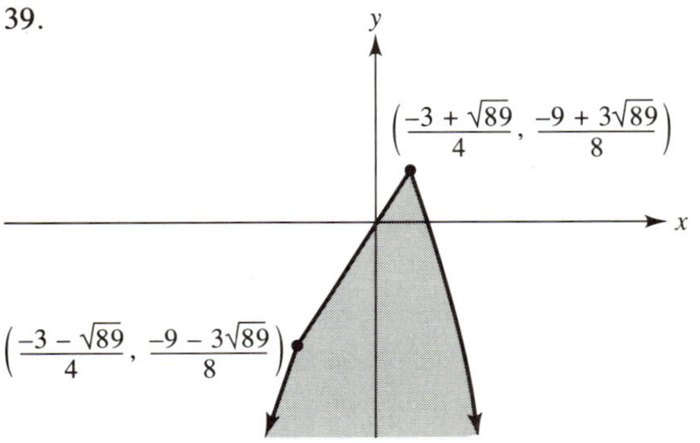

41.

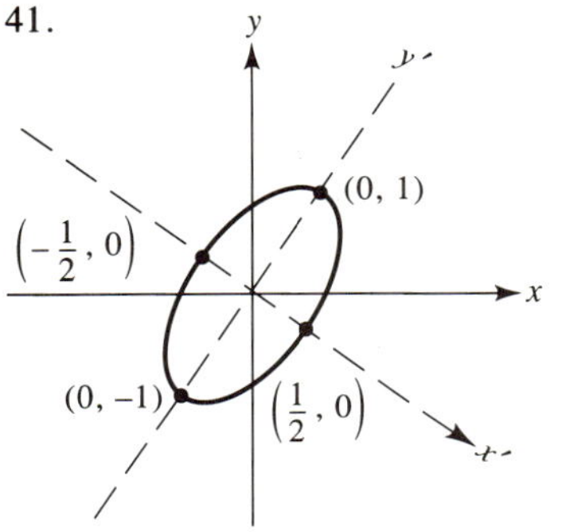

θ is $-\pi/6$; $16(x')^2 + 4(y')^2 = 4$.

Chapter Ten Test

3. 30,240
6. 196
9. 715
12. 32,432,400
15. 1440
18. The basis step in exercise 17 is true. For the induction step,

$$\sum_{i=1}^{k+1} i^3 = \sum_{i=1}^{k} i^3 + (k+1)^3, \text{ so } P(k+1):$$

$$\sum_{i=1}^{k+1} i^3 = \tfrac{1}{4}\cdot(k+1)^2\cdot(k+2)^2$$

$$\Leftrightarrow \quad \tfrac{1}{4}k^2(k+1)^2 + (k+1)^3 = \tfrac{1}{4}(k+1)^2(k+2)^2 \qquad \text{Using } P(k)$$

$$\Leftrightarrow \quad \tfrac{1}{4}(k+1)^2[k^2 + 4(k+1)] = \tfrac{1}{4}(k+1)^2(k+2)^2$$

$$\Leftrightarrow \quad \tfrac{1}{4}(k+1)^2(k+2)^2 = \tfrac{1}{4}(k+1)^2(k+2)^2.$$

21. −56,229,888
24. Arithmetic
27. None of these (power)
30. $6, 9, \frac{27}{2}, \frac{81}{4}, \frac{243}{8}$
33. a_n is $a_{n-1} + a_{n-2}$ and $a_1 = 3$, $a_2 = 4$.
36. a_n is $0.12n - 0.32$.
39. $2^n - 1$
42. $2n - \frac{3}{2}(n-1)n$
45. 171
48. 143

Appendix Test

3. {2, 3, 8, 13, 18, . . .}
6. {1, 3, 5, 7, . . .}
9.

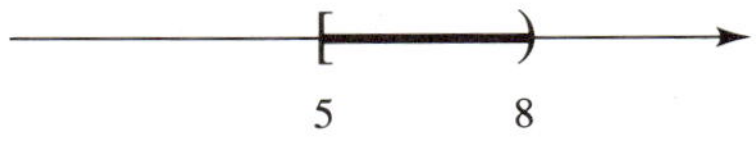

[5, 8)

12.

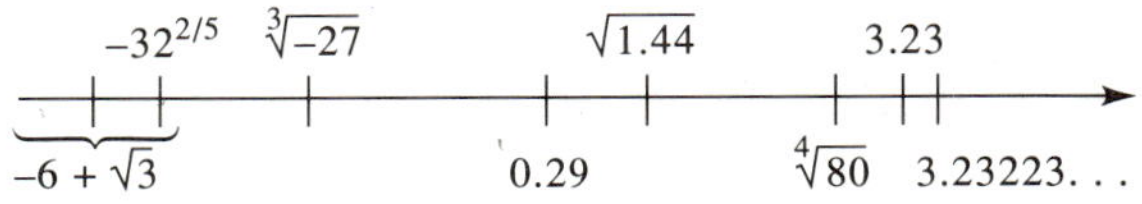

$[-6 + \sqrt{3}, 3.23223 \ldots]$

15. Distributive property
18. <
21. <
24. $-106 - 228i$
27. $\frac{40}{3} + \frac{16}{3}i$
30. $(2a-1)^{1/2}(a+4)^{-1/2}(8a+23)$
33. $y^{-1/2}(y-2)(y^2+2y+4)(y+2)(y^2-2y+4)$
36. $x^3 - x^2 + 3x - 2 + \dfrac{3x+3}{x^2+x+1}$
39. $\dfrac{x^{-2/5} - 1}{x^{-2/5} + 5}$; $x \neq 0, 5^{-5/2}$
42. $\dfrac{6x^{1/3} - 18}{x + 27}$, $x \neq -27$

Index

Note: Boldface type indicates where definitions may be found.

Absolute value, 36, **394**
  properties, 394
Acia, 41, 324, 335
Addition principle of counting, **358**
Additive identity, 393
Additive inverse, 393
Algebraic expression, **402.** *See also* Expression
Ambiguous case, 244
Amplitude, **210**
Angle(s), 227
  complementary, 232
  initial side, 227
  of rotation, 337–338
  standard position, 227
  terminal side, 227
Application problems, guidelines for working, 10
Area:
  circular sector, 240
  rectangle, 13
  trapezoid, 13
  triangle, 236, 239
Argument of a function, **115**
Arithmetic sequence, **372,** 375
  common difference, 372
Arithmetic series, 377–379
Associative property:
  of addition, 392
  of function addition, 154
  for function composition, 168
  of function multiplication, 156
  of multiplication, 392
Asymptote, 67
  horizontal, 129, 131–133, 135, 299
  oblique, 302
  vertical, 129, 135, 299
Asymptotic curve, **300**
Axis, 35

Base, **398**
Basis step, 361
Binomial, **402**
Binomial coefficients, 366, 367
Binomial theorem, **367**
Boolean matrix, **335**
Boolean product, 335
Boundary line, 98
Boundary number, 86, 95, 130

Characteristic operation, 75, 125
Circle, **59,** 59–60, 70, 340
  center of, 59
  radius of, 59, 340
  unit, 60
$C(n,r)$, 357, 367
Coefficient, **402**
  leading, 276
Cofunction, 232
Combination(s), **357**
  $C(n,r)$, 357, 367
  without repeats, 357–358
  with unlimited repeats, 358
Commutative property:
  of addition, 393
  of function addition, 156
  of function multiplication, 156
  of multiplication, 393
Complex number(s), **394**
  conjugate, 285, 397
  imaginary part, 396
  modulus, 265
  power of, 267
  principal square root, 270, 272
  product, 265–266, 395
  quotient, 266, 396–397
  real part, 396
  root of, 268–269
  standard form, 396
  sum, 272, 395
  trigonometric representation, 265
Composition of functions, **157,** 157–161. *See also* Decomposition
  graphing, 158–160
  identity for, 161

Composition of functions *(continued)*
inverse for, 161–167
Compound interest, 12, 176
continuous, 170, 179
Compound statement, 103
Conic sections, **69,** 69–71, 337–339. *See also* Circle, Ellipse, Hyperbola, Parabola
degenerate, 72
invariant measurement, 340–341
Conjugates, 397, 401
Conjugate complex numbers, 284–285, **397**
Constant, **386**
Constant of proportionality, 182
Coordinates, 35
polar, 264–265
cosecant, **197,** 231, 232
domain, 197
range, 198
cosine, **195,** 230, 232
Cos, **217**
domain, 195, 197
graph, 202–204
partial inverse, 217
range, 198
cotangent, **197,** 231, 232
domain, 197
range, 198
Cramer's rule, 323
Cubic formula, 32
Cusp, 139, 143
Cycle, **141**

Decomposition, 167
Decreasing function, **136**
Degenerate conic, **72**
Degree:
as angle measure, 231
equivalence with radians, 231
of a polynomial, **402**
De Moivre's theorem, 267, 363–364
Dependent variable, **113**
Descartes, René, 35
rule of signs, 289
Dimension of a matrix, **325**
Discriminant, **17**
Distance:
from a to b, **36**
in the plane, 36–37
from a point to a line, **58**
Distributive property, 154, 393
div_p, **120**
Division:
of polynomials, 403–404
synthetic, 279, 282
Domain, **113**
Double angle formulas, 251–252

e, 134
Element:
matrix, **325**
set, 388, 390

Ellipse, **61,** 61–64, 71, 341
center, 61
foci, 61
general form, 62
maximum chord length, 341
Equation(s), **2.** *See also* System of linear equations
absolute value, 6–7, 48
linear, **3,** 3–4, 42
quadratic, **14,** 14–17, 271
radical, **22**
rules for, 393
to solve, 2, 388
Equivalent equations, **3**
rules to produce, 3–4, 166
Equivalent systems, 310
Expansion, 337
contraction factor, 75–76
vertical scale, 76, 125, 209
Exponent(s), **398**
integer, 398, 399
irrational, 134, 402
rational, 399–400
rules for, 398
Exponential function(s), 134, **170,** 172–173
properties, 173
Expression, 402
fractional, 403, 407–409
quadratic in form, 30, 406
rational, 403, 407–409
Extrapolation, 181–182

Factoring, 404–406
Fibonacci sequence, **374**
type, 374
Fibonacci-type series, 380–381
Finite differences, method of, 180–181
Fractional expression, **403,** 407–409
Function(s), **113**
composition, 157
constant, **143**
decomposition, 167
decreasing, 136
difference, 153
domain, 113
even, **138**
exponential, 170
fractional, 133
identity for composition, 161
increasing, 136
inverse, 161–166, 216–218
logarithmic, 172
multiplicative identity, 155
notation, 114
odd, **138**
one-to-one, 137
periodic, **140,** 364–365
piecewise defined, 116
polynomial, 143–145, 276
product, 153
quotient, 153
range, 113
rational, **129,** 129–133

Function(s) *(continued)*
reciprocal, 155
sum, 152
transcendental, 188
trigonometric, 195–197, 230–232
wrapping, 189–193
zero of, 143
Fundamental theorem of algebra, 284

Gauss-Jordan elimination, **318,** 320, 321
Geometric sequence, **373,** 375
common ratio, 373
Geometric series, 379–380
Golden ratio, 33
Golden rectangle, 33
Graph, two-dimensional, **39**
Graph, **335**
incidence matrix, 335
Graphing techniques, 145
for trigonometric functions, 211–212

Half-angle formulas, 253, 332
Half-life, 176
Harmonic motion, simple, 202
Harmonic sequence, **374**
Harmonic series, 380
Horizontal asymptote, 129, 131–133, **135,** 299
Hyperbola, **65,** 65–68, 71, 341
asymptotes, 67
foci, 65
general forms, 66
minimum distance between arms, 341

i, **394**
Identity, **246**
additive, 393
function, 161
matrix, 328
multiplicative, 393
properties, 393
trigonometric, 247, 251, 252, 253, 259
Index, **399**
Index of summation, 377
Increasing function, **136**
Independent variable, **113**
Indirect proof, 314
Induction. *See* Mathematical induction
Induction step, 361
Inequality(s), 86
rules for, 393
solving, 109
solving one-variable, 97, 105
system of, 347
Integers, **392**
Intermediate value theorem, 280
Interpolation, 181–182
linear, 117–118, 121
Intersection of sets, **390**
Interval, 389
notation, 389
Inverse function(s), 161–166
trigonometric, 216–218

Inverse properties, 393
Irrational numbers, **392**

Law of Cosines, 237
Law of Sines, 241
lg, 172
Like terms, **403**
Linear factor theorem, 284
Linear-in-form system, **321**
pad to form, 345
Linear interpolation, 117–118, 121
Linear programming, 347–348
ln, **172**
Logarithmic function(s), 171–173, **172,** 175
change of base, 175
common, 172
natural, 171
properties, 173
Lower bound on real zeros. *See* Zero(s) of a function

Mapping, 113
Mathematical induction, **361**
Matrix (matrices), **325**
augmented, 315
Boolean, 335
difference, 326
dimension of, 325
element of, 325
equal, 325
equation, 332–333
identity, 327–328
incidence, 335
inverse, 328–329
product, 327
scalar multiplication, 326
sum, 326
transpose of, 330
zero, 326
m by *n* system, **311**
Midpoint, **38,** 38–39, 41
Mil, 235
Modeling, 177–182
mod_p, **118**
Modulus of a complex number, **265**
Monomial, **402**
Multiple root, 284
Multiplication principle of counting, **355**
Multiplicative identity, 393
function, 155
Multiplicative inverse, 393

Natural logarithm, **171**
Natural numbers, **392**
n factorial, **355**

Oblique asymptote, **302**
One-to-one function. *See* Function(s)
Opposite of a number, 393
Order of operations, 115, 386–387
Origin, 35
Oscillation, 260, 263

Parallel lines, 45
Parabola, 50, **58,** 58–59, 70, 340–341
  directrix, 58
  focus, 58
  vertex, 50, 58
  width at focus, 341
Partial sum, 376
Pascal's triangle, 367, 369–370
pc, **128**
Perfect square, **15**
  creation of, 15
Period, **140,** 364–365
  trigonometric function, 211
  wrapping function, 190
Permutation, **355**
  $P(n,r)$, 355
  with limited repeats, 356–357
  without repeats, 356
  with unlimited repeats, 356
Perpendicular lines, 45
Phase shift, **208**
Plane, real-number, 35
Point-slope form, **46**
Polar coordinates, 264–265
  principal value of angle, 265
Polynomial(s), 276, **402**
  degree of, 402
  equal, 277
  function, 276
  in one variable, **276**
Polynomial principle for graphing, **143,** 280, 287
Power series, **381**
Principal square root of a complex number, **270**
Principal value, 265
Proof techniques:
  counterexample, 155
  direct, 155
  indirect, 314
  mathematical induction, 361
  verification of an identity, 253, 255
ps, **146**
Pythagorean theorem, 35, 188

Quadrant, 35
Quadrant number, **190**
Quadratic formula, 16, 19, 30

Radian, **228**
Radical, **399**
Radicand, **399**
Range, **113**
  of a polynomial function, 289, 293
Rational numbers, **392**
Rational expression, **403,** 407–409
  proper, 298
Real numbers, **392**
Reciprocal, 393
  function, 155
Reference number, **194**
Reflection, 77–78, 124, 337
Relation, **113**
Restriction:
  explicit, 130, 407
  implicit, 130, 407
Root of a number, 399
Roots of a polynomial, 282
  multiple, 284
  of multiplicity m, 284
Roots of unity, **270**
Rotation by Θ radians, 337
Rules for equations, 393
Rules for inequalities, 393

Scalar multiplication, **326**
secant, **197,** 231, 232
  domain, 197
  graph, 206–207
  range, 198
Sequence, **372**
  arithmetic, 372, 375
  base, 376
  Fibonacci, 374–375
  geometric, 373, 375
  harmonic, 374
  series, 376
Series, **376**
  arithmetic, 377–378
  base sequence, 376
  Fibonacci-type, 380–381
  geometric, 379–380
  harmonic, 380
  partial sum, 376
  power, 381
Set(s), **388**
  element of, 388, 390
  enumeration, 389
  intersection, 390
  subset, 390
  union, 390
Set-builder notation, **389**
Shift. *See* Translation
Sigma notation, 377
Sine, **196,** 201, 231, 232
  domain, 196, 197
  graph, 204–205
  partial inverse, 216
  range, 198
  Sine, **216**
Slope, 43, **44,** 45
Slope-intercept form, **45**
Solution, **2,** 388
  of a system, 308
Solving:
  an equation, 2, 388
  an inequality, 388
  a triangle, 231
Spike. *See* Cusp
Standard form of a complex number, **396**
Standard graph, **47**
  1 (line), 47
  2 (parabola), 50
  3 (circle), 60
  4 (ellipse), 63

Standard graph (*continued*)
5 (reciprocal), 79
6 (cubic), 122
7 (absolute value), 123
8 (square root), 123
9 (natural exponential), 134
10 (natural logarithm), 171
11 (sine), 204
12 (tangent), 206
13 (secant), 207
Standard labeling of a right triangle, 231
Subset(s), 362, **390**
Sum and difference formulas, 251
Sum and product formulas, 259
Symbols of grouping, 115, 386, 394, 400
Symmetry:
line, **137**
origin, **138**
y-axis, 138
$y = x$, **140**
Synthetic division, **279,** 282
System of linear equations, **308**
m by *n*, 311
solution, 308

Tangent, **197,** 231, 232
domain, 197
graph, 205–206
partial inverse, 218
range, 198
Tangent, **218**
Term(s), 116, **402**
generalized, **116**
like, 403
of a sequence, 372
of a series, 376
Theorem:
binomial, 367
De Moivre's, 267, 363–364
fundamental, of algebra, 284

Theorem *(continued)*
intermediate value, 280
linear factor, 284
Pythagorean, 35, 188
Toolkit, **2**
Translation, **74,** 75, 139, 337
horizontal, 74, 123, 208
vertical, 74, 123, 208, 292–293
Transpose of a matrix, **330**
Trapezoid area formula, 13
Trinomial, **402**

Union of sets, **390**
Unit box, **91**
Unit circle, 60, 189
Upper bound on real zeros. *See* Zero(s) of a function

Variable, **386**
Variations, 82
Vector(s), **273**
in the plane, 273
sum, 273
Vertical asymptote, 129, **135,** 299
Vertical line test, 114
Vinculum, 386

Whole numbers, **392**
Wrapping function, 189–193

x-axis, 35
x-intercept, **42,** 143

y-axis, 35
y-intercept, **42**

Zero(s) of a function, **143,** 282
lower bound on real, **290,** 292
upper bound on real, **290,** 292
Zwab, 33